AF846467

Nano-Biomaterials For Ophthalmic Drug Delivery

Yashwant Pathak • Vijaykumar Sutariya
Anjali A. Hirani
Editors

Nano-Biomaterials For Ophthalmic Drug Delivery

Springer

Editors
Yashwant Pathak
College of Pharmacy
University of South Florida Health
Tampa, FL, USA

Vijaykumar Sutariya
College of Pharmacy
University of South Florida Health
Tampa, FL, USA

Anjali A. Hirani
College of Pharmacy
University of South Florida
Tampa, FL, USA

ISBN 978-3-319-29344-8 ISBN 978-3-319-29346-2 (eBook)
DOI 10.1007/978-3-319-29346-2

Library of Congress Control Number: 2016937003

© Springer International Publishing Switzerland 2016
This work is subject to copyright. All rights are reserved by the Publisher, whether the whole or part of the material is concerned, specifically the rights of translation, reprinting, reuse of illustrations, recitation, broadcasting, reproduction on microfilms or in any other physical way, and transmission or information storage and retrieval, electronic adaptation, computer software, or by similar or dissimilar methodology now known or hereafter developed.
The use of general descriptive names, registered names, trademarks, service marks, etc. in this publication does not imply, even in the absence of a specific statement, that such names are exempt from the relevant protective laws and regulations and therefore free for general use.
The publisher, the authors and the editors are safe to assume that the advice and information in this book are believed to be true and accurate at the date of publication. Neither the publisher nor the authors or the editors give a warranty, express or implied, with respect to the material contained herein or for any errors or omissions that may have been made.

Printed on acid-free paper

This Springer imprint is published by Springer Nature
The registered company is Springer International Publishing AG Switzerland

To the loving memories of my parents and Dr. Keshav Baliram Hedgewar, who gave a proper direction; my wife Seema, who gave a positive meaning; and my son Sarvadaman, who gave a golden lining to my life.

Yashwant Pathak

Dedicated to my mother and family. I have achieved much in my life because of their blessings.

Vijaykumar Sutariya

To my sister and parents for their love and support, my family in India for their prayers, and my mentors, Dr. Pathak and Dr. Sutariya, for being a beacon of light to help direct me to my goals.

Anjali A. Hirani

Foreword

A rapid expansion of new technologies in ocular drug delivery and new drug candidates, including biologics, to treat challenging diseases of the eye have recently emerged. These approaches are necessary because the eye has many unique barriers that block drug delivery. In the last decade, a significant growth in polymer science, nanotechnology, and biotechnology have been observed. This has led to the development of newly engineered nano-biomaterials, extensively explored as drug delivery carriers, implantable devices, and scaffolds.

This book is a collaborative effort of the editors Yashwant Pathak, Anjali Hirani, and Vijaykumar Sutariya as well as the numerous contributors who are leading scientists in this field. The subject matter is of prime importance in the area of nanotechnology and its application in ophthalmic drug delivery. All of the authors elucidate in their chapters, the potential interface between nanomaterials and the ophthalmic environment. This book presents a variety of nanomaterials and their applications in the treatment of ocular disease. I would like to commend Dr. Yashwant Pathak, Dr. Anjali Hirani, and Dr. Vijaykumar Sutariya at the University of South Florida for editing this important and timely issue.

It is my great pleasure to present to you *Nano-Biomaterials for Ophthalmic Drug Delivery*. I hope you will gain as much insight as I did from these chapters.

Kakarla V. Chalam
Department of Ophthalmology
UF-College of Medicine
Gainesville, FL, USA

Preface

This book is edited with a focus on nano-biomaterials for ophthalmic drug delivery. In recent years, the development of different nano-biomaterials has given a boost to drug delivery systems, and many products approved by the US FDA are in the market. The application of these nano-biomaterials, especially in the area of ophthalmic drug delivery, is receiving significant attention from the scientific community. It really created a need for an extensive study of these materials and for an excellent reference book, as these topics were previously addressed in detail. We think this book will fill the gaps of the knowledge presently available and provide detailed coverage of the potential applications of nano-biomaterials in ophthalmic drug delivery.

We also tried to cover various assessment and characterization techniques that have been developed to evaluate systems in the ophthalmic environment. The safety of the interaction of nano-biomaterials in the physiological environment is a concern for the health community which has also been covered well in this edited volume.

This book is targeted toward academic institutions, especially those working in the field of polymeric materials for drug delivery and ophthalmic drug delivery systems. Additionally, this will be a useful resource for pharmaceutical, medical, and other healthcare professionals. We envision this book will be a reference material and resource for researchers investigating technology to update understanding of drug delivery systems for the eye.

The volume comprises 26 chapters written by leading scientists in this field. The first chapter covers an introduction to nanotechnology with a special reference to ophthalmology. The next section addresses the ophthalmic system from the physiological and pharmacological point of view.

The next group of chapters covers nanoscale materials and their applications in ophthalmic drug delivery, including implantable and non-implantable systems. This is followed by the assessment and characterization of nano systems. The final group of chapters covers nano safety concerns and solutions.

We sincerely hope this book will be well-received by the scientific communities in the fields of nanotechnology and ophthalmology.

Thanks and acknowledgments to our families, the publishers, and our contributing authors. A special thanks to Aum Solanki and Rohini Nimbalkar for their help in the final compilation. We would like to acknowledge Ms. Carolyn Honour and all the other staff at Springer who helped to create this wonderful contribution to the field of ophthalmic drug delivery.

University of South Florida, Tampa, FL, USA

Yashwant Pathak
Anjali A. Hirani
Vijaykumar Sutariya

Contents

Contributors

Sridhar Anand Department of Pharmaceutical Sciences, Wegmans School of Pharmacy, St. John Fisher College, Rochester, NY, USA

Orhan E. Arslan Department of Pathology and Cell Biology, University of South Florida Morsani College of Medicine, Tampa, FL, USA

Paromita Bakshi Gupta College of Technological Sciences (College of Pharmacy), Asansol, West Bengal, India

Suraj G. Bhansali Novartis Pharmaceuticals, East Hanover, NJ, USA

Vanildo Martins Lima Braga Rua Sebastião Gonçalves Coelho, Divinópolis, MG, Brazil

Amy Broadwater College of Pharmacy, University of South Florida, Tampa, FL, USA

Lipika Chablani Department of Pharmaceutical Sciences, Wegmans School of Pharmacy, St. John Fisher College, Rochester, NY, USA

Harsh Chauhan Pharmacy Sciences, Creighton University, Omaha, NE, USA

Smita Chavan Department of Pharmaceutics, Bombay College of Pharmacy, Mumbai, India

Yahya E. Choonara Wits Advanced Drug Delivery Platform Research Unit, Department of Pharmacy and Pharmacology, School of Therapeutic Sciences, Faculty of Health Sciences, University of the Witwatersrand, Johannesburg, Parktown, South Africa

Mahavir B. Chougule The Daniel K. Inouye College of Pharmacy, University of Hawaii at Hilo, Hilo, HI, USA

Vivek S. Dave Pharmaceutical Sciences, St. John Fisher College, Wegmans School of Pharmacy, Rochester, NY, USA

Sumir Desai College of Pharmacy, University of South Florida, Tampa, FL, USA

Fahima Dilnawaz Laboratory for Nanomedicine, Institute of Life Sciences, Bhubaneswar, OR, India

Alhasan Elghouche Department of Ophthalmology, Indiana University School of Medicine, Indianapolis, IN, USA

Morgan V. Fedorchak Departments of Ophthalmology and Chemical Engineering, University of Pittsburgh and the Louis J. Fox Center for Vision Restoration, Pittsburgh, PA, USA

Willy Gama Department of Ophthalmology, Indiana University School of Medicine, Indianapolis, IN, USA

Rutika Godse Bombay College of Pharmacy, Mumbai, India

Aditya Grover College of Medicine, University of South Florida, Tampa, FL, USA

Frank X. Gu Department of Chemical Engineering, Waterloo Institute for Nanotechnology, University of Waterloo, Waterloo, Canada

Deepak Gupta Pharmaceutical Sciences, L|E|C|O|M Bradenton School of Pharmacy, Bradenton, FL, USA

Sheeba Varghese Gupta College of Pharmacy, University of South Florida, Tampa, FL, USA

Alon Harris Department of Ophthalmology, Indiana University School of Medicine, Indianapolis, IN, USA

Shruti Hazare Department of Pharmaceutics, Bombay College of Pharmacy, Mumbai, India

Anjali Hirani Department of Pharmaceutical Sciences, USF College of Pharmacy, University of South Florida, Tampa, FL, USA

School of Biomedical Engineering and Sciences, Virginia Tech-Wake Forest University, Blacksburg, VA, USA

Lyndon W. Jones Centre for Contact Lens Research, School of Optometry, University of Waterloo, Waterloo, Canada

Bibhuti B. Kakoti Department of Pharmaceutical Sciences, Dibrugarh University, Dibrugarh, India

Manjir Sarma Kataki Department of Pharmaceutical Sciences, Dibrugarh University, Dibrugarh, India

Nathaniel J. Kim Department of Ophthalmology, Indiana University School of Medicine, Indianapolis, IN, USA

Vijay Kumar Unither Pharma Ceuticals, Rochester, NY, USA

Yong W. Lee School of Biomedical Engineering and Sciences, Virginia Tech-Wake Forest University, Blacksburg, VA, USA

Shengyan Liu Department of Chemical Engineering, Waterloo Institute for Nanotechnology, University of Waterloo, Waterloo, Canada

Joyce S. Macwan Simulations Plus, Inc., Lancaster, CA, USA

Sabyasachi Maiti Gupta College of Technological Sciences (College of Pharmacy), Asansol, West Bengal, India

Sharad Malavade Department of Global Health, College of Public Health, University of South Florida, Tampa, FL, USA

Mala D. Menon Department of Pharmaceutics, Bombay College of Pharmacy, Mumbai, India

Preya Patel College of Arts and Sciences, University of South Florida, Tampa, FL, USA

Department of Pharmaceutical Sciences, College of Pharmacy, University of South Florida, Tampa, FL, USA

Yashwant Pathak Department of Pharmaceutical Sciences, USF College of Pharmacy, University of South Florida, Tampa, FL, USA

Viness Pillay Wits Advanced Drug Delivery Platform Research Unit, Department of Pharmacy and Pharmacology, School of Therapeutic Sciences, Faculty of Health Sciences, University of the Witwatersrand, Johannesburg, Parktown, South Africa

Dev Prasad School of Pharmacy, Massachusetts College of Pharmacy and Health Sciences, Boston, MA, USA

Charles Preuss Department of Molecular Pharmacology and Physiology, Morsani College of Medicine, University of South Florida, Tampa, FL, USA

Sayantan Sadhukhan Gupta College of Technological Sciences (College of Pharmacy), Asansol, West Bengal, India

Sanjeeb Kumar Sahoo Laboratory for Nanomedicine, Institute of Life Sciences, Bhubaneswar, Orissa, India

Nikhil A. Sangave MCPHS University, University of Southern California, Los Angeles, USA

Ankit Shah Department of Ophthalmology, University of South Florida, Tampa, FL, USA

Ujwala Shinde Bombay College of Pharmacy, Mumbai, India

Agnivesh Shrivastava Bombay College of Pharmacy, Mumbai, India

Brent Siesky Department of Ophthalmology, Indiana University School of Medicine, Indianapolis, IN, USA

Kavita Singh Bombay College of Pharmacy, Mumbai, India

Aum Solanki Honors College, University of South Florida, Tampa, FL, USA

Vijaykumar Sutariya Department of Pharmaceutical Sciences, College of Pharmacy, University of South Florida, Tampa, FL, USA

Muktika Tekade TIT College of Pharmacy, Technocrats Institute of Technology Campus, Anand Nagar, MP, India

Rakesh K. Tekade Preclinical Nuclear Imaging Laboratory, The University of Texas Southwestern Medical Center, Dallas, TX, USA

Department of Pharmaceutical Technology, School of Pharmacy, International Medical University, Kuala Lumpur, Malaysia

Lisa C. du Toit Wits Advanced Drug Delivery Platform Research Unit, Department of Pharmacy and Pharmacology, School of Therapeutic Sciences, Faculty of Health Sciences, University of the Witwatersrand, Johannesburg, Parktown, South Africa

Rongbing Yang The Daniel K. Inouye College of Pharmacy, University of Hawaii at Hilo, Hilo, HI, USA

About the Editors

Yashwant Pathak, M.Pharm., E.M.B.A., M.S. (Conflict Management), Ph.D.
Dr. Yashwant Pathak completed his M.S. and Ph.D. in pharmaceutical technology from Nagpur University, India, and E.M.B.A. and M.S. in conflict management from Sullivan University, Louisville, KY. He is the associate dean for Faculty Affairs at the College of Pharmacy, University of South Florida in Tampa, FL. With an extensive experience in academia as well as industry, he has more than 100 publications, 2 patent applications, 5 books in nanotechnology and 4 in nutraceuticals, and several books in cultural studies. His areas of research include drug delivery systems and its characterization in animal models.

Vijaykumar Sutariya, B.Pharm., M.Pharm., Ph.D.
Dr. Vijaykumar Sutariya received his bachelor's and master's degree in pharmacy from L. M. College of Pharmacy, Gujarat University, Ahmedabad, India, and Ph.D. in pharmacy from the M.S. University of Baroda, Vadodara, India. He did his postdoctoral training in pharmaceutics and drug delivery at Butler University, Indianapolis, IN. Dr. Sutariya is an assistant professor in the Department of Pharmaceutical Sciences at the USF College of Pharmacy. He has a joint appointment with the Department of Internal Medicine, Division of Translational Medicine, at USF. Dr. Sutariya's research is focused on the development of novel drug delivery systems such as nanoparticles, liposomes, and thermoreversible gels.

Anjali Hirani, Ph.D.
Dr. Anjali Hirani received her M.S. and Ph.D. in biomedical engineering from Virginia Tech, Blacksburg, VA. She is currently a postdoctoral fellow at the University of South Florida in the College of Pharmacy. Her research is focused on the development of sustained drug delivery systems for ocular disease.

Chapter 1
Introduction to Nanotechnology with Special Reference to Ophthalmic Delivery

Anjali Hirani and Yashwant Pathak

Abstract The capabilities of nanotechnology are leading to widespread use of nanomaterials in biotechnology and medicine. Nanomaterials exist in various sizes, compositions, and morphology. Due to their large numbers, classification systems should be implemented for proper evaluation of the nanomaterials for efficacy and prevention of toxicity. Many nanomaterials are being studied for potential therapeutic applications in ocular disease. Conventional treatments still possess side effects and limitations and the use of nano-biomaterials for ophthalmic drug delivery can allow for design and administration of delivery systems with targeted and sustained release capabilities. Additionally, off-target activities and toxicity can be reduced.

Keywords Nanotechnology/nanomaterials • Biodistribution • Bottom-up fabrication • Top-down fabrication • Dendrimers • Composites • Nanofibers • Nanotoxicity • Nanoemulsions • Ophthalmic drug delivery

1.1 Introduction to Nanotechnology

Nanotechnology involves the use of materials sized under 100 nm. For decades, it was mainly used in fields such as engineering, electronics, and physics; however, other avenues for nanotechnology in pharmaceuticals and biomedical areas are being explored due to its incredible potential in improving the treatment of a wide

---

A. Hirani
Department of Pharmaceutical Sciences, USF College of Pharmacy, University of South Florida, Tampa, FL 33612, USA

School of Biomedical Engineering and Sciences, Virginia Tech-Wake Forest University, Blacksburg, VA 24061, USA

Y. Pathak (✉)
Department of Pharmaceutical Sciences, USF College of Pharmacy, University of South Florida, Tampa, FL 33612, USA
e-mail: ypathak1@health.usf.edu

© Springer International Publishing Switzerland 2016

Y. Pathak et al. (eds.), *Nano-Biomaterials For Ophthalmic Drug Delivery*,
DOI 10.1007/978-3-319-29346-2_1

variety of medical disorders. This stems from the ability of nanomaterials to be modified to meet the specific needs of each application. The molecular and supramolecular structures can be manipulated for developing innovative devices programmed with specific functions.

Nanomaterials exist in many sizes, forms, and compositions. These properties are key in developing novel pharmaceuticals and biomedical products. The conventional forms of nanomaterials include liposomes, polymeric micelles, and fluorescent crystals, each of which has properties that are suitable for specific applications. Currently, there is a trend toward improving conventional pharmaceutical drugs (active pharmaceutical ingredients) with nanomaterials. These pharmacological ingredients have applications in treating severe illnesses but have limited benefits for patients due to severe side effects. Nanostructures comprised of atoms or molecules, ranging in size and with different morphologies such as amorphous, crystalline, or spherical, are used to entrap active pharmacological ingredients [1]. These nanoparticulate drug delivery systems offer many benefits over conventional pharmaceutical drugs such as improved therapeutic efficacy, reduced toxicity, improved biodistribution, and improved patient compliance.

The formulation of nanomaterials is developed by two fabrication methods: bottom up or top down. The bottom-up method involves controlling the thermodynamic properties such as self-assembly, precipitation, and crystallization of the nanomaterials to piece together molecules by molecules to build the nanostructures. The top-down approach involves the use of nanofabrication technologies, such as photolithography, nanomolding, and nanofluidics [1, 2]. These two fabrication methods are used to make nanostructures of various forms and properties. For instance, drug-encapsulated nanospheres or nanocapsules can be made using bottom-up approaches. In manufacturing, however, the top-down approach is a more cost-effective approach to make nanostructures in large quantities.

1.2 Classification of Nanomaterials

The use of nanomaterials is growing exponentially and already has a large commercial impact. Some known commercial uses of nanomaterials include cosmetics, electronics, textiles, and varnishes. Since nanomaterials are widely used and their numbers are growing, a classification system is needed to group the nanomaterials. However, the classification of nanomaterials is not simple, since they may possess overlapping characteristics, sizes, composition, and functionalities.

A multiple classification system should be considered for nanomaterials depending on the purpose of the system. One simple classification is to categorize the nanomaterials by the process through which they are generated. Nanomaterials are generated through natural or anthropogenic processes. Naturally made nanomaterials can be generated by living organisms, soil, air, and heat and can be categorized into biogenic, geogenic, atmospheric, and pyrogenic categories [3]. For instance, a study identified that *Arthrobotrys oligospora*, a flesh-eating fungus, produce natural-based nanomaterials consisting of glycosaminoglycan and protein from their trap to capture and digest nematodes. The organic nanoparticles were found to induce tumor necrosis factor which suggests that the nanoparticles have immunostimulatory properties [4]. Nanomaterials were also found geochemically: a study found volcano ash-derived nanoparticles and colloids in glacial-fed rivers after the eruption of Eyjafjallajokull [5].

As for anthropogenic processes, these nanomaterials are generated through human activity. They can either be made intentionally or unintentionally. For nanomaterials that are generated intentionally, they can be classified into five categories: carbon-based materials, metal-based materials, dendrimers, polymeric particles, and composites. These nanoparticles are made through chemical synthesis and/or self-assembly. For instance, FePt nanoparticles can be generated by both methods [6]. FePt can be generated using thermal decomposition of iron pentacarbonyl, $Fe(CO)_5$, and reduction of platinum acetylacetonate, Pt $(acac)_2$. Another method for generating FePt is to use metal salts. Metal salts of platinum and iron can be combined with polyalcohol to generate the FePt nanoparticles. The self-assembly method is mainly used to produce nanofilm FePt. The films are developed with the use of polymer, such as PEI, phospholipids, and oleic acid. Nanofilms can also be developed by using a silicon wafer with photoresistant film, where dispersion of the FePt nanoparticles is used to develop patterned nanofilms.

Nanomaterials that are generated unintentionally are usually by-products of combustion. Combustion has been known to generate harmful particles. Engine-derived particles, also known as diesel exhaust particles, are generated by nucleation of hydrocarbon from dilution and cooling of the exhaust [7]. These particles can agglomerate directly from the combustion. Other types of combustion-derived particles include welding fume and carbon black nanoparticles. These are generated in the workplace and are occupational hazards.

Size and morphology are important characteristics of nanomaterials. The morphology of nanomaterials can be classified based on their dimension. Nanomaterials that are considered to have a nanometer size in one dimension are referred to as surface films; nanomaterials in two dimensions are called strands or fiber; nanomaterials in the three dimensions are referred to as particles. Nanomaterials can be classified based on their forms, such as single, fused, aggregated, or agglomerated, and their shape, such as spherical, tubular, and irregular. For instance, ZnO particles with various sizes and shapes are developed using organometallic synthesis methods [8]. By varying the solvents, ligands, and time of incubation, nanorods and spherical particles with various sizes are generated. This classification is useful in identifying the biological effects of nanomaterials. For instance, tubular nanomaterials, such as carbon nanotubes, are found to be more cytotoxic in contrast to spherical carbon black nanomaterials and crystal zinc oxide nanomaterials [9].

In manufacturing, phase composition properties and the method of production of nanomaterials are also useful classifications. Nanomaterials can be categorized as single-phase solids (crystalline, amorphous particles and layer), multiphase solids (matrix composites and coated particles), and multiphase systems (colloids, aerogels, and ferrofluids). Their production can be classified into three methods: gas phase reactions (flame synthesis, condensation, etc.), liquid phase reactions (sol-gel, precipitation, and hydrothermal processing), and mechanical procedures (ball milling, plastic deformation). There are many ways to classify nanomaterials; however, due to the complexity and number of nanomaterials, it is still difficult to group all particles in a distinct, data table.

1.3 Nanotoxicity

Despite the potential applications of nanotechnology, environmental and health risks of nanomaterials are under constant scrutiny. Nanotoxicology emerged as a discipline to study potential environmental, health, and safety risks that come with the applications of nanomaterials [10–12]. This discipline examines the physicochemical determinants, routes of exposure, biodistribution, molecular determinants, and genotoxicities of nanomaterials. These risks are of concern for scientific, governmental, and nongovernmental agencies [13, 14]. Multidisciplinary research programs that examine the impact of nanomaterials on human health and the environment have been initiated by the National Center for Environmental Research of the United States Environmental Protection Agency, National Toxicology Program, National Institute of Environmental Health Sciences, and National Institutes of Health [14]. Since the 1980s, epidemiological and toxicological studies were conducted to evaluate the respiratory and pulmonary toxicity of ambient, ultrafine particles. Studies have shown that TiO_2 and aluminum oxide (Al_2O_3) nanoparticle exposure induced inflammatory responses in the lungs of rats compared to larger particles of the same compositions [15, 16]. It was also identified that engineered nanomaterials, such as carbon nanotubes, present an inhalational risk in rodent toxicity studies [17, 18]. Intratracheal dosing of single- or multiwalled carbon nanotubes causes a more pronounced inflammatory pulmonary response in mice as compared to rats [17–19].

Since 2000, research has focused on toxicokinetics and toxicodynamics of nanomaterials, the human exposure of nanomaterials, and the use of nanomaterials in medical devices, diagnostics, and therapeutics [14]. To examine the life cycle of nanomaterials, it is important to study both human and environmental exposure and determine the risk of each.

Nanomaterials impact the human body at the cellular and molecular levels, where they can interact with biomolecules, such as proteins and lipids. These interactions account for most of the toxicological effects. For example, cellular internalization of nanoparticles can cause organelle damage, as they bind to DNA and activate the ROS system, leading to mitochondrial damage. Cellular damage can also be caused by the generation of membrane, nanoscale holes, or depolarization of the membrane. Nanomaterials interact with the membrane through hydrophobic forces, electrostatic forces, van der Waals forces, hydrogen bonding, or receptor interactions. Constant generation of ROS leads to the activation or proinflammatory cytokines and induces toxicity in organs such as the liver, spleens, lungs, and others.

There are several mechanisms through which nanomaterials induce toxicity in the body. The most common mechanism is via oxidative stress through free radical generations [20, 21]. Free radicals are known to cause cellular damages by oxidizing lipids, proteins, and DNA. Oxidation results in the upregulation of inflammatory factors, such as redox-sensitive transcription factors (e.g., NF-κB), activator protein-1, chemokines, and cytokines [20, 22, 23]. In addition to generating free radicals, nanomaterials interact with cellular components, such as the mitochondria and nucleus, which causes a cascade of events leading to apoptosis, cell arrest, and nuclear DNA damage [21, 24].

Nanoparticles can complex with proteins, forming the nanoparticle-protein complex or the "protein corona." The protein coronae are comprised of hard layers (protein that strongly attached to the nanomaterials) and dynamic, soft layers (protein that weakly adsorbed onto the nanomaterials) [25]. The interaction of nanomaterials with proteins can block the protein's active site, alter the protein's conformation, or denature the protein. This can have devastating effects on the protein's normal biological function [25, 26].

Pharmacokinetics is important for understanding the safety and toxicity of nanomaterials. Examining the mechanisms of absorption, distribution, metabolism, and excretion will reveal the time of exposure and dose of the nanomaterials required for toxicity. Information on the pharmacokinetics will help assess those who are exposed and determine the necessary approach for treatment.

1.4 Applications in Ophthalmic Drug Delivery

Nano-biomaterials are gaining new ground in the development of novel drug delivery systems. These systems are divided into five categories: (1) sustained and controlled delivery systems, (2) stimuli-sensitive delivery systems, (3) functionalized systems for bioactive delivery, (4) multifunctional systems for combined delivery of therapeutics and diagnostics, and (5) site-specific drug delivery systems. The use of nanocarriers, such as polymeric nanoparticles, dendrimer, and antibody-drug conjugates, can facilitate the efficient delivery of active pharmacological ingredients and minimize drug peak plasma levels.

These nanoparticle systems are useful for improving ocular treatment. The anterior segment of the eye can be readily accessed; however, clearance mechanisms and physical barriers prevent access to the posterior portion of the eye, which contains the retina, vitreous, choroid, and sclera. There are several conventional treatments for the eyes. The topical route is a convenient method of drug delivery for treating ocular diseases; however, there is poor bioavailability due to nasolacrimal drainage and systemic absorption [27]. A model of transient diffusion has shown that less than 5 % of a lipophilic drug and 0.5 % of a hydrophilic drug reach the anterior chamber [28]. The amount of available drug transported further decreases across the sclera, choroid, and retinal pigment epithelium [29]. Permeability via sclera is reduced with cationic and lipophilic solutes, and the retinal pigment epithelium has tight intercellular junctions for hydrophilic molecules [29]. Additionally, the lymphatic system, blood vessels, and active transporters all work to clear drugs administered through transscleral routes. The eye can also be treated through systemic routes; however, this route requires high doses to obtain a therapeutic concentration in the posterior eye due to the tight barrier of the RPE. Another approach is through intravitreal injections. This method circumvents physiological barriers and maintains therapeutic doses without damage to bystander tissues. However, due to the liquefaction of the vitreous body related to aging, drug delivery can be nonuniform across different retinal areas [30]. Furthermore, frequent injections can lead to sight-threatening complications like retinal detachment, increase in intraocular pressure, hemorrhage, and endophthalmitis [31, 32]. Given the presence of these physiological barriers, the development of therapies that efficiently deliver drugs and extend drug release to the posterior segment of the eye would be beneficial to the progression of ocular disease treatment.

The use of drug delivery systems with nano-biomaterials provides an effective method to deliver drugs to the different regions of the eye for extended periods of time. Drug delivery systems can provide strategies to circumvent physiological barriers and provide sustained release with minimal systemic side effects, thereby expanding current disease therapy and repurposing presently used drugs and extending their patent life.

Biodegradable nanomaterials were found to reduce complications from conventional ocular disease treatment. Biodegradable and polymeric materials are used to entrap and adsorb drugs [33]. These biodegradable materials were found to significantly enhance the quality of life for patients with degenerative eye disorders. A recent study showed that intravitreal injection of 20 nm particles localized in the retinal pigment epithelium within 6 h and cytoplasmic concentration of the nanoparticles remained elevated up to 4 months [34].

Polysaccharide nanomaterials have shown to be effective at treating ocular diseases. For instance, chitosan is hydrophilic and biodegradable that is cleaved by lipases and proteases. It is found that chitosan loaded with cyclosporine A showed promise in treating in conjunctivitis and dry eye disease, since the cyclosporine A remained elevated in the cornea for 48 h [35]. Incorporation of chitosan in other nanomaterials will increase the zeta potentials, membrane permeability, and ocular bioavailability [36].

Along with polysaccharide, nanoemulsions and polymeric micelles are currently being explored for ocular disease treatments. Nanoemulsions can be incorporated with anti-VEGF bevacizumab for sustained drug release delivery. Micelles encapsulated with pDNA are found to ameliorate choroidal neovascularization in age-related macular degeneration [37]. These nanomaterials offer effective drug delivery and maintain therapeutic drug concentration for ocular disease treatment.

1.5 Future Considerations

Research and implementation of nanomaterials for ophthalmic drug delivery is still in the initial phases of development. It has the potential to increase the efficacy and longevity of treatment in numerous ocular disorders, which will be discussed in greater detail in the following chapters.

References

1. Sahoo SK, Labhasetwar V (2003) Nanotech approaches to drug delivery and imaging. Drug Discov Today 8(24):1112–20
2. Peppas NA (2004) Intelligent therapeutics: biomimetic systems and nanotechnology in drug delivery. Adv Drug Deliv Rev 56(11):1529–31
3. Nowack B, Bucheli TD (2007) Occurrence, behavior and effects of nanoparticles in the environment. Environ Pollut 150(1):5–22
4. Wang Y et al (2013) Naturally occurring nanoparticles from arthrobotrys oligospora as a potential immunostimulatory and antitumor agent. Adv Funct Mater 23(17):2175–2184
5. Tepe N, Bau M (2010) Importance of nanoparticles and colloids from volcanic ash for riverine transport of trace elements to the ocean: evidence from glacial-fed rivers after the, eruption of Eyjafjallajökull Volcano, Iceland. Sci Total Environ 2014(S488–489):243–251
6. Sun S (2006) Recent advances in chemical synthesis, self-assembly, and applications of FePt nanoparticles. Adv Mater 18:393–403
7. Kittelson D (1998) Engines and nanoparticles: a review. J Aerosol Sci 29(5/6):575–588
8. Kahn ML et al (2005) Size- and shape-control of crystalline zinc oxide nanoparticles: a new organometallic synthetic method. Adv Funct Mater 15(3):458–468
9. Yang H et al (2008) Comparative study of cytotoxicity, oxidative stress and genotoxicity induced by four typical nanomaterials: the role of particle size, shape, and composition. J Appl Toxicol 29(1):69–78
10. Fischer HC, Chan WCW (2007) Nanotoxicity: the growing need for in vivo study. Curr Opin Biotechnol 18(6):565–571
11. Donaldson K et al (2004) Nanotoxicology. Occup Environ Med 61(9):727–728
12. Arora S, Rajwade JM, Paknikar KM (2012) Nanotoxicology and in vitro studies: the need of the hour. Toxicol Appl Pharmacol 258(2):151–65
13. Colvin VL (2003) The potential environmental impact of engineered nanomaterials. Nat Biotechnol 21(10):1166–1170
14. Santamaria A (2012) Historical overview of nanotechnology and nanotoxicology. Methods Mol Biol 926:1–12
15. Ferin J et al (1990) Increased pulmonary toxicity of ultrafine particles.1. Particle clearance, translocation, morphology. J Aerosol Sci 21(3):381–384
16. Oberdorster G et al (1990) Increased pulmonary toxicity of ultrafine particles.2. Lung lavage studies. J Aerosol Sci 21(3):384–387
17. Warheit DB et al (2004) Comparative pulmonary toxicity assessment of single-wall carbon nanotubes in rats. Toxicol Sci 77(1):117–125
18. Shvedova AA et al (2005) Unusual inflammatory and fibrogenic pulmonary responses to single-walled carbon nanotubes in mice. Am J Physiol Lung Cell Mol Physiol 289(5):L698–L708
19. Lam CW et al (2004) Pulmonary toxicity of single-wall carbon nanotubes in mice 7 and 90 days after intratracheal instillation. Toxicol Sci 77(1):126–134
20. Lanone S, Boczkowski J (2006) Biomedical applications and potential health risks of nanomaterials: molecular mechanisms. Curr Mol Med 6(6):651–663
21. Aillon KL et al (2009) Effects of nanomaterial physicochemical properties on in vivo toxicity. Adv Drug Deliv Rev 61(6):457–466
22. Rahman I (2000) Regulation of nuclear factor-kappa B, activator protein-1, and glutathione levels by tumor necrosis factor-alpha and dexamethasone in alveolar epithelial cells. Biochem Pharmacol 60(8):1041–1049
23. Rahman I et al (2005) Glutathione, stress responses, and redox signaling in lung inflammation. Antioxid Redox Signal 7(1–2):42–59
24. Unfried K et al (2007) Cellular responses to nanoparticles: target structures and mechanisms. Nanotoxicology 1(1):52–71
25. Mahmoudi M et al (2011) Protein-nanoparticle interactions: opportunities and challenges. Chem Rev 111(9):5610–37

26. Lynch I, Dawson KA (2008) Protein-nanoparticle interactions. Nano Today 3(1–2):40–47
27. Gambhire S, Bhalerao K, Singh S (2013) In situ hydrogel: different approaches to ocular drug delivery. Int J Pharm Sci 5(2):27–36
28. Zhang W, Prausnitz MR, Edwards A (2004) Model of transient drug diffusion across cornea. J Control Release 99(2):241–58
29. Urtti A (2006) Challenges and obstacles of ocular pharmacokinetics and drug delivery. Adv Drug Deliv Rev 58(11):1131–5
30. Kleinberg TT et al (2011) Vitreous substitutes: a comprehensive review. Surv Ophthalmol 56(4):300–23
31. Christoforidis JB et al (2012) Intravitreal devices for the treatment of vitreous inflammation. Mediat Inflamm 2012
32. Peyman GA, Lad EM, Moshfeghi DM (2009) Intravitreal injection of therapeutic agents. Retina 29(7):875–912
33. Bu HZ et al (2007) Ocular disposition, pharmacokinetics, efficacy and safety of nanoparticle-formulated ophthalmic drugs. Curr Drug Metab 8(2):91–107
34. Bourges JL et al (2003) Ocular drug delivery targeting the retina and retinal pigment epithelium using polylactide nanoparticles. Invest Ophthalmol Vis Sci 44(8):3562–9
35. Singh AV (2011) Biopolymers in drug delivery: a review. Pharmacologyonline 1:666–674
36. Vaishya R et al (2014) Controlled ocular drug delivery with nanomicelles. WIREs Nanomed Nanobiotechnol 6:422–437
37. Ong FS et al (2013) Personalized medicine in ophthalmology: from pharmacogenetic biomarkers to therapeutic and dosage optimization. J Pers Med 3(1):40–69

Chapter 2
Overview of the Ophthalmic System

Sharad Malavade

Abstract Drug delivery to the eye is a challenging task given the inherent structural and physiological uniqueness of the eye. The eye is a composite organ comprised of unique structures with distinct structure and functions that work together for the perception of sight. Structures like the cornea, the crystalline lens, and the vitreous humor are avascular structures, while the uveal tract is primarily the vascular layer of the eye. The retinal neuro-circuitry is complex and situated deep inside the eye and is a derivative of neuroectoderm. Similar to the blood–brain barrier, there is the blood–retina barrier in the eye. The limited external surface area and the deep isolated location of the different structures of the eye along with avascularity of some of the structures and the blood–retina barrier pose unique challenges for delivery of drugs to intraocular structures.

Keywords Ocular drug delivery • Ocular anatomy • Cornea • Conjunctiva • Sclera • Uveal tract • Retina

2.1 Introduction

The eye is probably the most important sensory organ of the body. It is responsible for sight and also visual cues for maintaining balance. The eye is also an immune-privileged site, and being derived from the forebrain during embryological development, it retains the challenges of the blood–brain barrier for drug delivery in the form of the blood–retina barrier among others. This chapter briefly reviews the anatomy and physiology of the eye with the aim of understanding the ophthalmic system from the drug delivery perspective. It is not meant to be the most comprehensive review of the ophthalmic system and ocular anatomy and physiology as there are works that delve deep into the subjects, and the reader is encouraged to refer to those works for further detailed information (Fig. 2.1).

S. Malavade (✉)
Department of Global Health, College of Public Health, University of South Florida, Tampa, FL, USA
e-mail: smalavad@health.usf.edu

© Springer International Publishing Switzerland 2016

Y. Pathak et al. (eds.), *Nano-Biomaterials For Ophthalmic Drug Delivery*,
DOI 10.1007/978-3-319-29346-2_2

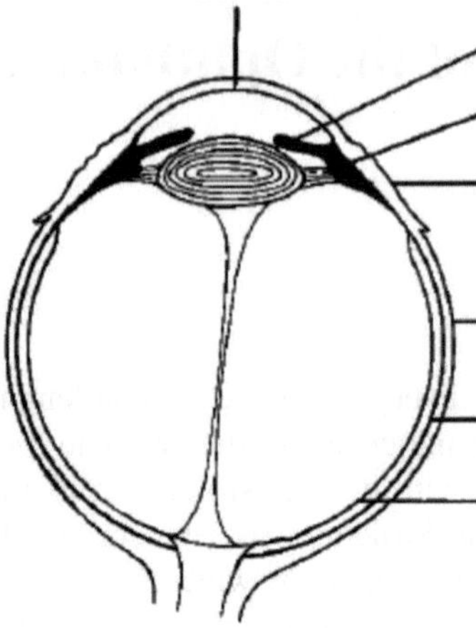

Fig. 2.1 Schematic cross section of the human eye [1]

2.2 Cornea

The cornea is the clear and transparent avascular part of the eye that forms the anteriormost coat of the eye. The cornea is anatomically comprised of five main layers. The first layer is the corneal epithelium, which is the most exterior. The next is the Bowman's membrane followed by the corneal stroma, both of which are formed by collagen fibers. The fourth layer is Descemet's membrane. It is the basement membrane that supports the corneal endothelium, which is the innermost layer of the cornea. The cornea is a prolate structure and measures about 11.5–12.00 mm in diameter with the horizontal diameter being greater than the vertical diameter by about 1.00 mm (Fig. 2.2).

2.2.1 Corneal Epithelium

The corneal epithelium is a 4–6 cell thick stratified squamous epithelium and varies from 40 to 50 μ in thickness. It is devoid of keratin. The outermost layer of the corneal epithelium is irregular, being characterized by apical microvilli. The microvillus surface is covered by a glycocalyx that binds the mucus secreted by the conjunctival goblet cells. This helps neutralize any irregularities of the anterior corneal epithelium by smoothening it out and prevents irregular refraction of light rays from entering the eye. It also prevents the drying up of the corneal epithelium. The anterior 2–3 layers of cells are polygonal in shape. These cells are joined to each other laterally by tight junctional complexes that do not allow any tears to penetrate through them. The middle 2–3 cell layers are composed of wing cells also known as suprabasal cells. They have extensive tight junctions along their lateral surfaces that hinder penetration of drugs into the substance of the cornea. The innermost layer of

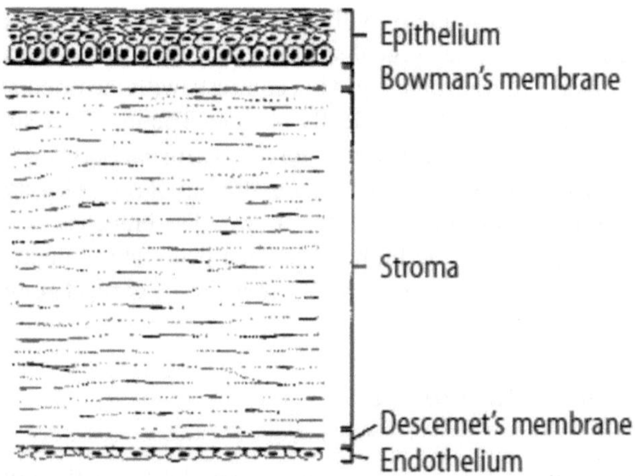

Fig. 2.2 Schematic section through human cornea [1]

the corneal epithelium is called the basal layer, and it is formed of columnar epithelial cells, which are mitotically active. The basal cells are concentrated in the limbal region in crypts, and as they divide and spread anteriorly, they transiently form amplifying cells, which also have mitotic abilities. There are no amplifying cells at the anterior, most central portion of the corneal epithelium. The basal cells also have extensive lateral and basal hemidesmosomal tight junctions. The basement membrane of the corneal epithelium is a thin layer of extracellular matrix and is attached to the underlying Bowman's membrane.

2.2.2 Bowman's Membrane

It is the next prominent structure in the cornea. It is considered to be the anteriormost region of condensed corneal stromal collagen fibers. For a very long time, the acellular Bowman's layer was thought to be the basement membrane for the corneal epithelium.

2.2.3 Corneal Stroma

The corneal stroma is formed from neural crest cells early in embryonic development. It is responsible for the mechanical strength of the cornea. It is about 500 μ in the center and about 700 μ in the periphery. It is mainly comprised of type I collagen fibers that are arranged as tight bundles by the heterodimeric complexes with type V collagen. There are multiple layers of these collagen fiber bundles. The collagen fibers in one layer are parallel to each other but set at an angle to those in the other layers.

This parallelism of collagen fibers within the same layer and angular orientation between different layers of the corneal stroma gives the cornea mechanical strength while maintaining optical transparency. This contrasts with the orientation of the collagen fibers in the sclera, which is optically opaque. The interfiber stromal matrix is comprised of proteoglycans like keratin sulfate, dermatan sulfate, and chondroitin sulfate that play a role in maintaining corneal hydration. Stromal keratocytes maintain the stromal milieu being concentrated in the stroma. The branches of the anterior ciliary nerve enter the stroma at the limbus and then penetrate through the Bowman's membrane to form a plexus of sensory nerve fibers in the corneal epithelium.

2.2.4 Descemet's Membrane

Descemet's membrane (DM) is the basement membrane of the corneal endothelium, and it is secreted by the corneal endothelium itself. It is a compact structure that is attached to the endothelial cells by strong numerous hemidesmosomes. The peripheral DM is a little thicker, and its junction with the trabecular meshwork marks an important anatomical and gonioscopic landmark called Schwalbe's line.

2.2.5 Corneal Endothelium

The corneal endothelium is a single cell layer. It is formed of cuboidal epithelial cells and is about 20 μ thick. The endothelial cells have extremely limited mitotic capacity, and progressive stresses and injuries lead to declining numbers of the endothelial cells. However, the remaining endothelial cells expand to cover the DM by a process called polymegathism. Endothelial cells are critical for maintaining the hydration level of the corneal stroma at optimal levels and thereby the transparency of the corneal stroma and cornea. They have extensive lateral interdigitations along with gap and tight junctions. The lateral surfaces have a high concentration of ATP-dependent Na^+ and K^+ channels that help in the movement of the water and ions from the corneal stroma toward the aqueous humor in the anterior chamber of the eye. The number of endothelial cells progressively decreases over the lifetime, and in critical cases, deturgescence function is compromised, resulting in corneal hydration that causes visual impairment. These cells are metabolically active and thus have numerous mitochondria and Golgi apparatus.

2.3 Conjunctiva

This is probably the most important structure of the eye from a drug delivery standpoint. It forms the exposed surface of the eye and is the site for administration of local drugs for treatment of ocular diseases on the surface (e.g., conjunctivitis) as well as inside the eye (e.g., glaucoma, iritis). The conjunctiva is a mucous membrane that covers the anterior surface of the eyeball. Anteriorly, the conjunctival epithelium is continuous with the corneal epithelium. Anatomically, the conjunctiva can be divided into the palpebral and the bulbar parts (Fig. 2.3).

2.3.1 Palpebral Conjunctiva

Palpebral conjunctiva is the conjunctiva that lines the posterior surfaces of the eyelids (palpebrum = eyelid). It can be further divided into three parts as marginal, tarsal, and orbital conjunctiva.

2.3.1.1 Marginal Conjunctiva

This is a 2–3 mm narrow strip of conjunctiva, as the mucous membrane becomes continuous with the keratinized epithelium of the skin. The conjunctival epithelium is a nonkeratinized, stratified, squamous epithelium. The superficial layer is comprised of squamous cells, the middle three layers are comprised of polyhedral cells, and the basal layer is made up of cylindrical epithelial cells.

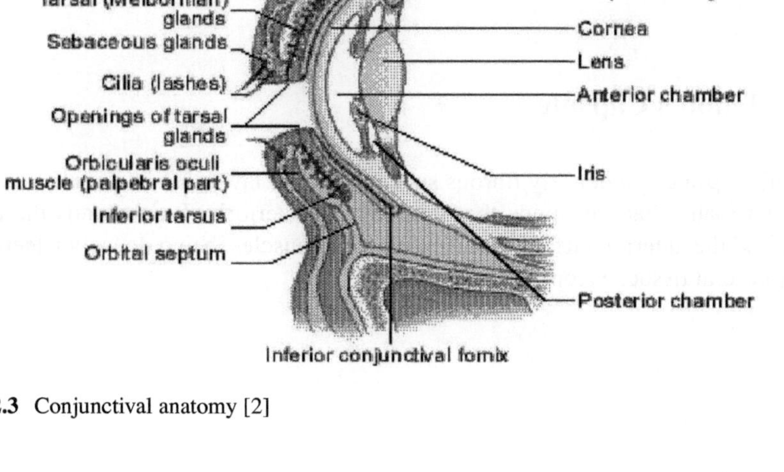

Fig. 2.3 Conjunctival anatomy [2]

2.3.1.2 Tarsal Conjunctiva

This part of the palpebral conjunctiva is only comprised of two layers of cuboidal epithelial cells and closely adheres to the underlying cartilage called the tarsal plate.

2.3.1.3 Orbital Conjunctiva

The remaining portion of the palpebral conjunctiva that is against the orbital wall is called the orbital conjunctiva.

2.3.2 Bulbar Conjunctiva

The part of conjunctiva that lines the surface of the eyeball or that which covers the sclera is called the bulbar conjunctiva. It is separated from the sclera by a fascial sheath called Tenon's capsule.

2.3.3 Forniceal Conjunctiva

The part of the conjunctiva that folds in the orbital recesses to reflect and come anteriorly again is called the forniceal conjunctiva. The conjunctival fornices are superior (deepest), inferior, medial, and lateral. The medial fornix is the shallowest and has the caruncle. A study found the average normal depths of the superonasal and temporal, inferonasal and temporal, and nasal and temporal canthal fornices to be about 11.5–16.6 mm, 11.5–16.6 mm, 7.9–12.1 mm, 8.6–12.2 mm, 2.1–3.9 mm, and 4.0–6.4 mm, respectively. The mean overall area of the conjunctival fornix in normal subjects was 737.4–1071.8 mm^2 [3].

2.4 Tenon's Capsule

Tenon's capsule is a densely fibrous subconjunctival layer that is a continuation of the dura mater that surrounds the optic nerve posteriorly. It surrounds the entire globe and the anterior surface of the extraocular muscles, being densely adherent to the episcleral tissue, except over the cornea.

2.5 Lacrimal Gland and Tears

2.5.1 The Lacrimal Gland Is a Serous Gland

It is a bilobed structure that is located in the upper and outer quadrant of the orbit. It can be seen upon eversion of the lateral third of the upper eyelid as a pinkish mass. It has a smaller palpebral lobe and a larger orbital lobe. It secretes tears dependent on stimulation from the parasympathetic nervous system. It is covered by the palpebral and the orbital conjunctiva. It opens on the ocular surface through multiple openings in the conjunctival fornix. Glands of Krause and glands of Wolfring comprise the accessory lacrimal gland system. They are located in the lamina propria of the conjunctiva in the fornix and along the edge of the tarsus, respectively. Tear secretion is peripheral sensory, central, and baseline. The cells of the lacrimal gland are pyramid-shaped apocrine cells with a concentration of secretory granules toward the apex. The gland consists of multiple lobules in a connective tissue framework. The lobules are drained by intralobular ducts. Multiple intralobular ducts join to form the interlobular ducts, which ultimately form multiple secretory ducts that open in the conjunctival fornix [4].

2.5.2 Tears

Tear fluid is the liquid lubricant that constantly bathes the ocular surface and ensures smooth movement of the eyelids over the surface of the eye and also washes away particulates and microorganisms from the ocular surface. The tear fluid nourishes the epithelial cells, forms a barrier between the epithelium and the external environment, and with help of lysozymes protects the ocular surface from bacterial cells [5, 6]. From a drug delivery perspective, this is important because we want the surface of the eye to be healthy and unencumbered with maximum possible surface available for drug delivery.

The tear fluid is composed of three layers. The outermost is the lipid layer. It is derived from the oil glands (meibomian glands) that line the margins of the eyelids as we discussed earlier. It covers the aqueous layer and retards its evaporation from the ocular surface. The second layer is the aqueous layer. As the name suggests, it is primarily water with some electrolytes and other proteins in it. This is secreted by the lacrimal glands. Its osmolality is critical in maintaining the function of ocular surface homeostasis. The change in the osmolality can affect the retention and delivery of drugs administered on the ocular surface. In a condition called "dry eye," the increased watering of the eyes can lead to lowered retention time on the ocular surface for the eye drops and increased drainage and loss through the tear drainage system. The third layer of the tear fluid is the mucinous layer that is comprised of the glycoproteins secreted by the conjunctival epithelial cells and small goblet cells on the conjunctival surface. It helps smoothen the ocular surface cellular surfaces and also acts as bed for the aqueous layer. It is considered to be the thickest component of the tear film complex [7].

Disturbances of any of the three components lead to instability of the tear film. Tear film instability can also be caused by inappropriate lid position and movement. Dry eye is characterized by redness, watering, and gritty foreign body sensation in the eye primarily due to the disturbance of the tear film stability (Fig. 2.4).

2.6 Eyelids

The eyelids are composed of skin, muscles, fibrous septae, and cartilage with glandular structures and hair that form eyelashes. The upper lid is larger than the lower lid and more mobile. The upper lid functions more to spread the tear film and any drug on the surface of the eye, and the lower lid forms support for the inferior fornix in which the drug delivered on the ocular surface can be retained. Problems with lid

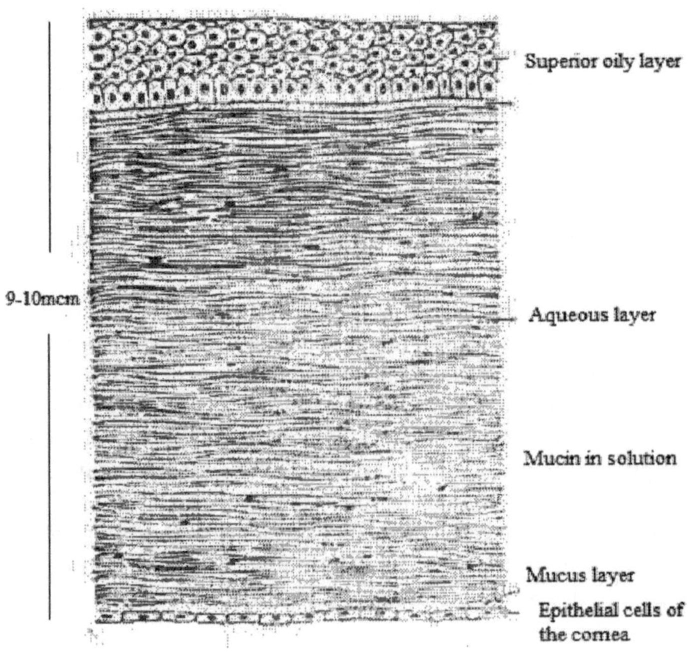

Fig. 2.4 Schematic of tear film structure [2]

function and structure affect the dispersion of the surface delivered drugs and their retention for a period of time optimal for drug action. Also, from a drug delivery perspective, muscles of the eyelids are the site of injection for botulinum toxin in oculoplastics and in treatment of certain conditions such as blepharospasm (Fig. 2.5).

The upper lid is formed by skin, subcutaneous tissue with slips of levator palpebrae superioris muscle tendon insertions into the skin, tarsal plate, and palpebral conjunctiva. Another small muscle that works as an elevator of the upper lid is called Muller's muscle. This is supplied by the sympathetic nervous system and causes lid retraction in conditions with increased sympathetic tone such as hyperthyroidism.

The lower eyelid is formed of skin, subcutaneous tissue, inferior tarsal plate, orbital septum, and palpebral conjunctiva. The tarsal plate in the lower lid is smaller than in the upper lid [8].

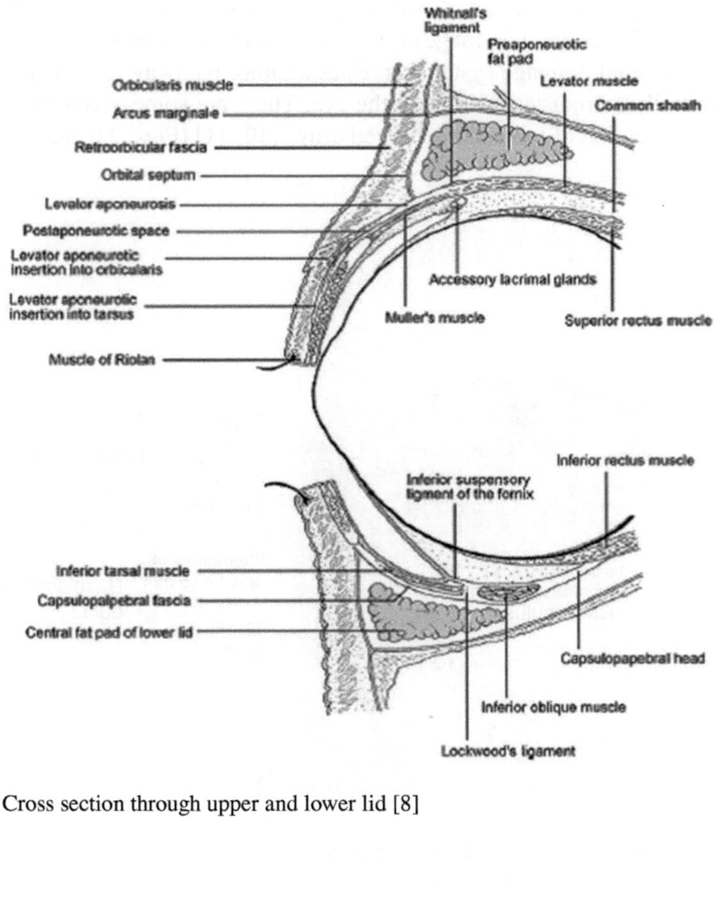

Fig. 2.5 Cross section through upper and lower lid [8]

2.7 Lacrimal Drainage System

The lacrimal drainage system is comprised of the lacrimal puncta, lacrimal canaliculi, lacrimal sac, and nasolacrimal duct. The epithelium lining the lacrimal system is non-ciliated columnar epithelium.

The lacrimal puncta are small openings located near the medial end of each lid about 0.8 mm in diameter. The lower punctum is a little more laterally placed than the upper one so they do not appose on lid closure. The punctum leads into the ampulla which is placed perpendicular to the lid margin. This portion of the lacrimal canaliculus is 2 mm long. It then bends medially and continues as the canalicular duct and is about 10 mm long and about 1 mm in diameter. In about 90 % of individuals, the upper and lower canaliculi meet and form a common canaliculus before joining the lacrimal sac. The lacrimal sac is situated in the groove formed by the lacrimal bone and the frontal process of the maxillary bone. It continues downward as the nasolacrimal duct. The nasolacrimal duct opens into the inferior turbinate. The sac measures about 3–5 mm above its junction with the common canaliculus (fundus of the sac) and about 10 mm until it joins the nasolacrimal duct. The nasolacrimal duct is about 12–15 mm in length and opens into the nasal cavity at the inferior turbinate [9]. Obstruction at any part along the lacrimal drainage system leads to epiphora (overflow of tears from the eyes). Lacrimal drainage system serves as a minor pathway for systemic absorption of topically administered drugs to the eye. There are some excellent reviews on the ocular absorption of topically delivered drugs [10, 11] (Figs. 2.6 and 2.7).

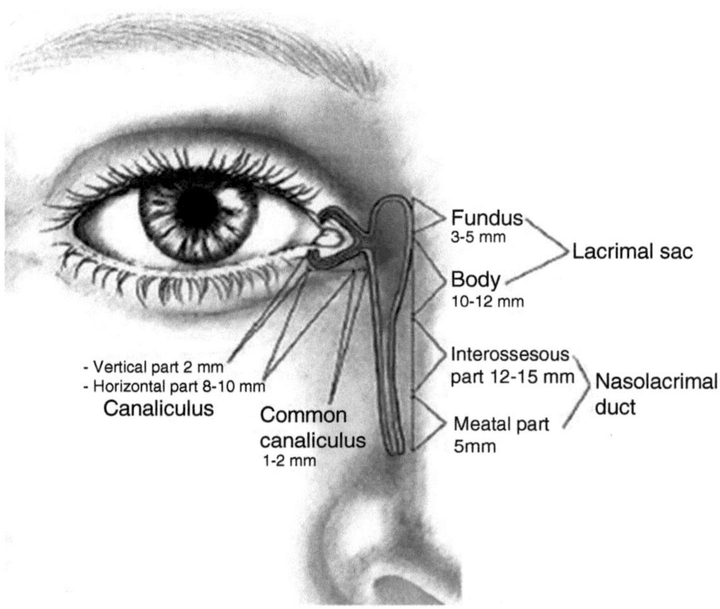

Fig. 2.6 Schematic of the lacrimal drainage system [8]

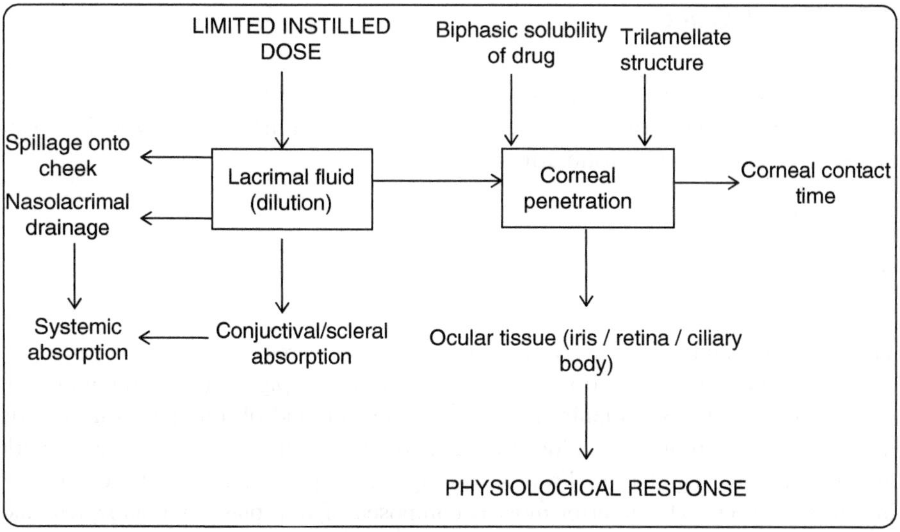

Fig. 2.7 Fate of ocular drug delivery systems [2]

2.8 Sclera

Sclera is the white of the eye. It is composed of fibrous tissue that is composed of collagen fibers. It is the same as the collagen that forms the cornea. However, there are significant differences. While the corneal collagen fibers are all arranged in uniform diameter bundle of about 25 nm each, the scleral collagen bundles vary in size from 25 to 250 nm. Also, the corneal collagen bundles are arranged in lamellae, and all are parallel to each other with inter-bundle communicating fibrils. In the sclera, the bundled collagen fibers are intertwined with each other and with those from other bundles and are randomly oriented. The sclera is quite avascular [12].

From a drug delivery perspective, certain aspects of the human sclera merit consideration. The scleral surface available for potential drug diffusion or transport is higher than that of the cornea. Also, once the drug is in the episcleral space, the drug has a potential to achieve higher intraocular concentrations especially since the scleral permeability is higher than corneal permeability [13]. Molecular characteristics such as weight, diameter, hydrophilicity, and charge determine the transscleral permeability. Additional biological considerations such as blood flow and lymphatic drainage influence the drug residence time in the episcleral site [14].

2.9 Uveal Tract

Uvea is the vascular layer of the eye. Its name is derived from Latin (uva = grape). It receives its blood supply from the branches of the ophthalmic artery which itself arises from the internal carotid artery.

2.9.1 Iris

The iris is the shutter diaphragm of the eye and measures 12 mm in diameter and 37.5 mm in circumference. It is made up of a bilayer of pigmented epithelium. The ratio of the amount of melanin pigments (eumelanin and pheomelanin) in the iris melanocytes determines the color of the eye. Blue irises have lowest, while the dark brown irises have a high ratio. From a histological perspective, the iris can be divided into three regions. The anterior most is composed of fibroblasts, melanocytes, and collagen with numerous undulations on the surface. The middle layer is similarly composed without any folds. Also, in the stroma are nerves, blood vessels, melanocytes, and clump cells. The blood vessels of the iris have a non-fenestrated endothelium and therefore are impermeable. There are two important muscles. The first is the sphincter pupillae which is a concentric muscle around the pupil. It is responsible for decreasing the size of the pupil to help focus on nearby objects. The other is the dilator pupillae, which is a radial muscle, and its function is to dilate the pupil, like at night. Both the muscles lie in the posterior stroma. The posterior layer is comprised of a double layer of pigment epithelium that has cells arranged apex to apex [15, 16].

The iris is a watertight structure, and therefore aqueous produced by the ciliary body behind the iris has to traverse through the pupil to drain via the trabecular meshwork located in the angle of anterior chamber.

2.9.2 Ciliary Body

The ciliary body is a pigmented band-like structure and extends from the ora serrata posteriorly to the scleral spur anteriorly for 3.5 mm to about 5.5 mm posteriorly from the limbus. It has two parts—the pars plicata and pars plana. As the name indicates, the pars plicata, which is the anterior portion, is seen as having multiple radial fingerlike folded processes, and it is the main secretory region of the ciliary body. There are about 70–75 processes in each eye. The pars plana is the posterior flat portion. Posteriorly, the pars plana shows the attachment of the retina at ora serrata. The ciliary body is mainly comprised of ciliary muscles, pigmented ciliary epithelium that is secretory in function, and stroma that has the blood vessels. These blood vessels have fenestrations, unlike the iris vessels, and so are permeable, which is important since the ciliary body secretes the aqueous humor. The ciliary muscles are arranged in three distinct bundles as circular, oblique, and longitudinal fibers from inside out. These are responsible for the change in the lens shape and thereby in focusing of images on the retina. The longitudinal fibers affect aqueous outflow by their effect on the trabecular outflow track. The innermost layer of the ciliary body is an epithelial bilayer that has cells arranged apex to apex. The outer layer is pigmented and the inner layer is nonpigmented. There are tight zonulae adherens at the apical surface of the nonpigmented layer, and they comprise the blood aqueous barrier. The nonpigmented epithelium of the ciliary body is responsible for the secretion of aqueous and the vitreous mucopolysaccharides. This epithelium continues posteriorly as the neuro-retina bearing a common embryologic origin, and the pigmented layer continues as the retinal pigment epithelium(RPE) [15, 16].

2.9.3 Choroid

The choroid is the vascular layer of the eye. It receives blood supply from the short posterior ciliary arteries and the recurrent branches of the anterior ciliary arteries. The choroid extends from the optic nerve head posterior until the ora serrata anteriorly. The choroid is a thin structure that is about 0.2 mm posteriorly and thins to about 0.1 mm anteriorly toward the ora serrata. The choroid is held against the sclera by the intraocular pressure so a potential space exists between the choroid and the sclera. The points of anchorage of the choroid to the sclera are at the points of exit of the vortex vein which drain the choroid. The choroid is divided into four layers. The outermost layer is Haller's layer, followed by Sattler's layer, then the choriocapillaris, and finally the Bruch's membrane.

2.9.3.1 Haller's Layer

This is the outermost layer of the choroid. It is composed of large-diameter vessels. The arteries have a tunica adventitia, muscular layer, and inner elastic lamina. These are not fenestrated and are impermeable. Among the blood vessel are melanocytes, nevus cells, and nerves.

2.9.3.2 Sattler's Layer

This is the second outer layer of the choroid. It is comprised of mainly the medium-sized blood vessels along with fibroblasts, mast cells, melanocytes, lymphocytes, and collagen fibers. The blood vessels, as they progress deeper, gradually decrease in their diameter, losing the muscular and elastic layers forming arterioles. The vessels are un-fenestrated and impermeable as in Haller's layer.

2.9.3.3 Choriocapillaris

The next layer of the choroid is choriocapillaris. This is the layer of capillaries in the choroid. The vessels are smaller about 25–50 μ in diameter. Unlike the blood vessels in the outer two layers of the choroid, the vessels of the choriocapillaris are fenestrated and help nourish the RPE and the outer layers of the retina (photoreceptor layer, outer plexiform layer, and the outer aspects of the inner nuclear layer). The endothelial cells are joined together by gap junctions. The organization of the choriocapillaris is thought to be in polygonal lobules that range in diameter from 200 to 250 μ with a central afferent arteriole and multiple efferent venules.

2.9.3.4 Bruch's Membrane

Bruch's membrane is the acellular layer that functions as the basement membrane of the choriocapillaris on the outer side and the RPE on the inner side. It is composed of five layers. The outermost is the basement membrane of the choriocapillaris, followed by the outer collagenous layer. The middle layer is the elastic layer. The fourth layer is the inner collagenous layer, and the innermost layer is the basement membrane of the RPE. All layers of the Bruch's membrane are normally permeable to fluorescein.

2.9.3.4.1 Basement Membrane of Choriocapillaris

It is the thinnest layer of the Bruch's membrane and is only about 0.14 μ in thickness being composed of fine collagen fibers.

2.9.3.4.2 Outer Collagen Layer

This is the second outer layer of the Bruch's membrane. It is about 0.7 μ thick and is composed of loose collagen that is randomly arranged. Some of these fibers penetrate the elastic layer and join the inner collagen layer.

2.9.3.4.3 Elastic Layer

This is the middle layer of the Bruch's membrane. It is about 0.8 μ in thickness and has openings through which the collagen layers on its either side exchange collagen fibers.

2.9.3.4.4 Inner Collagen Layer

This is the fourth layer of the Bruch's membrane and is 1.0 μ in thickness and is similar in composition to the outer collagen layer.

2.9.3.4.5 Basement Membrane of the Retinal Pigment Epithelium

This is the innermost layer of the Bruch's membrane. Being 0.3 μ thick, it extends margins of the optic disk to the margin of the pupil [15, 16].

2.10 Angle of Anterior Chamber

The angle of anterior chamber is the most important part of the ocular anatomy in terms of the outflow of aqueous and glaucoma. Anteroposteriorly it is comprised of the following structures. Anteriormost is the Schwalbe's line followed in order by trabecular meshwork, scleral spur, and ciliary body band. The point of attachment or the iris to the ciliary body marks the end of the angle posteriorly in the widest possible configuration.

2.10.1 Schwalbe's Line

Schwalbe's line is the peripheral thickening of the Descemet's membrane where it joins the Bowman's membrane and is an important landmark to assess the openness of the angle by gonioscopy (Fig. 2.8).

2.10.2 Trabecular Meshwork

Trabecular meshwork is divided into anterior and posterior trabecular meshwork and extends from the peripheral cornea to the scleral spur. It is divided into uveal, corneoscleral, and juxtacanalicular meshwork as we go deeper in the tissue toward the Schlemm's canal. The uveal meshwork is the deepest part of the trabecular meshwork. It is this portion of the trabecular meshwork that the aqueous traverses on its path of egress from the eye in the conventional pathway. It is composed of thick trabeculae that extend from the iris root and ciliary body to the peripheral cornea. The pore size is largest in this portion and measures 25–75 μ in size. The corneoscleral meshwork extends from the Schwalbe's line anteriorly to the anterior edge of the scleral sulcus. The trabecular arrangement becomes progressively more compact toward the Schlemm's canal, resulting in the pore size that ranges from 5 to 50 μ. The trabeculae themselves are made up of a tetra-layer of collagen core surrounded by elastin layer. A clear layer surrounds the elastic layer. This clear layer is covered by trabecular endothelial cells with their basement membrane. The trabecular endothelial cells are irregular and larger than corneal endothelial cells. Their borders are less distinct and have gap junctions and desmosomes. The innermost portion of the trabecular meshwork is called the juxtacanalicular meshwork.

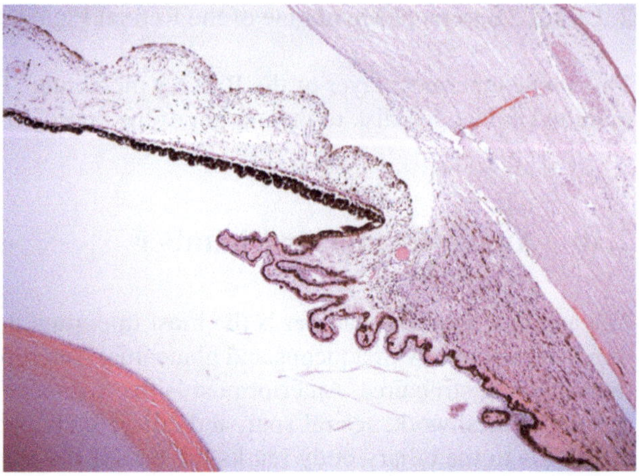

Fig. 2.8 Microanatomy of angle of anterior chamber showing iris root, trabecular meshwork, and ciliary body microanatomy [17]

The collagen trabeculae are more compactly arranged. The juxtacanalicular trabecular meshwork opens into the Schlemm's canal through pores that range in size from 0.5 to 2 μ. Schlemm's canal is an annular drainage channel that drains the aqueous filtered through the trabecular meshwork. It is lined with endothelial cells that have pores and giant vacuoles that facilitate aqueous drainage.

2.10.3 Scleral Spur

Scleral spur is the site where the cornea and sclera join to form a continuous coat. It forms the site of attachment of the ciliary muscles. As discussed earlier, the ciliary muscle regulates aqueous outflow through their action on the scleral spur.

2.10.4 Ciliary Body Band

Ciliary body band is the anteriormost part of the ciliary body that can be seen in some individuals. When the ciliary body band is seen, the angle configuration is said to be open wide.

Depending on the structures seen on visualization of the angle by gonioscopy (gonio = angle), the angle is classified into various grades (I–IV) with grade IV being an open angle and grade I being a narrow angle. When the peripheral iris covers the trabecular meshwork and no structures are seen in the angle, then it is termed as closed angle. Closed angle gives rise to angle closure glaucoma where the intraocular pressure rises due to impedance to aqueous drainage.

2.11 Aqueous Humor Formation, Circulation, and Drainage

Aqueous humor or aqueous is the secretion of the ciliary epithelium of the ciliary body. It is considered to be protein-free plasma. Its pH is 7.2, and it has a very high level of ascorbate in it which is about 15 times the level in the plasma. Also, it has very little protein (0.02 %) compared to plasma (15 %). It is formed in the posterior chamber and bathes the crystalline lens to nourish it. It then flows anteriorly through the pupil to drain peripherally through the trabecular meshwork in the angle of the anterior chamber.

2.11.1 Formation of Aqueous

The formation of aqueous has the following steps.

2.11.1.1 Uptake of Fluid from Stroma

This occurs as transport of Na^+ and Cl^- ions from the ciliary body stroma across the ciliary epithelial bilayer. As discussed earlier in the chapter, the ciliary epithelial bilayer is comprised of the outer pigmented epithelium (PE) and the inner nonpigmented epithelium (NPE). The cells of these two are arranged apex to apex. Gap junctions connect cells of PE with those of NPE. Gaps junctions also connect cells of the PE and NPE with surrounding cells in the same layer. However, the number of gap junctions connecting a PE cell to an NPE is higher than those that connect a PE cell to another PE cell or an NPE cell to another NPE cell. PE cells take up Na^+ and Cl^- ions by two mechanisms. The first is by utilizing a Na^+–H^+ exchange protein to take up Na^+ ions and give out H^+ ions and by exchanging HCO_3^- ions for Cl^- ions. This is facilitated by the carbonic anhydrase enzyme that generates the H^+ and HCO_3^- ions necessary for the exchange. The second method is by using a Na^+–K^+–$2Cl^-$ cotransporter. This is dependent on the concentration gradient of the three electrolytes and particularly the ratio of extracellular concentration of Cl^- to the intracellular concentration.

2.11.1.2 Fluid Transfer Through Gap Junctions

This likely occurs through the PE–NPE gap junctions formed by connexins as water follows the osmotic gradient.

2.11.1.3 Fluid Transfer into the Aqueous Humor

This occurs as a two-step process. The first is by an ATP-dependent active transport of Na^+ ions into the posterior chamber. The Cl^- ions follow an electrochemical gradient. The water then follows the osmotic gradient and aqueous humor is formed. Some of the Cl^- ions are exchanged for HCO_3^- ions using an exchange transporter.

2.11.2 Circadian Rhythm of Aqueous Flow

The flow during the evening is about half of during the morning. The exact mechanism of how the circadian rhythm is regulated is not completely understood.

2.11.3 Circulation of Aqueous Humor

The aqueous humor after formation bathes the lens providing nourishment and then flows into the anterior chamber through the pupil. In the anterior chamber, it nourishes the other avascular structure, i.e., the cornea. It then drains through the trabecular meshwork into the Schlemm's canal and out through the aqueous veins of Ascher into the episcleral and conjunctival veins to the systemic circulation.

2.11.4 Drainage of Aqueous Humor

There are two different mechanisms of aqueous egress from the eye.

2.11.4.1 Conventional Pathway

The first is called the conventional pathway and is the major pathway for aqueous drainage and is responsible for 70–95 % of aqueous outflow. It is through the trabecular meshwork. The trabecular meshwork is a sievelike filter. The columns and beams are made up of a framework of collagen and elastin fibers with other connective tissue and are lined by endothelial cells. It is divided into three zones as one goes from the anterior chamber toward the Schlemm's canal. The part of trabecular meshwork toward the anterior chamber has larger-sized pores, the middle zone has medium-sized pores, and the portion adjacent to the Schlemm's canal has the smallest-diameter pores. Schlemm's canal is a continuous channel that goes all around the limbus. It collects the aqueous as it is filtered through the trabecular meshwork. The aqueous humor egresses the eye through the veins of Ascher that drain the Schlemm's canal. The veins of Ascher drain into the episcleral veins.

2.11.4.2 Uveoscleral Pathway

The minor pathway is the uveoscleral pathway and is responsible for 5–30 % of the aqueous outflow. In this pathway, the aqueous flows out of the eye in the potential space between the choroid and the sclera by passing through the root of the iris between the ciliary muscle bundles and thence to the supraciliary and suprachoroidal space [18].

2.12 The Lens

The lens is a biconvex transparent and avascular structure in the normal eye. It is divided into the lens capsule, the lens cortex, and lens nucleus. The lens nucleus is the central compact aggregation of oldest lens fibers. The lens fibers are lens epithelial cells which arise from the anterior capsule. The lens capsule is actually the basement membrane of the lens secreted by the epithelial cells anteriorly and the superficial cells posteriorly. However, the epithelial cells continue to accumulate within the capsule of the lens over their lifetime and get compacted over time and sclerosed. The periphery of the lens is called the lens equator. During embryological development, the lens is supplied by the hyaloid artery which atrophies and disappears. The lens is then nourished by the aqueous humor. The most important function of the lens is to focus the light rays entering the eye through the pupil to form a sharp image on the retina, as this lens needs to be clear and transparent. In disease conditions and with age, the lens can lose its transparency due to opacity of the lens fibers or the capsule. This is called a cataract. The lens hydration is maintained by active and passive transportation of substance across the lens capsule, and water follows passively in or out of the cell.

The anterior lens capsule is the site of the mitotically active lens epithelial cells. It is also the site where ATP-dependent active transport of K^+ into the lens and Na^+ out of the lens occurs on this surface. Also, amino acids are actively transported inward on this surface. Water and other solutes passively follow. On the posterior surface, there is only passive exchange of water and solutes between the lens and vitreous humor [19]. In diabetes mellitus, glucose enters the cell and is diverted to the sorbitol pathway to form sorbitol. The lens capsule is impermeable to sorbitol, and, thus, the oncotic pressure in the lens increases, leading to hydration and formation of cataract.

The lens is suspended in the visual axis by a system of fibers called the zonules of Zinn or simply as lens zonules. The zonules are filamentous structures about 60 μ in diameter and composed of non-collagenous acidic glycoproteins. They are arranged in three bundles: anterior, equatorial, and posterior. The anterior and posterior zonules are longer than the equatorial [20]. Currently we do not have any drugs that are specifically directed at treating lens or zonular pathologies.

2.13 Vitreous Humor

The vitreous humor or vitreous is a clear gel that fills the globe of the eye. It is formed of mainly type II collagen and hyaluronan. Other components of vitreous include collagens (type IX, type V/XI, type VI), glycosaminoglycans (chondroitin sulfate and heparin sulfate), miscellaneous structural proteins (fibrillins, opticin, and VIT1), and miscellaneous molecules (amino acids and proteins, glycoproteins, lipids, and ascorbic acid). Vitreous can be described as framework or scaffold of collagen that is filled with diffuse arrays of hyaluronan. The vitreous is secreted by the pars plana of the ciliary epithelium, hyalocytes of vitreous, and the Muller cells of the retina.

The vitreous is in multiple layers like the peels of an onion. Anatomically it is divided into three parts. The first is the vitreous base. This is a three-dimensional structure that is about 2.5–5 mm anteriorly where the vitreous is attached to the ora serrata and pars plana of the ciliary body, and, anteriorly, the vitreous fibers delve into the body of the vitreous. This adhesion of vitreous base to the retina is quite strong. The remaining posterior part of the vitreous that lies adjacent to the retina and optic disk and behind the lens anteriorly is called the cortical vitreous. This has strong attachments to the retina at the margins of optic disk, fovea, and the retinal blood vessels. Hyalocytes secrete hyaluronan and are located in the cortical vitreous about 20–50 μ from the internal limiting membrane of the retina and the basement membrane of the nonpigmented epithelium of the pars plana of ciliary body. Fibroblasts, which comprise less than 10 % of the cells in the vitreous, are concentrated anteriorly near the ciliary body in the cortical vitreous. The central part of the vitreous is called the corpus vitreous [21]. There are potential spaces between the vitreous and the retina, between vitreous and the lens, and between the layers of the vitreous. As a result, hemorrhage from the retina can be seen isolated and contained in the sub-hyaloid epiretinal space. Occasionally a hemorrhage can be seen between the cortical vitreous and body of vitreous. In disease conditions, the cortical vitreous can act as a scaffold for the growth and proliferation of cells, as in proliferative diabetic retinopathy. Drug delivery to the retina and/or vitreous is mainly through intravitreal injections such as in endophthalmitis or age-related macular degeneration. This is because the systemically administered drugs do not achieve significant levels in the vitreous. Also, intravitreal injections and drug depot implants ensure persistent drug concentrations in the vitreous.

2.14 Retinal Pigment Epithelium

Retinal pigment epithelium (RPE) is the pigmented layer that separates the retina from the choroid. It rests on Bruch's membrane. There is a growing realization over the last few years of the immense role played by the RPE in the pathogenesis of retinal diseases, e.g., age-related macular degeneration (AMD). The RPE, although single layered, functions like an organ by itself. The apices of the RPE are all oriented to the center of the eye and their bases oriented centrifugally toward the choroid. The RPE cells are cuboidal epithelial cells arranged in a horizontal manner. They have tight junctions with the each other and form the blood–retina barrier.

The apical membranes of RPE show microvilli, which are in intimate apposition with the bodies of rod and cone photoreceptors. However, there are no junctional complexes between the photoreceptor cells and the RPE. The glue that holds the photoreceptors and the RPE together is called the inter-photoreceptor matrix. It is composed of proteins and proteoglycans containing an admixture of enzymes, growth factors, carrier protein, and other substances. Also, there are Na^+–K^+ ATPases along with the Na^+–K^+–$2Cl^-$ cotransporter which help maintain the extracellular milieu around the photoreceptors and the electrolyte gradient inward for maintenance of the fluid flux. A potential space exists between the retina and the RPE that is significantly manifest in retinal detachment. Also, on the apical surface are aquaporin channels for transit of fluid to the RPE from the sub-retinal space that help to preserve the RPE–retina attachment.

Blood–retina barrier is formed at the lateral membrane surface of the RPE cells. The lateral membrane is flat unlike the apical and the basal membrane. It bears tight junctions (zonula occludens) composed of occludins, adherens junctions (zonula adherens), and desmosomal attachments made of transmembrane proteins belonging to the cadherin family. The adjacent RPE cells are also connected by gap junctions composed of connexin proteins. They allow electric coupling and transfer of Ca^{2+}, H^+, cyclic adenosine monophosphate (cAMP), and other solutes.

The basal membrane of the RPE cells shows folds greatly increasing the surface area as is required for absorptive and secretory activity. Here numerous protein channels responsible for the transport and uptake of nutrients across the Bruch's membrane from the choriocapillaris are located. The basal surface also has focal adhesions to the Bruch's membrane formed by integrins.

The RPE functions in supporting vision in numerous ways. The melanin pigment absorbs the light incident on it through the retina, preventing backscatter and enhancing the quality of the stimulus. Also, the melanin serves as a protective factor against oxidative stress and in inactivation and clearance of toxic metabolites. RPE participates in the visual cycle by converting the all-*trans* retinol, which is generated after photic stimulus and bound to the inter-photoreceptor binding protein (IRBP), into 11-*cis* retinol and binding it to the IRBP for transport to and use by the retinal photoreceptors. RPE also functions in clearing the debris generated by shedding of outer photoreceptor membrane disks. These are phagocytized by the RPE and digested in lysosomes [22].

In view of the important function of the RPE for retinal function, newer therapeutic modalities aim at transplantation of genetically engineered RPE in patients, but the best modality would be to genetically engineer the RPE in situ in vivo and avoid the complications and risks associated with the procedure of RPE culture and transplantation.

2.15 Retina

The retina is the neurosensory portion of the eye. It is directly connected to the brain through the optic nerve. Retina is a very thin flimsy structure composed neurons, nerve fibers, and glial cells. It extends from the optic disk margin to the ora serrata. The total surface area of the retina is estimated to be approximately 1200 mm^2.

The retinal blood supply is by the retinal artery, which is derived from the ophthalmic artery. Upon entering the eye at the optic disk, the retinal artery divides into superior and inferior trunks, and then each of those divides into superior and inferior arteries forming arcades. The veins have a similar configuration and the medial and temporal veins converge at the optic disk to form the central retinal vein. Branches and tributaries of the arterial and venous arcades, respectively, supply and drain the retina. The nourishment of the retina is from the retinal blood vessels for the inner two thirds of its thickness and through the choriocapillaris for the outer third.

From a clinical standpoint, the retina can be divided into central retina, equatorial retina, and peripheral or anterior retina. The central retina has two important landmarks. These are the optic disk and the macula. The optic disk is also called the optic nerve head, and it is the point of exit of the optic nerve from the eye. The macula is the central portion of the eye that is responsible for acuity of vision and color vision. The macula is generally thought to be the area between the two temporal vascular arcades. At the center of the macula slightly below the horizontal is an area called fovea. It has more than one layer of ganglion cells. At the center of the fovea is a small area called the foveola. This is the point on the retina where there are only cone photoreceptors. The central 0.27 mm^2 area of the fovea is devoid of any retinal capillaries and is known as the foveal avascular zone [23]. Equatorial retina is about 15 mm behind the limbus and about 3 mm behind the anatomical equator of the eye. Anteriorly, the retina terminates at the ora serrata. The vitreous base is attached to the retina at the ora serrata.

The retina is a highly complex structure with multiple layers that consist of vascular, neural, and glial cells and nerve fibers.

The layers of the retina from the vitreous toward the RPE are:

- Internal limiting membrane (ILM)
- Nerve fiber layer (NFL)
- Ganglion cell layer (GCL)
- Inner plexiform layer (IPL)
- Inner nuclear layer (INL)
- Outer plexiform layer (OPL)
- Outer nuclear layer (ONL)
- The inner and outer segments of photoreceptors

The ILM is formed by the basement membrane of the Muller cells, which are the major supporting glial cells in the retina. The ILM is thinnest at the fovea. The vitreous is attached to the ILM by insertion of the collagen fibers and, thus, is said to be prone to traction causing cystic maculopathy. The drug ocriplasmin, which is a recombinant protease, is directed toward the laminin and fibronectin and disrupts the attachment of the vitreous at the fovea to the ILM and relieves traction stress on the macula. Under the ILM is the nerve fiber layer (NFL). This is comprised of the axons of the ganglion cells of the retina. The NFL fibers all come together at the optic disk to exit the eye as the optic nerve. The ganglion cell layer (GCL) is composed of ganglion cells. They are one of the main neural cells of the retina along with photoreceptors, bipolar cells, amacrine cells, and horizontal cells. The next layer is the inner plexiform layer. This is the layer where the dendrites of the ganglion cells synapse with the axons of cells in the inner nuclear layer. The inner nuclear layer is comprised of bipolar cells. The next layer is the outer plexiform layer, which is essential for the synapses of the bipolar cells with the photoreceptor cells. The outer nuclear layer is basically the layer that has the cell bodies of the photoreceptor cells. The outermost layer of the retina is the IS/OS layer, which has the inner and outer segments of the photoreceptors. These are in very close apposition to the Muller cells and the microvilli of the RPE.

The signal transduction occurs in the photoreceptor layers where the light stimulus generates an electrical signal. This is communicated as an electrical signal to bipolar cells. Horizontal cells help in gathering and compiling the signals from adjacent bipolar cells to form a composite signal. The bipolar cells further transmit the signal to the ganglion cells. Amacrine cells further modulate the signal in the retina prior to the transmission of the electrical signal to the brain by the nerve fibers of the ganglion cells in the optic nerve to the brain. The signal conveys information of size, form, color, contrast, depth, and field of vision. In the brain, the nerve fibers synapse in the lateral geniculate body and are transmitted to the visual cortex in the optic radiation (Fig. 2.9).

In the retina, the pigment is of two types and is mainly found in the macular region. The two types of pigments are zeaxanthin and lutein, and they are structural isomers of dihydroxyxanthophyll. Both of these are potent antioxidants with zeaxanthin being localized at the fovea, while lutein is more diffusely spread over the macular region [25].

The drug delivery to the retina is by the use of systemic drugs intravenously (antibiotics), orally (steroids and antibiotics, vitamins), or locally by injections (antibiotics, steroids) or eye drops (NSAID like flurbiprofen and ketorolac tromethamine).

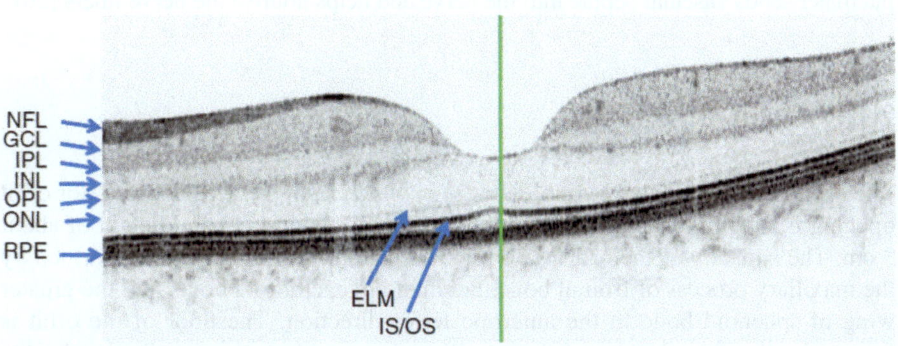

Fig. 2.9 Structural details of retina in a section taken through the macula as seen in ultrahigh-resolution optical coherence tomography (OCT) with 7 μ axial resolution in tissue and 14 μ transverse resolution in tissue. *NFL* nerve fiber layer, *GCL* ganglion cell layer, *IPL* inner plexiform layer, *INL* inner nuclear layer, *OPL* outer plexiform layer, *ONL* outer nuclear layer, *ELM* external limiting membrane, *IS/OS* inner/outer segment junction, *RPE* retinal pigment epithelium [24]

2.16 Optic Nerve

The optic nerve carries the visual signal from the retina to the brain. It is approximately 7 cm in length. It can be divided anatomically into four portions as intraocular, orbital, intracanalicular, and intracranial.

The intraocular portion is also called the optic nerve head. It is about 1.5 mm in diameter and about 1–1.5 mm long and exits the eye through the sclera. The sclera forms a meshwork with round and oval openings called the lamina cribrosa. It is made up of compact collagenous sheets of scleral trabeculae alternating with glial sheets along with elastic tissue. Astrocytes keep the nerve fibers separated from connective tissue. The blood supply of the optic nerve head is from the short posterior ciliary arteries, central retinal artery, and the pial branches of the ophthalmic artery. Compromise of the blood supply to the optic nerve head results in visually disabling conditions called ischemic optic neuropathy, e.g., in giant cell arteritis. There are no meningeal coverings for the intraocular portion of the optic nerve unlike in the other parts.

The intra-orbital portion is about 3–4 cm in length and is S shaped, which allows significant excursion along with ocular movements. The central retinal artery enters the optic nerve in this region about 8–15 mm behind the eyeball. From the intra-orbital portion until the optic chiasm, the optic nerve is now surrounded by the meningeal sheaths.

The intracanalicular portion is about 5–8 mm long. The outermost meningeal sheath, i.e., the dura mater, is fused to the periosteum and anchors the optic nerve in the canal. This portion receives its blood supply from branches of the internal carotid artery, anterior communicating artery, and ophthalmic arteries.

The intracranial portion of the optic nerve is about 1 cm long. It meets its counterpart from the other side to form the optic chiasm.

The extraocular portion of the optic nerve is comprised of myelinated nerve fibers with a supporting framework of astrocytes, microglial cells, and oligodendrocytes. The pia mater sends vascular septae into the nerve and helps nourish the nerve fibers [26].

2.17 Orbit

The orbit is the four-sided pyramidal cage in which rests the eye. The anterior orbit opening is roughly quadrangular with the width of about 4 cm and height of about 3 cm. The capacity of the orbit is 30 ml. The medial wall of the orbit is formed by the maxillary process of frontal bone, lacrimal bone, ethmoid bone, and the greater wing of sphenoid bone in the anteroposterior direction. The floor of the orbit is formed by the orbital plate of the maxillary bone with some contribution by the palatine and the zygomatic bone. The lateral wall is incomplete and formed by the zygomatic process of the frontal bone and the frontal process of the zygomatic bone anteriorly and the greater wing of sphenoid bone posteriorly. It is also the strongest wall of the orbit. The superior wall of the orbit is formed by the frontal bone anteriorly and the lesser wing of sphenoid bone posteriorly (Figs. 2.10 and 2.11).

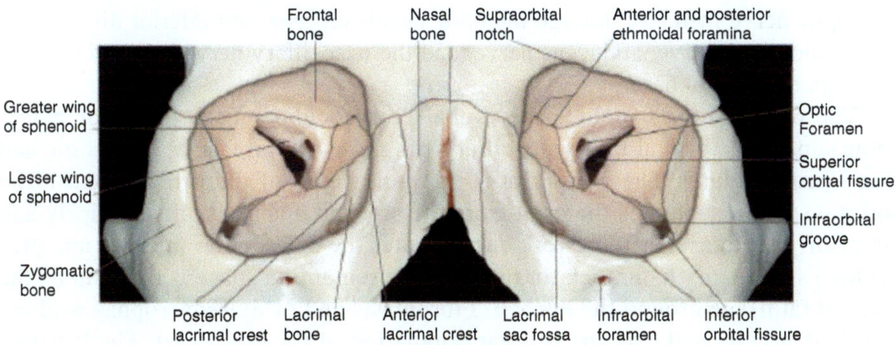

Fig. 2.10 Anatomy of bony orbit [8]

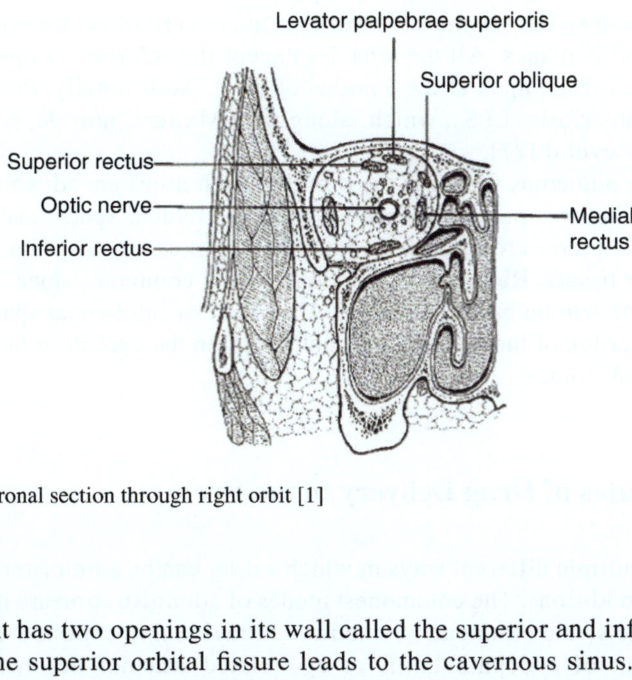

Fig. 2.11 Coronal section through right orbit [1]

The orbit has two openings in its wall called the superior and inferior orbital fissures. The superior orbital fissure leads to the cavernous sinus. The orbital and ophthalmic veins drain into the cavernous sinus through the superior orbital fissure. Also, other blood vessels traverse the opening like meningeal artery anastomosis with the ophthalmic artery, and also the lacrimal, frontal, and trochlear nerve pass through it. The inferior orbital fissure transmits the second division of the trigeminal nerve, the zygomatic nerve, and small branches of the sphenopalatine ganglion.

The apex of the orbit has the annulus of Zinn, which is a fibrous ring that straddles both the superior and inferior orbital fissures posteriorly. The apex transmits the optic nerve medially through the optic canal, superior and inferior divisions of the oculomotor nerve, abducens nerve, and the nasociliary nerve laterally through the superior orbital fissure.

The orbit contents include the eyeball, extraocular muscles, levator palpebrae superioris, orbital fat tissue, ciliary ganglion, orbital fascia, periorbita, and the lacrimal gland. The periorbita is the periosteum of the orbital bones. The orbital septum separates the orbit from the preorbital space both superiorly and inferiorly. The fascial framework encapsulates the extraocular muscles and provides a suspensory hammock for the eye. The remaining space is filled up by the orbital fat tissue around the eyeball. Fibroblasts and orbital macrophages along with other cells make up the cellular population of the orbital fat. The lacrimal gland is situated in the lacrimal fossa of the orbit in the supero-temporal quadrant.

There are six extraocular muscles. They include four recti and two oblique muscles. The recti muscles are superior rectus, medial rectus, inferior rectus, and lateral rectus. The two oblique muscles are the superior and inferior oblique muscles. The six muscles working in concert allow the eye to move in the X, Y, and Z planes. All the muscles except the inferior oblique have their origin at the orbital apex at the annulus of Zinn. Additionally, there is levator palpebrae superioris (LPS), which, along with Muller's muscle, is an elevator of the upper eyelid [27].

There are numerous spaces in the orbit in which drugs are administered. They are subconjunctival space, sub-Tenon's space, peribulbar space, and retrobulbar space. The first three are employed commonly for injecting drugs for better delivery to ocular tissues. Retrobulbar injections are not commonly done. They have a higher risk of retrobulbar hemorrhage. Also, there is intra-conal space, which is from the insertion of the recti muscles anteriorly on the eyeball to their origins at the annulus of Zinn.

2.18 Routes of Drug Delivery to the Eye

There are multiple different ways in which a drug can be administered for treating ocular conditions. The commonest modes of administration are by the use of eye drops, gels, and ointments. Sometimes intravitreal injections are used, as in the treatment of endophthalmitis or age-related macular degeneration. Oral treatment using pills and capsules is also used along with systemic intravenous therapy. Anesthetic drugs can be administered to the eye either topically or by local injection in skin, subconjunctival space, peribulbar space, or retrobulbar space (Fig. 2.12).

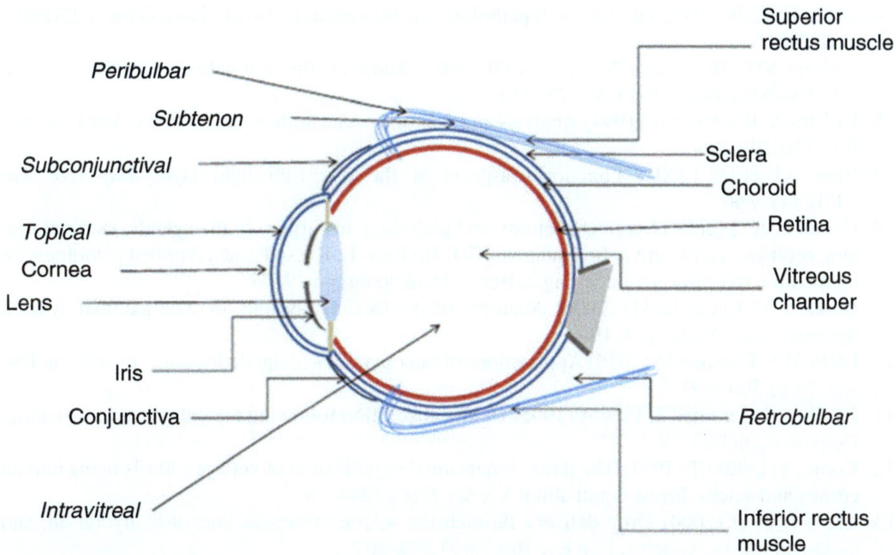

Fig. 2.12 Schematic showing route of drug delivery to the eye [14]

2.19 Conclusion

The eye is a fascinating sensory organ with quite some interesting qualities. It is an energy transducer and is immune privileged by the blood–retina barrier. It has the only optically transparent tissues of the human body visible externally and, also, the only place in the human body where systemic disease processes can be seen as they evolve like in diabetes mellitus and hypertension. Yet, drug delivery to ocular tissues presents challenges due to the inherent necessity to have tight junctions, circulation of tear fluid, windshield wiper effects of the eyelids, avascularity, and compartmentalization of structures. The existing modes of drug delivery are not optimal given the challenges inherent in the ocular structure and physiology, and newer modes of drug delivery that achieve significant levels of drug molecules in target tissues of the eye such as nano-biomaterials are desirable.

References

1. Galloway NR, Amoaku WMK, Galloway PH, Browning AC (2006) Basic anatomy and physiology of the eye. In: Common eye diseases and their management. Springer, London, pp 7–15
2. Rathore K, Nema R (2009) An insight into ophthalmic drug delivery system. Int J Pharm Sci Drug Res 1(1):1–5
3. Kawakita T et al (2008) Measurement of fornix depth and area: a novel method of determining the severity of fornix shortening. Eye 23(5):1115–1119

4. Obata H (2006) Anatomy and histopathology of the human lacrimal gland. Cornea 25(Supp. 1):S82–S89
5. Fleiszig SM, McNamara NA, Evans DJ (2002) The tear film and defense against infection. Adv Exp Med Biol 506(Pt A):523–530
6. Kijlstra A, Kuizenga A (1994) Analysis and function of the human tear proteins. Adv Exp Med Biol 350:299–308
7. Bron AJ et al (2004) Functional aspects of the tear film lipid layer. Exp Eye Res 78(3):347–360
8. Heindl L et al (2008) Surgical anatomy and pathology in surgery of the eyelids, lacrimal system, orbit and conjunctiva. In: Naumann GH, Holbach L, Kruse F (eds) Applied pathology for ophthalmic microsurgeons. Springer, Berlin/Heidelberg, pp 29–75
9. Burkat CN, Lucarelli MJ (2006) Anatomy of the lacrimal system. In: The lacrimal system. Springer, New York, pp 3–19
10. Diebold Y, Calonge M (2010) Applications of nanoparticles in ophthalmology. Prog Retin Eye Res 29(6):596–609
11. Järvinen K, Järvinen T, Urtti A (1995) Ocular absorption following topical delivery. Adv Drug Deliv Rev 16(1):3–19
12. Komai Y, Ushiki T (1991) The three-dimensional organization of collagen fibrils in the human cornea and sclera. Invest Ophthalmol Vis Sci 32(8):2244–58
13. Lee S-B et al (2004) Drug delivery through the sclera: effects of thickness, hydration, and sustained release systems. Exp Eye Res 78(3):599–607
14. Gaudana R et al (2010) Ocular drug delivery. AAPS J 12(3):348–360
15. Evans M (2014) Anatomy of the uvea. In: Yanoff M, Duker JS (eds) Ophthalmology. Elsevier Saunders, Philadelphia, pp 687–689
16. Shiuey Y, Jakobiec FA, Friedman E (2000) Uveal physiology and circulatory abnormalities. In A.a.J.A. Gragoudas (ed) Principles and practice of ophthalmology. W.B. Saunders, Philadelphia, pp 1175–1185
17. Giampani, A.S.B.-. and J.G. Junior, Anatomy of Ciliary Body, Ciliary Processes, Anterior Chamber Angle and Collector Vessels. Glaucoma - Basic and Clinical Aspects. 2013. InTech, DOI: 10.5772/52780. Available from: http://www.intechopen.com/books/glaucoma-basic-and-clinical-aspects/anatomy-of-ciliary-body-ciliary-processes-anterior-chamber-angle-and-collector-vessels
18. Allingham RR (2010) Cellular and molecular biology of aqueous humor dynamics. In: Shields textbook of glaucoma. Wolters Kluver Health, Philadelphia, pp 1–23
19. Delamere NA (2006) The lens. In: Tasman W, Jaeger EA (eds) Duane's foundations of ophthalmology. Lippincott Williams & Wilkins Publishers, Inc, Philadelphia
20. Worgul BV, Kleiman NJ (2006) Lens. In: Tasman W, Jaeger E (eds) Duane's foundations of clinical ophthalmology. Lippincott Williams and Wilkins, Philadelphia
21. Sebag J (2006) Vitreous biochemistry, structure, and clinical evaluation. In: Tasman W, Jaeger E (eds) Duane's clinical ophthalmology. Lippincott Williams and Wilkins, Philadelphia
22. Grunwald GB (2006) Structure and function of the retinal pigment epithelium. In: Tasman W, Jaeger E (eds) Duane's foundations of clinical ophthalmology. Lippincott Williams and Wilkins, Philadelphia
23. John D et al (2011) Dimensions of the foveal avascular zone using the Heidelberg retinal angiogram-2 in normal eyes. Indian J Ophthalmol 59(1):9–11
24. Göran Darius Hildebrand ARF (2011) Anatomy and physiology of the retina. In: Reynolds J, Olitsky S (eds) Springer/Berlin Heidelberg, New York
25. Park SS (2006) The anatomy and cell biology of the retina. In: Tasman W, Jaeger E (eds) Duane's foundations of clinical ophthalmology. Lippincott Williams and Wilkins, Philadelphia
26. Jie Zhang, Rubin RM, Rao NA (2006) Anatomy and embryology of the optic nerve. In: Tasman W, Jaeger EA (eds) Duane's foundations of clinical ophthalmology. Lippincott Williams and Wilkins, Philadelphia
27. Sherman DD, Burkat CN, Lemke BN (2006) Orbital anatomy and its clinical applications. In: Tasman W, Jaeger EA (eds) Duane's clinical ophthalmology. Lippincott Williams and Wilkins, Philadelphia

Chapter 3
Pharmacological Considerations in Ophthalmic Drug Delivery

Nikhil A. Sangave, Charles Preuss, and Yashwant Pathak

Abstract The eye is housed in a bony orbit, anchored by extraocular muscles and multiple layers of soft tissue. The globe is comprised of several layers including the sclera, uvea, and retina. The refractive elements of the eye include the lens and cornea, which provide refractive power necessary to focus light on the retina.

On a micro-molecular level, the intraocular environment is immunologically naïve. This sequestration is maintained by a complex series of active and passive transporters. Other biochemical elements, namely, enzymatic reactions and pH, dictate the level of metabolism in the eye. The unique nature of ocular anatomy and physiology presents similarly unique challenges for ophthalmic drug delivery. Thorough understanding of these clinically relevant facets may be important in developing solutions for drug delivery.

Keywords Drug delivery • Ophthalmology • Ocular anatomy • Ocular physiology • Ocular transporters • Ocular pharmacology

3.1 Introduction

Novel forms of drug delivery hold therapeutic promise for various disease states once thought to be untreatable. This is especially true when it comes to ocular pharmacotherapy. Although, ocular drug delivery has faced its share of hurdles, it has proven to be effective in treating inflammatory and infectious afflictions of the eye.

N.A. Sangave (✉)
MCPHS University, University of Southern California, Los Angeles, USA
e-mail: nsangave@gmail.com

C. Preuss
Department of Molecular Pharmacology and Physiology, Morsani College of Medicine, University of South Florida, Tampa, FL, USA

Y. Pathak
Department of Pharmaceutical Sciences, USF College of Pharmacy, University of South Florida, Tampa, FL 33612, USA
e-mail: ypathak1@health.usf.edu

© Springer International Publishing Switzerland 2016

Y. Pathak et al. (eds.), *Nano-Biomaterials For Ophthalmic Drug Delivery*,
DOI 10.1007/978-3-319-29346-2_3

In order to fully understand the limitations, capabilities, and goals of ocular drug treatment, it is important to fully grasp the anatomy and physiology of the eye.

Currently, there are significant barriers that create problems for ophthalmic pharmacotherapy. Most of the problems arise from a lack of distribution of the therapeutic agent due to factors such as low absorption, nasolacrimal drainage, or enzyme neutralization.

Knowledge of ocular structures is crucial to an understanding of the pharmacokinetics and pharmacodynamics that are and will be quintessential to the development of new therapeutic managements for ophthalmic diseases.

3.2 Extraocular Structures

The eyes are encased in a bony orbit of the skull. The bony orbit consists of seven bones: frontal, zygomatic, maxilla, ethmoid, sphenoid, lacrimal, and palatine [1].

3.3 Aqueous Layer

The aqueous layer is a protective layer, secreted by the lacrimal glands, which consists of electrolytes, water, and proteins that help maintain homeostasis on the ocular surface. The lacrimal glands are heavily mediated by the parasympathetic nervous system. The main neurotransmitters include norepinephrine, neuropeptide Y, and to a much less extent substance P and CGRP [2].

Osmotic pressure is created by the electrolytes between the corneal epithelial tissue and the tear film, thereby controlling the membrane permeability. Electrolytes including bicarbonate buffer the tear pH. Iron, magnesium, copper, calcium, and phosphate all serve as cofactors for enzymatic reactions. Sodium, potassium, and chloride regulate the osmotic pressure ([2]).

The outermost layer is the superficial oily layer. The lipids in this section are secreted by the meibomian glands and protect against evaporation. In drug delivery, this is the first barrier that the drug must penetrate in order to reach its target ([2]).

Tear film consists of glucose, urea, lactate, citrate, and amino acids, among other organic compounds. The proteins that make up tear film are immunoglobulins (IgA and sIgA). These antibodies help with the immune response in the eye and can trigger inflammation. The tear film also consists of lysozymes, phospholipase II, and other proteins that are used for defense against microbes ([2]).

In summary, the aqueous layer oxygenates the corneal epithelial cells and is the regulatory center for electrolytes and antimicrobial factors.

The innermost layer of the tear film consists of a mucin component secreted by conjunctival goblet cells. This component contains mucin, proteins, electrolytes, and water. This layer acts as yet another protective barrier for the corneal epithelium ([2]).

3.4 Precorneal Tear Film

The precorneal tear film is a highly complex amalgam of substances that protect, nourish, and maintain the cornea. From anterior to the posterior, the tear film consists of an oily layer, a middle aqueous layer, and a deep mucin layer. The tear film whose dysfunction can result in severe ocular disease distributes crucial agents like lactoferrin, lysozymes, and immunoglobulins [1].

3.5 Tear Film

The meibomian (tarsal) gland helps to secrete both polar and nonpolar lipids to make up the anterior section of the tear film. The polar lipids are usually phospholipids that are located at the boundary of the aqueous layer. However, the majority of the nonpolar lipids are hydrophobic. The glands are located at the tarsal plate of both the upper and lower eyelids. They are innervated by the parasympathetic nervous system. The sebaceous glands of Zeis also secrete lipids into the tear film. They are located near the eyelash roots. The lipid layer serves to fulfill functions like preventing evaporation, maintaining surface tension and tear overflow, and inhibit damage to the lid margin [1].

3.6 Cornea

The cornea is the most anterior portion of the globe and consists of five classically described layers: epithelium, Bowman's membrane, stroma, Descemet's membrane, and endothelium [1].

The corneal epithelium is itself made up of several layers – most anteriorly a nonkeratinized squamous epithelium whose basal layer is attached to a basal lamina by hemidesmosomes. Between superficial epithelium and the basal lamina are 2–3 layers of polygonal "wing cells" that attach to each other by occluding zonules. These attachments form the semipermeable membrane that is characteristic to the epithelium [3].

Deep to the epithelium is Bowman's membrane though this is a misnomer as it is not a true basement membrane. This layer is composed of randomly dispersed collagen fibrils approximately 8–14 μm thick. Damage to Bowman's membrane results in permanent scarring [1].

The thickest part of the cornea (~90 % by volume) is the hydrophilic stroma that consists of collagen lamellae and collagen-producing keratocytes. Woven between collagen fibrils is ground substance – a mix of proteoglycans like keratan sulfate. It is the exquisite organization of the mentioned contents that provides optical clarity that is essential for light transmission [1].

A single layer of hexagonal cells derived from the neural crest makes up the corneal endothelium. Crucial to the maintenance of corneal clarity, the endothelial cells actively transport ions and pump water from the corneal stroma. Endothelial cell dysfunction, as seen in Fuch's dystrophy, can lead to marked corneal clouding. Since these cells don't replicate, their number declines with age. Additionally, they are attached to one another by a variety of convoluted, but adherent, junctions consisting of complexes that may impede the absorption of water or ions [1].

3.7 Uveal Tract

The vascular component of the eye consists of the iris, ciliary body, and choroid. Together, these three structures make up the uveal tract.

3.7.1 Iris

The iris is the most anterior segment of the uveal tract. The iris contains a significant amount of blood vessels as well as the area known as the stroma. The stroma contains the melanocytes that give the iris its unique color. The iris contains smooth muscle, which controls the size of the pupil and how much light is allowed to pass through. During mydriasis, or pupil dilation, the muscles appear convoluted, and during miosis, or when the pupils constrict, the muscle is smooth [1].

3.7.2 Stroma

The stroma is made up of nonpigmented and pigmented cells. The nonpigmented area consists of a region of hyaluronic acid and collagen, while the pigmented cells contain abundant melanocytes. Aqueous humor runs along the anterior side of the iris and through the stroma. A person's unique eye color is attributed to the amount of pigmentation in the anterior border and deep stroma. The stroma is a highly vascularized area. The blood vessels originate from the major arterial circle and travel to the pupil. The apical end of the ciliary body is where the major arterial circle lies. The minor vascular circle of the iris – made of anastomoses – makes up the collarette, the thickest portion of the iris. Nerve fibers pass through the stroma as well for sensory and muscular functions [1].

The posterior part of the stroma consists of the pigmented cells. It runs continuously with the nonpigmented portion of the ciliary body and the polarities of the cells are kept constant. The posterior chamber borders the posterior pigmented layer at the basal surface. The apex faces the stroma and merges into the dilator muscle [1].

3.7.3 Dilator Muscle

The dilator muscle stems from the neuroectoderm. Myofilaments and melanosomes make up the smooth muscle. The basal part of the cells contains mostly myofibrils and these extend to the stroma. The muscle contracts when alpha-1 adrenergic receptors (in the sympathetic nervous system) are stimulated, while stimulating the cholinergic receptors (parasympathetic nervous system) does the opposite [1].

3.7.4 Sphincter Muscle

The sphincter muscle also stems from the neuroectoderm and consists of circular bands of smooth muscle fibers. The muscle is located anterior to the pigmented epithelium in the deep stroma. The sphincter muscle is predominantly innervated by the parasympathetic nervous system; stimulating cholinergic receptors contracts the sphincter muscle. Conversely, stimulating adrenergic receptors does the opposite by relaxing the sphincter muscle [1].

3.7.5 Ciliary Body

The foundation of the ciliary body forms the iris. Longitudinal muscle fibers attach the ciliary body to the scleral spur. The apical end of the ciliary body is oriented to the ora serrata. The main function of the ciliary body is to augment aqueous humor production and aid in lens adjustment, which is also known as accommodation [1].

3.7.6 Ciliary Epithelium and Stroma

The pars plana and the pars plicata make up the ciliary body and stretch roughly 6–7 mm. The pars plana is a smooth and pigmented region with little vasculature. It is about 4 mm wide and spans from the ora serrata to the ciliary processes. The pars plicata, contrary to the aforementioned region, has much more vasculature and contains about 70 radial folds, otherwise known as ciliary processes. This region is primarily where zonular fibers attach the lens [1].

The nerves of the ciliary processes – the nexus – are accessed by arterioles from the major arterial circle. There are at least 1–2 larger venules that circulate fluid from the capillary plexus. Arteriolar smooth muscle and sphincter tone affect the hydrostatic pressure gradient of the capillaries and determine whether the blood will flow to the plexus or completely bypass it and pass into the choroidal vein only. These processes, along with neuronal innervation, and innate vasoactive agents all impact the production of aqueous humor [1].

The ciliary body consists a layer of pigmented and a layer of nonpigmented epithelial cells. The nonpigmented layer is located on the interior bordering the aqueous humor and the outer-pigmented epithelial layer. Complex junctions bridge the apical segments of the two epithelial layers. More specifically, the lateral intercellular spaces near the border of the nonpigmented layer are fused via tight junctions. These junctions preserve the integrity of the blood-aqueous barrier. The basal lamina spans the basal surface of the nonpigmented layer. The thick basal lamina resides in the valleys of the ciliary processes and borders the stroma of the iris. The pigmented layer is known to have a large nucleus and multiple mitochondria. This layer has more melanosomes that are located in close proximity to the iris. The endoplasmic reticulum and Golgi complex are two organelles that play pivotal roles when it comes to the upkeep of aqueous humor. The pigmented area of the ciliary body has larger capillaries. It consists of many collagen fibers and fibroblasts. The ciliary arteries merge and provide nourishment for a deep, intramuscular, multilayered arterial plexus. The veins help circulate the blood through a vortex system and through episcleral veins into the limbal region [1].

3.7.7 *Choroid*

The choroid, the final portion of the uveal tract, which is located in the posterior, serves to provide nourishment to the retina. It contains bodies of vasculature: the choriocapillaris, the middle layer of smaller vessels, and the outer layer of larger vessels. The choroid is perfused through the long and short ciliary arteries and the vortex system removes the blood from venules. This area is highly vascularized and both the arterial and the venous blood consist of high levels of oxygen [1].

3.7.8 *Bruch's Membrane*

Bruch's membrane spans the distance from the optic disk to the ora serrata. This membrane is periodic acid-Schiff (PAS) positive due to the joining of the basal laminae of the retinal pigment epithelial layer and the choriocapillaris. It has five components: the basal lamina of the retinal pigment layer, the inner collagenous layer, the porous elastic fibers, the outer collagen layer, and the basal lamina of the choriocapillaris. These layers make up connective tissue that is highly permeable to smaller molecules [1].

3.7.9 *Choriocapillaris*

The choriocapillaris is a bed of large capillaries underneath the retinal-pigmented epithelial layer. The capillary walls become thin on the surface when approaching the retina, and pericytes are found on the outer wall. The blood vessels have many fenestrations. The middle and outer choroidal vessels, however, are not fenestrated. As a result, drug delivery to the surrounding areas proves somewhat more difficult. The larger vessels have an elastic layer and smooth muscle cells, and thus, small molecules can diffuse through the endothelial layer but not the middle and larger choroidal capillaries. The stroma contains many melanocytes, but macrophages, lymphocytes, and plasma cells have also been seen in this region as well. Nerves and collagen primarily make up the intercellular space [1].

3.8 Lens

The convex structure that is located directly behind the pupil is called the lens, and it accounts for about a third of the focusing power in the adult eye. The lens capsule surrounds this structure. At birth, the equatorial diameter is approximately 6.5 mm, but this grows to about 9–10 mm. The width, however, starts at about 3 mm and increases to 6 mm in the later stages of life [1].

Ciliary muscle contraction causes the muscle to move forward; this process leads to lens accommodation. Ciliary muscle contraction releases zonular tension on the lens and allows the lens to conform to a rounder shape. This change in shape is what causes the lens to thicken. As we age, the lens' ability to accommodate wanes because of the lens becoming more stiff and increasing in size [1].

The lens is unique in that it is entirely dependent on the aqueous and vitreous humors for sustenance. It is not innervated and has no vasculature.

3.8.1 Lens Capsule

The basal lamina that encloses the lens is called the lens capsule. Collagen fibers are consistently added to the anterior segment of the lens capsule, while the posterior zone of the capsule remains somewhat constant throughout life. The anterior capsule contains a unique material called laminin, which adds to the thickness of this segment. The filaments of the capsule are organized in a lamellar fashion [1].

3.8.2 Lens Epithelium

Underneath the capsule lies the lens epithelium. The apical end of this layer faces the lens, and borders are so tight that there is little to no intercellular space. The central region of the epithelial layer is made up of cells that decline as we age. In an intermediate region, there is evidence that smaller cells do go through a degree of cell replication. The germinative zone of the lens is made of cuboidal cells that are organized in meridional rows. These cells replicate throughout the lifetime, they stretch, and they compose the lens fibers [1].

3.8.3 Lens Fibers

The lens contains an inner nucleus that is a component of the fibers, which is formed at birth. The lens also has an outer cortex component, which assists in the creation of new fibers itself. The cells of the fibers are spindle shaped, have many projections, and have only a small amount of organelles. Lens crystallins in the cytoplasm of fiber cells account for the refractive index of the lens [1].

3.9 Aqueous Layer

Osmotic pressure is created by the electrolytes between the corneal epithelial tissue and the tear film, thereby controlling the membrane permeability. Electrolytes like bicarbonate buffer the tear pH. Iron, magnesium, copper, calcium, and phosphate all serve as cofactors for certain enzymatic reactions. Sodium, potassium, and chloride regulate the osmotic pressure ([2]).

The outermost layer is the superficial oily layer. The lipids in this section are secreted by the meibomian glands and serve to protect against evaporation. In drug delivery, this is the first barrier that the drug must penetrate in order to reach its target ([2]).

Tear film consists of glucose, urea, lactate, citrate, and amino acids, among other organic compounds. The proteins that make up tear film are immunoglobulins (IgA and sIgA). These antibodies help with the immune response in the eye and can trigger inflammation. The tear film also consists of lysozymes, phospholipase II, and other proteins that are used for defense against microbes ([2]).

In summary, the aqueous layer oxygenates the corneal epithelial cells and is the regulatory center for electrolytes and antimicrobial factors.

The innermost layer of the tear film consists of a mucin component secreted by conjunctival goblet cells. This component contains mucin, proteins, electrolytes, and water. This layer acts as yet another protective barrier for the corneal epithelium ([2]).

Underneath the basal lamina lies the Bowman's layer. This layer is composed of randomly dispersed collagen fibrils approximately 8–14 μm thick. If these cells are destroyed, scar tissue replaces them [1].

Descemet's membrane is the basal lamina of the endothelial cells that increases in thickness with age. At birth, the layer starts out at 3–4 μm thick and grows with type IV collagen fibers anywhere from 10–12 μm thick by adulthood. This layer is periodic acid-Schiff (PAS) positive. The membrane has two stages of development. In utero, it is composed of an anterior banded zone, but as life progresses, the corneal endothelial cells add a posterior nonbanded zone [1].

A single layer of hexagonal cells derived from the neural crest makes up the corneal endothelium. This layer lies between anterior chamber and Descemet's membrane. The cells usually have large nuclei and many mitochondria. The endothelial cells actively transport ions and pump water from the corneal stroma in order to maintain proper osmotic pressure. Endothelial cells rarely undergo replication and the number of cells decreases with aging. Endothelial cells junctions are usually never joined by desmosomes. They do, however, form a variety of convoluted, but adherent, junctions consisting of complexes that may impede the absorption of water or ions. Endothelial cell damage may ultimately lead to visual dysfunction [1].

3.10 Sclera

The sclera is the majority of the posterior surface of the globe. The anterior part opens to the cornea, while the posterior part opens to the optic nerve. The rectus muscles flow into the superficial collagen fibers of the sclera. Anteriorly, the Tenon's capsule encompasses the sclera and rectus muscles. Both structures lie below the bulbar conjunctiva. The Tenon's capsule and bulbar conjunctiva meet at the limbus. The sclera is the thickest (1.0 mm) near the optic nerve head at the posterior pole and thinnest (0.3 mm) behind the rectus muscles. The sclera is virtually avascular but does contain vessels, called the emissaria, that allow for passage of arteries, veins, and nerves into the sclera. The episclera is a network of vascular connective tissue that meets with superficial sclera, the Tenon's capsule, and conjunctiva. Collagen and fibroblasts make up the scleral stroma. Usually, the outer scleral collagen fibers have a larger diameter than the inner collagen fibers. The collagen fibers in the inner layer of the sclera, the lamina fusca, meet with the suprachoroidal and supraciliary lamellae of the uveal tract. The fibers also mix with the ciliary body and the choroid. The lamina fusca consist of fibroblasts, melanocytes, and electron-dense bodies [1].

3.11 Limbus

The peripheral cornea and the anterior sclera meet at an area known as the limbus. The limbus consists of the conjunctiva and the limbal palisades as well as the aforementioned Tenon's capsule. Additionally, the limbus consists of the episclera, corneoscleral stroma, and aqueous outflow apparatus.

3.12 Anterior Chamber

The layer behind the corneal stroma is called the anterior chamber. This ciliary body produces aqueous humor, which is then stored in this compartment. This fluid is crucial in removing waste and nourishing the anterior segment of the eye [4].

3.13 Trabecular Meshwork

This region of the eye consists of spongy, connective tissue that is lined by trabeculocytes. These cells are both contractile and phagocytic. The apical part of the meshwork is at the Schwalbe line and the foundation is at the scleral spur and the ciliary body. The meshwork has three components: the uveal meshwork, the corneoscleral meshwork, and the juxtacanalicular tissue [1].

3.14 Uveal Trabecular Meshwork

The uveal meshwork has trabeculae that are less circular and less elastic than the corneoscleral meshwork. They usually contain some pigment and are cord shaped.

3.15 Corneoscleral Meshwork

The corneoscleral meshwork is a series of thin connective tissue covered by a single layer of pinocytic vesicles. The inner core is primarily composed of collagen and elastin. The outer cortex of the trabecular beam is made up of the basal lamina of this meshwork [1].

3.16 Schlemm Canal

Made of nonfenestrated endothelium and thin connective tissue, this circular tube helps with the outflow of aqueous humor. Tight junctions join the walls of the endothelial cells and the surfaces of the cells are lined with pinocytic vesicles. The size and number of these vesicles are directly proportional to the intraocular pressure [1].

3.17 Collector Channels

About 25–30 collector channels drain into the deep and midscleral venous plexus. These channels originate from the Schlemm canal. A maximum of 8 channels can meet at the episcleral venous plexus as aqueous veins [1].

3.18 Retina

The retina is the translucent inner lining of the back of the eye that collects and transmits light signals to the brain. It is a multilaminar structure comprised of the following layers (from external to internal): rod and cone segments, external limiting membrane, outer nuclear layer, outer plexiform layer, inner nuclear layer, ganglion cell layer, nerve fiber layer, and the internal limiting membrane.

3.18.1 Retinal Pigment Epithelium (RPE)

The retinal pigment epithelium carries out a number of vital functions for the eye. It maintains the outer blood-retina barrier, assists in the metabolism of vitamin A, and aids in the active transport of various agents in and out of this layer. The characteristics are important considerations when formulating ophthalmic medications. Additionally, the RPE has been known to help with the absorption of light, heat exchange, and phagocytosis of photoreceptor outer regions. This layer converges with the aforementioned RPE of the ciliary body. It is a monolayer that spans from the optic disk to the ora serrata [1].

3.18.2 Retinal Photoreceptor Layer

This layer, considered part of the neurosensory retina, contains rods and cones. The outer portions of these cells are encapsulated by a mucopolysaccharide matrix, which lies between the photoreceptor layer and the RPE.

The rod photoreceptors have both an outer and inner segment. The outer segment looks similar to a stack of coins. The outer layer also contains a central cilium. The central cilium is made up of microtubules that are organized in a "9 plus 0" cross-sectional configuration. Additionally, there is the rod inner section. This part contains an outer ellipsoid and an inner myoid. The outer ellipsoid holds abundant mitochondria, while the inner myoid contains a large storage of glycogen. The synaptic body, or spherule, is located in the inner portion of the rod cell.

The next type of photoreceptor is the cone. The extrafoveal cone photoreceptor has conical ellipsoids and myoids. Unlike rods, cone structures are actually attached to the cell membrane. The synaptic body of the cone is called the pedicle. The cone pedicle shares synapses with other rods and cones. Foveal cones are essentially identical to extrafoveal cones. However, these cones contain cylindrical inner segments, a characteristic also found in rod photoreceptors. Horizontal cells not only synapse at pedicules and spherules, but they also reach to the outer plexiform layer. Vertical bipolar cells have dendrites that synapse with either another pedicule or spherule. The vertical bipolar cells also have axons that can reach the ganglion and the amacrine cells of the inner plexiform.

The ganglionic axons run parallel to the inner surface of the retina. There, they form the nerve fiber layer, which then becomes the beginnings of the axons of the optic nerve. Each of the optic nerves has about one million nerve fibers. The fibers reach the superior and inferior poles of the optic disk while circumventing the macula. The papillomacular fibers begin at the fovea and travel directly to the optic nerve [1].

3.18.3 Components of Retina

Glial cells extend from the external limiting membrane to the internal limiting membrane and are called Müller cells. Astrocytes, microglia, and Müller cells feed and support the retina.

The central retinal artery and its capillaries are the chief blood supply to the inner parts of the retina. In some cases, however, the cilioretinal artery has been seen to supply parts of the inner retina as well. The blood-retina barrier is maintained by the blood vessels. The barrier is present because of a nonfenestrated monolayer of endothelial cells with tight junctions. A basal lamina encompasses the endothelial cells, and Müller cells usually surround this layer. The basement membrane that anchors these cells contains pericytes, or mural cells. These retinal blood vessels lack smooth muscle cells except near the optic nerve head. Usually, the blood vessels reach to the level of middle limiting membrane [1].

3.19 Macula

Today, there are two definitions as to where the macula actually begins and what its boundaries are. Histologically, the macula is the layer that starts with more than one layer of ganglion cell nuclei. However, most clinicians consider the region within the temporal vascular arcades as the macula. The macula lutea or "yellow spot" gets its name from the distinct yellow color that carotenoid pigments create. Aside from carotenoids, other pigments are lutein and zeaxanthin. Lutein is mostly found in areas where rods are abundant – in the periphery, further from the fovea. Zeaxanthin is mostly found in areas where cones dominate in number – closer to the fovea [1].

In the macula exists a concave structure called the fovea. The fovea is about 1.5 mm in diameter. At this structure, the thickness of the basal lamina decreases.

The parafovea is a structure that surrounds the fovea. The ganglion cell layer is the thickest here. Around this region lies the perifovea.

Inside the fovea lies a central depression called the foveola. Photoreceptors and Müller cells are the predominant types of cells that are present in this area. One unique property of foveola is the fact that the photoreceptors are all cones. This high density of cones accounts for high visual perception. These cones resemble rods but contain all the same properties as extramacular cones [1].

3.20 Ora Serrata

The ora serrata divides the retina from the pars plana. The diameter of the eye is about 20 mm and the circumference is about 63 mm at the ora serrata. The retinal blood vessels appear looped before they meet the ora serrata [1].

3.21 Vitreous

Eighty percent of the volume of the globe is the vitreous cavity. The vitreous humor plays an integral role in metabolic processes. It gives the metabolites of the lens, retina, and ciliary body a pathway. Despite the vitreous humor's gel-like consistency, it predominantly consists of water. Hyaluronic acid, a mucopolysaccharide, provides the viscosity of the vitreous humor. The vitreous humor contains type II collagen fibrils, hyalocytes, with an unknown function. These collagen fibrils begin at the basal lamina of the nonpigmented epithelium of the pars plana and the internal limiting membrane (ILM) of the retina. The vitreous base, which extends 2.0 mm anterior to the ora serrata to about 4.0 mm posterior to the ora serrata, anchors the vitreous. The vitreous cavity is also maintained by attachments at the disk margin, at the perimacular region, near the retinal vessels, and at the periphery of the posterior lens capsule. With age, the vitreous humor loses its viscosity [1].

3.22 Optic Nerve

The optic nerve is the gateway into the forebrain. It consists of an optic stalk that is made up of neuroectodermal cells and undifferentiated neural crest cells. Normal adult optic nerves consist of about 1.1 million axons that meet at the lateral geniculate nucleus (LGN). Crossover occurs at the optic chiasm. During development, the optic nerve grows as a part of the brain. Its fibers average about 40 mm in length and are encased by glial cells. The optic nerve can be broken down into four different regions: the intraocular region, the intraorbital region, the intracanalicular region, and the intracranial region [1].

3.22.1 Intraocular Region

The anterior surface of the optic nerve head is known as the optic disk. A depression, the physiologic cup, resides in the optic disk. This feature is crucial for circulation as the central retinal artery and central retinal vein pass through the area. The optic nerve itself can be broken down into four subdivisions: the superficial nerve fibers, the prelaminar area, laminar area, and retrolaminar area [1].

3.22.2 Superficial Nerve Fibers

This layer consists of fibers from the retina entering the nerve head.

3.22.3 Prelaminar Area

This region consists of astrocytic glial cells, fascicles, which are used to bolster the ganglionic axons that enter the nerve head. These fascicles extend to the lamina cribrosa and create a circular enclosing for the nerve fibers in the intraocular and intraorbital course. This in turn separates the nerve fibers from the connective tissue. The astrocytic glial cells are continuous with the retina and completely cover the surface of the head, creating an internal limiting membrane (ILM). The ILM consists of Müller cells, but toward the end of the retina, astrocytes become the main component [1].

3.22.4 Laminar Area

The laminar area, otherwise known as the lamina cribrosa, is integrated with the sclera. It consists of connective tissue plates that have pores through which the axons may travel. The laminar area consists of types I and III collagen, elastin, laminin, and fibronectin. This area serves to act as support for axons and posterior segment of the globe. Astrocytes enclose the axons and smaller blood vessels are present in the area. The area also serves as a target for the central retinal artery (CRA) and the central retinal vein (CRV) [1].

3.22.5 Intraorbital Region

A component that forms from the rectus muscles, known as the annulus of Zinn, encloses the optic nerve in the intraorbital region. The dural sheath of the nerve meets at the periosteum and thus renders the optic nerve stationary.

The intraorbital region also contains meningeal sheaths. The pia mater is the innermost connective tissue layer, containing meningothelial cells. These cells create septa that separate axons into bundles. The septa terminate just before the chiasm. The septa predominantly consist of fibroblasts, nerves, elastin, collagen, arterioles, and venules and serve to nourish the axons and glia. A layer of astrocytic glial cells separates the layer from directly touching the axons [1]

The arachnoid mater, consisting of collagen, some elastin, and meningothelial cells, lines the dura mater. The vascular trabeculae connect this region to the pia mater across the subarachnoid space. This subarachnoid space terminates at the lamina cribrosa. The CRA and CRV travel through this space [1].

The dura mater is a thick layer that encloses the brain. It makes up the outermost portion of the meningeal sheath. It is heavily reinforced with collagen and elastin fibers that fuse with the sclera [1].

3.22.6 Intracanalicular Region

In the optic canal, the pial vessels, which begin at the ophthalmic artery, perfuse the optic nerve. The periosteum of the bony canal anchors the optic nerve in this area [1].

3.22.7 Intracranial Region

Two optic nerves that reside above the ophthalmic arteries emerge after passage through the optic canals. The optic nerve passes over the cavernous sinus to the optic chiasm. The chiasm bifurcates into separate right and left optic tracts. These tracts terminate in their own lateral geniculate nuclei [1].

3.23 Pharmacokinetics

Ocular pharmacokinetics is a complicated set of interactions. Several modes of ocular drug delivery exist: topical, intravitreal, intravenous, and oral. Agents can also be injected subconjunctivally, underneath the Tenon's capsule and through the retrobulbar routes. Injection of anesthetics, glucocorticoids, and antibiotics provides high local concentration and perhaps greater therapeutic efficacy [1].

The absorption of a drug administered topically depends on a number of factors like the residence time of the drug in the tear film, drug-protein binding and metabolism, diffusion through the cornea and conjunctiva, and elimination of the drug through nasolacrimal drainage. Drugs being administered topically that are absorbed through the nasal mucosa bypass first-pass metabolism in the liver and have higher concentrations in serum. Therefore, use of these topical drugs for extended periods of time may lead to undesired systemic toxicities. When ophthalmic drugs are administered topically, there are several routes in which they could travel, but ultimately they end up in systemic circulation. For instance, from the tears, the drug could travel to the cornea through the aqueous humor and to the iris [1].

Evidence has shown that the best results for local absorption are when the drug penetrates the cornea or the conjunctiva via passive diffusion. Besides drug concentrations, there are several other factors that affect the diffusion across the cornea or conjunctiva. These factors may include particle size and chemical structure. Lag time is the term applied to the time at which the drug is administered to when it is first found in aqueous humor [1].

Drugs that are intended to cross the cornea have to overcome unique obstacles. The drug must penetrate through a labyrinth of layers that have different physicochemical properties. The drug must pass through the epithelial layer, the stroma, and then the endothelial layer. Thus, the drug must pass through a hydrophobic, a hydrophilic, and then another hydrophobic layer again. Therefore, it is imperative that when drugs are being designed for the transcorneal route, they have both lipophilic and hydrophilic characteristics in order to pass through these compartments and reach therapeutic concentrations in the eye [1].

Drug absorption may also be affected by a physical barrier or deficiency in the epithelium or elsewhere in the cornea. Rectifying these defects may lead to an improvement in absorption. Additionally, other factors like nasolacrimal drainage, blink rate, and tear dilution must all be considered when it comes to absorption of ophthalmic drugs [1].

Topical drug administration results in systemic uptake via the nasolacrimal system as well as, to a lesser extent, through the cornea and conjunctiva. As mentioned earlier, many regions in the eye are pigmented and, thus, contain melanocytes. The melanocytes produce melanin, which can bind to drugs and cause the delayed distribution through tissues. Therefore, it is often seen that people with darker eyes (and therefore, more melanin) may experience a longer time for the drug to take effect in the eye. This may not necessarily pose as a detriment because this property could also be applied to sustained-release formulations. After absorption, the drug tends to deposit in the aqueous humor, which can subsequently distribute the drug to the other parts of the eye or to systemic circulation [1].

Metabolism in the eye is carried out extensively by a wide number of enzymes. These include esterases, lysozymes, oxidoreductases, and monoamine oxidases, among many others [1]. Although metabolism in the eye has not been thoroughly understood, it is well established that corneal esterases play an integral role in the metabolism of ophthalmic medications. Most of these enzymes are located in the corneal epithelium, the iris, and the corneal endothelium. Most ophthalmic drugs given topically are not capable of penetrating through to the retina. Therefore, the enzymes located in the retina itself are not heavily considered unless the drug is being administered systemically. In order to design drugs effectively, it is imperative to comprehend the full range of metabolism of a particular medication as well as an individual's specific reaction to drugs based on the amount of enzymes dictated by his or her genetic makeup [5].

pH plays a crucial role in delivery of drugs to the eye. The environment in which drugs are usually instilled is slightly alkali. Drugs that are formulated as weak bases prove advantageous in penetration of the eye. Instilling slightly basic drops causes a change in pH. This means that the drug will convert into its more non-ionized form and is henceforth more lipid soluble [5].

Viscosity is another property that is important when delivering ophthalmic drugs. Nasolacrimal drainage is somewhat impeded when the viscosity of medication is increased. Viscosity-adding agents, like polyvinyl alcohol (PVA) and methylcellulose (MC), also help increase residence time in the conjunctiva and therefore increase the penetration and efficacy of the ophthalmic agent. One study suggests that the optimal viscosity of the ophthalmic agents should be between 12 and 15 cps

OPHTHALMIC AGENTS AVAILABLE IN THE U.S.			
Medication	**Brand Name**	**Pharmacological Class**	**Indication**
Ketorlac Tromethamine	Acuvail®	NSAID	Allergic Conjunctivitis, Post-Operative Inflammation
Diclofenac Sodium	Voltaren®	NSAID	Corneal Sensitivity, Post-Operative Inflammation
Timolol maleate	Timoptic®	Beta Blocker	Glaucoma
Levobunolol HCl	Betagan®	Beta Blocker	Glaucoma
Bimatoprost	Lumigan®	Prostaglandin Analog	Glaucoma
Latanoprost	Xalatan®	Prostaglandin Analog	Glaucoma
Difluprednate	Durezol®	Corticosteroid	Iritis, Episcleritis
Lotprednol Etabonate	Lotemax ®	Corticosteroid	Dry eye, Uveitis, other inflammatory disease
Moxifloxacin	Vigamox ®	Fluoroquinolone Antibiotic	Bacterial conjunctivitis
Azithromycin	AzaSite®	Macrolide Antibiotic	Bacterial conjunctivitis
Ganciclovir	Zirgan®	Antiviral	Cytomegalovirus
Trifluridine	Viroptic®	Antiviral	Keratoconjunctivits caused by Herpes Simplex Virus I & II
Olopatadine HCl	Pataday®	H1 Antagonist & Mast Cell Stabilizer	Allergic Conjunctivitis
Bepotastine	Bepreve®	H1 Antagonist & Mast-Cell Stabilizer	Allergic Conjunctivitis

Fig. 3.1 Indicates a few of the many ophthalmic agents on the market in the United States. The chart includes generic and proprietary names, pharmacological class, and a few indicated uses [7]

[6]. If the viscosity exceeds the upper limit, the medication may cause irritation or reduce absorption [5] (Fig. 3.1).

Another physicochemical factor that must be considered when creating ophthalmic medications is tonicity. Most drugs are created isotonic with tears in order to minimize irritation. The eye, however, can tolerate a significant range of tonicities without immediately feeling discomfort. Usually, the normal tonicity of ophthalmic medications may extend anywhere from 220 to 640 mOsM [5].

3.24 Tear Film

The meibomian (tarsal) gland helps to secrete both polar and nonpolar lipids to make up the anterior section of the tear film. The polar lipids are usually phospholipids that are located at the boundary of the aqueous layer. However, the majority of the nonpolar lipids are hydrophobic. The glands are located at the tarsal plate of both the upper and lower eyelids. They are innervated by the parasympathetic nervous system. The tear film also accepts lipids secreted by the sebaceous glands of

Zeis. They are located near the eyelash roots. The lipid layer serves to fulfill functions like preventing evaporation, maintaining surface tension and tear overflow, and inhibit damage to the lid margin [1].

3.25 Biochemical Transporters

Biochemical transporters in the eye are essential in maintaining the structural, functional, and metabolic processes of the eyes. These transporters are essential to the study of ocular pharmacology because they may help with drug delivery. If a drug has low bioavailability in a specific part of the eye, it can be formulated with the same properties so that it can pass through one of these transporters. This method of drug design holds promise in the future.

In the eye, glucose is primarily transported via the GLUT-1 transporter [8]. The GLUT-1 transporters are located on the retinal pigment epithelium as well as the retinal capillary epithelium. However, GLUT-1 transporters are also commonly seen in the corneal epithelium, the iris, the photoreceptor cells, the ganglion, and the nerve fiber layer. Glucose is transported into each layer via facilitated diffusion in order to carry out normal metabolic processes. These transporters have shown significance when it comes to diabetic retinopathy. One study suggests that the absence of the GLUT-1 transporters in the neovascular endothelium caused by retinopathy leads to hindered glucose uptake in the eye in patients with diabetes mellitus [9].

Peptides are also commonly transported across the epithelium. Three main transporters are known as the PepT1, PepT2, and the peptide-histidine (PHT) transporters. According to a study, these transporters commonly accept drugs as their substrates. These drugs may include renin inhibitors or ACE inhibitors. PHT1 was found on neural retina but was not found to have any significant function. PEPT2 and PHT2 were found to be in the retina. This study also found that peptide prodrugs were endocytosed by carrier proteins [10].

ASCT1, LAT1, and ATB0+ are all examples of amino acid transporters that have been found on the cornea. These transporters are used to extract amino acids, like L-alanine and L-arginine, from the blood supply and distribute them to the necessary tissue. Amino acids are essential for the creation of proteins in both the conjunctiva and the RPE. LAT1 mostly transports larger, more neutral amino acids [10].

Transporters for vitamin C (ascorbic acid) also exist. Vitamin C protects corneal tissue with its antioxidant properties. The two main transporters are the SVCT1 and SVCT2. These transporters shuttle L-ascorbic acid across membranes. Hexose transporters (GLUT) have also been known to help transport vitamin C in the form of dehydroascorbic acid. The vitamin C subsequently is supplied to the cornea and lens via the aqueous humor [10].

The eye also has many efflux proteins. One of these proteins is P-glycoprotein (P-gp). This efflux pump prevents certain compounds from passing through the membranes of certain cells. It is primarily meant to keep out toxins and other foreign materials from entering the eye. However, P-gp may also stop therapeutic agents from being absorbed in the cell. Therefore, this biochemical transporter serves as another significant barrier to drug delivery [11].

Based on the complex nature of ocular anatomy, treatment of ophthalmic diseases continues to pose a significant hurdle for drug developers. Typically, ophthalmic medications are delivered topically, via drops (e.g., ocular solutions or suspensions). This allows for an easy route of administration, thereby increasing patient adherence. Although these benefits are seen, there are many problems with the current forms of drug delivery. The corneal epithelium is predominantly a lipophilic membrane that allows for little passage of hydrophilic drugs. Underneath this layer is the stroma, which is mostly a hydrophilic layer, and the same type physicochemical barrier can be seen here again, but this time it is for lipophilic drug particles. Aside from these chemical mechanisms, natural processes like nasolacrimal drainage dilute the drug to further reduce the bioavailability. Additionally, the eye contains many enzymes and transporters that render the drug particles inactive or unable to reach the target tissue. This means that drugs need to be administered in larger quantities and frequencies which can lead to increased risk of toxicity and decreased medication adherence [5].

Some progress has been made with drug design when it comes to these problems. For instance, many ophthalmic agents have been formulated as prodrugs. Prodrugs are chemicals that must first be broken down by metabolic processes in order to be activated. These metabolites are either more potent or more bioavailable than the prodrug itself. One example of an ophthalmic prodrug includes latanoprost (Xalatan). Esterases in the eye hydrolyze part of this prodrug for the treatment of glaucoma. In turn, the active metabolite now has a heightened ability to penetrate the corneal epithelium, increase the therapeutic efficacy of the drug, and decrease chances of adverse effects [12]. Many ophthalmic prodrugs, however, have failed to be marketed for reasons such as instability and difficulty maintaining sterility with the drug formulation [5].

References

1. Chalam KV (2011) Fundamentals and principles of ophthalmology: section 2, 2011–2012. American Academy of Ophthalmology, San Francisco, Print
2. Henderer J, Rapuano C (2011) Ocular pharmacology. In: Goodman and Gilman's the pharmacological basis of therapeutics, 12th edn. McGraw-Hill, New York, chapter 64
3. Yanoff M, Duker JS (2014) Ophthalmology, 4th edn. Elsevier Saunders, Philadelphia/Pennsylvania, Print
4. Bores LD (2008) Ocular anatomy – anterior segment. Ocular anatomy, 15 Jan 2008. Web. 08 Oct 2014
5. Zimmerman TJ (1997) Textbook of ocular pharmacology. Lippincott-Raven, Philadelphia, Print
6. Netland PA (2007) Ocular pharmacology. In: Glaucoma medical therapy principles and management, 2nd edn. Oxford UP In Cooperation with the American Academy of Ophthalmology, New York, p 13, Print

7. Ron Melton OD, Randall Thomas OD (2014) A clinical guide to ophthalmic drugs. Rev Optom:1A–34A. *Review of Optometry*. Bausch and Lomb. Web. 10 Aug. 2014 https://www.reviewofoptometry.com/publications/clinical-guide-to-ophthalmic-drugs2014
8. Mantych GJ (1993) Characterization of glucose transporter isoforms in the adult and developing human eye. Endocrinology 133(2):600–607, *PubMed*. Web. 5 May 2014
9. Kumagai AK, Glasgow BJ, Pardridge WM (1994) GLUT1 glucose transporter expression in the diabetic and nondiabetic human eye. Invest Ophthalmol Vis Sci 35(6):2887–894, Web. 7 July 2014
10. Gaudana R, Jwala J, Boddu SHS, Mitra AK (2009) Recent perspectives in ocular drug delivery. Pharm Res 26(5):1197–216, Web
11. Levin LA, Francis Heed A (2011) Adler's physiology of the eye, 11th edn. Saunders/Elsevier, Edingburg, Print
12. Molokhia SA, Thomas SC, Garff KJ, Mandell KJ, Wirostko BM (2013) Anterior eye segment drug delivery systems: current treatments and future challenges. J Ocul Pharmacol Ther 29(2):92–105, Web. 17 July 2014

Chapter 4
Pathophysiology of Vision

Orhan E. Arslan

Abstract The visual system is tasked with the function of processing visual impulses and recognizing visual images that incorporate associated memories. This system also mediates associated visual reflexes. To attain proper visual images, several neuronal stations at various areas of the central nervous system are utilized. Non-neuronal structures, such as the cornea, iris, anterior and posterior chambers of the eye, and the vitreous body, are also involved in directing visual images to the retina. Images that project to the retina are further transmitted through the optic nerve and optic tract and then through specific thalamic neurons to the visual cortex.

Despite the dependence of the visual system on the integrity of the cerebral cortex and other neuronal chains, vestibular neurons and gaze centers in the brain and brain stem remain important elements in securing coordinated eye movements and thus maintaining accurate binocular vision. Because of the diversity of the neuronal chain involved in transmission and construction of the visual image, lesions of one or more of these neuronal elements can result in partial or complete disorders that may be unilateral or bilateral.

Keywords Pupillary • Refractive • Drusen • Retina • Visual • Optic • Anopsia • Radiation • LGN • Striate • Brodmann • Columns • Vertebrobasilar

Through an elaborate peripheral apparatus and a highly organized conducting and integrating system, a binocular vision with associated memories is constructed. In order to achieve a precise binocular vision, the role of the visual cortex, gaze centers, and other neuronal chains in the brain stem that maintain coordinated eye movements must be not overlooked [1]. Examination of the eyeball reveals an external fibrous layer, a middle vascular layer, and a primarily inner neuronal layer. The

O.E. Arslan (✉)
Department of Pathology and Cell Biology, University of South Florida Morsani College of Medicine, Tampa, FL 33612, USA
e-mail: oarslan@health.usf.edu

© Springer International Publishing Switzerland 2016

Y. Pathak et al. (eds.), *Nano-Biomaterials For Ophthalmic Drug Delivery*,
DOI 10.1007/978-3-319-29346-2_4

external fibrous layer, which is also known as the tunica fibrosa, is comprised of the cornea and sclera. The cornea, which represents 1/6 of the fibrous layer, is an avascular structure that forms the main refractive medium of the eyeball. It receives abundant innervation from the long ciliary nerves. Despite the lack of lymphatics, the cornea remains highly resistant to infection. The cornea constitutes the anterior boundary of the anterior chamber. Discoloration of the cornea may reveal pathological conditions that have affected other organs. For instance, in hepatolenticular degeneration (Wilson's disease), greenish gray discoloration of the cornea at the limbus (Kayser-Fleischer ring) appears due to deposition of excess of copper in the Descemet's membrane of the cornea [2, 3]. The circulation of the aqueous humor is maintained by the Schlemm's canal, a venous channel at the limbus. Narrowing or occlusion at this site may impede the drainage of the aqueous humor into systemic circulation and lead to an increase in intraocular pressure and the development of glaucoma.

The ability of the eyeball to resist intraocular pressure, preserve its shape, provide attachments for eye muscles, and form a smooth articular surface for eye movements is afforded by the sclera, which forms 5/6 of the tunica fibrosa. The sclera continues posteriorly with the dural sheath of the optic nerve and allows passage of the axons that form the optic nerve, as well as the central retinal and the posterior ciliary vessels through the lamina cribrosa sclera.

The intermediate layer of the eyeball is formed by the tunica vasculosa, which consists of the iris anteriorly, the ciliary body posteriorly, and the choroid layer, which is interposed between the two previous parts. The choroid layer consists primarily of veins and arteries with the veins joining together to form 4–5 vorticose veins that eventually drain into the systemic circulation via anterior ciliary veins. Quantitative assessment of blood flow in these vessels can be determined in two dimensions through the utilization of color Doppler noninvasive techniques [4]. The long and short ciliary that form the arterial component of the choroid layer arteries forms the major and minor arterial iridal circles through branches that bypass the ciliary body to the iris. In addition to the central retinal artery, these vessels provide a supplementary source of blood to the retina.

The ciliary body consists of muscles and processes. It is sandwiched between the lateral end of the iris and the ora serrata, a transitioning zone between the optic and non-optic parts of the retina. The ciliary processes secrete the aqueous humor via diffusion or active transport into the posterior chamber of the eye. An increase in lens curvature when viewing near objects can be achieved by contraction of the ciliary muscles and movement of the ciliary body toward the iris, accompanied by relaxation of the suspensory ligament of the lens. On the contrary, looking at a distant visual field requires a decrease in curvature (flattening) of the lens and increase in tension of the suspensory ligament with relaxation of the ciliary muscles. Sustained contraction of the ciliary muscles renders the lens in both eyes in a state of short sightedness (pseudomyopia). Patients exhibit blurring of distance vision, orbital pain, and fluctuating visual acuity [5].

The lens, an integral part of the eye chambers, is located posterior to the iris and anterior eye chamber. It consists of an outer elastic capsule, a central nucleus, and an intermediate cortical zone. An imaginary line that connects the anterior and posterior poles of the lens signifies the most convex parts of the lens. This colorless, biconvex, avascular, and lymphatic structure possesses less refractive power than the cornea. As indicated earlier, contraction and relaxation of the ciliary muscles modify lens curvature through changes in the degree of tension of the suspensory ligaments.

The absence of the lens due to developmental anomaly is known as primary aphakia; if it is a result of degeneration, it is secondary aphakia. In congenital anomaly of Peter's, a gradual impairment of vision (misty appearance of visual objects) and diplopia occur as a result of corneal opacification (cloudy cornea) and lens cataracts [6].

When the opacity is confined to the periphery of the lens (peripheral cataract), poor vision in dim light and better vision in bright daylight develop. In contrast, cataract of the lens nucleus (central cataract) causes myopia and poor vision during the day but clearer vision in the dim light. Chromosomal abnormalities, metabolic disorders, infection, or maternal diseases may cause congenital cataract. Other disorders that affect the lens include presbyopia and sunflower cataract (chalcosis lentis). In presbyopia, commonly seen in aging, the lens becomes less pliable and thus less reactive to contraction of the ciliary muscles. Patients with presbyopia are hyperopic (farsighted) and their ability to read fine print becomes limited. To attain optimum vision, patients usually hold reading materials at a distance. Sunflower cataract, seen in Wilson's disease, results from impregnation and then metallic green grayish opacification of the subcapsular area of the lens [7].

Miosis, which denotes constriction of the pupil, is commonly seen ipsilaterally in Horner's syndrome and bilaterally in metabolic encephalopathies, destructive pontine lesions, or as a result of opiate use. Horner's syndrome encompasses anhydrosis, ptosis in addition to miosis. Miosis of pontine or medullary origin may occur as a result of disruption of the descending sympathetic pathways from the hypothalamus. Disruption of the efferent sympathetic fibers in the carotid sheath or the stellate ganglion posterior to the apex of the lung or base of the neck may also produce unilateral miosis [8].

Mydriasis (dilatation of the pupil) occurs when the parasympathetic fibers of the oculomotor nerve to the constrictor pupillae are disrupted, and as a result the dilator pupillae muscle becomes unopposed. It can be unilateral or bilateral, and it can be caused by vascular or traumatic conditions. It may be seen subsequent to aneurysms of arteries that bear close relationship to the oculomotor nerve, such as the posterior communicating, superior cerebellar, and posterior cerebral arteries. Mydriasis and miosis may alternate in patients with Cheyne-Stokes respiration exhibiting mydriasis in hyperventilation stage and miosis in apneustic stage [9].

Traumatic mydriasis is usually on the same side and occurs in response to direct trauma and may not be accompanied by ocular muscle palsy. Traumatic conditions that cause rupture of the middle meningeal artery and formation of epidural hematoma or development of a brain-shifting mass can displace the uncus through the tentorial notch and produce oculomotor palsy and the resultant fixed dilated pupil (mydriasis). Poor vascular perfusion subsequent to increased intracranial pressure or hypotension in trauma patients may eventually cause reversible bilateral mydriasis. That can be reversed with adequate vascular perfusion.

Pupils may not be of equal sizes in both eyes. It may occur in darkness and/or in illumination. This condition is known as anisocoria. Determining whether the pupil is abnormally small or large may require that the pupillary diameter be assessed alternately in bright and dark light. The inability of the pupil to constrict on the side of the sympathetic dysfunction produces sympathetic anisocoria that becomes more pronounced in dim light. This type of anisocoria is associated with iritis, disruption of the cervical sympathetic ganglia, or as a result of application of miotic (miosis inducing) medications. Parasympathetic anisocoria may result from conditions that disrupt the parasympathetic fibers to the constrictor pupillae muscle, such as oculomotor palsy, glaucoma, or subsequent application of mydriatic (mydriasis inducing) drugs such as atropine.

The combination of light-near dissociation becomes evident when the pupil shows constriction in response but not when viewing near objects which is known as Argyll Robertson pupil. This pupil, which responds to light but not to accommodation, is seen in diabetes mellitus, severe vitamin B deficiency, and tertiary syphilis [10]. The lesion that produces this type of pupil is usually located medial to the lateral geniculate nucleus, disrupting the afferent limb of the pupillary light reflex, while preserving the afferent limb of the accommodation reflex. Reverse Argyll Robertson pupil is seen in Parkinson's disease.

The pupil may display constrictions and dilatations that are spontaneous, intermittent, and rhythmical. This condition, which is known as hippus, may be associated with Cheyne-Stokes respiration, cerebral abscess, hysteria, and multiple sclerosis [11].

Adie (tonic) pupil or syndrome exhibits ipsilateral tonic dilation that fails to constrict in accommodation but without other signs of oculomotor palsy. Random regeneration of the injured parasympathetic fibers produces a dilated pupil, which constricts poorly and slowly in accommodation. This syndrome, mainly seen in young females, is associated with other manifestations, such as hyperreflexia, sweating disorders, photophobia, and hypermetropia.

A slow reacting pupil to light or its inability to constrict in response to direct light (appears to be dilated) is known as Marcus Gunn pupil or relative afferent pupillary defect (RAPD). It occurs in patients with ipsilateral retrobulbar neuropathy and patients with lesions of the optic nerve or retina. In this defective pupil, the consensual reflex indicating normal oculomotor nerve. However, when the light is quickly passed from the intact to the affected eye, both eyes show dilatation (positive swinging flashlight test). The depth of perception of moving and colored objects is lost. Exacerbation of the visual deficits subsequent to exercise or any physical activity that increases the body temperature by possibly changing the nerve conduction or producing fluctuating sodium and potassium concentration around the myelin of the optic nerve is known as Uhthoff's sign [12]. The Gunn pupil is a congenital anomaly that is characterized by ipsilateral ptosis (drooping of the upper eyelid) and retraction of the eyelid as a reflex response to deviation of the jaw or depression of the mandible.

The anterior chamber of the eye is bounded anteriorly by the cornea and posteriorly by the iris and the lens. The aqueous humor within this chamber drains into the systemic circulation passing through a trabecular network in its way to the irideocorneal angle and the canal of Schlemm. The posterior chamber is bounded anteriorly by the iris and the lens and posteriorly by the zonular fibers. The pupil enables the aqueous humor to circulate between these two chambers and, thus, help maintain the intraocular pressure. Impairment or obstruction of the aqueous humor circulation may lead to increased intraocular pressure (glaucoma). It transports nutrients, plays a role in respiratory gaseous exchange, and due to its close proximity to the cornea and lens, the aqueous humor could serve as an avenue for removal of metabolites from these refractive media.

Glaucoma is classified into primary and secondary glaucoma. In primary glaucoma, the increase in intraocular pressure occurs independent from any other ocular diseases, while secondary glaucoma develops secondary to other ocular conditions. Primary glaucoma, which is categorized into chronic or open angle, may be congenital or acquired. Acquired primary glaucoma may be associated with obstruction at the irideocorneal angle, sinus venosus sclera (canal of Schlemm), or aqueous veins that drain eventually into the systemic circulation.

In open-angle glaucoma (develops slowly), a gradual increase in intraocular pressure, patients experience a gradual loss of peripheral vision that may end in total blindness. Patients with familial history of glaucoma are prone to develop this most common type of glaucoma in the fourth or fifth decades of life. Patients are asymptomatic at the beginning, and the diagnosis may be confirmed by funduscopic examination. This necessitates regular ocular examination for individuals over the age of forty. As the disease progresses, the optic cup abnormally deepens and permanently impaired. These structural changes are accompanied over time by reductions in retrobulbar and retinal blood circulation [13]. This condition may be treated topically by the administration of pilocarpine, which produces constriction of the pupil and increases the outflow of the aqueous humor through the Schlemm canal. Timolol (Timoptic) can be used for the treatment of this condition. However, side effects on the cardiovascular system limit its usage. Patients with this type of glaucoma may benefit from the use of marijuana to decrease the intraocular pressure.

Adhesion of the iris to the cornea and obstruction of the irideocorneal angle produce increased intraocular pressure and development of closed-angle glaucoma. Patients are usually older than 40 years of age with family history of glaucoma. This condition may be spontaneous or the result of administration of pupillary dilators that cause blockage of the irideocorneal (filtration) angle. Additionally, tricyclic antidepressants with anticholinergic properties may also precipitate this condition. This condition may be acute or chronic. The acute (closed)-angle glaucoma, as the name indicates, develops upon sudden obstruction of the path of the aqueous humor. Due to the sudden nature of the onset, patients experience pain and visual impairment of the affected eye. In this condition, the eye appears red due to dilated blood vessels, and the cornea appears hazy.

A gradual impairment of the aqueous humor circulation at the irideocorneal angle leads to the development of the chronic (angle-closure) glaucoma in which patients exhibit signs similar to that of the acute closed-angle glaucoma. Reduction of production of aqueous humor in this condition may require topical and systemic treatments and even laser iridectomy. Peripheral scotoma may be associated with this type of glaucoma when the etiologies include lesions involving the peripheral retina or the optic nerve. It has been suggested that angle-closure glaucoma is linked to progression of Alzheimer's disease. However, this claim has not been fully investigated and warrants large and adequately designed studies due to heterogeneity of this type of glaucoma [14].

Posterior to the lens and surrounded by the hyaloid membrane, the vitreous body can be seen as a clear and colorless jellylike substance. This membrane is pierced centrally by the hyaloid canal that extends from the optic disc to the posterior pole of the lens. The main constituent of this body is water, but it also contains hyaluronic acid and trace amounts of salts and mucoproteins. The fibrils contained within the vitreous body may occasionally be visible as floating objects. Thickening of the hyaloid membrane at the ora serrata results in the formation of the ciliary zonules; extension of these constitutes the suspensory ligaments of the lens.

The refractive media discussed above may undergo changes that translate into defective vision. In a normal (emmetropic) eye, the lens is flat and the suspensory ligaments of the lens are taut and, therefore, no optical defects exist. However, accommodative changes will be needed, such as an increase in lens curvature, constriction of the pupil, and convergence when viewing near objects. An increase in lens curvature shortens the focal distance and allows images from closer objects to fall on the same point of the retina in both eyes. However, these changes become effective and properly functional when the anteroposterior dimension of the eyeball and refractive media is within normal range. A variety of visual defects that range from myopia, hyperopia, and astigmatism to presbyopia, and anisometropia, may ensue if the anteroposterior dimensions of the eyeball or refractive media are out of normal range.

The most common optical disorder is hyperopia (farsightedness). In this condition, which is corrected by a convex lens, the image from a distant visual field falls behind the retina in the relaxed eye due to an unusually short axis of the eyeball or weak refractive power of the lens or cornea. Another optical disorder is myopia (nearsightedness) in which the focal point of an image from a distant visual field falls anterior to the retina instead of the retina, restricting the ability to visualize far objects. However, as the object moves closer, the focal point moves closer until the image is marked on the retina and a clear vision is attained. This disorder, which may be corrected by a concave lens, may occur as a result of an abnormally long eyeball axis or an unusually strong refractive power. It may be associated with rhegmatogenous retinal detachment, which is characterized by detached retina which breaks up into pieces.

An irregular surface (egg shaped) of the cornea and/or lens results in a change in the angle of refraction of the light rays and subsequent of funneling these rays to different focal points on the retina and the resultant astigmatism. Patients with this condition experience blurred vision, which can be corrected by a cylindrical lens that has a flat surface on one axis and concave or convex on the other.

Another disorder of the refractive media is presbyopia, which is characterized by the conversion of the lens into an inelastic and hard structure that lacks or has limited power of accommodation. Patients with this condition are hyperopes (farsighted). Despite the perceived inflexibility by presbyopic patients, multifocal intraocular lenses appear to provide the steadiest and dependable surgical treatment compared to contact lenses [15]. When the refractive powers between both eyes are different, the condition is known as anisometropia.

The innermost layer of the eyeball is the tunica nervosa, which consists of the pigment epithelial layer and retina. The pigment epithelium is loosely bound to the retina that contains the photoreceptor layer. It nourishes the photoreceptors, offers physical support, and absorbs excess light. The retina, a derivative of the diencephalic optic vesicle, is comprised of a visual (optic) part, ciliary part that covers the ciliary body, and iridal part that covers the iris. The optic part contains photoreceptors and is lightly connected to the pigment epithelium from which it derives nourishment and mechanical support. Congenital hypertrophy of retinal pigment epithelium (CHRPE) seen in early postnatal life is considered a commonly occurring non-digestive system manifestations (90 %) of patients with familial adenomatous polyposis (FAP). The latter is an autosomal dominant disease in which patients exhibit multiple benign adenomatous colic polyps that may become malignant later [16].

The site of junction of the optic and ciliary parts of the retina is marked by a notched area, the ora serrata, where the photosensitive layer of the retina and the pigment epithelium unite to form a barrier restricting the spread of pathological processes and subretinal exudate.

Separation of the optic part of the retina from the adjoining pigment epithelium is known as detachment of the retina. Patients with this condition may see light flickers or moving bodies in the visual filed. This separation may be complete or partial, occurring as a result of traumatic conditions or pathologic mechanisms. Direct traumatic condition may induce a rhegmatogenous retinal detachment. Chronicity of this condition may produce, although rarely, peripheral retinal sea-fan neovascularization. Regression of the new vessels can be arrested by cryotherapy and a segmental buckle limited to the sites of retinal break [17].

Disease processes that result in the formation of exudate from the choroid layer to sweep into the subretinal space produce exudative retinal detachment. Nontraumatic processes such as diabetic vitreoretinopathy can produce retinal detachment in which the retina is left intact but endures undue traction by the fibrovascular bundles that stretch between the vitreous body and the retina.

Histologic examination of the optic part of the retina reveals ten layers with nine types of cellular elements. The tunica nervosa contains photoreceptors, which are categorized into cones and rods. The total number of the cones varies from 6 to 7 million and is centrally located but also peripherally scattered among the rods. The site of acute vision, the fovea centralis, consists only of cones, lies inside the macula lutea and on the lateral to the optic disc. Cones, which are designed to facilitate daylight (photopic) vision and color discrimination, exhibit higher threshold of excitability, maintaining 1:1 synapse ratio with the bipolar neurons at the outer plexiform layer.

In contrast to the cones, rods are peripherally scattered and activated by lower illumination (dim light) and are the most abundant, averaging between 100 and 120 million. Due to low threshold of excitability, the rods are specialized for scotopic (achromatic) or twilight vision, enabling visualization of black, white, and gray colors. Cones and rods maintain synaptic connections through the dendrites of horizontal cells in the inner nuclear layer. Retinal layers are glued together by the processes of the large glial (Müller) cells that form the outer limiting membrane. A unique characteristic of the rods is the ability to contract in dim light, maximizing the surface area exposed to light. Proximity of the pigment epithelium to the rods allows these epithelial cells to remove the sloughed discs from the outer segments of the rods.

Accumulation of pigment cells and floating discs of the outer segments of the rods is seen in retinitis pigmentosa (RP). This is a genetic disease in which mutation of peripherin and rhodopsin in the outer segments causes disc fragments and clusters of black pigment "bone spicules" to be deposited in the periphery of the retina. The presence of the bone spicules and disc fragments impairs the dissemination of nutrients from the capillary network of the choroid layer to the cones and rods and the resultant retinal degeneration and the formation of funduscopically visible retinal black-colored lesions. In this blinding disease, advancing nyctalopia (night blindness) and focal blindness in the form of ring scotoma are seen. Interestingly, the midportion of the periphery of the ring scotoma is affected largely compared to the center and outermost part of the retina that remain intact. The degree of severity of RP can be correlated with quantitative attenuation in the retinal vessels and intensity of oxygen saturations. This fact can serve as a possible clinical marker in the evaluation of RP and extent of its progression [18].

Mediation of color vision is accomplished by the cones, which are centrally located, and, thus, transmissions of color-related visual impulses are uniquely segregated from other modalities. The 1:1 synapse ratio of the cones continues to the lateral geniculate nucleus and through the optic radiation to the visual cortex. All three-color cone genes maintain a common ancestral red gene. The red gene may have given rise to the blue cone pigment, which, in turn, may have given origin to the red and green cone pigments. Color blindness may be universal across all colors (achromatopsia) or toward two colors (dichromatopsia) or may be confined to a single color (monochromatopsia). This disability, which affects predominantly males, can be inherited (sex linked) or acquired as a result of chemical or physical injury. The prevalence of this condition in males may be correlated with the variation in the amount of photopigments in the blue, green, and red cones. This is particularly true with regard to the red and green genes, which exist as a recessive trait on the male X chromosome. One percent of males are protanopes (lacking the red gene and associated long wave mechanism), while 2–3 % are deuteranopes, because of the deficiency of the green gene and the related medium-length mechanism. The rarity with which the blue color gene can be affected by mutation is attributed to the fact that this gene exists on an autosome on the seventh chromosome [19, 20].

Reduction in the amount of cone pigment and not the neuronal circuitry associated with processing of the color vision can explain protanomaly in which trichromats with normal three-color vision or with one normal color vision and two feeble red color vision. The same fact accounts for deuteranomaly and tritanomaly in which trichromats may also have three feeble green vision or four weak blue vision, respectively. Dichromats, with two-color vision, lack one of the pigments. They may be protanopes deficient of red color perception due to the absence of erythrolabe, or deuteranopes deficient of green color perception subsequent to the absence of chlorolabe, or tritanopes, who are unable to see the blue color, lacking the cyanolabe and the associated short wavelength mechanism. The ratio of deuteranopia to protanopia among color blind individuals is 3:1. Individuals with red-green color blindness may have a hybrid gene on the X chromosome due to gene loss or recombination between genes. Color blindness may also be acquired. Tritanopia may also develop as a result of pathological conditions that affect the outer layer of the retina producing loss of the processing mechanism of short wavelength. Similarly, blindness to red-green color may occur subsequent to pathological processes that affect the inner retinal layer or the optic nerve.

Constellation of color blindness, blurred vision, loss of contrast sensitivity that result in indistinct contours and shadows, metamorphopsia, central scotomas, and/or drastic reduction in central vision may be seen in macular degeneration, which is a pain-free condition that affects individuals over the age of 50 years [21]. Contrary to past reports, a rigorous meta-analysis indicates an inconsistent but statistically significant link between the risk of age-related macular degeneration and the use of aspirin [22].

This condition is categorized into nonexudative (dry) or exudative (wet) type. In the nonexudative (dry) type, distinct yellowish clusters of cellular remains (drusen) deposit around the macula lutea causing funduscopically visible lesions. Drusen can be associated with high levels of cholesterol deposits. As the condition advances, these lesions unite and produce death of the retinal pigment epithelial cells compromising the flow of nutrients and mechanical support and the atrophy of the cones and central vision. In addition, cellular debris may also deposit between the retinal pigment epithelium and choroid layer and impair the nourishment of the photoreceptors leading to retinal detachment. Despite the lack of clarity with regard to the pathogenesis of dry AMD, the presence of β-amyloid in the drusen points to the similarities between AMD and other degenerative diseases. The role of the inflammatory process in the development of this disease is based on the discovery of Nod-like receptors containing pyrin domain 3 inflammasome and dysfunction of complement system. Oxidative stress, mitochondrial disorders, and reduction in autophagic capacity in the aging retinal pigment epithelial cells symbolize metabolic basis of the disease [23].

In the exudative (wet) type, a less common form, the release of blood and protein under the macula occurs through the Bruch's membrane from the newly formed abnormal capillaries as a result of the neovascularization of the choriocapillaris in the choroid layer. Leakage of blood leads to scar formation and induces irreversible retinal dysfunction. The main initial deficit in the exudative macular degeneration is the inability to identify misalignment of visual objects (visual hyperacuity). This deficit can be evaluated by the preferential hyperactivity perimetry. Blindness may occur in advanced stages of the disease, though rarely, when the photoreceptors are irreversibly damaged. In addition to older age, family history of the disease and the presence of the genes for factor H (CFH), factor B (CFB), and factor 3 (C3) play important roles in the pathogenesis of this disease. Prevention of blindness may require the early administration of angiogenesis inhibitors, and anti-VEGF-A factors, such as ranibizumab (Lucentis) or bevacizumab (Avastin). Studies have shown that reduction of the vitreal VEGF-A levels in exudative age-related macular degeneration can be achieved by the combined administration of omega-3 supplementation with anti-VEGF treatment [24].

The bipolar neurons, which represent the first-order neuron of the visual pathway, are classified into depolarizing (invaginating) neurons and hyperpolarizing (flat) neurons. Depolarizing neurons activate the “on” type ganglionic neurons and are inhibited by darkness and released from inhibition by illumination. Hyperpolarizing neurons excite the “off” type multipolar ganglionic neurons and are inhibited by light. The nuclei of these neurons are located in the inner nuclear layer, while the axons extend into the inner plexiform layer and form synaptic contacts with the dendrites of the multipolar ganglion cells.

The amacrine, the only excitable neurons of the retina, facilitate majority of visual processing. They resemble the granule cells of the olfactory bulb and establish inhibitory synaptic connections with the dendrites of the multipolar ganglion cells. The type of neurotransmitter they utilize and the morphology of their dendritic arborization enable classification of these diverse retinal cells. They form reciprocal connections with the bipolar neurons [25]. The horizontal cells intensify contrast by establishing inhibitory dendrodendritic synapses with the bipolar and multipolar ganglionic neurons.

The ganglionic multipolar neurons establish dendroaxonic synaptic contacts with the bipolar neurons in the inner plexiform layer. Some ganglionic neurons in the medial halves of the retina play a role in maintaining the fixation of the visual field image on the fovea centralis. Fixation of the image on the fovea centralis occurs by projection of the ganglionic neuronal axons to the inferior olivary nucleus via the midbrain reticular formation and by the inferior olivary nucleus projection via the olivocerebellar fibers to the Purkinje neurons that receive afferents from the medial vestibular nucleus. These multipolar neurons possess the ability to fire at a fairly steady rate irrespective of the presence or absence of visual stimuli. These neurons, which form the second-order neurons in the visual pathway, exhibit “on” or “off” characteristics relative to their synaptic contacts with the bipolar neurons. They are classified into sustained X, transient Y, and intermediate W cells. Sustained X cells explore the shapes and colors of objects in the visual field, exhibiting constant on or off response. Despite their limited number, transient Y cells are important for responding to and detecting rapidly moving objects through projection to the superior colliculus and thalamus. W cells mediate the pupillary light reflex through projection to the pretectum, which contains the efferent neurons of the reflex.

The retina can be partitioned by the optic axis into medial (nasal) and lateral (temporal) halves; a horizontal plane further divides each half of the retina into upper and lower quadrants. The nasal half of the retina of each eye receives visual impulses from the temporal half of the visual field and vice versa. The upper quadrant of the retina for each eye visualizes the lower quadrant of the contralateral visual field and vice versa.

The site of emerging axons of the ganglionic neurons of the retina is known as the optic disc. Lack of photoreceptors at this site renders it a blind spot of the retina. In a healthy individual, the optic disc exhibits a more sharply defined lateral edge than the medial edge. Further, the optic disc appears pinkish in fair-colored and yellowish-orange in dark-skinned individuals. Central retinal vessels course in the center of the optic disc after piercing the optic nerve and then branch in a radial pattern supplying the four quadrants of the retina. These central retinal artery branches show morphologic changes in hypertension and patients with arteriosclerosis. Retinal exudate, bleeding into the plexiform layer, stenosis of the retinal arteries, cotton wool spots, and lipid in the macula lutea may accompany essential or malignant hypertension.

The physiologic cup is the central lighter-colored part of the disc, which is penetrated by central retinal vessels. The normal cup to disc (C/D) ratio of 1:5 may be lost in late stages of glaucoma. In the initial stages of glaucoma, the increase in the dimensions of the cup may not be perceived as pathological due to the fact that even with this change in the cup size, it remains smaller than the physiological cup. Based on this gauging, the C/D ratio may not by itself bear a meaningful diagnostic significance unless the change is major. A dramatic alteration of the C/D ratio becomes significant because the enlargement of the optic cup will superimpose upon the physiological cupping that preceded the onset of the glaucoma pressure. Glaucomatous cups are usually larger than physiological cups, although a large cup may not always mean it is pathological. Only 2 % of healthy eyes have a ratio more than 0.7. In nearly two-thirds of patients with glaucoma, the difference in the C/D ratios is more than 0.1, though the same difference exists among one-tenth of healthy patients.

The depths of the optic cup and the lamina cribrosa can help in the differentiation between compressive optic neuropathy (CON) with glaucomatous optic neuropathy (GON). Studies have shown the lack of laminar remodeling apparent from the absence of laminar surface position in the patients with CON compared to those with GON [26].

Scotoma refers to deficient or absent vision in a specific area of the visual field in the form of the dark or colored area that the patient is aware of them (positive scotoma) or oblivious to the deficit (negative scotoma). This is commonly seen after stroke, multiple sclerosis, hypertension, and metabolic or nutritional disorders. Central scotoma interferes with acute central vision and is associated with a small lesion or petechial hemorrhage in the retina near the optic disc, whereas peripheral scotoma produces deficient or depressed vision in the outer segment of the visual field distant from the gaze fixation point. Physiological scotoma is an optic disc-induced condition, while arcuate scotoma, a glaucoma-induced condition, is the result of damage to the retina or optic nerve. This arch-shaped scotoma curves superiorly or inferiorly toward the nasal field of the retina. Scintillating or flittering scotoma is a form of central scotoma in which transient irregular sparks and flickering flashes and lucid geometric forms, sometimes with a zigzag (fortification) or city wall-like pattern, hover in the visual filed for a period of time. It is usually seen secondary to occipital lobe lesion, but may also be associated with migraine (migraine aura).

Vitamin A encompasses the retinoic acid, retinol, retinal, and provitamin-A carotenoids. Retinal form, 11-cis-retinal, represents the light-absorbing molecule that mediates visualization of objects in dim light and illumination. It binds to the opsin to form rhodopsin in rods and iodopsin in cones at lysine residue. Illumination isomerizes the 11-cis-retinal to the retinal form that detaches from the opsin. Later, the retinal form is recycled and converted back to the 11-"cis"-retinal form. Despite this conversion, some of the retinal form may be transformed to retinol form and then conveyed via a specific receptor known as interphotoreceptor retinol-binding protein (IRBP) to the pigment epithelium. Esterification of the retinol to retinyl esters enables storage of retinol for later usage. Darkness enables the reverse changing process of rebinding 11-cis-retinal to the opsin, leading to the formation of rhodopsin essential for night and dim light. A corollary to this is the fact that retinal binding and the formation of the rhodopsin can be impeded when vitamin A is deficient leading to nyctalopia (night blindness). This may occur secondary to gastric bypass surgery as this operation produces deficiency of the fat-soluble vitamin A [27]. Tay-Sachs disease shows cherry red spots on the retina that can be seen through funduscopic examination.

The retina receives blood supply through branches of the central retinal artery, which arises from the ophthalmic artery, a branch of the internal carotid artery. The ophthalmic artery arises from the internal carotid artery medial to the anterior clinoid process and then gains access to the orbit through the optic canal accompanied by the optic nerve. In addition to the central retinal artery, the ophthalmic artery also gives rise to the long and short posterior ciliary arteries that provide additional blood supply to the retina via the pigment epithelium.

Prior to its division to end arterial branches that supply the four quadrants of the retina, the central retinal artery pierces the optic nerve as it courses in the orbit. The branches of the central retinal artery lie superficial to the corresponding veins appear thinner and brighter red and less in number compared to the veins (ratio of 2:3). Arteriovenous nicking in hypertensive individuals is accompanied by narrowing or spasm and thickening of these vessels or assumption of orange metallic color and appearance of a discontinuous venous column. Incomplete occlusion of the internal carotid artery or compression of the ophthalmic artery by an increased intraocular pressure on one side may produce amaurosis fugax, which is characterized by transient blindness. This type of blindness appears as a sudden blackout or cloudy vision appearing as a vertical or horizontal shade across the visual field lasting from seconds to minutes on the affected side. Reduction of the blood pressure in both ophthalmic arteries relative to the pressure in the brachial arteries may point to possible occlusion of both carotid arteries. Occlusion of one of the four branches of the central retinal artery may produce quadranopsia.

As the unmyelinated axons of the ganglionic layer of the retina leave the eyeball through the lamina cribrosa sclera, they join to form the optic nerve. The retrobulbar part of the optic nerve acquires myelin and courses partly inferior to the frontal lobe. The optic nerve also forms the common afferent limb for both the pupillary light and accommodation reflexes. This nerve develops with the retina from the telencephalon, a fact that renders this nerve a CNS structure. As it runs in the orbit and optic canal, it is invested by the meninges and the intervening space and the CSF.

A unique arrangement of the optic nerve fibers is displaced in which those arising from the fovea centralis pursue a straight course to the temporal quadrant of the optic disc, resulting in the formation of a centrally positioned and spindle-shaped papillomacular bundle. Additionally, the fibers that arise from upper and lower parts of the macula meet at the upper and lower poles of the optic disc, respectively. While those that originate from the medial retina assume a radial course compared to the fibers of the lateral retina that maintain an arcuate path around the papillomacular bundle. The upper and lower medial fibers also contain fibers from the peripheral lateral parts of the retina. The papillomacular bundle is more prone to metabolic, toxic, and compressive injuries and is, thus, predisposed to the development of central scotomata that may expand to the optic disc due to its central location within the optic nerve, high metabolic activity, and reliance on the blood supply of the ciliary arteries. Despite this fact, the papillomacular bundle remains most resistant to compressive effect of glaucoma compared to the more susceptible arcuate fibers to this type of injury that reach the upper and lower temporal aspects of the optic disc. Multiple sclerosis, which affects the temporal part of the optic disc, disrupts this bundle producing central scotoma. Paget's disease, suprasellar tumors, fractures involving the optic canal, glaucoma, optic neuritis, ischemic neuropathy, papilledema, optic nerve drusen, and congenital dysplasia are conditions that may cause damage to the optic nerve. A slowly progressive optic neuropathy associated with "cupping" atrophy of the optic disc may be induced by glaucoma. Patients with this condition initially experience scotomata, which appear as circles or arches in the peripheral visual field that may advance to tunnel vision.

Inflammation of the optic nerve and the development of retrobulbar demyelinating patches is known as optic neuritis. When the inflammation is confined to the intraocular part of the optic nerve, it is known as papillitis. These pathologic changes may be associated with diabetes mellitus, herpes zoster, multiple sclerosis, syphilis, Lyme disease, and vasculitis. It also occurs in measles, mumps, or infection with varicella viruses. Optic neuritis is seen in nearly 50 % of patients with multiple sclerosis (MS), and one-third of patients with optic neuritis can develop other signs of MS. Optic neuritis in MS patients produces pallor of the temporal quadrant of the optic disc and accompanied by defects in temporal halves of the visual fields. MS commonly affects individuals between the ages of 20–40. Examination may be accompanied by demyelination in the white matter of the brain and papillomacular bundle of the optic nerve fibers. Multiple sclerosis may also be associated with degeneration of the foveal part of the retina and the corresponding part of the optic nerve as well as optic neuritis or retrobulbar neuritis. Episodes of eye pain triggered by ocular movements followed by loss of acute vision and color blindness, particularly to red, may also be manifested in patients with MS. This disease shows bilateral manifestations in pediatric patients as compared to adults, who usually show unilateral symptoms. The recovery of vision may occur spontaneously after few months irrespective of medical therapy.

Anterior ischemic optic neuropathy, a condition induced by an unusually small optic disc due to optic ischemia, is characterized by early morning sudden, painless, and unilateral visual loss and nocturnal hypotension. This ischemia may be associated with giant cell (temporal) arteritis, a systemic vasculitis that affects the superficial temporal artery and other branches of the external and internal carotid arteries including the ophthalmic artery. This condition is seen at the age of 60 years or older, presenting with ipsilateral or bilateral visual loss, accompanied by diplopia, tinnitus, temporal scalp tenderness, weak pulsation, fever, and jaw claudication.

Polymyalgia rheumatica is a preexisting condition in 50 % of patients with temporal arteritis. Due to the fact that the presence of giant cells in the biopsy specimens of affected arteries is not a universal finding in this disease due to segmental pattern of vasculitis, multiple biopsies from different segments of the vessel may be needed to confirm the diagnosis. Temperate elevation of the optic disc margin, exudate, and bleeding observed upon funduscopic examination may accompany inflammatory conditions of the optic disc.

Patients with increased intracranial pressure usually manifest papilledema associated with ipsilateral pulsatile synchronous tinnitus, transient blurred vision (visual obscuration), or even hemianopsia. Increased intracranial pressure may also produce horizontal diplopia due to abducens nerve palsy. Papilledema (choked disc) refers to the bilateral passive elevation of the margins of the optic discs, dilation of the subarachnoid space around the optic nerve, and impairment of the venous return from the retina as a result of increased intracranial pressure. It may be detected through funduscopic examination of the retina and the observation of the dilated central retinal veins. Papilledema may also result from optic nerve sheath meningioma or gliomas, tectal gliomas, cerebellar medulloblastoma, ependymoma of the fourth ventricle, and cerebral and corpus callosal tumors. Medulloblastoma, a cerebellar tumor, may produce papilledema earlier than any other CNS tumors by invading the fourth ventricle, obstructing the CSF circulation, and causing increased intracranial pressure. On the contrary, pontine or medullary tumors do not generally interfere with CSF circulation and, thus, do not usually produce papilledema. A major danger in these patients is the fact that death may ensue from brain stem compression before any detectable sign of papilledema. Pseudotumor cerebri (idiopathic intracranial hypertension), as the name indicates, is another condition that produces increased intracranial pressure in obese females of childbearing age. Cavernous sinus thrombosis and ethmoidal cellulitis may contribute to papilledema by impairing the venous blood flow in the central retinal vein. Papilledema is rarely seen in congenital cyanotic conditions such as tetralogy of Fallot, persistent truncus arteriosus or tricuspid atresia, or in inflammatory conditions that produces polyneuropathy, such as Guillain-Barre syndrome (an idiopathic acute febrile inflammatory disease).

Bilateral progressive accumulation of globules of mucopolysaccharides and proteinaceous material (hyaline bodies or optic nerve drusen) anterior to the lamina cribrosa may lead to compression of the optic nerve axons and associated central retinal vessels causing scar formation. The development of abnormal neovascularization and choroidal neovascular membrane and the growth of abnormal blood vessel beneath the retina may cause the loss of central acute vision. These changes, which affect males and females to an equal degree, result from axonal degeneration of the retinal multipolar neurons or inertia of the axoplasmic transport. It may be associated with retinitis pigmentosa and Noonan syndrome and can mimic the manifestations associated with papilledema and increased intracranial pressure. Fortunately, several examination techniques are currently available to diagnose optic nerve drusen including ultrasound, fundus photography, fluorescein angiography, computerized perimetry, autofluorescence funduscopic examination, optical coherence tomography (OCT), or nerve fiber layer analyzer [28].

Confrontational visual field testing can be utilized to check intactness of the optic nerve through examination of visual acuity and visual fields. This test starts with the examiner and the patient standing at eye level opposite each other followed by closure of the left eye of the patient and the right eye of the examiner. Then, the examiner shows one or two fingers in each hand concurrently, asking the patient to confirm the number of fingers he/she has seen. This test will be repeated in the same manner with the other eye from superior to inferior quadrants. The fingers should be seen at the same time by the examiner and patient in healthy individuals. Widening of the visual field of a patient by pulling the examiner's hand away from the patient may help detect glaucoma-induced scotoma and possibly tumors of the central nervous system. Specifying the timing of the appearance and the direction of the examiner's flashing light beam or pencil by the patient may also be used for this purpose.

The Snellen eye chart may be used to evaluate visual acuity. This chart is positioned approximately 20 ft from the patient and each eye is tested independently. The first number in the 20/20 standard ratio shows the actual distance of the patient from the chart, while the second number in the ratio is the distance at which a healthy eye can read the Snellen chart. In order to accurately determine the visual acuity and thus macular function, each eye should be tested separately with and without the patient's glasses. For this purpose, the examiner may present a newspaper article, a picture, or small objects and request the patient to identify them. Visual acuity may vary by environmental factors such as illumination and degree of contrast.

The swinging light test is another method by which the difference in response between both eyes to afferent stimuli may be determined. This requires the patient to look at a distant object while the examiner swings a light beam from one eye to the other in a very rapid fashion. Due to the lack of afferents to the retina and optic nerve, neither eye will constrict when directing the light into the blind eye. However, when moving the light back to the unaffected eye, the blind eye shows apparent pupillary dilatation. This is known as Marcus Gunn pupil [29].

Rostral to the infundibulum and tuber cinereum, the multipolar neuronal axons from the nasal part of the retina and macula lutea that run in the optic nerve decussate to form the optic chiasma. The optic chiasma lays superior and rostral to the pituitary gland and inferior and anterior to the third ventricle. The presence of the internal carotid at or near the site of its bifurcation into the anterior and middle cerebral arteries on the medial side of the optic chiasma explains the contralateral monocular nasal hemianopsia seen in patients with internal carotid artery aneurysm. It also lies caudal to the anterior communicating and medial to the posterior communicating arteries. The arterial blood supply to the optic chiasma is maintained through this vascular network. Similarly, the venous blood drains into the adjacent anterior cerebral vein and also to the basal vein of Rosenthal.

Chiasmal fibers are arranged in a manner in which the inferior nasal fibers occupy a rostral position to that of the superior fibers, whereas fibers from the superior nasal quadrant form a short loop (anterior knee fibers of von Willebrand) that curves into the medial part of the contralateral optic nerve before joining the optic tract. The presence of the anterior knee fibers accounts for the contralateral superior temporal quadranopsia, which accompanies optic nerve lesion. Likewise, the fibers from the superior nasal quadrant of the retina form a short loop that curves backward into the ipsilateral optic tract known as the posterior knee fibers of von Willebrand before joining the optic chiasma. Macular fibers and nearby central retinal area occupy the center of the optic chiasma. Beyond the optic chiasma, the temporal fibers continue caudally to join the ipsilateral optic tract, whereas the crossed fibers join the contralateral optic tract.

The anterior pituitary (adenohypophysis) location posterior and inferior to the optic chiasma has clinical implication. Adenoma, a benign tumor of the pituitary gland, and craniopharyngioma can disrupt the crossed nasal fibers that form the optic chiasma and produce bitemporal heteronymous hemianopsia (tunnel vision) in which pedestrian patients report bumping into people or stationary objects, such as electric poles during walking. Patients frequently state that they can only read left-hand letters only with the right eye and the right-hand letters with the left eye. Initially, the visual deficit associated with the benign tumor is confined to the upper temporal quadrant but with no deficits in central vision, visual acuity, and color vision. This is due to the fact that pituitary adenoma is slowly expanding the tumor size and disruption of the crossed nasal fibers occurs gradually. As the tumor becomes larger, the nasal fibers from the upper medial quadrant of the retina will disrupt, leading to inferior temporal quadranopsia. At later stages complete visual loss in the temporal visual halves of the visual field will be apparent. A pituitary tumor may be accompanied by radiographically detectable sellar enlargement. This type of tumor can be surgically removed by trans-sphenoidal guided by ultrasonography [30].

A variety of deficits may occur including superior or inferior bitemporal quadranopsia or monocular temporal hemianopsia in patients with chiasmal lesion due to pattern of chiasmal fiber arrangement. Disruption of the fibers emanating from the inferior nasal part of the retina located in the rostral part of the optic chiasma near its transition with the optic nerve produces junctional scotoma in the form of superior temporal quadranopsia. Patients also experience ipsilateral central blindness. This deficit may be seen in pituicytoma, a rare benign glioma, which is thought to derive from the pituicytes of the posterior pituitary or the folliculostellate cells of the anterior pituitary [31].

Posterior junctional scotoma, which is characterized by incongruous contralateral hemianopsia and ipsilateral inferior temporal quadranopsia, is the result of a lesion that affects the posterior chiasma and beginning of the optic tract. Incongruous contralateral hemianopsia occurs due to separation of the crossed and uncrossed fibers at the site of the lesion and disruption of the nascent optic tract fibers, whereas damage to the crossing superior nasal fibers of the optic chiasma in this lesion is responsible for the ipsilateral inferior temporal quadranopsia.

Due to the close proximity of the internal carotid arteries to the ipsilateral temporal fibers of the optic chiasma, aneurysms of these vessels produce binasal heteronymous hemianopsia. Ipsilateral nasal hemianopsia is associated with the aneurysm of the internal carotid artery on the side of deficit.

The optic tract is the result of the union of the decussated nasal fibers and ipsilateral temporal fibers of the optic nerve that convey visual impulses from the opposite visual field. It pursues a course between the tuber cinereum of the hypothalamus and the anterior perforated substance in close proximity to the internal carotid artery and then the crus cerebri before terminating largely in the lateral geniculate nucleus (LGN). The fibers that originate from the macula lutea occupy an intermediate position, and the fibers from the upper quadrant of the retina reside in an anterior and medial location, while the fibers from the lower quadrant of the retina occupy a more lateral and posterior position in the optic tract.

Most of the fibers of the optic tract project and establish synaptic linkage with the neurons of the lateral geniculate nucleus (LGN), a visual relay nucleus of the thalamus. Some fibers of the optic tract bypass the LGN and terminate in the pretectum that contains the efferent neurons of the pupillary light reflex. These pretectal efferent neurons activate the Edinger-Westphal nuclei bilaterally through the crossed fibers of the posterior commissure. Presynaptic parasympathetic axons of the Edinger-Westphal neurons project to the neurons of the ciliary ganglion located between the optic nerve and the lateral rectus. Activation of the ciliary ganglion neurons causes contraction of the sphincter pupillae and ciliary muscles through the short ciliary nerves, mediating light and accommodation reflexes, respectively. The blood supply of the anterior one-third the optic tract is provided by branches of the internal carotid, middle cerebral, and posterior communicating arteries, whereas a single anterior choroidal artery is responsible for the vascularization of the posterior two-thirds.

Lesions of the optic tract can produce deficits in other areas of the visual pathway. For instance, disruption of the ipsilateral temporal retinal fibers causes upper and lower pole optic disc atrophy characterized by disc pallor, while atrophy of the temporal and nasal poles of the optic disc produces "band or bow tie," subsequent to disruption of the contralateral nasal fibers and the contralateral nasal papillomacular bundle. Partial damage to the optic tract produces characteristic configuration of wedge-shaped loss of visual field. A compressive lesion that completely disrupts the fibers of the optic tract produces contralateral homonymous hemianopsia accompanied by mild afferent pupillary defect on the contralateral side due to the fact that more fibers enter the optic tract from the contralateral side than the ipsilateral side.

The lateral geniculate nucleus (LGN) is the third-order neuron in the visual pathway that receives retinal fibers contained in the optic tract. Despite the fact that LGN is a visual synaptic station, retinal input constitutes only 20 % afferents, while the posterior parietal, occipital lobes and the midbrain reticular formation form the rest of the input to LGN. Fibers that bypass the LGN synapse in the pretectal area and the superior colliculus to mediate light and accommodation reflexes. This six-layered thalamic nucleus exhibit 1:1 synapse ration enabling the creation of the contralateral half of the visual field. In addition to these synaptic characteristics, the LGN also shows retinotopic arrangement in which layers 1, 4, and 6 receive fibers from the contralateral retina through the crossed nasal fibers, whereas layers 2, 3, and 5 establish synaptic linkage with fibers from the ipsilateral temporal fibers of the retina. In addition to the retinotopic arrangement, the lateral geniculate nucleus exhibits somatotopic synaptic arrangement with the incoming fibers of the optic tract. This arrangement entails that the optic tract fibers from the upper retinal quadrant terminate in the medial part of the lateral geniculate nucleus, whereas fibers from the lower retinal quadrant are received by the lateral part of the LGN. Macular fibers that represent the central acute vision retain its central position in the LGN.

Morphologically, layers 1 and 2 constitute the ventral (magnocellular) subnucleus, whereas layers 3–6 form the dorsal (parvicellular) subnucleus of the lateral geniculate nucleus. The magnocellular subnucleus contains neurons that quickly process and transmit visual input from the rods, and the parvicellular subnucleus conveys detailed visual information, though slowly, from the red and green cones. A less distinct layer with very small neuronal bodies sandwiched between the previous subnuclei is termed the koniocellular (interlaminar) subnucleus. It is thought to enhance integration of proprioceptive signals with visual perception and possibly transmitting visual input from blue cones of the retina [32].

Neuronal axons of both dorsal (parvicellular) and magnocellular subnuclei of LGN form the geniculocalcarine tract (optic radiation) in which the lateral and medial parts of this radiation establish synaptic connections with neurons of the inferior and superior banks of the visual cortex, respectively. Prior to their termination in the visual cortex, the LGN neurons occupy the retrolenticular portion of the internal capsule. LGN receives its primary blood supply from the branches of the posterior cerebral and posterior communicating arteries.

A lesion, which causes degeneration of the area medial to the lateral geniculate body (as is seen in tertiary syphilis), may selectively disrupt the fibers that mediate light reflex, sparing the accommodation reflex. Patients with this type of lesion may exhibit Argyll Robertson pupil that accommodates (constricts) when viewing near objects but does not respond to light.

The optic radiation (geniculocalcarine tract) represents the myelinated postsynaptic fibers of neurons from the dorsal (parvicellular) and magnocellular subnuclei of the lateral geniculate nucleus. This part of the visual pathway is crescent shaped with superior and inferior parts that course within the retrolenticular part of the internal capsule en route to the visual cortex. The upper and lower halves of each optic radiation represent one-fourth of the contralateral visual field. Superiorly positioned groups of optic radiation fibers, which are derived from the upper retinal quadrant, follow a straightforward path through the parietal lobe to the upper bank of the calcarine fissure represented by the cuneus. The inferiorly located group of fibers, which originates from the lower retinal quadrant, forms a curve into the temporal lobe, as the Meyer's loop, running initially adjacent to the amygdala and the tip of the inferior horn and then on the lateral side of the posterior horn of the lateral ventricle. This loop rejoins the rest of the optic radiation, transmitting visual impulses from the contralateral upper visual field and terminating in the lower bank of the calcarine fissure within the lingual gyrus. The foveal fibers occupy the most lateral position within the optic radiation.

The rostral part of the optic radiation is principally supplied by the anterior choroidal branch of the internal carotid artery and, secondarily, by the posterior choroidal artery, a branch of the posterior cerebral artery. The latter also supplies the posterior part of the optic radiation through the calcarine branch of the posterior cerebral artery. The central part is the middle cerebral artery. Transient homonymous hemianopsia may occur upon compression of the optic radiation by lateral striate artery bleeding-induced edema. This deficit will last until the edema subsides. Likewise, occlusion of the anterior choroidal and posterior cerebral arteries can also compromise the function of the optic radiation.

Contralateral homonymous hemianopsia due to disruption of the optic radiation may also be seen as a sequel to parietal or occipital lesions or from lesions of the optic tract and LGN. However, the presence of relative afferent pupillary defect and "bow tie" optic disc pallor in patients with optic tract damage may serve as distinguishing features. In parietal lobe lesion that disrupts the optic radiation, the visual field defects are bilateral and homonymous, accompanied by the loss of ipsilateral smooth pursuit ocular movement. Bilateral occipital lobe lesion and optic radiation damage may occur due to ischemia induced by hypotension, occlusion of the posterior cerebral or basilar arteries, contrecoup head trauma, or cardiac arrest. The contralateral homonymous hemianopsia that occurs in this lesion exhibits a distinguishing defect in the shape of a keyhole in the vertical meridian.

Mildly incongruous contralateral upper quadranopsia occurs as a result of selective damage to the optic radiation in the temporal lobe. This frequently occurring lesion involves the Meyer's loop that follows a separate course within the temporal lobe before joining the upper fibers of the geniculocalcarine tract. This visual field defect appears as a wedge or "slice" removed from the upper visual field and is commonly termed a "pie in the sky." An abscess that develops in the temporal lobe above the level of the auditory meatus may also disrupt the fibers of Meyer's loop and cause contralateral quadranopsia. Incongruous contralateral homonymous hemianopsia (edges of the visual field defect in each eye are not identical in shape) is usually seen with lesions of Meyer's loop.

Congruous contralateral homonymous hemianopsia (edges of the visual field defect in each eye is identical in shape) occurs as a result of damage to the optic radiation near the visual cortex. Damage to the upper fibers of the optic radiation, although rare, occurs in parietal lobe lesion and produces contralateral inferior quadranopsia. A lesion of the most rostral part of the calcarine fissure is most likely to produce monocular homonymous defect. This occurs due to the fact that the most lateral part of the temporal visual field, which measure 25°, is uniquely seen only by the ipsilateral eye and uniquely conveyed to the rostral most part of the striate cortex adjacent to the parieto-occipital cortex. Unlike the optic tract, a lesion of the primary visual cortex produces congruent visual deficits but without relative afferent pupillary defect, optic disc atrophy, or loss of central vision loss.

The primary visual (striate cortex) represented by Brodmann area 17 is the cortical area located on the medial surface of the occipital lobe on both banks of the calcarine fissure extending slightly to the lateral surface around the occipital pole. It includes portions of the cuneate and lingual gyri located on the superior and inferior banks of the calcarine fissure, respectively. The primary visual cortex consists of a very thin granular type or koniocortex. However, layer IV of this six-layered cortex is cytoarchitecturally divided into densely packed upper and lower sub-layers and a lighter less packed middle layer with fewer small cells between the giant stellate cells. A characteristic of the light middle layer is the presence of a thickened external myelin-dense band, the band of Gennari, which is visible to the naked eye in fresh brain sections. This six-layered isocortex exhibits a point-to-point precise synaptic connection with the neurons of the lateral geniculate nucleus, projecting to Brodmann areas 18 and 19. Due to this specific synaptic linkage, a lesion in the visual cortex, even if it is very small, may produce scotoma. It receives information from all neurons of the lateral geniculate body and projects to Brodmann areas 18 and 19 that represent the secondary and tertiary visual cortices.

The commissural fibers that establish interconnection between areas 17 of both cerebral hemispheres are not well developed. Pulvinar has an important role in mediating part of visual information. Pulvinar receives visual input from the contralateral visual field and projects to the supragranular layers of the primary visual cortex (Brodmann area 17) and to layers I, III, and IV of the secondary (area 18) and the tertiary (area 19) visual cortices. The projection of the pulvinar to supragranular layers of the primary visual cortex establishes the extrageniculate visual pathway.

Perception and recognition of visual images and the ability to read fine print if brought into focus remain unaffected even with bilateral damage to the striate cortex, as long as both or one occipital pole is spared [33]. This confirms the occipital poles' role in acute central vision. It may also explain the phenomenon of macular sparing in which a lesion involving the occipital lobe or occipital pole on one hemisphere produces no detectable deficit of the central vision. It has been reported that sparing of the central vision may either be due to incomplete lesion sparing of the posterolateral part of the striate cortex, particularly the area around the occipital pole, or possibly due to the efficient arterial anastomosis and collateral circulation between the middle and posterior cerebral arteries when vascular lesion affects this area. One of the other reported views is that bilaterality in the representation of the macular area in both cerebral hemispheres accounts for macular sparing when one hemisphere is affected. Cortical blindness and contralateral incongruous homonymous hemianopsia usually accompany macular sparing. However, bilateral destruction of the occipital poles, on the other hand, markedly impairs the ability to clearly and accurately observe visual fields and discern colors.

The primary visual cortex consists of cellular columns with diverse receptive fields. These functional columns or units of the cerebral cortex consist of ocular dominance and orientation columns that extend perpendicular to the cortical surface. The ocular dominance columns, partially formed at birth, which run at a right angle to the cortical surface, receive visual input from both eyes. However, due to specific arrangement of these columns, the visual input to a particular ocular dominance column originates from one (dominant) eye. Selective disruption of the input from a single eye to these columns becomes almost impossible, given the close proximity of the column that receives primary input from the right eye to that of the left eye. Segregation of the visual impulses into the right and left laminae of the dominance column occurs in layer IV of the visual cortex, although parts of the striate cortex which receive impulses from the optic disc and from the peripheral most temporal visual field of the ipsilateral eye lack ocular dominance columns. Experimental monocular strabismus during postnatal development may not hinder growth of the associated dominance column as reported in earlier investigation [34].

The orientation columns make up the second set of functional columns. They appear smaller than the dominance columns, extending vertically to the pial surface of the cerebral cortex. Their name is derived from the fact that these columns consist of cells that possess the same receptive field axis of orientation and have "on" and "off" centers.

The study of the coupled development of ocular dominance and orientation columns using an analytical approach shows that segregation of ocular dominance can induce the stabilization and the eventual production of pinwheel crystallization, a process linked to the overall dominance of one which occurs during early development of the cortex [35]. The striate cortex consists of simple, complex, and hypercomplex cells. Simple cells have similar characteristics to that of the retina and LGN. Dark bars against a light background and straight edges separating areas of different degrees of brightness effectively stimulate the visual cortex.

Amblyopia (lazy eye) is a disorder that develops from a prolonged suppression of an image in one eye between the second and fourth year of life. It may be the result of congenital strabismus and the inadequate stimulation of one eye by visual image. It occurs in children who exhibit diplopia as a sequel to functional imbalance between the extraocular muscles and subsequent attempts to eliminate the image in one eye by constantly utilizing the other eye. As the crossed-eyed child favors one eye over another, the unused eye eventually loses visual acuity and may permanently be blind. In this condition no deficits are recorded in the refractive media or ocular apparatus. Amblyopia may also occur as a result of nutritional deficiency and in alcoholics. This condition may be associated with damage to the optic nerves and bilateral scotoma. Blurred vision and optic atrophy may also occur in this condition. The risks of developing amblyopia appear to be higher in patients who undergo probing for congenital nasolacrimal duct obstruction and also among those who experience nasolacrimal duct irrigation [35].

Patients with Anton's syndrome exhibit normal pupils that show constriction in response to light but, interestingly, remain oblivious or ignore half of the visual field on the lesion side. They deny and are generally unaware of their deficits, attributing them to poor lighting or lack of personal interest in the surrounding. They attempt to identify objects or describe the items in the surrounding visual field to prove that they are indeed healthy. Ironically, they fail to distinguish illuminated from non-illuminated areas. This syndrome is the result of disruption of the corticothalamic connection between Brodmann area 17 and the LGN. It is also seen in patients with nondominant hemispheric injury and in cases of bilateral occlusion of the posterior cerebral arteries. Clinical data indicates recurrence of bilateral occipital lobe infarcts in patients with Anton's syndrome [37].

A lesion of the upper part of Brodmann area 17 and the occipital association cortex adjacent to the angular gyrus of the parietal lobe in the dominant hemisphere causes finger agnosia, left right disorientation, and the inability to identify fingers, name objects, or write.

The secondary visual cortex (Brodmann area 18), which adjoins the primary (striate) cortex, mediates visual memories, receiving input from Brodmann area 17. It is a mirror image representation of the Brodmann area 17 to which it is interconnected, though lacking the characteristic band of Gennari. This cortex, as the case with area 17, responds best to dark bars and edges. The majority of cells in Brodmann area 18 are complex cells arranged in columns. In order for the visual object to be recognized and undergo fine feature analysis, visual input must project via the splenium of the corpus callosum to Brodmann area 18 of the dominant hemisphere.

Unilateral visual agnosia, which is characterized by the inability to recognize images received by the right (nondominant) cerebral hemisphere, results from excision of the corpus callosum and disruption of the connection between Brodmann areas 18 in both cerebral hemispheres. Bilateral visual agnosia results from a lesion of Brodmann area 18 in the dominant cerebral hemisphere. Visually agnostic patients are unable to identify objects in the visual field without utilizing gustatory, tactile, auditory, or olfactory senses. Autotopagnosia, which refers to the failure to distinguish living people from inanimate objects, is a deficit produced by damage to the upper lateral or lower medial parts of the secondary visual cortex in the left dominant hemisphere. This is due to the fact that the upper lateral portion of Brodmann area 18 mediates memories of inanimate objects, while the lower medial portion is mostly concerned with memories of living parts or individuals.

Achromatopsia, the inability to recognize color in only one-half of the visual field, is associated with amblyopia, nystagmus, photophobia, and hemeralopia (day blindness). It may occur independently as a result of a lesion in the ventral occipitotemporal cortex or subsequent to congenital anomaly. The congenital type is an autosomal recessive condition in which the cones fail to perceive color even in a highly illuminated area. This type does not progress with age. Acquired achromatopsia shows progressive worsening of symptoms. Stimulation of Brodmann area 18 results in visual hallucinations in the form of sparkling lights.

The tertiary visual cortex (Brodmann area 19) is responsible for recalling (revisualizing) previously seen images (visual memory). It is considered a mirror image of and is located on the lateral side of the secondary visual cortex. In order for a visual image to be recalled, visual information must be conveyed between Brodmann areas 18 and 19, and integrated with stimuli from the auditory, tactile, and olfactory systems. Recalling symbols must be distinguished here, as it pertains to a function of the angular gyrus. Stimulation of this area produces colorful visual images of objects and poignant events. The primary neuronal elements in this area that receive visual input from the left and right eye are hypercomplex cells. However, there is a clear specificity in this area regarding the source and type of input. The inferior part of this area responds exclusively to color, the middle part to information derived from the macula and object sizes, while movement stimulates a small area anterior to the macular zone. An interesting fact to correlate with the above is the significant role that Brodmann areas 18 and 19 play in stereopsis. This mechanism, which becomes functional after 5–6 months of postnatal development, enables the brain to accurately determine the incongruity between the retinal images received by both eyes and construct a complete three-dimensional image.

Lesions involving Brodmann area 19 and the parieto-occipital cortex may produce dysfunction that mimic astereognosis in which patients lose the ability to recall objects using touch sensation. The inferior temporal gyrus, which serves as a visual association cortex, encodes visual memory through integration of input from the Brodmann area 22 (entorhinal cortex) and Brodmann areas 7, 18, and 19. The latter connections may account for the visual hallucination experienced by patients with temporal lobe epilepsy or vivid scenes reported by patients enduring surgical brain operation. Visual changes in migraine headaches include blurred vision, flashing lights, wavy lines, and scotoma. Hemiparesis, ophthalmoplegia, or aphasia may accompany these symptoms.

The visual cortex is primarily supplied by the posterior (vertebrobasilar) circulation through the posterior cerebral artery, a branch of the basilar artery. The middle cerebral artery also contributes blood supply, particularly to the occipital pole and the lateral surface of the occipital lobe through anastomotic connections with the posterior cerebral artery. Visual deficit of the occipital lobe origin is most commonly attributed to infarction of the posterior cerebral artery. Cervical spondylosis-induced transient bilateral occlusion of the vertebral arteries due to narrowing of the associated transverse foramina may dramatically reduce the blood flow in the posterior cerebral and labyrinthine branches of the basilar artery. As a result, an individual with this condition may experience vertigo (sense of rotation of self and/or environment) and transient blindness, which last for a few seconds. The patient maintains no memory that these disorders ever happened.

Occlusion of the posterior cerebral arteries can cause bilateral degeneration of the parieto-occipital cortex between Brodmann areas 19 and 7, producing signs and symptoms of Balint syndrome. This syndrome is characterized by the impairment of fixation or inability to scan the peripheral visual field (due to lack of coordination with the oculomotor system) or use visual cues to grasp an object [38].

Infarction of the posterior cerebral artery may also produce a combination of hemianopsia or quadranopsia, macular sparing, and hemianesthesia with no muscle paralysis. If the infarct involves the dominant hemisphere, Charcot-Wilbrand syndrome may develop which is characterized by visual agnosia. Gerstmann's syndrome, transcortical sensory aphasia, and alexia without agraphia are also seen in posterior cerebral artery infarcts.

Anterograde visual amnesia and difficulty adapting to new and unaccustomed territory despite intactness of the visual apparatus are attributed to bilateral disruption of the connections between the occipitotemporal and entorhinal cortices subsequent to basilar or possibly posterior cerebral artery insufficiency. Patients with this disconnection exhibit anterograde visual amnesia and difficulty adapting to new and unaccustomed territory despite intactness of the visual apparatus.

References

1. Hansson EE, Beckman A, Håkansson A (2010) Effect of vision, proprioception, and the position of the vestibular organ on postural sway. Acta Otolaryngol 130(12):1358–1363
2. Walshe JM (2014) The Kayser-Fleischer ring. Br J Hosp Med (Lond) 75(3):C38–C39
3. Mantas AM, Wells J, Trotter J (2013) Kayser-Fleischer rings of acute Wilson's disease. Proc (Bayl Univ Med Cent) 26(2):166–167
4. Mendívil A, Cuartero V, Mendívil MP (1995) Color Doppler imaging of the ocular vessels. Graefes Arch Clin Exp Ophthalmol 233(3):135–139
5. Lindberg L (2014) Spasm of accommodation. Duodecim 130(2):168–173
6. Meyer I, Rolim H, Medeiros A, Paiva L, Galvão FR (2010) Peter's anomaly, clinical and therapeutic aspects: case report. Arq Bras Oftalmol 73(4):367–369
7. Ram J, Gupta A (2014) Kayser-Fleischer ring and sunflower cataract in Wilson disease. JAMA Ophthalmol 132(7):873
8. Walker L, French S (2014) Horner's syndrome: a case report and review of pathophysiology and clinical features. West Indian Med J 63(3):282–284
9. Werner I (1951) Cheyne-Stokes respiration associated with rhythmic pupillary changes a case report. Acta Psychiatr Neurol Scand 26(2):213–217
10. Draper EM, Malloy KA (2012) Progressive visual and hearing loss secondary to neurosyphilis. Optom Vis Sci 89(11):e65–e71
11. Sullivan KN, Manfredi F, Behnke RH (1968) Hippus in Cheyne-Stokes respiration. Arch Intern Med 122(2):116–121
12. Rae-Grant AD (2013) Unusual symptoms and syndromes in multiple sclerosis. Continuum (Minneap Min) 19(4 Multiple Sclerosis):992–1006.
13. Tobe LA, Harris A, Hussain RM, Eckert G, Huck A, Park J, Egan P, Kim NJ, Siesky B (2015) The role of retrobulbar and retinal circulation on optic nerve head and retinal nerve fibre layer structure in patients with open-angle glaucoma over an 18-month period. Br J Ophthalmol 99(5):609–612, pii: bjophthalmol-2014-305780
14. Tsilis AG, Tsilidis KK, Pelidou SH, Kitsos G (2014) Systematic review of the association between Alzheimer's disease and chronic glaucoma. Clin Ophthalmol 8:2095–2104
15. Charman WN (2014) Developments in the correction of presbyopia II: surgical approaches. Ophthalmic Physiol Opt 34(4):397–426
16. Nusliha A, Dalpatadu U, Amarasinghe B, Chandrasinghe PC, Deen KI (2014) Congenital hypertrophy of retinal pigment epithelium (CHRPE) in patients with familial adenomatous polyposis (FAP); a polyposis registry experience. BMC Res Notes 7:734
17. Georgalas I, Paraskevopoulos T, Symmeonidis C, Petrou P, Koutsandrea C (2014) Peripheral sea-fan retinal neovascularization as a manifestation of chronic rhegmatogenous retinal detachment and surgical management. BMC Ophthalmol 14:112
18. Ma Y, Kawasaki R, Dobson LP, Ruddle JB, Kearns LS, Wong TY, Mackey DA (2012) Quantitative analysis of retinal vessel attenuation in eyes with retinitis pigmentosa. Invest Ophthalmol Vis Sci 53(7):4306–4314
19. Randolph SA (2013 Jun) Color vision deficiency. Workplace Health Saf 61(6):280

20. Yoshida M (2013) Neural mechanism of blindsight. Brain Nerve 65(6):671–677
21. Heider B, Spillmann L, Peterhans E (2002) Stereoscopic illusory contours—cortical neuron responses and human perception. J Cogn Neurosci 14(7):1018–1029
22. Li L, Li W, Chen CZ, Yi ZH, Zhou YY (2014) Is aspirin use associated with age-related macular degeneration? A meta-analysis. J Clin Pharm Ther 55(4):2687–2696
23. Zhang GM (2014) Chen Y Pathogenesis of dry age-related macular degeneration: updates. Zhonghua Yan Ke Za Zhi 50(6):464–470
24. Rezende FA, Lapalme E, Qian CX, Smith LE, SanGiovanni JP, Sapieha P (2014) Omega-3 supplementation combined with anti-vascular endothelial growth factor lowers vitreal levels of vascular endothelial growth factor in wet age-related macular degeneration. Am J Ophthalmol 158(5):1071–1078
25. Development of Retinal Amacrine Cells and their dendritic stratification 2014 Curr Ophthalmol Rep 2(3):100–106.
26. Hata M, Miyamoto K, Oishi A, Makiyama Y, Gotoh N, Kimura Y, Akagi T, Yoshimura N (2014) Comparison of optic disc morphology of optic nerve atrophy between compressive optic neuropathy and glaucomatous optic neuropathy. PLoS One 9(11):e112403
27. AlHassany AA 2014 Night blindness due to vitamin A deficiency associated with copper deficiency myelopathy secondary to bowel bypass surgery. BMJ Case Rep. pii: bcr2013202478. doi: 10.1136/bcr-2013-202478.
28. Ziak P, Jarabáková K, Koyšová M (2014) Optic disc drusen – current diagnostic possibilities. Cesk Slov Oftalmol 70(1):30–35
29. Levenson JH, Kozarsky A (1990) In: Walker HK, Hall WD, Hurst JW (eds) Clinical methods: the history, physical, and laboratory examinations, 3rd edn. Butterworths, Boston, Chapter. 111. ISBN 3
30. Ota Y, Mami L (2013) Ultrasonography imaging during nasal endoscopic trans-sphenoidal surgery. ORL J Otorhinolaryngol Relat Spec 75(1):27–31
31. Huynh N, Stemmer-Rachamimov AO, Swearingen B, Cestari DM (2012) Decreased vision and junctional scotoma from pituicytoma. Case Rep Ophthalmol 3(2):190–196
32. Hendry SHC, Reid RC (2000) The Koniocellular pathway in primate vision. Annu Rev Neurosci 23:127–153
33. Farivar R (2009) Dorsal-ventral integration in object recognition. Brain Res Rev 61(2):144–153
34. Adams DL, Horton JC (2006) Ocular dominance columns in strabismus. Vis Neurosci 23(5):795–805
35. Reichel LS, Wolf F (2009) Pinwheel stabilization by ocular dominance segregation. Phys Rev Lett 102(20):208101
36. Ozgur OR, Sayman IB, Oral Y, Akmaz B (2013) Prevalence of amblyopia in children undergoing nasolacrimal duct irrigation and probing. Indian J Ophthalmol 61(12):698–700
37. Kwong Yew K, Abdul Halim S, Liza-Sharmini AT, Tharakan J (2014) Recurrent bilateral occipital infarct with cortical blindness and anton syndrome. Case Rep Ophthalmol Med 2014:795837
38. Kerkhoff G, Heldmann B (1999) Balint syndrome and associated disorders. Anamnesis – diagnosis – approaches to treatment. Nervenarzt 70(10):859–869

Chapter 5
Nanotechnology and Nanomaterials in Ophthalmic Drug Delivery

Shengyan Liu, Lyndon W. Jones, and Frank X. Gu

Abstract The anatomical barriers and physiological clearance mechanisms on the ocular surface have presented enormous challenges for development of ocular drug delivery devices. More invasive methods, such as intravitreal injections, can improve the ocular bioavailability of therapeutic agents but often result in vision-threatening side effects. Recently, an increasing number of scientists have turned to nanomaterial-based drug delivery systems to address the challenges faced by conventional methods. This chapter highlights the recent applications of various nanomaterials, such as polymeric micelles, hydrogels, liposomes, niosomes, dendrimers, and cyclodextrins as ophthalmic drug delivery systems to enhance the ocular bioavailability of therapeutic agents.

Keywords Nanotechnology • Ocular drug delivery • Ophthalmic drug delivery • Nanomaterials

5.1 Introduction

Ocular drug delivery remains among the most challenging approaches to the administration of therapeutic agents to the human body. The eye is a small and complex organ that is separated from the rest of the body by multiple layers of biological barriers. Moreover, the internal ocular structures and tissue are protected from the external environment by the tight junctions of the corneal epithelium and the mucosal surface.

S. Liu • F.X. Gu (✉)
Department of Chemical Engineering, Waterloo Institute for Nanotechnology, University of Waterloo, Waterloo, Canada
e-mail: sandy.liu@uwaterloo.ca; frank.gu@uwaterloo.ca

L.W. Jones
Centre for Contact Lens Research, School of Optometry, University of Waterloo, Waterloo, Canada
e-mail: lwjones@uwaterloo.ca

© Springer International Publishing Switzerland 2016

Y. Pathak et al. (eds.), *Nano-Biomaterials For Ophthalmic Drug Delivery*,
DOI 10.1007/978-3-319-29346-2_5

Therefore, the primary challenge of ocular drug delivery is to circumvent these protective barriers in order to achieve therapeutically effective concentrations of drugs in the intraocular tissues. Thus, the objective of ocular drug administration is to treat ophthalmic diseases in a localized manner, as opposed to serving as an intermediate route to achieve systemic drug activity. This is an advantage for ocular drug delivery in that once the drug is successfully delivered to the intraocular tissues, the drug is not likely to be cleared to the systemic circulation, which may cause undesirable adverse effects.

There are a plethora of vision-threatening ocular diseases that require therapeutic treatment, and most of these diseases are associated with the intraocular structures, particularly the retina, macula, retinal pigment epithelium (RPE), or choroid [1]. Typically, anti-inflammatory drugs, antibacterial agents, or even angiogenesis inhibitors are used to treat various types of local infections or blinding disorders such as chronic uveitis, glaucoma, and choroidal neovascularization; nevertheless, the efficiency with which these substances are successfully delivered has largely been limited by the presence of a variety of ocular barriers. To date, most common ophthalmic drugs are administered topically in the form of eye drops to the corneal surface. However, topically administered drugs suffer from low bioavailability due to clearance mechanisms such as tear turnover in which the drugs are rapidly drained away from the ocular surface through the puncta to the nasolacrimal duct. Alternative delivery methods such as intravitreal or periocular injections have been developed to improve the bioavailability of the therapeutic agents, but due to the invasive nature of these methods, side effects such as retinal detachment and intravitreal hemorrhage have been observed. The challenges affiliated with these conventional methods of ocular drug delivery have led scientists to contribute significant effort into developing advanced drug delivery systems that provide targeted therapy with increased bioavailability [2, 3].

An ideal ocular drug delivery system should possess key properties, including: (1) a controlled and sustained-release profile to maintain therapeutic concentration of the drug for a prolonged period of time to reduce the frequency of administration; (2) specific targeting and prolonged retention in the diseased tissues to improve therapeutic efficiency and mitigate side effects; and (3) patient-friendly delivery routes that minimize or eliminate side effects resulting directly from these administration methods. At present, nanocarrier-based ocular drug delivery systems appear to be the most promising tool to meet the primary requirements of an ideal ocular delivery system. Nano-drug carriers, due to their small sizes, are likely to have high diffusivity across membranes such as the corneal epithelium. A significant number of studies have already demonstrated that the use of such nanomaterials via topical administration improved the corneal permeability of drugs [2–5]. Similarly, due to their high surface-area-to-volume ratio, nano-drug carriers may also show improved interaction with the outer mucous membrane of the corneal surface, prolonging the retention of the topically administered drug. Ongoing advances in developing improved nanocarrier-based ocular drug delivery systems may not only serve to enhance the delivery of drugs to the ocular surface but also provide new possibilities of effectively delivering therapeutic agents to intraocular tissues such as the retina or choroid using noninvasive delivery methods. The utilization of nanomaterials for enhanced ocular drug delivery applications has been reviewed in our previous publication [6] (Fig. 5.1), and this chapter expands on that review by including the most recent findings since the publication of the review.

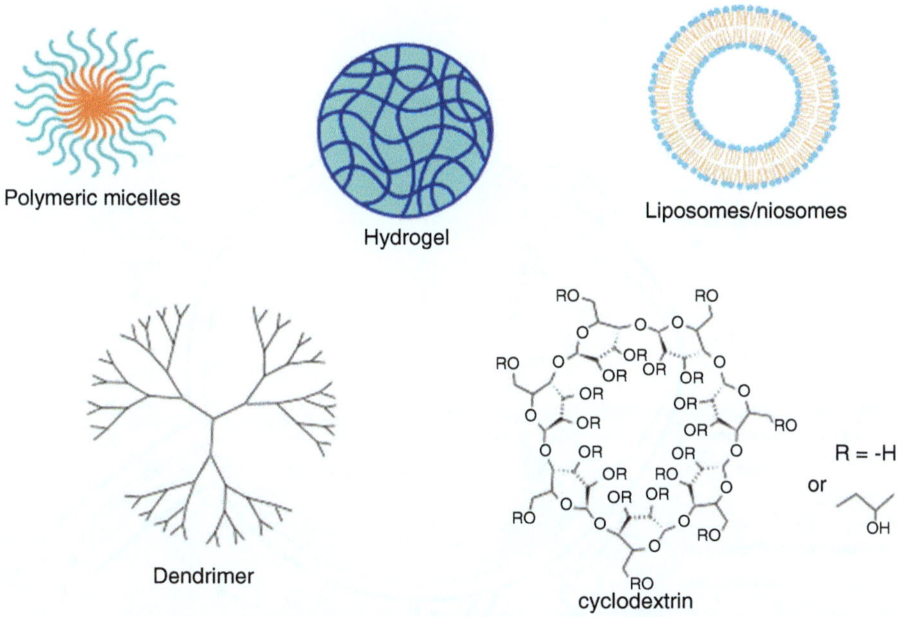

Fig. 5.1 A schematic illustration of different nanomaterial-based ocular drug delivery systems. Polymeric micelles are nanoparticles self-assembled from amphiphilic copolymers with the ability to encapsulate hydrophobic drugs in the core of the particles. Hydrogel colloids are 3D networks of water-soluble polymers which have the ability to incorporate hydrophilic drugs in the 3D network. Liposomes/niosomes are vesicle structures with both lipophilic and hydrophilic phases and therefore can encapsulate both hydrophobic and hydrophilic drug components. Dendrimers are 3D highly branched tree-structured macromolecules that can encapsulate hydrophobic drug molecules since they possess internal empty cavities and open conformations for low-generation dendrimers. Cyclodextrins are a family of cyclic oligosaccharides, composed of 6–8 glucose units which have been shown to improve pharmacokinetic properties of many drugs through the formation of an inclusion complex (Reproduced with permission from [6])

5.2 Challenges of Conventional Ophthalmic Drug Delivery

Difficulties in delivering drugs through the ophthalmic route arise from the anatomy and the physiology of the eye (Fig. 5.2), which is divided into anterior and posterior segments. Topical drug administrations, such as eye drops, suspensions, or ointments, are the most preferred route of delivery to the anterior segment of the eye due to the ease of administration and low cost. However, local drug administration to the anterior portion of the eye by topical application is significantly limited by the clearance mechanisms of the corneal surface, which include lacrimation, tear dilution, and tear turnover [7]. In addition, topically administered drugs are generally absorbed either through the corneal route (cornea → aqueous humor → intraocular tissues) or non-corneal route (conjunctiva → sclera → choroid/RPE) [7], which further limits the amount of drug that is ultimately absorbed in the intraocular tissues. Due to these clearance mechanisms and ocular barriers, less than 5 % of the total

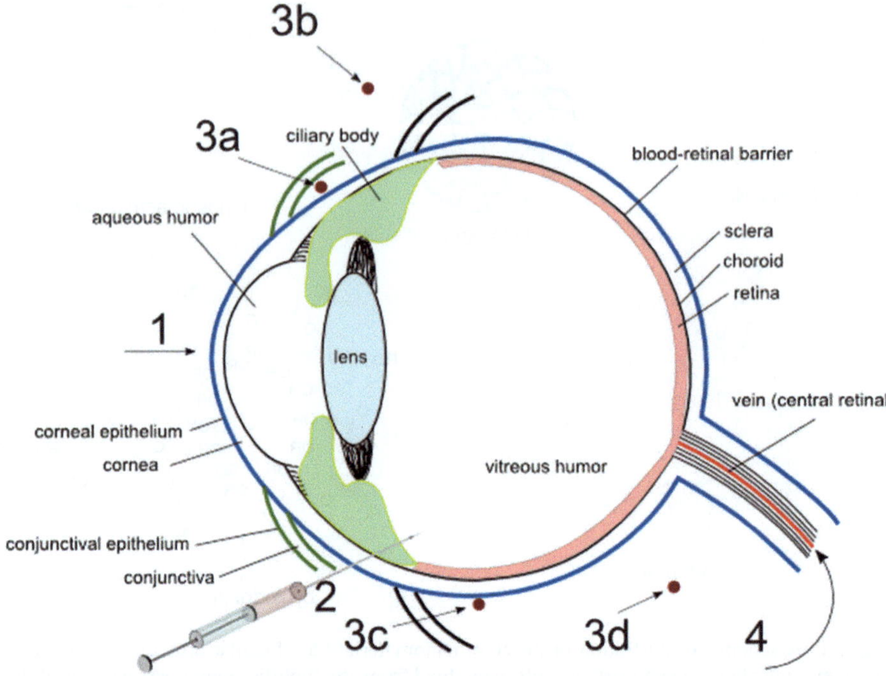

Fig. 5.2 The anatomy of the eye illustrating various ocular barriers and various mechanisms of drug delivery: *1*. topical administration; *2*. intravitreal injection; *3*. periocular injections (*3a*. subconjunctival, *3b*. peribulbar, *3c*. sub-tenon, and *3d*. retrobulbar); and *4*. systemic administration (Reproduced with permission from [6])

administered drug reaches the aqueous humor [8]. For example, after administration of a common eye-drop solution containing 0.05 % cyclosporine to treat dry eye syndrome, more than 95 % of the drug reaches systemic circulation through transnasal or conjunctival absorption [9].

The administration of drugs to the posterior portion of the eye is even more challenging, due to the lack of cellular components in the vitreous body, which reduces the convection of molecules to the posterior segment. Efficient drug delivery to the posterior segment of the eye is a challenge faced by many pharmaceutical researchers, as most blinding diseases are associated with the posterior segment [10]. Systemic administration has also been used to deliver therapeutic agents to the posterior segment of the eye; however, this route of administration requires large administration doses because of the inner and outer blood–retinal barriers that separate the retina and the vitreous humor from the systemic circulation [11] Studies to date have shown that less than 2 % of systemically administered drugs reaches the vitreous cavity [7]. High administration doses or frequent administration directly translates to poor patient compliance and increased risk of systemic side effects.

Recently, intravitreal injections and vitreal implants have been investigated in order to achieve therapeutic concentrations at the posterior segment, although both of these administration options are very invasive methods with high degrees of risk [2]. Currently, intravitreal injections of drugs such as bevacizumab or ranibizumab are being tested to treat vascular diseases associated with the posterior segment of the eye, including retinopathy of prematurity [12] or choroidal neovascularization [13], respectively. Frequent administration through the intravitreal route is associated with short-term adverse effects such as retinal detachment, endophthalmitis, intravitreal hemorrhage, and increased risk of cataract development [4, 7]. More recently, periocular drug delivery has been investigated as a less invasive method, compared to intravitreal administration, in order to achieve high drug concentrations in the vitreous cavity [14]. Periocular administration refers to the injection of the drug in the vicinity of the ocular organ, such as the subconjunctival, sub-tenon, or peribulbar regions, so that the drugs can reach the vitreous cavity by crossing the sclera, choroid, and RPE barriers [14]. However, periocular drug administration is also not without side effects, including a rise in intraocular pressure, cataract development, hyphema, strabismus, and corneal decompensation [7]. As a result, topical drug delivery carriers that possess enhanced permeability across corneal membranes and prolonged retention of the carriers in the ocular moiety are still largely preferred over intravitreal or periocular injections.

5.3 Contact Lenses in Ocular Drug Delivery

One of the earliest advancements in ocular drug delivery came from the idea of using contact lenses as a sustained drug delivery system [7, 15–18]. In 2007, there were about 35 million contact lens wearers in North America alone, and the number is expected to have risen continuously to date [19]. The idea of using contact lenses as a drug delivery tool, therefore, has been considered very practical. The contact lens drug delivery system is typically prepared by soaking commercial products in a drug solution, allowing retention of drugs in the hydrogel matrix. Subsequently, the drug is released at the corneal surface upon application of the lens on the eye (Fig. 5.3). The purpose of using the contact lens drug delivery system is to increase the residence time of drugs on the corneal surface, compared to topical application, to increase the amount of corneal absorption of these drugs. A mathematical model was used to postulate that as much as 50 % of the drug released from contact lens could be absorbed by the cornea, but in vivo studies have yet to confirm this hypothesis [21]. The drug release from the lens is governed by a simple diffusion mechanism, which typically results in a burst release of drugs for the initial few hours [22]. Therefore, several studies have focused on using additional components in the contact lenses to control the release rate of the drugs.

Ciolino et al. used the formulation of a poly(lactic-co-glycolic) acid (PLGA) film, which is an FDA-approved biocompatible polymer that has shown to be able to control drug-release kinetics, coated with poly(hydroxyethyl methacrylate) (pHEMA) hydrogel layers to control the release of ciprofloxacin [22]. The formulation resulted in steady zero-order release of ciprofloxacin for up to 4

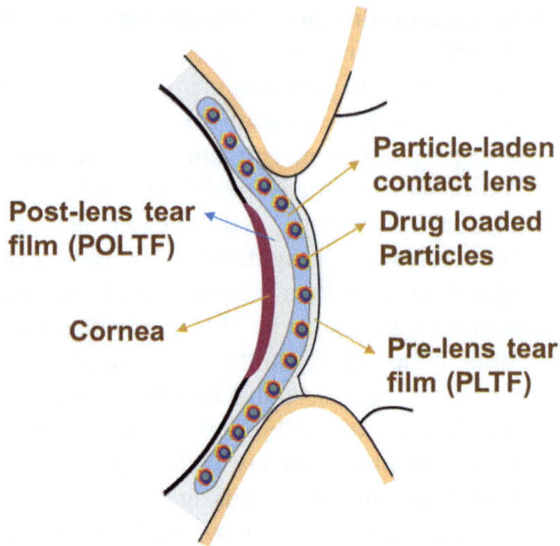

Fig. 5.3 Schematic illustration of the particle-laden lens inserted in the eye (Reproduced with permission from [20])

weeks. When the PLGA film was not present, ciprofloxacin released at a rate more than three times faster from the pHEMA hydrogel than from the PLGA–pHEMA combination. Several more recent studies examined the use of nanomaterial-laden hydrogel contact lenses to achieve controlled drug-release phenomena [20, 23–25]. For example, Phan et al. incorporated the poly(D,L-lactide)-b-dextran nanoparticles into the contact lens materials by soaking method and showed that the total drug uptake by the contact lens and the duration of the release were both improved [26]. Nevertheless, a nanomaterial-laden contact lens drug delivery system still faces substantial challenges. The obvious challenge is that with this controlled release mechanism the duration of the therapeutic window is also prolonged, resulting in the patient having to wear the lenses for a longer period of time, potentially inducing discomfort. Another challenge is that the diffusional-based drug-release mechanism still causes some of the drugs to be released in the pre-lens tear film, as opposed to the post-lens tear film (i.e., between the cornea and the lens), which are also exposed to the same clearance mechanisms that occur with topical administration. Lastly, the nanomaterials that allow a slow release mechanism may also trap the drugs permanently in the hybrid hydrogel matrix, which is shown in PLGA–pHEMA layers: only about 10 % of the total drug from the contact lens was released during the sustained-release phase [22]. This implies that the drug shelf life must be long enough so that an efficient amount of drug will be absorbed by the cornea before losing its therapeutic efficacy. Finally, the incorporation of nanomaterials into a contact lens must also not compromise key properties of the lens, such as biocompatibility, modulus (or stiffness), transparency, comfort, and wettability. Despite these issues, much progress has been made in utilizing contact lenses as drug delivery devices. With the advent of nanomaterials, contact lenses are potentially one of the more convenient ways of delivering ocular drugs.

5.4 Nanomaterials in Ocular Drug Delivery

Much research has been carried out on using different types of nanomaterials for the ophthalmic delivery of drugs. Sahoo et al. have outlined three major goals for ocular drug delivery nano-drug carriers: (1) enhancing drug permeation, (2) controlling the release mechanisms of drugs, and (3) modifying the surfaces of the nano-drug carriers with specific targeting moieties [27]. Due to the small size of these nano-drug carriers, the permeability of these drug carriers through the vitreous barrier is greatly enhanced, which in turn improves the delivery rate of the drugs to the posterior tissues. Moreover, with the mucoadhesive properties exhibited by certain types of nanoparticles, drug carrier retention may be significantly prolonged in the targeted tissues. The nanocarrier-drug complex can be administrated as eye-drop solutions, requiring less frequent administration due to the higher retention of the drugs, reducing the cost of administration, and increasing patient compliance. Zimmer et al. have suggested that the sizes of administered particles for ophthalmic applications must be less than 10 μm in order to avoid the sensation of scratching upon administration [28]. The potential advantages of using nanoparticles (NPs) for ocular drug delivery have been reviewed [29].

The field of ocular drug delivery has been significantly impacted in the past decade by advances in technology, especially the development of nanostructured drug carriers. It was proposed that microspheres and nanoparticles used for the encapsulation of drugs would enhance the overall performance of ocular drug delivery systems [28]. The application of nanotechnology in the field of ocular drug delivery has been extensively reviewed by several groups. Gaudana et al. have highlighted recent advances in nanotechnology applications for ocular drug delivery, listing numerous nanostructure carriers as potential ocular drug carriers [7]. Ludwig has discussed the use of mucoadhesive polymer-based nanostructures for the ophthalmic delivery of drugs, reporting the enhancement of targeting and retention of the drugs at the tissue site of interest, such as the corneal surface [30]. Nagarwal et al. have focused more on polymer-based nanoparticles, exploiting the application of the block copolymer micelles in the field of ocular drug delivery [2].

5.5 Nanomaterial-Based Ophthalmic Formulation Currently in the Clinical Stages

The studies discussed thus far have focused on the initial proof-of-concept stage of nanomaterials in the field of ocular drug delivery. Table 5.1 summarizes the progress of nanomaterial-based ocular drug delivery system in the clinical stages. The past few decades have seen a significant increase in the number of journal articles published dealing with developing advanced ocular drug delivery

Table 5.1 List of preclinical and clinical studies using nanomaterials in ocular drug delivery system (Reproduced with permission from [6])

Nanomaterial	Drug	Formulation	Administration	Treatment	Clinical stage	References
Microsphere	Triamcinolone acetonide	PLGA (RETAAC)	Intravitreal injection	Diabetic macular edema	Launched	Cardillo et al. [31]
Micelles	Dexamethasone	NIPAAM–VP–MAA	Eye drops	Inflammation	Preclinical	Rafie et al. [32]
	Dexamethasone	Pluronic F127–chitosan	Eye drops	Ocular hypertension	Preclinical	Pepic et al. [33]
	Cyclosporin A	MPEG-hexPLA	Eye drops	Dry eye, autoimmune uveitis	Preclinical	Di Tommaso et al. [34]
	Dexamethasone	PHEA–PEG	Eye drops	Ocular hypertension	Preclinical	Civiale et al. [35]
Hydrogel	Timolol		Topical	Ocular hypertension	Randomized human	Volotinen et al. [36]
			Trans-scleral iontophoresis		Randomized human	Parkinson et al. [37]
Liposomes	Flurbiprofen	Steric acid + castor oil	Eye drops	Inflammation	Preclinical	Gonzalez-Mira et al. [38]
	Ciprofloxacin HCl	Chitosan-coated liposome	Eye drops	Conjunctivitis	Preclinical	Abdelbary [39]
	Fluconazole	Liposome	Eye drops	Fungal infection	Preclinical	Habib et al. [40]

Nanomaterial	Drug	Formulation	Administration	Treatment	Clinical stage	References
	Ciprofloxacin HCl	Chitosan-coated liposome	Eye drops	Bacterial growth of *P. aeruginosa*	Preclinical	Mehanna et al. [41]
	Bevacizumab	Liposome	Intravitreal injection	Ocular neovascular activity	Preclinical	Abrishami et al. [42]
	Verteporfin	Liposome (Visudyne®)	Intravitreal injection	Classic subfoveal choroidal neovascularization	Launched	[43])
Dendrimer	Pilocarpine nitrate and tropicamide	PAMAM	Eye drops	Myosis and mydriasis	Preclinical	Vandamme and Brobeck [44]
	Carteolol	Phosphorus-containing dendrimers	Eye drops	Glaucoma	Preclinical	Spataro et al. [45]
Cyclodextrin	Methazolamide	HPβCD and HPMC	Eye drops	Ocular hypertension	Randomized human	Gudmundsdottir et al. [46]
	Latanoprost		Eye drops	Open-angle glaucoma	Randomized human	Gonzalez et al. [47]

systems. Among these delivery systems, eye-drop formulations incorporated with various drug-release-controlling excipients (i.e., polycarbophil) are currently available in the market [43]. A number of products in the form of intravitreal implants, inserts, or puctal plugs in the micro- or milli-scale are also currently being developed by various pharmaceutical companies [43]. Although numerous nanomaterial-based ocular drug delivery systems have already progressed onto preclinical studies (i.e., in vivo animal studies) to characterize their pharmacokinetic properties in a physiological environment, very few have progressed to full clinical studies. Hydrogel carriers of timolol showed improved control over the individual variation of timolol concentration in the aqueous humor among human patients, compared to the pure drug suspension [36]. The lowered variation in drug concentration reduces the probability of having insufficient local drug concentration, and it also reduces the risk of toxic effects caused by the presence of excessive amounts of drug. Another interesting delivery method using transscleral iontophoresis was also studied in human patients [37]. Iontophoresis uses a small electric current to deliver charged therapeutic agents across biological membranes, and it has been extensively studied in ocular drug delivery as it overcomes issues relating to the various ocular barriers [17, 48–52]. The hydrogel-iontophoresis delivery method in a randomized human study demonstrated that a total charge of less than 60 mA-min were well tolerated by patients, rendering itself as a promising ocular drug delivery method for targeting both anterior and posterior segments of the eye.

In addition to hydrogel ophthalmic delivery systems, incorporation of ophthalmic drugs with cyclodextrin has shown efficacy in lowering intraocular pressure in human patients [46, 47]. To date, evaluations of nanomaterials such as micelles, liposomes/niosomes, and dendrimers are currently limited to preclinical animal studies. However, these nanomaterials are also expected to expand to human clinical phase studies in the near future, as more preclinical studies demonstrate their enhanced pharmacokinetic properties.

5.6 Recent Developments of Nanomaterials in Ophthalmic Formulations

5.6.1 Polymeric Micelles

Polymeric micelles are core–shell structured nanoparticles formed by the self-assembly of amphiphilic copolymers. Polymeric micelles have been investigated extensively in the field of drug delivery as they present numerous advantages. Polymeric micelles can be fabricated through simple self-assembly process, usually involving techniques such as nanoprecipitation, emulsion, or dialysis, in aqueous medium. In addition, the core–shell structure enables encapsulation of hydrophobic drugs in its hydrophobic core. Because the hydrophobic core is shielded by the hydrophilic corona, the stability and the half-life of the drug is significantly prolonged in the vascular circulation. Moreover, biodegradable and biocompatible polymers are selected for formulation of micellar carriers, which would prevent any adverse effects caused in the physiological system by the carriers. With these benefits, polymeric micelles have attracted great interest in the field of drug delivery over the past few decades including drug delivery for various ocular routes [51, 53–59].

Several groups have used polylactide (PLA) or poly(lactide-co-glycolide) (PLGA) drug carriers to enhance the delivery of different types of drugs [60–65]. PLGA copolymers form micro- or nanoparticles in aqueous medium by self-assembly phenomena and can be used to encapsulate the drugs. However, the challenge with using PLGA carriers is that they are unstable in aqueous medium: the ester backbone of the PLGA polymer is prone to hydrolytic degradation, which ultimately lowers the shelf life of PLGA-encapsulated drug complexes. Thus, many scientists have turned to investigating copolymers with hydrophilic chains, such as polyethylene glycol (PEG) or polyethylene oxide (PEO), in order to increase the stability of the nanoparticles by shielding the hydrophobic cores from hydrolytic degradation. In one study, the effect of PEG versus chitosan surface modification on polymeric micelles was compared in terms of the ability to penetrate and deliver drugs across the corneal membrane [66]. Chitosan was electrostatically anchored to the surface of PCL nanocapsules, whereas the block copolymer Poly(ε-caprolactone)-b-PEG (PCL-b-PEG) was synthesized. The results showed both nanocapsules increased the permeability of encapsulated drugs through the corneal epithelium compared to free drug suspension, with chitosan-coated nanocapsules increasing the amount of drug permeation up to threefold after 4 h. However, the mechanism of transport enhancement remains to be completely dissected. A number of studies also showed that the PEG-based micellar carriers were safe in both intracameral and intravitreal injections [67] and effective in increasing the ocular bioavailability of the delivered drugs [68]. A triblock copolymer poly(oxyethylene)-poly(oxypropylene)-poly(oxyethylene) (PEO-PPO-PEO, trade name: Pluronic F127) was suggested by Pepic et al. for encapsulation and delivery of pilocarpine [69]. The triblock copolymer is comprised of a central hydrophobic chain (PPO) with two hydrophilic chains on either side (PEO) which was shown to self-assemble into a micellar structure in an aqueous environment, with sizes ranging from 16 to 30 nm having pilocarpine encapsulated. Although the drug encapsulation efficiency was very low (maximal 1.9 wt%), the authors demonstrated that the miotic response (area under the curve of miotic response vs. time) using this micellar solution of pilocarpine base showed a 64 % increase compared to a standard pilocarpine eye-drop solution. A PEO-PPO-PEO triblock copolymer was also used in two other studies. Liaw et al. developed an eye-drop solution containing the polymeric micelles of Pluronic F127 to deliver DNA plasmids for gene therapy and to increase the stability of DNA plasmids in vivo [70]. More recently, Kadam et al. used the micellization of Pluronic F127 in an aqueous environment to encapsulate the anti-epileptic drug carbamazepine to increase the water solubility of the drug [71].

A stimuli-responsive micellar drug delivery system has also been developed. A triblock copolymer comprising *N*-isopropylacrylamide (NIPAAM), vinylpyrrolidone

(VP), and acrylic acid (AA) were cross-linked with N,N-methylene-bis-acrylamide (MBA) to form NPs [72]. The micellar NPs were analyzed for the delivery of the anti-inflammatory drug ketorolac. The polymeric micellar system possesses a unique thermosensitivity, since the lower critical solution temperature (LCST) of the tri-block copolymer was found near the physiological temperature. Therefore, the micellar system showed a temperature-dependent release phenomena, which can be utilized to control the release of the drugs to specific targeting tissues.

Other types of block copolymers were also exploited by various research groups. Due to the small sizes and the stable suspension in aqueous environment, polymeric micelles have also been injected into the vitreous using an intravitreal delivery route. Roy et al. were able to demonstrate intravitreal delivery of antisense oligonucleotides to the retina of a rat using a polyoxyethylene-polyspermine (PEO-*b*-PSP) carrier to decrease the fibronectin expression in retinal vascular cells [73]. The block copolymer formed an effective diameter of 12 nm with the oligonucleotides embedded in the matrix and the NPs persisted up to 6 days in the retinal microvessels without showing any toxic effects. The study proved that the NP delivery method successfully reduced the expression of fibronectin mRNA level to 86.7 % 2 days after injection and down to 46.7 % at day 6. An in vivo study carried out by Bourges et al. demonstrated that nanoparticles injected through an intravitreal route means showed a tendency to migrate through the retinal layers and accumulate in RPE cells. They observed the presence of the nanoparticles in RPE cells up to 4 months after the initial single intravitreal injection, which provides a promising long-term drug delivery tool to the posterior segment of the eye [74].

Various types of amphiphilic copolymers have also demonstrated their ability to improve the in vivo bioavailability of the encapsulated drugs. Qu et al. synthesized chitosan-based amphiphilic copolymer, quaternary ammonium palmitoyl glycol chitosan (GCPQ) [75]. The modified amphiphilic copolymer encapsulated hydrophobic drugs by forming micellar clusters and showed encapsulation of up to 20–200 times more hydrophobic drugs compared to that of Pluronic block copolymers. Upon both intravenous injection and topical ocular application of GCPQ with prednisolone, the bioavailability of prednisolone increased tenfold compared to a commercial emulsion formulation of the same drug. A significant number of recent studies focused on developing mucoadhesive drug carriers to improve their corneal retention by reducing rapid clearance by tear turnover [76–78]. Chitosan is most commonly used material for the mucoadhesion property facilitated by the electrostatic interaction between positively charged chitosan and the negatively charged mucous membrane [79]. Recently, Lin et al. developed a micellar delivery system based on Pluronic–chitosan that showed enhanced pharmacological behavior compared to the commercial drug form [80]. Shen et al. utilized thiolated PEG stearate as a surfactant to modify the surface of nanostructured lipid carrier (NLC) in order to render mucoadhesiveness for the nano-drug carriers [81]. The mucoadhesion was achieved by the disulfide linkage between the thiol groups of the nanocarrier and the mucin particles (Fig. 5.4). When topically administered to the rabbit eye, the thiolated NLC improved the drug concentration and the precorneal retention time of the encapsulated cyclosporine, which remained in the cul-de-sac for up

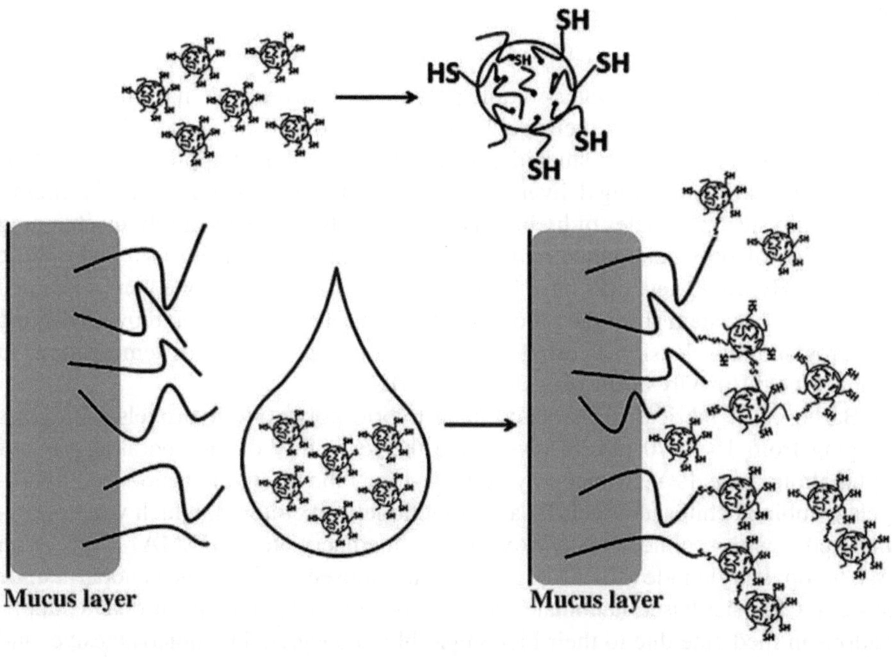

Fig. 5.4 Schematic representation of the interaction between the thiolated NLCs and mucus (Reproduced with permission from [81])

to 6 h. Several groups have also started modifying the surface of the drug carriers with phenylboronic acid molecules for mucoadhesive targeting. Phenylboronic acid molecules bind covalently with the *cis*-diol groups of sialic acids that are abundant on mucous membrane at physiological pH [82]. Liu et al. modified the dextran surface of the poly(D,L-lactide)-b-dextran nanoparticles with phenylboronic acid and showed that the extent of mucoadhesion was improved with increased density of phenylboronic acid molecules on the surface [83]. The same nanoparticles showed controlled release of cyclosporine A for up to 5 days in vitro. A number of studies have also focused on using prodrugs for enhanced penetration through ocular barriers [84, 85]. An amphiphilic prodrug of tilisolol, *O*-palmitoyl tilisolol, was developed by Kawakimi et al. in order to improve the ocular adsorption of the drug through the cornea [86]. Upon topical instillation, the prodrug form of tilisolol showed prolonged retention and high drug concentration in both tear fluid and cornea compared to the tilisolol suspension.

Micellar drug carriers have also shown increased ocular permeability of various drugs, when compared to their pure drug solution forms [35, 87]. From these studies, it is evident that polymeric micellar drug carriers have great potential not only in improving the drug stability and release mechanism to prolong the therapeutic activity of the encapsulated drugs, but they also show promise in increasing the permeability of the drugs through various ocular barriers.

5.6.2 *Hydrogels*

Hydrogels are defined as water-soluble polymeric networks that have the ability to absorb more than 20 % of their weight of water and still main their 3D structure [88]. Hydrogels have been studied extensively in medicinal applications, such as tissue engineering and drug delivery, due to their unique properties [89]. The chemical and physical properties of hydrogels can be easily modified simply by changing the choice of polymers, since hydrogels can be fabricated from any hydrophilic polymer. Hydrogel networks have been broadly studied as controlled and sustained drug delivery system since the porosity of their matrix can be tailored by modifying the cross-linking density or using external stimuli such as pH or temperature, to control the diffusivity of the drugs in the matrix.

Barbu et al. fabricated nanoparticulate hybrid polymeric hydrogels with sizes ranging from 10 to 70 nm, to investigate the feasibility of incorporating various ophthalmic drugs [90]. The copolymers used for fabricating the nano-sized hydrogels combined chitosan or acrylic acid-functionalized chitosan, which was copolymerized with either 2-hydroxyethyl methacrylate (HEMA) or with *N*-isopropylacrylamide (NIPAM), to impart thermosensitivity to the nanoparticulate system. Chitosan-based nanomaterials have been widely studied for diverse applications in medicine due to their biocompatible, biodegradable, mucoadhesive, and nontoxic properties [91]. Chitosan's mucoadhesive property is especially important in topical ocular drug delivery systems, since this prolongs the retention of the delivery system within the mucin layer, further increasing the duration of drug activity. The in vitro drug-release results demonstrated that the nanoparticulate system showed sustained-release phenomena, and the release rate of various drugs depended on the % content of chitosan materials in the hydrogels. More recent studies have focused on using in situ gelation properties of some of the hydrogel materials, to enhance the ocular bioavailability of the drugs and the in vivo drug activity. Pluronic F127, in addition to forming micellar structures due to its amphiphilic nature, has been used to fabricate hydrogel network systems. Due to its temperature-sensitive nature, Pluronic F127 was studied by various groups as an in situ gelling agent for ocular delivery applications [92, 93]. The sol–gel transition temperature of Pluronic F127 was below the human physiological temperature of 37 °C. This property was utilized by preparing a liquid solution of Pluronic F127 with drugs of interest at low temperatures. Subsequently, upon administration into the physiological environment, the solution undergoes gellation to create a polymeric matrix, where the drug is entrapped and released slowly through the gel by diffusion.

A handful of recent studies focused on incorporating micellar nanoparticles within the gel network of hydrogels to further improve the pharmacokinetic properties of the delivery system. Lu et al. demonstrated that the incorporation of PEG-b-PCL block copolymer micelles into the gel matrix of pHEMA achieved controlled release of the drug for up to 14 days, whereas the micelles themselves completed the release of all the drugs within 20 h without the gel matrix [94]. Li et al. developed an in situ gelling solution with nano-micelles incorporated [95]. The in situ gelled delivery system was nonirritating to the eye and maintained the minimal effective concentration of the drug for at least 24 h after administration. Because of the ability of the gel matrix to further delay the drug release from the nanocomposites within, more and more research are focused on developing this type of hybrid system for enhanced drug delivery.

5.6.3 Liposomes

Liposomes are biocompatible and biodegradable particles consisting of membrane-like lipid bilayers composed mostly of phospholipids [76]. Formed by amphiphilic phospholipids, liposomes have hollow spherical structures (vesicles) with both lipophilic (lipid bilayer) and hydrophilic (aqueous compartment) phases. Due to this biphasic nature, liposomes can encapsulate both hydrophilic and/or lipophilic therapeutic agents in each compartment. The properties of liposomes, such as surface charge, can be tuned substantially by varying the lipid composition. Since the sialic acid moieties in the mucous membrane of the corneal surface is negatively charged; positively charged liposomes are preferentially captured at the corneal surface, as compared to neutral or negatively charged liposomes [27]. In addition, a recent study showed that positively charged liposomes also increased the extent of absorption of the encapsulated drugs across corneal membranes [96]. To achieve this characteristic, numerous studies have focused on coating the liposomal surface with the mucoadhesive polymer chitosan.

Diebold at el. investigated the cellular uptake effects of liposome–chitosan nanoparticle complexes (LCS-NP) in the precorneal moieties. Strong cellular uptake of LCS-NP in the conjunctiva, and less intense uptake by the corneal epithelium, suggested that the NPs were first retained by the mucus layer and subsequently enter through the conjunctival cells. They also showed that ocular drug delivery system liposome–chitosan nanoparticle complexes (LCS-NP) showed negligible in vitro toxicity and acceptable in vivo tolerance [97]. Li et al. proposed using low molecular weight chitosan (LCH) coating on liposomes and investigated their in vitro and in vivo properties [98]. The coating of LCH not only modified the surface to have a positive charge to enhance its interaction with the mucous membrane of the corneal surface, but it also showed increased transcorneal penetration of the LCH-coated liposomes. The effect of surface coating the nanoparticles subsequently led to prolonged retention of the encapsulated drugs at the corneal surface and suggested the potential for using the liposomal carriers as a transcorneal drug delivery system, to increase bioavailability of the drugs in the aqueous compartment. However, chitosan has shown precipitation phenomena near physiological pH, and therefore, some studies used a modified version of chitosan to solve the precipitation issue. Wang et al. also explored using *N*-trimethyl chitosan (TMC)-coated liposomes to deliver coenzyme Q_{10} to the human lens epithelial cells to protect against oxidative damages [99]. Chitosan chains were modified to the quaternized derivative TMC to overcome issues such as precipitation of chitosan at neutral pH. This study also demonstrated that the TMC-coated liposomes exhibited excellent corneal permeation, increasing the permeability coefficient more than two times in comparison with values obtained from a control study. Poly(L-lysine), a cationic polymer, modified sub-100 nm liposomes were administered as eye drops to target the retina: the modified liposomes enhanced the penetration of the drugs across the ocular barriers in the anterior segment of the eye [100].

Other studies have used different coating materials on the liposomes to achieve enhanced pharmacokinetic properties. Hosny developed ofloxacin-encapsulated liposomes, which were further incorporated into an in situ thermosensitive hydrogel system composed of chitosan/β-glycerophosphate, to enhance the transcorneal permeation of ofloxacin [101]. The liposomal hydrogel system improved the transcorneal permeability of ofloxacin sevenfold compared to that of the aqueous solution. In addition, the drug delivery system ensured steady and prolonged permeation of the drugs, improving the bioavailability. Bochot et al. used a PEG coating on the surface of a liposomal formulation, to sterically stabilize the liposomes [102]. The liposomal formulation was injected into the vitreous humor, where the liposomes showed favored biodistribution at retina–choroid moieties, as compared with non-target tissues such as the sclera or lens. The administration of liposome-encapsulated phosphodiester (16-mer oligothymidylate) oligonucleotides into the vitreous humor also showed sustained release, which offers the prospect that the liposomes can be used as an intravitreal delivery system, targeting the retina and choroid, where the majority of ocular diseases occur. Hironaka et al. studied the effect of size and rigidity of the liposomes on the delivery performance of the liposomes to the posterior segment of the eye [103]. It was shown that rigid liposomes with size ranges in the nanoscale showed significant potential as an ocular delivery system, targeting the posterior segment of the eye. Another study used systemic intravenous injection of liposome delivery vehicles to target the posterior segment of the eye, and they showed that the paclitaxel-carrying liposomes had significant effect in reducing the size of the choroidal neovascularization [104].

Despite the recent reports of using liposomal formulations for ocular drug delivery and the fact that they demonstrated improvements in precorneal retention, sustained drug release, and transcorneal permeation, liposomes still face challenges in their limited long-term structural stability and drug-loading capacity due to the inherent complex nature of their structures.

5.6.4 Niosomes

Niosomes, a specific type of liposome, are comprised of amphiphilic nonionic surfactants which form vesicle structures. Niosomes, like liposomes, can entrap both hydrophilic and hydrophobic drugs. Niosomes are generally preferred over other vesicular systems for topical ocular drug delivery systems for several reasons: they are chemically more stable, they are less toxic because of the nonionic nature of the surfactants, they are easier to handle without special precautions, they are able to improve the performance of the drug via better bioavailability and controlled delivery at a specific site, and they are also biodegradable, biocompatible, and non-immunogenic [105–108].

Abdelbary et al. investigated niosomes as a potential topical ocular drug delivery system for the antibiotic gentamicin [105]. The in vitro results showed that the encapsulation efficiency was as high as 92 %, and the encapsulation of gentamicin showed prolonged release compared to the control. The encapsulation efficiency and the release rate varied depending on the cholesterol content, the type of surfactant used, and the presence of the charge-inducer dicetyl phosphate. An ocular irritancy test of niosomes demonstrated no sign of redness, inflammation, or increased tear production, suggesting that niosomes can be used as an enhanced and safe topical ocular drug delivery system. Prabhu et al. studied the feasibility of using both liposome and niosome systems as a drug delivery system for brimonidine tartrate, to improve its intraocular pressure (IOP) lowering activity for the treatment of glaucoma [109]. The vesicle formulations were in the size range of 210–245 nm, with drug payload up to 42 w/w%. Although the vesicular formulations did not show as high an IOP-lowering activity as that of the marketed formulation of pure drug solution, they extended the duration of activity sixfold to eightfold compared to the marketed formulation. Niosomes coated with the mucoadhesive material chitosan were developed by Kaur at el. to investigate its potential as a drug delivery system for timolol maleate (TM), which also lowers IOP in the treatment of glaucoma [110]. The chitosan-coated niosome showed higher and prolonged corneal permeation of TM into the aqueous humor compared to the pure drug solution. Consequently, the IOP-lowering effect of the niosomal formulation of TM was maintained up to 8 h, as compared to 1.5 h for the drug solution. Most recent study proposed using "proniosomes" to further improve the shelf life of the niosome drug formulation [111]. Proniosomes are liquid crystalline-compact niosome hybrids which will turn into niosomes upon hydration prior to administration to patients. The study utilized this method to demonstrate that the treatment using tacrolimus-loaded proniosomes delayed the corneal allograft rejection significantly and improved the median survival times of corneal allografts.

5.6.5 Dendrimers

Dendrimers are three-dimensional, highly branched, and tree-structured macromolecules which have nanoscale sizes due to their well-organized synthesis strategy. In addition, because of their highly branched structure, the surface of the dendritic macromolecules can be easily functionalized with a variety of desired properties. Dendrimers can encapsulate hydrophobic drug molecules since they possess internal empty cavities and open conformations, for low-generation dendrimers. Due to these specific combinations of properties, dendrimers offer some attractions for drug delivery applications [112, 113].

One of the most widely studied dendrimers for use as an ocular drug delivery system is poly(amidoamine) (PAMAM). Vandamme and Brobeck used PAMAM dendrimers to improve ocular residence time [44]. They have demonstrated that a 0.25 w/v% PAMAM solution mixture with pilocarpine prolonged the ocular residence time of pilocarpine for up to 5 h, which is comparable to Carbopol® (a commercial bioadhesive polymer for ophthalmic dosage forms) but without the irritation that was observed with Carbopol® administration. The study also showed that residence time was longer for dendrimers with hydroxyl or carboxyl surface groups. Puerarin–PAMAM complexes were developed by Yao et al. through hydrogen bond interactions [114]. Similar to the previous study, they have demonstrated prolonged ocular residence time of the puerarin complex formulation, as compared to puerarin eye drops. It was also reported that although the puerarin–PAMAM complex did not increase the corneal permeability coefficient, the physical mixture of cationic G4 PAMAM dendrimer with puerarin was found to enhance the corneal permeability coefficient and increased the cumulative amount of puerarin permeating across the cornea. Durairaj et al. used a complex formation between dendrimeric polyguanidilyated translocators (DPTs) and gatifloxacin (GFX) antibiotic to improve the permeation across sclera–choroid–retinal pigment epithelium (SCRPE) membranes [115]. The complex improved the permeation of GFX across SCRPE by 40 % in 6 h. In vivo studies showed that the topical administration of the complex was able to maintain higher drug concentration for up to 24 h, compared to pure drug solution. Novel phosphorus-containing dendrimers, with a quaternary ammonium salt as the core and carboxylic acid terminal groups, have also been synthesized [45]. These cationic dendrimers were able to physically associate with amino groups of carteolol (an ocular antihypertensive drug) by forming ion pairs. The generation 2 dendrimer-carteolol formulation improved the corneal penetration by 2.5 times when compared to carteolol alone. However, the generation 2 dendrimer showed poor water solubility, and therefore the amount of instilled carteolol was limited. Utilizing the ease of functionalization of dendrimers, Yandrapu et al. modified the dendrimers with thiol groups to achieve enhanced mucoadhesion [116]. Similar to micelle/hydrogel hybrid structures, several studies also developed dendrimers/hydrogel hybrid structures to improve the total uptake of the drug in the delivery system and also to prolong the total release time of the drugs [117, 118].

The ease of surface functionalization and the ability to encapsulate hydrophobic drugs render dendrimers an attractive ocular drug delivery system. In some animal studies however, a vision-blurring effect was observed after the administration of dendrimer on the ocular surface [4]. This issue must be addressed in future studies using dendrimers, before proceeding to full clinical trial studies.

5.6.6 Cyclodextrins

Cyclodextrins (CDs) are a family of cyclic oligosaccharides, composed of 6–8 glucose units which have been shown to improve the pharmacokinetic properties of many drugs, through the formation of an inclusion complex [119]. The application of cyclodextrins to ophthalmic formulations has been reviewed extensively over the past decade [119–122]. The complex inclusion of hydrophobic drugs with CDs can increase the aqueous solubility of those drugs. Toxicity studies have shown that orally administered CDs are practically nontoxic, due to the lack of absorption from the gastrointestinal tract [123]. One of the most extensively researched cyclodextrins used as an ocular drug delivery enhancement is 2-hydroxypropyl β-CD (HPβCD), which is modified with a hydroxylpropyl derivative to increase its water solubility. Topical administration of CDs have also shown no toxic effects, and eye-drop solutions containing 45 % of HPβCD showed no irritation effects in rabbits [124]. Similar to other types of nanomaterials described in this report, CDs also act as a permeation enhancer at biological barriers such as the cornea, by increasing the drug retention at the surface of the corneal epithelium.

Wang et al. used an inclusion complex of disulfiram (DSF) and HPβCD in an eye-drop formulation to increase the bioavailability of the drug across the corneal membrane [125]. Along with the addition of the penetration enhancer hydroxypropylmethylcellulose (HPMC), the eye-drop formulation successfully suppressed the cataract effect (i.e., lens opacity) for up to 48 h in rabbits. Similarly, Zhang et al. also used HPβCD with the complex inclusion of ketoconazole (KET) in a topical eye-drop formulation [126]. The incorporation of HPβCD increased the aqueous solubility of KET. Compared to the KET suspension, the HPβCD/KET complex significantly increased the bioavailability of KET, over eightfold in the aqueous humor and 12-fold in the cornea. More recently, another group has investigated the complex between HPβCD and indomethacin and found delayed release and a high drug stability effect [127]. While showing no signs of irritation after administration to rabbits, an HPβCD–indomethacin complex showed significant improvement in therapeutic efficacy in healing corneal wounds. Mahmoud et al. developed ionically cross-linked chitosan/sulfobutylether-β-cyclodextrin nanoparticles as an ocular drug delivery system [128]. The nanoparticle system showed sustained-release behavior of the encapsulated econazole nitrate. Furthermore, the incorporation of chitosan on the surface of the nanoparticles also significantly improved the ocular mucoadhesiveness, which supports nanoparticles as being a promising carrier for controlled ocular drug delivery systems. Similarly, many studies have also incorporated CDs into hydrogel matrix to utilize both the increased drug solubility using CDs and the slow release provided by the hydrogel matrix to improve the overall pharmacokinetic properties of the delivery system [129–133].

5.7 Future Research

The field of ocular drug delivery has taken a significant stride forward with the advent of nanotechnology. Numerous recent studies have focused on using various types of nanomaterials, such as polymeric micelles, hydrogels, liposomes, niosomes, dendrimers, and cyclodextrins as drug carriers, to increase the ocular bioavailability of various therapeutic agents. Topical administration of the drugs associated with the nanomaterials showed sustained release of the drug which increased the duration of therapeutic activity, consequently reducing the need for frequent administration. To date, demonstrating an enhancement in the extent of absorption into the vitreous cavity across corneal membranes, the topical administration of nanomaterial drug complexes to target the posterior segment of the eye has yet to be proven to be therapeutically effective compared to intravitreal or periocular injections. However, due to the invasive nature of intravitreal and periocular injections, further studies utilizing nanomaterials to overcome the various ocular barriers are expected. Although several in vivo studies have shown that the nano-drug carriers did not induce any signs of irritation or inflammatory responses, the long-term effect of these nanomaterials in both the ocular region and the systemic circulation must be rigorously analyzed. Most recently, many of the studies incorporated hybrid system with hydrogel using micelles, dendrimers, and cyclodextrins to further improve the pharmacokinetic properties of these delivery systems. There is no doubt that pharmaceutical researchers will continue improving the performance of ocular drug delivery systems with the help of nanotechnology, to achieve therapeutically effective, patient-compliant, and low-cost systems with negligible side effects.

References

1. Bejjani RA, Behar-Cohen F, Benezra D, Gurny R, Delie F (2005) Polymeric nanoparticles for drug delivery to the posterior segment of the eye. Chimia 59(6):344–347
2. Nagarwal RC, Kant S, Singh PN, Maiti P, Pandit JK (2009) Polymeric nanoparticulate system: a potential approach for ocular drug delivery. J Control Release 136(1):2–13
3. Tong Y, Chang S, Liu C, Kao WW, Huang CH, Liaw J (2007) Eye drop delivery of nano-polymeric micelle formulated genes with cornea-specific promoters. J Gene Med 9(11):956–966
4. Wadhwa S, Paliwal R, Paliwal SR, Vyas SP (2009) Nanocarriers in ocular drug delivery: an update review. Curr Pharm Des 15(23):2724–2750
5. Zarbin MA, Montemagno C, Leary JF, Ritch R (2010) Nanomedicine in ophthalmology: the new frontier. Am J Ophthalmol 150(2):144–162
6. Liu S, Jones L, Gu FX (2012) Nanomaterials for ocular drug delivery. Macromol Biosci 12(5):608–620
7. Gaudana R, Jwala J, Boddu SHS, Mitra AK (2009) Recent perspectives in ocular drug delivery. Pharm Res 26(5):1197–1216
8. Hughes PM, Olejnik O, Chang-Lin JE, Wilson CG (2005) Topical and systemic drug delivery to the posterior segments. Adv Drug Deliv Rev 57(14):2010–2032

9. Peng CC, Bengani LC, Jung HJ, Leclerc J, Gupta C, Chauhan A (2011) Emulsions and microemulsions for ocular drug delivery. J Drug Delivery Sci Technol 21(1):111–121
10. del Amo EM, Urtti A (2008) Current and future ophthalmic drug delivery systems. A shift to the posterior segment. Drug Discov Today 13(3–4):135–143
11. Yasukawa T, Ogura Y, Tabata Y, Kimura H, Wiedemann P, Honda Y (2004) Drug delivery systems for vitreoretinal diseases. Prog Retin Eye Res 23(3):253–281
12. Harder BC, Baltz S, Jonas JB, Schlichtenbrede FC (2011) Intravitreal bevacizumab for retinopathy of prematurity. J Ocul Pharmacol Ther 27(6):623–627
13. Kolomeyer AM, Roy MS, Chu DS (2011) The use of intravitreal ranibizumab for choroidal neovascularization associated with vogt-koyanagi-harada syndrome. Case Rep Med 2011:747648
14. Ranta V, Mannermaa E, Lummepuro K, Subrizi A, Laukkanen A, Antopolsky M, Murtomaki L, Hornof M, Urtti A (2010) Barrier analysis of periocular drug delivery to the posterior segment. J Control Release 148(1):42–48
15. Ciolino JB, Dohlman CH, Kohane DS (2009) Contact lenses for drug delivery. Semin Ophthalmol 24(3):156–160
16. Lavik E, Kuehn MH, Kwon YH (2011) Novel drug delivery systems for glaucoma. Eye 25(5):578–586
17. Weiner AL, Gilger BC (2010) Advancements in ocular drug delivery. Vet Ophthalmol 13(6): 395–406
18. Phan C-M, Subbaraman L, Jones L (2014) Contact lenses for antifungal ocular drug delivery: a review. Expert Opin Drug Deliv 11(4):537–546
19. Fonn D (2007) Targeting contact lens induced dryness and discomfort: what properties will make lenses more comfortable. Optom Vis Sci 84(4):279–285
20. Gulsen D, Chauhan A (2005) Dispersion of microemulsion drops in HEMA hydrogel: a potential ophthalmic drug delivery vehicle. Int J Pharm 292(1–2):95–117
21. Li CC, Chauhan A (2006) Modeling ophthalmic drug delivery by soaked contact lenses. Ind Eng Chem Res 45(10):3718–3734
22. Ciolino JB, Hoare TR, Iwata NG, Behlau I, Dohlman CH, Langer R, Kohane DS (2009) A drug-eluting contact lens. Invest Ophthalmol Vis Sci 50(7):3346–3352
23. Gulsen D, Li CC, Chauhan A (2005) Dispersion of DMPC liposomes in contact lenses for ophthalmic drug delivery. Curr Eye Res 30(12):1071–1080
24. Jimenez N, Galan J, Vallet A, Egea MA, Garcia ML (2010) Methyl trypsin loaded poly(D,L-lactide-coglycolide) nanoparticles for contact lens care. J Pharm Sci 99(3):1414–1426
25. Kapoor Y, Chauhan A (2008) Drug and surfactant transport in Cyclosporine A and Brij 98 laden p-HEMA hydrogels. J Colloid Interface Sci 322(2):624–633
26. Phan C-M, Subbaraman L, Liu S, Gu F, Jones L (2014) In vitro uptake and release of natamycin Dex-b-PLA nanoparticles from model contact lens materials. J Biomater Sci Polym Ed 25(1):18–31
27. Sahoo SK, Diinawaz F, Krishnakumar S (2008) Nanotechnology in ocular drug delivery. Drug Discov Today 13(3–4):144–151
28. Zimmer A, Kreuter J (1995) Microspheres and nanoparticles used in ocular delivery systems. Adv Drug Deliv Rev 16(1):61–73
29. Hamidi M, Azadi A, Rafiei P (2008) Hydrogel nanoparticles in drug delivery. Adv Drug Deliv Rev 60(15):1638–1649
30. Ludwig A (2005) The use of mucoadhesive polymers in ocular drug delivery. Adv Drug Deliv Rev 57(11):1595–1639
31. Cardillo JA, Souza-Filho AA, Oliveira AG (2006) Intravitreal bioerudivel sustained-release triamcinolone microspheres system (RETAAC). Preliminary report of its potential usefulnes for the treatment of diabetic macular edema. Arch Soc Esp Oftalmol 81(12):675
32. Rafie F, Javadzadeh Y, Javadzadeh AR, Ghavidel LA, Jafari B, Moogooee M, Davaran S (2010) In vivo evaluation of novel nanoparticles containing dexamethasone for ocular drug delivery on rabbit eye. Curr Eye Res 35(12):1081–1089

33. Pepic I, Hafner A, Lovric J, Pirkic B, Filipovic-Grcic J (2010) A nonionic surfactant/chitosan micelle system in an innovative eye drop formulation. J Pharm Sci 99(10):4317–4325
34. Di Tommaso C, Torriglia A, Furrer P, Behar-Cohen F, Gurny R, Moller M (2011) Ocular biocompatibility of novel cyclosporin a formulations based on methoxy poly(ethylene glycol)-hexylsubstituted poly(lactide) micelle carriers. Int J Pharm 416(2):515–524
35. Civiale C, Licciardi M, Cavallaro G, Giammona G, Mazzone MG (2009) Polyhydroxyethylaspartamide-based micelles for ocular drug delivery. Int J Pharm 378(1–2):177–186
36. Volotinen M, Maenpaa J, Kautiainen H, Tolonen A, Uusitalo J, Ropo A, Vapaatalo H, Aine E (2009) Ophthalmic timolol in a hydrogel vehicle leads to minor inter-individual variation in timolol concentration in aqueous humor. Eur J Pharm Sci 36(2–3):292–296
37. Parkinson TM, Ferguson E, Febbraro S, Bakhtyari A, King M, Mundasad M (2003) Tolerance of ocular iontophoresis in healthy volunteers. J Ocul Pharmacol Ther 19(2):145–151
38. Gonzalez-Mira E, Egea MA, Garcia ML, Souto EB (2010) Design and ocular tolerance of flurbiprofen loaded ultrasound-engineered NLC. Colloids Surf B Biointerfaces 81(2):412–421
39. Abdelbary G (2011) Ocular ciprofloxacin hydrochloride mucoadhesive chitosan-coated liposomes. Pharm Dev Technol 16(1):44–56
40. Habib FS, Fouad EA, Abdel-Rhaman MS, Fathalla D (2010) Liposomes as an ocular delivery system of fluconazole: in-vitro studies. Acta Ophthalmol 88(8):901–904
41. Mehanna MM, Elmaradny HA, Samaha MW (2010) Mucoadhesive liposomes as ocular delivery system: physical, microbiological, and in vivo assessment. Drug Dev Ind Pharm 36(1):108–118
42. Abrishami M, Zarei-Ghanavati S, Soroush D, Rouhbakhsh M, Jaafari MR, Malaekeh-Nikouei B (2011) Preparation, characterization, and in vivo evaluation of nanoliposomes-encapsulated bevacizumab (Avastin) for intravitreal administration (29:699, 2009). Retina J Retin Vitreous Dis 31(1):205
43. Kuno N, Fujii S (2011) Recent advances in ocular drug delivery systems. Polymers 3(1):193–221
44. Vandamme TF, Brobeck L (2005) Poly(amidoamine) dendrimers as ophthalmic vehicles for ocular delivery of pilocarpine nitrate and tropicamide. J Control Release 102(1):23–38
45. Spataro G, Malecaze F, Turrin C, Soler V, Duhayon C, Elena P, Majoral J, Caminade A (2010) Designing dendrimers for ocular drug delivery. Eur J Med Chem 45(1):326–334
46. Gudmundsdottir E, Stefansson E, Bjarnadottir G, Sigurjonsdottir JF, Gudmundsdottir G, Masson M, Loftsson T (2000) Methazolamide 1 % in cyclodextrin solution lowers IOP in human ocular hypertension. Invest Ophthalmol Vis Sci 41(11):3552–3554
47. Gonzalez JR, Baiza-Duran L, Quintana-Hau J, Tornero-Montano R, Castaneda-Hernandez G, Ortiz M, Alarcon-Oceguera F, Beltran-Loustaunau M, Cortez-Gastelum M, Garciduenas-Mejia J, Gomez-Bastar P, Jimenez-Roman J, Korder-Ortega V, Paczka-Zapata J, Torres-Segura M, Velasco-Gallegos G (2007) Comparison of the stability, efficacy, and adverse effect profile of the innovator 0.005 % latanoprost ophthalmic solution and a novel cyclodextrin-containing formulation. J Clin Pharmacol 47(1):121–126
48. Eljarrat-Binstock E, Domb AJ (2006) Iontophoresis: a non-invasive ocular drug delivery. J Control Release 110(3):479–489
49. Eljarrat-Binstock E, Orucov F, Aldouby Y, Frucht-Pery J, Domb AJ (2008) Charged nanoparticles delivery to the eye using hydrogel iontophoresis. J Control Release 126(2):156–161
50. Eljarrat-Binstock E, Orucov F, Frucht-Pery J, Pe'er J, Domb AJ (2008) Methylprednisolone delivery to the back of the eye using hydrogel iontophoresis. J Ocul Pharmacol Ther 24(3):344–350
51. Gaucher G, Marchessault RH, Leroux J (2010) Polyester-based micelles and nanoparticles for the parenteral delivery of taxanes. J Control Release 143(1):2–12

52. Molokhia SA, Jeong E, Higuchi WI, Li SK (2009) Transscleral iontophoretic and intravitreal delivery of a macromolecule: study of ocular distribution in vivo and postmortem with MRI. Exp Eye Res 88(3):418–425
53. Cho HK, Cheong IW, Lee JM, Kim JH (2010) Polymeric nanoparticles, micelles and polymersomes from amphiphilic block copolymer. Korean J Chem Eng 27(3):731–740
54. Gaucher G, Dufresne MH, Sant VP, Kang N, Maysinger D, Leroux JC (2005) Block copolymer micelles: preparation, characterization and application in drug delivery. J Control Release 109(1–3):169–188
55. Jagur-Grodzinski J (2009) Polymers for targeted and/or sustained drug delivery. Polym Adv Technol 20(7):595–606
56. Kim DK, Dobson J (2009) Nanomedicine for targeted drug delivery. J Mater Chem 19(35):6294–6307
57. Onaca O, Enea R, Hughes DW, Meier W (2009) Stimuli-responsive polymersomes as nanocarriers for drug and gene delivery. Macromol Biosci 9(2):129–139
58. Pepic I, Lovric J, Filipovic-Grcic J (2012) Polymeric micelles in ocular drug delivery: rationale, strategies and challenges. Chem Biochem Eng Q 26(4):365–377
59. Soppimath KS, Aminabhavi TM, Kulkarni AR, Rudzinski WE (2001) Biodegradable polymeric nanoparticles as drug delivery devices. J Control Release 70(1–2):1–20
60. Ayalasomayajula SP, Kompella UB (2005) Subconjunctivally administered celecoxib-PLGA microparticles sustain retinal drug levels and alleviate diabetes-induced oxidative stress in a rat model. Eur J Pharmacol 511(2–3):191–198
61. Barcia E, Herrero-Vanrell R, Diez A, Alvarez-Santiago C, Lopez I, Calonge M (2009) Downregulation of endotoxin-induced uveitis by intravitreal injection of polylactic-glycolic acid (PLGA) microspheres loaded with dexamethasone. Exp Eye Res 89(2):238–245
62. Cleland JL, Duenas ET, Park A, Daugherty A, Kahn J, Kowalski J, Cuthbertson A (2001) Development of poly-(D,L-lactide-coglycolide) microsphere formulations containing recombinant human vascular endothelial growth factor to promote local angiogenesis. J Control Release 72(1–3):13–24
63. Gavini E, Chetoni P, Cossu M, Alvarez MG, Saettone MF, Giunchedi P (2004) PLGA microspheres for the ocular delivery of a peptide drug, vancomycin using emulsification/spray-drying as the preparation method: in vitro/in vivo studies. Eur J Pharm Biopharm 57(2):207–212
64. Gupta H, Aqil M, Khar RK, Ali A, Bhatnagar A, Mittal G (2010) Sparfloxacin-loaded PLGA nanoparticles for sustained ocular drug delivery. Nanomed Nanotechnol Biol Med 6(2):324–333
65. Qaddoumi MG, Ueda H, Yang J, Davda J, Labhasetwar V, Lee VHL (2004) The characteristics and mechanisms of uptake of PLGA nanoparticles in rabbit conjunctival epithelial cell layers. Pharm Res 21(4):641–648
66. De Campos AM, Sanchez A, Gref R, Calvo P, Alonso MJ (2003) The effect of a PEG versus a chitosan coating on the interaction of drug colloidal carriers with the ocular mucosa. Eur J Pharm Sci 20(1):73–81
67. Xu L, Xu X, Chen H, Li X (2013) Ocular biocompatibility and tolerance study of biodegradable polymeric micelles in the rabbit eye. Colloids Surf B Biointerfaces 112:30–34
68. Di Tommaso C, Bourges J, Valamanesh F, Trubitsyn G, Torriglia A, Jeanny J, Behar-Cohen F, Gurny R, Moeller M (2012) Novel micelle carriers for cyclosporin a topical ocular delivery: in vivo cornea penetration, ocular distribution and efficacy studies. Eur J Pharm Biopharm 81(2):257–264
69. Pepic I, Jalsenjak N, Jalsenjak I (2004) Micellar solutions of triblock copolymer surfactants with pilocarpine. Int J Pharm 272(1–2):57–64
70. Liaw J, Chang SF, Hsiao FC (2001) In vivo gene delivery into ocular tissues by eye drops of poly(ethylene oxide)-poly(propylene oxide)-poly(ethylene oxide) (PEO-PPO-PEO) polymeric micelles. Gene Ther 8(13):999–1004

71. Kadam Y, Yerramilli U, Bahadur A (2009) Solubilization of poorly water-soluble drug carbamazepine in Pluronic (R) micelles: effect of molecular characteristics, temperature and added salt on the solubilizing capacity. Colloids Surf B Biointerfaces 72(1):141–147
72. Gupta AK, Madan S, Majumdar DK, Maitra A (2000) Ketorolac entrapped in polymeric micelles: preparation, characterisation and ocular anti-inflammatory studies. Int J Pharm 209(1–2):1–14
73. Roy S, Zhang K, Roth T, Vinogradov S, Kao RS, Kabanov A (1999) Reduction of fibronectin expression by intravitreal administration of antisense oligonucleotides. Nat Biotechnol 17(5):476–479
74. Bourges JL, Gautier SE, Delie F, Bejjani RA, Jeanny JC, Gurny R, BenEzra D, Behar-Cohen FF (2003) Ocular drug delivery targeting the retina and retinal pigment epithelium using polylactide nanoparticles. Invest Ophthalmol Vis Sci 44(8):3562–3569
75. Qu X, Khutoryanskiy VV, Stewart A, Rahman S, Papahadjopoulos-Sternberg B, Dufes C, McCarthy D, Wilson CG, Lyons R, Carter KC, Schatzlein A, Uchegbu IF (2006) Carbohydrate-based micelle clusters which enhance hydrophobic drug bioavailability by up to 1 order of magnitude. Biomacromolecules 7(12):3452–3459
76. du Toit LC, Pillay V, Choonara YE, Govender T, Carmichael T (2011) Ocular drug delivery – a look towards nanobioadhesives. Expert Opin Drug Deliv 8(1):71–94
77. Khutoryanskiy VV (2011) Advances in mucoadhesion and mucoadhesive polymers. Macromol Biosci 11(6):748–764
78. Shaikh R, Raj Singh TR, Garland MJ, Woolfson AD, Donnelly RF (2011) Mucoadhesive drug delivery systems. J Pharm Bioallied Sci 3(1):89–100
79. Sogias IA, Williams AC, Khutoryanskiy VV (2008) Why is chitosan mucoadhesive? Biomacromolecules 9(7):1837–1842
80. Lin H, Chang P (2013) Novel pluronic-chitosan micelle as an ocular delivery system. J Biomed Mater Res B Appl Biomater 101B(5):689–699
81. Shen J, Wang Y, Ping Q, Xiao Y, Huang X (2009) Mucoadhesive effect of thiolated PEG stearate and its modified NLC for ocular drug delivery. J Control Release 137(3–4):217–223
82. Otsuka H, Uchimura E, Koshino H, Okano T, Kataoka K (2003) Anomalous binding profile of phenylboronic acid with *N*-acetylneuraminic acid (Neu5Ac) in aqueous solution with varying pH. J Am Chem Soc 125(12):3493–3502
83. Liu S, Jones L, Gu FX (2012) Development of mucoadhesive drug delivery system using phenylboronic acid functionalized poly(D,L-lactide)-b-Dextran nanoparticles. Macromol Biosci 12(12):1622–1626
84. Saarinen-Savolainen P, Jarvinen T, Suhonen P, Urtti A (1996) Amphiphilic properties of pilocarpine prodrugs. Int J Pharm 133(1–2):171–178
85. Sasaki H, Yamamura K, Mukai T, Nishida K, Nakamura J, Nakashima M, Ichikawa M (1999) Enhancement of ocular drug penetration. Crit Rev Ther Drug Carrier Syst 16(1):85–146
86. Kawakami S, Nishida K, Mukai T, Yamamura K, Kobayashi K, Sakaeda T, Nakamura J, Nakashima M, Sasaki H (2001) Ocular absorption behavior of palmitoyl tilisolol, an amphiphilic prodrug of tilisolol, for ocular drug delivery. J Pharm Sci 90(12):2113–2120
87. Tu J, Pang H, Yan Z, Li P (2007) Ocular permeability of pirenzepine hydrochloride enhanced by methoxy poly(ethylene glycol)-poly(D,L-lactide) block copolymer. Drug Dev Ind Pharm 33(10):1142–1150
88. Gupta P, Vermani K, Garg S (2002) Hydrogels: from controlled release to pH-responsive drug delivery. Drug Discov Today 7(10):569–579
89. Hoare TR, Kohane DS (2008) Hydrogels in drug delivery: progress and challenges. Polymer 49(8):1993–2007
90. Barbu E, Verestiuc L, Iancu M, Jatariu A, Lungu A, Tsibouklis J (2009) Hybrid polymeric hydrogels for ocular drug delivery: nanoparticulate systems from copolymers of acrylic acid-functionalized chitosan and *N*-isopropylacrylamide or 2-hydroxyethyl methacrylate. Nanotechnology 20(22):225108

91. Peniche H, Peniche C (2011) Chitosan nanoparticles: a contribution to nanomedicine. Polym Int 60(6):883–889
92. El-Kamel AH (2002) In vitro and in vivo evaluation of Pluronic F127-based ocular delivery system for timolol maleate. Int J Pharm 241(1):47–55
93. Ma W, Xu H, Wang C, Nie S, Pan W (2008) Pluronic F127-g-poly(acrylic acid) copolymers as in situ gelling vehicle for ophthalmic drug delivery system. Int J Pharm 350(1–2):247–256
94. Lu C, Mikhail AS, Wang X, Brook MA, Allen C (2012) Hydrogels containing core cross-linked block co-polymer micelles. J Biomater Sci Polym Ed 23(8):1069–1090
95. Li X, Zhang Z, Chen H (2013) Development and evaluation of fast forming nano-composite hydrogel for ocular delivery of diclofenac. Int J Pharm 448(1):96–100
96. Law SL, Huang KJ, Chiang CH (2000) Acyclovir-containing liposomes for potential ocular delivery. Corneal penetration and absorption. J Control Release 63(1–2):135–140
97. Diebold Y, Jarrin M, Saez V, Carvalho ELS, Orea M, Calonge M, Seijo B, Alonso MJ (2007) Ocular drug delivery by liposome-chitosan nanoparticle complexes (LCS-NP). Biomaterials 28(8):1553–1564
98. Li N, Zhuang C, Wang M, Sun X, Nie S, Pan W (2009) Liposome coated with low molecular weight chitosan and its potential use in ocular drug delivery. Int J Pharm 379(1):131–138
99. Wang S, Zhang J, Jiang T, Zheng L, Wang Z, Zhang J, Yu P (2011) Protective effect of Coenzyme Q(10) against oxidative damage in human lens epithelial cells by novel ocular drug carriers. Int J Pharm 403(1–2):219–229
100. Sasaki H, Karasawa K, Hironaka K, Tahara K, Tozuka Y, Takeuchi H (2013) Retinal drug delivery using eyedrop preparations of poly-L-lysine-modified liposomes. Eur J Pharm Biopharm 83(3):364–369
101. Hosny KM (2009) Preparation and evaluation of thermosensitive liposomal hydrogel for enhanced transcorneal permeation of ofloxacin. AAPS PharmSciTech 10(4):1336–1342
102. Bochot A, Fattal E, Boutet V, Deverre JR, Jeanny JC, Chacun H, Couvreur P (2002) Intravitreal delivery of oligonucleotides by sterically stabilized liposomes. Invest Ophthalmol Vis Sci 43(1):253–259
103. Hironaka K, Inokuchi Y, Tozuka Y, Shimazawa M, Hara H, Takeuchi H (2009) Design and evaluation of a liposomal delivery system targeting the posterior segment of the eye. J Control Release 136(3):247–253
104. Gross N, Ranjbar M, Evers C, Hua J, Martin G, Schulze B, Michaelis U, Hansen LL, Agostini HT (2013) Choroidal neovascularization reduced by targeted drug delivery with cationic liposome-encapsulated paclitaxel or targeted photodynamic therapy with verteporfin encapsulated in cationic liposomes. Mol Vis 19:54–61
105. Abdelbary G, El-gendy N (2008) Niosome-encapsulated gentamicin for ophthalmic controlled delivery. AAPS PharmSciTech 9(3):740–747
106. Abdelkader H, Alani AWG, Alany RG (2014) Recent advances in non-ionic surfactant vesicles (niosomes): self-assembly, fabrication, characterization, drug delivery applications and limitations. Drug Deliv 21(2):87–100
107. Mahale NB, Thakkar PD, Mali RG, Walunj DR, Chaudhari SR (2012) Niosomes: novel sustained release nonionic stable vesicular systems – an overview. Adv Colloid Interface Sci 183:46–54
108. Vyas SP, Mysore N, Jaitely V, Venkatesan N (1998) Discoidal niosome based controlled ocular delivery of timolol maleate. Pharmazie 53(7):466–469
109. Prabu P, Chaudhari AA, Aryal S, Dharmaraj N, Park SY, Kim WD, Kim HY (2008) In vitro evaluation of poly(caprolactone) grafted dextran (PGD) nanoparticles with cancer cell. J Mater Sci Mater Med 19(5):2157–2163
110. Kaur IP, Aggarwal D, Singh H, Kakkar S (2010) Improved ocular absorption kinetics of timolol maleate loaded into a bioadhesive niosomal delivery system. Graefes Arch Clin Exp Ophthalmol 248(10):1467–1472
111. Li Q, Li Z, Zeng W, Ge S, Lu H, Wu C, Ge L, Liang D, Xu Y (2014) Proniosome-derived niosomes for tacrolimus topical ocular delivery: In vitro cornea permeation, ocular irritation, and in vivo anti-allograft rejection. Eur J Pharm Sci 62:115–123

112. Cheng Y, Xu Z, Ma M, Xu T (2008) Dendrimers as drug carriers: applications in different routes of drug administration. J Pharm Sci 97(1):123–143
113. Mignani S, El Kazzouli S, Bousmina M, Majoral J (2013) Expand classical drug administration ways by emerging routes using dendrimer drug delivery systems: a concise overview. Adv Drug Deliv Rev 64(10):1316–1330
114. Yao W, Sun K, Mu H, Liang N, Liu Y, Yao C, Liang R, Wang A (2010) Preparation and characterization of puerarin-dendrimer complexes as an ocular drug delivery system. Drug Dev Ind Pharm 36(9):1027–1035
115. Durairaj C, Kadam RS, Chandler JW, Hutcherson SL, Kompella UB (2010) Nanosized dendritic polyguanidilyated translocators for enhanced solubility, permeability, and delivery of gatifloxacin. Invest Ophthalmol Vis Sci 51(11):5804–5816
116. Yandrapu SK, Kanujia P, Chalasani KB, Mangamoori L, Kolapalli RV, Chauhan A (2013) Development and optimization of thiolated dendrimer as a viable mucoadhesive excipient for the controlled drug delivery: an acyclovir model formulation. Nanomed Nanotechnol Biol Med 9(4):514–522
117. Holden CA, Tyagi P, Thakur A, Kadam R, Jadhav G, Kompella UB, Yang H (2012) Polyamidoamine dendrimer hydrogel for enhanced delivery of antiglaucoma drugs. Nanomed Nanotechnol Biol Med 8(5):776–783
118. Yang H, Leffler CT (2013) Hybrid dendrimer hydrogel/poly(lactic-co-glycolic acid) nanoparticle platform: an advanced vehicle for topical delivery of antiglaucoma drugs and a likely solution to improving compliance and adherence in glaucoma management. J Ocul Pharmacol Ther 29(2):166–172
119. Loftssona T, Jarvinen T (1999) Cyclodextrins in ophthalmic drug delivery. Adv Drug Deliv Rev 36(1):59–79
120. Conway BR (2008) Recent patents on ocular drug delivery systems. Recent Pat Drug Deliv Formul 2(1):1–8
121. Kaur IP, Chhabra S, Aggarwal D (2004) Role of cyclodextrins in ophthalmics. Curr Drug Deliv 1(4):351–360
122. Malaekeh-Nikouei B, Bazzaz BSF, Soheili V, Mohammadian K (2013) Problems in ophthalmic drug delivery: evaluation of the interaction between preservatives and cyclodextrins. Jundishapur J Microbiol 6(5), UNSP e6333
123. Totterman AM, Schipper NGM, Thompson DO, Mannermaa JP (1997) Intestinal safety of water-soluble beta-cyclodextrins in paediatric oral solutions of spironolactone: effects on human intestinal epithelial Caco-2 cells. J Pharm Pharmacol 49(1):43–48
124. Loftsson T, Stefansson E (1997) Effect of cyclodextrins on topical drug delivery to the eye. Drug Dev Ind Pharm 23(5):473–481
125. Wang SL, Li DX, Ito Y, Nabekura T, Wang SJ, Zhang JH, Wu CF (2004) Bioavailability and anticataract effects of a topical ocular drug delivery system containing disulfiram and hydroxypropyl-betacyclodextrin on selenite-treated rats. Curr Eye Res 29(1):51–58
126. Zhang J, Wang L, Gao C, Zhang L, Xia H (2008) Ocular pharmacokinetics of topically-applied ketoconazole solution containing hydroxypropyl beta-cyclodextrin to rabbits. J Ocul Pharmacol Ther 24(5):501–506
127. Halim Mohamed MA, Mahmoud AA (2011) Formulation of indomethacin eye drops via complexation with cyclodextrins. Curr Eye Res 36(3):208–216
128. Mahmoud AA, El-Feky GS, Kamel R, Awad GEA (2011) Chitosan/sulfobutylether-beta-cyclodextrin nanoparticles as a potential approach for ocular drug delivery. Int J Pharm 413(1–2):229–236
129. Chen J, Zhou R, Li L, Li B, Zhang X, Su J (2013) Mechanical, rheological and release behaviors of a poloxamer 407/poloxamer 188/carbopol 940 thermosensitive composite hydrogel. Molecules 18(10):12415–12425

130. Glisoni RJ, Garcia-Fernandez MJ, Pino M, Gutkind G, Moglioni AG, Alvarez-Lorenzo C, Concheiro A, Sosnik A (2013) beta-Cyclodextrin hydrogels for the ocular release of antibacterial thiosemicarbazones. Carbohydr Polym 93(2):449–457
131. Hu X, Qiu J, Tan H, Li D, Ma X (2013) Synthesis and characterization of cyclodextrin-containing hydrogel for ophthalmic drugs delivery. J Macromol Sci Part A Pure Appl Chem 50(9):983–990
132. Moya-Ortega MD, Alves TFG, Alvarez-Lorenzo C, Concheiro A, Stefansson E, Thorsteinsdottir M, Loftsson T (2013) Dexamethasone eye drops containing gamma-cyclodextrin-based nanogels. Int J Pharm 441(1–2):507–515
133. Pawar P, Kashyap H, Malhotra S, Sindhu R (2013) Hp-beta-CD-voriconazole in situ gelling system for ocular drug delivery: in vitro, stability, and antifungal activities assessment. BioMed Res Int 2013:341218

Chapter 6
Pharmacokinetics and Pharmacodynamics of Ocular Drugs

Vivek S. Dave and Suraj G. Bhansali

Abstract This chapter aims to provide the readers a systematic overview of the pharmacokinetics and pharmacodynamics of the drugs intended for ophthalmic use. The concepts of ocular pharmacokinetics and pharmacodynamics are briefly discussed in the introduction. The chapter begins with a discussion on the common anatomical and physiological factors such as blood–ocular and tear fluid–corneal barriers, as well as anterior segment drug loss; and the challenges these factor pose in describing ocular pharmacokinetics and pharmacodynamics. The biopharmaceutics of the ocular drugs describes common pathways of ocular drug absorption. Further, commonly employed routes of administration for ocular drugs are discussed with respect to the choice of the route, properties of the drug, the nature of the ocular disease, the targeted ocular tissue, and the pharmacokinetic behavior of the drugs administered through the route. The pharmacokinetic–pharmacodynamic models that describe the fate of ocular drugs are further reviewed. Finally, recent advances and current trends in understanding of the pharmacokinetics/pharmacodynamics of ocular drugs are discussed based on the reported findings of the scientific and medical community.

Keywords Ocular pharmacokinetics • Ocular pharmacodynamics • Biopharmaceutics • Ocular routes • Compartment models

6.1 Introduction

The treatment of ocular diseases by the means of drugs mainly involves optimizing the bioavailability of drugs in the ocular tissues. Thus, achieving and maintaining an optimum drug concentration at a specific ocular site requires a clear understanding of

V.S. Dave (✉)
Pharmaceutical Sciences, St. John Fisher College, Wegmans School of Pharmacy, Rochester, NY, USA
e-mail: viveksdave@gmail.com

S.G. Bhansali (✉)
Novartis Pharmaceuticals, East Hanover, NJ, USA
e-mail: suraj.bhansali@novartis.com

© Springer International Publishing Switzerland 2016
Y. Pathak et al. (eds.), *Nano-Biomaterials For Ophthalmic Drug Delivery*,
DOI 10.1007/978-3-319-29346-2_6

the anatomy/physiology of the eye, pathophysiology of the ocular disease, as well as the pharmacokinetics and pharmacodynamics of the drug. *Ocular pharmacokinetics* has been defined earlier as the study of mechanisms and extent of drug absorption, distribution, metabolism, and excretion in the eye [1]. The pharmacokinetics of a drug is influenced by several factors including patient-specific factors (age, gender, race, genetic makeup, disease condition, etc.) and drug-specific factors [2]. Ocular pharmacodynamics has been defined as the study of the biochemical and physiological effects of a drug in the eye, including mechanisms of action; and ocular toxicology has been defined as the study of unwanted, mild, or severe adverse effects of drugs on one or more ocular tissues [1]. Generally, there is a correlation between concentrations of a drug at the site of action and its effects, which can be driven by binding with receptors. Hence the pharmacodynamics of a drug is influenced by receptor density in addition to its pharmacokinetics. This chapter will provide a brief overview of the common challenges in ocular pharmacokinetics/pharmacodynamics, typical biopharmaceutical fate of ocular drugs, pharmacokinetic models used in ocular systems, and modern approaches utilized in studying ocular pharmacokinetics and pharmacodynamics.

6.2 Challenges in Ocular Pharmacokinetics and Pharmacodynamics

Characterization of pharmacokinetics and pharmacodynamics of ocular drugs is challenging due to the existence of anatomical and physiological barriers that are unique to ocular system. The main factors that influence the ocular bioavailability of drugs (e.g., instilled dose, tear turnover, drug absorption, metabolism, elimination, etc.) are shown in Fig. 6.1. Similarly, the typical fate of drugs delivered via ophthalmic routes is briefly represented in Fig. 6.2. These factors influencing the availability of the drugs to the ocular tissues can be broadly classified into three categories.

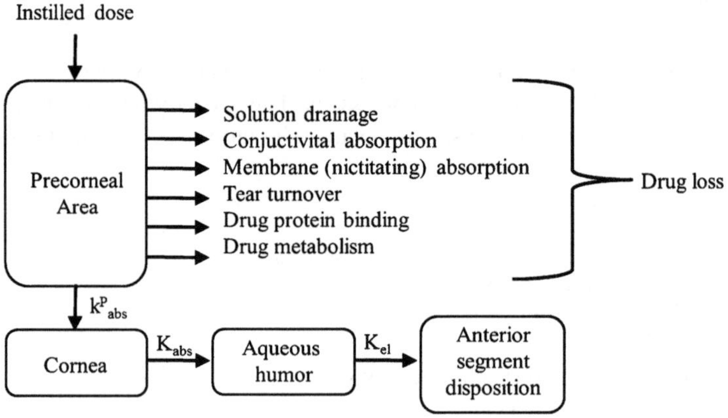

Fig. 6.1 Precorneal and intraocular drug movement following topical administration (Adapted from [3])

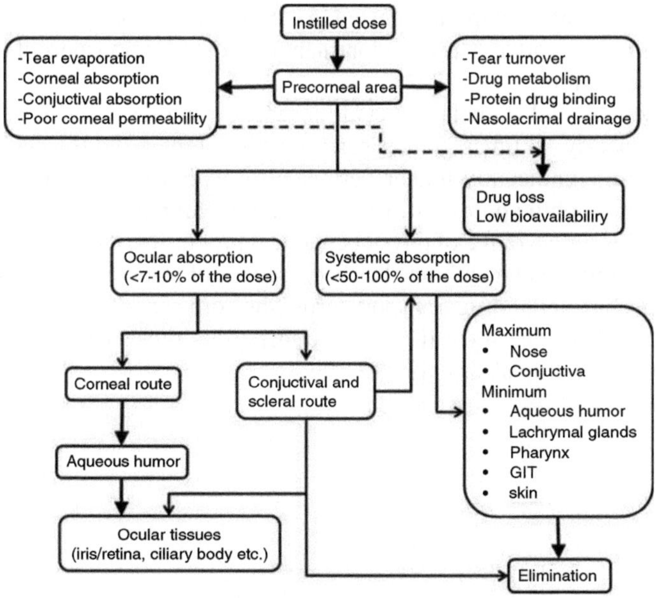

Fig. 6.2 Fate of a drug delivered via ophthalmic route (Adapted from Patel et al. [4])

6.2.1 *Anterior Segment Drug Loss*

After topical administration, ocular drugs are rapidly removed from the surface of the eye by the lacrimal fluid. Within minutes a large fraction of the instilled formulation is transported to the nasolacrimal duct [5]. Such precorneal removal of the drugs is often referred to as *nonproductive* drug loss [4, 6]. In addition to the drug loss due to lacrimal fluid, instilled doses of ocular drugs may also be removed from the anterior segment due to systemic absorption. Systemic absorption of topically administered ocular drugs may occur either directly in the conjunctival sac from the local blood capillaries or from the nasolacrimal cavity [7, 8]. Thus, nonproductive drug loss as well as loss due to systemic absorption can result in a significant lowering, i.e., less than 10 % of ocular bioavailability of drugs [9].

6.2.2 *Tear Fluid–Corneal Barriers*

Corneal epithelium is a primary barrier to the transport of drugs from the lacrimal fluid into the eye [10]. The apical cells of the corneal epithelium form tight junctions and restrict the permeation of ocular drugs across the epithelial membrane [11]. Huang et al. studied the permeability characteristics of a group of beta-blocking agents on rabbit corneas [12]. The studies found that the lipophilic compounds penetrated the cornea more rapidly, whereas the corneal epithelium was rate limiting for the most lipophilic compounds tested. As shown in Fig. 6.3, transcorneal permeation can still be considered as a predominant route for the transport of ocular drugs to the aqueous humor.

Prausnitz et al. and Hamalainen et al. in separate studies described that the conjunctiva has nearly 20 times greater surface area compared to the cornea and also demonstrates a significantly higher permeability to most ocular drugs due to the presence of hydrophilic pores [14, 15]. Owing to its high permeability to small and large molecules with a range of physicochemical properties, the study of drug transport across conjunctival membrane has been a focus of research in recent years [16]. The main mechanism of drug transport across the cornea and conjunctiva is thought to be passive diffusion, albeit the active transporters present in these membranes may play a role in drug absorption [13].

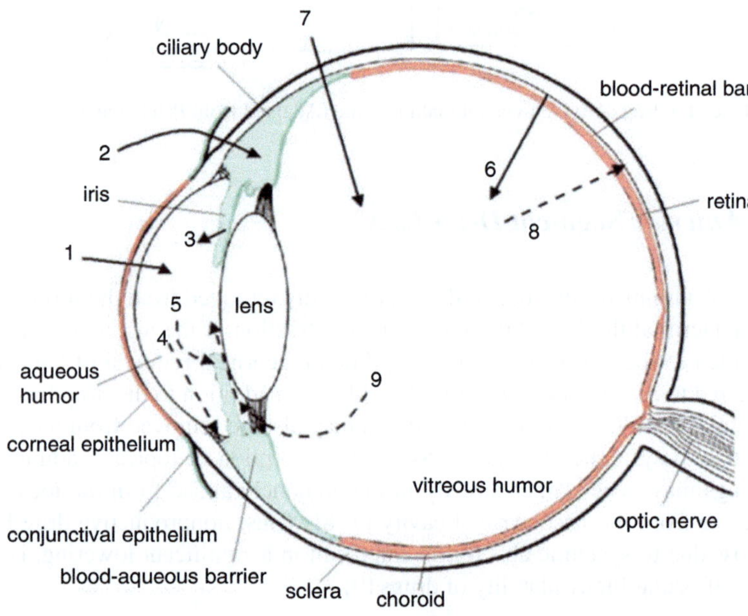

Fig. 6.3 Schematic of the ocular routes of drug transport: (*1*) transcorneal permeation from the lacrimal fluid into the anterior chamber, (*2*) non-corneal permeation across the conjunctiva and sclera into the anterior uvea, (*3*) distribution from the blood stream via blood–aqueous barrier into the anterior chamber, (*4*) elimination from the anterior chamber by the aqueous humor turnover to the trabecular meshwork and Schlemm's canal, (*5*) elimination from the aqueous humor into the systemic circulation across the blood–aqueous barrier, (*6*) distribution from the blood into the posterior eye across the blood–retina barrier, (*7*) intravitreal administration, (*8*) elimination from the vitreous via posterior route across the blood–retina barrier, and (*9*) elimination from the vitreous via anterior route to the posterior chamber (Adapted from Urtti et al. [13])

6.2.3 Blood–Ocular Barriers

Blood–ocular barriers exist to protect the ocular tissues from foreign, harmful molecules that may be present in the systemic circulation. As shown in Fig. 6.3, two types of blood–ocular barriers are present in the eye: (1) blood–aqueous barrier and (2) blood–retinal barrier. The blood–aqueous barrier is present in the anterior segment of the eye and consists of endothelial cells in the uvea. This barrier restricts the entry of albumin and other similar large biomolecules from the plasma into the aqueous humor. The blood–retinal barrier is present in the posterior segment of the eye and consists of retinal pigment epithelium (RPE). This barrier restricts the transport of drugs, ions, proteins, and water flux into and out of the retina and the choroid. Alterations of the blood–retinal barrier are known to play a crucial role in the development of retinal diseases [17]. For example, diabetic retinopathy and age-related macular degeneration (AMD) are known to be directly associated with alterations of the blood–retinal barrier [18–20].

6.3 Biopharmaceutics of Ocular Drug Delivery

Due to the abovementioned barriers, ophthalmic drug delivery is challenging. Since the administration of drugs via systemic routes does not achieve an acceptable ocular bioavailability, the topical route remains the most commonly used approach for ocular drug delivery. Ocular drug absorption occurs mainly via two pathways: corneal and non-corneal. Corneal absorption is a major route of drug absorption. The transport of drugs across corneal epithelium is known to be the rate-limiting step for ocular bioavailability and is highly dependent on the physicochemical properties of the drugs [12]. The non-corneal route of absorption includes transport of drugs across the sclera and conjunctiva into the intraocular tissues. The drugs absorbed by this pathway reach the aqueous humor and may enter the systemic circulation via local capillaries and show higher clearance.

After transcorneal absorption, the drug is typically accumulated in the aqueous humor and then distributed to the other intraocular tissues via passive diffusion. The distribution of drugs to these tissues depends on the extent of drug–protein binding in the iris, ciliary body, and the aqueous humor and the elimination of the drugs due to the aqueous humor turnover. The ocular tissues express a wide range of enzymes; and these enzymes play a major role in the metabolism of ocular drugs. Among the commonly expressed ocular metabolic enzymes include esterases, oxidoreductases, lysosomal enzymes, peptidases, glucuronide-o-methyl-transferase, monoamine oxidase, and corticosteroid β-hydroxylase [21]. These enzymes are mainly found in the cornea, the iris–ciliary body, and the retina. Topically administered drugs are metabolized by the enzymes in the cornea and the ciliary body, whereas drugs administered by systemic or periocular routes are metabolized by the retinal enzymes. After absorption, most drugs are eliminated by the aqueous humor turnover. Other drugs are eliminated by metabolism and uptake by the blood vessels present in the anterior uvea or iris [21].

6.4 Ocular Routes of Drug Delivery

Drugs used for the treatment of ophthalmic diseases can be administered via different routes. The choice of a route for ocular drug delivery depends on a number of factors. These include physicochemical properties of the drug, the nature of the disease, and the specific ocular tissue to be targeted. Among the most commonly utilized routes for ocular drug delivery are topical, subconjunctival, and intravitreal. Conventionally topical and subconjunctival routes are used to treat the diseases of anterior segment of the eye. The intravitreal route is used mainly for the diseases of the posterior segment.

6.4.1 Topical Administration

Most topically administered ocular drugs are available in the form of ophthalmic solutions (e.g., eye drops). Due to several anatomical/physiological factors mentioned above and elsewhere in this book, the residence time of these dosage forms at the site of action is short. Several formulation approaches (e.g., gels, ointments, insert, etc.) have been explored to increase the ocular residence time of topically administered drugs [6, 22–26]. These approaches are described in detail in Chap. 7 of this book. As illustrated in Fig. 6.3, after topical administration to the corneal surface, the drug begins partitioning in the corneal epithelium. Depending on its lipophilicity, it either remains in the epithelium for an extended period or is slowly absorbed into the corneal stroma followed by its release in the anterior chamber [27]. The peak concentration in the anterior chamber after topical administration is typically reached in less than 30 min; however, only a small fraction of the administered dose is absorbed [9]. A fraction of topically administered ocular drugs may be absorbed from the conjunctival surface into the sclera, followed by the uvea and the posterior segment of the eye (Fig. 6.3). Drug absorption from the conjunctival surface is a slow process and dependent upon the physicochemical properties of the drugs. In the anterior chamber, the drug binds to the ocular protein *melanin*, thus forming a complex which may serve as a reservoir and provide a sustained drug release. The drug is then distributed to the uvea and the lens; the distribution to the lens is slower due to its tightly packed protein-rich composition [10].

The drug absorbed in the aqueous humor is mainly eliminated via two pathways. The first pathway is by the aqueous turnover through the chamber angle and Schlemm's canal (Fig. 6.3). The approximate elimination rate via this pathway is 3 μL/min and is independent of the nature of the drug [10]. The second pathway is via systemic circulation in the anterior uvea (Fig. 6.3). Elimination via this pathway is slower and depends on the ability of the drug to cross the endothelial walls of the blood vessels [10]. Thus, typically lipophilic drugs tend to clear faster from the anterior segment compared to hydrophilic drugs. In addition, due to the slow equilibration of the drugs in ocular tissues, determination of the volume of distribution is challenging. Animal studies have demonstrated a wide range (250 μL to 2 mL) of volume of distribution in aqueous humor [11].

6.4.2 Subconjunctival Administration

Ocular drugs are administered via subconjunctival route to improve their bioavailability in the uvea. The main drivers for the use of this route in recent years are (1) the need to deliver newly developed drugs or dosage forms to the retina and choroid in the treatment of macular degenerative diseases and (2) the exponential progress made in fields of ophthalmic formulations and drug delivery systems to deliver the drugs to the posterior segment of the eye [28–30].

The sclera is known to have a higher permeability compared to the cornea, and its permeability to the drugs is independent of the drug lipophilic character [15, 31]. Thus, compared to the cornea and conjunctiva, sclera demonstrates a higher permeability to a wide range of molecules. This makes subconjunctival administration of drugs a more feasible approach to deliver the drugs to the choroid [32]. However, due to the presence of blood vessels, there is a significantly high systemic clearance of the drugs from the choroid. Pitkanen et al. also demonstrated that retinal pigment epithelium may be the rate-limiting factor in the retinal delivery of hydrophilic drugs and macromolecules through the transscleral route [31]. A comprehensive understanding of the scleral and choroidal pharmacokinetics is essential to optimize the drug bioavailability in the posterior segment of the eye.

Several researchers have reported the advantages of subconjunctival administration of ocular drugs over other routes. In a clinical study, Behren-Baumann et al. carried out a comparative investigation of the ocular pharmacokinetics of azlocillin between intravenous and subconjunctival routes of administration [33]. The study results revealed a superior bioavailability of azlocillin in aqueous humor with subconjunctival route compared to intravenous application. Moreover, the half-life of azlocillin in aqueous humor was also found to be longer with subconjunctival administration compared to intravenous route. In a series of studies following this study, Behren-Baumann et al. evaluated the ocular pharmacokinetics of a variety of drugs in animals and humans and demonstrated the superiority of subconjunctival route compared to other modes of drug administration in improving ocular drug bioavailability [34–37].

6.4.3 Intravitreal Administration

Intravitreal drug administration involves injecting a drug solution directly into the vitreous via pars plana to increase the vitreal bioavailability of drugs compared to the drugs delivered via topical or systemic routes [4]. This approach of drug delivery has an inherent advantage of ensuring highest vitreal and retinal bioavailability while avoiding systemic toxicity, since the drug is delivered directly into the posterior segment of the eye. The reduced systemic toxicity is attributed to the presence of retinal pigment epithelium (RPE), which serves as a barrier to the transport of drugs, particularly large, positively charged molecules to the choroid [38, 39]. In addition to passive diffusion, convection resulting from eye movements is known to play a role in the drug transport process across the vitreous [40]. As shown in Fig. 6.3, drugs administered via intravitreal route may be eliminated in the anterior segment and/or the posterior segment [10]. Elimination in the anterior segment occurs via aqueous turnover and uveal blood flow for most drugs. Elimination in the posterior segment is restricted to small, lipophilic molecules and occurs via passive diffusion across the blood–retinal barrier (BRB). Thus, an extended half-life is observed for large, hydrophilic drugs in the vitreous.

The pharmacokinetics and the pharmacodynamics of the drugs administered via intravitreal routes have been well researched. Shen et al. investigated the pharmacokinetics and safety of caspofungin, a potent antifungal agent, after intravitreal injection (50 μg/0.1 mL) in rabbits [67]. The results showed that caspofungin had a half-life of 6.28 h with the mean vitreous concentration of 6.06 ± 1.76 μg/mL. Detectable concentration of caspofungin was found up to 24 h. In addition, intravitreal injection of caspofungin was found to be safe and devoid of any focal necrosis or other abnormalities in retinal histology. Haller et al. studied the pharmacodynamics of intravitreal ocriplasmin in the treatment of vitreomacular adhesion and vitreomacular traction in human patients [41]. This multicenter, randomized, placebo-controlled, double-masked, 6-month clinical trial consisting of 652 randomized patients confirmed the positive effects of ocriplasmin across relevant subpopulations. Inoue et al. evaluated the pharmacodynamics of intravitreal injection of ranibizumab for the treatment of age-related macular degeneration in humans, as measured by the visual function and vision-related quality of life [42]. The study results demonstrated the tolerability, efficacy, and compliance with intravitreal ranibizumab in the test patients. Intravitreal injections are known to be associated with some specific drawbacks. Ausayakhun reported several short-term complications including retinal detachment, endophthalmitis, and intravitreal hemorrhages after intravitreal injection of ganciclovir in AIDS patients when used to treat cytomegalovirus retinitis [43]. Ornek et al. evaluated the corneal and the conjunctival sensitivity changes measured using the Cochet-Bonnet esthesiometer, following intravitreal ranibizumab (IVR) injection in patients with diabetic retinopathy [44]. The study found significantly increased sensitivities in the central, temporal, and nasal corneas after a single intravitreal ranibizumab injection; and the corneas remained sensitive up to a week. Aslan et al. reported hypersensitivity reaction in the form of marginal keratitis in a single patient after receiving an intravitreal injection of ranibizumab for the treatment of diffuse diabetic macular edema [45].

6.5 Ocular Pharmacokinetic–Pharmacodynamic Models

Pharmacokinetic–pharmacodynamic models consist of compartments that represent physiology of various tissues/organs to characterize and predict drug disposition and pharmacological response using mathematical expressions. Pharmacokinetic models describe relationship between administered dose and drug concentrations in blood/plasma and or tissues/organs. Pharmacodynamic models describe dose–/concentration–response relationships. Combined, pharmacokinetic–pharmacodynamic models can establish and predict effect-time course resulting from a dose of drug. The application of pharmacokinetic–pharmacodynamic models in drug development has become very critical as they provide valuable information during each phase of development. Worakul et al. have extensively reviewed several pharmacokinetic–pharmacodynamic models for drugs administered via ocular route [46]. One of the basic models that describe precorneal and intracorneal drug movement following topical administration is shown in Fig. 6.1. In this model, as described by Lee et al., the dose is instilled in the precorneal area from where fraction of drug can be absorbed to the cornea then to aqueous humor from where it can be eliminated. Also, the drug may be directly eliminated or lost from the precorneal area as shown in Fig. 6.1 [3].

Schematic of another model describing the ocular absorption is shown in Fig. 6.2. For topically administered drugs or dosage form (e.g., eye drops), there is very limited ocular absorption (only 5–7 %), which is absorbed via the cornea, sclera, or aqueous humor to ocular tissues. Majority of drug is lost through conjunctiva of the eye and nose which results in systemic absorption [6]. Based on the mechanisms of ocular absorption described above, several ocular pharmacokinetic models have been reported in the literature [3, 27, 47–56].

One compartment model is the simplest model where the eye is represented as a single compartment (Fig. 6.4). The drug concentration (C) using this model can be estimated by the following equation, where F is fraction of absorbed dose (D), K_a and K_{el} are rate constants for absorption and elimination, and apparent distribution volume is described by V_d [3, 52]:

$$C=\left(\frac{FD}{V_d}\right)\left(\frac{K_{el}}{K_a-K_{el}}\right)\left(e^{-K_{el}t}-e^{-K_a t}\right)$$

Makoid et al. described a two-compartment model representing the precornea (compartment 1) and cornea (compartment 2) as shown in Fig. 6.5a [52]. The model depicted elimination from both the compartments. The concentration in the cornea was estimated using the following equation:

$$C=\left(\frac{DK_{12}}{V_d\left(K_{12}+K_{10}-K_{23}\right)}\right)\left(e^{-K_{23}t}-e^{-(K_{12}+K_{10})t}\right)$$

A model developed by Himelstein et al. for pilocarpine consisted tears and aqueous humor as two compartments (Fig. 6.5b), which incorporated input to and elimination

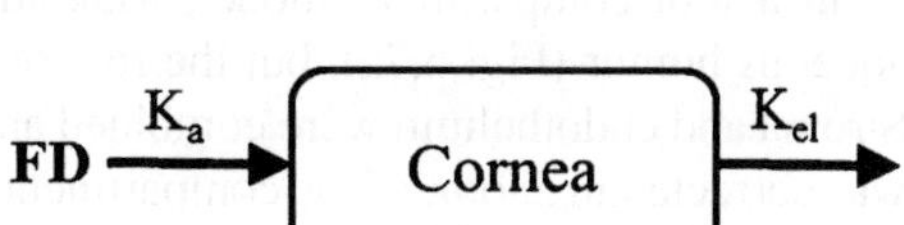

Fig. 6.4 Schematic of a one-compartment model to describe ocular pharmacokinetics

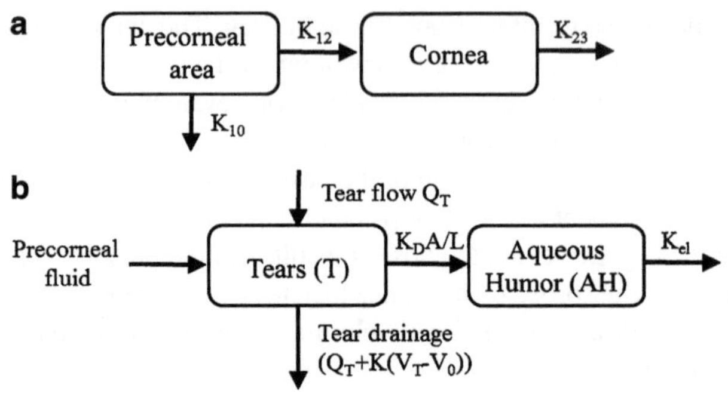

Fig. 6.5 Schematics of different types of two-compartment models ((**a**) Adapted from [52] (**b**) Adapted from [50])

from both the compartments [50]. The concentration in the tear fluid (T) was characterized by the following equation:

$$\frac{dC_T}{dt} = \frac{\left[-Q_T C_t - \left(\frac{K_D A}{L} \right) (C_T - C_{AH}) \right]}{\left(V_D e^{-Kt} + V_0 \right)}$$

The concentration in the aqueous humor (AH) was characterized by the following equation:

$$V_{AH} \left(\frac{dC_{AH}}{dt} \right) = \left(\frac{K_D A}{L} \right) (C_T - C_{AH}) - (K_{el} C_{AH})$$

where C_T represents the concentration in tear fluid; C_{AH}, the concentration in aqueous humor; Q_T, the normal rate of tear production; V_D, the volume of drop; V_0, the normal tear volume; V_{AH}, the aqueous humor volume; K_D, the coefficient of permeability; K_{el}, the first-order elimination constant from the aqueous humor compartment; and A and L, the area and thickness of the cornea, respectively. In this model, the cornea was assumed to be homogeneous membrane with no contribution to drug disposition. Also, the entire disposition from the precornea was explained by one rate constant.

A three-compartment model is described in Fig. 6.6a where anterior of the eye was divided into two compartments, the cornea and aqueous humor. Another three-compartment model with precorneal area, cornea, and aqueous humor is shown in Fig. 6.6b.

In a four-compartment model, Makoid et al. incorporated both the cornea and aqueous humor (Fig. 6.7a), but the role of stroma was not captured separately [52]. Stroma and endothelium were combined and described as a single compartment. This was corrected in another four-compartment model by Lee et al. as shown in Fig. 6.7b, with an assumption that instilled drug mixes completely and instantaneously and the

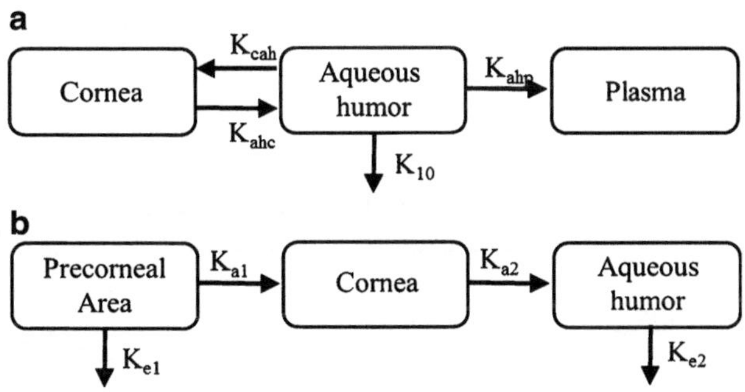

Fig. 6.6 Schematics of different types of three-compartment models ((**a**) Adapted from [51] (**b**) Adapted from [53])

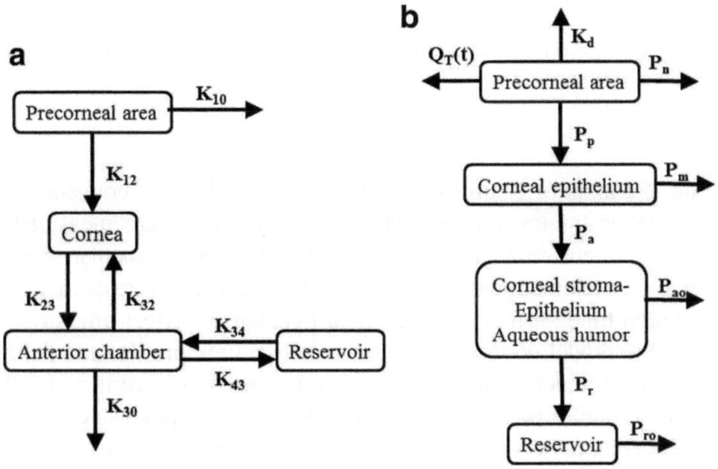

Fig. 6.7 Schematics of different types of four-compartment models ((**a**) Adapted from [52] (**b**) Adapted from [3])

reservoir compartment contains the iris, lens, vitreous humor, and ciliary body [3]. The group extended the model to separate stroma and aqueous humor to make it a five-compartment model (Fig. 6.8a), where drug loss or transfer occurred from each of the compartments. This model was used for a lipophilic drug, fluorometholone [56]. A five-compartment model was also used to describe the pharmacokinetics of pilocarpine as shown in Fig. 6.8b, where the five compartments were precorneal area, cornea, aqueous humor, lens, and iris [54]. The movement of drug to/from each compartment was considered reversible. In this model, elimination of drug was considered to be only from the aqueous humor and iris.

Sakanaka et al. developed one of the first ocular pharmacokinetic–pharmacodynamic (PK/PD) models for timolol (a beta-blocker), after topical administration or

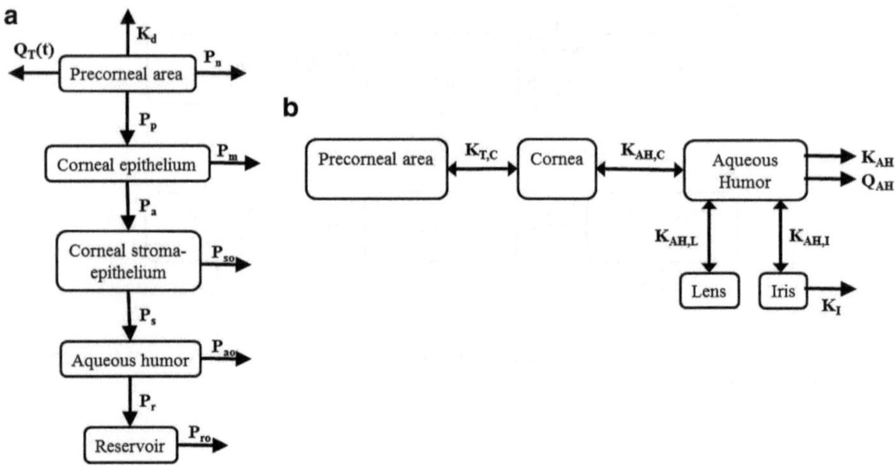

Fig. 6.8 Schematics of different types of five-compartment models ((**a**) Adapted from [56] (**b**) Adapted from [54])

injection into the anterior chamber in rabbit eyes [57]. Timolol concentrations in the tear fluid, aqueous humor, cornea, and iris–ciliary body were measured by high-performance liquid chromatography (HPLC). In addition, intraocular pressure (IOP) was measured by a telemetry system. The PK/PD parameters were estimated by fitting the concentration-time profiles and the ocular hypotensive effect-time profiles. A six-compartment PK model described the concentration-time profile; and the PD model which consisted of aqueous humor dynamics, i.e., timolol lowering of IOP by suppressing aqueous humor production, described the ocular hypotensive effect-time profiles of timolol (Fig. 6.9). Similarly, a model for bunazosin with seven compartments where cornea compartment was divided into two compartments, corneal epithelium and stroma, was developed [58]. All the other processes and parameters were similar to timolol model. Sakanaka et al. were also the first to develop a combined ocular PK/PD model for multidrug therapy (Fig. 6.10) [59]. This model considered mechanism of action of two drugs, timolol and bunazosin. The model was used to simulate the concentrations in aqueous humor and intraocular pressure-lowering effects following instillation of combination of timolol and bunazosin. The reliability of this model was verified with observed drug concentrations in aqueous humor and measurement of ocular hypotensive effects using telemetry. For further details including derivations of equations, readers are encouraged to refer to the original publication [59].

In pharmacodynamic models, measurements of pharmacological responses like miosis and mydriasis and intraocular pressures have been considered [46]. For cholinergic drugs, miotic response m_r was used to measure the effect at time t and was expressed as shown below [60]:

$$m_r = -k_{dm}t + m_0$$

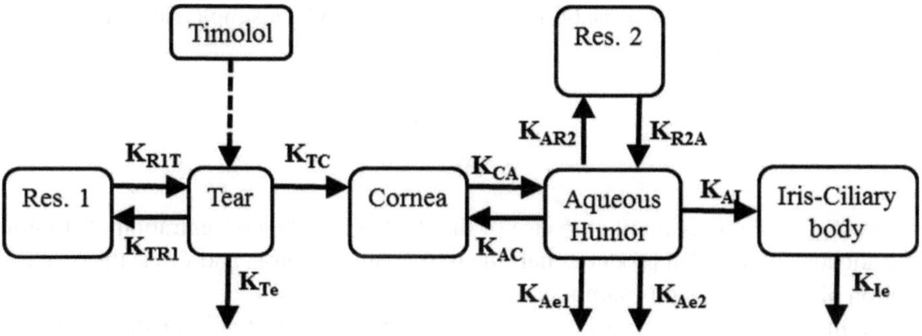

Fig. 6.9 Schematic of a six-compartment model for the ocular pharmacokinetics of timolol [57]

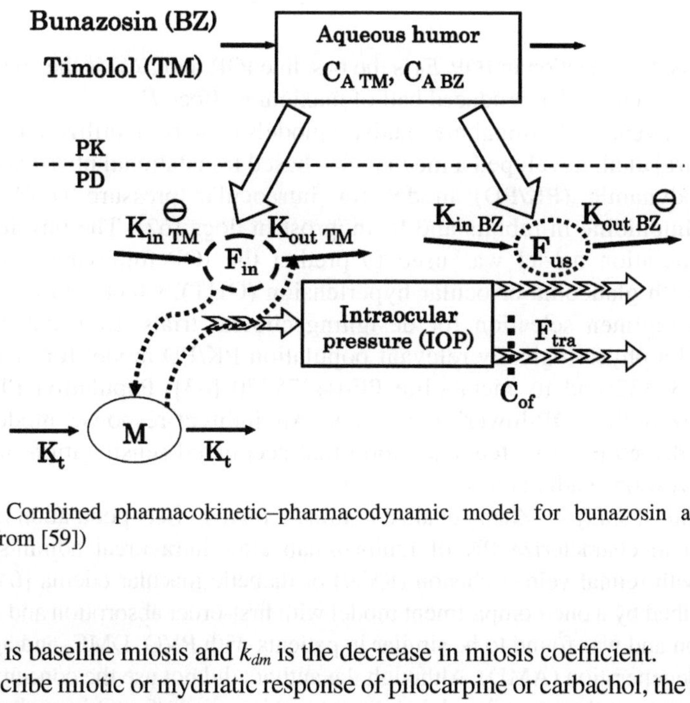

Fig. 6.10 Combined pharmacokinetic–pharmacodynamic model for bunazosin and timolol (Adapted from [59])

where m_0 is baseline miosis and k_{dm} is the decrease in miosis coefficient.

To describe miotic or mydriatic response of pilocarpine or carbachol, the following general model was used [46]:

$$R_1 = \frac{R}{(R_{max} - R)}$$

For the miotic response, $R = D_0 - D$ and $R_{max} = D_0 - D_{min}$, and for the mydriatic response, $R = D - D_0$ and $R_{max} = D_{max} - D_0$, where D_0 is baseline pupil diameter prior to treatment and D is pupil diameter after drug administration.

For measurement of mydriatic response at time t, ΔE_t, a simple Emax model has been used for phenylephrine:

$$\Delta E_t = \left(\frac{E_{\max} Ca_t}{K_m + Ca_t} \right)$$

where $E_{\max}$ is the maximum response of the drug, K_m is the concentration of drug in the aqueous humor that produces half the maximum response, and Ca_t is the concentration of drug in aqueous humor at time t.

Further, modified $E_{\max}$ model was used for the measurement of intraocular pressure (IOP) reduction [61]:

$$\Delta E = E - E_0 \left(\frac{E_{\max} C}{EC_{50} + C} \right)$$

Here, ΔE is the reduction in IOP, E_0 is the baseline IOP, and EC_{50} is the concentration C in aqueous humor that produced half of maximum effect, $E_{\max}$.

In recent years, additional mechanistic models have been utilized by researchers. Durairaj et al. developed a mechanism-based translational pharmacokinetic–pharmacodynamic (PK/PD) model for intraocular pressure (IOP)-lowering drugs, brimonidine in rabbits and latanoprost in dogs [62]. The physiologically based population model was used to predict the IOP following treatment in patients with glaucoma or ocular hypertension (OHT), which can be useful for dose and regimen selection for designing clinical trials. Luu and coworkers developed a physiologically relevant population PK/PD model for an EP4 agonist CP-734432 and its metabolite PF-04475270 [63]. Population PD models characterized the IOP-lowering profiles. An indirect-response model with a response-driven positive feedback loop that accounted sensitization of PD captured the response adequately.

In a clinical study by Zhang et al., a nonlinear mixed-effect population model was developed to characterize PK of ranibizumab after intravitreal administration in patients with retinal vein occlusion (RVO) or diabetic macular edema [64, 65]. PK was described by a one-compartment model with first-order absorption and first-order elimination and was found to be similar in patients with RVO, DME, and age-related macular degeneration (AMD). Although the authors did not use the concentrations in the ocular tissues, they postulated that intravitreal injection of ranibizumab may result in similar intraocular concentrations and eventually similar concentrations in the serum irrespective of the disease.

Most of the comprehensive modeling work is done in preclinical species based on animal data as it is challenging to obtain drug concentrations in the human eye or ocular tissues. Due to simplification of the processes, the models have certain limitations but are adequate to characterize and describe the data. The application of appropriate model depends on drug properties, mode of application, and available information (observed data).

6.6 Current Trends in the Pharmacokinetics and Pharmacodynamics (PD) of Ocular Drugs

The studies of pharmacokinetics and pharmacodynamics of drugs used to treat ophthalmic diseases have gained a lot of interest among researchers in recent years [9]. Due to the uniqueness of the drug delivery systems as well as the therapeutic targets, these studies bring significant and synergistic contributions to the knowledge base in ocular therapeutics. Several research studies focusing on PK/PD of ocular drugs have been carried out using a variety of biological models including cells, tissues, animals, and humans.

Tang-Liu et al. carried out a comprehensive investigation of the ocular pharmacokinetics and safety of 0.05 % cyclosporine ophthalmic emulsion, approved by the US FDA for treatment of keratoconjunctivitis sicca (dry-eye disease) [66]. The initial results from the cell culture studies showed that cyclosporine had no adverse effects on human corneal endothelial and stromal cells in vitro. Furthermore, long-term ocular administration of cyclosporine (up to 0.4 %) in rabbits (6 months) and dogs (1 year) was also found to be devoid of any systemic or ocular toxicity. The formulation was also found to be safe in the safety studies in human patients with dry eye. Bucolo et al. evaluated the ocular pharmacokinetics of two marketed ocular formulations, i.e., a suspension containing hydroxypropyl methylcellulose (IND-HPMC) and a solution with hydroxypropyl-beta-cyclodextrin (IND-CD) of indomethacin following topical administration in rabbits [67]. Indomethacin levels were measured in aqueous humor, vitreous humor, and the retina of the animals at regular intervals between 0 and 240 min by liquid chromatography–mass spectrometry (LC-MS). The results showed that the peak concentrations in the vitreous and humor were achieved within 30 and 60 min for IND-HPMC and IND-CD, respectively. The total bioavailability in the retina, vitreous, and humor as obtained from the area under the curve (AUC) was found to be significantly higher (threefold to fourfold) with IND-HPMC compared to that with IND-CD. Overall, the data indicated the importance of formulation composition in achieving therapeutically relevant pharmacokinetics of the drug.

Yuan et al. in a series of studies investigated the safety and pharmacokinetics of FK506 (*tacrolimus hydrate*, a macrolide immunosuppressant used to prevent graft rejection after organ transplantation) following topical administration in rabbits [68, 69]. The preliminary results indicated that FK506 had acceptable ocular safety profile. Moreover, the drug concentration in the cornea and aqueous humor measured after single-dose and multiple-dose administration at various time points indicated that the therapeutic concentration of FK506 required for treating corneal allograft rejection was achieved. Asena et al. studied the ocular pharmacokinetics, efficacy, and endothelial toxicity of moxifloxacin after a single intracameral bolus injection of 500 μg/0.1 ml in rabbit eyes [70]. Moxifloxacin concentrations in aqueous humor and vitreous samples were determined at 0.5, 1, 3, 6, 12, and 24 h by HPLC and compared with the minimum inhibitory concentrations and mutant prevention concentrations for endophthalmitis pathogens. The results showed that moxifloxacin exceeded the minimum inhibitory concentrations and mutant prevention concentrations of all common endophthalmitis pathogens in the aqueous humor for 6 h. The half-life of moxifloxacin in the aqueous humor was found to be 2.2 h.

In a recent study, Lin et al. investigated the ocular pharmacokinetics of 1 % naringenin (4′, 5, 7-trihydroxy flavanone) eye drops following topical administration to rabbits [71]. The authors analyzed the concentrations of naringenin in the ocular tissues and plasma using specific electrospray ionization liquid chromatography–tandem mass spectrometry method. The results found highest ocular bioavailability of naringenin in the cornea, followed by the aqueous humor, retina, and vitreous body, as measured by the area under the curve ($AUC_{0\text{-}t}$). The half-lives of naringenin were found to be 9.37 h (cornea), 0.65 h (aqueous humor), 1.17 h (vitreous), and 4.62 h (retina). The plasma levels of free naringenin and total naringenin were found to be similar based on the C_{max} and T_{max}. The study revealed the utility of naringenin in the treatment of several ocular diseases like age-related macular degeneration and retinitis pigmentosa. In another study, Shen et al. compared ocular and systemic pharmacokinetics of brimonidine and dexamethasone following a single intravitreal dose in rabbits and monkeys [72]. The results were also compared between healthy animals and those with chemical- or laser-induced breakdown of the blood–retinal barrier (BRB). The results showed that in both animals, the ocular bioavailability (as measured by AUC) was found to be lower in animals with damaged BRB compared to healthy animals. This was attributed to increased systemic clearance of the drugs due to the absence of BRB. There was no significant difference in the plasma concentrations of the drugs. The study emphasized the importance of the PK/PD data obtained and analyzed from healthy vs. disease models and suggested caution when extrapolating PK data from these models.

6.7 Summary

Optimizing ocular therapeutics via targeted drug delivery to specific ocular tissues is a challenge due to the unique anatomical and physiological constraints present in the eye. The rate and extent of absorption, distribution, metabolism, and elimination of ocular drugs may vary significantly depending on the physicochemical properties of the drugs, the route of drug administration, as well as the pathophysiology of the ocular disease state. The design and development of pharmacokinetic and pharmacodynamic models is of great help in evaluating the bioavailability, efficacy, and safety of ocular drugs. A robust ocular PK/PD model should include all relevant physiological/pharmacological processes and should have reliable predictive value. These models then form the basis of the theoretical understanding of the pharmacokinetics, pharmacodynamics, and toxicokinetics of the ocular drugs. Such an understanding is essential in identifying the challenges, predicting drug response, as well as implementing appropriate formulation and other interventional strategies to optimize ocular drug therapy.

References

1. Gilger BC, Gilger B (2013) Challenges in ocular pharmacokinetics, pharmacodynamics, and toxicology. In: Ocular pharmacology and toxicology. Humana Press, New York, pp 1–6
2. Ammar HO et al (2010) Development of dorzolamide hydrochloride in situ gel nanoemulsion for ocular delivery. Drug Dev Ind Pharm 36(11):1330–1339
3. Lee VH-L, Robinson JR (1979) Mechanistic and quantitative evaluation of precorneal pilocarpine disposition in albino rabbits. J Pharm Sci 68(6):673–684
4. Patel PB et al (2010) Ophthalmic drug delivery system: challenges and approaches. Syst Rev Pharm 1(2):113–120
5. Urtti A, Salminen L (1993) Minimizing systemic absorption of topically administered ophthalmic drugs. Surv Ophthalmol 37(6):435–456
6. Lee VH, Robinson JR (1986) Topical ocular drug delivery: recent developments and future challenges. J Ocul Pharmacol 2(1):67–108
7. Urtti A et al (1994) Controlled ocular timolol delivery: systemic absorption and intraocular pressure effects in humans. Pharm Res 11(9):1278–1282
8. Urtti A, Salminen L, Miinalainen O (1985) Systemic absorption of ocular pilocarpine is modified by polymer matrices. Int J Pharm 23(2):147–161
9. Urtti A et al (1990) Controlled drug delivery devices for experimental ocular studies with timolol 2. Ocular and systemic absorption in rabbits. Int J Pharm 61(3):241–249
10. Maurice DM, Mishima S (1984) Ocular pharmacokinetics. In: Sears M (ed) Pharmacology of the eye. Springer, Berlin/Heidelberg, pp 19–116
11. Hornof M, Toropainen E, Urtti A (2005) Cell culture models of the ocular barriers. Eur J Pharm Biopharm 60(2):207–225
12. Huang HS, Schoenwald RD, Lach JL (1983) Corneal penetration behavior of beta-blocking agents II: assessment of barrier contributions. J Pharm Sci 72(11):1272–1279
13. Urtti A (2006) Challenges and obstacles of ocular pharmacokinetics and drug delivery. Adv Drug Deliv Rev 58(11):1131–1135
14. HÃmÃlÃinen KM et al (1997) Estimation of pore size and pore density of biomembranes from permeability measurements of polyethylene glycols using an effusion-like approach. J Control Release 49(2–3):97–104
15. Prausnitz MR, Noonan JS (1998) Permeability of cornea, sclera, and conjunctiva: a literature analysis for drug delivery to the eye. J Pharm Sci 87(12):1479–1488
16. Geroski DH, Edelhauser HF (2001) Transscleral drug delivery for posterior segment disease. Adv Drug Deliv Rev 52(1):37–48
17. Cunha-Vaz J, Bernardes R, Lobo C (2011) Blood-retinal barrier. Eur J Ophthalmol 21(6):S3–S9
18. Xu HZ, Le YZ (2011) Significance of outer blood-retina barrier breakdown in diabetes and ischemia. Invest Ophthalmol Vis Sci 52(5):2160–2164
19. Murata T et al (1996) The relation between expression of vascular endothelial growth factor and breakdown of the blood-retinal barrier in diabetic rat retinas. Lab Invest 74(4):819–825
20. Cunha-Vaz J, Faria de Abreu JR, Campos AJ (1975) Early breakdown of the blood-retinal barrier in diabetes. Br J Ophthalmol 59(11):649–656
21. Yu D (2004) Pharmacokinetics in ocular drug development. In: Bonate PL, Howard DR (eds) Pharmacokinetics in drug development. AAPS Press, Arlington, pp 381–398
22. Gaudana R et al (2009) Recent perspectives in ocular drug delivery. Pharm Res 26(5):1197–1216
23. Kaur IP, Kanwar M (2002) Ocular preparations: the formulation approach. Drug Dev Ind Pharm 28(5):473–493
24. Kuno N, Fujii S (2011) Recent advances in ocular drug delivery systems. Polymers 3:193–221
25. Liaw J, Robinson JR (1993) Ocular penetration enhancers. In: Mitra AK (ed) Ophthalmic drug delivery systems. Marcel Dekker, New York, pp 369–381
26. Rawas-Qalaji M, Williams CA (2012) Advances in ocular drug delivery. Curr Eye Res 37(5):345–356

27. Sieg JW, Robinson JR (1976) Mechanistic studies on transcorneal permeation of pilocarpine. J Pharm Sci 65(12):1816–1822
28. Gomes dos Santos AL et al (2006) Sustained release of nanosized complexes of polyethylenimine and anti-TGF-β2 oligonucleotide improves the outcome of glaucoma surgery. J Control Release 112(3):369–381
29. Bashshur ZF et al (2006) Intravitreal bevacizumab for the management of choroidal neovascularization in age-related macular degeneration. Am J Ophthalmol 142(1):1–9
30. Zhou B, Wang B (2006) Pegaptanib for the treatment of age-related macular degeneration. Exp Eye Res 83(3):615–619
31. Pitkanen L et al (2005) Permeability of retinal pigment epithelium: effects of permanent molecular weight and lipophilicity. Invest Ophthalmol Vis Sci 46(2):641–646
32. Ambati J et al (2000) Transscleral delivery of bioactive protein to the choroid and retina. Invest Ophthalmol Vis Sci 41(5):1186–1191
33. Behrens-Baumann W, Ansorg R (1983) Azlocillin concentrations in human aqueous humor after intravenous and subconjunctival administration. Graefes Arch Clin Exp Ophthalmol 220(6): 292–293
34. Behrens-Baumann W, Ansorg R (1985) Mezlocillin concentrations in human aqueous humour after intravenous and subconjunctival administration. Chemotherapy 31(3):169–172
35. Behrens-Baumann W, Martell J (1987) Ciprofloxacin concentrations in human aqueous humor following intravenous administration. Chemotherapy 33(5):328–330
36. Behrens-Baumann W, Martell J (1988) Ciprofloxacin concentration in the rabbit aqueous humor and vitreous following intravenous and subconjunctival administration. Infection 16(1):54–57
37. Behrens-Baumann W et al (1986) Ciclosporin concentration in the rabbit aqueous humor and cornea following subconjunctival administration. Graefes Arch Clin Exp Ophthalmol 224(4): 368–370
38. Peeters L et al (2005) Vitreous: a barrier to nonviral ocular gene therapy. Invest Ophthalmol Vis Sci 46(10):3553–3561
39. Pitkanen L et al (2003) Vitreous is a barrier in nonviral gene transfer by cationic lipids and polymers. Pharm Res 20(4):576–583
40. Park J et al (2005) Evaluation of coupled convective-diffusive transport of drugs administered by intravitreal injection and controlled release implant. J Control Release 105(3):279–295
41. Haller JA et al (2014) Efficacy of intravitreal ocriplasmin for treatment of vitreomacular adhesion: subgroup analyses from two randomized trials. Ophthalmology 6420(14):00689-7
42. Inoue M et al (2014) Intravitreal injection of ranibizumab using a pro re nata regimen for age-related macular degeneration and vision-related quality of life. Clin Ophthalmol 8:1711–1716
43. Ausayakhun S et al (2005) Treatment of cytomegalovirus retinitis in AIDS patients with intravitreal ganciclovir. J Med Assoc Thai 88(9):S15–S20
44. Ornek N, Ornek K, Erbahceci IE (2014) Corneal and conjunctival sensitivity changes following intravitreal ranibizumab injection in diabetic retinopathy. J Ocul Pharmacol Ther 2014:22
45. Aslan Bayhan S et al (2014) Marginal keratitis after intravitreal injection of ranibizumab. Cornea 2014:12
46. Worakul N, Robinson JR (1997) Ocular pharmacokinetics/pharmacodynamics. Eur J Pharm Biopharm 44(1):71–83
47. Ahmed I, Patton TF (1987) Disposition of timolol and inulin in the rabbit eye following corneal versus non-corneal absorption. Int J Pharm 38(1–3):9–21
48. Chien DS et al (1990) Corneal and conjunctival/scleral penetration of p-aminoclonidine, AGN 190342, and clonidine in rabbit eyes. Curr Eye Res 9(11):1051–1059
49. Eller MG et al (1985) Topical carbonic anhydrase inhibitors IV: relationship between excised corneal permeability and pharmacokinetic factors. J Pharm Sci 74(5):525–529
50. Himmelstein KJ, Guvenir I, Patton TF (1978) Preliminary pharmacokinetic model of pilocarpine uptake and distribution in the eye. J Pharm Sci 67(5):603–606
51. Jones RF, Maurice DM (1966) New methods of measuring the rate of aqueous flow in man with fluorescein. Exp Eye Res 5(3):208–220

52. Makoid MC, Robinson JR (1979) Pharmacokinetics of topically applied pilocarpine in the albino rabbit eye. J Pharm Sci 68(4):435–443
53. Makoid MC, Sieg JW, Robinson JR (1976) Corneal drug absorption: an illustration of parallel first-order absorption and rapid loss of drug from absorption depot. J Pharm Sci 65(1):150–153
54. Miller SC, Himmelstein KJ, Patton TF (1981) A physiologically based pharmacokinetic model for the intraocular distribution of pilocarpine in rabbits. J Pharmacokinet Biopharm 9(6): 653–677
55. Rao CS et al (1992) Biopharmaceutical evaluation of ibufenac, ibuprofen, and their hydroxyethoxy analogs in the rabbit eye. J Pharmacokinet Biopharm 20(4):357–388
56. Sieg JW, Robinson JR (1981) Mechanistic studies on transcorneal permeation of fluorometholone. J Pharm Sci 70(9):1026–1029
57. Sakanaka K et al (2008) Ocular pharmacokinetic/pharmacodynamic modeling for timolol in rabbits using a telemetry system. Biol Pharm Bull 31(5):970–975
58. Sakanaka K et al (2004) Ocular pharmacokinetic/pharmacodynamic modeling for bunazosin after instillation into rabbits. Pharm Res 21(5):770–776
59. Sakanaka K et al (2008) Ocular pharmacokinetic/pharmacodynamic modeling for multiple antiglaucoma drugs. Biol Pharm Bull 31(8):1590–1595
60. Zimmer A et al (1994) Pharmacokinetic and pharmacodynamic aspects of an ophthalmic pilocarpine nanoparticle-delivery-system. Pharm Res 11(10):1435–1442
61. Tang-Liu D et al (1996) Pharmacokinetic and pharmacodynamic correlation of ophthalmic drugs. In: Reddy IK (ed) Ocular therapeutics and drug delivery: a multi-disciplinary approach. Technomic Publishing Co., Inc., Lancaster
62. Durairaj C, Shen J, Cherukury M (2014) Mechanism – based translational pharmacokinetic – pharmacodynamic model to predict intraocular pressure lowering effect of drugs in patients with glaucoma or ocular hypertension. Pharm Res 31(8):2095–2106
63. Luu KT et al (2009) Pharmacokinetic-pharmacodynamic and response sensitization modeling of the intraocular pressure-lowering effect of the EP4 Agonist 5-{3-[(2S)-2-{(3R)-3-hydroxy-4-[3-(trifluoromethyl)phenyl]butyl}-5-oxopyrrolidin- 1-yl]propyl}thiophene-2-carboxylate (PF-04475270). J Pharmacol Exp Ther 331(2):627–635
64. Siepmann J et al (1999) HPMC-matrices for controlled drug delivery: a new model combining diffusion, swelling, and dissolution mechanisms and predicting the release kinetics. Pharm Res 16(11):1748–1756
65. Zhang Y et al (2014) Pharmacokinetics of ranibizumab after intravitreal administration in patients with retinal vein occlusion or diabetic macular edema. Ophthalmology 6420(14): 00432-1
66. Tang-Liu DD, Acheampong A (2005) Ocular pharmacokinetics and safety of ciclosporin, a novel topical treatment for dry eye. Clin Pharmacokinet 44(3):247–261
67. Bucolo C, Melilli B, Piazza C, Zurria M, Drago F. (2011). Ocular Pharmacokinetics Profile of Different Indomethacin Topical Formulations. J Ocul Pharmacol Ther, 2011/12/01, 27(6): 571–576
68. Yuan J et al (2009) Preparation of 0.05 % FK506 suspension eyedrops and its pharmacokinetics after topical ocular administration. J Ocul Pharmacol Ther 25(4):345–50
69. Yuan J et al (2012) Ocular safety and pharmacokinetics study of FK506 suspension eye drops after corneal transplantation. J Ocul Pharmacol Ther 28(2):153–158
70. Asena L et al (2013) Ocular pharmacokinetics, safety and efficacy of intracameral moxifloxacin 0.5 % solution in a rabbit model. Curr Eye Res 38(4):472–479
71. Lin J et al (2014) Ocular pharmacokinetics of naringenin eye drops following topical administration to rabbits. J Ocul Pharmacol Ther 2014:17
72. Shen J et al (2014) Ocular pharmacokinetics of intravitreally administered brimonidine and dexamethasone in animal models with and without blood-retinal barrier breakdown. Invest Ophthalmol Vis Sci 55(2):1056–1066

Chapter 7
Physicochemical Requirements for Polymers and Polymer-Based Nanomaterial for Ophthalmic Drug Delivery

Sheeba Varghese Gupta

Abstract Polymers used for constructing ophthalmic nanodelivery systems play a crucial role in determining the drug entrapment capacity, biodegradation, and residence time of the nanoparticles. Polymers used for ophthalmic nanodelivery systems are biodegradable; the biodegradation may be enzymatically or chemically mediated. The physicochemical properties of the polymers such as molecular weight, hydrophobicity/hydrophilicity, polymer/copolymer composition, crystallinity, and glass transition temperature affect particle size, entrapment efficiency, adsorption/absorption pattern, degradation kinetics, and mechanical strength of the nanoparticles. PLGA is the most widely used polymer for ophthalmic delivery because of its biodegradability and flexibility in alteration of the physicochemical properties by altering the copolymer composition. Physicochemical properties of a polymer can be altered by chemical modifications. In-depth understanding of the physicochemical properties of the polymers is important to design a nanodelivery system with optimum drug encapsulation, degradation, and residence time.

Keywords Physicochemical properties • Molecular weight • Crystallinity • Glass transition temperature • Biodegradation • PLGA

Polymeric materials play a significant role in controlled drug delivery. Polymers, in general, are large molecules made up of small repeating units or monomers. The syntheses of polymers include covalently attaching monomers to form a long chain like structure. When a polymer consists of repetition of same type of monomer, it is known as homopolymer. If the polymer is a result of repetition of different types of monomer, then it is known as copolymer. Both homopolymers

S.V. Gupta (✉)
College of Pharmacy, University of South Florida, Tampa, FL 33612, USA
e-mail: svarghes@health.usf.edu

© Springer International Publishing Switzerland 2016

Y. Pathak et al. (eds.), *Nano-Biomaterials For Ophthalmic Drug Delivery*,
DOI 10.1007/978-3-319-29346-2_7

and copolymers are used in pharmaceutical preparations. Molecular weight of the polymers depends on the number of monomers present; the molecular weight increases with the number of monomers in the polymer structure. The physical and chemical properties of polymers are heavily influenced by its molecular weight and how the monomers in the chain bond and interact with each other. The interaction and the labiality of the covalent bonding dictate if the polymer can be degraded by the biological systems; in other words, the composition of the polymers dictates their biodegradability. Orally given pharmaceutical preparations containing polymers do not require the polymer to be metabolized in the gastrointestinal tract and are eliminated from the body unchanged. However, polymers intended for direct application to the site of action, such as the eye, are required to be biocompatible and biodegradable. Biocompatibility is important for the polymer interacting with the biological environment such as the ocular system. The polymer should not cause injury, damage, or any other adverse event while interacting with the biological system [2]. Biodegradable polymers are degraded into smaller nontoxic units by the biological systems. Major mode of biodegradation is through hydrolysis. As mentioned earlier, the structure of the polymer determines the biodegradability of the polymer; therefore, the presence of hydrolysis labile functional groups in the polymers structure will make them biodegradable. A variety of natural, synthetic, and biosynthetic polymers are biodegradable. In order for a polymer to be biodegradable, the polymer structure should have chemical linkages such as anhydride, ester, or amides among other labile linkages. In addition to biodegradability and biocompatibility, the polymers are required to have mucoadhesive properties if the formulation is intended for sustained release.

Polymers are used in both conventional ocular dosage forms such as liquids and semisolids as well as in ocular delivery systems such as ophthalmic inserts, microspheres, and nanoparticles. In both cases, the requirements of ideal physicochemical properties are more or less the same. Nanoparticles are built using biodegradable, biocompatible, natural, or synthetic polymers with some mucoadhesive properties [3, 4]. Commonly used polymers for making nanoparticles include poly(alkyl cyanoacrylate) (PAC), polylactic acid (PLA), poly(epsilon-caprolactone) (PCL), pol(lactide-co-glycolide) (PLGA), chitosan, gelatin, sodium alginate, and albumin [5]. Nanocarriers for ophthalmic delivery should meet three major goals for proper drug delivery: (i) enhance drug permeation, (ii) control the release mechanism of the drugs, and (iii) modify the surface of nanocarrier with specific target molecule [4]. In order to meet the abovementioned goals, the nanomaterial should be able to encapsulate the drug efficiently, should release the drug at a constant rate for the required amount of time, and should be lodged at the site of action. All these are possible by choosing the right polymer for building the nanocarrier. Here, only the physicochemical properties of the polymers affecting the nanoparticle-based drug delivery will be addressed. As the polymers used in ophthalmic delivery are biodegradable and biocompatible, this chapter will only address the examples of such polymers in detail.

7.1 Biodegradable Polymers

The biodegradable polymeric nanoparticles system ranging in size from 10 to 1000 nm are most commonly used for ophthalmic delivery. Biodegradable polymers can modify the surface characteristics of drug-encapsulated polymer carrier to improve cellular uptake, cell targeting, and immune system interactions. These polymeric systems can degrade enzymatically or nonenzymatically to yield in vivo biocompatible by-products that are safe and can be eliminated by normal metabolic pathways [6]. The biodegradable polymers can be classified based on their source, i.e., natural or synthetic.

7.1.1 Natural Biodegradable Polymers

Polysaccharides such as cyclodextrin, chitosan, alginate, and starches; proteins such as albumin, gelatin, fibrins, and collagen are the examples of natural biodegradable polymers. Mucoadhesive polymer such as chitosan and its derivative carboxymethyl chitosan has attracted much attention for their ability to prolong ophthalmic drug retention [7]. Sodium alginate is used in combination with chitosan and poloxamers to enhance the release and bioavailability of drug in aqueous humor of eye [8]. Cyclodextrins are used to solubilize water-insoluble drugs; cyclodextrin is usually used in combination with a mucoadhesive polymer to ensure sustained release of the drug in ocular system [9].

7.1.2 Synthetic Biodegradable Polymer

Most commonly used synthetic biodegradable polymers for ophthalmic delivery include poly(alkyl cyanoacrylate) (PAC), polylactic acid (PLA), poly(epsilon-caprolactone) (PCL), and pol(lactide-co-glycolide) (PLGA). Out of all the biodegradable polymers of synthetic origin, PLGA is one of the well-known carriers employed in ophthalmic drug delivery. Biocompatibility, ease of formulation, and predictable and versatile degradation kinetics make PLGA an excellent choice for polymeric carrier to achieve sustained and targeted drug delivery [10]. PLGA is a copolymer of polylactide (PLA, hydrophobic and crystalline polymer) and polyglycolide (PGA, hydrophilic polymer). The copolymer constituents of PLGA dictate its overall hydrophobicity/hydrophilicity.

7.2 Biodegradation

As mentioned earlier, the biodegradation process of polymers could be enzymatic or nonenzymatic hydrolysis of the polymer backbone in to water-soluble or insoluble products. Degradation and erosion constitutes to the process of biodegradation.

Degradation or depolymerization involves cleavage of polymer backbone, and erosion involves diffusion and dissolution of low molecular weight fractions from the polymer matrix. The products of polymer degradation are eliminated from body by normal metabolic pathways [11]. Rate of biodegradation of pharmaceutical polymers depends on the site of exposure in the human body. Chemical degradation is influenced by composition, molecular structure, polydispersity, crystallinity, surface area, and hydrophilic and hydrophobic characteristics. Chemical degradation causes the breaking of polymer chains by random cleavage of covalent bonds, depolymerization of linear polymers, interfering with the order of chain and crystallinity, thereby decreasing mechanical properties. The polymer can undergo surface or bulk degradation; in case of bulk degradation, the rate of water uptake by hydrophilic polymers is faster than the conversion of polymers to water-soluble derivatives. Bulk degradation occurs throughout the volume of the polymer and can lead to collapse of the entire structure. Surface degradation is commonly seen with hydrophobic polymers, where the inner surface is intact and the surface undergoes slow erosion [12]. Biodegradation in the body can happen by following three mechanisms.

Hydrolysis: Hydrolytic degradation of the polymer involves cleaving of covalent bonds in polymer backbone in the presence of water to form oligomers and monomers. Biodegradable polymers contain groups such as glycosides, ester, anhydrides, carbonates, amides, and urethane, which undergo hydrolysis. The first step of hydrolysis involves attack of water molecule on the water-labile bond in the polymer backbone followed by cleaving of the labile bonds. Acids and bases can catalyze the hydrolysis reactions. During the hydrolysis, the polymer chains are cleaved to smaller molecular weight fragments, which get further reduced in molecular weight with hydrolysis, allowing them to diffuse from the bulk material to the surface and into solution [13]. Hydrophilicity of the polymer increases the rate of hydrolysis. In PLGA-like polymers, the acid by-products can accelerate biodegradation by acid autocatalyzed hydrolysis.

Oxidation: Polymers can be degraded by oxidation when exposed to body fluids and tissues. The oxidation is mainly catalyzed by reactive oxygen species such as superoxide (O_2^-), hydrogen peroxide (H_2O_2), and nitric oxide (NO). Macrophages and leukocytes are able to produce reactive oxidative species and can cause polymers to degrade by the oxidation [13].

Enzymatic Degradation: The most commonly seen enzymatic degradation is hydrolysis reactions catalyzed by hydrolytic enzymes such as proteases, esterases, glycosidases, and phosphatases. The mechanism of enzyme degradation involves diffusion of the enzyme from the bulk solution, adsorption of the enzyme on the substrate forming enzyme–substrate complex, catalysis of hydrolysis, and ultimately diffusion of degradation product from solid substrate to the solution [13].

The adsorption of the enzyme and rate of hydrolysis is affected by the physicochemical properties of the polymer and by the inherent characteristics of the enzyme involved. Conditions of the reaction medium such as temperature and pH also affect the rate of enzyme-catalyzed hydrolysis. Chemical modifications such as cross-linking and removal and/or introduction of chemical groups in polymer chain also affect the rate of reaction by affecting the affinity of the enzyme for the substrate. It was reported that the lysozyme (an enzyme responsible for degradation of chitin materials and peptidoglycan) showed decreased activity toward chitosan with high degrees of deacetylation or cross-linking [14–16].

There are three types of polymer biodegradations as shown in Fig. 7.1.

Type I: This type of biodegradation is generally observed in cross-linked water-soluble polymers such as gelatin, collagen, polyacrylamides, poly(vinyl alcohol (PVA), and poly(N-vinyl pyrrolidone) (PVP). These polymers form hydrogels upon cross-linking to produce water-insoluble three-dimensional structures. Hydrolysis of the cross-linked polymers can produce high or low molecular weight fragments based on the type of parent polymer. Polymers used in implants generally undergo degradation by generating low molecular weight fragments, whereas polymers used for topical applications produces high molecular weight fragments. Polymers undergoing the type I degradation are highly permeable to water, making this degradation unsuitable for low molecular weight molecules with good water solubility. However, such a system is more suitable for delivering macromolecules as these molecules get entangled in the cross-linked polymeric system restricting their movement from the polymer matrix. Hence, the cross-linked polymers are best used as a delivery vehicle for macromolecules such as enzymes, antigens, or molecules with low water solubility.

Type II: This type of degradation converts water-insoluble linear polymers into water-soluble smaller units through the process of hydrolysis, ionization, or

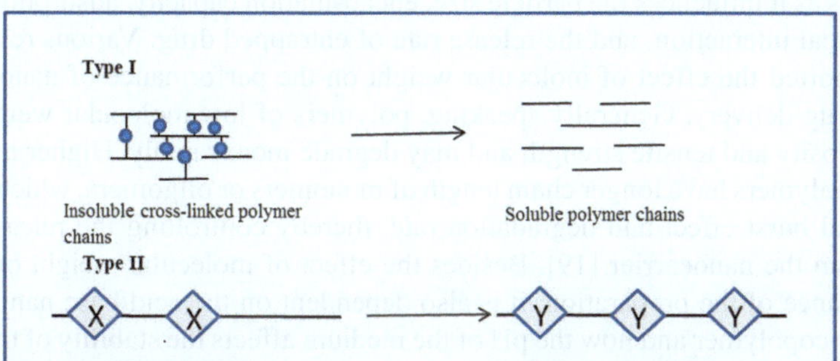

Fig. 7.1 Schematic representation of different types of polymer degradation [1]

protonation. These types of polymers are generally intended for topical applications. Copolymers of alkyl vinyl ether and maleic anhydride undergo degradation by type II mechanism [17].

Type III: This process involves hydrolytic cleavage of water-insoluble polymers into low molecular weight, water-soluble molecules. The by-products of the cleavage are nontoxic; hence, these polymers can be used for both topical and systemic applications. Polymers undergoing type III degradation have different physicochemical properties and a wide range of molecular weight, which makes them amenable for various drug delivery applications. Polymers such as PLA, PGA, PLGA, etc. that possess a hydrolytically unstable backbone undergo type III degradation [18].

7.3 Physicochemical Properties

The following physicochemical properties of polymers largely influence the biodegradation and performance of nanoparticles for ophthalmic delivery.

7.3.1 Molecular Weight

Choosing the right molecular weight for the polymer is important in making nanomaterials as it influences the particle size, encapsulation capacity, adsorption, physicochemical interaction, and the release rate of entrapped drug. Various researchers have reported the effect of molecular weight on the performance of nanoparticle-based drug delivery. Generally speaking, polymers of low molecular weight show low viscosity and tensile strength and may degrade more rapidly. Higher molecular weight polymers have longer chain length of monomers or oligomers, which reduces the initial burst effect and degradation rate, thereby controlling the release of the drug from the nanocarrier [19]. Besides the effect of molecular weight on overall performance of the preparation, it is also dependent on the acid/base nature of the polymer/copolymer and how the pH of the medium affects the stability of the preparation. Ideal value of the molecular weight depends on the type of polymer molecule and the type of entrapped drug. As seen in the study with chitosan-based nanoparticles, lower molecular weight chitosan polymers offered maximum encapsulation efficiency for the loaded drug [20]. Chitosan is a cationic, pH-responsive polymer (undergoes sol–gel transition at pH7.4) rich in primary amino groups. The reasoning for the increase in encapsulation efficiency was that in case of the lower molecular weight fragments of chitosan, the free amino groups were easily accessible for protonation and, therefore, were able to adsorb greater amount of negatively charged drug through ionic interactions [20]. Molecular weight of the polymer affects the stability of the nanoparticles. In case of PLGA-containing nanoparticles, it has been shown that the amount of carboxylic acid functional group available for peptide coupling was dependent on the molecular weight of the polymer. It was also shown that the glass transition temperature for the polymer increased with increase in molecular weight. The resulting microspheres had increased particle size with increase in molecular weight of the PLGA polymer [21]. However, in case of PLA polymers, lower molecular weight polymers produced nanoparticles with a tad bigger particle size compared to nanoparticles produced from a high molecular weight polymer. It was observed that zeta potential of the particles rises slightly with the rise in molecular weight of the polymer. The zeta potential is related to the electric charge on the surface of nanoparticles, and therefore, is also related to the carboxylate end groups of PLA. The number of carboxylate end group decreases with increase in molecular weight of PLA. The effect of molecular weight on the size of the nanoparticles, in case of PLA, can be minimized by the selection of right solvent and right surfactant while preparing the nanoparticles [22]. As mentioned above, the molecular weight of the polymer affects the pharmacokinetic and pharmacodynamics properties of the drug delivered via nanoparticles.

7.3.2 Hydrophobicity/Hydrophilicity

Hydrophobicity of the polymer affects both absorption pattern and degradation kinetics of the nanoparticles. Increase in polymer chain increases the hydrophobicity of a polymer; as molecular weight increases with increase in polymer chain length, it can be assumed that with increase in molecular weight, hydrophobicity of the polymer will increase [23]. Increase in hydrophobicity decreases the polymer degradation and, hence, can prolong the drug release from the nanoparticulate system. The bioavailability of ophthalmic drugs in aqueous solutions is low due to rapid elimination after mucosal instillation, as a result of constant blinking and tear drainage, as well as presence of corneal barrier [24]. Therefore, nanoparticles offer an advantage of offering controlled release of drug. As nanoparticles can be removed from precorneal area, utilization of mucoadhesive polymers can increase their residence time and increase ocular drug bioavailability. Mucoadhesiveness of a polymer is the function of its hydrophilic nature. Presence of hydrophilic groups in polymers promotes hydrogen bonding with mucin, a component of the cornea. Mucin is a hydrophilic polymer rich in sialic acid side chains. Because of its high sialic acid content, mucin assumes negative charge at higher pH. Mucin is capable of interacting with various polymers due to its hydrophilicity and negative charge. Polymers used in ophthalmic delivery swell and get entangled in mucin [25]. For ophthalmic delivery, hydrophilic polymers with good hydration tendency are preferred, as these types of polymers prolong the residence time in precorneal area. Hydrophilic and ionic polymers such as chitosan offer mucoadhesive properties by virtue of electrostatic interaction with charged mucin [26]. Hydrophobic polymers do not possess any affinity toward mucus membrane and are cleared by the lymphatic system. Highly hydrophobic polymers are seldom used for making nanoparticles for ophthalmic delivery. Nanoparticles made with hydrophilic polymers have high encapsulation efficiency for hydrophilic drugs [27]. Chitosan can be utilized to coat the nanoparticles to enhance their affinity for corneal and conjunctival epithelial cells.

7.3.3 Polymer/Copolymer Composition

Polymer composition affects mainly the degradation kinetics of polymer matrix by affecting the hydrophilicity of the system. The degradation of PLGA-containing nanoparticles can be well programmed by increasing the lactide composition of the polymer [23]. The effect of PLGA composition on the rate of degradation is faster in PLGA with equal ratio of PLA and PGA, as compared to PLGA with higher concentration of PLA. As mentioned in previous section, PGA contributes to the overall hydrophilicity of PLGA; decrease in proportion PGA increases hydrophobicity of PLGA, slowing its degradation kinetics. Conversely, increase in proportion of PGA in PLGA will increase the degradation rate of polymer, thereby increasing the drug release [28].

The end group of the polymer also determines the fate of polymer carrier systems. As in case of PLGA, the presence of ester end groups decreases the amount of drug release as compared to polymer with acidic end groups. The type of end group has shown to have a pronounced effect on drug loading and zeta potential. Zeta potential exhibited by ester end groups is lower than the acid end group, indicating enhanced stability of the resultant nanoparticles in the suspension form [29]. The degradation kinetics of the delivery system can be modified by using different copolymer composition.

7.3.4 Glass Transition Temperature (T_g)

Glass transition temperature is the temperature or range of temperature at which the polymer starts transformation from a glassy state to a viscous liquid phase. T_g depends on the structure and stereochemistry of the polymer chains; the presence of large population of flexible chains decreases T_g, whereas increase in proportion of inflexible polymer chains increases T_g. There is a decrease in T_g with an increase in flexibility of polymer chains because these structures can easily gain enough thermal energy to initiate movements with each other. PLGA, which is a common polymer for ophthalmic delivery, has T_g higher than physiological temperature due to rigidity in its structure [30]. The rigidity contributes to the mechanical strength of PLGA polymer, thereby increasing the robustness of the resulting nanoparticles. T_g is dependent on the molecular weight and polymer/copolymer composition; low molecular weight polymers allow rapid hydration due to hydrophilicity which leads to increased movement of the polymer chains and degradation [23]. Decrease in lactide content and molecular weight of PLGA decreases its T_g due to high hydrophilicity of the PLGA copolymer [30]. T_g is important in determining the release kinetics of the drug. Polymers with T_g lower than body temperature will exist in molten state, facilitating rapid absorption of water and creating channels for diffusion. Therefore, it leads to sudden degradation of the nanocarrier, resulting in rapid and increased rate of drug release.

7.3.5 *Polymer Crystallinity*

The crystalline nature of the polymer significantly affects the entrapment efficiency (EE) and degradation kinetics of nanoparticles. The crystalline nature of the polymer increases its barrier properties and is impermeable to water. The increase in barrier property also prevents encapsulation of drugs; conversely, the amorphousness of the polymer enhances EE of the nanoparticles. Degradation is inversely proportional to the crystallinity of the polymer; the more crystalline the polymer, the slower will be the degradation of the nanoparticles, and hence, the slower will be the drug release [30]. Crystallinity is largely dependent on the molecular weight of the polymer; low molecular weight polymer possesses high degree of crystallinity and high molecular weight polymer has low crystallinity. As high molecular weight polymer will have high proportion of amorphousness, the EE would be better with faster drug release rate. As mentioned earlier, crystalline form is less conducive to offering optimum EE; therefore, a proper balance between amorphousness/crystallinity and molecular weight should be maintained to achieve optimum EE [31].

In case of PLGA polymer, the crystallinity is mainly affected by the type and the molar ratio of the copolymers (PLA and PGA). PLA exists as optically active L-isomer and as a racemic mixture of D/L forms. The racemic mixture is amorphous, whereas the L-form is semicrystalline. Therefore, the type and proportion of PLA isomer will determine the overall crystalline nature of PLGA polymer. PLGA containing less than 70 % of PGA are generally amorphous in nature, as PLGA used in formulating nanoparticles contain PLA in the racemic form. With increase in crystallinity, drug release will decrease because a polymer in crystalline form creates microchannels in their structure, which allows slow passage of the dissolved drug across the channels, resulting in slower release [30].

7.4 Examples of Polymers Used in Ophthalmic Delivery

Following sections will address the physicochemical properties of the most commonly used polymers in ophthalmic delivery.

7.4.1 Chitosan

Chitosan is a natural polymer obtained by deacetylation of chitin, a (1→4)2-amino-2-deoxy-β-D-glucan. Chitin is found in the exoskeleton of crustaceans, crab and shrimp shells, and insects [32]. Chitosan's physicochemical properties strongly influence its biological properties. Its ability to associate with water and (poly) electrolytes is responsible for formation of colloidal systems trapping water molecules within itself; these are also known as hydrogels. The degree of deacetylation (DA) and molecular weight (M_w) of the chitosan affects its biological properties [33–35]. Chitosan is a weak base with a pKa of 6, dependent on the DA and M_w [36]. Chitosan is soluble between pH values of 2 and 6, if the DA is below 25 %. As the value of DA increases to 50 %, chitosan becomes soluble at all pH. However, if the degree of DA further increase and crosses over 60 %, then chitosan becomes completely insoluble at pH above 6 due to change in hydrogen bonding between amide hydroxyl and ether groups. Salts of chitosan formed with organic or inorganic acids upon dissolution in water give solution with low pH. Chloride and glutamate salts of chitosan are commonly used in pharmaceutical preparations [36].

Increase in molecular weight of polymeric chitosan results in decreased solubility of the polymer as a result of intrachain hydrogen bonding. However, the oligomeric chitosan with a degree of polymerization is soluble at any pH and DA [37].

Chitosan can undergo enzymatic degradation in vivo in the presence of lysozyme enzyme. The rate of degradation decreases with increase in degree of DA and crystallinity [38]. In addition to mucoadhesiveness mentioned in earlier section of this chapter, chitosan also has penetration enhancing nature, which makes it an ideal polymer for application in ophthalmic formulations. The permeating enhancing property of chitosan is results from its ability to modulate the tight junctions of the corneal epithelial cells in a reversible manner. This mechanism can affect the paracellular and transcellular transport of the drug molecules without affecting the cellular integrity [39, 40]. Pseudoelastic and viscoelastic rheological properties of chitosan allow its prolonged residence time in cornea. It has been shown by multiple researchers that the chitosan nanoparticles and chitosan-coated nanoparticles have high affinity for the corneal and conjunctival epithelial cells [41, 42] (Fig. 7.2).

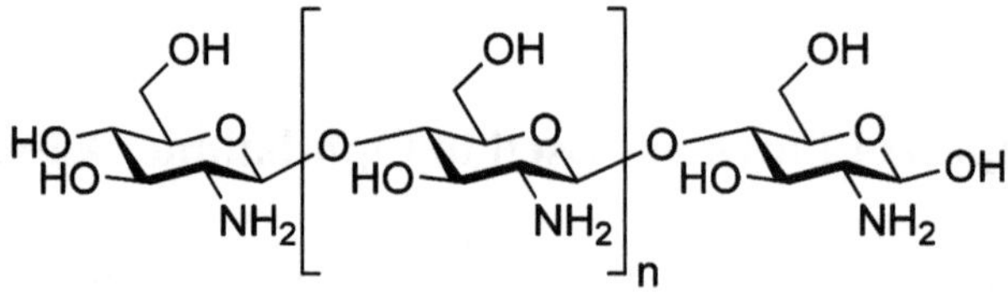

Fig. 7.2 Structure of chitosan

7.4.2 Gelatin

Gelatin is a collagen-derived protein with good gelling properties; gelatin is liquid above the gelling temperature (t_{gel}) and a gel below t_{gel}. Gelatin is amphipathic and bioresorbable, and its physicochemical properties can be easily modified [43]. An absorbable cross-linked hydrogel of gelatin can be formed by thermal and chemical modification. Hydrogel formed by the above said modifications does not dissolve in water, but swells up on contact with water [44]. Gelatin used for ocular delivery systems is cross-linked to improve the stability and thermal degradation of the hydrogels. The cross-linked gelatin also offers slow drug release from the delivery system [45].

7.4.3 Poly(N-vinyl pyrrolidone) (PVP)

PVP is a synthetic and biocompatible polymer widely used for preparation of hydrogels for ophthalmic delivery. Cross-linking of PVP prolongs its degradation and drug release ability [46]. Degradation kinetics of PVP can be modified by altering the cross-linking density. It has also been shown that the PVP-based hydrogels can cause intravitreal opacity, hazy corneas, and inflammation [47]. Various researchers have reported applications of PVP cross-linked polymers as artificial vitreous substance, scaffold for lens regeneration and long-term drug delivery system for antiglaucoma drugs [48, 49].

7.4.4 Poly(lactide) (PLA), Polyglycolide (PGA), and Polylactide-co-glycolide (PLGA)

As mentioned in the previous sections, PLGA is the most widely used polymer for ophthalmic dug delivery. PLA and PLGA are used in ocular delivery systems such as implants, injectable microspheres, and nanoparticles [1]. PLGA is a copolymer of PLA and PGA; the physicochemical properties of PLGA can be modulated by altering the proportions of PLA and PGA. The degradation kinetics of PLGA can be tailored by changing molecular weight, composition confirmation, and crystallinity of the polymers [50]. The effect of varying proportions of PLA and PLGA on physicochemical properties of PLGA is addressed in a previous section under respective properties. Drug release from PLGA delivery systems depends on the proportion of the two polymers used, porosity, surface area of the carrier, and physicochemical properties of the drug [51]. Most of the PLGA-based drug delivery system undergoes an initial burst phase. This initial burst can release particulates, which may cause visual obstruction, or retinal irritation. It has been shown that the initial burst phase of PLGA delivery system can be prevented by using PLGA-PEG-PLGA triblock thermogel to suspend the particulate system [52, 53]. The thermoreversible gel forms a gel on contact with eye tissue and holds particles at the site of administration, preventing vision interference. Additionally, the gel nature of the delivery system offers slow release of the incorporated drug by protecting the delivery system against cellular and enzymatic degradation [54] (Fig. 7.3).

7.4.5 *Poly-ε-caprolactone (PCL)*

PCL is a rigid semicrystalline polymer with melting point of 59–64 °C and a T_g of −60 °C [1]. Permeability and crystallinity of PCL can be altered by copolymerization with PLA and PGA [55]. PCL is an attractive choice for sustained drug delivery because of its excellent permeability to drug molecule and resistance to degradation. There reports of utilization of PCL-based ophthalmic drug delivery systems capable of releasing drug in therapeutic range over a period of 1 year in animal model [56]. PCL-based triblock polymer (PEG-PCL-PEG) has also been evaluated for ophthalmic applications and has been found to be biocompatible with ocular tissues and offered controlled release of the drug [57] (Fig. 7.4 and Table 7.1).

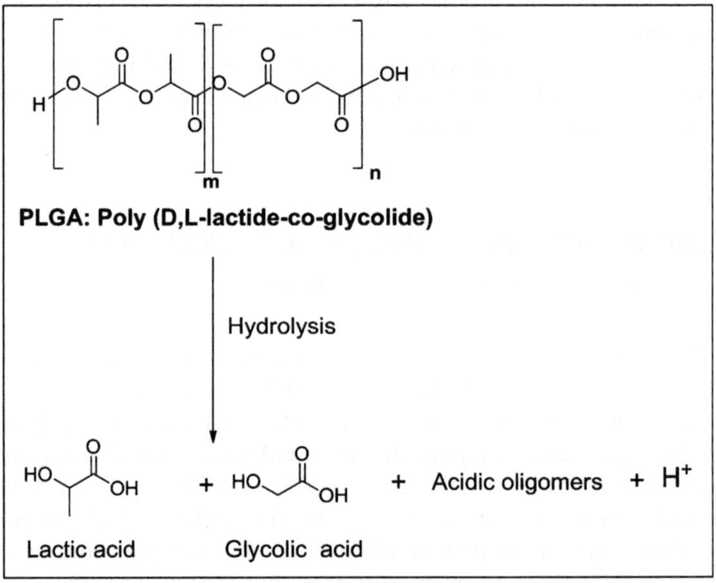

Fig. 7.3 Structure of PLGA and its hydrolysis products

Fig. 7.4 Chemical structure of polycaprolactone

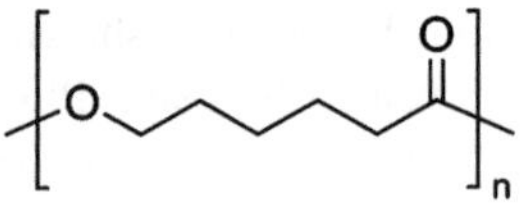

Table 7.1 Summarizes various polymers and the examples of type of delivery systems used in ophthalmic research (this list does not include all examples)

Polymer	Drug	Dosage form	Animal model
Collagen	Mitomycin-C	Implant	Rabbits [58]
Gelatin	Pilocarpine	In-situ gel	Rabbits [59]
	Timolol Maleate	Microsphere	Rabbits [60]
Chitosan	Amphotericin	Nanoparticles	Rabbits [61]
	Methazolamide	Nanoparticles	Rabbits [62]
PLA	Ganciclovir	Implant	Rabbits [63]
PLGA	Clindamycin	Implant	Rabbits [64]
	Vancomycin	Microsphere	Rabbits [65]
PCL	Triamcinolone	Episcleral plaques	Rabbits [66]

7.5 Conclusion

The physicochemical properties of polymers have a great effect on the performance of drug delivery system in general. The polymers used for ophthalmic delivery are generally biodegradable and biocompatible. The biodegradability is attributed to tissue and enzyme degradation at the site of application. Physicochemical properties of the polymer affect the drug release from the delivery system by altering the degradation kinetics and residence time at the site of application. Knowledge of physicochemical properties of the polymers is important in designing a delivery system with optimum drug encapsulation, degradation, and residence time. Physicochemical properties of the polymer system can be modulated by varying the proportion of chemical modifications or the proportion of individual polymers in case of copolymer. Therefore, not only choosing the right polymer but also using the right blend of polymers is important for affording delivery systems with optimum drug encapsulation and release properties.

References

1. Tamboli V, Mishra GP, Mitra AK (2012) Biodegradable polymers for ocular drug delivery. In: Mitra AK (ed) Advances in ocular drug delivery. Research Signpost, Trivandrum, pp 65–86
2. Ahmed F, Pakunlu RI, Brannan A, Bates F, Minko T, Discher DE (2006) Biodegradable polymersomes loaded with both paclitaxel and doxorubicin permeate and shrink tumors, inducing apoptosis in proportion to accumulated drug. J Control Release 116(2):150–158, Epub 2006/09/01
3. Nagarwal RC, Kant S, Singh PN, Maiti P, Pandit JK (2009) Polymeric nanoparticulate system: a potential approach for ocular drug delivery. J Control Release 136(1):2–13, Epub 2009/04/01
4. Sahoo SK, Dilnawaz F, Krishnakumar S (2008) Nanotechnology in ocular drug delivery. Drug Discov Today 13(3–4):144–151, Epub 2008/02/16
5. Mudgil M, Gupta N, Nagpal M, Pawar P (2012) Nanotechnology: a new approach for ocular drug delivery system. Int J Pharm Sci 4(2):105–112

6. Makadia HK, Siegel SJ (2011) Poly lactic-co-glycolic acid (PLGA) as biodegradable controlled drug delivery carrier. Polymers 3(3):1377–1397, Epub 2012/05/12
7. Di Colo G, Zambito Y, Burgalassi S, Nardini I, Saettone MF (2004) Effect of chitosan and of N-carboxymethylchitosan on intraocular penetration of topically applied ofloxacin. Int J Pharm 273(1–2):37–44
8. Nagarwal RC, Kumar R, Pandit JK (2012) Chitosan coated sodium alginate-chitosan nanoparticles loaded with 5-FU for ocular delivery: in vitro characterization and in vivo study in rabbit eye. Eur J Pharm Sci 47(4):678–685, Epub 2012/08/28
9. Mahmoud AA, El-Feky GS, Kamel R, Awad GE (2011) Chitosan/sulfobutylether-beta-cyclodextrin nanoparticles as a potential approach for ocular drug delivery. Int J Pharm 413(1–2):229–236, Epub 2011/05/05
10. Joshi SA, Chavhan SS, Sawant KK (2010) Rivastigmine-loaded PLGA and PBCA nanoparticles: preparation, optimization, characterization, in vitro and pharmacodynamic studies. Eur J Pharm Biopharm 76(2):189–199, Epub 2010/07/20
11. Katti DS, Lakshmi S, Langer R, Laurencin CT (2002) Toxicity, biodegradation and elimination of polyanhydrides. Adv Drug Deliv Rev 54(7):933–961, Epub 2002/10/18
12. Pak J, Lakes RS (2007) Biomaterials an introduction, 3rd edn. Springer, New York
13. Azevedo HS, Reis RL (2004) Understanding the enzymatic degradation of biodegradable polymers and strategies to control their degradation rate. In: Reis RL (ed) Biodegradable systems in tissue engineering and regenerative medicine. CRC Press, Boca Raton, pp 177–201
14. Nordtveit RJ, Vårum KM, Smidsrød O (1996) Degradation of partially N-acetylated chitosans with hen egg white and human lysozyme. Carbohydr Polym 29:163–167
15. Tomihata K, Ikada Y (1997) In vitro and in vivo degradation of films of chitin and its deacetylated derivatives. Biomaterials 18(7):567–575, Epub 1997/04/01
16. Mi FL, Tan YC, Liang HC, Huang RN, Sung HW (2001) In vitro evaluation of a chitosan membrane cross-linked with genipin. J Biomater Sci Polym Ed 12(8):835–850, Epub 2001/11/23
17. Woodruff CW, Peck GE, Banker GS (1972) Dissolution of alkyl vinyl ether-maleic anhydride copolymers and ester derivatives. J Pharm Sci 61(12):1912–1916, Epub 1972/12/01
18. Kimura H, Ogura Y (2001) Biodegradable polymers for ocular drug delivery. Ophthalmologica 215(3):143–155, Epub 2001/05/08
19. Matsumoto J, Nakada Y, Sakurai K, Nakamura T, Takahashi Y (1999) Preparation of nanoparticles consisted of poly(L-lactide)-poly(ethylene glycol)-poly(L-lactide) and their evaluation in vitro. Int J Pharm 185(1):93–101, Epub 1999/07/30
20. Yang H-C, Hon M-H (2009) The effect of the molecular weight of chitosan nanoparticles and its application on drug delivery. Microchem J 92(1):87–91
21. Mehta RC, Thanoo BC, Deluca PP (1996) Peptide containing microspheres from low molecular weight and hydrophilic poly(d, l-lactide-co-glycolide). J Control Release 41(3):249–257
22. Zambaux MF, Bonneaux F, Gref R, Maincent P, Dellacherie E, Alonso MJ et al (1998) Influence of experimental parameters on the characteristics of poly(lactic acid) nanoparticles prepared by a double emulsion method. J Control Release 50(1–3):31–40
23. Mittal G, Sahana DK, Bhardwaj V, Ravi Kumar MN (2007) Estradiol loaded PLGA nanoparticles for oral administration: effect of polymer molecular weight and copolymer composition on release behavior in vitro and in vivo. J Control Release 119(1):77–85, Epub 2007/03/14
24. Araújo J, Vega E, Lopes C, Egea MA, Garcia ML, Souto EB (2009) Effect of polymer viscosity on physicochemical properties and ocular tolerance of FB-loaded PLGA nanospheres. Colloids Surf B Biointerfaces 72(1):48–56
25. Mikos AG, Peppas NA (1986) Systems for controlled release of drugs v. bioadhesive systems. STP Pharm 2:705–716
26. Dash TK, Konkimalla VB (2012) Poly-small je, Ukrainian-caprolactone based formulations for drug delivery and tissue engineering: a review. J Control Release 158(1):15–33, Epub 2011/10/04
27. Bilensoy E, Sarisozen C, Esendagli G, Dogan AL, Aktas Y, Sen M et al (2009) Intravesical cationic nanoparticles of chitosan and polycaprolactone for the delivery of Mitomycin C to bladder tumors. Int J Pharm 371(1–2):170–176, Epub 2009/01/13

28. Park TG (1995) Degradation of poly(lactic-co-glycolic acid) microspheres: effect of copolymer composition. Biomaterials 16(15):1123–1130, Epub 1995/10/01
29. Panyam J, Williams D, Dash A, Leslie-Pelecky D, Labhasetwar V (2004) Solid-state solubility influences encapsulation and release of hydrophobic drugs from PLGA/PLA nanoparticles. J Pharm Sci 93(7):1804–1814, Epub 2004/06/04
30. Sonam HC, Arora V, Koli K, Kumar V (2013) Effect of physicochemical properties of biodegradable polymers on nano drug delivery. Polymer Rev 53(4):546–567
31. Youan BB, Benoit MA, Baras B, Gillard J (1999) Protein-loaded poly(epsilon-caprolactone) microparticles. I. Optimization of the preparation by (water-in-oil)-in water emulsion solvent evaporation. J Microencapsul 16(5):587–599, Epub 1999/09/28
32. Kumar M (2000) A review of chitin and chitosan applications. React Funct Polym 46:1–27
33. Sabnis S, Block LH (2000) Chitosan as an enabling excipient for drug delivery systems. I. Molecular modifications. Int J Biol Macromol 27(3):181–186, Epub 2000/06/01
34. Benesch J, Tengvall P (2002) Blood protein adsorption onto chitosan. Biomaterials 23(12):2561–2568, Epub 2002/05/30
35. Chatelet C, Damour O, Domard A (2001) Influence of the degree of acetylation on some biological properties of chitosan films. Biomaterials 22(3):261–268, Epub 2001/02/24
36. Ottoy MHV, Varum KM, Smidsrod O (1996) Compositional heterogeneity of heterogeneously deacetylated chitosans. Carbohydr Polym 29:17–24
37. Anthonsen MWV, Varum KM, Smidsrod O (1993) Solution properties of chitosans-confirmation and chain stiffness of chitosans with different degrees. Carbohydr Polym 22:193–201
38. Shi C, Zhu Y, Ran X, Wang M, Su Y, Cheng T (2006) Therapeutic potential of chitosan and its derivatives in regenerative medicine. J Surg Res 133(2):185–192, Epub 2006/02/07
39. Badawi AA, El-Laithy HM, El Qidra RK, El Mofty H, El dally M (2008) Chitosan based nanocarriers for indomethacin ocular delivery. Arch Pharm Res 31(8):1040–1049, Epub 2008/09/13
40. De Campos AM, Sanchez A, Alonso MJ (2001) Chitosan nanoparticles: a new vehicle for the improvement of the delivery of drugs to the ocular surface. Application to cyclosporin A. Int J Pharm 224(1–2):159–168, Epub 2001/07/27
41. de Campos AM, Diebold Y, Carvalho EL, Sanchez A, Alonso MJ (2004) Chitosan nanoparticles as new ocular drug delivery systems: in vitro stability, in vivo fate, and cellular toxicity. Pharm Res 21(5):803–810, Epub 2004/06/08
42. De Campos AM, Sanchez A, Gref R, Calvo P, Alonso MJ (2003) The effect of a PEG versus a chitosan coating on the interaction of drug colloidal carriers with the ocular mucosa. Eur J Pharm Sci 20(1):73–81, Epub 2003/09/19
43. Gaspard S, Oujja M, Abrusci C, Catalina F, Lazare S, Desvergne JP, Castillejo M (2008) Laser induced foaming and chemical modifications of gelatin films. J Photochem Photobiol A 193:187–192
44. Young S, Wong M, Tabata Y, Mikos AG (2005) Gelatin as a delivery vehicle for the controlled release of bioactive molecules. J Control Release 109(1–3):256–274, Epub 2005/11/04
45. Natu MV, Sardinha JP, Correia IJ, Gil MH (2007) Controlled release gelatin hydrogels and lyophilisates with potential application as ocular inserts. Biomed Mater 2(4):241–249, Epub 2008/05/07
46. Hong Y, Chirila TV, Vijayasekaran S, Shen W, Lou X, Dalton PD (1998) Biodegradation in vitro and retention in the rabbit eye of crosslinked poly(1-vinyl-2-pyrrolidinone) hydrogel as a vitreous substitute. J Biomed Mater Res 39(4):650–659, Epub 1998/03/10
47. Colthurst MJ, Williams RL, Hiscott PS, Grierson I (2000) Biomaterials used in the posterior segment of the eye. Biomaterials 21(7):649–665, Epub 2000/03/11
48. Niu G, Yang Y, Zhang H, Yang J, Song L, Kashima M et al (2009) Synthesis and characterization of acrylamide/N-vinylpyrrolidone copolymer with pendent thiol groups for ophthalmic applications. Acta Biomater 5(4):1056–1063, Epub 2008/12/17
49. Hacker MC, Haesslein A, Ueda H, Foster WJ, Garcia CA, Ammon DM et al (2009) Biodegradable fumarate-based drug-delivery systems for ophthalmic applications. J Biomed Mater Res A 88(4):976–989, Epub 2008/04/04

50. Yasukawa T, Ogura Y, Kimura H, Sakurai E, Tabata Y (2006) Drug delivery from ocular implants. Expert Opin Drug Deliv 3(2):261–273, Epub 2006/03/02
51. Vega E, Gamisans F, Garcia ML, Chauvet A, Lacoulonche F, Egea MA (2008) PLGA nanospheres for the ocular delivery of flurbiprofen: drug release and interactions. J Pharm Sci 97(12):5306–5317, Epub 2008/04/22
52. Zentner GM, Rathi R, Shih C, McRea JC, Seo MH, Oh H et al (2001) Biodegradable block copolymers for delivery of proteins and water-insoluble drugs. J Control Release 72(1–3):203–215, Epub 2001/06/08
53. Duvvuri S, Janoria KG, Mitra AK (2005) Development of a novel formulation containing poly(d, l-lactide-co-glycolide) microspheres dispersed in PLGA-PEG-PLGA gel for sustained delivery of ganciclovir. J Control Release 108(2–3):282–293, Epub 2005/10/19
54. Duvvuri S, Janoria KG, Pal D, Mitra AK (2007) Controlled delivery of ganciclovir to the retina with drug-loaded Poly(d, L-lactide-co-glycolide) (PLGA) microspheres dispersed in PLGA-PEG-PLGA Gel: a novel intravitreal delivery system for the treatment of cytomegalovirus retinitis. J Ocul Pharm Ther 23(3):264–274, Epub 2007/06/27
55. Sinha VR, Bansal K, Kaushik R, Kumria R, Trehan A (2004) Poly-epsilon-caprolactone microspheres and nanospheres: an overview. Int J Pharm 278(1):1–23, Epub 2004/05/26
56. Fialho SL, Behar-Cohen F, Silva-Cunha A (2008) Dexamethasone-loaded poly(epsilon-caprolactone) intravitreal implants: a pilot study. Eur J Pharm Biopharm 68(3):637–646, Epub 2007/09/14
57. Yin H, Gong C, Shi S, Liu X, Wei Y, Qian Z (2010) Toxicity evaluation of biodegradable and thermosensitive PEG-PCL-PEG hydrogel as a potential in situ sustained ophthalmic drug delivery system. J Biomed Mater Res B Appl Biomater 92(1):129–137, Epub 2009/10/06
58. Adjadj E, Roy S, Zimmermann C, Shaarawy T, Flammer J, Mermoud A et al (2006) Dosage et cinetique de liberation de mitomycine C d'un implant de collagene utilise comme moyen d'administration lors d'une chirurgie filtrante chez le lapin. J Fr Ophtalmol 29(9):1042–1046, Epub 2006/11/23. Dosage and kinetics of MMC release of a collagen implant used as a delivery device in glaucoma surgery in the rabbit eye
59. Lai J-Y, Hsieh A-C (2012) A gelatin-g-poly(N-isopropylacrylamide) biodegradable in situ gelling delivery system for the intracameral administration of pilocarpine. Biomaterials 33(7):2372–2387
60. Bonferoni MC, Chetoni P, Giunchedi P, Rossi S, Ferrari F, Burgalassi S et al (2004) Carrageenan-gelatin mucoadhesive systems for ion-exchange based ophthalmic delivery: in vitro and preliminary in vivo studies. Eur J Pharm Biopharm 57(3):465–472, Epub 2004/04/20
61. Chhonker YS, Prasad YD, Chandasana H, Vishvkarma A, Mitra K, Shukla PK et al (2014) Amphotericin-B entrapped lecithin/chitosan nanoparticles for prolonged ocular application. Int J Biol Macromol 72C:1451–1458, Epub 2014/12/03
62. Wang F, Chen L, Zhang D, Jiang S, Shi K, Huang Y et al (2014) Methazolamide-loaded solid lipid nanoparticles modified with low-molecular weight chitosan for the treatment of glaucoma: vitro and vivo study. J Drug Target 22(9):849–858, Epub 2014/07/22
63. Yasukawa T, Kimura H, Kunou N, Miyamoto H, Honda Y, Ogura Y, Ikada Y (2000) Biodegradable scleral implant for intravitreal controlled release of ganciclovir. Graefes Arch Clin Exp Ophthalmol 238(2):186–190
64. Fernandes-Cunha GM, Gouvea DR, Fulgencio GD, Rezende CM, da Silva GR, Bretas JM et al (2014) Development of a method to quantify clindamycin in vitreous humor of rabbits' eyes by UPLC-MS/MS: application to a comparative pharmacokinetic study and in vivo ocular biocompatibility evaluation. J Pharm Biomed Anal 102:346–352, Epub 2014/12/03
65. Gavini E, Chetoni P, Cossu M, Alvarez MG, Saettone MF, Giunchedi P (2004) PLGA microspheres for the ocular delivery of a peptide drug, vancomycin using emulsification/spray-drying as the preparation method: in vitro/in vivo studies. Eur J Pharm Biopharm 57(2):207–212, Epub 2004/03/17
66. Meng Y, Sun S, Li J, Nan K, Lan B, Jin Y et al (2014) Sustained release of triamcinolone acetonide from an episcleral plaque of multilayered poly-ε-caprolactone matrix. Acta Biomater 10(1):126–133

Chapter 8
Formulation Approaches for Ocular Drug Delivery

Vivek S. Dave

Abstract This chapter aims to provide the readers a comprehensive review of the current trends and approaches used in the development of ocular drug delivery systems. After the introduction to the topic, the chapter begins with a brief description of the common anatomical and physiological factors that are barriers to ocular drug transport and presents a persistent challenge to the development of formulations targeted to ophthalmic tissues. Further, traditional as well as relatively newer approaches used in the development of ocular formulations are discussed. Approaches such as viscosity modifiers, penetration enhancers, liposomal delivery, niosomes, microemulsions, prodrugs, and nanotechnology-derived approaches are extensively reviewed. These approaches are described with respect to the materials, formulation development, formulation performance, relative advantages and drawbacks, and recent advances in the technology. Additionally, the design approaches for the modified ocular drug delivery such as ocular implants, inserts, lenses, and ocular shields are described with respect to the formulation development. Finally, ocular drug delivery systems that are relatively newer or employed less frequently such as intravitreal injections, ocular iontophoresis, etc. are discussed. All approaches described are supported with relevant examples and up-to-date literature reports.

Keywords Ocular drug delivery • Formulation approaches • Nanotechnology • Implants • Inserts

8.1 Introduction

The main goal of ocular therapeutics with respect to drug delivery is to achieve an optimal bioavailability of an active pharmaceutical ingredient in specific ocular tissues. Unique anatomical and physiological barriers exist, which makes this goal

V.S. Dave (✉)
Pharmaceutical Sciences, St. John Fisher College, Wegmans School of Pharmacy, Rochester, NY, USA
e-mail: viveksdave@gmail.com

© Springer International Publishing Switzerland 2016

Y. Pathak et al. (eds.), *Nano-Biomaterials For Ophthalmic Drug Delivery*,
DOI 10.1007/978-3-319-29346-2_8

challenging. In recent years, the focus of research has been to understand the extent of influence these barriers have in ocular drug delivery and to explore formulation strategies that could potentially circumvent these barriers. This chapter aims to provide a comprehensive overview of the current formulation approaches researched to optimize ocular delivery of drugs.

8.2 Ocular Drug Delivery: Common Challenges

The predominant challenge in designing a drug delivery system for ophthalmic use is to attain and maintain an optimal therapeutic concentration at the site of action [1]. Figure 8.1 shows a schematic representation of the predominant routes of ocular drug absorption, distribution, metabolism, and elimination. Several physiological and precorneal factors influence the bioavailability of topically administered ophthalmic drugs. These include instilled dose/volume, nasolacrimal drainage, tear turnover, tear evaporation, tear dilution, tear pH, drug protein interaction, drug metabolism, blood-retinal barrier (BRB), influx/efflux transporters, etc. These factors have been broadly categorized into those that influence the delivery to the anterior segment and those that influence the delivery to the posterior segment of the eye [1]. As shown in Figs. 8.2 and 8.3 below, the precorneal loss factors are the main determinants of the bioavailability and the ultimate fate of ocular drugs in the anterior segment of the eye. Moreover, due to the presence of blood-retinal barrier

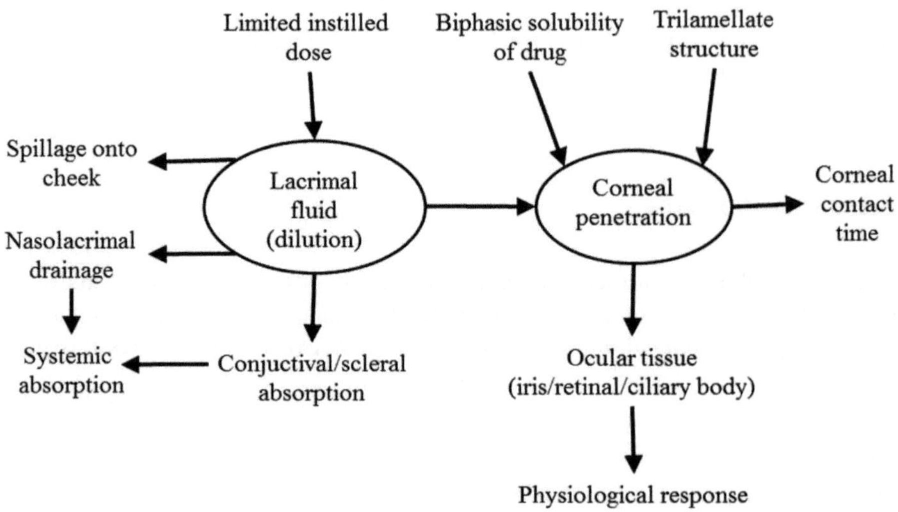

Fig. 8.1 Distribution of drug within ocular tissues after ophthalmic drug delivery (Adopted from Patel et al. [1])

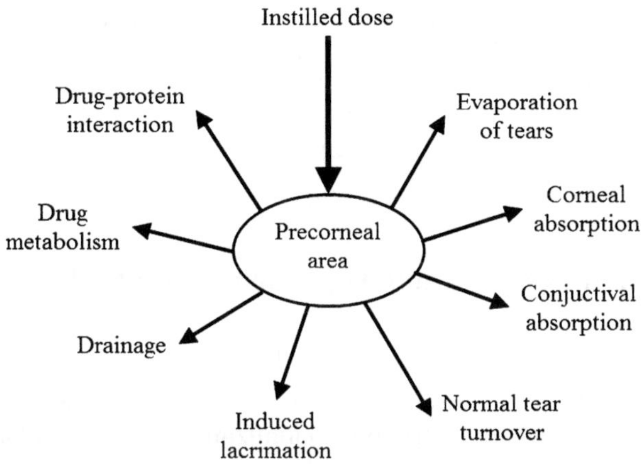

Fig. 8.2 Precorneal factors influencing the bioavailability of topical ocular drugs (Adopted from Patel et al. [1])

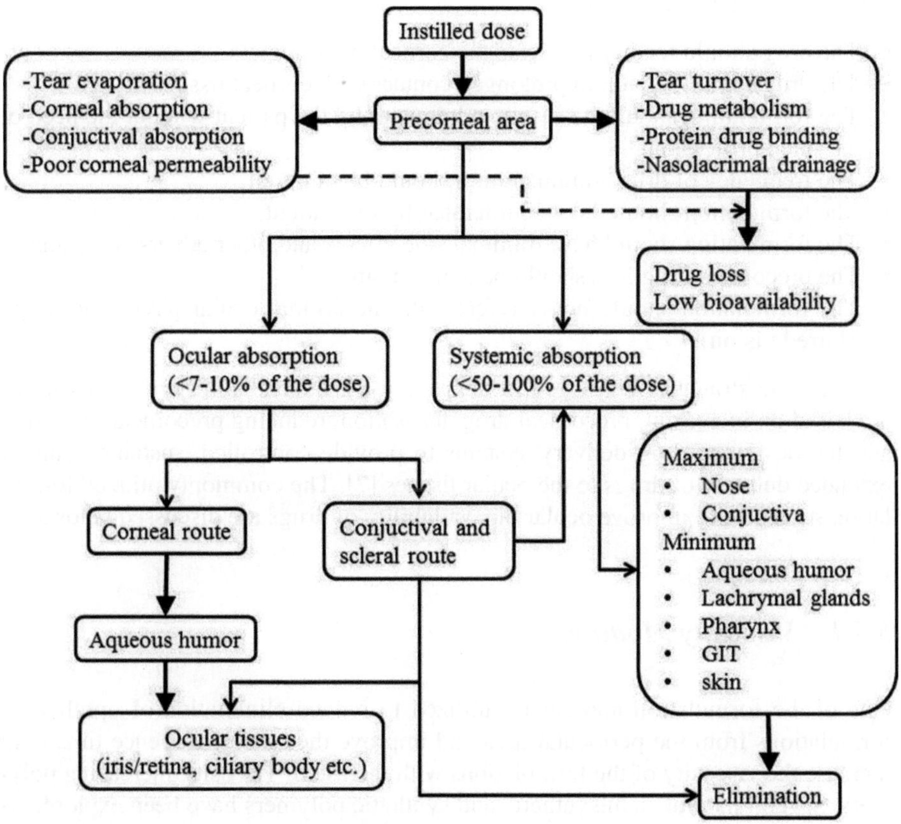

Fig. 8.3 Fate of a drug delivered via ophthalmic route (Adopted from Patel et al. [1])

(BRB), most topically administered ocular drugs are unable to reach the posterior segment of the eye [2]. In addition, the tight junctions of BRB prevent the crossing of systemically administered drugs into the retina [3]. It has been reported that due to these anatomical and physiological constraints, only a fraction (≤1 %) of the administered dose is effectively absorbed [4].

8.3 Common Formulation Approaches for Ocular Drug Delivery

Considering the inherent constraints of the anatomy and physiology of the eye mentioned above and elsewhere in the book, optimizing ocular bioavailability via topically applied drugs is a formulation challenge. The main goal of a formulation scientist is to improve the ocular bioavailability of drugs from 1 % to 3 % (currently achieved with conventional formulations) to at least 15–20 % (therapeutically relevant) [5]. An ideal drug delivery system for ophthalmic application is expected to have following characteristics [6]:

- The drug should readily penetrate the corneal membrane.
- The drug should remain in prolonged contact with corneal tissue.
- The formulation should be easy to administer by the patient without the need of a trained professional.
- The frequency of drug administration should be minimal.
- The formulation should be well tolerated by the patient.
- The formulation should have minimal side effects and other adverse reactions.
- The precorneal drug loss should be at minimum.
- The formulation should not interfere with the normal ocular physiology (e.g., blurred vision).

Thus, the strategies employed in improving the bioavailability of ocular drugs are based on increasing precorneal drug absorption, reducing precorneal drug loss, and the design of drug delivery systems to provide controlled, sustained, and/or extended delivery of drugs to the ocular tissues [7]. The commonly utilized formulation strategies to improve ocular bioavailability of drugs are discussed below.

8.3.1 Viscosity Modifiers

One of the formulation approaches utilized to reduce elimination of ophthalmic formulations from the periocular area and improve the ocular residence time is to increase the viscosity of the formulations with the use of viscosity-increasing polymers. Several natural, semisynthetic and synthetic polymers have been explored as viscosity modifiers in ophthalmic formulations. These include xanthan gum, polyvinyl alcohol (PVA), polyvinylpyrrolidone (PVP), methyl cellulose, hydroxyethyl cellulose, hydroxypropyl methyl cellulose, hydroxyl propyl cellulose, etc. Meseguer et al. compared, in healthy human volunteers, the corneal contact time of various formulations, each containing one viscosity enhancer including gellan gum, xanthan gum, hydroxyethyl cellulose, hydroxypropyl methylcellulose, or polyvinyl alcohol. They observed that formulations containing gellan gum or xanthan gum were superior in delaying the clearance of the instilled solution by the tear flow [8]. Zaki et al. evaluated the precorneal drainage of radio-labeled solutions containing different concentrations of PVA or HPMC in rabbits and humans using gamma scintigraphy. They noted that reducing precorneal drainage in humans required inclusion of higher concentration of polymers than in rabbits [9]. Saettone et al. compared the effects of carboxymethylcellulose (CMC), low-molecular-weight hydroxypropylcellulose (HPCL), medium-molecular-weight hydroxypropylcellulose (HPCM), poly(vinylpyrrolidone) (PVP), and polyvinyl alcohol (PVA) at a constant viscosity of 70 cps on ocular bioavailability of tropicamide solution (0.2 %) in humans and rabbits. They found PVA to be the most effective viscosity modifier for the formulation. They indicated that apart from viscosity modification, bioavailability can be improved by influencing the spreading characteristics and the thickness of the medication layer over the precorneal area. The adhesive properties of PVA and its capacity to increase the thickness of the precorneal tear film have also been attributed to its superior functionality in ocular formulations [10].

Another aspect regarding ophthalmic formulations containing viscosity modifiers is the in vitro evaluation of the drug release pattern from these preparations. Several approaches such as diffusion cells, USP apparatus I and II, and dialysis tubes have been used to evaluate the drug release behavior of viscous ophthalmic preparations. Diffusion cells have been commonly employed to study in vitro drug release from viscous and semisolid formulations [11]. Rozier et al. compared the in vitro release rates of timolol from HEC and Gelrite® gel using diffusion cell [12]. Hui et al. used the USP rotating basket method to evaluate progesterone release from a viscous formulation [13]. Stevens et al. described a novel in vitro method to measure time-release profiles of drugs from various ophthalmic dosage forms [14]. This method based on sample flow cells claimed to reproduce the small volume and slow the exchange rate of the human precorneal tear reservoir in vitro. Cohen et al. used dialysis tubing made of cellulose ester membrane to evaluate the release rate of pilocarpine from eyedrops [15].

8.3.2 Penetration Enhancers

A common approach to improve ocular bioavailability of drugs is to increase the permeability of corneal epithelial membrane [16–20]. The stratified corneal epithelium cell layer is known to be a highly resistive tissue that serves as a rate-limiting barrier to the uptake of drugs [21]. A class of substances commonly known as "permeation enhancers" increase the corneal uptake by temporarily modifying the integrity of corneal epithelium [22]. Agents acting as permeation enhancers belong to a group of chemicals known as "surfactants." Several agents evaluated for their ability to enhance transcorneal permeability include cetylpyridinium chloride, lasalocid, benzalkonium chloride, Tween 20, parabens, saponins, Brij® 35, Brij® 78, Brij® 98, bile salts and bile acids, capric acid, azone, fusidic acid, hexamethylene lauramide, hexamethylene octamide, and decylmethyl sulfoxide [23–31]. Due to their surface acting behavior, penetration enhancers have also known to decrease the drop size of ophthalmic liquid preparations [32]. Surfactants typically act by getting incorporated into the lipid bilayer of the corneal epithelium where they form mixed micelles resulting in the removal of cell membrane phospholipids and membrane solubilization. By inducing a transient modification in the structure and permeability of the membrane, they increase diffusion of molecules across corneal epithelium due to their mucolytic properties. These agents are also known to create intercellular channels or pores thus facilitating the transport of molecules across the membrane [28].

Apart from surfactants, other compounds such as preservatives and chelating agents have been explored for their permeation-enhancing capabilities [33, 34]. Higaki et al. demonstrated a 2.5-fold enhancement of in vitro corneal transport of an antiglaucoma agent with the addition of benzalkonium chloride (0.005 %) [25]. Pfister et al. showed the disruptive behavior of benzalkonium chloride (0.01 %) on corneal epithelium using scanning electron microscopy [35]. Chelating agents such as EDTA when used as permeation enhancers facilitate paracellular transport by depleting Ca^{2+} and thus disrupting tight junctions between epithelial cells [36, 37]. Grass et al. demonstrated a significant increase in the ocular absorption of topically applied glycerol and cromolyn sodium in the presence of 0.5 % EDTA [38]. Several researchers reported an increased transcorneal penetration of drugs with azone, a transdermal absorption promoter. Newton et al. showed that inclusion of azone resulted in an increased delivery of cyclosporine, an immunosuppesant [39]. Tang-Liu et al. demonstrated selective permeation-enhancing capabilities of azone for hydrophilic drugs [31]. The use of permeation enhancers is, thus, a highly effective approach to improve ocular bioavailability. However, due to the inherent mechanism of permeation enhancers being membrane disruptors, the extent of their use in ophthalmic formulation should precede great precautions. Extensive use of permeation enhancers may lead to their accumulation in the cornea and/or may cause potentially adverse effects or unknown toxicities. Green et al. found that anionic and cationic detergents are retained by the corneal epithelium for several days [40]. Grass et al. showed that EDTA in the ophthalmic formulations reached the iris-ciliary body in high concentrations and altered the permeability of the blood vessels in the uveal tract, indirectly speeding drug removal from the aqueous humor [41]. Ismail et al. demonstrated the irritating, discomforting, and/or toxic effects of azone (0.2 %) in rabbit eyes [42]. Chiou et al. studied the efficacy of seven compounds saponin, fusidic acid, BL-9, EDTA, glycocholate, decamethonium, and Tween 20 in enhancing insulin absorption [23]. They observed that saponins caused eye irritation at 0.5 % level and recommended less than 1 % of saponin to be used for enhancing peptide absorption through the eyes. Other researchers have also observed the irritating and cell membrane-damaging effects of bile salts and other enhancers [34, 43, 44].

8.3.3 Liposomes

Liposomes are generally defined as lipid-based microscopic vesicles with sizes ranging from 50 nm to over 100 μm. Liposomes may possess a negative, positive, or a neutral surface charge depending on their composition. In addition, depending on their size, liposomes may be categorized as small unilamellar vesicles (SUV, 10–100 nm), large unilamellar vesicles (LUV, 100–3000 nm), or, in the presence of more than one bilayer, multilamellar vesicles (MLV) [7]. Among the earliest studies reporting the applicability of a lipid microsphere as ophthalmic drug delivery system was carried out by Komatsu et al. in 1988 where they examined ophthalmic preparations using a lipid microsphere containing hydrocortisone 17-butyrate 21-propionate (HBP) [45]. In recent decades, liposomes have been explored as microencapsulating agents for several hydrophilic and lipophilic drugs [46, 47]. Liposome-based ocular delivery systems have shown to enhance corneal absorption of drugs such as idoxuridine compared to simple solution-based formulations in laboratory animals [48, 49]. Their activity is largely attributed to their ability to achieve intimate contact with the corneal and conjunctival membranes. Shaeffer et al. carried out a comparative in vitro liposome-corneal interaction study using SUV and MLV with neutral, positive, or negative surface charge [50]. They observed that positively charged liposomes were more readily taken up by the cornea compared to the ones with negative or neutral charge. This is generally attributed to the fact that the corneal epithelium is thinly coated with negatively charged mucin to which the positive surface charge of the liposomes may absorb more strongly [51]. This was confirmed by Al-Muhammed et al. who studied in vitro and in vivo efficacy of dexamethasone sodium phosphate liposomes as an ocular drug delivery system. They found that positively charged liposomal formulations achieved the highest bioavailability in the anterior segment of the eye [52]. Similarly El-Gazayerly et al. reported a significantly higher bioavailability of acetazolamide from positively charged liposomal preparations [53]. Apart from improving ocular bioavailability, liposomes have also been explored for their potential benefits in reducing ophthalmic organ graft rejections and improving graft survival rates [54]. The ease of preparation and the ability to customize the physical characteristics lend itself to liposomes being an ideal ocular drug delivery system. However, due to limited drug-loading capacity, chemical and oxidative instability, and difficulties in maintaining sterility, liposomal preparations are yet to achieve a broad acceptance [55]. Liposomes have additionally been explored for their potential sustained-release properties when used synergistically with other technologies. Durrani et al. showed that coating of reverse phase evaporation vesicles (REVs) by Carbopol 1342 resulted in a larger area under the miotic intensity curve [56] and a longer duration of action compared to uncoated REVs [57]. Pleyer et al. investigated the ability of liposomes to deliver the immunosuppressive agent cyclosporine A (CsA) to the cornea, anterior sclera, aqueous humor, and vitreous in rabbit eyes [58]. It was observed that the combination of liposome-encapsulated Cyclosporin A with collagen shields indicated improved corneal bioavailability and sustained-release properties of such formulations.

8.3.4 Niosomes

Another vesicular drug delivery systems used in ophthalmics include niosomes [59]. Niosomes are formed by self-assembly when a mixture of cholesterol and a single-alkyl chain, nonionic surfactant is added to aqueous media [60]. Niosomes can encapsulate both hydrophilic and lipophilic drugs either in an aqueous layer or in a vesicular membrane [61]. Niosomes offer several advantages such as chemical stability, lower toxicity, easier handling, improved bioavailability, etc. compared to liposomal delivery systems [62]. In addition niosomes are biodegradable, biocompatible, and non-immunogenic. Several studies have reported the utility of niosomes for the ocular delivery of a variety of drugs [63, 64]. A small group of active drug molecules with a free carboxyl group or active hydrogen can be esterified to the hydroxyl group of a lipid molecule to generate an amphiphilic prodrug. Such prodrugs are then diluted with water and are commonly known as *pharmacosomes* [7]. Pharmacosomes are currently being explored as drug delivery systems in ophthalmics [65].

8.3.5 Microemulsions

Microemulsions are dispersions of oil and water that are stabilized with the use of surfactants or co-surfactants. These systems are transparent, are thermodynamically stable, and are known to increase corneal membrane permeability [66]. Some of the critical parameters that influence the stability and drug-incorporating ability of microemulsions include selection of aqueous phase, organic phase, surfactants, and co-surfactants [67]. Studies have reported several aspects of microemulsions pertaining to ocular drug delivery. These include improving drug solubility, increasing corneal membrane permeability, providing sustained release of drugs, increasing hydrophilic and lipophilic drug loading, and increasing corneal residence time. Hasse et al. designed an oil-in-water system consisting of pilocarpine using lecithin, propylene glycol, PEG 200 as surfactant/co-surfactant, and isopropyl myristate as the oil phase [68]. Fialho et al. compared a dexamethasone (0.1 %) in a microemulsion system to a conventional eyedrop preparation. It was observed that microemulsion-based system had a twofold bioavailability and faster penetration rate compared to conventional eye drops [66]. The study also demonstrated that microemulsions were stable for at least 3 months and did not cause eye irritation in the tested animals. Studies such as these reveal the possibility of reformulating other ophthalmic drugs in microemulsion in order to improve their ocular bioavailability. Klang et al. showed that positively charged microemulsion containing indomethacin had increased bioavailability compared to either the commercially available solution or a negatively charged microemulsion [69]. Microemulsions can also be developed for sustained delivery of drugs, thus resulting in a decrease in dosing frequency. Chan et al. investigated the release and ocular retention of pilocarpine from phase transitioned water-in-oil microemulsions containing nonionic surfactants [70]. They found that changing water content resulted in a change in the rheological properties of the formulations. Increased viscosity results in retaining the formulation on the cornea for an extended period, thus, requiring lower dosing frequency. Kapoor et al. demonstrated an extended release (~20 days) of cyclosporine A from a microemulsion-filled poly(2-hydroxyethyl methacrylate) (P-HEMA) hydrogel contact lens [71]. Despite several advantages with the use of microemulsions as ocular drug delivery systems, these systems have not gained popularity due to several drawbacks. These drawbacks include limited choice of surfactants/co-surfactants as well as potential toxicities associated with their use in high concentrations [7].

8.4 Nanotechnology-Derived Formulations

In recent years, one of the most aggressively researched approaches for ocular drug delivery is the nanotechnology-based formulations. Due to the perceived advantage of small particle size (≦100 μm) in increasing transmembrane permeability, nanotechnology-based products are assumed to provide a significant improvement in the ocular bioavailability. Nanosuspensions and other nanomicellar formulations are among the widely studied approaches for ocular drug delivery.

8.4.1 Nanosuspensions

Nanosuspensions are submicron (particle size range from 10 to 1000 nm) colloidal systems consisting of drug entrapped, adsorbed, or encapsulated with an inert mucoadhesive polymer such as polyacrylic acid (PAA) or its derivatives [72]. Nanosuspensions offer several advantages such as improving drug solubility, improving bioavailability, and, unlike microemulsions and liposomes, negligible irritant properties. Pignatello et al. developed flurbiprofen-loaded polymeric nanoparticle suspensions using Eudragit® RS100R and RL100R polymer resins [73]. They observed an inhibition of the miotic response to the surgical trauma with flurbiprofen-loaded nanosuspension comparable to a control eyedrop formulation. They also found that the developed nanosuspension was devoid of any toxicity on ocular tissues. Adibkia et al. studied the anti-inflammatory effects of methylprednisolone acetate nanosuspensions in rabbits with endotoxin-induced uveitis [74]. Preparing nanoformulations of methylprednisolone acetate (MPA) using a copolymer of poly(ethylacrylate, methyl methacrylate, and chlorotrimethyl-ammonioethyl methacrylate), the authors showed that localized, controlled ocular delivery of methylprednisolone is achievable with nanosuspensions. In a similar study, Adibkia et al. showed significantly improved anti-inflammatory effects of a piroxicam nanosuspension compared to microsuspension in rabbits [75]. Nanosuspensions have also been explored for the ocular delivery of poorly water-soluble drugs such as glucocorticoids. Kassem et al. formulated hydrocortisone, prednisolone, and dexamethasone into high viscosity nanosuspensions [76]. These formulations showed extended drug release and a significantly improved bioavailability. Nanosuspensions are also reported to enhance the stability of drugs in a formulation. Pignatello et al. prepared cloricromene (AD6)-loaded polymeric nanoparticle suspensions using Eudragit® RS100 and RL100 to improve the stability of cloricromene in ophthalmic formulations and its drug availability at the ocular level [77]. Their study showed that nanosuspensions prepared with saline solution and no or lower amounts of surfactant (Tween 80) showed an enhanced stability of the drug for several months, compared to an AD6 aqueous solution.

8.4.2 Nanoparticles/Nanospheres

Nanoparticles are defined as particles with a diameter of less than 1000 nm, consisting of various biodegradable and non-biodegradable polymers, lipids, phospholipids, or metals [78]. They are further subclassified as nanospheres or nanocapsules depending on whether the drug is uniformly dispersed within a polymeric material (nanoparticles) or coated with a polymeric material (nanocapsules). The uptake and distribution of nanoparticles in the ocular tissues depend on the size of the nanoparticles. Sakurai et al. investigated the kinetics of fluorescein nanospheres (50, 200 and 2000 nm) following the intravitreal injection in rabbits [79]. The study found that the larger nanospheres distributed mainly in the vitreous cavity and trabecular meshwork. In addition, the nanospheres smaller than 200 nm are able to penetrate the retinal cells, thus, emphasizing the importance of particle size in ocular tissue distribution. Bourges et al. studied the kinetics of polylactide (PLA) nanoparticle [80] localization within the intraocular tissues and evaluated their potential to release encapsulated material [81]. They observed a transretinal movement of the nanoparticles, with a preferential localization in the retinal pigment epithelium (RPE) cells. This supported the idea that with the use of nanoparticles as ocular drug delivery systems, specific targeting of ocular tissues is feasible.

The release of drugs from nanoparticles depends on the surface charge interaction of the drug and the polymer. Li et al. demonstrated the incomplete release of progesterone from the polybutyl cyanoacrylate nanospheres due to the surface charge interaction between the drug and the polymer [82]. Cavalli et al. evaluated solid-lipid nanoparticles (SLN) as carriers for topical ocular delivery of tobramycin [83]. The study demonstrated an extended release (up to 6 h) of tobramycin from tobramycin-loaded solid-lipid nanoparticles. The study also showed that when compared with an equal dose of tobramycin from standard commercial eye drops, tobramycin-loaded solid-lipid nanoparticles produced a significantly higher bioavailability in the aqueous humor. In addition to formulation variables, physiological factors may also influence the bioavailability and distribution of nanoparticles in the ocular tissues. Amrite et al. studied the ocular disposition and distribution of nanoparticles of various sizes (20 and 200 nm) in rats [84]. They observed a significantly low transport of 20 nm particles across the sclera and sclera-choroid-RPE due to rapid clearance via periocular circulation. It was also observed that following periocular administration, a higher concentration of particles was seen in the ocular tissues of dead animals. This was attributed to the lack of physiological barrier, i.e., periocular clearance.

Several naturally occurring biodegradable polymeric materials have been evaluated for their applicability as carriers for the nanoparticle-based ocular drug delivery. Irache et al. prepared and evaluated albumin nanoparticles as carriers for a positively charged drug ganciclovir and a negatively charged phosphodiester oligonucleotide analog of fomivirsen [85]. Calvo et al. designed nanoparticles and nanocapsules made of poly-epsilon-caprolactone and submicron emulsions and investigated their capacity for increasing the corneal penetration of indomethacin [86]. It was found that the in vitro corneal penetration of the indomethacin from the nanoparticles, nanocapsules, and nanoemulsion was more than threefold that of the commercial eye drops. De Campos et al. observed that the PEG coating enhanced the permeation of the nanocapsules across the epithelium of the ocular mucosa, whereas the chitosan coating favors the retention of the nanocapsules in the superficial layers of the epithelium [87]. Xu et al. studied the inhibitory efficacy of intravitreal dexamethasone acetate (DA) loaded with (D, L-lactide-co-glycolide) (PLGA) nanoparticles on choroidal neovascularization (CNV) in a laser-induced rat model and evaluated its mode of drug release [88]. The results showed that about half of the dexamethasone acetate was released over two weeks followed by a controlled release over 40 days. Motwani et al. investigated mucoadhesive chitosan-sodium alginate nanoparticles as a carrier for the extended topical ocular delivery of gatifloxacin [89]. They found that gatifloxacin was rapidly released from the nanoparticles in the first hour and was subsequently released in controlled manner in the following 24 h period, indicating a possible reduction in dosing frequency. Arnedo et al. evaluated the potential of albumin nanoparticles as a delivery system for antisense oligonucleotides (ODN) adsorbed or incorporated in the nanoparticles [90]. The in vivo studies showed that ODNs entrapped in the nanoparticles revealed a higher antiviral activity compared to ODNs adsorbed on the surface. This was attributed to the protection of nanoparticle-entrapped ODNs from enzymatic activity. Nanoparticles, thus, have emerged in recent years as potential ocular drug delivery systems.

8.4.3 Prodrugs

Prodrugs are broadly defined as biologically inactive compounds that can be metabolized in the body to produce a pharmacologically active drug. Prodrug approach has been explored for improving the therapeutic efficacy of several drug molecules [91]. This approach offers several advantages for efficient drug delivery including improved solubility, stability, bioavailability, as well as resisting efflux mechanisms. Prodrug approaches have also been studied for ocular drug delivery. Gunda et al. prepared dipeptide monoester prodrugs of ganciclovir (GCV) and evaluated their corneal absorption against L-valine-GCV and GCV [92]. Additionally they evaluated the pharmacokinetics of these prodrugs in aqueous humor. The study found that dipeptide (glycine-valine and tyrosine-valine) monoester prodrugs of GCV showed a superior corneal absorption and bioavailability compared to the active drug alone. In a similar study, Majumdar et al. evaluated a series of dipeptide monoester ganciclovir (GCV) prodrugs with the aim of improving ocular bioavailability of GCV from topical ophthalmic solutions [93]. The results showed that one of the prodrugs of ganciclovir, valine-valine-GCV, demonstrated excellent corneal permeability, chemical stability, high aqueous solubility, and a significant in vivo antiviral activity against the HSV-1. Kansara et al. showed that the dipeptide monoester prodrugs of ganciclovir, valine-GCV, tyrosine-valine-GCV, and glycine-valine-GCV, had significantly higher transcorneal permeability in rabbits compared to GCV alone [94]. This was attributed to the interaction of prodrugs with the oligopeptide transporter present on the retinal pigment epithelium. Janoria et al. developed sustained-release microsphere formulations of ganciclovir and its lipophilic prodrug (GCV monobutyrate) and evaluated their in vitro release behavior [95]. The results demonstrated a higher efficiency of drug entrapment for the drug and a longer residence time at the site of administration with the prodrug. Using a similar strategy, Dias et al. prepared and evaluated L-valyl ester of acyclovir (ACV) and found that the prodrug revealed a higher corneal permeability due to its interaction with oligopeptide transporter [96]. Anand et al. prepared a dipeptide prodrug of acyclovir (ACV), valine-valine-ACV, and evaluated it in vivo as a potential drug candidate for improving antiviral efficacy against herpetic epithelial and stromal keratitis [97]. The results showed that the developed prodrug was highly water soluble, less cytotoxic, and possessed improved bioavailability excellent in vivo activity against HSV-1. Katragadda et al. investigated the modulation of efflux mechanisms using transporter-targeted prodrug derivatization of quinidine and demonstrated that the prodrug approach can effectively overcome P-gp-mediated efflux of drug across the cornea [98].

Doukas et al. investigated the effects of the topical delivery of TG100801, an inactive prodrug of TG100572 (an inhibitor of Src kinases and selected receptor tyrosine kinases) in a murine model of laser-induced choroidal neovascularization (CNV) [99]. It was observed that topical delivery TG100801 significantly suppressed laser-induced CNV in mice and avoided systemic toxicity observed with the parent drug. Al-Ghananeem et al. in a review justified the development of prodrugs for ocular drug delivery based on the role of phase I and phase II ocular metabolic activities [100]. Lallemand et al. prepared UNIL088, a water-soluble prodrug of cyclosporine A (CsA), for topical ocular delivery and elucidated the conversion mechanisms by which UNIL088 generates CsA [101]. Juntunen et al. improved the water solubility of lipophilic cannabinoids (arachidonylethanolamide, R-methanandamide, and noladin ether) by making their phosphate ester prodrugs and demonstrated a significant increase in the corneal permeability of these cannabinoids in the treatment of glaucoma [102]. Nambu et al. studied the effect of a prodrug of combretastatin A-4 (CA-4) and combretastatin A-4-phosphate (CA-4-P) in mice models of ocular neovascularization [103]. The results showed a potential for CA-4-P in both prevention and treatment of ocular neovascularization. In a similar study, Takahashi et al. demonstrated that topical nepafenac, a prodrug of amfenac, inhibited choroidal neovascularization and ischemia-induced retinal neovascularization in murine models [104]. Nepafenac is known to readily penetrate the cornea and get metabolized to amfenac, a potent cyclooxygenase [105]-1 and COX-2 inhibitor.

8.5 Systems for Modified Ocular Drug Delivery

Most traditional ocular drug delivery systems are "immediate-release" type of formulations. These include solutions, suspensions, emulsions, and various forms of semisolid preparations. Due to the inherent anatomical constraints of the site of administration, these dosage forms require multiple and frequent dosing. In recent years the focus has shifted toward development of dosage forms that could provide a sustained, extended, or controlled delivery of drugs to the ocular tissues. Several such systems that provide a modified ocular delivery of drug are currently being developed and marketed. These include implants, inserts, lenses, etc. developed using one or more biodegradable or non-biodegradable materials.

8.5.1 Ocular Implants

Ocular implants are devices prepared using biodegradable or non-biodegradable materials and need to be surgically inserted in the vitreal compartment. Implants provide several advantages over conventional ocular drug delivery systems in that they reduce potential for systemic side effects, avoid restrictive blood-ocular barrier, and provide a sustained release of medications. This makes them immensely useful for use in chronic diseases such as uveitis, cytomegalovirus retinitis, and age-related macular degeneration [106].

Biodegradable poly(lactic acid) and poly(lactide-co-glycolide) polymers are among the most widely studied materials for their applications in implantable ocular drug delivery [107]. Hashizoe et al. designed a scleral plug made of poly(lactic-glycolic acid) containing 1 % doxorubicin and evaluated the drug release into the vitreous after being implanted and fixed at the pars plana [108]. The in vivo studies demonstrated that the device maintained the therapeutic drug concentration in the vitreous humor over 4 weeks. Kimura et al. similarly prepared ocular implantable devices based on poly(dl-lactic acid) and evaluated the feasibility of using these devices to deliver drugs into the vitreous humor [109]. The results showed an extended release (25–45 days) and an extended bioavailability (17–28 days) of drug in the vitreous from these devices. Kunou et al. prepared and evaluated biodegradable scleral implants made of poly(dl-lactide) and poly(dl-lactide-co-glycolide) for the intraocular delivery of ganciclovir in the treatment of cytomegalovirus retinitis in rabbits [110]. They reported a 5-month, controlled delivery of ganciclovir from devices. Sanborn et al. prepared a surgically implantable ganciclovir ocular device based on polyvinyl alcohol (PVA) and ethylene vinyl acetate (EVA) and evaluated them in patients with acquired immunodeficiency syndrome (AIDS) and associated cytomegalovirus (CMV) retinitis [111]. They observed a controlled release of ganciclovir from this device over a period of 4–5 months resulting in complete treatment of CMV retinitis.

8.5.1.1 Insoluble Inserts

Insoluble inserts are ocular drug delivery systems prepared using non-biodegradable polymeric materials. Depending on the shape and size, these devices are meant to be placed in the lower fornix, the upper fornix, or on the cornea to deliver drugs for topical or systemic effects [7, 112]. Typically, these systems comprise of drug in a liquid or a semisolid form encapsulated in a polymeric system. The common advantages and drawbacks of ocular inserts are well summarized in literature [113–117] and include the following:

- Improve ocular residence time, hence higher bioavailability
- Feasibility of sustained, extended, or controlled release of medication
- Relative control over dosing accuracy (avoiding product loss seen with conventional dosage forms)
- Improved local action (avoiding systemic drainage via the nasolacrimal duct and nasal mucosa)
- Lower dosing frequency and, hence, improved patient compliance
- Feasibility of drug targeting to specific ocular tissues
- Reduced need for preservatives (avoiding sensitivity reactions associated with preservatives)
- Improved overall stability of the formulation

In addition to the abovementioned advantages, ocular inserts are not devoid of drawbacks. The development and application of inserts present several challenges. The most common drawbacks include the following:

- Patient compliance: due to the "extraneous solid" nature of the delivery system, inserts constitute a physical and a psychological challenge to patient acceptance.
- Unrestricted movement: eyeball movements may cause discomfort.
- Product dislodgement: may happen during sleep or washing of eye, face, etc.
- Vision interference: depending on the size, may interfere with normal vision. It may also restrict wearing of contact lenses.
- Application: correct application may require training.

Depending on the design and the mechanism of drug release, inserts are broadly classified as membrane-controlled systems, minidisc, contact lenses, etc.

8.5.1.1.1 Membrane-Controlled Reservoir Inserts

As the name suggests, membrane-controlled reservoir inserts, also known as "ocuserts," are delivery systems wherein a reservoir of drug is incorporated in a release rate-controlling polymeric membrane. Over the years membrane-controlled inserts have proven to be significantly better than conventional ocular drug delivery systems [118, 119]. The Ocusert pilocarpine system, developed by the Alza Corporation (USA), was among the earliest ocular inserts designed to release the drug in a controlled manner over 7 days [120]. This system consists of a mixture of pilocarpine and alginic acid incorporated between layers of ethylene vinyl acetate [121] rate-controlling membrane. Urtti et al. studied the systemic absorption of pilocarpine from matrix-based ophthalmic inserts prepared using hydroxypropylcellulose (HPC) or n-butyl half-ester of poly(methyl vinyl ether/maleic anhydride) (PVM/MA) [122]. The in vitro and in vivo results demonstrated a sustained-release behavior of these matrices. In a similar study, Urtti et al. studied pilocarpine release from cast plasticized hydroxypropylcellulose (HPC) and HPC-polyvinylpyrrolidone (PVP) matrices [123]. The results showed that increased molecular weight and concentration of HPC resulted in the decreased rate of pilocarpine release due to the decreased rate of drug diffusion from the matrix. Vasantha et al. developed and evaluated three types of modified collagen-based pilocarpine nitrate drug delivery systems [124]. The studies showed that depending on the type of modification made on the collagen carrier, the release of pilocarpine could be varied between 5 and 15 days. Attia et al. developed ocular inserts based on Eudragit® and cellulose acetate phthalate films and evaluated the absorption of dexamethasone form these inserts in different eye tissues [125]. The study indicated that film-based ocular inserts can be designed to achieve therapeutic concentrations of drugs in specific targets within the ocular tissues. A patent by Darougar S. describes a silicone elastomer-based insert (Ocufit® SR) to fit the human conjunctival fornix [126]. This system demonstrates a sustained release of drugs from the inserts over a period of 2 weeks. Urtti et al. compared the pharmacokinetics and the intraocular pressure-lowering effects of silicone tubing-based timolol inserts to those of 0.5 % timolol eyedrops in glaucoma patients [127]. The results showed that while the inserts were therapeutically as effective as eyedrops, they afforded a decrease in the systemic absorption and subsequent side effects. Gurtler et al. described several osmotic-controlled inserts for the controlled delivery of drugs to the ocular tissues [112]. Dumitriu et al. demonstrated the usefulness of modified xanthan-based ocular inserts prepared by ionic bonding and gel entrapment for the delivery of pilocarpine, adrenaline, and hydrocortisone for the treatment of glaucoma [128].

Despite the potential benefits of ocular inserts such as controlled delivery, reduction in systemic side effects, targeted delivery, etc., there are limitations to their use, and caution has to be observed when doing so. Kushnick et al. and others have reported potential for pilocarpine toxicity with symptoms such as miosis, retinal detachment, headache, nausea, vomiting, and diarrhea due to the leakage from ocusert [129, 130].

8.5.1.1.2 Minidisc and Soft Contact Lenses

Minidiscs have been described as matrix-based devices developed for controlled delivery of ocular drugs [131]. These are small, disk-shaped delivery systems with a convex-concave shape designed to be placed on the eyeball [7]. Various methacrylic acid-based polymeric materials have been used to fabricate minidiscs. Hitoshi et al. prepared and evaluated ophthalmic inserts of tilisolol based on poly(2-hydroxypropyl methacrylate) and poly(2-hydroxyethyl methacrylate) polymers [131]. The study results indicated that the prepared inserts resulted in a controlled release and an improved ocular bioavailability of tilisolol. Soft contact lenses are similar in design, fabrication, and applicability to minidiscs. These systems are also prepared using non-biodegradable polymers such as poly(2-hydroxyethyl methacrylate). The copolymers of poly(2-hydroxyethyl methacrylate) with polyvinylpyrrolidone (PVP) are reported to provide a controlled release of drugs [7]. In a recent paper, Fazly Bazzaz et al. prepared hydrogel-based soft contact lenses for the ocular delivery of silver nanoparticles and evaluated their antimicrobial activity against *Pseudomonas aeruginosa* and *Staphylococcus aureus* [132]. The delivery of silver nanoparticles exhibited desirable antimicrobial effect for a period of 6 h.

8.5.1.2 Soluble Inserts

As the name suggests, soluble inserts are devices prepared using biodegradable polymeric materials that either dissolve or erode in the ocular tissues after the drug is released. The inherent advantage of soluble inserts is that they are not required to be removed after application. Depending on the nature of material used to prepare soluble inserts and the mechanism of drug delivery, these systems can be broadly categorized as soluble ophthalmic drug inserts (SODI), bioadhesive ophthalmic drug inserts (BODI), and collagen shields.

8.5.1.2.1 Soluble Ophthalmic Drug Inserts (SODI)

Soluble ophthalmic drug inserts (SODI) are sterile thin ocular drug delivery devices prepared from soluble copolymer of acrylamide, N-vinylpyrrolidone, and ethyl acrylate [7]. They are oval shaped and weigh between 10 and 20 mg. The mechanism of drug release from a SODI is fairly simple. A SODI after insertion in the conjunctival sac immediately softens and acquires the shape of the eyeball. It then slowly starts to dissolve while simultaneously releasing the drug. Within a few hours, a SODI completely disappears from the eye. Several researchers have formulated and evaluated the performance of SODI. Maichuk et al. prepared a pilocarpine SODI and compared it with conventional pilocarpine preparations for the treatment of glaucoma [133]. The study concluded that a single application of SODI containing 2.7 mg pilocarpine can replace 4–12 eyedrop instillations or 3–6 ointment applications and can be used as a once-a-day treatment for glaucoma. Baeyens et al. developed a soluble insert based on cellulose acetate phthalate (CAP) containing gentamicin sulfate and dexamethasone phosphate as a combination therapy for external ophthalmic infections [134]. This newly developed insert demonstrated a concomitant release of both drugs for a period of 10 h. This was followed by an extended release of gentamicin over a period of 50 h. Thus, the system was thought to provide a dual advantage: (1) feasibility of a combination therapy and (2) sustained release of drugs (thus significantly reducing the dosing frequency). The only drawback reported was the temporary blurring of vision which occurs during the insert solubilization. Aiache et al. in a review described the mechanism of drug release from a SODI [135]. They proposed that the drug release from a SODI follows a two-step process: (1) interaction of the hydrophilic matrix material with the aqueous surroundings and formation of a hydrogel and (2) diffusion of the dissolved drug across the hydrogel matrix. Kamath et al. designed a film-based SODI for ocular delivery of ciprofloxacin hydrochloride [136]. The authors reported the feasibility of once-a-day application of ciprofloxacin for ocular infections with this product. Mundada et al. in a series of publications reported preparation and evaluation of natural polymer-based ciprofloxacin SODI [137, 138]. The results from the studies indicated that with proper optimization of formulation and process variables, SODI may offer several advantages like increased residence time, prolonged drug release, reduction in dosing frequency, and improving the patient compliance.

8.5.1.2.2 Bioadhesive Ophthalmic Drug Inserts (BODI)

Bioadhesive ophthalmic drug inserts (BODI) are drug delivery systems similar to SODI in that they are soluble inserts. However, one of the major drawbacks of SODI is the risk associated with the expulsion from the site of application [7]. The formulation of BODI involves incorporation of a water-soluble bioadhesive material in the design thus reducing the risk of expulsion and increasing the residence time on the ocular tissues [112, 119, 139]. Due to the soluble nature of the bioadhesive component, the insert is expected to disappear as the drug is released. Saettone et al. prepared cylindrical pilocarpine BODI based on mixtures of PVA, glyceryl behenate, and different polymers (xanthan gum, jota-carrageenan, hydroxypropyl methylcellulose, hyaluronic acid) by extrusion and subsequent coating with a mixture of Eudragit RL and RS [140]. The study showed that the BODI released pilocarpine over a period of 3–5 h. Gurtler et al. developed a long-acting BODI based on mixtures of hydroxypropyl cellulose, ethyl cellulose, poly(acrylic) acid, and 25.0 % (w/w) of gentamicin sulfate with and without cellulose acetate phthalate for veterinary use [119]. The study results indicated that compared to an eyedrop solution of gentamicin (~15 min), gentamicin BODI showed an extended release of gentamicin (~72 h), thus offering the feasibility of soluble BODI as promising drug delivery concept for animal health care.

8.5.1.2.3 Collagen Corneal Shields

Collagen shields are ocular lens-like devices designed from the collagen obtained from porcine and bovine sources [141]. Collagen shields were earlier developed as ocular surface protecting devices following surgery or other traumatic and nontraumatic corneal conditions [141]. Drug delivery via collagen shields typically involves saturating collagen shield with drug by soaking it in a drug solution. This drug-impregnated collagen shield upon application dissolves on the surface of the eye, forms a lubricating viscous layer, and along with drug release provides a protective epithelial healing [142]. Kaufman et al. described the benefits of collagen matrix as a drug carrier for ocular delivery [143, 144]. The collagen matrix acts as a drug reservoir, reversibly binds with the drug molecules to provide a delayed release, and prevents the likelihood of potential systemic toxicity. Willoughby et al. extensively reviewed literature that supports the role of collagen shields in drug delivery and corneal wound healing and recommended a reevaluation of the application and benefits of corneal collagen shields to clinical practice [141]. Bloomfield et al. prepared gentamicin collagen shields and compared its performance against conventional eyedrop and an eye ointment [145]. It was observed that collagen shield provided the highest tear film and ocular tissue bioavailability of gentamicin. Punch et al. studied the release rate of procaine penicillin, erythromycin, and erythromycin estolate from soluble and insoluble collagen films in an attempt to develop a collagen shield-based ocular insert for the treatment of infectious bovine keratoconjunctivitis [105]. Their results revealed that due to the poor retention of collagen shields in the conjunctival sacs, the development of these inserts for veterinary use was challenging. Sawusch et al. compared the treatment of *Pseudomonas keratitis* by topical tobramycin with and without the presence of a commercially available collagen corneal shield [146]. The studies showed a 30-fold increase in the bioavailability of tobramycin into the anterior chamber in the eyes in the presence of a collagen shield. In similar studies, Clinch et al., Assil et al., and O'Brien et al. also observed the effectiveness of collagen corneal shields of tobramycin in sustained treatment of Pseudomonas keratitis in animal models [147–149].

Several collagen shield-based products are now commercially available on the market. A patent by McDonnell describes MediLens® (Chiron Ophthalmics Inc., Irvine, CA) as a collagen shield-based lens used for the prevention of keratolytic loss [150]. A similar product, ProShield® (Alcon Laboratories, Inc., Fort Worth, TX), is prepared from bovine corium tissue and releases the drug over a period of 24 or 48 h [151]. SOFT-SHIELD™ (Oasis® Medical Inc., Glendora, CA) is a collagen corneal shield with dissolution times ranging from 12 to 72 h [152].

Apart from a small drawback of blurred vision due to lack of transparency, collagen shields have proven to be a promising ocular drug delivery system. The common advantages seen with collagen shields include pharmacological inertness, physical stability, compatibility with ocular tissues, and economical to produce.

8.5.2 New Ophthalmic Delivery Systems (NODS)

Several new ophthalmic drug delivery systems or NODS that cannot be classified as the ones described above have been developed for ocular drug delivery. Lawrenson et al. reported the development and evaluation of NODS based on polyvinyl alcohol (PVA) film to investigate sustained corneal anesthesia from different concentrations of proxymetacaine in humans [153]. The study results demonstrated an improved bioavailability of the drug and an increased duration of anesthesia with NODS. Greaves and Wilson emphasized the importance of the properties of the polymeric materials used in the formulation of NODS, especially the mucoadhesive properties of the excipient in improving the drug retention in the cornea and conjunctiva [154]. Greaves et al. also evaluated the precorneal bioavailability and pharmacological activity of a radiolabelled NODS containing pilocarpine nitrate in healthy human volunteers [155]. The study showed that compared to a 2 % w/v pilocarpine nitrate solution, pilocarpine NODS demonstrated eight times increase in corneal bioavailability and a higher compliance among the test subjects. Bentley reported a PVA-based NODS for precise and sustained delivery of ocular drugs [156]. Kelly et al. also developed a pilocarpine NODS based on PVA film and reported approximately eightfold greater bioavailability from NODS than from a conventional eyedrop formulation [157]. Diestelhorst et al. compared the ocular tolerability of NODS vs. Isopto-Naturale R eyedrops in healthy volunteers in a two-period crossover study [158]. The study results found significantly lower ocular discomfort with NODS compared to the eyedrops. Alani et al. reported the development and evaluation of "ophthalmic rod" (OR), prepared from nontoxic plastic as a new ocular drug delivery system [159]. This OR is reported to be a single-dose sterile applicator with the drug deposited on it in the form of a thin film. The drug is delivered to the conjunctiva by the means of rubbing the rod against the palpebral conjunctiva of the lower lid. The main advantage reported with this delivery system is the elimination of the need of preservatives and sterility and avoidance of cross contamination. Using clonidine ORs and fluorescein ORs, the authors also demonstrated the tolerability and efficacy of ORs in human patients. In a similar study, Gwon et al. evaluated the safety, comfort, and ease of use of ophthalmic rods coated with fluorescein (30 μg) in healthy human volunteers [160]. A majority of test subjects rated the ORs as better than eyedrops due to perceived cleanliness, comfort, and ease of application. Eller et al. described the fabrication of a NODS for a constant infusion of drugs to the specific restricted area of ocular tissues [161]. This NODS takes the shape of a plastic well containing the drug solution and is fixed on the cornea with silicone grease.

Several analytical techniques have been studied to evaluate the precorneal pharmacokinetics of drug administered via NODS. These include traditional pharmacokinetic techniques involving plasma and tissue sampling as well as newer techniques such as gamma scintigraphy and slit-lamp fluorophotometry [162, 163]. Wilson et al. developed a gamma scintigraphy-based analytical technique for monitoring the ocular distribution of solutions for ophthalmic use and used this technique to monitor the effects of polyvinyl alcohol on the precorneal drainage of (99 mTc) pertechnetate in lab animals [164]. Ludwig et al. evaluated the precorneal kinetics of viscous eyedrops in humans using split-lamp fluorophotometry [165, 166].

8.5.3 Intravitreal Injections

As the name suggests, intravitreal injection is a process of injecting a drug solution directly into the vitreous via pars plana to improve bioavailability of drugs compared to the drugs delivered via topical or systemic routes [1]. Since the drug is delivered directly into the posterior segment of the eye, this approach of drug delivery has an inherent advantage of avoiding systemic toxicity. This approach also ensures the highest vitreal and retinal bioavailability. Shen et al. evaluated the pharmacokinetics and safety of intravitreal caspofungin in rabbits [167]. The study results found intravitreal injection of caspofungin to be safe and recommended further studies. Haller et al. studied the efficacy of intravitreal ocriplasmin for the treatment of vitreomacular adhesion and vitreomacular traction in human patients and confirmed the positive effect of ocriplasmin across relevant subpopulations [168]. Inoue et al. assessed the visual function and vision-related quality of life after intravitreal injection of ranibizumab for the treatment of age-related macular degeneration in a prospective clinical study [169]. The study results demonstrated the tolerability, efficacy, and compliance of intravitreal ranibizumab in test subjects.

Intravitreal injections are not devoid of drawbacks. Several short-term complications including retinal detachment, endophthalmitis, and intravitreal hemorrhages have been reported when intravitreal injection of ganciclovir was administered in AIDS patients to treat cytomegalovirus retinitis [170]. Ornek et al. reported an increase in corneal and conjunctival sensitivity after intravitreal ranibizumab injection in patients with diabetic retinopathy [171]. Aslan et al. reported hypersensitivity reaction in the form of marginal keratitis in a 56-year-old man after receiving an intravitreal injection of ranibizumab for the treatment of diffuse diabetic macular edema [172].

8.5.4 Ocular Iontophoresis

Ocular iontophoresis is described as an ocular drug delivery system wherein a topically applied ionized drug is transported across the cornea or the sclera with the assistance of a mild electric current [106]. For a drug to be delivered to the ocular tissues via iontophoresis mechanism, it has to fulfill two criteria: (1) the pH of the drug formulation should be below 6 or above 8 – this facilitates ionization of the drug at physiological pH – and (2) the salt form of the drug is preferred and the dissociated form will have higher aqueous solubility [106]. The noninvasive nature of drug delivery, combined with minimization of potential unwanted effects associated with intraocular injections and implants, ocular iontophoresis have gained a significant popularity among both patients and health-care providers. Several researchers have explored the applicability and efficacy of iontophoresis for the drug delivery to the ocular tissues.

Parkinson et al. studied the ocular tolerance with transscleral administration of fluorescein via iontophoresis using OcuPhor™, a novel hydrogel drug delivery applicator filled with balanced salt solution in healthy human volunteers [173]. The test subjects found iontophoresis technique to be tolerable up to 4.0 mA for 40 min. Vollmer et al. studied the bioavailability of the aminoglycoside antibiotic amikacin following transscleral iontophoresis in rabbits [174]. The study carried out in triplicate found that therapeutically relevant bioavailability amikacin was achieved in the ocular tissues with iontophoresis. Frucht-Pery et al. evaluated the efficacy of gentamicin iontophoresis for the treatment of *Pseudomonas keratitis* in the rabbits [175]. The study results showed an applied-current-dependent efficacy of gentamicin iontophoretic treatment in reducing the pseudomonas colony count. In similar studies, Hobden et al. investigated the applicability of iontophoresis for the delivery of tobramycin and ciprofloxacin for the treatment of tobramycin-resistant and aminoglycoside-resistant keratitis caused by *Pseudomonas aeruginosa*, respectively, in rabbits [176, 177]. Eljarrat-Binstock et al. evaluated the transcorneal and transscleral iontophoretic penetration of dexamethasone from a drug-loaded hydrogel into rabbit eyes [178]. The study found a 30-fold increase in the transcorneal bioavailability of dexamethasone after a single iontophoretic administration compared to multiple eyedrop instillations. Berdugo et al. investigated the feasibility of delivering antisense oligonucleotides to the vascular endothelial growth factor receptor in the cornea of the rat eye [179]. Barza et al. studied the transcorneal penetration of cefazolin, ticarcillin, and gentamicin via iontophoresis and demonstrated a correlation of drug penetration with strength of current and with duration of iontophoresis [180]. Behar-Cohen et al. evaluated the pharmacokinetics of intraocular penetration of hemisuccinate methylprednisolone via the transscleral coulomb-controlled iontophoresis system in rabbit eyes and compared it with that from i.v. injection [181]. The study data revealed that a significantly higher bioavailability can be safely achieved in intraocular tissues using coulomb-controlled iontophoresis. Grossman et al. compared transcorneal and transscleral iontophoresis to subconjunctival injection in the delivery of gentamicin into rabbit eyes and observed a significantly higher bioavailability of gentamicin via either modes compared to injection [182]. Sarraf et al. investigated transscleral iontophoresis for vitreous delivery of foscarnet for local treatments of cytomegalovirus retinopathy [183]. The study demonstrated the vitreous bioavailability of foscarnet for 60 h after iontophoretic application. Similarly Lam et al. demonstrated a 72-h bioavailability of ganciclovir after transscleral iontophoresis in rabbit eyes [184].

8.5.5 Periocular Administration

Periocular administration involves placing drug solutions in close proximity to the sclera, resulting in increased retinal and vitreal bioavailability [1]. This approach of ocular drug delivery has several advantages over systemic or topically administered routes including delivery to the posterior segment (avoiding potential systemic toxicities), higher bioavailability, and targeted drug delivery. Hayden et al. characterized pharmacology and toxicity of intravenous versus periocular carboplatin delivery in the rabbit eye [185]. The study showed that periocular administration of carboplatin resulted in an increase in intraorbital carboplatin concentration while decreasing its systemic exposure. Ghate et al. compared the pharmacokinetics of periocular injections of sodium fluorescein with those from retrobulbar and subconjunctival routes [186]. The results indicated a higher vitreous bioavailability of sodium fluorescein via periocular route compared to others.

8.6 Conclusions

Development of an ideal ocular drug delivery system is a formulation challenge. Over the last three decades, a plethora of formulation approaches have been explored in much detail in order to optimize ocular drug delivery. All of these approaches offer unique advantages in terms of improving ocular bioavailability of drugs. These approaches are also associated with one or more formulation, application, and/or therapeutic drawbacks. Nevertheless, most of the novel formulation approaches seem to be significantly better compared to conventional ocular formulations such as simple ophthalmic solutions. Moreover, combining multiple formulation approaches may provide synergies in achieving an optimal bioavailability of ocular drugs while minimizing the drawbacks associated with individual approach. Thus, the choice of formulation approach/approaches for the development of an ocular drug delivery system should be based on a sound scientific understanding of the anatomy/physiology of the eye, pathophysiology of ocular diseases, physicochemical properties of the drug and the delivery system, and the expected therapeutic outcomes.

References

1. Patel PB et al (2010) Ophthalmic drug delivery system: challenges and approaches. Syst Rev Pharm 1(2):113–120
2. Peyman GA, Ganiban GJ (1995) Delivery systems for intraocular routes. Adv Drug Deliv Rev 16(1):107–123
3. Janoria KG et al (2007) Novel approaches to retinal drug delivery. Expert Opin Drug Deliv 4(4):371–388
4. Pal Kaur I, Kanwar M (2002) Ocular preparations: the formulation approach. Drug Dev Ind Pharm 28(5):473–493
5. Prausnitz MR, Noonan JS (1998) Permeability of cornea, sclera, and conjunctiva: a literature analysis for drug delivery to the eye. J Pharm Sci 87(12):1479–1488
6. Keister JC et al (1991) Limits on optimizing ocular drug delivery. J Pharm Sci 80(1):50–53
7. Kaur IP, Kanwar M (2002) Ocular preparations: the formulation approach. Drug Dev Ind Pharm 28(5):473–493
8. Meseguer G et al (1996) Gamma scintigraphic comparison of eyedrops containing pilocarpine in healthy volunteers. J Ocul Pharmacol Ther 12(4):481–488
9. Zaki I et al (1986) A comparison of the effect of viscosity on the precorneal residence of solutions in rabbit and man. J Pharm Pharmacol 38(6):463–466
10. Deshpande AA, Heller J, Gurny R (1998) Bioerodible polymers for ocular drug delivery. Crit Rev Ther Drug Carrier Syst 15(4):40
11. Bottari F et al (1974) Influence of drug concentration on in vitro release of salicylic acid from ointment bases. J Pharm Sci 63(11):1779–1783
12. Rozier A et al (1989) Gelrite®: a novel, ion-activated, in-situ gelling polymer for ophthalmic vehicles. Effect on bioavailability of timolol. Int J Pharm 57(2):163–168
13. Hui H-W, Robinson JR (1985) Ocular delivery of progesterone using a bioadhesive polymer. Int J Pharm 26(3):203–213
14. Stevens LE, Missel PJ, Lang JC (1992) Drug release profiles of ophthalmic formulations. 1. Instrumentation. Anal Chem 64(7):715–723
15. Cohen S et al (1997) A novel in situ-forming ophthalmic drug delivery system from alginates undergoing gelation in the eye. J Control Release 44(2â€"3):201–208

16. Lee VHL (1993) Ophthalmic drug delivery systems. Precorneal Corneal and Postcorneal factors. In: Mitra AK (ed) Marcel Dekker, New York, pp 59–82
17. Lee VHL (1993) Biopharmaceutics of ocular drug delivery. Formulation Approaches to Improve Ocular Bioavailability. In: Edman P (ed) CRC Press, Boca Raton, pp 121–143
18. Liaw J, Robinson JR (1992) The effect of polyethylene glycol molecular weight on corneal transport and the related influence of penetration enhancers. Int J Pharm 88(1–3):125–140
19. Sasaki H et al (1995) Different effects of absorption promoters on corneal and conjunctival penetration of ophthalmic beta-blockers. Pharm Res 12(8):1146–1150
20. Sasaki H et al (1995) Penetration of β-blockers through ocular membranes in albino rabbits. J Pharm Pharmacol 47(1):17–21
21. Marshall W, Klyce D (1983) Cellular and paracellular pathway resistances in the 'tight' Cl^--secreting epithelium of rabbit cornea. J Membr Biol 73(3):275–282
22. Lee VHL (1990) Mechanisms and facilitation of corneal drug penetration. J Control Release 11(1–3):79–90
23. Chiou GCY, Chuang CY (1989) Improvement of systemic absorption of insulin through eyes with absorption enhancers. J Pharm Sci 78(10):815–818
24. Godbey REW, Green K, Hull DS (1979) Influence of cetylpyridinium chloride on corneal permeability to penicillin. J Pharm Sci 68(9):1176–1178
25. Higaki K, Takeuchi M, Nakano M (1996) Estimation and enhancement of in vitro corneal transport of S-1033, a novel antiglaucoma medication. Int J Pharm 132(1–2):165–173
26. Marsh RJ, Maurice DM (1971) The influence of non-ionic detergents and other surfactants on human corneal permeability. Exp Eye Res 11(1):43–48
27. Mikkelson TJ, Chrai SS, Robinson JR (1973) Competitive inhibition of drug-protein interaction in eye fluids and tissues. J Pharm Sci 62(12):1942–1945
28. Morimoto K, Nakai T, Morisaka K (1987) Evaluation of permeability enhancement of hydrophilic compounds and macromolecular compounds by bile salts through rabbit corneas in-vitro. J Pharm Pharmacol 39(2):124–126
29. Sasaki H et al (1995) Ophthalmic preservatives as absorption promoters for ocular drug delivery. J Pharm Pharmacol 47(9):703–707
30. Sasaki H et al (1994) Effect of preservatives on systemic delivery of insulin by ocular instillation in rabbits. J Pharm Pharmacol 46(11):871–875
31. Tang-Liu DDS et al (1994) Effects of four penetration enhancers on corneal permeability of drugs in vitro. J Pharm Sci 83(1):85–90
32. Van Santvliet L, Ludwig A (1998) The influence of penetration enhancers on the volume instilled of eye drops. Eur J Pharm Biopharm 45(2):189–198
33. Lee VH, Robinson JR (1986) Topical ocular drug delivery: recent developments and future challenges. J Ocul Pharmacol 2(1):67–108
34. Green K (1993) Biopharmaceutics in ocular drug delivery. The effects of preservatives on corneal permeability of drugs. In: Edman P (ed) CRC PRess, Boca Raton, pp 43–49
35. Pfister RR, Burstein N (1976) The effects of ophthalmic drugs, vehicles, and preservatives on corneal epithelium: a scanning electron microscope study. Invest Ophthalmol 15(4):246–259
36. Hochman J, Artursson P (1994) Mechanisms of absorption enhancement and tight junction regulation. J Control Release 29(3):253–267
37. Saettone MF et al (1996) Evaluation of ocular permeation enhancers: in vitro effects on corneal transport of four Î²-blockers, and in vitro/in vivo toxic activity. Int J Pharm 142(1):103–113
38. Grass GM, Robinson JR (1988) Mechanisms of corneal drug penetration I: in vivo and in vitro kinetics. J Pharm Sci 77(1):3–14
39. Newton C, Gebhardt BM, Kaufman HE (1988) Topically applied cyclosporine in azone prolongs corneal allograft survival. Invest Ophthalmol Vis Sci 29(2):208–215
40. Green K et al (1987) Detergent penetration into young and adult rabbit eyes: comparative pharmacokinetics. Cutan Ocul Toxicol 6(2):89–107
41. Grass GM, Robinson JR (1984) Relationship of chemical structure to corneal penetration and influence of low-viscosity solution on ocular bioavailability. J Pharm Sci 73(8):1021–1027

42. Ismail IM et al (1992) Comparison of azone and hexamethylene lauramide in toxicologic effects and penetration enhancement of cimetidine in rabbit eyes. Pharm Res 9(6):817–821
43. Merkus FWHM et al (1993) Absorption enhancers in nasal drug delivery: efficacy and safety. J Control Release 24(1–3):201–208
44. Rojanasakul Y, Liaw J, Robinson JR (1990) Mechanisms of action of some penetration enhancers in the cornea: laser scanning confocal microscopic and electrophysiology studies. Int J Pharm 66(1–3):131–142
45. Komatsu A et al (1988) Application of lipid microsphere drug delivery system to steroidal ophthalmic preparation. Jpn J Ophthalmol 32(1):41–43
46. Oppenheim RC (1980) Drug delivery systems characteristics and biomedical applications. Nanoparticle. In: Juliano RJ (ed) Oxford University Press, New York, pp 177–188
47. Meisner D, Mezei M (1995) Liposome ocular delivery systems. Adv Drug Deliv Rev 16(1):75–93
48. Dharma SK, Fishman PH, Peyman GA (1986) A preliminary study of corneal penetration of 125l-labelled idoxuridine liposome. Acta Ophthalmol (Copenh) 64(3):298–301
49. Smolin G et al (1981) Idoxuridine-liposome therapy for herpes simplex keratitis. Am J Ophthalmol 91(2):220–225
50. Schaeffer HE, Krohn DL (1982) Liposomes in topical drug delivery. Invest Ophthalmol Vis Sci 22(2):220–227
51. Shek PN, Barber RF (1987) Liposomes are effective carriers for the ocular delivery of prophylactics. Biochim Biophys Acta Biomembr 902(2):229–236
52. Al-Muhammed J et al (1996) In-vivo studies on dexamethasone sodium phosphate liposomes. J Microencapsul 13(3):293–305
53. El-Gazayerly ON, Hikal AH (1997) Preparation and evaluation of acetazolamide liposomes as an ocular delivery system. Int J Pharm 158(2):121–127
54. Milani JK et al (1993) Prolongation of corneal allograft survival with liposome-encapsulated cyclosporine in the rat eye. Ophthalmology 100(6):890–896
55. Lee VHL et al (1985) Ocular drug bioavailability from topically applied liposomes. Surv Ophthalmol 29(5):335–348
56. Liddell MR et al (2007) Evaluation of glass dissolution vessel dimensions and irregularities. Dissolution Technol 14(1):28–33
57. Durrani AM et al (1992) Pilocarpine bioavailability from a mucoadhesive liposomal ophthalmic drug delivery system. Int J Pharm 88(1–3):409–415
58. Pleyer U et al (1994) Ocular absorption of cyclosporine A from liposomes incorporated into collagen shields. Curr Eye Res 13(3):177–181
59. Kaur IP et al (2004) Vesicular systems in ocular drug delivery: an overview. Int J Pharm 269(1):1–14
60. Uchegbu IF, Vyas SP (1998) Non-ionic surfactant based vesicles (niosomes) in drug delivery. Int J Pharm 172(1–2):33–70
61. Carafa M et al (1998) Preparation and properties of new unilamellar non-ionic/ionic surfactant vesicles. Int J Pharm 160(1):51–59
62. Carafa M, Santucci E, Lucania G (2002) Lidocaine-loaded non-ionic surfactant vesicles: characterization and in vitro permeation studies. Int J Pharm 231(1):21–32
63. Abdelbary G, El-gendy N (2008) Niosome-encapsulated gentamicin for ophthalmic controlled delivery. AAPS PharmSciTech 9(3):740–747
64. Vyas SP et al (1998) Discoidal niosome based controlled ocular delivery of timolol maleate. Pharmazie 53(7):466–469
65. Pandita A, Sharma P (2013) Pharmacosomes: an emerging novel vesicular drug delivery system for poorly soluble synthetic and herbal drugs. ISRN Pharm 2013:1–10
66. Fialho SL, da Silva-Cunha A (2004) New vehicle based on a microemulsion for topical ocular administration of dexamethasone. Clin Experiment Ophthalmol 32(6):626–632
67. Vandamme TF (2002) Microemulsions as ocular drug delivery systems: recent developments and future challenges. Prog Retin Eye Res 21(1):15–34
68. Hasse A, Keipert S (1997) Development and characterization of microemulsions for ocular application. Eur J Pharm Biopharm 43(2):179–183

69. Klang S, Abdulrazik M, Benita S (2000) Influence of emulsion droplet surface charge on indomethacin ocular tissue distribution. Pharm Dev Technol 5(4):521–532
70. Chan J et al (2007) Phase transition water-in-oil microemulsions as ocular drug delivery systems: in vitro and in vivo evaluation. Int J Pharm 328(1):65–71
71. Kapoor Y, Chauhan A (2008) Ophthalmic delivery of cyclosporine A from Brij-97 microemulsion and surfactant-laden p-HEMA hydrogels. Int J Pharm 361(1–2):222–229
72. G.Wilson C, Y.P.Zhu, P.Kurmala, L.S.Rao, B.Dhillon. Ophthalmic Drug Delivery. In: M.Hillery A, W.Lloyd A, Swarbrick J, editors. Drug Delivery and Targeting for Pharmacists and Pharmaceutical Scientists. London: Taylor & Francis; 2005. p. 298–319
73. Pignatello R et al (2002) Flurbiprofen-loaded acrylate polymer nanosuspensions for ophthalmic application. Biomaterials 23(15):3247–3255
74. Adibkia K et al (2007) Inhibition of endotoxin-induced uveitis by methylprednisolone acetate nanosuspension in rabbits. J Ocul Pharmacol Ther 23(5):421–432
75. Adibkia K et al (2007) Piroxicam nanoparticles for ocular delivery: physicochemical characterization and implementation in endotoxin-induced uveitis. J Drug Target 15(6):407–416
76. Kassem MA et al (2007) Nanosuspension as an ophthalmic delivery system for certain glucocorticoid drugs. Int J Pharm 340(1–2):126–133
77. Pignatello R et al (2006) Preparation and characterization of eudragit retard nanosuspensions for the ocular delivery of cloricromene. AAPS PharmSciTech 7(1):E192–E198
78. Sahoo SK, Dilnawaz F, Krishnakumar S (2008) Nanotechnology in ocular drug delivery. Drug Discov Today 13(3–4):144–151
79. Sakurai E et al (2001) Effect of particle size of polymeric nanospheres on intravitreal kinetics. Ophthalmic Res 33(1):31–36
80. Gan L et al (2013) Recent advances in topical ophthalmic drug delivery with lipid-based nanocarriers. Drug Discov Today 18(5–6):290–297
81. Bourges JL et al (2003) Ocular drug delivery targeting the retina and retinal pigment epithelium using polylactide nanoparticles. Invest Ophthalmol Vis Sci 44(8):3562–3569
82. Li VH et al (1986) Ocular drug delivery of progesterone using nanoparticles. J Microencapsul 3(3):213–218
83. Cavalli R et al (2002) Solid lipid nanoparticles (SLN) as ocular delivery system for tobramycin. Int J Pharm 238(1–2):241–245
84. Amrite AC et al (2008) Effect of circulation on the disposition and ocular tissue distribution of 20 nm nanoparticles after periocular administration. Mol Vis 14:150–160
85. Irache JM et al (2005) Albumin nanoparticles for the intravitreal delivery of anticytomegaloviral drugs. Mini Rev Med Chem 5(3):293–305
86. Calvo P, Vila-Jato JL, Alonso MJ (1996) Comparative in vitro evaluation of several colloidal systems, nanoparticles, nanocapsules, and nanoemulsions, as ocular drug carriers. J Pharm Sci 85(5):530–536
87. De Campos AM et al (2003) The effect of a PEG versus a chitosan coating on the interaction of drug colloidal carriers with the ocular mucosa. Eur J Pharm Sci 20(1):73–81
88. Xu J et al (2007) Inhibitory efficacy of intravitreal dexamethasone acetate-loaded PLGA nanoparticles on choroidal neovascularization in a laser-induced rat model. J Ocul Pharmacol Ther 23(6):527–540
89. Motwani SK et al (2008) Chitosan-sodium alginate nanoparticles as submicroscopic reservoirs for ocular delivery: formulation, optimisation and in vitro characterisation. Eur J Pharm Biopharm 68(3):513–525
90. Arnedo A et al (2004) Albumin nanoparticles improved the stability, nuclear accumulation and anticytomegaloviral activity of a phosphodiester oligonucleotide. J Control Release 94(1):217–227
91. Gaudana R et al (2009) Recent perspectives in ocular drug delivery. Pharm Res 26(5): 1197–1216
92. Gunda S, Hariharan S, Mitra AK (2006) Corneal absorption and anterior chamber pharmacokinetics of dipeptide monoester prodrugs of ganciclovir (GCV): in vivo comparative evaluation of these prodrugs with Val-GCV and GCV in rabbits. J Ocul Pharmacol Ther 22(6):465–476

93. Majumdar S et al (2005) Dipeptide monoester ganciclovir prodrugs for treating HSV-1-induced corneal epithelial and stromal keratitis: in vitro and in vivo evaluations. J Ocul Pharmacol Ther 21(6):463–474
94. Kansara V, Hao Y, Mitra AK (2007) Dipeptide monoester ganciclovir prodrugs for transscleral drug delivery: targeting the oligopeptide transporter on rabbit retina. J Ocul Pharmacol Ther 23(4):321–334
95. Janoria KG, Mitra AK (2007) Effect of lactide/glycolide ratio on the in vitro release of ganciclovir and its lipophilic prodrug (GCV-monobutyrate) from PLGA microspheres. Int J Pharm 338(1–2):133–141
96. Dias C et al (2002) Ocular penetration of acyclovir and its peptide prodrugs valacyclovir and val-valacyclovir following systemic administration in rabbits: an evaluation using ocular microdialysis and LC-MS. Curr Eye Res 25(4):243–252
97. Anand BS et al (2003) In vivo antiviral efficacy of a dipeptide acyclovir prodrug, val-val-acyclovir, against HSV-1 epithelial and stromal keratitis in the rabbit eye model. Invest Ophthalmol Vis Sci 44(6):2529–2534
98. Katragadda S, Talluri RS, Mitra AK (2006) Modulation of P-glycoprotein-mediated efflux by prodrug derivatization: an approach involving peptide transporter-mediated influx across rabbit cornea. J Ocul Pharmacol Ther 22(2):110–120
99. Doukas J et al (2008) Topical administration of a multi-targeted kinase inhibitor suppresses choroidal neovascularization and retinal edema. J Cell Physiol 216(1):29–37
100. Al-Ghananeem AM, Crooks PA (2007) Phase I and phase II ocular metabolic activities and the role of metabolism in ophthalmic prodrug and codrug design and delivery. Molecules 12(3):373–388
101. Lallemand F et al (2007) Biological conversion of a water-soluble prodrug of cyclosporine A. Eur J Pharm Biopharm 67(2):555–561
102. Juntunen J, Jarvinen T, Niemi R (2005) In-vitro corneal permeation of cannabinoids and their water-soluble phosphate ester prodrugs. J Pharm Pharmacol 57(9):1153–1157
103. Nambu H et al (2003) Combretastatin A-4 phosphate suppresses development and induces regression of choroidal neovascularization. Invest Ophthalmol Vis Sci 44(8):3650–3655
104. Takahashi K et al (2003) Topical nepafenac inhibits ocular neovascularization. Invest Ophthalmol Vis Sci 44(1):409–415
105. Punch PI et al (1987) The release of insoluble antibiotics from collagen ocular inserts in vitro and their insertion into the conjunctival sac of cattle. J Vet Pharmacol Ther 10(1):37–42
106. Rawas-Qalaji M, Williams CA (2012) Advances in ocular drug delivery. Curr Eye Res 37(5):345–356
107. Jalil R-U (1990) Biodegradable poly(lactic acid) and poly (lactide-co-glycolide) polymers in sustained drug delivery. Drug Dev Ind Pharm 16(16):2353–2367
108. Hashizoe M et al (1994) SCleral plug of biodegradable polymers for controlled drug release in the vitreous. Arch Ophthalmol 112(10):1380–1384
109. Kimura H et al (1994) A new vitreal drug delivery system using an implantable biodegradable polymeric device. Invest Ophthalmol Vis Sci 35(6):2815–2819
110. Kunou N et al (1995) Controlled intraocular delivery of ganciclovir with use of biodegradable scleral implant in rabbits. J Control Release 37(1–2):143–150
111. Sanborn GE et al (1992) Sustained-release ganciclovir therapy for treatment of cytomegalovirus retinitis: use of an intravitreal device. Arch Ophthalmol 110(2):188–195
112. Gurtler F, Gurny R (1995) Patent literature review of ophthalmic inserts. Drug Dev Ind Pharm 21(1):1–18
113. Chien YW (1992) Ocular drug delivery and delivery systems. In: Chien YW (ed) Novel drug delivery systems, 2nd edn. Taylor and Francis, New York
114. Karthikeyan D et al (2008) The concept of ocular inserts as drug delivery systems: an overview. Asian J Pharm 2(4):192–200
115. Khar RK, Vyas SP (2002) Targeted and controlled drug delivery novel carrier systems. In: Khar RK, Vyas SP (eds), 1st edn. CBS Publishers and Distributors, New Delhi
116. Saettone MF, Salminen L (1995) Ocular inserts for topical delivery. Adv Drug Deliv Rev 16(1):95–106

117. Sahane NK et al (2010) Ocular inserts: a review. J Pharm Res 3(1):57–64
118. El-Shanawany S (1992) Ocular delivery of pilocarpine from ocular inserts. Pharma Sci 2(4):337–341
119. Gurtler F et al (1995) Long-acting soluble bioadhesive ophthalmic drug insert (BODI) containing gentamicin for veterinary use: optimization and clinical investigation. J Control Release 33(2):231–236
120. Pollack IP, Quigley HA, Harbin TS (1976) The Ocusert pilocarpine system: advantages and disadvantages. South Med J 69(10):1296–1298
121. Johansen S, Prause JU, Rask-Pedersen E (1996) I: a bioavailability comparison in rabbits after a single topical ocular application of prednisolone acetate formulated as a high-viscosity gel and as an aqueous suspension. Acta Ophthalmol Scand 74(3):253–258
122. Urtti A, Salminen L, Miinalainen O (1985) Systemic absorption of ocular pilocarpine is modified by polymer matrices. Int J Pharm 23(2):147–161
123. Urtti A, Juslin M, Miinalainen O (1985) Pilocarpine release from hydroxypropyl-cellulose-polyvinylpyrrolidone matrices. Int J Pharm 25(2):165–178
124. Vasantha V, Sehgal PK, Rao KP (1988) Collagen ophthalmic inserts for pilocarpine drug delivery system. Int J Pharm 47(1–3):95–102
125. Attia MA, Kassem MA, Safwat SM (1988) In vivo performance of [3H]dexamethasone ophthalmic film delivery systems in the rabbit eye. Int J Pharm 47(1–3):21–30
126. Darougar S (1992) Ocular insert for the fornix. In: U.S.P.A.T. Office (ed) Google. Google Patents, USA
127. Urtti A et al (1994) Controlled ocular timolol delivery: systemic absorption and intraocular pressure effects in humans. Pharm Res 11(9):1278–1282
128. Dumitriu S et al (1988) Polycomponent ophthalmic inserts with polysaccharide support. J Bioact Compat Polym 3(4):370–389
129. Kalenak JW, Zakov ZN (1994) Presumed sudden leakage of a pilocarpine ocusert and rhegmatogenous retinal detachment. J Glaucoma 3(2):152–153
130. Kushnick H, Liebmann JM, Ritch R (1996) Systemic pilocarpine toxicity from ocusert leakage. Arch Ophthalmol 114(11):1432–1432
131. Hitoshi S et al (1993) Drug release from an ophthalmic insert of a beta-blocker as an ocular drug delivery system. J Control Release 27(2):127–137
132. Fazly Bazzaz BS et al (2014) Preparation, characterization and antimicrobial study of a hydrogel (soft contact lens) material impregnated with silver nanoparticles. Cont Lens Anterior Eye 37(3):149–152
133. Maichuk YF, Erichev VP (1981) Soluble ophthalmic drug inserts with pilocarpine: experimental and clinical study. Glaucoma 3:239–242
134. Baeyens V et al (1998) Optimized release of dexamethasone and gentamicin from a soluble ocular insert for the treatment of external ophthalmic infections. J Control Release 52(1–2):215–220
135. Aiache JM et al (1997) The formulation of drug for ocular administration. J Biomater Appl 11(3):329–348
136. Kamath U, Singh U, Udupa N (1993) Evaluation of ciprofloxacin hydrochloride ocular preparations. Ind J Pharm Sci 55(4):148–150
137. Mundada AS, Shrikhande BK (2006) Design and evaluation of soluble ocular drug insert for controlled release of ciprofloxacin hydrochloride. Drug Dev Ind Pharm 32(4):443–448
138. Mundada AS, Shrikhande BK (2008) Formulation and evaluation of ciprofloxacin hydrochloride soluble ocular drug insert. Curr Eye Res 33(5):469–475
139. Gurtler F, Gurny R (1998) Bioadhesive ophthalmic insert. In: U.S.P.A.T. Office, Vetoquinol SA (ed) Google patents, Lure, USA
140. Saettone MF et al (1992) Controlled release of pilocarpine from coated polymeric ophthalmic inserts prepared by extrusion. Int J Pharm 86(2–3):159–166
141. Willoughby CE, Batterbury M, Kaye SB (2002) Collagen corneal shields. Surv Ophthalmol 47(2):174–182

142. Poland DE, Kaufman HE (1988) Clinical uses of collagen shields. J Cataract Refract Surg 14(5):489–491
143. Kaufman HE (1988) Collagen shield symposium. J Cataract Refract Surg 14(5):487–488
144. Kaufman HE et al (1994) Collagen-based drug delivery and artificial tears. J Ocul Pharmacol 10(1):17–27
145. Bloomfield SE et al (1978) Soluble gentamicin ophthalmic inserts as a drug delivery system. Arch Ophthalmol 96(5):885–887
146. Sawusch MR et al (1988) Use of collagen corneal shields in the treatment of bacterial keratitis. Am J Ophthalmol 106(3):279–281
147. Assil KK et al (1992) Efficacy of tobramycin-soaked collagen shields vs tobramycin eyedrop loading dose for sustained treatment of experimental Pseudomonas aeruginosa-induced keratitis in rabbits. Am J Ophthalmol 113(4):418–423
148. Clinch TE et al (1992) Collagen shields containing tobramycin for sustained therapy (24 hours) of experimental Pseudomonas keratitis. CLAO J 18(4):245–247
149. O'Brien TP et al (1988) Use of collagen corneal shields versus soft contact lenses to enhance penetration of topical tobramycin. J Cataract Refract Surg 14(5):505–507
150. McDonnell PJ (1998) Method for preventing keratocyte loss. In: U.S.P.A.T. Office (ed) Google patents. University of Southern California (Los Angeles), USA
151. Sintzel MB et al (1996) Biomaterials in ophthalmic drug delivery. Eur J Pharm Biopharm 42(6):358–374
152. OASIS® Medical, I. SOFT SHIELD collagen corneal shield 2014 [cited 2014 September 22, 2014]; Widest range of dissolution rates available. Thin, purified collagen. 14.5-mm diameter shield. Accurate and consistent dissolution times]. Available from: http://oasismedical.com/soft-shield-collagen-corneal-shield.html
153. Lawrenson JG et al (1993) A comparison of the efficacy and duration of action of topically applied proxymetacaine using a novel ophthalmic delivery system versus eye drops in healthy young volunteers. Br J Ophthalmol 77(11):713–715
154. Greaves JL, Wilson CG (1993) Treatment of diseases of the eye with mucoadhesive delivery systems. Adv Drug Deliv Rev 11(3):349–383
155. Greaves JL et al (1992) Scintigraphic studies on the corneal residence of a new ophthalmic delivery system (NODS): rate of clearance of a soluble marker in relation to duration of pharmacological action of pilocarpine. Br J Clin Pharmacol 33(6):603–609
156. Bentley PH (1990) A new ophthalmic delivery system (NODS). In: 9th pharmaceutical technology conference. Veldhoven, Netherlands
157. Kelly JA et al (1989) Relative bioavailability of pilocarpine from a novel ophthalmic delivery system and conventional eyedrop formulations. Br J Ophthalmol 73(5):360–362
158. Diestelhorst M, Krieglstein GK (1994) The ocular tolerability of a new ophthalmic drug delivery system (NODS). Int Ophthalmol 18(1):1–4
159. Alani S, Hammerstein W (1990) The ophthalmic rod – a new drug-delivery system II. Graefes Arch Clin Exp Ophthalmol 228(4):302–304
160. Gwon A et al (1986) Ophthalmic rods. Ophthalmology 93(9):82–85
161. Eller MG et al (1985) Topical carbonic anhydrase inhibitors IV: relationship between excised corneal permeability and pharmacokinetic factors. J Pharm Sci 74(5):525–529
162. Maren TH et al (1987) Ocular pharmacology of methazolamide analogs: distribution in the eye and effects on pressure after topical application. J Pharmacol Exp Ther 241(1):56–63
163. Putnam ML et al (1987) Ocular disposition of aminozolamide in the rabbit eye. Invest Ophthalmol Vis Sci 28(8):1373–1382
164. Wilson CG, Olejnik O, Hardy JG (1983) Precorneal drainage of polyvinyl alcohol solutions in the rabbit assessed by gamma scintigraphy. J Pharm Pharmacol 35(7):451–454
165. Ludwig A, Ooteghem MV (1986) The study of the precorneal dynamics of ophthalmic solutions by fluorophotometry. Pharm Acta Helv 61(8):236–240
166. Ludwig A, Van Ooteghem M (1989) The evaluation of viscous ophthalmic vehicles by slit lamp fluorophotometry in humans. Int J Pharm 54(2):95–102

167. Shen YC et al (2014) Pharmacokinetics and safety of intravitreal caspofungin. Antimicrob Agents Chemother 2014(22):03324–14
168. Haller JA et al (2014) Efficacy of intravitreal ocriplasmin for treatment of vitreomacular adhesion: subgroup analyses from two randomized trials. Ophthalmology 6420(14):00689–7
169. Inoue M et al (2014) Intravitreal injection of ranibizumab using a pro re nata regimen for age-related macular degeneration and vision-related quality of life. Clin Ophthalmol 8:1711–1716
170. Ausayakhun S et al (2005) Treatment of cytomegalovirus retinitis in AIDS patients with intravitreal ganciclovir. J Med Assoc Thai 88(9):S15–S20
171. Ornek N, Ornek K, Erbahceci IE (2014) Corneal and conjunctival sensitivity changes following intravitreal ranibizumab injection in diabetic retinopathy. J Ocul Pharmacol Ther 2014:22
172. Aslan Bayhan S et al (2014) Marginal keratitis after intravitreal injection of ranibizumab. Cornea 2014:12
173. Parkinson TM et al (2003) Tolerance of ocular iontophoresis in healthy volunteers. J Ocul Pharmacol Ther 19(2):145–151
174. Vollmer DL et al (2002) In vivo transscleral iontophoresis of amikacin to rabbit eyes. J Ocul Pharmacol Ther 18(6):549–558
175. Frucht-Pery J et al (2006) Iontophoretic treatment of experimental pseudomonas keratitis in rabbit eyes using gentamicin-loaded hydrogels. Cornea 25(10):1182–1186
176. Hobden JA et al (1989) Tobramycin iontophoresis into corneas infected with drug-resistant Pseudomonas aeruginosa. Curr Eye Res 8(11):1163–1169
177. Hobden JA et al (1990) Ciprofloxacin iontophoresis for aminoglycoside-resistant pseudomonal keratitis. Invest Ophthalmol Vis Sci 31(10):1940–1944
178. Eljarrat-Binstock E et al (2005) Transcorneal and transscleral iontophoresis of dexamethasone phosphate using drug loaded hydrogel. J Control Release 106(3):386–390
179. Berdugo M et al (2003) Delivery of antisense oligonucleotide to the cornea by iontophoresis. Antisense Nucleic Acid Drug Dev 13(2):107–114
180. Barza M, Peckman C, Baum J (1986) Transscleral iontophoresis of cefazolin, ticarcillin, and gentamicin in the rabbit. Ophthalmology 93(1):133–139
181. Behar-Cohen FF et al (2002) Transscleral Coulomb-controlled iontophoresis of methylprednisolone into the rabbit eye: influence of duration of treatment, current intensity and drug concentration on ocular tissue and fluid levels. Exp Eye Res 74(1):51–59
182. Grossman RE, Chu DF, Lee DA (1990) Regional ocular gentamicin levels after transcorneal and transscleral iontophoresis. Invest Ophthalmol Vis Sci 31(5):909–916
183. Sarraf D et al (1993) Transscleral iontophoresis of foscarnet. Am J Ophthalmol 115(6):748–754
184. Lam TT et al (1994) Intravitreal delivery of ganciclovir in rabbits by transscleral iontophoresis. J Ocul Pharmacol 10(3):571–575
185. Hayden BC et al (2004) Pharmacokinetics of systemic versus focal Carboplatin chemotherapy in the rabbit eye: possible implication in the treatment of retinoblastoma. Invest Ophthalmol Vis Sci 45(10):3644–3649
186. Ghate D et al (2007) Pharmacokinetics of intraocular drug delivery by periocular injections using ocular fluorophotometry. Invest Ophthalmol Vis Sci 48(5):2230–2237

Chapter 9
Ocular Permeation Enhancers

Nathaniel J. Kim, Alon Harris, Alhasan Elghouche, Willy Gama, and Brent Siesky

Abstract Nanotechnology-based penetration enhancers provide novel mechanisms to improve current drug delivery methodologies. New developments in this field have been shown to overcome previous difficulties of ocular therapeutics like anatomical barriers and filtration mechanisms. Furthermore, nanoparticles exhibit additional benefits as a drug carrier, such as protection of drugs against degradation, biocompatible carrier structures, or reduction of adverse or immunological effects. Penetration enhancers exhibit great promise in ophthalmology. While much of the current research in the field surrounds lipid-based nanotechnologies, other innovative works have expanded the spectrum of ocular drug delivery.

This chapter investigates the most current work in the field of nanotechnology-based ocular penetration enhancers that have laid the foundation for future investigations. In addition to nanoparticle constructs, drug delivery techniques that may be used in conjunction with nanoparticles are also explored.

Keywords Drug delivery • Ocular barrier • Penetration enhancer • Nanobiomaterials • Iontophoresis • Implants • Microsphere • Microneedle

9.1 Introduction

Drug penetration is a vital feature that affects the efficacy of the therapeutic molecule. While drug compounds can be designed for target sites to elicit a specific reaction, there are physical and chemical challenges that the drug encounters en route to the target. Within the eye, barriers fortify target sites and impede drug delivery and action, subsequently interfering with desired effects. Penetration enhancement,

N.J. Kim, PhD • A. Elghouche • W. Gama • B. Siesky, PhD
Department of Ophthalmology, Indiana University School of Medicine, Indianapolis, IN, USA

A. Harris, MS, PhD, FARVO, (✉)
Department of Ophthalmology, Indiana University School of Medicine, 1160 West Michigan Street, Indianapolis, IN 46202, USA
e-mail: alharris@indiana.edu

© Springer International Publishing Switzerland 2016
Y. Pathak et al. (eds.), *Nano-Biomaterials For Ophthalmic Drug Delivery*,
DOI 10.1007/978-3-319-29346-2_9

therefore, is the facilitation of a compound's path to a target site by modifying the drug with a carrier. This chapter describes ocular permeation pathways and the barriers therein, followed by an overview of various nano-biomaterial technology and instrumentation that may enhance penetration through those barriers.

There is vast potential for the use of nanoparticle drug delivery in patients with different ophthalmic diseases. Nanomaterials demonstrate important properties as delivery agents that can overcome current challenges in ocular drug delivery. Delivering sufficient amounts of drug to target tissues and enhancing its penetration into the eye is challenging due to ocular protective and filtration mechanisms [1]. Penetration enhancers may facilitate the transport of molecules, such as pharmaceutical compounds, nucleic acids, or antibodies, into target sites while maintaining structural integrity and functionality and protecting the compound of interest from degradation [2]. Depending on the type of nanomaterial and its particular composition, the drug may be incorporated as a component of, or encapsulated within, the enhancer. There are various methods used to synthesize hollow, solid, or porous nanoparticles in a variety of shapes and sizes [3].

The pathology of certain clinical conditions may take place within difficult-to-access sites of the eye, necessitating enhanced penetration to achieve greater therapeutic effects. Conditions such as glaucoma, diabetic retinopathy, and age-related macular degeneration (ARMD) are among the most significant causes of visual loss and encompass processes within insulated ocular tissues. For example, glaucoma drugs modulating aqueous humor production or drainage must interact with receptors present on the ciliary body; if applied topically, the most common method of ocular drug delivery, these compounds must penetrate the overlying structures (i.e., conjunctiva, cornea, sclera) before such an interaction may take place [4]. In such topical drop formulations, less than 10 % of the drug may be absorbed into the eye. Furthermore, approximately 1 % may enter the aqueous humor [5]. Despite patient confidence, studies have shown that complications of improper eye drop technique, patient noncompliance, and dispenser contamination continue to persist [6, 7]. Gupta et al. found that nine out of ten glaucoma patients were unable to apply their drop medications properly [7]. Vitreal injections are an alternative to topical administration and are used for direct administration of a compound into the posterior segment. Whereas this method certainly enhances penetration, its invasiveness can result in significant complications (e.g., endophthalmitis, retinal detachment, cataract formation) and negatively impacts patient compliance [8].

9.2 The Barriers

9.2.1 Precorneal Barriers

Permeation of a compound via the conventional mode of ocular drug administration, the topical drop, is opposed by precorneal and corneal barriers. Upon proper eye drop application, a compound first encounters the lacrimal fluid overlying the cornea and conjunctiva. This tear film is normally in constant flux, being primarily supplied by lacrimal glands, and travels toward the lacrimal lake in the conjunctival cul-de-sac, which has a maximum capacity of 30 μL, before drainage via the nasolacrimal duct at a rate of approximately 1 μL/min [9, 10]. Compounds may enter the nasolacrimal duct and go on to be systemically absorbed via the nasal mucosa, potentially contributing to adverse side effects. These fluid mechanics are such that the instilled compound is quick to lose contact with the already limited absorptive surface of the eye, particularly when blinking and tear production are increased secondary to the irritative effects of the instilled compound, causing spillage of tears and drug onto the face [4, 11].

9.2.2 *Corneal Barriers*

If the compound successfully contacts the corneal surface, it is yet again faced with an array of barriers to permeation. These barriers are contained within the structural organization of the cornea. The most anterior aspect of the cornea comprises five to seven layers of cells joined via occluding junctions, which limit interstitial drug permeation [12, 13]. Due to the impaired paracellular approach, hydrophobic compounds that penetrate the cell membranes by crossing the epithelium intracellularly are favorable. This concept is illustrated by the compound benzalkonium chloride which, in addition to being a preservative widely used in ophthalmic formulations, enhances corneal permeability by disrupting epithelial tight junctions and increasing the space between cells [14–16].

Once a substance traverses the corneal epithelium, it is then opposed by the hydrophilic corneal stroma, with a thickness of approximately 450 μm [17]. The corneal stroma is largely acellular, being comprised of collagen fibrils and glycosaminoglycans, accounting for its highly aqueous nature. The high water content may potentially impede passage of significantly hydrophobic compounds, though there appears to be no significant correlation between stromal permeability and a compound's partition coefficient [13]. The fibrous framework of the stroma may impact permeability of larger compounds, posing a barrier when molecular radius exceeds one angstrom [13].

The internal corneal layers consist of Bowman's and Descemet's membranes, followed by the corneal endothelium. The membranes do not pose a significant obstacle to drug penetration [4]. The endothelium, with its single cell layer and significant interstitial space, may be significantly rate limiting only for hydrophobic molecules [13, 18]. While the epithelium is the preeminent corneal barrier, particularly for hydrophilic compounds, the stroma and endothelium may pose as minor obstacles to hydrophobic compounds. Given the tight barricade of cell membranes, which shape the corneal epithelium, transcorneal passage is more amenable to hydrophobic or electrically neutral compounds.

9.2.3 *Aqueous Humor*

The anterior chamber is the space between the iris, chamber angle, and cornea. It is filled with aqueous humor, which is a fluid that provides nutrients and oxygen to the surrounding tissues and carries waste out of the eye. The aqueous humor maintains an avascular cornea and an optimal lens transparency for light transmission. The liquid is secreted by ciliary processes into the posterior chamber and exits the eye through the trabecular meshwork and Schlemm's canal at a flow rate of 2–5 uL/min [19]. While there is some contribution to drug clearance from the turnover of aqueous humor, most of the clearance is due to the metabolism and systemic uptake by the vascular tissues of the uvea.

9.2.4 Conjunctival-Scleral Barriers

The conjunctiva comprises of a stratified squamous epithelium with underlying neurovasculature. A topically applied compound may be redirected from its intended course by absorption into the conjunctival vasculature, which drains into the systemic circulation. This phenomenon, like the nasal diversion previously described, impairs ocular permeation and promotes adverse systemic effects. If a compound permeates the tear film and avoids uptake into the vessels of the conjunctiva, it may succeed in traversing the conjunctival membrane itself. This membrane is relatively more permeable than the corneal apex, particularly with regards to hydrophilic compounds—this enhanced permeability may be attributed to the comparatively larger absorptive surface and "leakier" tight junctions, which better facilitate a paracellular approach [13, 20, 21]. The compound then converges on the sclera which, compared to the conjunctiva and cornea, is more permissive of larger compounds and where passage is independent of hydrophobicity [13, 22]. Mechanisms do exist for subconjunctival drug delivery, which involves placing a compound in direct contact with the scleral surface. Following scleral clearance, the molecules can gain access to the uvea.

9.2.5 Lens Barriers

The lens with its pigmented tissues may act as a drug reservoir for the anterior chamber. Approximately one-third of the lens is made of crystallin, while the lens nucleus is made of a hard condensed cellular material that resists drug penetration. However, drug accumulation in the lens is known to result in cataract formation. Ahmed et al. showed that timolol penetration into the vitreous body is via diffusion around the epithelial capsule and peripheral cortical layers of the lens [23].

9.2.6 Vitreal Barriers

A well-established mechanism of enhancing permeation is direct injection into the vitreous. The vitreous body is a hydrogel of collagen, hyaluronic acid, and water, which has a turnover of approximately 10–50 min. This turnover quickly eliminates drugs in the vitreous body. The invasive nature of this approach is associated with undesirable effects such as endophthalmitis, retinal detachment, or cataract formation [24].

9.2.7 *Blood–Ocular Barrier*

The blood–ocular barrier is comprised of the blood–aqueous barrier and the blood–retinal barrier. These prevent toxic substances from entering the eye and preserve physiological homeostasis [25]. The blood–aqueous barrier consists of the endothelium of Schlemm's canal, the endothelium of the iris vessels with non-fenestrated tight junctions, the posterior iris epithelium, and the nonpigmented epithelium of the ciliary body. The blood–retinal barrier is made up of the endothelial membrane of the retinal vessels and the retinal pigment epithelium, both of which contain non-fenestrated tight junctions [26]. Due to these junctions, orally or intravenously administered drugs have a limited ability to pass through the vasculature into the intraocular chambers to reach therapeutic levels [27]. Small protein tracers like microperoxidase (19 kDa) have been demonstrated to access the periaxonal space but not the retina [27].

As compared to the blood–retinal barrier, the blood–aqueous barrier does allow for some amount of molecular diffusion. After intravenous (IV) injection, test substances such as phosphate, inulin, sucrose, urea, proteins, potassium, sodium, chloride, and some antibiotics were found on the anterior side of the vitreous humor from the ciliary circulation. However, the substances were unable to reach the retina due to the blood–retinal barrier [28].

9.3 The Penetration Enhancers

9.3.1 *Liposomes*

Liposomes are artificial phospholipid bilayers that are biocompatible with the human body [29]. They have demonstrated applicability across all modes of ocular drug delivery (topical, intravitreal, subconjunctival, IV, and implantable devices) [30]. These vesicles comprise an aqueous compartment encapsulated by a hydrophobic shell. This configuration and the many ways it may be adapted make liposomes promising penetration enhancers because chemical modifications facilitate expression of desired properties [31]. Studies have shown that designed liposomes can increase bioavailability, bioefficacy, and sustained release of drug molecules [29, 32, 33]. The shell can be further altered to change the construct's hydrophobicity, fluidity, circumference, and electrical charge [30, 34, 35]. Most liposomal formulations comprise of phosphatidylcholine membrane components along with cholesterols and other lipid-associated constituents [30]. Additionally, various compounds (e.g., antibodies, peptides, etc.) may be integrated into the outer membrane, which aid in guiding the contents to desired targets [36, 37]. Similarly, polymers and bioadhesive components may be incorporated into the liposome for further penetration enhancement [30]. Surface modifications may also enhance the integrity of delivered compounds. In one study, polyethylene glycol (PEG) was integrated into the surface of a liposome, which induced an aqueous film around the liposome as a means of steric stabilization within the vitreous body [38].

Liposomes measure from 10 nm to 10 μm in diameter, and their hydrophobic exterior enables passage through cell membranes (such as those of the corneal epithelium) along with the enclosed hydrophilic components [4]. It should be noted that liposomes may also deliver hydrophobic components by encapsulation or integration into its lipid phase [30, 39]. Liposomes can be designed to respond to different triggers such as temperature changes, electromagnetic waves, and pH environment [40]. Liposomes could also be designed to degrade over time by variations in shell width, composition, and particle size, thus facilitating controlled release of entrapped molecules [40].

As explained previously, the corneal membrane typically bars penetration of unencapsulated hydrophilic substances. Liposomes present an efficient method for topical ocular drug delivery to entrap and protect solutes with varying sizes, properties, and levels of hydrophobicity and hydrophilicity en route to target structures through otherwise formidable barriers [31, 32, 41–43]. In rabbits, liposomal pilocarpine formulations have been demonstrated to sustain efficacy in comparison to standard eye drops and subsequently halved the number of required dosing intervals [44]. Annexin A5, a phospholipid binding protein, has been shown to enhance penetration of liposomes containing the anti-vascular endothelial growth factor (VEGF) antibody bevacizumab into the posterior segment of the eye [45]. Anti-VEGF antibodies are increasingly used in the treatment of ARMD, but intravitreal complications make alternative topical application an appealing prospect. Annexin A5 is believed to enhance liposomal uptake and transcytosis across the corneal epithelium, which may be due to formation of endocytic vesicles upon binding phosphatidylserine residues on the surface of epithelial cells [46, 47]. Compared to topical administration of free bevacizumab, encapsulation in annexin-A5-containing liposomes achieved significantly higher concentrations in the posterior segment of both rat and rabbit eyes [45].

The aforementioned properties of liposomes may also promote the efficiency and efficacy of intravitreal injection [31]. For instance, in treating cytomegalovirus (CMV) retinitis, liposomal formulations of the antiviral agent ganciclovir injected into rabbit vitreous demonstrated enhanced therapeutic concentrations lasting up to two weeks postinjection and no retinopathic effects as compared to free ganciclovir [48]. Liposomal bevacizumab injected into rabbit eyes achieved a greater mean vitreal concentration after intravitreal injection at both 28 days (48 μg/mL vs 28 μg/mL) and 42 days (16 μg/mL vs 3.3 μg/mL) as compared to free bevacizumab [49]. In rats with endotoxin-induced uveitis, intravitreal injections of liposomal vasoactive intestinal peptide (VIP) demonstrated 15 times greater concentration in both aqueous humor and vitreous body as compared to VIP/saline injection [50]. In a rat model for choroidal neovascularization, liposomes incorporating an angiogenic vessel-homing peptide and encapsulating an angiogenesis inhibitor (SU5416) led to significant reductions in neovascularization as compared to intravitreal injection with soluble SU5416 [37].

Subconjunctival introduction of liposomal formulations may act to attenuate lymphovascular drainage of the desired compound, decrease systemic effects, increase retention, and facilitate steady release, thereby increasing the opportunity for drug interaction with the target site [30]. Various liposomal formulations, including low-molecular-weight heparin, streptokinase, and clodronate, have been investigated [42, 51–53]. Natarajan and colleagues constructed latanoprost-loaded liposomes to evaluate therapeutic evaluation of the nanomaterial in rabbit eyes. The group developed latanoprost-loaded egg-phosphatidylcholine (EggPC) liposomes using a thin film hydration technique [54]. The size of the liposomes averaged 109 ± 18 nm with a narrow polydispersity index (PDI). The latanoprost-loaded EggPC liposomes were stable with respect to aggregation for at least 6 months stored at 4 °C and for at least 1 month at 25 °C. Furthermore, a single subconjunctival injection of the latanoprost-loaded EggPC liposomes effectively lowered intraocular pressure (IOP) in rabbit eyes for at least 90 days. The effect was significantly greater than once-daily topical latanoprost controls [54].

With regard to IV applications, a liposomal formulation containing verteporfin is currently being marketed for treatment of ARMD. Following IV injection, the diseased retina is irradiated via laser, thereby activating the liposomal formulation within the targeted region [31]. Rostaporfin is another agent with a similar administration and mechanism of action that is being investigated for treatment of ARMD [30].

9.3.2 Niosomes

Niosomes are structurally similar to liposomes in that they are also vesicular bilayers. However, niosomes differ in their specific bilayer components. The outer layer of a niosome comprises combinations of surfactant and cholesterol, which forms an anionic exterior as opposed to the hydrophobic phospholipid shell found in liposomes. These distinguishing membrane characteristics reportedly give niosomes several advantages over liposomal formulations: increased chemical stability, decreased toxicity due to the anionic outer shell, and lower production and storage costs [55, 56]. When topically applied, the lipid bilayer protects the internal contents from enzymatic action while simultaneously engaging the corneal surface and facilitating controlled release of the contained compound [57]. As with the liposome, the inner portion of the niosome is an aqueous compartment.

In the context of enhanced ocular drug delivery, there are several examples of niosomal formulations incorporating various combinations of cholesterols and surfactants. They have mostly been applied to the enhancement of corneal penetration in the context of topical administration and have, thus far, demonstrated minimal irritation to ocular tissue. Saettone et al. constructed polysorbate niosomes encapsulating the muscarinic antagonist cyclopentolate. The niosomal form demonstrated enhanced in vivo penetration as compared to a control solution containing cyclopentolate [58]. Increased incorporation of surfactant may increase dispersion within the niosome membrane with subsequent enlargement and flattening into a disk-shaped capsule, termed a discome [56, 59]. The increased volume within these larger, modified niosomes facilitates encapsulation and transport of larger amounts of a compound. Discomes were shown to carry 11 % more timolol maleate as compared to niosomes [60]. Characteristics of these larger formulations may also include enhanced conformation within the ocular cul-de-sac and better resistance to outflow into the nasolacrimal duct and/or systemic vasculature, resulting in enhanced retention and bioavailability [56].

Abdelkader et al. developed niosomal and discomal formulations of naltrexone, an opioid antagonist used in the treatment of diabetic keratopathy [57, 61]. The niosomal formulation demonstrated a 1.6-fold increase in permeation across excised bovine corneas. No irritative effects were observed upon instilling the niosomes onto hen chorioallantoic membrane [60].

Timolol maleate was incorporated within a niosomal formulation. Following topical administration to rabbit eyes, the niosomal form demonstrated almost 1.7 times greater penetration into the aqueous humor as measured via a microdialysis technique. Furthermore, these higher concentrations were sustained for nearly 2 h. Interestingly, timolol release from the niosomes into the aqueous humor was so sustained that it effectively caused a second elevation in drug concentration, presumably via concentration secondary to the inhibitory effects of timolol on aqueous humor production [62]. Reductions in intraocular pressure were noted for 8 h after administration, 6 h longer than with free timolol maleate [62]. This highlights that enhanced penetration and retention decreases the required dosages of a drug and reduces the frequency of administration, thus increasing patient compliance and decreasing toxicity and adverse systemic effects.

Maiti et al. showed that variations in the cholesterol–surfactant ratio within the niosome surface provide a mechanism to modulate the amount of drug contained within the aqueous cavity along with the kinetics of drug release. Using a film hydration technique, they created niosomes encapsulating brimonidine tartrate, an anti-glaucoma medication acting as an alpha-agonist to increase aqueous humor outflow through the trabecular meshwork. In vivo administration to rabbit eyes demonstrated enhanced efficacy that was sustained for nearly 7.5 h. Additionally, no irritating effects were noted [63].

Another study incorporated the antimicrobial agent gentamicin within niosomes whose membranes varied in the types of surfactant and the levels of surfactant and cholesterol. In vitro testing revealed that the niosomal formulations demonstrated variations in how much gentamicin was contained and how slowly it was subsequently released, based on the particular membrane composition. In vivo analysis demonstrated that administration was well tolerated and without overt irritant effects [64].

9.3.3 Cyclodextrins

Cyclodextrins (CDs), or cyclic oligosaccharides, are a class of molecules made up of six to eight sugar molecules arranged in a ring [65]. They display an outer hydrophilic surface surrounding an inner, hydrophobic chamber. This composition has found application in drug delivery as CDs can form inclusion complexes with water-insoluble compounds, ensure compound stability, and effectively transport the molecule through an aqueous medium [65–68]. Various types of CDs have been studied with respect to enhanced ocular penetration, with an emphasis on enhanced retention at the corneal surface and subsequent penetration. 2-hydryoxypropyl-β-CD (HPβCD) is a cyclodextrin with a complexed hydroxypropyl moiety designed to augment the characteristic water solubility [65].

Inclusion of the antifungal agent ketoconazole within HPβCD complexes demonstrated 12-fold and eightfold greater enhanced corneal penetration within the cornea and aqueous humor, respectively, as compared to standard suspension [69]. Administration of HPβCD to rabbit eyes has been shown to cause no tissue irritation [70]. Formulations of the anti-inflammatory agent indomethacin complexed with HPβCD demonstrated enhanced stability and prolonged release in vitro when compared to commercially available formulations. The cyclodextrin–indomethacin composite also demonstrated improved efficacy upon topical administration as compared to established eye drop formulations [71].

In vitro studies have shown enhanced penetration of dexamethasone through a semipermeable membrane and through porcine sclera with increased HPβCD inclusion [72]. Furthermore, inclusion within both HPβCD and γCD facilitated a synergistic effect with increased dexamethasone solubility and penetration [72]. This highlights the ability to further control the rate and extent of dissolution and penetration by employing differing formulations in combination.

Another study also analyzing dexamethasone penetration incorporated the compound with a CD nanogel composed of cross-linked γCD units [73]. These nanogels were further incorporated within an HPβCD-containing medium administered to rabbits. As compared to a commercially available eye drop, the cyclodextrin formulation was present within the tear fluid for at least four additional hours and achieved a threefold increase in aqueous humor concentration. No overt signs of irritation or toxicity were noted [73].

Morrison et al. demonstrated enhanced solubility and diffusion of riboflavin upon inclusion with two separate CDs [74]. Incorporation with αCD and βCD resulted in a 1.4-fold and 2.4-fold increase in solubility, respectively [74]. In vitro studies showed enhanced permeation through bovine cornea with βCD (threefold) and HPβCD (1.5-fold) incorporation as compared to free riboflavin [74].

The anti-glaucoma cannabinoid delta-8-tetrahydrocannabinol (Δ^8-THC) has also been studied as a CD, inclusion. Incorporation with HPβCD, randomly methylated βCD (RMβCD), and sulfobutyl ether-βCD (SβCD) showed remarkably increased aqueous solubility by up to 12,000-fold [75]. In vitro transcorneal permeation of Δ^8-THC was also increased upon CD incorporation from previously near-undetectable levels [75].

Zhang et al. generated doxycycline–HPβCD complexes which demonstrated enhanced stability of the doxycycline molecule with decreased formation of degradation products, which may be a consequence of increased protection of vulnerable sites by inclusion within the hydrophobic CD cavity [76].

Of note, incorporation of CDs , with other penetration enhancers such as the previously mentioned liposomes or niosomes has been proposed as a potential avenue for further magnified permeation across ocular barriers [66, 71].

9.3.4 Dendrimers

Dendrimers are synthetic, spherical polymer compounds composed of a core cavity surrounded by side chain portions called dendrons [31]. The core acts to contain an active compound (e.g. drug) and thereby increase its solubility and retention, effectively enhancing penetration [77]. The compound is retained within the core cavity by physical, non-covalent encapsulation [77]. The core is especially conducive to transporting hydrophobic molecules and may act to enhance transit through aqueous barriers [52, 78]. Alternatively, dendrimers may also enhance penetration of compounds via covalent attachment to the outer surface [77]. The branching structure formed by the dendrons, which act as reactive moieties, enables synthesis of dendrimers of differing sizes and with various physical and biochemical properties [52, 79]. Smaller dendrimers may demonstrate greater transport between cells [80]. The terminal components of the dendrons may be modified to elicit desired biological characteristics that enhance penetration by increased cellular uptake and distribution throughout the desired tissue [81]. Furthermore, addition of neutral or anionic components to dendrons increases biocompatibility and reduces toxicity by attenuating deleterious interactions with negatively charged cell membranes [77, 82–84].

Vandamme and colleagues utilized poly(amidoamine) (PAMAM) dendrimers with amine and carboxylic acid terminal ends to deliver pilocarpine and tropicamide more efficiently than commercially available Carbopol eye drop formulations [85]. While the PAMAM dendrimers had similar physical properties to Carbopol, the dendrimers exhibited significantly lower ocular irritation and greater bioavailability. Although initial lacrimal drainage washed out small amounts of the dendrimers, the dendrimers also exhibited improved mucoadhesive properties [85].

Holden and colleagues also developed a dendrimer hydrogel using amine PAMAM dendrimers with PEG acrylate chains as cross-linkers on the surface and brimonidine or timolol maleate stored in the hydrophobic cores [86]. When the gel was exposed to light, the PEG acrylate chains cross-linked to form a water-soluble viscous solution. Topical application of the gel increased the drug molecule concentration in the epithelium, stroma, and endothelium. Furthermore, the gel was nontoxic and adhered to the cornea [86]. Lastly, a carboxylic terminal group dendrimer with a quaternary ammonium salt was designed by Spataro and colleagues to stabilize carteolol, an antihypertensive agent, and it was topically delivered to rabbit eyes [87]. The carteolol–dendrimer conjugate showed no irritation after several hours and increased precorneal residence time compared to solely carboxylic terminal group dendrimers, and the carteolol concentration delivered by the carteolol–dendrimer conjugates in the aqueous humor was approximately 2.5 times greater than that delivered by carteolol solutions [87]. Yang et al. topically administered a hybrid dendrimer hydrogel/PLGA nanoparticle platform modified with an acrylate-carrying PEG network to rabbit eyes. They observed an effective and sustained IOP reduction over a period of time [88]. Moreover, there were significantly greater local concentrations of brimonidine (aqueous humor and cornea) and timolol maleate (aqueous humor, cornea, and conjunctiva) for 7 days as compared to a saline control [89].

Dendrimers have also been utilized in a process termed photochemical internalization, acting as photosensitizing agents to enhance penetration of gene therapy [90]. The dendrimer core is comprised of the photosensitizer phthalocyanine encapsulating a DNA/polycation complex, surrounded by associated dendrons. The complex is incorporated intracellularly via an endosome, at which point photosensitization promotes disruption of the endosome membrane and release of its contents. In vitro, this dendrimer complex was shown to enhance gene delivery by over 100 times when compared to administration of the DNA/polycation complex alone. Subconjunctival injection of the dendrimer complex in rat eyes demonstrated successful gene transfection localized to the desired sites that were subjected to laser irradiation [90].

Dendrimers were applied to intravitreal injections in treatment of ARMD and retinitis pigmentosa in a study that utilized the innate affinity of PAMAM dendrimers to the neuroinflammatory areas [91]. In a rat model, injection of dendrimers loaded with fluocinolone acetonide showed enhanced targeting of the retinal microglia that potentiate the inflammatory response when compared to injection of the free compound. The dendrimeric form also demonstrated enhanced retention within the target tissue [91]. Intravitreal injection of dendrimers also had proved effective in facilitating anti-VEGF penetration to target sites. By protecting the antibody against endogenous nucleases, the dendrimer–antibody complexes were capable of accessing the choroidal membrane and significantly decreased the extent of neovascularization as compared to exposed antibody injection [92].

9.3.5 Other Nanoparticles

Bhagav and colleagues used brimonidine tartrate (BRT)-encapsulated Eudragit nanoparticles to study the prolonged release of BRT [93]. The Eudragit nanoparticles are inert polymeric resin copolymers used to entrap lipophilic drugs. The IOP-lowering efficacy of the nanoparticles was evaluated in α-chymotrypsin-induced glaucomatous rabbits. BRT nanoparticles were prepared as an aqueous dispersion in phosphate-buffered saline (PBS) and compared to a conventional ophthalmic drop, Iobrim E/D. In one group, two drops of the nanoparticle solution were placed in the lower cul-de-sac of the left eye. In a second group, two drops of the Iobrim were placed in the same location. For both groups, normal saline was administered in the same location of the right eye. Results showed that the nanoparticles were well tolerated, with no signs of irritation or toxicity. Moreover, the BRT nanoparticles decreased the elevated IOP in rabbits with glaucoma for a longer period of time compared to the conventional drop formulation [93].

Tártara and colleagues similarly studied the performance of 6-O-lauryl-L-ascorbic acid (coagel) nanostructures as a carrier for acetazolamide (AZM) in ophthalmic administration in rabbit eyes [94]. Rabbits were randomly allocated into four treatment groups: 0.1 % AZM-loaded ASC12 coagel, 0.4 % AZM-loaded ASC12 coagel, AZM–Ringer solution, and Azopt (brinzolamide ophthalmic suspension). IOP was measured four times at 30, 60, 90, 120, 150, 180, and 240 min following drug application. In an ex vivo study, the transcorneal permeation of coagel–AZM 0.1 % and 0.4 % was approximately three and six times greater than the transcorneal permeation of AZM–Ringer and Azopt for the same drug concentration, respectively. The in vivo study revealed that the hypotensive effect (HE) for coagel–AZM 0.1 %, coagel–AZM 0.4 %, and Azopt each had a similar IOP endpoint after 4 h. However, the Azopt HE diminished after 2 h, while the coagel–AZM 0.4 % stayed at the maximum HE until 3 h. Moreover, the AZM concentration for the coagel–AZM 0.4 % was 2.5-fold lower than the Azopt concentration, suggesting that the nanocarrier had greater permeability efficiency. These results show that the coagels promoted increased AZM permeation, ocular bioavailability, and absorption [94].

9.3.6 *Natural Polymers*

Natural polymers are those found and extracted from the environment. These are often condensation polymers, which involve the loss of water as a by-product. The end behavior of natural polymers used for ocular drug delivery tends to be less predictable than synthetic polymers [95]. Some naturally occurring polymers include chitosan, gelatin, starch, collagen, sodium alginate, sodium hyaluronate, xanthan gum, gellan gum, and albumin [95]. Chitosan, gelatin, and collagen have been the most widely studied of all natural polymers for drug delivery [95].

Chitosan is a linear polysaccharide composed mostly of D-glucosamine in β(1,4) linkage. It is produced via the deacetylation of chitin, which is a component of the exoskeleton of arthropods. Chitin possesses many physical properties important for ocular drug delivery and, being a mucoadhesive polymer, possesses a penetration-enhancing nature and excellent ocular tolerance [96]. Its mucoadhesive quality improves ocular contact time and is attributed to its positive charge, which interacts with the ocular tissue's negative charge [97]. Chitosan has been shown to open the tight junctions between the epithelial cells without disrupting cellular integrity and result in increased paracellular absorption [97].

Collagen is the most abundant protein in the human body and is present in large quantities in the cornea, sclera, and stroma. Bovine and porcine collagen is utilized for ocular drug delivery due to its similarities to human collagen. Its physical and chemical properties can be easily optimized to adjust for different delivery methods. Collagen is widely used in corneal shields because of its absorption, biocompatibility, and low immunogenicity [98]. These shields improve drug contact time and are biocompatible. Adversely, collagen shields can interfere with vision and cause patient discomfort [98].

Gelatin is obtained by hydrolyzing collagen from animal by-products. It is a protein that is highly biocompatible and biodegradable. Gelatin is easily manipulated in order to change ocular delivery properties, and it has been shown to be an effective polymer for ocular drug delivery given the large amount of time it can stay at the ocular surface [99]. This can be accomplished by changing the electrical and physical properties of gelatin, thus creating an optimized ocular delivery system. Lai et al. were able to effectively deliver an ocular pilocarpine solution for the treatment of glaucoma [100].

9.3.7 Synthetic Polymers

Synthetic polymers can be produced with predictable properties such as permeability, molecule size, degradation, solubility, and rate of dissolution. Given these designable advantages, this type of molecule is incorporated into many ocular drug delivery nanoparticle constructs.

Polylactide (PLA) is a biodegradable polyester polymer synthesized from organic sources such as sugarcane and cornstarch. PLA can also be copolymerized with poly-glycolic acid to form PLGA. This copolymerization can be adjusted for different polymeric ratios, therefore allowing a great deal of control over physical and chemical properties [101]. These polyester polymers degrade though nonenzymatic and enzymatic hydrolysis into products that are not harmful to living organisms [102]. In a 2007 study, researchers were able to synthesize PLGA microspheres containing ganciclovir and intravitreously administer the medication to rabbit eyes with prolonged maintenance of concentrations [103].

Polyvinylpyrrolidone (PVP) is a class of synthetic polymers that is biocompatible with the vitreous and is employed in the production of hydrogels that can be used as vitreous substitutes. Polymer elimination from the vitreous occurs through phagocytosis, and the degraded products are not harmful to humans [104]. PVP-based degradation properties can be modulated by varying polymer cross-linking density [105]. Hacker et al. studied a poly(propylene fumarate)/PVP for the sustained release of acetazolamide, dichlorphenamide, and timolol maleate and were able to successfully attain steady drug release rates [100, 106]. It is important to note that PVP-based hydrogels have been associated with side effects, such as intravitreal opacity, hazy corneas, and certain degrees of inflammation [107].

Polyanhydrides are a biocompatible and biodegradable class of polymers joined together by anhydride bonds. This molecule degrades in vivo by surface erosion and not drug diffusion, which makes this class ideal for drug delivery devices [108]. Subclasses of polyanhydrides include aliphatic, unsaturated, and aromatic. These subclasses differ mostly on the rate of degradation. Aliphatic polyanhydrides take a few days to degrade, while aromatic ones may take a few years. Polyanhydride disks containing 5-flurouridine have been shown useful in a glaucoma filtration surgery model of primates [109].

Poly-ε-caprolactone (PCL) is a biocompatible, biodegradable, semicrystalline polymer [110]. This molecule's degradation occurs in two steps. The first is a cleavage of the ester linkage. The second involves chain scission of low-molecular-weight fragments from the larger initial polymer [111]. PCL has a slow degradation time making it quite suitable for long-term drug delivery systems in the form of microspheres, nanospheres, and implants [110]. The degradation rates of this molecule can be modified by copolymerizing with faster degrading polymers. PCL has been successfully studied in the ocular drug delivery of dexamethasone and triamcinolone [112, 113].

9.3.8 Nano-implants

Implants present a unique permeation enhancement technology for intraocular sustained-release drugs for the treatment of posterior segment diseases. The dimensions and material of the implant dictate the class and concentration of drug molecules that can be delivered to the target site. The molecule can be enclosed inside the core of the implant, which can be further encased by a polymeric layer. A concern for implants is that they may necessitate follow-up surgery for removal or that the implant will remain inside the patient indefinitely. Biodegradable implants, which have been approved for clinical use, could avoid this potential dilemma [114, 115].

Vitrasert®, a nonbiodegradable implant that contains the antiviral drug ganciclovir, is used in the treatment of patients with CMV retinitis and acquired immunodeficiency syndrome. Ganciclovir implants have been found to deliver drugs for approximately 8 months and achieve higher intravitreal levels than can be achieved by maintenance IV therapy [116–119]. Retisert™ and Medidur™ are nonbiodegradable implants that contain fluocinolone acetonide, an anti-inflammatory drug, for the treatment of noninfectious uveitis. Fluocinolone implants are also under investigation for applications in chronic inflammation, diabetic macular edema, and macular degeneration [120]. Ozurdex®, which is a matrix reservoir system containing dexamethasone (700 mg) in PLGA, is a biodegradable implant that is clinically approved for the treatment of macular edema following branch retinal vein occlusion or central retinal vein occlusion as well as for the treatment of noninfectious uveitis affecting the posterior segment of the eye [115].

Implants have also been used to deliver nanoparticles into the eye. Jung and Chauhan developed a fornix insert to continuously release timolol in vitro [121] to lower IOP. Timolol was cross-linked to propoxylated glycerol triacrylate nanoparticles by an ester bond. The nanoparticles were then mixed with hydroxyethyl methacrylate. From their design, timolol could be dispensed over a range of 10–30 days by adjusting the fraction of the particles in the insert or the amount of drug loaded in the insert [121].

9.3.9 Iontophoresis

Iontophoresis is a noninvasive physical approach that precisely controls the flux of drug molecules and nucleic acids by varying an applied electric current [122–124]. This technique minimizes trauma and risk of infection and serves as an alternative to invasive injections. Advantages of iontophoresis for treating local conditions are the high local drug concentration, avoiding repeated intravitreal injections, and minimal systemic uptake of drugs, thus decreasing the occurrence of adverse systemic side effects [125–127]. Iontophoresis accurately regulates drug delivery over a period of time and facilitates the transport of charged molecules of high molecular weight [128]. Inter- and intra-subject drug concentration variability is largely decreased because the rate of drug delivery is proportional to the applied current [129–131]. Thus, dosage regimens can be modified on an individual basis to transport drug at predetermined rates [132]. In ophthalmology, transcorneal and transscleral iontophoresis are the two main applications of this technique [24].

Transcorneal iontophoresis has been shown to supply and maintain high concentrations of drugs into the cornea and the aqueous humor [122, 133–137]. Studies have demonstrated that antibiotic transcorneal iontophoresis may be an effective method to manage corneal ulcers. Tobramycin iontophoresis in a rabbit model of corneal ulcers supplied a significantly greater corneal concentration than by topically applied drops [135].

Iontophoresis may also improve drug delivery for corneal infections. In a study to compare treatment efficacy by iontophoresis or topical instillation of tobramycin in rabbit eyes after intrastromal inoculation with *Pseudomonas aeruginosa*, Rootman et al. found that tobramycin iontophoresis resulted in significantly fewer bacterial colonies in the cornea as compared to topical instillation [138]. Frucht-Pery et al. examined gentamicin iontophoresis penetration efficiency into rat corneas by varying initial gentamicin concentration in agar, iontophoresis duration, and current densities. High concentrations of gentamicin were observed in all treated corneas. Increasing the gentamicin concentration of the agar or increasing the duration of iontophoresis was found to be associated with greater gentamicin concentrations in the cornea [139]. Frucht-Pery et al. also evaluated iontophoresis of a hydrogel gentamicin-containing probe into normal and *Pseudomonas aeruginosa*-infected rabbit corneas. As with the rat studies, corneal penetration increased with current intensity or duration of iontophoresis. While gentamicin concentration in the corneas could range from 88.60 ± 38.64 to 0.10 ± 0.89 mg/mL, therapeutic levels were achieved at current intensities of 0.1 mAmp or greater. The logarithmic value of *Pseudomonas* colony-forming units decreased in the iontophoretic treated group as compared to the control group [140, 141].

Circumventing the corneal–lens–iris barrier, transscleral iontophoresis is the application of current through the pars plana of the sclera. This method has been shown to successfully deliver effective doses of ganciclovir, foscarnet, gentamicin, cefazolin, vancomycin, bevacizumab, and ticarillin into the vitreous body for treatment of posterior segment diseases, such as CMV retinitis and endophthalmitis [122, 127, 136, 137, 142–145]. Barza et al. demonstrated that a combination therapy by iontophoresis and intravitreal injection of gentamicin in an endophthalmitis rabbit model resulted in significantly lower numbers of bacterial colonies when compared to an injection alone [146]. However, this process may also be associated with damage to the choroid and retina. Barza et al. reported that after 3 h of a transscleral iontophoresis of gentamicin and cefazolin in rabbits, side effects varied and included hemorrhagic necrosis, edema, or inflammatory infiltration of the retina, choroid, and ciliary body [144].

Davies et al. demonstrated the transport of dyes and DNA across full-thickness human scleral fragments, which were embedded vertically in an agarose gel and positioned to completely span individual gel lanes. Xylene cyanol, bromophenol blue, or DNA was loaded into the wells and electrophoresis was carried out at about 3.3 V/cm for approximately 2 h per run. Agarose gel electrophoresis showed that the two dyes passed through the sclera in the presence of an electric field as measured by digital time-lapse photography or DNA extraction and purification from the gel. Additionally, SYBR green staining showed that small DNA and linear double-stranded DNA successfully passed through the sclera [124].

Hayden et al. evaluated the transcorneoscleral Coulomb-controlled iontophoresis (CCI) efficacy and dose–response of carboplatin for a murine model of retinoblastoma. In 30 mice, six serial iontophoretic treatments were administered two times per week using a current density of 2.57 mA/cm^2 for 5 min. Fourteen animals received carboplatin treatments at concentrations of 1.4, 7.0, 10.0, or 14.0 mg/mL without current, while 10 control mice underwent treatment with balanced saline solution. A dose-dependent inhibition of intraocular tumors after repetitive iontophoretic treatment was found. 50 % of the treated eyes exhibited tumor control at carboplatin concentrations of 7 mg/mL. In carboplatin concentrations under 10 mg/mL, no corneal toxicity was observed [147]. In rabbit eyes, they also compared the pharmacology and toxicity of a single IV carboplatin infusion of 18.7 mg/kg of body weight, a single subconjunctival carboplatin injection of 5.0 mg/400 μL, and a single application of 14 mg/mL carboplatin delivered by CCI of 5.0 mA/cm^2 over 20 min. Animals were euthanatized after each treatment, and eyes were analyzed at 1, 2, 6, or 24 h for carboplatin concentration in ocular structures. The retina, choroid, vitreous humor, and optic nerve were found to have higher levels of carboplatin after subconjunctival injection or iontophoretic delivery than with IV administration. Carboplatin concentrations were also found to be significantly higher in blood plasma after IV delivery than after subconjunctival injection or CCI [148].

Voigt et al. compared the pharmacokinetics and safety of 10 mg/mL acetylsalicylic acid delivered by transscleral CCI and IV injection in forty-one rabbits. Blood and pharmacokinetic distribution was studied at 0.5, 1, 2, 4, 6, and 8 h after the treatments. Salicylic acid (SA) was found in all tissues and fluids throughout the study period. CCI resulted in the highest concentrations of SA in all tissues, whereas IV administration revealed a delayed tissue peak of salicylate at 2 h after administration. At 8 h, ocular SA concentration in both groups was approximately equal. However, the blood plasma level of salicylate in the IV group was 28 times greater than that of CCI group and remained significantly elevated until 8 h after the treatments. Thus, CCI may be advantageous to avoid adverse systemic side effects [149].

Behar-Cohen compared hemisuccinate–methylprednisolone (MP) pharmacokinetics, safety profile, and intraocular distribution using a transscleral CCI system or after IV injection. Six rabbit eyes were used for each trial arm. For the CCI system, 62.5 and 150 mg/mL of MP at various intensities of current and duration of treatment were analyzed. In rabbits serving as controls, the MP was infused in the CCI device but without the applied electric current. For the IV group, 10 mg/kg as a 62.5 mg/mL MP solution was used. Among groups that received CCI, the MP concentrations increased in all ocular tissues and fluids correspondingly to the intensities of current used and its duration. The choroid and the retina had sustained the highest levels of MP at the highest current and 10 min duration of CCI. Following CCI, no clinical toxicity or histological lesions were observed. Without the application of current, negligible amounts of MP were found in ocular tissues. CCI also achieved greater sustained tissue concentrations with negligible systemic absorption as compared to the IV group [150].

A clinical study of iontophoresis has shown promise. Cohen et al. delivered EGP-437, a dexamethasone phosphate formulated for iontophoresis, in patients with noninfectious anterior uveitis through both the conjunctiva and sclera. Forty patients received an iontophoresis treatment in 1 qualifying eye and were randomized into 1 of 4 iontophoresis dose groups: 1.6, 4.8, 10.0, or 14.0 mA-min. After a single EGP-437 treatment, 48 % of patients achieved an anterior chamber cell (ACC) score of zero at day 14. By day 28, 60 % achieved an ACC score of zero. Throughout the study, mean best-corrected visual acuity at 4 m remained relatively stable and mean intraocular pressure remained within normal range. This treatment method was well tolerated and no systemic corticosteroid-mediated effects were observed [151].

While iontophoresis may be effective for delivering ophthalmic drugs to the cornea, aqueous humor, and vitreous body [152], it involves the use of inconvenient equipment and has specific electrical requirements. To complete the circuit on the body, a drug must be applied with an electrode carrying the same charge as the drug while the ground electrode of the opposite charge is placed elsewhere [152]. Additionally, the charge of the drug can influence the effectiveness of iontophoresis. Yoshizumi et al. demonstrated that transscleral iontophoresis into the vitreous body had a higher drug concentration with a negatively charged form of ciprofloxacin than with the positively charged form [153]. Iontophoresis may also cause damage to ocular tissues. Damage can be lessened by using the lowest current densities with the shortest duration needed to ensure an acceptable amount of drug penetration [152]. Despite these difficulties, benefits of iontophoresis over other drug delivery techniques include drug delivery directly to the target site, improved consistency of drug penetration, and minimization of systemic exposure to the drug [133]. Recent advances in iontophoresis are improving the potential to treat ocular disease by providing a unique system that decreases systemic toxicity while improving the efficacy and delivery of the therapeutic molecule.

9.3.10 Microemulsions

Microemulsions are monodisperse, homogenous colloids of oil, water, and surfactant that have diameters ranging from 1 to 100 nm. They consist of one of three main compositions: direct (oil dispersed in water; "o/w"), reversed (water dispersed in oil; "w/o"), or bicontinuous. Boonme et al. showed that the percentage of water in the microemulsion composition influences the type of microemulsion that forms. In a system of isopropyl palmitate/polyoxyethylene (10) oleyl ether (Brij-97) +1-butanol (2:1)/water, systems with <15 % water resulted in reverse micelles. 15–30 % water produced the w/o type, and >35 % water developed into o/w microemulsions [154]. Microemulsions may be further classified as either ternary or pseudoternary, which refers to the absence or presence of cosurfactant in the system [155–157]. Often in microemulsion synthesis, medium-chain fatty acids and derivatives are used because they are less lipophilic than long-chain lipids and easier to emulsify. Medium-chain fatty acids have been also been shown to mediate destabilization and solubilization of membranes and open tight junctions, both mechanisms leading to increased transport of drugs. Lipid excipients have been shown to improve both paracellular and transcellular transport by different mechanisms of action [158].

The droplet size of the microemulsion dispersed phase is influenced by the composition of the microemulsion system. Attwood et al. studied the effect of varying levels of isopropyl myristate (IPM) as oil, polysorbate (P) as amphiphiles, sorbitol (S) as cosurfactant, and water for an o/w microemulsion. At fixed surfactant/cosurfactant concentrations, a linear increment of droplet size corresponded with an increase in oil content. An increase in volume fraction of IPM increased linearly with droplet radius, which was also influenced by the structure of the polysorbate, P/S ratio, and the total (P+S) content. Droplet size decreased with a change of surfactant at a constant P/S ratio [159, 160].

These colloids can be modified to specifically target sites by modifying microemulsion surfaces with specific hydrophilic polymers or by altering the microemulsion surface properties [43, 161–163]. Trotta et al. investigated the influence of modified phospholipids on the o/w region in microemulsions of soybean lecithin as oils, ethanol as cosurfactant, and water microemulsions and found that a decrease in acyl chain length corresponded with an increase in the microemulsion surface area [164].

Microemulsions are stable over a wide temperature range and can self-assemble into uniform-sized particulate products, which is beneficial for scaling up microemulsion mass production to an industrial level [165, 166]. Kahlweit et al. showed that nontoxic microemulsions of ethyl oleate as oil, soybean lecithin as amphiphile, and 1,2-alkane diols as cosurfactant were stable up to 40 °C [167–169]. Surfactants and cosurfactants can be used to facilitate microemulsion membrane permeability [170]. Drug molecules encapsulated or incorporated into microemulsion microdomains will also protect the drugs from enzymatic degradation. The polar, nonpolar, and interfacial microdomains in microemulsions can provide a range of encapsulation environments for a variety of drugs [171–178].

The encapsulation efficiency (EE), which is the percentage of the amount of drug incorporated into the microemulsion divided by the initial amount of drug used in the synthesis of the drug-loaded microemulsion, is affected by the characteristics of the drug, the components of the microemulsion, and the microemulsion synthesis procedure. Salmon calcitonin was shown to be loaded into the water phase of water/oil/water emulsion with an EE of 95 % [179]. Ma et al. demonstrated that a self-microemulsifying drug delivery system (SMEDDS) with a final high EE of 99.2 % was prepared by dispersing concentrated insulin solution (60 mg/mL) into the oil phase of a SMEDDS preconcentrate [180]. To optimize the EE for individualized microemulsion drug delivery, it will be vital to understand the drug microsomal interaction and ideal microemulsion parameters.

Studies have demonstrated that drug-encapsulated microemulsions can be successfully formulated for potential clinical application. Feng-Feng et al. hydrolyzed chloramphenicol into glycols and then encapsulated the drug in an o/w microemulsion of isopropyl myristate/(Span-20/80 + Tween-20/80 + n-butanol)/water to improve drug stability. Drug molecules were protected from bulk water in the oxyethylene hydrophilic shells, while the nitro groups of the drug remained near the surfactant molecules. Microemulsions free of alcohols also were shown to have the same characterization and effect [181, 182].

Fialho et al. developed a microemulsion-based dexamethasone eye drop that demonstrated a higher degree of bioavailability and penetration and was well tolerated by the eye [183]. Radomska et al. created a microemulsion system of retinol and its esters as the oil component, Tween-80, Tween-60, SbPC as surfactants, and n-butanol, triacetin, and propylene glycol as cosurfactants that were studied for ocular delivery. The drug-containing optimal composition proved suitable in regard to refractive index, viscosity, pH, and osmotic tension. No physical change was observed for 6 months at 20 °C, and the preparation was physiologically tolerated [184].

A microemulsion containing 1 % Brazilian propolis was shown to improve wound healing and inflammation in rats following silver nitrate-induced corneal burns. The group found that the rats treated with the microemulsion had smaller epithelial defect areas and fewer neutrophils within the cornea at 24 and 48 h after administration [185].

A potential limitation of microemulsions is their dependence on a specific ratio between its components. Dilution can result in instability and lead to disintegration. Additionally, shearing force and resulting heat required to form crude emulsions may destabilize and denature the protein during the emulsion preparation [186]. However, microemulsions represent a unique methodology to deliver hydrophilic or hydrophobic drug molecules and will require further studies to optimize drug delivery features.

9.3.11 Microneedle

Microneedles are needlelike structures with diameters on the order of microns and lengths between 100 and 1000 μm [187, 188]. These can deliver any macromolecule because the size of macromolecules is usually in the nanometer range [189]. Microneedles can be used to deliver macromolecules such as insulin, growth hormones, immunobiologicals, proteins, and peptides [189–191]. There are several categories of designable microneedle characteristics for a variety of purposes including solid or hollow and drug coated or drug encapsulating [188, 192, 193]. A recent advancement is dissolving or biodegradable microneedles for sustained release of drugs or vaccines [194]. These were developed to overcome the disadvantages of hypodermic needle and conventional drug delivery systems, such as sharp disposal, waste, needle fractures, and needle phobia experienced by patients [195, 196].

Conventional therapy for the treatment of eye disorders involved the use of eye drops, laser treatment, injections, and surgeries. While successful to some extent, these techniques are associated with unfavorable systemic side effects, risks of bleeding, loss of vision, and failure to achieve permanent treatment. Thus, there is a need to explore alternate treatments for ocular diseases. Microneedles could pierce the conjunctiva, sclera, or cornea for anterior or posterior segment drug delivery or fluid extraction [197]. Patal et al. applied minimally invasive hollow borosilicate microneedles to deliver sulforhodamine B solution and FluoSpheres suspensions into the suprachoroidal space in whole rabbit, pig, and human eyes. Optimal microneedle parameters for reliable drug delivery were determined to be a needle length of 800–1000 μm and pressure of 250–300 kPa [198].

Jiang et al. used pilocarpine-coated microneedles to administer the drug into the sclera and corneal stroma. This technique achieved a pilocarpine bioavailability almost two orders of magnitude greater in the corneal stroma than by topical delivery and induced significant and extended constriction of the pupil in the rabbit eye [199]. They also studied if inflammation or microchannel persistence occurred as a result from insertion of microneedles into the cornea of a rabbit. Using a slit lamp, they found that microchannels disappeared within 3 h, and there was no evidence of an inflammatory response [199]. A subsequent study used hollow microneedles to inject nano- and microparticles into the sclera [200]. Scleral thickness, infusion pressure, microneedle retraction depth, and concomitant use of enzymes were found to influence intrascleral injection. However, the fluid injection was successfully targeted to the suprachoroidal space by introducing the hollow microneedles at the base of the sclera [198]. Fluid injected by this method was found to flow circumferentially around the eye and could reach the macula after injections near the limbus [198].

In addition to drug injections, microneedles could be used as nonpermanent implants serving as a source for sustained drug delivery. Palakurthi et al. developed an implantable microneedle prepared from methotrexate and biodegradable polymer for the treatment of primary vitreoretinal lymphoma. In a preliminary toxicity test in rabbit eyes, no inflammatory response followed in vivo implantation [201].

Microneedles may find applications in the clinical or surgical setting. Ohashi et al. and Saeki et al. compared physiological and pharmacological changes in rat and mouse intraocular pressure using microneedle manometry, TonoLab, and TonoPen-XL. The microneedle and TonoLab provided similar readings and could detect diurnal variation and drug effects. The TonoPen-XL readings, however, were less reliable than the microneedle tonometer [202, 203].

A microfabricated needle-based cannulation system was demonstrated for application in retinal endovascular surgery. 40 retinal vein vessels in 40 porcine eyes were pierced and injected. Half were pierced with a microfabricated needle that had an outer diameter of 50 um, and the other half were pierced with a micropipette that had an outer diameter of 50 um. While the micropipette group had a 40 % and 25 % piercing and injection success rate, the microneedle group was successful in all 20 eyes. Micropipettes have been produced with fine tips and diameters and have been considered the most suitable surgical tool for retinal endovascular surgery [204–214]. However, they are fragile, delicate, and difficult to use during cannulation procedures. Microneedles may be applied as an improved surgical instrument to increase retinal endovascular surgical success [215].

Microneedles are an adaptable technology that can be designed for a variety of drug delivery purposes. Whether utilized as an implant or as a conduit for injecting solutions, microneedles may play a large role as a drug delivery technique.

9.3.12 Microspheres

Microspheres are another drug delivery system that may be an alternative to multiple intravitreal injections and surgical implantation of sustained-release devices. Microspheres are colloids of solid polymers within 1–1000 mm that can act as carriers for different antimicrobial, chemotherapeutic, and anti-inflammatory agents [216–227]. Drug molecules can be stored within the microsphere polymeric matrix or inside a reservoir system with a polymeric shell encapsulating the drug [228–230].

PLGA and PLA microspheres are the most popular due to their biodegradability and biocompatibility [231]. Soppimath et al. demonstrated that the PLGA degrades into glycolic and lactic acids that are cleared from the body [232]. In vivo studies have validated the biodegradability, safety, and feasibility of PLGA microsphere intravitreal injections [224, 233].

Drug release can be influenced by dimensional and structural microspherical features, such as polymer composition and molecular weight, drug hydrophilicity, and loading [234]. Smaller microspheres were found to have lower drug-loading and faster drug release profiles than larger microspheres [220]. Hydrophilic drugs loaded in large surface area microspheres tend to have an initial burst of drug release [218, 220, 221]. In a study by Nan et al., daunorubicin (DNR), a water-soluble antibiotic, was loaded into oxidized porous silicon (pSiO2), which was further encapsulated in a layer of PLGA. The PLGA-pSiO2-DNR model was small in size and exhibited sustained release of DNR over 74 days [235].

Microsphere size not only impacts drug release kinetics but can also cloud the vitreous. Sakurai et al. delivered nonbiodegradable fluorescein-loaded microspheres into rabbit eyes. By indirect ophthalmoscopy, they observed dense vitreous opacities with 2 μm microspheres, while mild vitreous opacities occurred when using nanospheres of 200 nm or smaller. However, microspheres of 2 μm or greater diameters were able to remain suspended in residual vitreous body or sink due to gravity [234, 236]. Future studies may need to monitor adverse visual effects of microsphere administration.

Microspheres offer advantages when compared to conventional dosage methods such as improved homogeneity, solubility, efficacy, biodegradability, and biocompatibility, as well as reduced toxicity [237–239]. Li et al. developed and characterized the therapeutic efficacy and sustained release of triamcinolone acetonide-loaded biodegradable PLGA microspheres (TA-MS) in a rabbit model of uveitis. As compared to TA suspensions, the TA-MS had a longer-term therapeutic effect on intraocular inflammation in LPS-induced uveitis. Furthermore, TA in the microsphere was released for up to 87 days as compared to only 46 days for a TA suspension [240].

Rong et al. explored neuroprotective effects of erythropoietin (EPO)-loaded PLGA/PLA microspheres on retinal ganglion cells (RGCs) in optic nerve crush rats. These microspheres sustained EPO release for at least 60 days. The microsphere-treated group exhibited reduction in RGC death and GFAP expression in the retina. This suggests that a single intravitreal injection of EPO–dextran PLGA/PLA microspheres may be a feasible protein delivery system for retinal degeneration diseases [241].

Carrasquillo et al. studied transscleral delivery of anti-$VEGF_{165}$ aptamer EYE001 (Eyetech Pharmaceuticals, New York, NY, USA) encapsulated into PLGA microspheres. Encapsulation efficiency of the drug was found to be 70–85 %. Studies in rabbit scleral tissue over 6 days showed a steady transfer rate of 2 μg/day of EYE001 across the scleral tissue regardless of the amount of drug encapsulated [22, 242].

Microspheres can protect drug molecules from metabolism. Kompella et al. assessed budesonide-loaded PLA microsphere sustained delivery to the retina. Corticosteroid budesonide could be a potential anti-VEGF treatment, but the liver can clear up to 85 % of budesonide on its first pass. The study showed that loaded microspheres were able to inhibit VEGF expression in cultured retinal pigment epithelium cells (ARPE-19) without any significant toxicity. In a rat study, the microspheres showed a superior sustained-release effect as compared to budesonide solution over 14 days [243].

This system may also serve as an important alternative to oral administration of drugs. PKC412 is a kinase inhibitor that blocks two VEGF receptors that, when administered orally, reduces retinal thickening attributed to diabetic macular edema [244]. A microsphere delivery system could prevent systemic adverse effects associated with oral administration, such as nausea and vomiting. PKC412 was encapsulated into PLGA glucose and delivered via periocular injection. Ten days after the injection into choroidal neovascularization (CNV)-induced porcine models, image analysis of the rupture sites showed smaller areas of CNV in the PKC412 microsphere-treated group as compared to control [245].

Additionally, Ayalasomayajula and Kompella demonstrated the improved delivery of subconjunctivally injected anti-VEGF celecoxib-loaded microspheres as compared with oral administration. Previous studies have shown that repeated oral dosages of celecoxib decreased diabetes-induced retinal VEGF expression and subconjunctival administration improved retinal drug delivery. Celecoxib-loaded PLGA microparticles had an encapsulation efficiency of 68.5 % and demonstrated sustained celecoxib release in vitro over a 7-week period. After posterior subconjunctival injection in rats, the microsphere group was found to have higher levels of drug in the retina, vitreous, and cornea by at least a factor of 10 after day 7 and in all tissues after day 14. The study suggests the potential for microspheres as a sustained-release system for celecoxib, to minimize drug concentration fluctuations and to improve patient compliance by decreasing adverse effects and administration frequency [246–248].

Local administration of biodegradable microspheres offers an excellent alternative to multiple drug administrations. Microspheres have desirable characteristics of controlled and sustained drug delivery, biocompatibility, biodegradability, and adjustable features for personalized therapy. Future development of microspherical formulations for different ophthalmic therapeutic uses may provide novel therapeutic systems for the benefit of patients.

9.4 Conclusion

Nanotechnology shows great promise in providing novel modifications for current and future therapies in ophthalmology. The advancement of nanoparticles and other drug delivery methodologies as penetration enhancers represents one of the most important far-reaching impact areas of medical research. Further work will be necessary for the characterization of nanotechnology-based delivery systems to refine penetration enhancers for clinical applications. These nanotechnology-based penetration enhancements signify a paradigm shift to overcome barriers and increase the efficacy of therapeutic molecules.

References

1. Cholkar K et al (2013) Novel strategies for anterior segment ocular drug delivery. J Ocul Pharmacol Ther 29:106–123
2. Wagner V et al (2006) The emerging nanomedicine landscape. Nat Biotechnol 24:1211–1217
3. Institute NN (2012) United States National Nanotechnology Initiative. What it is and how it works. Cited 10 Aug 2012
4. Ghate D, Edelhauser HF (2008) Barriers to glaucoma drug delivery. J Glaucoma 17(2):147–156
5. Macha S, Mitra AK, Hughes PM (2003) Overview of ocular drug delivery. In: Mitra AK (ed) Ophthalmic drug delivery systems. Marcel Dekker, Inc., New York, pp 1–12
6. Stone JL et al (2009) An objective evaluation of eyedrop instillation in patients with glaucoma. Arch Ophthalmol 127:732–736
7. Gupta R et al (2012) Evaluating eye drop instillation technique in glaucoma patients. J Glaucoma 21:189–192
8. Jager RD et al (2004) Risks of intravitreous injection: a comprehensive review. Retina 24(5):676–698
9. Urtti A, Salminen L (1993) Minimizing systemic absorption of topically administered ophthalmic drugs. Surv Ophthalmol 37:435–457
10. Mishima S et al (1966) Determination of tear volume and tear flow. Invest Ophthalmol 5:264–276

11. Patton TF, Francoeur M (1978) Ocular bioavailability and systemic loss of topically applied ophthalmic drugs. Am J Ophthalmol 85:225–229
12. Hornof M, Toropainen E, Urtti A (2005) Cell culture models of the ocular barriers. Eur J Pharm Biopharm 60:207–225
13. Prausnitz MR, Noonan JS (1998) Permeability of cornea, sclera, and conjunctiva: a literature analysis for drug delivery to the eye. J Pharm Sci 87:1479–1488
14. Green K, Tonjum A (1971) Influence of various agents on corneal permeability. Am J Ophthalmol 72:897–905
15. Gasset AR et al (1974) Cytotoxicity of ophthalmic preservatives. Am J Ophthalmol 78:98–105
16. Chang SW et al (2000) Benzalkonium chloride and gentamicin cause a leak in corneal epithelial cell membrane. Exp Eye Res 71:3–10
17. Robinson JR (1993) Ocular anatomy and physiology relevant to ocular drug delivery. In: Mitra AK (ed) Ophthalmic drug delivery systems. Marcel Dekker, New York, pp 29–57
18. Schoenwald RW (1997) Ocular pharmacokinetics. In: Zimmerman TJ (ed) Textbook of ocular pharmacology. Lippincott-Raven Publishers, Philadelphia, pp 119–138
19. Jones RF, Maurice DM (1966) New methods of measuring the rate of aqueous flow in man with fluorescein. Exp Eye Res 5(3):208–220
20. Hämäläinen KM et al (1997) Estimation of pore size and porosity of biomembranes from permeability measurements of polyethylene glycols using an effusion-like approach. J Control Release 49:97–104
21. Geroski DH, Edelhauser HF (2001) Transscleral drug delivery for posterior segment disease. Adv Drug Deliv Rev 52:37–48
22. Ambati J et al (2000) Transscleral delivery of bioactive protein to the choroid and retina. Invest Ophthalmol Vis Sci 41(5):1186–1191
23. Ahmed I et al (1989) The kinetics of timolol in the rabbit lens: implications for ocular drug delivery. Pharm Res 6(9):772–778
24. Sasaki H et al (1999) Enhancement of ocular drug penetration. Crit Rev Ther Drug Carrier Syst 16(1):85–146
25. Cunha-Vaz J (1979) The blood-ocular barriers. Surv Ophthalmol 23(5):279–296
26. Peyman GA, Bok D (1972) Peroxidase diffusion in the normal and laser-coagulated primate retina. Invest Ophthalmol 11(1):35–45
27. Smith RS, Rudt LA (1975) Ocular vascular and epithelial barriers to microperoxidase. Invest Ophthalmol 14(7):556–560
28. Pederson JE, Green K (1973) Aqueous humor dynamics: experimental studies. Exp Eye Res 15(3):277–297
29. Immordino ML, Dosio F, Cattel L (2006) Stealth liposomes: review of the basic science, rationale, and clinical applications, existing and potential. Int J Nanomedicine 1:297–315
30. Mishra GP et al (2011) Recent applications of liposomes in ophthalmic drug delivery. J Drug Deliv. doi:10.1155/2011/863734
31. Honda M et al (2013) Liposomes and nanotechnology in drug development: focus on ocular targets. Int J Nanomedicine 8:495–504. doi:10.2147/IJN.S30725
32. Mastrobattista E et al (2002) Functional characterization of an endosome-disruptive peptide and its application in cytosolic delivery of immunoliposome-entrapped proteins. J Biol Chem 277:27135–27143
33. Schnyder A, Huwyler J (2005) Drug transport to brain with targeted liposomes. NeuroRx 2:99–107
34. Bangham AD, Horne RW (1964) Negative staining of phospholipids and their structural modification by surface-active agents as observed in the electron microscope. J Mol Biol 8:660–668
35. Oku N (1999) Anticancer therapy using glucuronate modified long-circulating liposomes. Adv Drug Deliv Rev 40:63–73

36. Sapra P, Tyagi P, Allen TM (2005) Ligand-targeted liposomes for cancer treatment. Curr Drug Deliv 2(4):369–381
37. Honda M et al (2011) Suppression of choroidal neovascularization by intravitreal injection of liposomal SU5416. Arch Ophthalmol 129(3):317–321
38. Bochot A et al (2002) Intravitreal delivery of oligonucleotides by sterically stabilized liposomes. Invest Ophthalmol Vis Sci 43(1):253–259
39. Allen TM, Cullis PR (2004) Drug delivery systems: entering the mainstream. Science 303(5665):1818–1822
40. Puri A et al (2009) Lipid-based nanoparticles as pharmaceutical drug carriers: from concepts to clinic. Crit Rev Ther Drug Carrier Syst 26:523–580
41. Lopez-Berestein G et al (1983) Effects of sterols on the therapeutic efficacy of liposomal amphotericin B in murine candidiasis. Cancer Drug Deliv 1(1):37–42
42. Oku N, Nojima S, Inoue K (1980) Selective release of non-electrolytes from liposomes upon perturbation of bilayers by temperature change or polyene antibiotics. Biochim Biophys Acta 595(2):277–290
43. Klibanov AL et al (1990) Amphipathic poly-ethyleneglycols effectively prolong the circulation time of liposomes. FEBS Lett 268(1):235–237
44. Monem AS, Ali FM, Ismail MW (2000) Prolonged effect of liposomes encapsulating pilocarpine HCl in normal and glaucomatous rabbits. Int J Pharm 198(1):29–38
45. Davis BM et al (2014) Topical delivery of Avastin to the posterior segment of the eye in vivo using annexin A5-associated liposomes. Small 10(8):1575–1584
46. Kenis H et al (2004) Cell surface-expressed phosphatidylserine and annexin A5 open a novel portal of cell entry. J Biol Chem 279:52623–52629
47. Ungethüm L et al (2011) Engineered annexin A5 variants have impaired cell entry for molecular imaging of apoptosis using pretargeting strategies. J Biol Chem 286:1903–1910
48. Díaz-Llopis M et al (1992) Liposomally-entrapped ganciclovir for the treatment of cytomegalovirus retinitis in AIDS patients. Experimental toxicity and pharmacokinetics, and clinical trial. Doc Ophthalmol 82(4):297–305
49. Abrishami M et al (2009) Preparation, characterization, and in vivo evaluation of nanoliposomes-encapsulated bevacizumab (avastin) for intravitreal administration. Retina 29(5):699–703
50. Lajavardi L et al (2007) Downregulation of endotoxin-induced uveitis by intravitreal injection of vasoactive intestinal Peptide encapsulated in liposomes. Invest Ophthalmol Vis Sci 48(7):3230–3238
51. Moon JW et al (2006) Effect of subconjunctivally injected, liposome-bound, low-molecular-weight heparin on the absorption rate of subconjunctival hemorrhage in rabbits. Invest Ophthalmol Vis Sci 47(9):3968–3974
52. Baek SH et al (2009) Subconjunctivally injected, liposome-encapsulated streptokinase enhances the absorption rate of subconjunctival hemorrhages in rabbits. Eur J Pharm Biopharm 72(3):546–551
53. Fukushima A et al (2005) Suppression of macrophage infiltration into the conjunctiva by clodronate liposomes in experimental immune-mediated blepharoconjunctivitis. Cell Biol Int 29(4):277–286
54. Natarajan JV et al (2012) Nanomedicine for glaucoma: liposomes provide sustained release of latanoprost in the eye. Int J Nanomedicine 7:123–131
55. Carafa M, Santucci E, Lucania G (2002) Lidocaine-loaded non-ionic surfactant vesicles: characterization and in vitro permeation studies. Int J Pharm 231(1):21–32
56. Kaur IP et al (2004) Vesicular systems in ocular drug delivery: an overview. Int J Pharm 269(1):1–14
57. Abdelkader H et al (2011) Design and evaluation of controlled-release niosomes and discomes for naltrexone hydrochloride ocular delivery. J Pharm Sci 100(5):1833–1846
58. Saettone MF et al (1996) Non-ionic surfactant vesicles as ophthalmic carriers for cyclopentolate. A preliminary evaluation. STP Pharma Sci 6:94–98

59. Uchegbu IF, Bowstra JA, Florence AT (1992) Large disk-shaped structures (discomes) in nonionic surfactant vesicle to micelle transitions. J Phys Chem 96(25):10548–10553
60. Vyas SP et al (1998) Discoidal niosome based controlled ocular delivery of timolol maleate. Pharmazie 53(7):466–469
61. Abdelkader H et al (2010) Design and evaluation of controlled-release niosomes and discomes for naltrexone hydrochloride ocular delivery. J Pharm Sci 100(5):1833–1846
62. Kaur IP et al (2010) Improved ocular absorption kinetics of timolol maleate loaded into a bioadhesive niosomal delivery system. Graefes Arch Clin Exp Ophthalmol 248:1467–1472
63. Maiti S et al (2011) Nanovesicular formulation of brimonidine tartrate for the management of glaucoma: in vitro and in vivo evaluation. AAPS PharmSciTech 12(2):755–763
64. Abdelbary G, El-Gendy N (2008) Niosome-encapsulated gentamicin for ophthalmic controlled delivery. AAPS PharmSciTech 9(3):740–747
65. Liu S, Jones L, Gu FX (2012) Nanomaterials for ocular drug delivery. Macromol Biosci 12:608–620
66. Kaur IP, Chhabra S, Aggarwal D (2004) Role of cyclodextrins in opthalmics. Curr Drug Deliv 1(4):351–360
67. Brewster ME, Loftsson T (2007) Cyclodextrins as pharmaceutical solubilizers. Adv Drug Deliv Rev 59:645–666
68. Loftsson T, Duchêne D (2007) Cyclodextrins and their pharmaceutical applications. Int J Pharm 329:1–11
69. Zhang J et al (2008) Ocular pharmacokinetics of topically-applied ketoconazole solution containing hydroxypropyl beta-cyclodextrin to rabbits. J Ocul Pharmacol Ther 24(5):501–506
70. Loftsson T, Stefánsson E (1997) Effect of cyclodextrins on topical drug delivery to the eye. Drug Dev Ind Pharm 23(5):473–481
71. Halim Mohamed MA, Mahmoud AA (2011) Formulation of indomethacin eye drops via complexation with cyclodextrins. Curr Eye Res 36(3):208–216
72. Jansook P et al (2010) yCD/HPyCD mixtures as solubilizer: solid-state characterization and sample dexamethasone eye drop suspension. J Pharm Pharm Sci 13(3):336–350
73. Moya-Ortega MD et al (2013) Dexamethasone eye drops containing γ-cyclodextrin-based nanogels. Int J Pharm 441(1–2):507–515
74. Morrison PW, Connon CJ, Khutoryanskiy VV (2013) Cyclodextrin-mediated enhancement of riboflavin solubility and corneal permeability. Mol Pharm 10(2):756–762
75. Hippalgaonkar K et al (2011) Enhanced solubility, stability, and transcorneal permeability of delta-8-tetrahydrocannabinol in the presence of cyclodextrins. AAPS PharmSciTech 12(2):723–731
76. Zhang H et al (2013) Molecular modeling-based inclusion mechanism and stability studies of doxycycline and hydroxypropyl-β-cyclodextrin complex for ophthalmic delivery. AAPS PharmSciTech 14(1):10–18
77. Yavuz B, Pehlivan SB, Unlü N (2013) Dendrimeric systems and their applications in ocular drug delivery. Sci World J 2013
78. Cheng YY et al (2008) Dendrimers as drug carriers: applications in different routes of drug administration. J Pharm Sci 97(1):123–143
79. Gillies ER, Fréchet JM (2005) Dendrimers and dendritic polymers in drug delivery. Drug Discov Today 10(1):35–43
80. Kaminskas LM, Boyd BJ, Porter CJ (2011) Dendrimer pharmacokinetics: the effect of size, structure and surface characteristics on ADME properties. Nanomedicine 6(6):1063–1084
81. D'Emanuele A, Attwood D (2005) Dendrimer-drug interactions. Adv Drug Deliv Rev 57(15):2147–2162
82. Jevprasesphant R et al (2003) Engineering of dendrimer surfaces to enhance transepithelial transport and reduce cytotoxicity. Pharm Res 20(10):1543–1550
83. Chen HT et al (2004) Cytotoxicity, hemolysis, and acute in vivo toxicity of dendrimers based on melamine, candidate vehicles for drug delivery. J Am Chem Soc 126(32):10044–10048

84. Jain K et al (2010) Dendrimer toxicity: let's meet the challenge. Int J Pharm 394(1–2):122–142
85. Vandamme TF, Brobeck L (2005) Poly(amidoamine) dendrimers as ophthalmic vehicles for ocular delivery of pilocarpine nitrate and tropicamide. J Control Release 102(1):23–38
86. Holden CA et al (2012) Polyamidoamine dendrimer hydrogel for enhanced delivery of antiglaucoma drugs. Nanomedicine 8(5):776–783
87. Spataro G et al (2010) Designing dendrimers for ocular drug delivery. Eur J Med Chem 45:326–334
88. Yang H et al (2012) Hybrid dendrimer hydrogel/PLGA nanoparticle platform sustains drug delivery for one week and antiglaucoma effects for four days following one-time topical administration. ACS Nano 6(9):7595–7606
89. Yang H, Leffler CT (2013) Hybrid dendrimer hydrogel/poly(lactic-co-glycolic acid) nanoparticle platform: an advanced vehicle for topical delivery of antiglaucoma drugs and a likely solution to improving compliance and adherence in glaucoma management. J Ocul Pharmacol Ther 29(2):166–172
90. Nishiyama N et al (2005) Light-induced gene transfer from packaged DNA enveloped in a dendrimeric photosensitizer. Nat Mater 4(12):934–941
91. Iezzi R et al (2012) Dendrimer-based targeted intravitreal therapy for sustained attenuation of neuroinflammation in retinal degeneration. Biomaterials 33(3):979–988
92. Marano RJ et al (2005) Dendrimer delivery of an anti-VEGF oligonucleotide into the eye: a long-term study into inhibition of laser-induced CNV, distribution, uptake and toxicity. Gene Ther 12(21):1544–1550
93. Bhagav P, Upadhyay H, Chandran S (2011) Brimonidine tartrate-eudragit long-acting nanoparticles: formulation, optimization, in vitro and in vivo evaluation. AAPS PharmSciTech 12(4):1087–1101
94. Tártara LI et al (2012) Improvement of acetazolamide ocular permeation using ascorbyl laurate nanostructures as drug delivery system. J Ocul Pharmacol Ther 28(2):102–109
95. Pahuja P, Arora S, Pawar P (2012) Ocular drug delivery system: a reference to natural polymers. Expert Opin Drug Deliv 9(7):837–861
96. Paolicelli P et al (2009) Chitosan nanoparticles for drug delivery to the eye. Expert Opin Drug Deliv 6(3):239–253
97. de la Fuente M et al (2010) Chitosan-based nanostructures: a delivery platform for ocular therapeutics. Adv Drug Deliv Rev 62(1):100–117
98. Bourlais CL et al (1998) Ophthalmic drug delivery systems–recent advances. Prog Retin Eye Res 17(1):33–58
99. Gorgieva S, Kokol V (2014) Processing of gelatin-based cryogels with improved thermomechanical resistance, pore size gradient and high potential for sustainable protein-drug release. J Biomed Mater Res A 103(3):1119–1130
100. Lai JY (2013) Biodegradable in situ gelling delivery systems containing pilocarpine as new antiglaucoma formulations: effect of a mercaptoacetic acid/N-isopropylacrylamide molar ratio. Drug Des Devel Ther 7:1273–1285
101. Shive MS, Anderson JM (1997) Biodegradation and biocompatibility of PLA and PLGA microspheres. Adv Drug Deliv Rev 28(1):5–24
102. Cam D, Hyon SH, Ikada Y (1995) Degradation of high molecular weight poly(L-lactide) in alkaline medium. Biomaterials 16(11):833–843
103. Duvvuri S et al (2007) Controlled delivery of ganciclovir to the retina with drug-loaded Poly(d, L-lactide-co-glycolide) (PLGA) microspheres dispersed in PLGA-PEG-PLGA Gel: a novel intravitreal delivery system for the treatment of cytomegalovirus retinitis. J Ocul Pharmacol Ther 23(3):264–274
104. Bruining MJ et al (1999) New biodegradable networks of poly(N-vinylpyrrolidinone) designed for controlled nonburst degradation in the vitreous body. J Biomed Mater Res 47(2):189–197

105. Bruining MJ et al (2000) Biodegradable three-dimensional networks of poly(dimethylamino ethyl methacrylate). Synthesis, characterization and in vitro studies of structural degradation and cytotoxicity. Biomaterials 21(6):595–604
106. Hacker MC et al (2009) Biodegradable fumarate-based drug-delivery systems for ophthalmic applications. J Biomed Mater Res A 88(4):976–989
107. Colthurst MJ et al (2000) Biomaterials used in the posterior segment of the eye. Biomaterials 21(7):649–665
108. Jain JP et al (2005) Role of polyanhydrides as localized drug carriers. J Control Release 103(3):541–563
109. Jampel HD et al (1990) Glaucoma filtration surgery in monkeys using 5-fluorouridine in polyanhydride disks. Arch Ophthalmol 108(3):430–435
110. Sinha VR et al (2004) Poly-epsilon-caprolactone microspheres and nanospheres: an overview. Int J Pharm 278(1):1–23
111. Pitt CG et al (1981) Aliphatic polyesters II. The degradation of poly (DL-lactide), poly (epsilon-caprolactone), and their copolymers in vivo. Biomaterials 2(4):215–220
112. Beeley NR et al (2005) Fabrication, implantation, elution, and retrieval of a steroid-loaded polycaprolactone subretinal implant. J Biomed Mater Res A 73(4):437–444
113. Fialho SL, Behar-Cohen F, Silva-Cunha A (2008) Dexamethasone-loaded poly(epsilon-caprolactone) intravitreal implants: a pilot study. Eur J Pharm Biopharm 68(3):637–646
114. Yasukawa T et al (2001) Biodegradable scleral plugs for vitreoretinal drug delivery. Adv Drug Deliv Rev 52(1):25–36
115. Herrero-Vanrell R, Cardillo JA, Kuppermann BD (2011) Clinical applications of the sustained-release dexamethasone implant for treatment of macular edema. Clin Ophthalmol 5:139–146
116. Anand R et al (1993) Control of cytomegalovirus retinitis using sustained release of intraocular ganciclovir. Arch Ophthalmol 111(2):223–227
117. Kuppermann BD et al (1993) Intravitreal ganciclovir concentration after intravenous administration in AIDS patients with cytomegalovirus retinitis: implications for therapy. J Infect Dis 168(6):1506–1509
118. Martin DF et al (1994) Treatment of cytomegalovirus retinitis with an intraocular sustained-release ganciclovir implant. A randomized controlled clinical trial. Arch Ophthalmol 112(12):1531–1539
119. Sanborn GE et al (1992) Sustained-release ganciclovir therapy for treatment of cytomegalovirus retinitis. Use of an intravitreal device. Arch Ophthalmol 110(2):188–195
120. Kurz D, Ciulla TA (2002) Novel approaches for retinal drug delivery. Ophthalmol Clin North Am 15(3):405–410
121. Jung HJ, Chauhan A (2013) Extended release of timolol from nanoparticle-loaded fornix insert for glaucoma therapy. J Ocul Pharmacol Ther 29(2):229–235
122. Grossman RE, Chu DF, Lee DA (1990) Regional ocular gentamicin levels after transcorneal and transscleral iontophoresis. Invest Ophthalmol Vis Sci 31(5):909–916
123. Singh P, Maibach HI (1994) Iontophoresis in drug delivery: basic principles and applications. Crit Rev Ther Drug Carrier Syst 11(2–3):161–213
124. Davies JB et al (2003) Delivery of several forms of DNA, DNA-RNA hybrids, and dyes across human sclera by electrical fields. Mol Vis 9:569–578
125. Sloan JB, Soltani K (1986) Iontophoresis in dermatology. A review. J Am Acad Dermatol 15(4 Pt 1):671–684
126. Lark MR, Gangarosa LP Sr (1990) Iontophoresis: an effective modality for the treatment of inflammatory disorders of the temporomandibular joint and myofascial pain. Cranio 8(2):108–119
127. Pescina S et al (2010) In-vitro permeation of bevacizumab through human sclera: effect of iontophoresis application. J Pharm Pharmacol 62(9):1189–1194
128. Fay MF (1989) Indications and applications for iontophoresis. Todays OR Nurse 11(4):10–16

129. Burnette RR, Marrero D (1986) Comparison between the iontophoretic and passive transport of thyrotropin releasing hormone across excised nude mouse skin. J Pharm Sci 75(8):738–743
130. Burnette RR, Ongpipattanakul B (1987) Characterization of the permselective properties of excised human skin during iontophoresis. J Pharm Sci 76(10):765–773
131. Phipps JB, Padmanabhan RV, Lattin GA (1989) Iontophoretic delivery of model inorganic and drug ions. J Pharm Sci 78(5):365–369
132. Banga AK, Chien YW (1993) Hydrogel-based iontotherapeutic delivery devices for transdermal delivery of peptide/protein drugs. Pharm Res 10(5):697–702
133. Hughes L, Maurice DM (1984) A fresh look at iontophoresis. Arch Ophthalmol 102(12):1825–1829
134. Hill JM et al (1978) Iontophoresis of vidarabine monophosphate into rabbit eyes. Invest Ophthalmol Vis Sci 17(5):473–476
135. Hobden JA et al (1988) Iontophoretic application of tobramycin to uninfected and Pseudomonas aeruginosa-infected rabbit corneas. Antimicrob Agents Chemother 32(7):978–981
136. Choi TB, Lee DA (1988) Transscleral and transcorneal iontophoresis of vancomycin in rabbit eyes. J Ocul Pharmacol 4(2):153–164
137. Grossman R, Lee DA (1989) Transscleral and transcorneal iontophoresis of ketoconazole in the rabbit eye. Ophthalmology 96(5):724–729
138. Rootman DS et al (1988) Pharmacokinetics and safety of transcorneal iontophoresis of tobramycin in the rabbit. Invest Ophthalmol Vis Sci 29(9):1397–1401
139. Frucht-Pery J et al (1996) Efficacy of iontophoresis in the rat cornea. Graefes Arch Clin Exp Ophthalmol 234(12):765–769
140. Frucht-Pery J et al (2004) Iontophoresis-gentamicin delivery into the rabbit cornea, using a hydrogel delivery probe. Exp Eye Res 78(3):745–749
141. Frucht-Pery J et al (2006) Iontophoretic treatment of experimental pseudomonas keratitis in rabbit eyes using gentamicin-loaded hydrogels. Cornea 25(10):1182–1186
142. Sarraf D et al (1993) Transscleral iontophoresis of foscarnet. Am J Ophthalmol 115(6):748–754
143. Burstein NL, Leopold LH, Bernacchi DB (1985) Trans-scleral iontophoresis of gentamicin. J Ocul Pharmacol 1(4):363–368
144. Barza M, Peckman C, Baum J (1986) Transscleral iontophoresis of cefazolin, ticarcillin, and gentamicin in the rabbit. Ophthalmology 93(1):133–139
145. Church AL, Barza M, Baum J (1992) An improved apparatus for transscleral iontophoresis of gentamicin. Invest Ophthalmol Vis Sci 33(13):3543–3545
146. Barza M, Peckman C, Baum J (1987) Transscleral iontophoresis as an adjunctive treatment for experimental endophthalmitis. Arch Ophthalmol 105(10):1418–1420
147. Hayden B et al (2006) Iontophoretic delivery of carboplatin in a murine model of retinoblastoma. Invest Ophthalmol Vis Sci 47(9):3717–3721
148. Hayden BC et al (2004) Pharmacokinetics of systemic versus focal Carboplatin chemotherapy in the rabbit eye: possible implication in the treatment of retinoblastoma. Invest Ophthalmol Vis Sci 45(10):3644–3649
149. Voigt M et al (2002) Ocular aspirin distribution: a comparison of intravenous, topical, and coulomb-controlled iontophoresis administration. Invest Ophthalmol Vis Sci 43(10):3299–3306
150. Behar-Cohen FF et al (2002) Transscleral Coulomb-controlled iontophoresis of methylprednisolone into the rabbit eye: influence of duration of treatment, current intensity and drug concentration on ocular tissue and fluid levels. Exp Eye Res 74(1):51–59
151. Cohen AE et al (2012) Evaluation of dexamethasone phosphate delivered by ocular iontophoresis for treating noninfectious anterior uveitis. Ophthalmology 119(1):66–73
152. Sarraf D, Lee DA (1994) The role of iontophoresis in ocular drug delivery. J Ocul Pharmacol 10(1):69–81

153. Yoshizumi MO et al (1991) Experimental transscleral iontophoresis of ciprofloxacin. J Ocul Pharmacol 7(2):163–167
154. Boonme P et al (2006) Characterisation of microstructures formed in isopropyl palmitate/water/Aerosol OT:1-butanol (2:1) system. Pharmazie 61(11):927–932
155. Myers D (1988) Surfactant science and technology. VCH publications, New York
156. Danielsson T, Lindman B (1981) The definition of microemulsion. Colloids Surf 3:391–392
157. Hoar TP, Schulman JH (1943) Transparent water-in-oil dispersions the oleopathic hydromicelles. Nature 152:102–103
158. Soderholm JD et al (1998) Reversible increase in tight junction permeability to macromolecules in rat ileal mucosa in vitro by sodium caprate, a constituent of milk fat. Dig Dis Sci 43(7):1547–1552
159. Attwood D, Elworthy PH, Lawrence MJ (1989) Effect of structural variations of non-ionic surfactants on micellar properties and solubilization: surfactants with semi-polar hydrophobes. J Pharm Pharmacol 41(9):585–589
160. Attwood D et al (1992) A study on factors influencing the droplet size in nonionic oil-in-water microemulsions. Int J Pharm 88(1–3):417–422
161. Hallouard F et al (2010) Iodinated blood pool contrast media for preclinical X-ray imaging applications–a review. Biomaterials 31(24):6249–6268
162. Klibanov A (1998) Long circulating liposomes old drugs new therapeutics. Springer, Berlin
163. Torchilin VP, Trubetskoy VS (1995) Which polymers can make nanoparticulate drug carriers long-circulating? Adv Drug Deliv Rev 16(2–3):141–155
164. Trotta M et al (1999) Investigation of the phase behaviour of systems containing lecithin and 2-acyl lysolecithin derivatives. Int J Pharm 190(1):83–89
165. Anton N, Vandamme TF (2009) The universality of low-energy nano-emulsification. Int J Pharm 377(1–2):142–147
166. Anton N, Benoit JP, Saulnier P (2008) Design and production of nanoparticles formulated from nano-emulsion templates-a review. J Control Release 128(3):185–199
167. Kahlweit M, Busse G, Faulhaber B (1996) Preparing nontoxic microemulsions with alkyl monoglucosides and the role of alkane diols as cosolvents. Langmuir 12:861–862
168. Kahlweit M (1995) Preparing non-toxic microemulsion. Langmuir 11:4185–4187
169. Kahlweit M, Busse G, Faulhaber B (1995) Preparing microemulsions with lecithin. Langmuir 11:1576–1583
170. Baker R (1984) Investigations into the formation and characterization of microemulsions. J Coll Int Sci 100:311–331, 332–349
171. Sarciaux JM, Acar L, Sado PA (1995) Using microemulsion formulations for oral drug delivery of therapeutic peptides. Int J Pharm 120(2):127–136
172. Starker S (2002) Quercetin: critical evaluation as an antileishmanial agent in vivo in hamsters using different vesicular delivery modes. J Drug Target 10:573–578
173. Baroli B et al (2000) Microemulsions for topical delivery of 8-methoxsalen. J Control Release 69(1):209–218
174. Lianly IL (2002) Development of an ethyl laurate based microemulsion for rapid-onset of intranasal delivery of diazepam. Int J Pharm 237:77–85
175. Malcomson C (1998) Effect of oil on the level of solubilization of testosterone propionate into non-ionic oil-in-water microemulsion. J Pharm Sci 87:109–116
176. Changez M, Varshney M (2000) Aerosol-OT microemulsions as transdermal carriers of tetracaine hydrochloride. Drug Dev Ind Pharm 26(5):507–512
177. Trotta M, Morel S, Gasco MR (1997) Effect of oil phase composition on the skin permeation of felodipine from o/w microemulsions. Pharmazie 52(1):50–53
178. Malcolmson C et al (1998) Effect of oil on the level of solubilization of testosterone propionate into nonionic oil-in-water microemulsions. J Pharm Sci 87(1):109–116
179. Dogru ST, Calis S, Oner F (2000) Oral multiple w/o/w emulsion formulation of a peptide salmon calcitonin: in vitro-in vivo evaluation. J Clin Pharm Ther 25(6):435–443
180. Ma EL et al (2006) In vitro and in vivo evaluation of a novel oral insulin formulation. Acta Pharmacol Sin 27(10):1382–1388

181. Lv FF, Zheng LQ, Tung CH (2005) Phase behavior of the microemulsions and the stability of the chloramphenicol in the microemulsion-based ocular drug delivery system. Int J Pharm 301(1–2):237–246
182. Lv FF et al (2006) Studies on the stability of the chloramphenicol in the microemulsion free of alcohols. Eur J Pharm Biopharm 62(3):288–294
183. Fialho SL, da Silva-Cunha A (2004) New vehicle based on a microemulsion for topical ocular administration of dexamethasone. Clin Experiment Ophthalmol 32(6):626–632
184. Radomska A, Dobrucki R (2000) The use of some ingredients for microemulsion preparation containing retinol and its esters. Int J Pharm 196(2):131–134
185. Martin LF et al (2013) Topical Brazilian propolis improves corneal wound healing and inflammation in rats following alkali burns. BMC Complement Altern Med 13:337
186. Li G et al (2012) In vitro and in vivo evaluation of a simple microemulsion formulation for propofol. Int J Pharm 425(1–2):53–61
187. Donnelly RF et al (2010) Optical coherence tomography is a valuable tool in the study of the effects of microneedle geometry on skin penetration characteristics and in-skin dissolution. J Control Release 147(3):333–341
188. Pettis RJ, Harvey AJ (2012) Microneedle delivery: clinical studies and emerging medical applications. Ther Deliv 3(3):357–371
189. Felton L (2012) Transdermal and intradermal delivery of therapeutic agents application of physical technologies. Drug Dev Ind Pharm 38(4):513
190. Bariya SH et al (2012) Microneedles: an emerging transdermal drug delivery system. J Pharm Pharmacol 64(1):11–29
191. Chandrasekhar S et al (2013) Microarrays and microneedle arrays for delivery of peptides, proteins, vaccines and other applications. Expert Opin Drug Deliv 10(8):1155–1170
192. Gupta J et al (2011) Infusion pressure and pain during microneedle injection into skin of human subjects. Biomaterials 32(28):6823–6831
193. Hirobe S, Okada N, Nakagawa S (2013) Transcutaneous vaccines–current and emerging strategies. Expert Opin Drug Deliv 10(4):485–498
194. Hong X et al (2013) Dissolving and biodegradable microneedle technologies for transdermal sustained delivery of drug and vaccine. Drug Des Devel Ther 7:945–952
195. Nir Y et al (2003) Fear of injections in young adults: prevalence and associations. Am J Trop Med Hyg 68(3):341–344
196. Hamilton JG (1995) Needle phobia: a neglected diagnosis. J Fam Pract 41(2):169–175
197. McAllister DV et al (2003) Microfabricated needles for transdermal delivery of macromolecules and nanoparticles: fabrication methods and transport studies. Proc Natl Acad Sci USA 100(24):13755–13760
198. Patel SR et al (2011) Suprachoroidal drug delivery to the back of the eye using hollow microneedles. Pharm Res 28(1):166–176
199. Jiang J et al (2007) Coated microneedles for drug delivery to the eye. Invest Ophthalmol Vis Sci 48(9):4038–4043
200. Jiang J et al (2009) Intrascleral drug delivery to the eye using hollow microneedles. Pharm Res 26(2):395–403
201. Palakurthi NK et al (2011) Toxicity of a biodegradable microneedle implant loaded with methotrexate as a sustained release device in normal rabbit eye: a pilot study. J Ocul Pharmacol Ther 27(2):151–156
202. Ohashi M et al (2008) Efficacy of TonoLab in detecting physiological and pharmacological changes in rat intraocular pressure: comparison of TonoPen and microneedle manometry. Jpn J Ophthalmol 52(5):399–403
203. Saeki T et al (2008) The efficacy of TonoLab in detecting physiological and pharmacological changes of mouse intraocular pressure–comparison with TonoPen and microneedle manometry. Curr Eye Res 33(3):247–252
204. Allf BE, de Juan E Jr (1987) In vivo cannulation of retinal vessels. Graefes Arch Clin Exp Ophthalmol 225(3):221–225

205. Weiss JN (1992) Treatment of central retinal vein occlusion by injection of tissue plasminogen activator into a retinal vein. Am J Ophthalmol 113:429–434
206. Weiss JN, Bynoe LA (2001) Injection of tissue plasminogen activator into a branch retinal vein in eyes with central retinal vein occlusion. Ophthalmology 108(12):2249–2257
207. Glucksberg MR, Dunn R, Giebs CP (1993) In vivo micropuncture of retinal vessels. Graefes Arch Clin Exp Ophthalmol 231(7):405–407
208. Han SK, Kim SW, Kim WK (1998) Microvascular anastomosis with minimal suture and fibrin glue: experimental and clinical study. Microsurgery 18(5):306–311
209. Fekrat S, de Juan E Jr (1999) Chorioretinal venous anastomosis for central retinal vein occlusion: transvitreal venipuncture. Ophthalmic Surg Lasers 30(1):52–55
210. Tang WM, Han DP (2000) A study of surgical approaches to retinal vascular occlusions. Arch Ophthalmol 118(1):138–143
211. Weiss JN (2000) Retinal surgery for treatment of central retinal vein occlusion. Ophthalmic Surg Lasers 31(2):162–165
212. Suzuki Y, Matsuhashi H, Nakazawa M (2003) In vivo retinal vascular cannulation in rabbits. Graefes Arch Clin Exp Ophthalmol 241(7):585–588
213. Tsillimbaris M (2004) Retinal microvascular surgery: a feasibility study. Invest Ophthalmol Vis Sci 45:1963–1986
214. Tameesh MK et al (2004) Retinal vein cannulation with prolonged infusion of tissue plasminogen activator (t-PA) for the treatment of experimental retinal vein occlusion in dogs. Am J Ophthalmol 138(5):829–839
215. Kadonosono K et al (2011) An experimental study of retinal endovascular surgery with a microfabricated needle. Invest Ophthalmol Vis Sci 52(8):5790–5793
216. Mansoor S, Kuppermann BD, Kenney MC (2009) Intraocular sustained-release delivery systems for triamcinolone acetonide. Pharm Res 26(4):770–784
217. Herreo-Vanrell R (2001) Biodegradable micro-spheres for vitreoretinal drug delivery. Adv Drug Deliv Rev 52:5–16
218. Khoobehi B et al (1991) Clearance of sodium fluorescein incorporated into microspheres from the vitreous after intravitreal injection. Ophthalmic Surg 22(3):175–180
219. Moritera T et al (1992) Biodegradable microspheres containing adriamycin in the treatment of proliferative vitreoretinopathy. Invest Ophthalmol Vis Sci 33(11):3125–3130
220. Herrero-Vanrell R et al (2000) Biodegradable PLGA microspheres loaded with ganciclovir for intraocular administration. Encapsulation technique, in vitro release profiles, and sterilization process. Pharm Res 17(10):1323–1328
221. Kimura H et al (1994) In vitro phagocytosis of polylactide microspheres by retinal pigment epithelial cells and intracellular drug release. Curr Eye Res 13(5):353–360
222. Moritera T et al (1994) Feasibility of drug targeting to the retinal pigment epithelium with biodegradable microspheres. Curr Eye Res 13(3):171–176
223. Ogura Y, Kimura H (1995) Biodegradable polymer microspheres for targeted drug delivery to the retinal pigment epithelium. Surv Ophthalmol 39(Suppl 1):S17–S24
224. Veloso AA Jr et al (1997) Ganciclovir-loaded polymer microspheres in rabbit eyes inoculated with human cytomegalovirus. Invest Ophthalmol Vis Sci 38(3):665–675
225. Moritera T et al (1991) Microspheres of biodegradable polymers as a drug-delivery system in the vitreous. Invest Ophthalmol Vis Sci 32(6):1785–1790
226. Thies CA (1996) A survey of microencapsulation processes. In: Microencapsulation: methods and industrial applications. Marcel Dekker, New York
227. Delie F, Blanco-Prieto MJ (2005) Polymeric particulates to improve oral bioavailability of peptide drugs. Molecules 10(1):65–80
228. Barratt GM (2000) Therapeutic applications of colloidal drug carriers. Pharm Sci Technol Today 3(5):163–171
229. Washington C (1996) Drug release from microparticulate systems. In: Microencapsulation: methods and industrial applications. Marcel Decker, New York

230. Herrero-Vanrell R, Refojo MF (2001) Biodegradable microspheres for vitreoretinal drug delivery. Adv Drug Deliv Rev 52(1):5–16
231. Sinha VR, Trehan A (2003) Biodegradable microspheres for protein delivery. J Control Release 90(3):261–280
232. Soppimath KS et al (2001) Biodegradable polymeric nanoparticles as drug delivery devices. J Control Release 70(1–2):1–20
233. Giordano GG et al (1995) Biodegradation and tissue reaction to intravitreous biodegradable poly(D, L-lactic-co-glycolic)acid microspheres. Curr Eye Res 14(9):761–768
234. Yasukawa T et al (2004) Drug delivery systems for vitreoretinal diseases. Prog Retin Eye Res 23(3):253–281
235. Nan K et al (2014) Porous silicon oxide-PLGA composite microspheres for sustained ocular delivery of daunorubicin. Acta Biomater 10(8):3505–3512
236. Sakurai E et al (2001) Effect of particle size of polymeric nanospheres on intravitreal kinetics. Ophthalmic Res 33(1):31–36
237. Ravi Kumar MN (2000) Nano and microparticles as controlled drug delivery devices. J Pharm Pharm Sci 3(2):234–258
238. Siepmann J et al (2004) Effect of the size of biodegradable microparticles on drug release: experiment and theory. J Control Release 96(1):123–134
239. Wang SH et al (2005) Controlled release of levonorgestrel from biodegradable poly(D, L-lactide-co-glycolide) microspheres: in vitro and in vivo studies. Int J Pharm 301(1–2):217–225
240. Li W et al (2014) Evaluations of therapeutic efficacy of intravitreal injected polylactic-glycolic acid microspheres loaded with triamcinolone acetonide on a rabbit model of uveitis. Int Ophthalmol 34(3):465–476
241. Rong X et al (2012) Effects of erythropoietin-dextran microparticle-based PLGA/PLA microspheres on RGCs. Invest Ophthalmol Vis Sci 53(10):6025–6034
242. Carrasquillo KG et al (2003) Controlled delivery of the anti-VEGF aptamer EYE001 with poly(lactic-co-glycolic)acid microspheres. Invest Ophthalmol Vis Sci 44(1):290–299
243. Kompella UB, Bandi N, Ayalasomayajula SP (2003) Subconjunctival nano- and microparticles sustain retinal delivery of budesonide, a corticosteroid capable of inhibiting VEGF expression. Invest Ophthalmol Vis Sci 44(3):1192–1201
244. Moshfeghi AA, Peyman GA (2005) Micro- and nanoparticulates. Adv Drug Deliv Rev 57(14):2047–2052
245. Saishin Y et al (2003) Periocular injection of microspheres containing PKC412 inhibits choroidal neovascularization in a porcine model. Invest Ophthalmol Vis Sci 44(11):4989–4993
246. Ayalasomayajula SP, Kompella UB (2003) Celecoxib, a selective cyclooxygenase-2 inhibitor, inhibits retinal vascular endothelial growth factor expression and vascular leakage in a streptozotocin-induced diabetic rat model. Eur J Pharmacol 458(3):283–289
247. Ayalasomayajula SP, Kompella UB (2004) Retinal delivery of celecoxib is several-fold higher following subconjunctival administration compared to systemic administration. Pharm Res 21(10):1797–1804
248. Ayalasomayajula SP, Kompella UB (2005) Subconjunctivally administered celecoxib-PLGA microparticles sustain retinal drug levels and alleviate diabetes-induced oxidative stress in a rat model. Eur J Pharmacol 511(2–3):191–198

Chapter 10
Ocular Bioadhesives and Their Applications in Ophthalmic Drug Delivery

Rakesh K. Tekade and Muktika Tekade

Abstract Most ophthalmic illnesses are treated by topical administration of drug in the form of conventional dosage forms (solutions, suspensions, or ointments). In spite of remarkable patient compliance, these dosage forms suffer from the issues of poor ocular bioavailability due to pathophysiological barriers prevailing in the eye. In recent decades, numerous efforts have been made to advance ophthalmic drug delivery including the one adopted to prolong the residence time of drug in the conjunctival epithelium. The current approach is to develop biodegradable and bioadhesive ocular delivery systems that are capable of prolonging corneal residence time and sustaining the drug effect. The rate of release of drug from these ocular delivery systems is primarily governed by the molecular weight of the drug; mechanism of polymer degradation; shape, size, and surface charge of the polymeric architect; etc. The objective of this book chapter is to summarize various natural and synthetic ocular bioadhesives with their applications in the opthalmic drug delivery.

Keywords Bioadhesives • Natural and synthetic polymers • Conjunctival epithelium • Ophthalmic drug delivery

10.1 Introduction

Ocular drug delivery is one of the most standard and well-accepted routes for drug administration in case of eye disorders [1]. However, attaining as well as maintaining a therapeutic drug concentration following ocular drug delivery is highly

R.K. Tekade (✉)
Preclinical Nuclear Imaging Laboratory, The University of Texas Southwestern Medical Center, 5323 Harry Hines Boulevard, Dallas, TX 75390, USA

Department of Pharmaceutical Technology, School of Pharmacy, International Medical University, No. 126, Jln Jalil Perkasa 19, Bukit Jalil, 57000 Kuala Lumpur, Malaysia
e-mail: rakeshtekade@gmail.com; rakeshkumar@imu.edu.my

M. Tekade
TIT College of Pharmacy, Technocrats Institute of Technology Campus, Anand Nagar, Bhopal, MP 462021, India

© Springer International Publishing Switzerland 2016

Y. Pathak et al. (eds.), *Nano-Biomaterials For Ophthalmic Drug Delivery*,
DOI 10.1007/978-3-319-29346-2_10

challenging due to unique anatomy and physiology (protective mechanisms) of the eye [2]. The natural protective mechanisms of the eye render this organ inaccessible to foreign substances and drug molecules [3]. The precorneal factors such as blinking of the eye and impulsive lachrymation together with lachrymal drainage promptly remove the instilled substances leading to less ocular bioavailability (< 5 %) of topically applied drugs in the cul-de-sac [4, 5].

Further, the lipophilic environment of the corneal epithelium drastically limits the admission of hydrophilic bioactive molecules [6]. Likewise, the physicochemical behavior of bioactives also significantly regulates the permeability as well as relative diffusion resistance offered by ocular epithelia [2, 4] (Fig. 10.1). The occurrence of drug efflux pumps such as P-glycoproteins (P-gp), multidrug resistance-associated proteins on ocular tissues, presents further challenges toward the entry of the drugs into the site of action in the eye [7, 8].

This scenario mandates repeated instillations of ocular preparations to maintain a therapeutic drug concentration at the site of action, which in turn is promptly associated with induction of toxic side effects as well as injury at the ocular compartments [4, 7, 8]. Hence, to overcome the toxicity issue and augment the local therapeutic effect in the cul-de-sac, the residence time of the drug in the tear film should be lengthened. To fulfill this objective, absorption enhancers or viscosity enhancers are largely employed in the ocular preparations.

Alternatively, both systemic and local routes are also preferred in situations of poor corneal drug permeation [9]. However, systemic drug administration necessitates administration of higher and more frequent dosages that ultimately result in severe adverse effects [10]. Local injection into the ocular site is another approach to accomplish therapeutic concentrations in the retinal disorders [11]. However, this scenario also obligates repeated injections to retain the therapeutic concentration that may result in associated clinical complications.

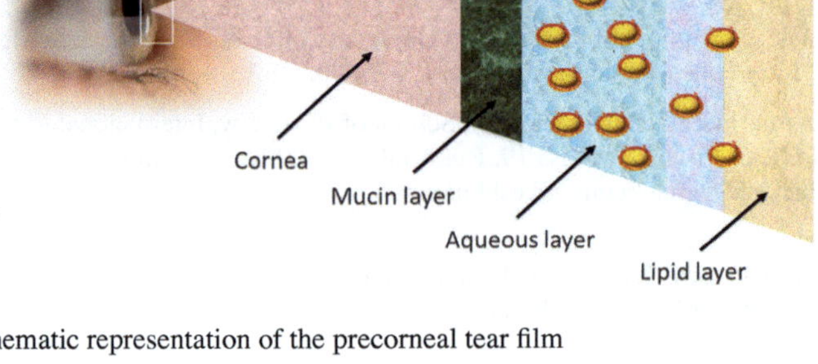

Fig. 10.1 Schematic representation of the precorneal tear film

In recent decades, several approaches have been appraised to improve ophthalmic drug delivery including the one adopted to prolong the residence time of drug formulation in the conjunctival epithelium [1, 2, 6]. Submission of a sustained, controlled, and bioadhesive drug formulation is cumulatively agreed to be an effective approach to circumvent all these limitations [12]. These strategies provide viable opportunities to evade recurrent administration, which is one of the prime problems related with ocular formulations [10]. With innovative approaches, the delivery of potential therapeutic biomolecules with very short half-lives could be conceivable. Viscous gel- or ointment-based preparations offer prolonged drug-eye contact; however, they elicit a tacky sensation, distorted vision, and eye-blinking reflex due to persistent discomfort in the eye [13]. The employment of a water-soluble polymer-based formulation to improve residence time of a bioactive was first reported in the literature, wherein a very promising improvement in corneal residence as well as drug bioavailability was observed in animal experiments [14].

In the past two decades, the development of synthetic biodegradable and bioadhesive polymers for ocular drug delivery has gained significant momentum. The rate of release of a drug from the polymeric architect is primarily governed by drug molecular weight, polymer degradation machineries, drug physicochemical properties, and shape, size, and surface charge of the polymeric architect [8]. The objective of this book chapter is to summarize various natural and synthetic bioadhesive polymers and their application in ocular drug delivery.

10.2 Anatomy and Physiology of Eye: Urgency of Developing Mucoadhesion Notion

The human eye consists of three layers, viz., an outer coat (known as sclera), a middle coat, and the inner coat (known as retina) [15]. Externally, the eye is covered by mobile tarsal eyelids, which can easily fold over the eyeball and permit rapid opening and closing movements of the palpebral fissures [16]. This movement produces a shear stress of approximately 2 K/s to distribute the tear fluid uniformly over the eye, providing an optically smooth surface over the cornea [5]. The shear rate during blinking directly affects the rheological characteristics of viscous ocular formulations administered into the eye [17].

The cornea is an avascular tissue compartment to which oxygen and required nutrients are supplied by lachrymal fluid and aqueous humor [18]. The low permeability of the cornea submits the presence of tight junctions between the ocular epithelial cells [19]. The cellular calcium levels and actin filaments of the cytoskeleton play a vital role in the integrity of the tight junctions. It may be noted that the squamous cells have microvilli on their surface, which enhance the cohesion and stability characteristics of the tear film [18].

The conjunctiva is a thin membrane that is composed of a highly vascularized submucosal epithelium [20]. It lines the inner surface of the eyelids and is connected by tight junctions, which makes the conjunctiva relatively impermeable. Notably, the conjunctival tissue is permeable to molecules up to ≈20 KDa, whereas the cornea is impermeable to molecules > 5 KDa [21].

The goblet cells are an important anatomical element of the conjunctiva and there are about 1.5 million goblet cells [22]. Goblet cells synthesize secretory mucins that contribute to the rheological properties of the tear film by specific non-covalent interactions with mucins forming an entangled network. Under normal circumstances, 2–3 μl of mucus is secreted daily by the goblet cells of the eye [23]. The gel-forming characteristics of the eye are an important phenomenon in entrapping foreign particles. During blinking, entrapped foreign particles are swept to the ocular puncta, which then discharge it to the drainage system [24]. The developed mucus layer lasts for more than 15 h, and it should be noted that the interaction of mucus with polymers and bioadhesive formulations is a major factor to be considered for the development of a sustained and controlled releasing ocular medication.

10.3 Considerations of Tear Film and Lachrymal Drainage for the Design of Ocular Drug Delivery System

The outer part of the eye is shielded by a thin layer of fluid (≈10 μm thickness) and is commonly referred to as "tear film." The mean pH value of normal tear film is approximately 7.4, while the same can be altered during pathological conditions and also be affected by the patient's age. In normal eyes, the osmolality of tear film ranges between 300 and 350 mOsm/kg (viscosity, 3 mPas; surface tension, 44×10^{-3} N/m) and is autonomously maintained by inorganic sodium, potassium, chlorine, or bicarbonate ions [25]. Interestingly, diurnal patterns of pH also exist in the eye, with a more pronounced shift of tear toward alkaline conditions during the day [26].

A superficial multimolecular lipid layer is also secreted by the meibomian glands during the blinking of the eye. Patients with dysfunctional meibomian glands display a high tear film osmolality as well as an elevated rate of tear film evaporation [27]. It plays an imperative role in dropping the evaporation rate to maintain normal tear osmolality especially when the tear production is reasonably short [28].

The mucus layer, which is secreted onto the eye surface by the goblet cells, is very sensitive to hydration and forms a gel layer with viscoelastic rheological properties. It protects the epithelia from damage and facilitates the movements of the eyelids (Fig. 10.1). It entraps bacteria, cell debris, and foreign bodies, forming mucous threads consisting of thick fibers arranged in bundles [29]. The large gel-forming mucins formed due to oligomerization through disulfide bonds are responsible for the rheological properties of mucus [30]. Due to their extended heavily glycosylated conformation, membrane-associated mucins in general have antiadhesive properties and provide a protective barrier for the cell membrane, which can prevent the interaction of ocular formulations with the epithelial cells. Several studies suggest that an increased intracellular Ca^{+} ion level also plays a central role to keep negatively charged mucins condensed and tightly packed (by cationic shielding) within the goblet cells [31].

The structure, composition, and physicochemical properties of the tear film are the prime factors that influence the bioadhesive behavior of ocular delivery systems [32, 33]. To be a noble bioadhesive agent, the polymeric assembly of the drug must make close contact with the mucus layer. The chains of the developed polymeric architect must be stretchy enough to make contact with the mucus and subsequently penetrate to an adequate depth. The polymeric assembly must be capable of eliciting hydrogen bond formation and electrostatic or hydrophobic interactions with mucins [14, 19].

On the contrary, these interaction parameters also depend on the ionic strength and pH of the vehicle, because changes in functional groups or charge shielding deeply impact the electrostatic repulsion and expansion of the mucus network [26]. An acidic environment promotes the mucoadhesion of mucin with polymers containing proton-donating groups (like –COOH), because uncharged carboxyl groups react with mucin molecules through hydrogen bonding. On the contrary, increasing the pH promotes electrostatic repulsion of –COOH anions, finally resulting in extension of the polymeric chains [34]. When the extent and number of available charged groups increases, the swelling pressure across the polymer also amplifies, which enlarges the void spaces within the polymeric complex. This reduction in density of the polymeric complex in turn increases the mobility of chains and enhances chances of mucoadhesion. Therefore, the precorneal residence time of polymeric formulation increases from a few hours to one day following its interaction with mucus [2, 12].

The naso-lachrymal drainage arrangement cumulatively comprises of (i) a secretory, (ii) distributive, and (iii) collection fragment, wherein the lacrimal gland serves as secretory unit [35]. During blinking, the eyelids uniformly distribute the tears secreted by lachrymal glands over the entire ocular surface. The basal tear flow is required to conserve a thin tear film on the corneal surface for optical, metabolic, and lubricant purposes [36].

The tear production by the lachrymal gland is largely a basic reflex, or an emotional tearing phenomenon [37]. The basal tear flow is of about 0.4–2.3 μl/min. It may be noted that no noteworthy difference exists between the tear production aptitudes in men and women; however, the tear production rate is also age-dependent, with elderly people suffering more from dry eyes [36]. Reflex tears are predominantly effected by peripheral stimuli (like chemical, mechanical, light, irritation, pathologic conditions, temperature, etc.). Existence of any of these reflex stimuli may enhance the tear formation by several folds, ultimately resulting in wash out of the foreign body or drug formulation from the ocular compartment [38].

These aforementioned factors must be taken into account during selection of polymer combination for the designing of a bioadhesive dosage formulation to attain a substantial mucus interaction. As discussed above, the anatomy and physiology of the eye render this organ exquisitely impervious to foreign substances [39]. Due to the existence of critical pharmacokinetic prerequisites, ophthalmic delivery of drug is one of the most challenging tasks confronted by the pharmaceutical investigators [40]. The innovation in drug delivery is, thus, increasingly becoming important in the treatment of ocular diseases, not only to improve drug efficacy but also to diminish the drug's side effects [1, 12]. The major goals of ocular drug targeting are to (i) increase the permeation of drug across deeper ocular tissues, (ii) sustain and prolong the drug release, (iii) attain site-specific drug delivery, (iv) diminish drug-related side effects, and (v) increase therapeutic bioavailability of drugs. The following sections of the chapter specifically deal with various bioadhesive polymeric approaches investigated for their possible ocular drug delivery applications.

10.4 Various Bioadhesive Approaches for Ocular Drug Delivery

10.4.1 Chitosan as Ocular Bioadhesive

Chitosan is a cationic biodegradable, biocompatible, and nontoxic polymer that furthermore bears antimicrobial and wound-healing properties [41]. Its excellent bioadhesiveness, hydrophilicity, corneal dispersion aptitude, and exceptional topical acceptance make it a unique polymer. It bears superior bioadhesive property owing to its ability to elicit molecular attraction by electrostatic interactions with the negatively charged mucus [42]. Moreover, chitosan also exhibits a pseudoplastic and viscoelastic behavior [43].

The bioadhesive characteristic of chitosan is governed by the formation of either secondary chemical bonds such as hydrogen bonds or ionic interactions between the positively charged $-NH_2$ groups of chitosan and the negatively charged sialic acid residues of mucins, depending on environmental pH [44]. The bioadhesive behavior of chitosan is expressively higher under neutral or alkaline pH similar to those that exist in the tear film [45]. It may be noted that a reinforcement of a strong bioadhesive force occurs only in the attendance of excess mucin-containing surrounding [46, 47].

The antibacterial action of chitosan is an additional plus especially concerning the topical formulation in case of ocular infections like conjunctivitis [48]. Several-fold rise in the precorneal residence time of tobramycin was achieved when adding chitosan to ocular formulations, in comparison to one other conventional commercial drug formulation [49]. A variety of chitosan derivatives have been synthesized with an aim to improve mucoadhesion as well as to enhance the penetration of the parent chitosan polymer [50]. The quaternized trimethyl and carboxyl methyl derivatives of chitosan were observed to bear potentially higher tissue penetration capability compared to that of the parent chitosan polymer [51]. Formulations based on the modified chitosan bear special interest in ocular drug delivery in situations when raised aqueous humor levels are required. In an experimental setting, the quaternized trimethyl and carboxyl methyl derivatives of chitosan drastically enhanced the corneal permeability of ofloxacin in the rabbit model [52]. The ocular formulation based on these polymers also improved the ocular pharmacokinetics of ofloxacin in the tear film [53].

The chitosan polymer has also attained consideration for the formulation of nanoparticle drug delivery systems (NDDS) owing to its useful bioadhesive, enhanced permeability, and exceptional biodegradability [42]. The degradation rate as well as the rate of release of drug from chitosan nanoparticles principally depends on the acetylation degree of the engaged chitosan [54]. Consequently, the attainment of sustained and controlled drug release kinetics is therefore highly feasible by adjusting the acetylation degree of the chitosan molecule [55].

Using cyclosporine as a model drug, the potential of loaded chitosan microspheres were assessed in rabbit models [56]. The outcome of this investigation infers the ability of the developed system to come into close contact with the corneal and conjunctiva epithelium layers of the eye. This attribute, in particular, increases the delivery of cyclosporine to ocular tissues without compromising inner ocular structures and systemic drug exposure. This feature greatly enhanced the drug residence and provides long-term drug levels in these target tissues [54, 56]. It may be noted that the chitosan-based microparticles and nanoparticles have a greater corneal retention than plain chitosan solution-based drug solutions [57]. Further, depending on the size, the nanoparticles may enter the corneal epithelium to a certain depth by paracellular or transcellular pathways. Numerous epithelial cells show a dissimilar affinity, which could explain the more imperative retention of chitosan nanoparticles in the conjunctiva compared to the cornea [58].

In a similar line, the chitosan hydrogels developed in association with other agents (such as β-glycerol PO_4, ethyl celluloses) and also showed immense potential as possible vehicles for drug delivery to the conjunctiva, wherein the release of therein employed drug (pilocarpine) showed a Fickian type of release behavior [59]. Notably, the pattern of drug release could be amended by adapting the composition as well as formulation conditions [60].

10.4.2 Hyaluronan as Ocular Bioadhesive

High polymer molecular weight is a prerequisite requirement for a polymer to function as machinery to prolong the corneal residence time of the therapeutics [61, 62]. High molecular weight hyaluronan polymers have a composition analogous to that of glycoproteins present in human tears, and hence it can easily cover the corneal epithelium. After adsorption at the mucin/aqueous interface, these polymers can extend the contiguous aqueous segment and, in doing so, stabilize the thick layer of water in the conjunctiva. It combines the benefit of high polymer viscosity at rest between blinks with an inimitable property of lowering inherent viscosity during the blinking process [62, 63]. In the case of hyaluronans, its bioadhesive behavior synergizes its shear-thinning characteristics to facilitate ocular drug delivery [64].

The bioadhesive and bioavailability enhancement effect of hyaluronan polymers have been successfully reported with the topical administration of tropicamide in the rabbit model [65]. Hyaluronic acid is another key constituent that plays an imperative role as an anti-inflammatory agent and hence may additionally promote healing of corneal epithelial cells [66]. Sodium hyaluronate, a class of hyaluronan, has been successfully employed as artificial tears for treatment of dry eye syndromes. It helps in overcoming dry eye syndrome due to its viscoelasticity, mucin-like biophysical behavior, long-lasting hydration, as well as ocular retention capacity. Additionally, it also offers an excellent lubrication sense to the ocular surface [66]. A formulation based on pilocarpine and hyaluronate showed a significantly improved bioavailability (approximately twofold increase) with an extended duration of miotic response in rabbits [67, 68].

Hyaluronan has also been successfully employed to prepare microparticles and nanoparticles [69]. For example, methyl prednisolone microparticles consisting of hyaluronic acid esters have shown a slower drug release with sustained drug concentration in the tear film in rabbits compared to a conventional drug suspension [70]. Biocompatibility is a key attribute for a formulation to be labeled as generally regarded as safe (GRAS) tag. In this direction, hyaluronic acid is devoid of any inherent cytotoxicity. It is also helpful in reducing the toxicity of the cytotoxic drug, benzalkonium chloride, due to its entrapment inside the sponge architect of the hyaluronan network, which diminishes the interaction of drug with ocular cell membranes [71]. Hence, hyaluronan represents an interesting class of polymers for delivery of drug to soft ocular tissues [72].

10.4.3 Poly(ethylene oxide) in Ocular Drug Delivery

Poly(ethylene oxide) has long chains of polyether that form strong hydrogen bonds with water to form hydrates and gels [73]. These hydrates and gel networks of poly(ethylene oxide) can load an exceptional quantity of drug, which can be released from the matrices in a controlled manner by swelling, followed by erosion and drug diffusion [74]. Engaging these properties of poly(ethylene oxide), sustained drug-releasing inserts were developed for the extended delivery of ofloxacin. Notably, such systems provide motifs to control the drug release by amending the formulation with different grades (400, 900) of poly(ethylene oxide) [75]. Further, innovative strategies were employed to enhance ocular absorption in rabbit models. For instance, chitosan microspheres containing drugs were amalgamated inside poly(ethylene oxide) to obtain delayed release as well as increase absorption [76].

It was noted that by increasing the proportion of microspheres, the erosion rate can be increased. The enhancement of hydration and swelling rate due to the presence of chitosan, and its subsequent interaction with poly(ethylene oxide), result in disentanglement in the long poly(ethylene oxide) chains. Further, the inclusion of chitosan in the formulation not only reduces the swelling time but also increases the precorneal residence time (increased blood concentration, C_{max} value) of the architect [52, 53, 55, 63, 75, 76].

10.4.4 Albumin

The word albumin is derived from a German word "albumen," generally indicating proteins. Albumen, on the other hand, derives from the Latin word albus (white) indicating the white part of the cooked egg surrounding the yolk. It is basically a polymer that consists of a single polypeptide chain and exists mainly in the α-helical form [77, 78]. Albumin can be found in egg whites and in blood. The three-dimensional structure of human serum albumin (HAS) has been determined by X-ray crystallography to a resolution of 2.5 Å and is comprised of three homologous domains that gather to form a heart-shaped molecule [79, 80]. Each domain of HAS is a product of two subdomains that bear universal structural design. The primary regions of ligand binding to HAS are located in hydrophobic cavities in subdomains IIA and IIIA, which exhibit similar chemistry. Structurally, the serum albumins are similar with each domain containing 5–6 internal disulfide bonds [80].

Albumin has numerous unique properties that make it an attractive drug vehicle in medicine. It is a versatile carrier for an array of hydrophobic drugs and biomolecules including hormones, vaccines, DNA, siRNA, and vitamins [81–83]. Nanoparticles made of HAS offer numerous advantages including ease of preparation, reproducibility, propensity to undergo surface modification (ligand binding), as well as its high tolerability and biodegradability in a given bioenvironment [84–86]. Drug molecules can be integrated in the particle matrix, covalently bound to surface functional groups, or adsorbed onto the particle. Albumin was well reported to facilitate endothelial transcytosis of bound constituents principally through binding to a cell-surface, glycoprotein receptor (albondin). This glycoprotein binds to caveolin-1 (an intracellular protein) with successive formation of transcytotic vesicles (known caveolae). Caveolin-1 is often present in some neoplasms, which could explain why albumin is known to accumulate in some tumors, and thus, facilitates intratumoral accumulation of albumin-bound drugs [81, 87]. In addition, surface modification and the introduction of targeting ligands can be attained with great ease to make the nanoparticle target-specific [85, 86].

Albumin-bound paclitaxel (Abraxane®; Abraxis BioScience and AstraZeneca) represents one of the strategies adopted to overcome the solvent-related problems of paclitaxel, and it has been recently approved by the US Food and Drug Administration. This novel, albumin-bound formulation of paclitaxel (size 130 nm) is used as a colloidal suspension derived from the lyophilized formulation of paclitaxel and HSA diluted in saline solution (0.9 % w/v NaCl) [88]. Preclinical studies, conducted in athymic mice, demonstrated that this product has a higher penetration into tumor cells with an increased antitumor activity, compared with an equal dose of standard paclitaxel [89, 90]. Similarly, albumin microspheres also represent an inimitable class of biodegradable colloidal particles, which have been used for lung scanning and circulating studies both in animal and human subjects. Bovine serum albumin (BSA) has also been successfully employed to prepare dry powder formulations for therapeutics [77]. BSA microspheres loaded with Ciprofloxacin were prepared by spray drying technique for lung delivery [78]. In another study, BSA was used to deliver and retain tetrandrine in the lung so as to induce any antisilicotic effects to other organs [91].

Numerous animal model-based investigations reported significant prolongation in the effect of drugs (like piroxicam, pilocarpine, etc.) incorporated inside albumin nanoparticles (ANP) compared to traditional formulations [92]. Ocular instillation of hydrocortisone via ANP in a rabbit model reduced the tissue concentration compared to a plain drug solution. This might be due to the sturdy linkage of the drug molecules with the ANPs [93]. It was also observed that coadministration of pilocarpine-loaded ANP in combination with other bioadhesive polymers provides an effective way to increase the intraocular pressure-lowering capacity of drug compared to the lone administration of ANPs [92].

10.4.5 Bioadhesive Acrylate Polymers

The mucoadhesive properties of poly(acrylic acid) (PAA, a long-used bioadhesive acrylate) are largely due to its hydrogen bond formation capacity [94]. Hydrophobic interaction with mucin is often proposed to be the mechanism behind its mucoadhesion, but available reports claim it to be an insignificant phenomenon [65]. This polymer after swelling gets entangled with mucin on the surface of conjunctiva via formation of a stable profuse hydrogel assembly [95].

The application of in-situ gel formation and drug complexation with acrylates for lengthening the residence time and improvement in ocular bioavailability drug in cornea was widely explored [96]. Acrylates such as polyacrylates and carbomers have already proven their efficacy as artificial tears for the relief of dry eye and ocular injury. Intrinsic mucus resembling, shear thinning behavior, good ocular retention, and lubricating characteristics form the rationale for the application of these high molecular weight polymers as artificial tears [97, 98]. However, it may be noted that with the usage of these polymers, a concentration-dependent blurring of vision and painful sensations were also reported in literature [99].

10.4.6 Alginate-Based Polymers

Alginate is a natural polymer containing β-mannuronic acid and α-guluronic acid monomers [100]. Hydrogels can be easily formed using alginates by interaction of its α-guluronic acid monomers with calcium ions. A significant prolongation in therapeutic efficacy of drugs like atropine, pilocarpine, and serine were observed upon using water-soluble alginates as drug carriers [101]. The in situ formation of calcium cross-linked alginate hydrogels showed sustained drug release and a prolonged therapeutic effect (intraocular pressure reducing capacity) [100].

It was proposed that the viscosity enhancement and interaction of alginate with mucus are responsible for prolonging the therapeutic effect of drugs [102], while the bioadhesive behavior was owing to the low surface tension of the alginate (approx. 30 mN/m), which is quite lower than the surface tension of cornea (approx. 40 mN/m). This difference in surface tension results in a uniform spreading and bioadhesion of the polymer. It elicits viscosity suitable for topical administration without causing any blurring effect [103, 104]. Importantly, the alginates were well tolerated in chronic glaucoma patients who require treatment for a prolonged period of time.

10.4.7 Particulate Vesicular Bioadhesives

Liposomes are particulate vesicular nanocarriers (multilamellar, small unilamellar, large unilamellar, oligolamellar vesicles) that are capable of improving the topical bioavailability of ocular drugs [105]. It represents a nanotechnology invention that has transformed from lab to clinic level in the form of nanoformulation (e.g., liposomal doxorubicin, Doxil®) because of its exceptional biocompatibility and biodegradability [106]. Positively charged liposomes prolong the corneal retention time of the loaded drug due to its interaction with the negatively charged corneal epithelial membrane, wherein it binds closely with the eye surface and thereby increases the residence time and enhances drug absorption [107].

Further, to enhance bioretention of liposomal formulation at corneal site, coating of liposomes with bioadhesive polymers [such as chitosan poly(acrylic acid), poloxamer hyaluronic acid, and chitosan] is a well-established and recommended strategy [105, 108]. The influence of Carbopol on the bioavailability of liposomes loaded with pilocarpine was also studied, wherein the Carbopol coating significantly reduced the drug release and prolonged the duration of the miotic activity as compared to plain drug solution and plain non-coated liposomes [109].

Poloxamer is yet another polymer solution that destabilizes the phospholipid lipid bilayer of liposomes due to its inherent surface-active characteristics. Conversely, poloxamer not only acts as a penetration enhancer but also could induce irritation and lachrymation in the eye [110]. This lachrymal secretion is the key reason behind the removal of the liposomal formulation from the ocular compartment, diluting and washing away the formula. Hence, it presents limitations for ocular delivery [111].

10.4.8 Bioadhesive Polycaprolactone Microspheres as Ophthalmic Delivery Device

Polycaprolactone (molecular weight 40 k) is a safe, biocompatible, and biodegradable bioadhesive polymer. It is reasonably more hydrophobic than polyalkyl cyanoacrylates [112], which helps in delaying its washing off from the ocular compartments [113]. Following its administration, polycaprolactone microspheres aggregate in the cul-de-sac of the eye from where drug is released in a slow and sustained fashion [114]. The application of microspheres in ophthalmic delivery of drugs is based on the principle of their interaction with the mucin of the eye to effect a delayed drainage as well as prolonged drug effect [22, 25–27, 31, 95]. For the preparation of bioadhesive formulations employing microsphere technology, the drug molecules are entrapped inside the bioadhesive polymeric architect, wherein the drug release profile can be further modulated by changing the composition, type of polymer, or the mode of manufacturing [14, 115–117].

Polycaprolactone microspheres can be coated with other polymers to further increase its bioadhesiveness [118]. For example, indomethacin microspheres when coated with chitosan showed net positive surface charge that significantly increased the adhesion of particles with mucus layer in the cul-de-sac of the eye. The chitosan coating enhanced the bioavailability of indomethacin (>2 folds) [112].

The amendment of microsphere periphery with polyethylene glycol changes the interfacial surface tension and bioadhesive behavior of the resultant formulation [119]. Further, it was observed that it increases the drug absorption, while causing no change in sight of the patient. However, the major challenge in this type of formulation lies in its manufacturing process, wherein achieving desired size of particles (for effective retention, sustaining drug effect, and avoiding ocular irritation) is quite a challenging task, which requires potential research and development. Further, to enhance the activity, polycaprolactone particles were coated with hyaluronic acid with a rationale to improve the electrostatic interactions with mucus in the eye [120, 121].

10.4.9 Polylactic Glycolic Acid (PLGA)-Based Ocular Bioadhesive Systems

PLGA is a copolymerization product of polylactic acid and polyglycolic acid, wherein the relative composition of both components influences the physicochemical properties (solubility, stability, ocular retention, bioadhesion, degradation, release pattern, etc.) of the copolymer [122]. Hence, it is highly conceivable to design an innovative PLGA polymer with desired physic-chemical properties for specific pharmaceutical application [123].

Additionally, the microspheres based on PLGA can entangle with the eye mucous layering to enhance residence time and prolonged drug release kinetics compared to plain drug solution [1]. In an investigation, the bioavailability and drug penetration of drug was observed to be promisingly increasing by coating of PLGA nanoparticles with polyethylene glycol [124]. In this investigation, almost 50 % of enhancement in bioavailability was noted upon polyethylene glycol coating of the nanoparticles due to its prompt interaction with mucin epithelium present in the eye [125].

A pronouncedly sustained and well-tolerated polymer was synthesized by conjugation of polylactic acid with PEGylated 1,2-distearoyl-3-phosphatidylethanolamine for the ocular delivery of acyclovir. The formulation was successfully tested for its bioadhesion, bioretention, and sustained/prolonged pharmacologic effect in the rabbit's eye [126]. Notably, the area under curve value was found to be significantly greater for the polyethylene glycol-coated polylactic acid nanoparticles (approx. 13 % enhancement) than that of uncoated nanoparticles (approx. 2 % enhancement). Interestingly, upon removal of eye mucus, an insignificant difference in the bioavailability was recorded between coated and non-coated nanoparticles [122, 126].

10.4.10 Microemulsion-Based Ocular Bioadhesive Formulations

Microemulsions represent another major class of architects that have immense application as ocular bioadhesives for topical drug delivery [127]. They are transparent, thermodynamically stable droplets that are of a size below 1 μ. These are basically micronized dispersions of oil and water, with an aid external of surfactant and cosurfactant to stabilize the interfacial surface tension [128]. These formulations offer exceptional alternatives for ocular drug delivery because of their inherent characteristic properties. Microemulsions are advantageous particularly with reference to their low viscosity, high loading, biocompatibility, penetration absorption ability, and formulation simplicity [129].

One type of microemulsion is an oil-in-water system, which is highly beneficial due to the existence of surfactant and cosurfactant in its architect that can increase corneal membrane permeability as well as drug uptake [130]. Microemulsion technology can penetrate into the deeper layers and may sustain the release of drug in the cornea. Additionally, because of these underlining features, microemulsions can significantly decrease the dosing frequency of eye drops [131]. An investigation in a rabbit model instilled with pilocarpine microemulsions leads to a substantial sustained effect of the drug activity. Notably, the microemulsion-based delivery system reduced the frequency of the instillation to twice daily irrespective of four equivalent instillations of conventional eye drop formulation. It was suggested that because of their low interfacial tension, these formulations demonstrate uniform spreading effect on the cornea [132, 133].

10.5 Conclusion

It may be noted that the key site for the treatment of ophthalmic disorders like ocular infections, keratitis sicca, and conjunctivitis is the ocular mucosa, whereas for intraocular disorders like glaucoma or uveitis, the key site for treatment is the cornea. In this context, bioadhesive polymers like chitosan, hyaluronan, poly(ethylene oxide), albumin, acrylates, alginates, polycaprolactone, and PLGA represent some of the innovative classes of polymers that have significantly enhanced the prospective of sustained and controlled ocular therapy. Bioadhesive ocular drug therapy has shown promising approaches to enhance retention, bioavailability, as well as therapeutic efficacy of the bioactive. It may be noted that the implementation of these polymers for the development of ocular drug delivery platforms requires a balance between adherence, prolonged residence time, low irritation potential, controlled drug release, drug tolerability, and patient acceptance. The complications of ocular disorders like conjunctivitis, inflammation, and dry eye syndromes have shown better therapeutic efficacy with bioadhesive polymer-based therapy. This strategy has been successfully applied to enhance the half-life of rapidly eliminating drugs. Further, drug entrapment by particulate delivery systems coated with bioadhesive polymer chains offers the opportunity of entanglement to the mucus layer of the tear film in the conjunctival sac. This strategy has immense potential to increase drug retention at the site of action by withstanding drainage. A controlled release behavior of the drug from particulate nanocarriers significantly increases its bioavailability. It is anticipated that research will continue with the development of more successful ocular formulations.

Acknowledgments The authors would like to acknowledge the International Medical University, Malaysia, for the support on cancer and arthritis researches. We also acknowledge internal grants from the IMU-JC [Project# BMSc I-01/2012(02)2014] for providing start-up financial support to our research group.

References

1. Al-Halafi AM (2014) Nanocarriers of nanotechnology in retinal diseases. Saudi J Ophthalmol 28:304–309
2. Gaudana R, Jwala J, Boddu SH, Mitra AK (2009) Recent perspectives in ocular drug delivery. Pharm Res 26:1197–1216
3. ElShaer A, Ghatora B, Mustafa S, Alany RG (2014) Contact lenses as drug reservoirs & delivery systems: the successes & challenges. Ther Deliv 5:1085–1100
4. Kaur IP, Kanwar M (2002) Ocular preparations: the formulation approach. Drug Dev Ind Pharm 28:473–493
5. Ahmed I, Patton TF (1985) Importance of the noncorneal absorption route in topical ophthalmic drug delivery. Invest Ophthalmol Vis Sci 26:584–587
6. Mun EA, Morrison PW, Williams AC, Khutoryanskiy VV (2014) On the barrier properties of the cornea: a microscopy study of the penetration of fluorescently labeled nanoparticles, polymers, and sodium fluorescein. Mol Pharm 11:3556–3564
7. Moosa RM, Choonara YE, du Toit LC, Kumar P, Carmichael T, Tomar LK et al (2014) A review of topically administered mini-tablets for drug delivery to the anterior segment of the eye. J Pharm Pharmacol 66:490–506
8. Chen P, Chen H, Zang X, Chen M, Jiang H, Han S et al (2013) Expression of efflux transporters in human ocular tissues. Drug Metab Dispos 41:1934–1948
9. Reichl S, Kolln C, Hahne M, Verstraelen J (2011) In vitro cell culture models to study the corneal drug absorption. Expert Opin Drug Metab Toxicol 7:559–578
10. Geamanu Panca A, Popa-Cherecheanu A, Marinescu B, Geamanu C, Voinea L (2014) Retinal toxicity associated with chronic exposure to hydroxychloroquine and its ocular screening. Review. J Med Life 7:322–326
11. Ibraheem D, Elaissari A, Fessi H (2014) Administration strategies for proteins and peptides. Int J Pharm 477:578–589
12. Ako-Adounvo AM, Nagarwal RC, Oliveira L, Boddu SH, Wang XS, Dey S et al (2014) Recent patents on ophthalmic nanoformulations and therapeutic implications. Recent Pat Drug Deliv Formul 8:193–201
13. Xu L, Xu X, Chen H, Li X (2013) Ocular biocompatibility and tolerance study of biodegradable polymeric micelles in the rabbit eye. Colloids Surf B Biointerfaces 112:30–34
14. Kimura H, Ogura Y (2001) Biodegradable polymers for ocular drug delivery. Ophthalmologica 215:143–155
15. Galicia-Carreon J, Santacruz C, Hong E, Jimenez-Martinez MC (2013) The ocular surface: from physiology to the ocular allergic diseases. Revista Alerg Mex 60:172–183 (Tecamachalco, Puebla, Mexico: 1993)
16. Raghava S, Hammond M, Kompella UB (2004) Periocular routes for retinal drug delivery. Expert Opin Drug Deliv 1:99–114

17. Fischer N, Narayanan R, Loewenstein A, Kuppermann BD (2011) Drug delivery to the posterior segment of the eye. Eur J Ophthalmol 21(Suppl 6):S20–S26
18. Versura P, Bonvicini F, Caramazza R, Laschi R (1985) Scanning electron microscopy of normal human corneal epithelium obtained by scraping-off in vivo. Acta Ophthalmol 63:361–365
19. Thrimawithana TR, Young S, Bunt CR, Green C, Alany RG (2011) Drug delivery to the posterior segment of the eye. Drug Discov Today 16:270–277
20. Huang AJ, Tseng SC, Kenyon KR (1990) Alteration of epithelial paracellular permeability during corneal epithelial wound healing. Invest Ophthalmol Vis Sci 31:429–435
21. Huang AJ, Tseng SC, Kenyon KR (1991) Change of paracellular permeability of ocular surface epithelium by vitamin A deficiency. Invest Ophthalmol Vis Sci 32:633–639
22. Kunert KS, Keane-Myers AM, Spurr-Michaud S, Tisdale AS, Gipson IK (2001) Alteration in goblet cell numbers and mucin gene expression in a mouse model of allergic conjunctivitis. Invest Ophthalmol Vis Sci 42:2483–2489
23. Roat MI, Ohji M, Hunt LE, Thoft RA (1993) Conjunctival epithelial cell hypermitosis and goblet cell hyperplasia in atopic keratoconjunctivitis. Am J Ophthalmol 116:456–463
24. Nichols BA, Chiappino ML, Dawson CR (1985) Demonstration of the mucous layer of the tear film by electron microscopy. Invest Ophthalmol Vis Sci 26:464–473
25. Dartt DA (2004) Control of mucin production by ocular surface epithelial cells. Exp Eye Res 78:173–185
26. Bron AJ, Tiffany JM, Gouveia SM, Yokoi N, Voon LW (2004) Functional aspects of the tear film lipid layer. Exp Eye Res 78:347–360
27. Pandit JC, Nagyova B, Bron AJ, Tiffany JM (1999) Physical properties of stimulated and unstimulated tears. Exp Eye Res 68:247–253
28. McCulley JP, Shine W (1997) A compositional based model for the tear film lipid layer. Trans Am Ophthalmol Soc 95:79–88, discussion 88-93
29. Lozato PA, Pisella PJ, Baudouin C (2001) The lipid layer of the lacrimal tear film: physiology and pathology. J Fr Ophtalmol 24:643–658
30. Frohlich E, Roblegg E (2014) Mucus as barrier for drug delivery by nanoparticles. J Nanosci Nanotechnol 14:126–136
31. Paz HB, Tisdale AS, Danjo Y, Spurr-Michaud SJ, Argueso P, Gipson IK (2003) The role of calcium in mucin packaging within goblet cells. Exp Eye Res 77:69–75
32. Smart JD (2004) Recent developments in the use of bioadhesive systems for delivery of drugs to the oral cavity. Crit Rev Ther Drug Carrier Syst 21:319–344
33. Kaur IP, Smitha R (2002) Penetration enhancers and ocular bioadhesives: two new avenues for ophthalmic drug delivery. Drug Dev Ind Pharm 28:353–369
34. Mortazavi SA, Smart JD (1994) Factors influencing gel-strengthening at the mucoadhesive-mucus interface. J Pharm Pharmacol 46:86–90
35. Paulsen F (2003) The human nasolacrimal ducts. Advances in anatomy, embryology, and cell biology, 170:III–XI. Springer, Berlin, Heidelberg/New York, pp 1–106
36. Murube J, Murube L, Murube A (1999) Origin and types of emotional tearing. Eur J Ophthalmol 9:77–84
37. Murube J (2009) Hypotheses on the development of psychoemotional tearing. Ocul Surf 7:171–175
38. Norn MS (1965) Tear secretion in normal eyes. Estimated by a new method: the lacrimal streak dilution test. Acta Ophthalmol 43:567–573
39. Lee SS, Hughes P, Ross AD, Robinson MR (2010) Biodegradable implants for sustained drug release in the eye. Pharm Res 27:2043–2053
40. Koevary SB (2003) Pharmacokinetics of topical ocular drug delivery: potential uses for the treatment of diseases of the posterior segment and beyond. Curr Drug Metab 4:213–222
41. Wang JJ, Zeng ZW, Xiao RZ, Xie T, Zhou GL, Zhan XR et al (2011) Recent advances of chitosan nanoparticles as drug carriers. Int J Nanomedicine 6:765–774
42. Nagpal K, Singh SK, Mishra DN (2010) Chitosan nanoparticles: a promising system in novel drug delivery. Chem Pharm Bull 58:1423–1430

43. Panos I, Acosta N, Heras A (2008) New drug delivery systems based on chitosan. Curr Drug Discov Technol 5:333–341
44. Kato Y, Onishi H, Machida Y (2003) Application of chitin and chitosan derivatives in the pharmaceutical field. Curr Pharm Biotechnol 4:303–309
45. Illum L (1998) Chitosan and its use as a pharmaceutical excipient. Pharm Res 15:1326–1331
46. Rossi S, Ferrari F, Bonferoni MC, Caramella C (2000) Characterization of chitosan hydrochloride-mucin interaction by means of viscosimetric and turbidimetric measurements. Eur J Pharm Sci 10:251–257
47. Sorlier P, Denuziere A, Viton C, Domard A (2001) Relation between the degree of acetylation and the electrostatic properties of chitin and chitosan. Biomacromolecules 2:765–772
48. Tavaria FK, Costa EM, Gens EJ, Malcata FX, Pintado ME (2013) Influence of abiotic factors on the antimicrobial activity of chitosan. J Dermatol 40:1014–1019
49. Kong M, Chen XG, Xing K, Park HJ (2010) Antimicrobial properties of chitosan and mode of action: a state of the art review. Int J Food Microbiol 144:51–63
50. Kotze AF, Luessen HL, de Leeuw BJ, de Boer AG, Verhoef JC, Junginger HE (1998) Comparison of the effect of different chitosan salts and N-trimethyl chitosan chloride on the permeability of intestinal epithelial cells (Caco-2). J Control Release 51:35–46
51. Kotze AF, Thanou MM, Luebetaen HL, de Boer AG, Verhoef JC, Junginger HE (1999) Enhancement of paracellular drug transport with highly quaternized N-trimethyl chitosan chloride in neutral environments: in vitro evaluation in intestinal epithelial cells (Caco-2). J Pharm Sci 88:253–257
52. Di Colo G, Zambito Y, Burgalassi S, Nardini I, Saettone MF (2004) Effect of chitosan and of N-carboxymethylchitosan on intraocular penetration of topically applied ofloxacin. Int J Pharm 273:37–44
53. Di Colo G, Burgalassi S, Zambito Y, Monti D, Chetoni P (2004) Effects of different N-trimethyl chitosans on in vitro/in vivo ofloxacin transcorneal permeation. J Pharm Sci 93:2851–2862
54. Kas HS (1997) Chitosan: properties, preparations and application to microparticulate systems. J Microencapsul 14:689–711
55. Sahariah P, Gaware VS, Lieder R, Jonsdottir S, Hjalmarsdottir MA, Sigurjonsson OE et al (2014) The effect of substituent, degree of acetylation and positioning of the cationic charge on the antibacterial activity of quaternary chitosan derivatives. Mar Drugs 12:4635–4658
56. De Campos AM, Sanchez A, Alonso MJ (2001) Chitosan nanoparticles: a new vehicle for the improvement of the delivery of drugs to the ocular surface. Application to cyclosporin A. Int J Pharm 224:159–168
57. Ameeduzzafar AJ, Bhatnagar A, Kumar N, Ali A (2014) Chitosan nanoparticles amplify the ocular hypotensive effect of cateolol in rabbits. Int J Biol Macromol 65:479–491
58. Zhou EH, Watson C, Pizzo R, Cohen J, Dang Q, Ferreira de Barros PM et al (2014) Assessing the impact of engineered nanoparticles on wound healing using a novel in vitro bioassay. Nanomedicine 9:2803–2815
59. Li J, Xu Z (2002) Physical characterization of a chitosan-based hydrogel delivery system. J Pharm Sci 91:1669–1677
60. Boucard N, Viton C, Domard A (2005) New aspects of the formation of physical hydrogels of chitosan in a hydroalcoholic medium. Biomacromolecules 6:3227–3237
61. Hartmann V, Keipert S (2000) Physico-chemical, in vitro and in vivo characterisation of polymers for ocular use. Pharmazie 55:440–443
62. Rupenthal ID, Green CR, Alany RG (2011) Comparison of ion-activated in situ gelling systems for ocular drug delivery. Part 2: precorneal retention and in vivo pharmacodynamic study. Int J Pharm 411:78–85
63. Ohguro N, Matsuda M, Kinoshita S (1991) The effects of denatured sodium hyaluronate on the corneal endothelium in cats. Am J Ophthalmol 112:424–430
64. Hamano T, Horimoto K, Lee M, Komemushi S (1996) Sodium hyaluronate eyedrops enhance tear film stability. Jpn J Ophthalmol 40:62–65

65. Sandri G, Bonferoni MC, Chetoni P, Rossi S, Ferrari F, Ronchi C et al (2006) Ophthalmic delivery systems based on drug-polymer-polymer ionic ternary interaction: in vitro and in vivo characterization. Eur J Pharm Biopharm 62:59–69
66. Aragona P, Di Stefano G, Ferreri F, Spinella R, Stilo A (2002) Sodium hyaluronate eye drops of different osmolarity for the treatment of dry eye in Sjogren's syndrome patients. Br J Ophthalmol 86:879–884
67. Bucolo C, Spadaro A, Mangiafico S (1998) Pharmacological evaluation of a new timolol/pilocarpine formulation. Ophthalmic Res 30:101–106
68. Bucolo C, Mangiafico P (1999) Pharmacological profile of a new topical pilocarpine formulation. J Ocul Pharmacol Ther 15:567–573
69. Wadhwa S, Paliwal R, Paliwal SR, Vyas SP (2010) Hyaluronic acid modified chitosan nanoparticles for effective management of glaucoma: development, characterization, and evaluation. J Drug Target 18:292–302
70. Avitabile T, Marano F, Castiglione F, Bucolo C, Cro M, Ambrosio L et al (2001) Biocompatibility and biodegradation of intravitreal hyaluronan implants in rabbits. Biomaterials 22:195–200
71. Contreras-Ruiz L, de la Fuente M, Garcia-Vazquez C, Saez V, Seijo B, Alonso MJ et al (2010) Ocular tolerance to a topical formulation of hyaluronic acid and chitosan-based nanoparticles. Cornea 29:550–558
72. Debbasch C, De La Salle SB, Brignole F, Rat P, Warnet JM, Baudouin C (2002) Cytoprotective effects of hyaluronic acid and Carbomer 934P in ocular surface epithelial cells. Invest Ophthalmol Vis Sci 43:3409–3415
73. Chiappetta DA, Sosnik A (2007) Poly(ethylene oxide)-poly(propylene oxide) block copolymer micelles as drug delivery agents: improved hydrosolubility, stability and bioavailability of drugs. Eur J Pharm Biopharm 66:303–317
74. Di Colo G, Zambito Y (2002) A study of release mechanisms of different ophthalmic drugs from erodible ocular inserts based on poly(ethylene oxide). Eur J Pharm Biopharm 54:193–199
75. Di Colo G, Zambito Y, Burgalassi S, Serafini A, Saettone MF (2002) Effect of chitosan on in vitro release and ocular delivery of ofloxacin from erodible inserts based on poly(ethylene oxide). Int J Pharm 248:115–122
76. Zambito Y, Zaino C, Di Colo G (2006) Effects of N-trimethylchitosan on transcellular and paracellular transcorneal drug transport. Eur J Pharm Biopharm 64:16–25
77. Moebus K, Siepmann J, Bodmeier R (2012) Cubic phase-forming dry powders for controlled drug delivery on mucosal surfaces. J Control Release 157:206–215
78. Li FQ, Hu JH, Lu B, Yao H, Zhang WG (2001) Ciprofloxacin-loaded bovine serum albumin microspheres: preparation and drug-release in vitro. J Microencapsul 18:825–829
79. Sugio S, Kashima A, Mochizuki S, Noda M, Kobayashi K (1999) Crystal structure of human serum albumin at 2.5 A resolution. Protein Eng 12:439–446
80. He XM, Carter DC (1992) Atomic structure and chemistry of human serum albumin. Nature 358:209–215
81. Hawkins MJ, Soon-Shiong P, Desai N (2008) Protein nanoparticles as drug carriers in clinical medicine. Adv Drug Deliv Rev 60:876–885
82. Paal K, Muller J, Hegedus L (2001) High affinity binding of paclitaxel to human serum albumin. Eur J Biochem/FEBS 268:2187–2191
83. Purcell M, Neault JF, Tajmir-Riahi HA (2000) Interaction of taxol with human serum albumin. Biochim Biophys Acta 1478:61–68
84. Wartlick H, Michaelis K, Balthasar S, Strebhardt K, Kreuter J, Langer K (2004) Highly specific HER2-mediated cellular uptake of antibody-modified nanoparticles in tumour cells. J Drug Target 12:461–471
85. Weber C, Coester C, Kreuter J, Langer K (2000) Desolvation process and surface characterisation of protein nanoparticles. Int J Pharm 194:91–102
86. Langer K, Balthasar S, Vogel V, Dinauer N, von Briesen H, Schubert D (2003) Optimization of the preparation process for human serum albumin (HSA) nanoparticles. Int J Pharm 257:169–180

87. Minshall RD, Sessa WC, Stan RV, Anderson RG, Malik AB (2003) Caveolin regulation of endothelial function. Am J Physiol Lung Cell Mol Physiol 285:L1179–L1183
88. Miele E, Spinelli GP, Miele E, Tomao F, Tomao S (2009) Albumin-bound formulation of paclitaxel (Abraxane ABI-007) in the treatment of breast cancer. Int J Nanomedicine 4:99–105
89. Desai N, Trieu V, Yao Z, Louie L, Ci S, Yang A et al (2006) Increased antitumor activity, intratumor paclitaxel concentrations, and endothelial cell transport of cremophor-free, albumin-bound paclitaxel, ABI-007, compared with cremophor-based paclitaxel. Clin Cancer Res 12:1317–1324
90. Sparreboom A, Scripture CD, Trieu V, Williams PJ, De T, Yang A et al (2005) Comparative preclinical and clinical pharmacokinetics of a cremophor-free, nanoparticle albumin-bound paclitaxel (ABI-007) and paclitaxel formulated in Cremophor (Taxol). Clin Cancer Res 11:4136–4143
91. Wang C, Wu QH, Li CR, Wang Z, Ma JJ, Zang XH et al (2007) Interaction of tetrandrine with human serum albumin: a fluorescence quenching study. Anal Sci 23:429–433
92. Rathod S, Deshpande SG (2008) Albumin microspheres as an ocular delivery system for pilocarpine nitrate. Indian J Pharm Sci 70:193–197
93. McCarty GR, Schwartz B (1999) Reduced plasma cortisol binding to albumin in ocular hypertension and primary open-angle glaucoma. Curr Eye Res 18:467–476
94. Bromberg L, Temchenko M, Alakhov V, Hatton TA (2004) Bioadhesive properties and rheology of polyether-modified poly(acrylic acid) hydrogels. Int J Pharm 282:45–60
95. Cleary J, Bromberg L, Magner E (2004) Adhesion of polyether-modified poly(acrylic acid) to mucin. Langmuir 20:9755–9762
96. Bonengel S, Haupstein S, Perera G, Bernkop-Schnurch A (2014) Thiolated and S-protected hydrophobically modified cross-linked poly(acrylic acid) - a new generation of multifunctional polymers. Eur J Pharm Biopharm 88:390–396
97. Xiao Q, Hu Y, Chen F, Chen X (2008) A comparative assessment of the efficacy of carbomer gel and carboxymethyl cellulose containing artificial tears in dry eyes. J Huazhong Univ Sci Technolog Med Sci 28:592–595
98. Springs CL (2010) Novel hydroxypropyl-guar gellable lubricant eye drops for treatment of dry eye. Adv Ther 27:681–690
99. Koffler BH, McDonald M, Nelinson DS (2010) Improved signs, symptoms, and quality of life associated with dry eye syndrome: hydroxypropyl cellulose ophthalmic insert patient registry. Eye Contact Lens 36:170–176
100. Giri TK, Thakur D, Alexander A, Ajazuddin BH, Tripathi DK (2012) Alginate based hydrogel as a potential biopolymeric carrier for drug delivery and cell delivery systems: present status and applications. Curr Drug Deliv 9:539–555
101. Pahuja P, Arora S, Pawar P (2012) Ocular drug delivery system: a reference to natural polymers. Expert Opin Drug Deliv 9:837–861
102. Tissie G, Sebastian C, Elena PP, Driot JY, Trinquand C (2002) Alginic acid effect on carteolol ocular pharmacokinetics in the pigmented rabbit. J Ocul Pharmacol Ther 18:65–73
103. Henness S, Swainston Harrison T, Keating GM (2007) Ocular carteolol: a review of its use in the management of glaucoma and ocular hypertension. Drugs Aging 24:509–528
104. Demailly P, Allaire C, Trinquand C (2001) Ocular hypotensive efficacy and safety of once daily carteolol alginate. Br J Ophthalmol 85:921–924
105. Honda M, Asai T, Oku N, Araki Y, Tanaka M, Ebihara N (2013) Liposomes and nanotechnology in drug development: focus on ocular targets. Int J Nanomedicine 8:495–503
106. Barenholz Y (2012) Doxil(R)–the first FDA-approved nano-drug: lessons learned. J Control Release 160:117–134
107. Eloy JO, Claro de Souza M, Petrilli R, Barcellos JP, Lee RJ, Marchetti JM (2014) Liposomes as carriers of hydrophilic small molecule drugs: strategies to enhance encapsulation and delivery. Colloids Surf B Biointerfaces 123:345–363
108. Li N, Zhuang CY, Wang M, Sui CG, Pan WS (2012) Low molecular weight chitosan-coated liposomes for ocular drug delivery: in vitro and in vivo studies. Drug Deliv 19:28–35

109. Davies NM, Farr SJ, Hadgraft J, Kellaway IW (1992) Evaluation of mucoadhesive polymers in ocular drug delivery. II. Polymer-coated vesicles. Pharm Res 9:1137–1144
110. Dumortier G, Grossiord JL, Agnely F, Chaumeil JC (2006) A review of poloxamer 407 pharmaceutical and pharmacological characteristics. Pharm Res 23:2709–2728
111. Bochot A, Fattal E, Boutet V, Deverre JR, Jeanny JC, Chacun H et al (2002) Intravitreal delivery of oligonucleotides by sterically stabilized liposomes. Invest Ophthalmol Vis Sci 43:253–259
112. Pohlmann AR, Fonseca FN, Paese K, Detoni CB, Coradini K, Beck RC et al (2013) Poly(-caprolactone) microcapsules and nanocapsules in drug delivery. Expert Opin Drug Deliv 10:623–638
113. Sharma S, Gupta D, Mohanty S, Jassal M, Agrawal AK, Tandon R (2014) Surface-modified electrospun poly(epsilon-caprolactone) scaffold with improved optical transparency and bioactivity for damaged ocular surface reconstruction. Invest Ophthalmol Vis Sci 55:899–907
114. Di Tommaso C, Behar-Cohen F, Gurny R, Moller M (2011) Colloidal systems for the delivery of cyclosporin A to the anterior segment of the eye. Ann Pharm Fr 69:116–123
115. Faisant N, Akiki J, Siepmann F, Benoit JP, Siepmann J (2006) Effects of the type of release medium on drug release from PLGA-based microparticles: experiment and theory. Int J Pharm 314:189–197
116. Grenha A, Remunan-Lopez C, Carvalho EL, Seijo B (2008) Microspheres containing lipid/chitosan nanoparticles complexes for pulmonary delivery of therapeutic proteins. Eur J Pharm Biopharm 69:83–93
117. Klose D, Siepmann F, Elkharraz K, Krenzlin S, Siepmann J (2006) How porosity and size affect the drug release mechanisms from PLGA-based microparticles. Int J Pharm 314:198–206
118. Bernards DA, Bhisitkul RB, Wynn P, Steedman MR, Lee OT, Wong F et al (2013) Ocular biocompatibility and structural integrity of micro- and nanostructured poly(caprolactone) films. J Ocul Pharmacol Ther 29:249–257
119. Yin H, Gong C, Shi S, Liu X, Wei Y, Qian Z (2010) Toxicity evaluation of biodegradable and thermosensitive PEG-PCL-PEG hydrogel as a potential in situ sustained ophthalmic drug delivery system. J Biomed Mater Res B Appl Biomater 92:129–137
120. Yenice I, Mocan MC, Palaska E, Bochot A, Bilensoy E, Vural I et al (2008) Hyaluronic acid coated poly-epsilon-caprolactone nanospheres deliver high concentrations of cyclosporine A into the cornea. Exp Eye Res 87:162–167
121. Barbault-Foucher S, Gref R, Russo P, Guechot J, Bochot A (2002) Design of poly-epsilon-caprolactone nanospheres coated with bioadhesive hyaluronic acid for ocular delivery. J Control Release 83:365–375
122. Pandita D, Kumar S, Lather V (2015) Hybrid poly(lactic-co-glycolic acid) nanoparticles: design and delivery prospectives. Drug Discov Today 20:95–104
123. Makadia HK, Siegel SJ (2011) Poly lactic-co-glycolic acid (PLGA) as biodegradable controlled drug delivery carrier. Polymers 3:1377–1397
124. Mathers WD (2000) Why the eye becomes dry: a cornea and lacrimal gland feedback model. CLAO J 26:159–165
125. Takeuchi H, Thongborisute J, Matsui Y, Sugihara H, Yamamoto H, Kawashima Y (2005) Novel mucoadhesion tests for polymers and polymer-coated particles to design optimal mucoadhesive drug delivery systems. Adv Drug Deliv Rev 57:1583–1594
126. Giannavola C, Bucolo C, Maltese A, Paolino D, Vandelli MA, Puglisi G et al (2003) Influence of preparation conditions on acyclovir-loaded poly-d, l-lactic acid nanospheres and effect of PEG coating on ocular drug bioavailability. Pharm Res 20:584–590
127. Kumar R, Sinha VR (2014) Preparation and optimization of voriconazole microemulsion for ocular delivery. Colloids Surf B Biointerfaces 117:82–88
128. Sapra B, Thatai P, Bhandari S, Sood J, Jindal M, Tiwary AK (2014) A critical appraisal of microemulsions for drug delivery: part II. Ther Deliv 5:83–94
129. Hegde RR, Verma A, Ghosh A (2013) Microemulsion: new insights into the ocular drug delivery. ISRN Pharm 2013:826798

130. Jadhav KR, Shaikh IM, Ambade KW, Kadam VJ (2006) Applications of microemulsion based drug delivery system. Curr Drug Deliv 3:267–273
131. Vandamme TF (2002) Microemulsions as ocular drug delivery systems: recent developments and future challenges. Prog Retin Eye Res 21:15–34
132. Fialho SL, da Silva-Cunha A (2004) New vehicle based on a microemulsion for topical ocular administration of dexamethasone. Clin Experiment Ophthalmol 32:626–632
133. Naveh N, Muchtar S, Benita S (1994) Pilocarpine incorporated into a submicron emulsion vehicle causes an unexpectedly prolonged ocular hypotensive effect in rabbits. J Ocul Pharmacol 10:509–520

Chapter 11
Nanovesicular Carrier Systems for Ophthalmic Drug Delivery

Lipika Chablani and Vijay Kumar

Abstract Ophthalmic drug delivery systems include a range of pharmaceutical dosage forms capable of either topical or systemic drug delivery. Nanovesicular drug delivery systems have been explored extensively for various such ophthalmic applications. Drug delivery using the nanovesicular systems requires a thorough understanding of the anatomy of the eye to determine the barriers and pathways. A good vesicular system is able to utilize these pathways and cross the barriers efficiently to deliver drugs to either topical or systemic tissues. Nanovesicular systems offer several advantages including good permeability, prolonged residence/contact time, sustained release profiles, easy administration, and are often patient compliant. Liposomes, niosomes, pharmacosomes, and spanlastics are some of the frequently studied nanovesicular drug delivery systems for ophthalmic applications. Multiple commercial ophthalmic liposomal products are available; however, many more are still being evaluated and are pending clinical trials. Formulation and development of these vesicular delivery systems has evolved in the past few decades to resolve issues related to vesicular aggregation, collapse of the vesicular system, and toxicity concerns. Approaches such as surface modification to avoid aggregation and use of cyclodextrin polymers for enhanced drug loading and stability are some of the examples. Overall, nanovesicles for ophthalmic drug delivery are a promising approach to deliver both hydrophilic and hydrophobic drug candidates efficiently.

Keywords Ophthalmic drug delivery • Nanovesicles • Liposomes • Niosomes and spanlastics

L. Chablani (✉)
Department of Pharmaceutical Sciences, Wegmans School of Pharmacy,
St. John Fisher College, Rochester, NY 14618, USA
e-mail: lchablani@sjfc.edu

V. Kumar
Unither Pharma, Rochester, NY 14623, USA

© Springer International Publishing Switzerland 2016

Y. Pathak et al. (eds.), *Nano-Biomaterials For Ophthalmic Drug Delivery*,
DOI 10.1007/978-3-319-29346-2_11

11.1 Introduction

Drug delivery to the eye has been a challenging task for pharmaceutical scientists for several decades. The complex anatomy of the eye, barriers to drug penetration, and absorption, and dynamic microenvironment of this unique organ limit the use of several dosage forms. Along with these limitations, formulators often come across issues pertaining to drug solubility and permeability. Thus, addressing issues related to the drug's physical properties and identifying a suitable formulation to ensure adequate drug delivery to the eye is a competent process. Nanotechnology has evolved from 1950s to date and offers several potential carriers for ophthalmic drug delivery (ODD). Nanotechnology has led to the development of various nanoscale drug delivery systems including polymeric nanoparticles, solid lipid nanoparticles, dendrimers, micelles, nanoemulsions, nanosuspensions, and nanovesicles (liposomes, niosomes, discomes, transfersomes, spanlastics, etc.). Each of these drug delivery systems offers the advantage of being smaller in size (10–1,000 nm), aiding the penetration and hence the absorption of the drug. However, these delivery systems have been uniquely designed to serve different purposes and should be carefully chosen depending upon the intended application. This chapter would address some of the key applications of nanovesicular systems so far in terms of ophthalmic drug delivery. Some of these nanovesicles have been comprehensively studied (such as liposomes), while others are still in their infancy (such as spanlastics) and are being explored further.

11.2 Structure of the Eye and Barriers to Nanovesicular Drug Delivery Systems

The anatomy, physiology, and biochemistry of the eye are very important to understand drug delivery via nanovesicles. This organ is highly specialized and serves the primary function of allowing external light to traverse through the pupils and form an image that is then processed by the brain. Often, eyes are subjected to external and internal assault in physical or biological form leading to significant damages, which may require medical intervention either surgically or pharmaceutically. Naturally, eyes are highly impervious to foreign substances and limit their exposure to avoid any damage; however, this also makes drug delivery a challenge [1].

Anatomically, the eye is comprised of anterior and posterior segments. The anterior segment consists of cornea iris, lens, ciliary body, and conjunctival epithelium and is lubricated with aqueous humor. On the other hand, the posterior segment is filled with vitreous humor composed of choroid, retinal pigment epithelium, neural retina, and optic nerves, primarily responsible for the sharp, colored images perceived by the organ. Disorders affecting the anterior segment, including conjunctivitis, glaucoma, fungal keratitis, etc., are commonly treated with conventional ophthalmic solutions, suspensions, or ointments. However, reaching the posterior

Fig. 11.1 Anatomical representation of the ocular structure. Anterior chamber of the eye is filled with aqueous humor and is separated by the lens from the posterior eye. Retina and retinal barriers limit the systemic exchange of drug molecules. Posterior segment of the eye contains vitreous gel; often intravitreal route of drug delivery is used to deliver drugs to this segment (Image source: National Eye Institute, National Institutes of Health)

segment for treatment of disorders such as macular degeneration, acute neurochorioretinitis, etc., is difficult and often requires intravitreal or periocular administration of drugs [1]. Drug delivery to the posterior segment of the eye has been investigated with the use of nanocarriers and nanovesicles in particular.

Along with the structural complexity, eye also presents ocular barriers and selective membranes to maintain the physiology of the organ (Fig. 11.1). The two blood-ocular barriers that are responsible for primary control of nutrients, solutes, cytokines, and growth factors are blood-aqueous barrier and blood-retinal barrier.

- Blood-aqueous barrier (BAR) is located in anterior segment of the eye and is responsible for controlling the exchange of solutes into the ocular fluids. Large molecules are often eliminated from aqueous humor to the systemic circulation to maintain vision acuity. Similarly, hydrophilic drug molecules are also eliminated as aqueous humor replenishes. This makes lipid-based nanovesicular systems a good candidate for anterior chamber drug delivery; due to their small size and hydrophobic exterior, these carrier systems are not primarily affected by the BAR.
- Blood-retinal barrier (BRB) is located in posterior chamber of the eye and allows selective movement of nutrients/drugs between the choroid and retina [2]. BRB is permeable to lipophilic molecules but has poor permeability for proteins and small hydrophilic molecules. It also restricts absorption of drug available in systemic circulation, thus limiting the entry of drug reaching the posterior chamber. Due to the presence of BRB, drug delivery to posterior segment of the eye has been challenging and often achieved by intravitreal injections. However, intravitreal injections pose risks of endophthalmitis, retinal detachment, and damage to the lens [2]. Also, this mode of drug delivery has poor patient compliance and alternative

approaches are always under investigation. Systemic route of drug delivery is also alternatively used for posterior eye diseases; however, it often results in systemic adverse effects and is capable of delivering only a minor fraction to the intended site of action. Nanovesicles have been studied as an alternative to address this unmet need and will be discussed further in this chapter.

11.3 Ophthalmic Drug Delivery Systems

Drug delivery to the eye involves various dosage forms ranging from conventional solutions, suspensions, and ointments to engineered implants, nanoparticles, dendrimers, micelles, nanoemulsions, nanosuspensions, and nanovesicles. Ophthalmic solutions and suspensions are one of the most common dosage forms. Over 90 % of ophthalmic formulations are marketed as eye drops and are intended for drug delivery to the anterior chamber of the eye. As the eye can accommodate only 30 μL of the liquid, the rest of the instilled volume (approximately 20 μL of a 50 μL average droplet) is swept away by the nasolacrimal duct entering the systemic circulation. This results in less than 5 % of drug being delivered to the eye, thus accounting for poor bioavailability and multiple dosing of the liquid formulation.

Ointments have also been used as an alternative to liquid dosage forms for ophthalmic drug delivery. These are often composed of hydrocarbon bases containing the therapeutic ingredient in the form of a solution or micronized powder. The physician or pharmacist would recommend applying a thin ribbon of the ointment onto the lower eyelid and gently flushing it with a blink of the eye. Ointments are mostly used at bedtime as they lead to blurry vision upon application and occasional transitory discomfort. Concerns of incorporating the drug in hot hydrocarbon base during formulation and patient compliance post application are often limitations for the use of this dosage form.

To address the unmet needs of drug delivery systems for ODD, the following criteria should be considered during formulation development:

- Good permeability and uptake from cornea.
- Encapsulate/entrap therapeutic drug concentrations.
- Balance of lipophilic and hydrophilic characteristics.
- Potential applications for posterior eye disorders.
- Prolonged residence or contact time.
- Safe and easy to use.
- Patient compliance.

Along with formulation approaches, drug delivery technologies (iontophoresis, microneedles, etc.) have been explored for management of ocular disorders. Novel drug delivery systems particularly offer advantages of being nano-sized, improved permeability and capability of surface modifications for enhanced ocular uptake. Here, we discuss the role of nanovesicles in ODD.

11.4 Nanovesicular Drug Delivery Systems

Nanovesicles are one of the most studied novel drug delivery systems. As the name indicates, the system is a nano-sized lipid vesicle encapsulating the active pharmaceutical ingredient. These vesicular drug delivery systems can further be categorized as liposomes, niosomes, transfersomes, discomes, pharmacosomes, emulsomes, and spanlastics. Essentially, these vesicles are phospholipid bilayers capable of encapsulating both hydrophilic and hydrophobic drug molecules. Each of the above-listed vesicular systems is unique based on its composition and applications. Liposomes are the original vesicular delivery systems, which have undergone several modifications based on the applications and unmet needs of drug delivery systems. Some of the common advantages of such vesicular systems could be summarized as:

- Ease of administration: viscous liquid administered via an eye dropper
- Encapsulated drug: avoids drug degradation and clearance
- Non-irritant, biodegradable, and biocompatible components
- Capable of targeted and sustained release
- Suitable for both hydrophilic and lipophilic drugs

11.4.1 Liposomes

Liposomes were first reported by Bangham et al. in 1965 and has evolved significantly since then. These are categorized based on their lipid bilayers and size. A vesicle can be unilamellar (UVs, one lipid bilayer) or multilamellar (MLVs, two or more lipid bilayers) based on concentric lipid bilayers and small (SUVS: 10–100 nm) or large (LUVs: 100–3,000 nm) based on the size (Fig. 11.2). As they have a lipophilic exterior but an aqueous core, these are ideal for loading both hydrophilic and hydrophobic drugs and also to permeate through lipophilic stromal layer of the cornea. Further, positively charged liposomes have been reported superior to the negatively charged or neutral liposomes for corneal uptake. This is due to the negatively charged mucin layer on cornea, which selectively prefers cationic particles [4–6]. Several pioneering studies of ocular drug delivery using liposomes were published in early 1980s and have led to further research [7, 8]. A wide range of therapeutic molecules have been delivered to the eye via liposomes. Table **11.1** provides a comprehensive summary of different studies attesting the use of liposomal ocular drug delivery. Vitamin A- and E-containing liposomes are commercially available for dry eye treatment (Lipo Nit, Optima Pharmazeutische GmbH, Wittibsmühle, Wang, Germany; OPTO lipo, Optox, Montegiorgio, AP, Italy; Tears Again, OCuSOFT, EUA, Rosenberg, TX, USA; Optrex Actimist Eye Spray, Optrex, Kingston-upon-Thames, UK) [16].

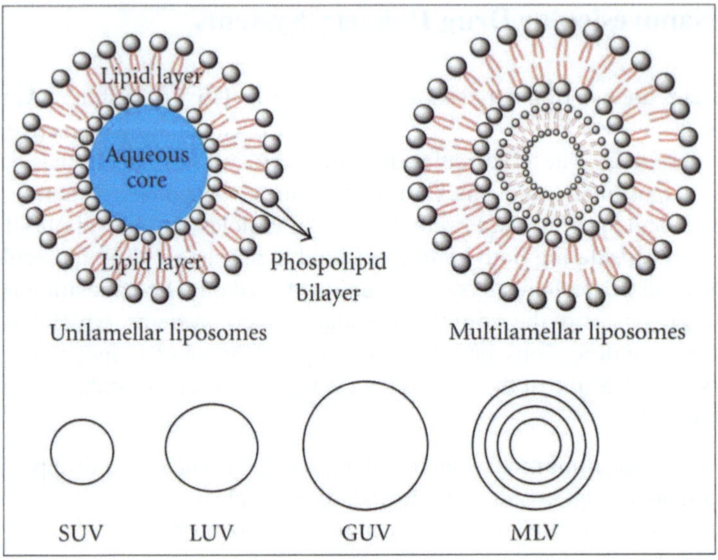

Fig. 11.2 Liposomal structure and types of liposomes based on size and lamellae (Image source: [3])

Table 11.1 Ocular drug delivery using liposomal carriers

Drug delivered	Therapeutic application	Results	References
Fluconazole	Antifungal	Reduced retinal toxicity of the intravitreal injections	[9]
Tacrolimus	Immunosuppressant	Improved safety and efficacy	[10]
Ciprofloxacin	Antibacterial	Enhanced bioavailability of the drug	[11]
Latanoprost	Anti-glaucomatic	Enhanced therapeutic effect and sustained delivery	[12]
Diltiazem HCl	Anti-glaucomatic	Efficacious delivery system for enhanced drug delivery and longer duration of action	[13]
Avastin (bevacizumab)	Anti-vascular endothelial growth factors	Topical delivery of the antibody can provide therapeutic concentrations to the posterior segment of the eye	[14]
Timolol maleate	Anti-glaucomatic	Liposomes incorporated in in situ gels provide rapid and effective delivery of the drug	[15]

11.4.2 Niosomes

Niosomes have been developed as efficient vesicular systems for ophthalmic drug delivery. They are usually within the size range of 10–1,000 nm. Niosomes are bilayered nonionic surfactant vesicles which can entrap hydrophilic and lipophilic drug molecules. Niosomes are chemically stable, biodegradable, biocompatible, and non-immunogenic and show low toxicity because of their nonionic nature. Having a flexible vesicular structure, they improve the performance of the drug via enhanced corneal uptake and controlled release at the targeted site. Niosomes also improve drug absorption by enhancing its pre-corneal residence time. A number of nonionic surfactants such as alkyl ethers, alkyl glyceryl ethers, polyoxyethylene 4 lauryl ether (Brij 30), polyoxyethylene acetyl ethers (Brij 58), Pluronic F127, and sorbitan fatty acid esters have been reported in various niosomal-based formulations [17]. Niosomes are usually prepared by hydration of nonionic surfactants using a number of fabrication methods like ether injection, transmembrane pH gradient, lipid layer hydration, reversed phase evaporation, microfluidization, bubbling of nitrogen, etc. [17].

Discomes are a modified nonconventional form of niosomes, made with the addition of nonionic surfactant like Solulan C24 (poly-24-oxyehtylene cholesteryl ether) [18]. They are disk shaped and larger in size than conventional niosomes. Discomes are reported to provide several advantages over niosomes, such as better fit in the cul-de-sac of the eye and improved ocular drug bioavailability due to slower nasolacrimal drainage [18, 19]. Also, high entrapment efficiency and bioavailability of discomes are more reported than of niosomes [19]. The increased corneal uptake by niosomes and discomes is attributed to many factors like their ability to disrupt the tight junctions of corneal epithelium, better spreadability on the lipophilic corneal surface, and suitable rheological properties. There are many reports of enhanced ocular drug absorption due to higher viscosity (leading to increased pre-corneal residence time) of ophthalmic solution [20]. However, this increase in bioavailability by increasing viscosity has its limitation. Application of conventional viscous formulation leads to blurry vision and poor patient compliance. Also, the drug bioavailability does not increase proportionally to the increase in viscosity, it reaches a plateau, and further increase in viscosity yields insignificant increase in the ocular bioavailability. Thus, niosomes and discomes are proposed as alternative vesicular carrier for ocular delivery.

A number of drugs for various therapeutic indications have been incorporated in these vesicles. Prabhu et al. have reported formulation and evaluation of niosomes containing anti-glaucoma drug Brimonidine tartrate [21]. They demonstrated significant reduction in intraocular pressure by this niosomal preparation on male albino rabbits. Similarly, Sathyavathi et al. have described improved bioavailability and increased pre-corneal residence time of Brimonidine tartrate in niosomal gel compared to marketed drops [22]. Abdelkader et al. have reported development of niosomes and discomes for naltrexone hydrochloride (NTX), an opioid antagonist for treatment of the diabetic complications affecting the cornea [20]. They suggest that the developed formulation provided better wetting properties, higher viscosity, and better protection to NTX against light-induced degradation and oxidizing agent than aqueous formulation. Akhtar et al. have reviewed a wide variety of drug molecules delivered to the eye through niosomal vesicular carrier [22]. Table **11.2** summarizes such drug molecules investigated for ocular delivery using niosomal carrier system.

Table 11.2 Ocular drug delivery using niosomes

Ocular delivery of drugs via niosomes	Therapeutic category	Conclusion	References
Brimonidine tartrate	Anti-glaucomatic	Improved drug absorption, significant reduction in intraocular pressure with no side effects	[23]
Ofloxacin	Antibacterial	Improved corneal penetration efficacy with no side effects	[24]
Diclofenac sodium	Nonsteroidal anti-inflammatory (NSAID)	Reduced irritation or redness	[25]
Fluconazole	Antifungal	Improved drug permeability, more stable	[26]
Flupirtine maleate	Analgesic	Improved delivery with reduced dosing frequency	[27]
Gentamicin	Antibacterial	Less toxicity	[28]
Chloramphenicol	Antibacterial	Less toxicity and improved patient compliance	[29]
Acetazolamide	Anti-glaucomatic	Enhanced bioavailability	[30]
Levofloxacin	Antibacterial	Prolonged drug release with reduced side effects	[31]
Cyclopentolate	Anticholinergic	Enhanced bioavailability	[32]
Timolol maleate	Anti-glaucomatic	Sustained controlled ocular drug delivery	[33]

11.4.3 Pharmacosomes

The term pharmacosome is derived from the word "Pharmakon," which translates to drug, and "soma" meaning carrier. These are vesicular drug delivery systems containing the drug as a colloidal dispersion, covalently bound to lipid, the carrier. Pharmacosomes have come up as potential alternative to conventional vesicles as they have potential to improve the bioavailability of hydrophilic as well as lipophilic drug molecules along with drug targeting and controlled release. The critical factor in pharmacosome development is surface and bulk interactions of lipids with drug [34]. Any drug molecule with free carboxylic or an active hydrogen atom like amino, hydroxyl groups, an active hydrogen atom can be converted to an ester using hydroxyl moiety of the lipid. This yields a prodrug, an amphiphilic compound, which can reduce interfacial tension, improve area of contact, facilitate membrane transport, and improve bioavailability. On dilution with water, the amphiphilic prodrug leads to formation of pharmacosomes.

Pharmacosomes have some unique advantages such as improved shelf life, due to the covalent linkage of drug to carrier and facilitated transport across the cornea. As drug release from pharmacosomes relies on hydrolysis of the covalent linkages (controlled by enzymes), this property can be used to achieve targeted and controlled release profile.

11.4.4 Spanlastics

These are Span (sorbitan)-based elastic vesicular drug delivery systems patented by Kaur et al. As transfersomes are modified liposomes, spanlastics are referred to as modified niosomes by their formulators. Spanlastics are prepared by ether injection method, where nonionic surfactants, such as Span 40/60/80, are solubilized in ether and injected in an aqueous phase containing edge activators (e.g., sodium taurocholate, sodium deoxycholate, Tween 80, etc.). The organic phase is injected slowly in the aqueous phase (maintained at 60 °C) via an 18-gauge needle, yielding bilamellar/multilamellar vesicles. The edge activators are surfactants used to enhance flexibility of these vesicles as they destabilize the lipid layers allowing the vesicles to permeate through pores smaller than their diameter. Kaur et al. conducted stability and safety studies on this nanovesicular systems and have reported that they are stable at refrigerated conditions comparatively to room temperature. Also, both in vitro and in vivo safety studies confirmed that the vesicles were nonirritant/noncorrosive and regarded safe for ocular applications. Spanlastics have been shown to be superior to niosomes in terms of enhanced deformability and thus bear the potential to deliver drugs to the posterior eye. These nanovesicles have been studied for antifungal (fluconazole) drug delivery for the treatment of ocular mycoses. The fluconazole-containing spanlastics were compared to commercial Zocon® (fluconazole) eye drops using ex vivo corneal permeability studies. Considering the hydrophilicity of fluconazole, drug permeation was least in case of Zocon® eye drops when compared to different types of spanlastic vesicles. Drug delivery to the posterior eye is limited to intravitreal injections, periocular routes, and systemic dosing. The use of spanlastics provides a unique drug delivery system that can be instilled topically as an eye drop and deliver drugs to the back of the eye. Several diseases such as fungal infections, uveitis, and age-related macular degeneration can benefit from such delivery systems. Spanlastics will also serve as a patient compliant drug delivery system as they replace intravitreal injections.

11.5 Other ODD Systems and Drug Delivery Techniques

Apart from nanovesicular ophthalmic drug delivery systems, various other dosage forms are available. This ranges from solutions, suspensions, emulsions, gels, ointments, ocular inserts, ocular implants, corneal shields, and contact lenses to nanoparticulate delivery systems (solid lipid nanoparticles, micelles, dendrimers, etc.). These formulation approaches have been discussed in detail elsewhere in this book for the readers. The use of technology including iontophoresis and microneedle-based drug delivery systems has also gained significance for ODD recently. Microneedles have been used to deliver drug to the posterior segment of the eye. Few hundred micron-sized needles penetrate the sclera to deliver the drug, avoiding damage to the ocular tissues. Such micropores are also efficient to deliver drug via nanocarriers to the posterior eye, thereby preventing vision threatening diseases.

11.6 Summary and Perspectives

Nanovesicular drug delivery systems serve as a unique carrier for ophthalmic diseases. These are a few of the most well-studied, characterized, and tested drug delivery systems. Their size, shape, charge, capability of hosting hydrophilic and hydrophobic drug molecules, and ease of administration clearly distinguish them from their counterparts. Several advances in vesicular drug delivery systems, leading to development of niosomes, discomes, pharmacosomes, and spanlastics, are evident that these have been modified over the years to meet unique unmet needs. As discussed earlier, 90 % of the ophthalmic dosage forms are administered as eye drops and lead to improved patient compliance and adherence to the therapy. Such vesicular drug delivery systems also meet this criterion of being incorporated into a viscous, buffered solution for ophthalmic administration. This maintains patient compliance and often leads to reduced multiple dosing of the drug as the nanovesicles can provide prolonged drug release. Further approaches, such as use of cyclodextrin matrix to enhance drug stability, enhance bioavailability, and avoid systemic toxic effect of the entrapped drug in the nanovesicles, have been studied [35–37]. Nanovesicle preparation is associated with drug aggregation, degradation, and poor drug loading issues. These problems have been successfully addressed over the years by surface modifications (use of PVA) to prevent aggregation and the dispersion of liposomes in a polymeric formulation (such as thermo-sensitive gels) to impart stability [38, 39]. Considering these developments and the success of vesicular systems for ophthalmic drug delivery, these nanocarriers remain an integral part of ODD systems.

References

1. Patel PB, Shastri D, Shelat PK, Shukla AK (2010) Ophthalmic drug delivery system: challenges and approaches. Syst Rev Pharm 1:8
2. Barar J, Asadi M, Mortazavi-Tabatabaei SA, Omidi Y (2009) Ocular drug delivery; Impact of in vitro cell culture models. J Ophthalmic Vis Res 4:238–252
3. Mishra GP, Bagui M, Tamboli V, Mitra AK (2011) Recent applications of liposomes in ophthalmic drug delivery. J Drug Deliv 2011:863734
4. Bourlais CL, Acar L, Zia H, Sado PA, Needham T, Leverge R (1998) Ophthalmic drug delivery systems – recent advances. Prog Retin Eye Res 17:33–58
5. Felt O, Furrer P, Mayer JM, Plazonnet B, BURI P, Gurny R (1999) Topical use of chitosan in ophthalmology: tolerance assessment and evaluation of precorneal retention. Int J Pharm 180:185–193
6. Sasaki H, Karasawa K, Hironaka K, Tahara K, Tozuka Y, Takeuchi H (2013) Retinal drug delivery using eyedrop preparations of poly-L-lysine-modified liposomes. Eur J Pharm Biopharm 83:364–369
7. Schaeffer HE, Krohn DL (1982) Liposomes in topical drug delivery. Invest Ophthalmol Vis Sci 22:220–227
8. Smolin G, Okumoto M, Feiler S, Condon D (1981) Idoxuridine-liposome therapy for herpes simplex keratitis. Am J Ophthalmol 91:220–225
9. Velpandian T, Narayanan K, Nag TC, Ravi AK, Gupta SK (2006) Retinal toxicity of intravitreally injected plain and liposome formulation of fluconazole in rabbit eye. Indian J Ophthalmol 54:237–240

10. Zhang R, He R, Qian J, Guo J, Xue K, Yuan YF (2010) Treatment of experimental autoimmune uveoretinitis with intravitreal injection of tacrolimus (FK506) encapsulated in liposomes. Invest Ophthalmol Vis Sci 51:3575–3582
11. Abdelbary G (2011) Ocular ciprofloxacin hydrochloride mucoadhesive chitosan-coated liposomes. Pharm Dev Technol 16:44–56
12. Natarajan JV, Ang M, Darwitan A, Chattopadhyay S, Wong TT, Venkatraman SS (2012) Nanomedicine for glaucoma: liposomes provide sustained release of latanoprost in the eye. Int J Nanomedicine 7:123–131
13. Mokhtar Ibrahim M, Tawfique SA, Mahdy MM (2014) Liposomal diltiazem HCl as ocular drug delivery system for glaucoma. Drug Dev Ind Pharm 40:765–773
14. Davis BM, Normando EM, Guo L, Turner LA, Nizari S, O'shea P, Moss SE, Somavarapu S, Cordeiro MF (2014) Topical delivery of Avastin to the posterior segment of the eye in vivo using annexin A5-associated liposomes. Small 10:1575–1584
15. Yu S, Wang QM, Wang X, Liu D, Zhang W, Ye T, Yang X, Pan W (2015) Liposome incorporated ion sensitive in situ gels for opthalmic delivery of timolol maleate. Int J Pharm 480:128–136
16. Souza JG, Dias K, Pereira TA, Bernardi DS, Lopez RF (2014) Topical delivery of ocular therapeutics: carrier systems and physical methods. J Pharm Pharmacol 66:507–530
17. Sahoo RK, Biswas N, Guha A, Sahoo N, Kuotsu K (2014) Nonionic surfactant vesicles in ocular delivery: innovative approaches and perspectives. Biomed Res Int 2014:263604
18. Abdelkader H, Ismail S, Kamal A, Alany RG (2011) Design and evaluation of controlled-release niosomes and discomes for naltrexone hydrochloride ocular delivery. J Pharm Sci 100:1833–1846
19. Vyas SP, Mysore N, Jaitely V, Venkatesan N (1998) Discoidal niosome based controlled ocular delivery of timolol maleate. Pharmazie 53:466–469
20. Abdelkader H, Wu Z, Al-Kassas R, Alany RG (2012) Niosomes and discomes for ocular delivery of naltrexone hydrochloride: morphological, rheological, spreading properties and photo-protective effects. Int J Pharm 433:142–148
21. Prabhu P, Nitish KR, Koland M, Harish N, Vijayanarayan K, Dhondge G, Charyulu R (2010) Preparation and evaluation of nano-vesicles of brimonidine tartrate as an ocular drug delivery system. J Young Pharm 2:356–361
22. Akhtar N (2013) Vesicular ocular drug delivery system: preclinical and clinical perspective of drugs delivered via niosomes. Int J Biopharm 4:11
23. Maiti S, Paul S, Mondol R, Ray S, Sa B (2011) Nanovesicular formulation of brimonidine tartrate for the management of glaucoma: in vitro and in vivo evaluation. AAPS PharmSciTech 12:755–763
24. Gupta N, Shrivastava V, Saxena S, Pandey A (2010) Formulation and evaluation of non-ionic surfactant vesicles (niosomes) for ocular delivery of ofloxacin. Int J Pharm Life Sci 1:413–418
25. Karthikeyan D, Pandey VP (2009) Study on ocular absorption of diclofenac sodium niosome in rabbits eye. Pharmacologyonline 1:769–779
26. Kaur IP, Garg A, Singla AK, Aggarwal D (2004) Vesicular systems in ocular drug delivery: an overview. Int J Pharm 269:1–14
27. Patidar S, Jain S (2012) Non-ionic surfactant based vesicles (Niosomes) containing Flupirtine Maleate as an ocular drug delivery system. J Chem Pharm Res 4:4495–4500
28. Abdelbary G, El-Gendy N (2008) Niosome-encapsulated gentamicin for ophthalmic controlled delivery. AAPS PharmSciTech 9:740–747
29. Yasin MN, Hussain S, Malik F, Hameed A, Sultan T, Qureshi F, Riaz H, Perveen G, Wajid A (2012) Preparation and characterization of chloramphenicol niosomes and comparison with chloramphenicol eye drops (0.5%w/v) in experimental conjunctivitis in albino rabbits. Pak J Pharm Sci 25:117–121
30. Aggarwal D, Pal D, Mitra AK, Kaur IP (2007) Study of the extent of ocular absorption of acetazolamide from a developed niosomal formulation, by microdialysis sampling of aqueous humor. Int J Pharm 338:21–26

31. Raghuwanshi N, Dikshit S, Sharma A, Upamanyu N, Dubey A, Pathak A (2010) Formulation and evaluation of niosome- encapsulated levofloxacin for ophthalmic controlled delivery. Int J Adv Pharm Res 3:901–906
32. Saettone MF, Perini G, Carafa M, Santucci E, Alhaique F (1996) Non-ionic surfactant vesicles as ophthalmic carriers for cyclopentolate. A preliminary evaluation. S T P Pharm Sci 6:94–98
33. Kaur IP, Aggarwal D, Singh H, Kakkar S (2010) Improved ocular absorption kinetics of timolol maleate loaded into a bioadhesive niosomal delivery system. Graefes Arch Clin Exp Ophthalmol 248:1467–1472
34. Pandita A, Sharma P (2013) Pharmacosomes: an emerging novel vesicular drug delivery system for poorly soluble synthetic and herbal drugs. ISRN Pharm 2013:348186
35. Ioele G, de Luca M, Ragno G (2014) Photostability of barnidipine in combined cyclodextrin-in-liposome matrices. Future Med Chem 6:35–43
36. Kaur IP, Chhabra S, Aggarwal D (2004) Role of cyclodextrins in ophthalmics. Curr Drug Deliv 1:351–360
37. Mccormack B, Gregoriadis G (1994) Entrapment of cyclodextrin-drug complexes into liposomes: potential advantages in drug delivery. J Drug Target 2:449–454
38. Fujisawa T, Miyai H, Hironaka K, Tsukamoto T, Tahara K, Tozuka Y, Ito M, Takeuchi H (2012) Liposomal diclofenac eye drop formulations targeting the retina: formulation stability improvement using surface modification of liposomes. Int J Pharm 436:564–567
39. Morand K, Bartoletti AC, Bochot A, Barratt G, Brandely ML, Chast F (2007) Liposomal amphotericin B eye drops to treat fungal keratitis: physico-chemical and formulation stability. Int J Pharm 344:150–153

Chapter 12
Cyclodextrins in Ocular Drug Delivery

Sridhar Anand and Vanildo Martins Lima Braga

Abstract Cyclodextrins are a group of water-insoluble, donut-shaped, naturally occurring cyclic oligosaccharides that are produced as a result of the bacterial breakdown of α-D-glucose polymers, such as cellulose. They are inactive molecules that do not display any innate biological activity. The ocular delivery of certain drugs is affected not only by the drug's physicochemical properties but also by the anatomical barriers of the eye. Drug-cyclodextrin complexes, formed by the physical occlusion of the torus cavity by the drug molecule, redress this issue. Thus, the ocular delivery of glaucoma drugs (e.g., carbonic anhydrase inhibitors, prostaglandin derivatives, pilocarpine), nonsteroidal anti-inflammatory drugs (e.g., indomethacin), and antifungal drugs (e.g., voriconazole) is enhanced by such inclusion complexes. Additionally, specific drug delivery systems (e.g., hydrogels, mucoadhesives, ocular gels) are harnessed to deliver the drug payload from the drug-cyclodextrin complex in a tunable fashion. This chapter seeks to elaborate on the chemical, biological, and pharmaceutical aspects governing cyclodextrins in the context of ocular drug delivery.

Keywords Cyclodextrin • Inclusion complexes • Hydrogels • Mucoadhesives • Glaucoma

12.1 Introduction

The principal aim of any drug delivery system is the precise and efficient delivery of a necessary amount of drug to the site of action for a specific duration of time. To achieve this aim, appropriate carrier materials are chosen to deliver the drug while simultaneously addressing certain inherently undesirable properties of the same drug molecule.

S. Anand (✉)
Department of Pharmaceutical Sciences, Wegmans School of Pharmacy, St. John Fisher College, Rochester, NY, USA
e-mail: sanand@sjfc.edu

V.M.L. Braga
Rua Sebastião Gonçalves Coelho, 400 Chanadour, Divinópolis, MG, Brazil

© Springer International Publishing Switzerland 2016

Y. Pathak et al. (eds.), *Nano-Biomaterials For Ophthalmic Drug Delivery*,
DOI 10.1007/978-3-319-29346-2_12

Cyclodextrins (CDs) are a group of naturally occurring cyclic oligosaccharides resulting from the bacterial breakdown of cellulose, an α-D-glucose polymer. CDs are toroidal (donut-shaped) molecules structurally related as a family on account of their (α-1,4)-linked α-D-glucopyranose units. The chair conformations of the six-membered glucopyranose ring, with the carbon skeleton and anomeric oxygen atom, create the hydrophobic wall in the interior, while the pendant hydroxyl groups create the outer hydrophilic edge. Consequently, the CD torus, which exists in the form of a truncated cone, possesses a lipophilic central cavity and relatively more hydrophilic outer surface.

α-CDs, β-CDs, and γ-CDs consist of six, seven, and eight (α-1,4)-linked α-D-glucopyranose units, respectively (Fig. 12.1). These natural CDs possess limited aqueous solubility compared to their acyclic saccharide counterparts. Improvement of aqueous solubility can be achieved by the random substitution of hydroxyl groups, even as methoxy ethers. CD derivatives of pharmaceutical interest include the hydroxypropyl derivatives of β-cyclodextrin and γ-cyclodextrin (HPβCD and HPγCD), the randomly methylated β-cyclodextrin (RMβCD), sulfobutylether β-cyclodextrin (SBEβCD), and glucosyl-β-cyclodextrin (GlcβCD).

12.2 Complex Formation and Solubility of Drug Molecules

Complexation is a molecular phenomenon where a molecule of drug comes into associative contact with a molecule of cyclodextrin. Water bridges the hydroxyl functional groups on adjacent D-glucose molecules to create a cage which is essential to maintaining the stability of the drug-CD complex. In aqueous solutions, CDs may form inclusion complexes by accommodating the drug molecule into the central lipophilic cavity. The molecule is held in place by non-covalent interactions, like hydrogen bonding and van der Waals interactions, as well as electrostatic interactions [1]. Additionally, favorable enthalpy changes caused by the displacement of water molecules from the cavity, relief of conformational strain, etc. contribute to enhanced binding of the drug in the cyclodextrin cavity [1]. Stable drug-CD complexes can be prepared by a variety of methods, viz., dissolution, coprecipitation, neutralization, slurry, kneading, and grinding. In most cases, the presence of at least some water is essential for successful complex formation [2]. The complexation of a drug molecule in the cavity of the CD affects the solubility of the drug. The functional groups of the drug interact with the hydroxyl groups and water molecules within the inside of the cavity, while the outer surface of the CD interacts with the solvent. Since the solubility of the drug-CD complex depends on the hydroxyl groups on the outer surface, chemical modifications, including capping of hydroxyls with small alkyl groups, affect the overall solubility of the drug-CD complex.

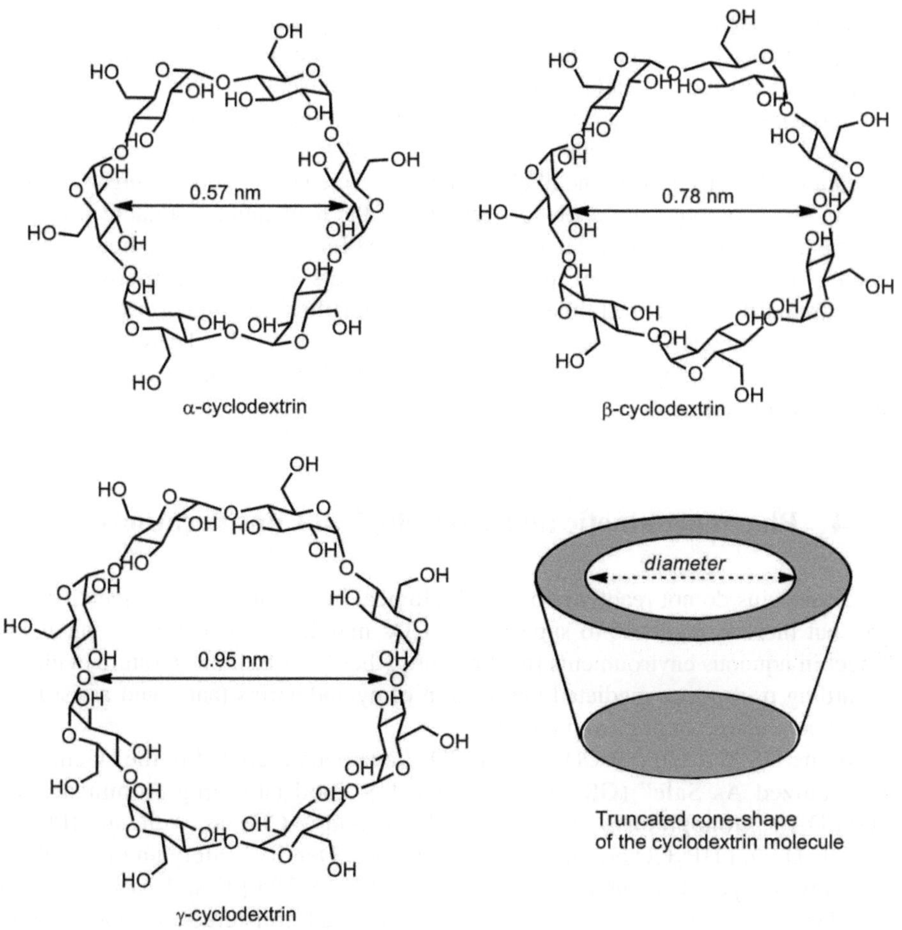

Fig. 12.1 Structures of cyclodextrins

12.3 Anatomical and Physiological Barriers in the Human Eye

Ophthalmic drugs are usually administered topically in the form of low-viscosity aqueous eye drops as solutions, suspensions, ointments and gels, and even solid inserts. However, the ocular bioavailability of such topically delivered drugs is low, and administration is usually accompanied by eye irritation and blurred vision.

The cornea, the conjunctiva, and the sclera are the primary barriers to ocular drug absorption. The corneal and conjunctival epithelia are made up of layers of cells with tight junctions that prevent the entry of drug molecules. The conjunctival epithelium is relatively less tight than the corneal epithelium which allows the permeation of the drug into the conjunctival vasculature in order to reach the sclera.

The surface of the eye is cleansed continually by tears and the turnover rate is approximately 15 % of the total tear volume per minute. This rapid clearing of tears directly affects the bioavailability of drugs as the contact time of drug with the ocular surface is reduced as a result of blinking and tearing. As the tears drain into the lacrimal duct, the drug is also washed away and may potentially enter systemic circulation [3]. To prevent the likelihood of systemic side effects arising as a result and in order to optimize the oculospecificity of the medications, suitable strategies can be used. These include the reduction of the volume of eye drops, increased viscosity, and the addition of permeation enhancers or vasoconstrictive agents. Increased drug-eye contact time is possible through the formation of polymeric bioadhesive hydrogels, *in situ* activated gel-forming systems, colloidal systems such as liposomes and nanoparticles, collagen shields (soluble ophthalmic inserts), and cyclodextrins [4].

12.4 Pharmacokinetic and Toxicological Considerations

Cyclodextrins do not readily permeate biological membranes via passive diffusion [5], but there is evidence to suggest that CDs may have a penetration enhancing effect in aqueous environments [6]. Presently, there is a lack of literature evidence regarding transporter-mediated permeation of cyclodextrins that could assist their movement across biological membranes.

In the USA, α-CDs, β-CDs, and γ-CDs have been included in the "Generally Recognized As Safe" (GRAS) list of the US Food and Drug Administration (USFDA). Monographs for α-CDs, β-CDs, and γ-CDs as well as HPβCD, SBEβCD, and HPγCD are listed in the USP/NF. Detailed safety and toxicology features of cyclodextrins have been reviewed by Stella [7] and Arima [8]. By and large, these studies have shown that orally administered cyclodextrins of pharmaceutical interest are nontoxic due to their lack of absorption. HPβCD, SBEβCD, γCD, and HPγCD can all be found in marketed parenteral formulations. Parenteral administration of αCD, βCD, or methylated βCD can result in renal toxicity and, consequently, do not find application in parenteral formulations. Although γCD can be found in parenteral formulations, its concentrations are very low. Randomly methylated βCD (RMβCD) is currently only used in topical or nasal drug formulations and that too only at relatively low concentrations. Among ophthalmic preparations, CDs are components of the marketed pharmaceutical products Indocid® eye drop solution (contains indomethacin-HPβCD complex) and Voltaren Ophtha® eye drop solution (diclofenac sodium-HPγCD) (Fig. 12.2).

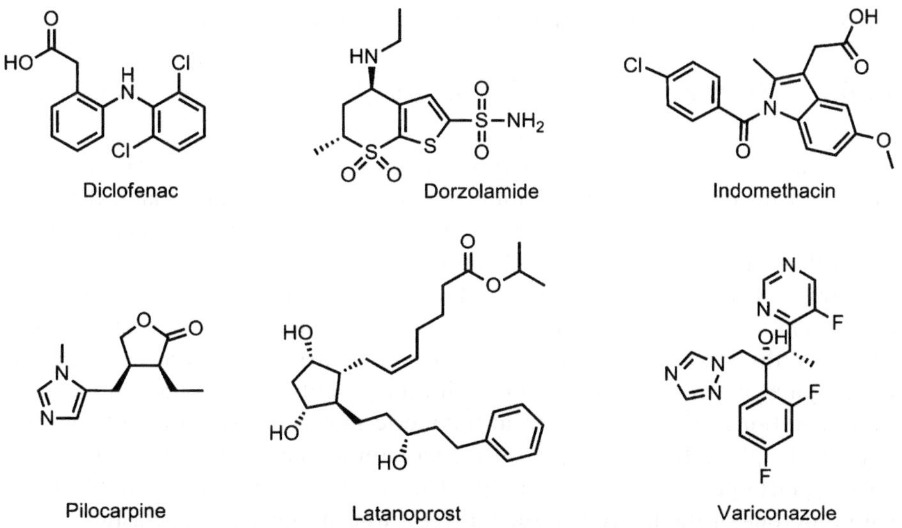

Fig. 12.2 Examples of drugs used in CD-drug complexes in ocular drug delivery

12.5 Applications of Cyclodextrins in Ocular Drug Delivery

The most common applications of CDs in ocular drug delivery are generally, associated with one or more of the following aspects: improvement of solubilization of the drug or of the drug-CD complex, chemical stabilization of the drug, reduction of ocular irritation, and/or enhancement of ocular drug permeability. Simultaneous enhancements of solubilization and chemical stabilization are achieved by various techniques described by Hedges [2]. As described earlier, the complexation of a drug molecule in the cavity of the CD affects the solubility of the drug. Since the solubility of the drug-CD complex depends on the hydroxyl groups on the outer surface, chemical modifications, including capping of hydroxyls with small alkyl groups, affect the overall solubility of the drug-CD complex.

Another important factor is the reduction of irritation, which results in rapid reflex blinking and causing a fast washout of the drug. This results in an overall loss in the available drug concentration, even before it passes the corneal barrier. It has been reported that drug loss is nearly 95 % of the administered dose [9, 10].

The primary goals of the development of ophthalmic drug delivery systems (ODDS) are to increase the amount of the drug penetration through the cornea, to extend its contact time in the eye structures, to reduce the required dose, and to lower the frequency of administration, eventually resulting in increased patient compliance [10, 11] (Fig. 12.2).

12.6 CDs in Ocular Drug Delivery Systems

12.6.1 Hydrogels: Drug-CD Complex for Soft Contact Lenses

Researchers have developed acrylic hydrogels with high proportions of cyclodextrins that assist in maintaining biocompatibility and mechanical properties of hydrogels while also improving their drug-loading ability and controlling their release rate from the drug-CD complex. Santos et al. [12] describe a procedure where glycidyl methacrylate (GMA) was copolymerized with hydrogels of acetazolamide, hydrocortisone [13] at various proportions to create poly (hydroxyethylmethacrylate) hydrogels. β-cyclodextrin (βCD) was embedded in the polymer network by conjugation with the glycidyl groups under mild reaction conditions [12]. This led to a polymeric matrix in which diclofenac was loaded; the βCDs formed no part of the structural chains but were connected to the drug through their hydroxyl groups. The pendant βCDs did not modify physical properties of the hydrogels such as light transmittance, viscoelasticity, oxygen permeability, or surface contact angle of the hydrogels but reduced the friction coefficient by half. The addition of βCDs improved diclofenac loading 13-fold leading to 15-fold enhancement of drug affinity. As an overall result, these hydrogels with pendant βCDs were able to sustain drug delivery in lacrimal fluid for 2 weeks. This finds application in the development of cytocompatible medicated implants or biomedical devices. Antibacterial thiosemicarbazones have also been extensively studied [14–17] for coupling with βCDs for loading on hydrogels of use in SCLs.

12.6.2 Ocular Mucoadhesives: Drug-CD Complex for the Resolution of Glaucoma

Carbonic anhydrase inhibitors (CAIs) used to treat glaucoma [18] (e.g., dorzolamide (Trusopt®) eye drops) unfortunately cause eye irritation when used topically, which greatly limits their clinical usefulness. While pH adjustment addressed the aspect of irritation, and the use of RMβCD [19] and γCD [20, 21] as ocular penetration enhancers improved aqueous solubility of dorzolamide (Fig. 12.2) in eye drop solutions, the ocular permeation of the drug was still poor. Consequently, formulation of mucoadhesive aqueous microparticle suspension of dorzolamide complexed with γCD resulted in enhanced ocular bioavailability [22]. A single administration of this aqueous dorzolamide-γCD eye drop microsuspension ensures a high concentration of dorzolamide in various eye tissues for extended time period with improved ocular permeation and extended ocular bioavailability. Ocular delivery of other CAIs as drug HPβCD complexes involving acetazolamide [23, 24], ethoxzolamide [25], and brinzolamide [26] have also been investigated.

12.6.3 Eye Drops: Drug-CD Complex for the Treatment of Eye Inflammation

Indomethacin finds application in the management and prevention of ocular inflammation and in the maintenance of mydriasis during cataract surgery. Inclusion complexes of indomethacin with various cyclodextrins (βCD, HPβCD, SBEβCD) were compared to the drug solution alone in to order to enable the formulation of stable and effective aqueous indomethacin eye drops with improved drug solubility and lowered aqueous instability [27]. Of the compared CDs, HPβCD was found to have the highest solubilizing effect toward indomethacin and that the formulated indomethacin-HPβCD eye drops exhibited delayed drug release and high drug stability. Anti-inflammatory studies [28] indicate that the indomethacin-HPβCD eye drops significantly improved the therapeutic efficacy of the indomethacin as compared to Indocollyre® and Voltaren® eye drops.

12.6.4 Eye Drops: Drug-CD Complex in the Treatment of Glaucoma

Pilocarpine (Fig. 12.2) has been used for open-angle and acute angle-closure glaucoma for more than 100 years. However, it was only recently that the first studies of pilocarpine with β-CDs (more specifically HP-β-CD) were published [29], discussing the enhancement of bioavailability by CDs. Further studies [30, 31] showed that pilocarpine had not only increased ocular absorption but also possessed extended shelf life when associated with SBE-β-CD and HP-β-CDs. More recent reviews [32, 33] have also mentioned the reduction of irritation of classical prodrugs of pilocarpine (such as bispilocarpic acid diesters) when co-administered with CDs.

Latanoprost (Fig. 12.2), an F2α prostaglandin analogue has been used with success in the ophthalmic therapeutics for ocular hypertension and closed-angle glaucoma [34]. The prodrug is absorbed by the cornea and later hydrolyzed by esterases to the biological active compound. It acts by increasing the outflow of aqueous fluid from the eyes through the uveoscleral tract. However, it has been found to be unstable in aqueous solution at temperatures higher than 40 °C making refrigeration necessary to maintain its therapeutic value. This factor is extremely important, mainly in countries with warm climates and where optimal storage facilities are not available. Recently, a novel solution of CD-complexed latanoprost has been introduced in Latin America named GAAP Ofteno® from Laboratorios Sophia, Mexico. The product uses (2-hydroxypropyl)-β-cyclodextrin (HPβCD) to enhance the thermal stability of latanoprost. A stability study [35] has shown that the formulation with HPβCD was stable at temperatures up to 40 °C. Although stability enhancement is not clearly linked to the CD component, it nevertheless plays an important role on preserving the integrity of the prodrug, which is necessary for the drug's permeation through the corneal barrier. The role of HPβCD on the formulation remains an area of investigation.

12.6.5 Ocular Gel: Drug-CD Complexes for the Treatment of Fungal Infections

Recent studies [36, 37] indicate that second-generation fluconazole derivatives, viz., triazole antifungal drugs like voriconazole and ravuconazole show much better outcomes on fungal keratitis than do amphotericin B and natamycin. Voriconazole (Fig. 12.2), a lipophilic drug with poor aqueous solubility, forms an inclusion complex with HPβCD which results in increased solubility and stability in aqueous solutions while simultaneously maintaining its lipophilicity and high corneal permeability [38]. Pawar et al. [39] describe the synthesis and evaluation of a voriconazole ocular gel. In this work, voriconazole-HPβCD complex is dispersed in a mixture of Pluronics®, F-127 and F68, and sodium alginate to form a gel. Drug release kinetics further demonstrated appropriate levels of corneal hydration, excellent viability of the ocular gel in releasing the drug over an extended duration of time, low degradation, and elevated stability. These results suggest that this formulation is a suitable vehicle for ocular administration of antifungal triazoles in the treatment of fungal keratitis.

12.7 Summary

Although cyclodextrins (CDs) have been studied for over 100 years, they are still regarded as novel pharmaceutical excipients. CDs are oligosaccharides that possess biological properties similar to their linear counterparts, but some of their physicochemical properties differ. CDs form water-soluble inclusion complexes with many poorly soluble lipophilic drugs, which enhance the aqueous solubility and to drug bioavailability. Through CD complexation, poorly soluble drugs can be formulated as aqueous eye drop solutions, suspensions, and gels. Their properties have changed the way we perform CD research and have given rise to new CD formulation opportunities. The pharmaceutical applications of CDs are reviewed with an emphasis on their presence in ophthalmic drug delivery.

References

1. Liu L, Guo Q-X (2002) The driving forces in the inclusion complexation of cyclodextrins. J Incl Phenom Macrocycl Chem 42(1–2):1–14
2. Hedges AR (1998) Industrial applications of cyclodextrins. Chem Rev 98(5):2035–2044
3. Urtti A, Salminen L (1993) Minimizing systemic absorption of topically administered ophthalmic drugs. Surv Ophthalmol 37(6):435–456
4. Le Bourlais C, Acar L et al (1998) Ophthalmic drug delivery systems—recent advances. Prog Retin Eye Res 17(1):33–58

5. Loftsson T, Brewster ME (2011) Pharmaceutical applications of cyclodextrins: effects on drug permeation through biological membranes. J Pharm Pharmacol 63(9):1119–1135
6. Uekama K, Adachi H et al (1992) Improved transdermal delivery of prostaglandin e1 through hairless mouse skin: combined use of carboxymethyl-ethyl-β-cyclodextrin and penetration enhancers. J Pharm Pharmacol 44(2):119–121
7. Stella VJ, He Q (2008) Cyclodextrins. Toxicol Pathol 36:30–42
8. Arima H, Motoyama K, Irie T (eds) (2011) Recent findings on safety profiles of cyclodextrins, cyclodextrin conjugates, and polypseudorotaxanes. Cyclodextrins in pharmaceutics, cosmetics, and biomedicine: current and future industrial applications. Wiley, Hoboken
9. Alvarez-Lorenzo C, Yañez F et al (2010) Ocular drug delivery from molecularly-imprinted contact lenses. J drug deliv sci technol 20(4):237–248
10. Tieppo A, White C et al (2012) Sustained in vivo release from imprinted therapeutic contact lenses. J Control Release 157(3):391–397
11. Alvarez-Lorenzo C, Hiratani H et al (2006) Contact lenses for drug delivery. Am J Drug Deliv 4(3):131–151
12. Rosa dos Santos J-F, Alvarez-Lorenzo C et al (2009) Soft contact lenses functionalized with pendant cyclodextrins for controlled drug delivery. Biomaterials 30(7):1348–1355
13. dos Santos J-FR, Couceiro R et al (2008) Poly(hydroxyethyl methacrylate-co-methacrylated-β-cyclodextrin) hydrogels: synthesis, cytocompatibility, mechanical properties and drug loading/release properties. Acta Biomater 4(3):745–755
14. Glisoni RJ, García-Fernández MJ et al (2013) B-cyclodextrin hydrogels for the ocular release of antibacterial thiosemicarbazones. Carbohydr Polym 93(2):449–457
15. Moglioni AG, Alvarez-Lorenzo C et al (2013) Romina j. Glisoni, maría j. García-fernández c, marylú pino d, gabriel gutkind b, d. Carbohydr Polym 93:449–457
16. Glisoni RJ, Chiappetta DA et al (2012) Novel 1-indanone thiosemicarbazone antiviral candidates: aqueous solubilization and physical stabilization by means of cyclodextrins. Pharm Res 29(3):739–755
17. Glisoni RJ, Cuestas ML et al (2012) Antiviral activity against the hepatitis c virus (hcv) of 1-indanone thiosemicarbazones and their inclusion complexes with hydroxypropyl-β-cyclodextrin. Eur J Pharm Sci 47(3):596–603
18. Supuran CT, Scozzafava A (2007) Carbonic anhydrases as targets for medicinal chemistry. Bioorg Med Chem 15(13):4336–4350
19. Sigurdsson HH, Stefánsson E et al (2005) Cyclodextrin formulation of dorzolamide and its distribution in the eye after topical administration. J Control Release 102(1):255–262
20. Gudmundsdottir BS, Petursdottir D et al (2014) Γ-cyclodextrin nanoparticle eye drops with dorzolamide: effect on intraocular pressure in man. J Ocul Pharmacol Ther 30(1):35–41
21. Jóhannesson G, Moya-Ortega MD et al (2014) Dorzolamide cyclodextrin nanoparticle suspension eye drops and trusopt in rabbit. J Ocul Pharmacol Ther 30(6):464–467
22. Loftsson T, Jansook P et al (2012) Topical drug delivery to the eye: dorzolamide. Acta Ophthalmol 90(7):603–608
23. Loftsson T, Friðriksdóttir H et al (1994) 2-hydroxypropyl-β-cyclodextrin in topical carbonic anhydrase inhibitor formulations. Eur J Pharm Sci 1(4):175–180
24. Mora MJ, Tártara LI et al (2013) Characterization, dissolution and in vivo evaluation of solid acetazolamide complexes. Carbohydr Polym 98(1):380–390
25. García-Fernández MJ, Tabary N et al (2013) Poly-(cyclo)dextrins as ethoxzolamide carriers in ophthalmic solutions and in contact lenses. Carbohydr Polym 98(2):1343–1352
26. Zhang Y, Ren K et al (2013) Development of inclusion complex of brinzolamide with hydroxypropyl-β-cyclodextrin. Carbohydr Polym 98(1):638–643
27. Halim Mohamed MA, Mahmoud AA (2011) Formulation of indomethacin eye drops via complexation with cyclodextrins. Curr Eye Res 36(3):208–216
28. Bucolo C, Melilli B et al (2011) Ocular pharmacokinetics profile of different indomethacin topical formulations. J Ocul Pharmacol Ther 27(6):571–576
29. Freedman KA, Klein JW et al (1993) Beta-cyclodextrins enhance bioavailability of pilocarpine. Curr Eye Res 12(7):641–647

30. Loftsson T (1995) The effect of polymers on cyclodextrin complexation. Abstracts of papers, The American Chemical Society, ACS, PO BOX 57136, Washington, DC 20037-0136
31. Suhonen P, Järvinen T et al (1995) Ocular absorption and irritation of pilocarpine prodrug is modified with buffer, polymer, and cyclodextrin in the eyedrop. Pharm Res 12(4):529–533
32. Loftsson T, Stefánsson E (1997) Effect of cyclodextrins on topical drug delivery to the eye. Drug Dev Ind Pharm 23(5):473–481
33. Loftssona T, Järvinen T (1999) Cyclodextrins in ophthalmic drug delivery. Adv Drug Deliv Rev 36(1):59–79
34. Ishikawa H, Yoshitomi T et al (2002) Pharmacological effects of latanoprost, prostaglandin e2, and f2 α on isolated rabbit ciliary artery. Graefes Arch Clin Exp Ophthalmol 240(2):120–125
35. Gonzalez JR, Baiza-Duran L et al (2007) Comparison of the stability, efficacy, and adverse effect profile of the innovator 0.005 % latanoprost ophthalmic solution and a novel cyclodextrin-containing formulation. J Clin Pharmacol 47(1):121–126
36. Hariprasad SM, Mieler WF et al (2008) Voriconazole in the treatment of fungal eye infections: a review of current literature. Br J Ophthalmol 92(7):871–878
37. Prajna NV, Mascarenhas J et al (2010) Comparison of natamycin and voriconazole for the treatment of fungal keratitis. Arch Ophthalmol 128(6):672–678
38. Dupuis A, Tournier N et al (2009) Preparation and stability of voriconazole eye drop solution. Antimicrob Agents Chemother 53(2):798–799
39. Pawar P, Kashyap H et al (2013) Hpβcd-voriconazole *in situ* gelling system for ocular drug delivery: in vitro, stability, and antifungal activities assessment. BioMed Res Int 2013(73):1–9

Chapter 13
Ocular Preservatives: Risks and Recent Trends in Its Application in Ocular Drug Delivery (ODD)

Sabyasachi Maiti, Sayantan Sadhukhan, and Paromita Bakshi

Abstract Topical application of eye drops, ointments, gels, or emulsions to treat ocular anterior segment diseases is preferred due to the ease of administration and low cost. Preservatives are the most important ingredients in ocular formulations next to active ingredient. These are often included to prevent the chances of inadvertent contamination of the ocular formulations preserved in multidose containers, most frequently during opening and application to the infected eyes. Indeed preservatives that kill or damage growing microbial cells may also be toxic to growing cells of the ocular tissues. Clinical studies suggest that the long-term use of ophthalmic preparations for effective therapy may induce major and frequent ocular surface changes causing allergic or inflammatory reactions such as redness, stinging, burning, irritation, eye dryness, or less frequently conjunctivitis or corneal damage. With the aging of population, the need for treatments of ocular diseases has become more important than ever. Increasingly high incidences of ocular disorders demand better, more effective, and innovative treatments without appreciable toxic manifestations because the preservation of vision is critical for improving the quality of life. This chapter addresses various commonly used ocular preservatives along with their ocular toxicity, newer preservatives, or other options coming up in this arena.

Keywords Ophthalmic preparations • Ocular preservatives • Organomercurials • Stabilized oxychloro complex (SOC) • Polyquaternium-1 • Ionic-buffered preservative • Mucoadhesive ophthalmics • Ocular toxicity • Corneal damages

S. Maiti (✉) • S. Sadhukhan • P. Bakshi
Gupta College of Technological Sciences (College of Pharmacy),
Ashram More, G.T Road, Asansol 713301, West Bengal, India
e-mail: sabya245@rediffmail.com

© Springer International Publishing Switzerland 2016

Y. Pathak et al. (eds.), *Nano-Biomaterials For Ophthalmic Drug Delivery*,
DOI 10.1007/978-3-319-29346-2_13

13.1 Introduction

Ophthalmic drug delivery is used only for the treatment of local conditions of the eye and cannot be used as a portal of drug entry to the systemic circulation. For most of the topically applied drugs, the site of action is usually different layers of the cornea, conjunctiva, sclera, and the other tissues of the anterior segment such as the iris and ciliary body (anterior uvea). Topical applications of eye drops, suspension, contact lens solutions, gels, and ointments are the mainstay of treatment for various ocular diseases including infections, uveitic diseases, dry eye, ocular hypertension, and glaucoma. To treat such chronic eye diseases, ophthalmic medications are recommended for longer durations on regular basis than usual dosing regimens [1]. US Food and Drug Administration recommends that ophthalmic preparations must contain an antimicrobial agent (preservative) to avoid or to limit microbial proliferation after opening the bottle. Further, they prevent physicochemical deterioration and biodegradation due to their bactericidal and/or bacteriostatic effect and prolong the shelf life of ophthalmic medications. Therefore, it becomes clear that they are not there to prevent infection in the eye but just to maintain sterility of the bottle against microbes (viruses, bacteria, and fungi or molds).

Ophthalmic medications are unique mixtures of the active drug, preservative, viscosity-increasing agents, buffers and stabilizers, and a vehicle by which all the above ingredients are "carried." They are contaminated essentially by the hands when handling the bottle or by contact of the tip touching the eyelids, lashes, conjunctiva, or tears. There is also a risk of cross-transmission when the same eye drops are shared by several patients, especially in a hospital environment or within the same family. Approximately 29 % of ophthalmic solutions are contaminated by microorganisms [2]. Further, multidose containers without ocular preservative have been often found to be contaminated with bacteria after 1–2 weeks of twice-daily usage [3]. Besides physical contamination, such a contamination may occur during manufacture. The main sources of contamination are the raw materials (especially water), the air, the personnel, and the packing material. As disposable single-dose units are less cost-effective than multidose bottles, the greatest progress has been made in the field of preservative development [4].

Although topically administered medications are increasingly used with apparent safety and good tolerance, there is growing evidence that long-term use of topical drugs can induce changes in the ocular surface and may often produce damage to conjunctival and corneal epithelial cells and the preservatives become a "necessary evil." There have been several reports of the toxic effects of prolonged topical treatments, partly due to the preservatives associated with the formulation of such treatments [5]. The repeated doses of preserved eye drops can have a cumulative effect, because the preservatives are in prolonged contact with the epithelium and may lead to disruption of the glycocalyx, when ocular preparations are used beyond the recommended dosing. Several studies have confirmed the participation of preservatives in induction of ocular surface inflammation, allergy, fibrosis, and dry eye syndrome and others. The sequence of repeated dosing leaves the epithelium unable to keep the tear film in place and can lead to ocular surface disease. This is especially true for patients who overuse their artificial tear products, use multiple ocular medications, suffer from chronic eye diseases like dry eye or glaucoma, or require postsurgery dosing of medication [6].

The adverse ocular effects could be due to the topically applied drug or the non-active ingredients (preservatives) of the formulations [7]. The infection of the weakened eye can have serious consequences leading even to ocular perforations. In addition to these serious diseases, the deterioration of the contaminated ophthalmic solutions may alter their efficacy, thus vitiating any beneficial effects on the primary disease and compromising the success of the medical treatment.

An ideal ocular preservative should possess the following characteristics:

(a) Rapidly effective against a wide range of bacteria (including *Pseudomonas aeruginosa*) and fungi at normal storage temperatures. A satisfactory eye drop preservative system should be capable of sterilizing the solution within 1 h of adding large inoculums of a test organism.
(b) Harmless to the eye and should not cause either pain or irritation.
(c) Compatible with a wide range of medicaments and other ingredients such as stabilizers and thickeners.
(d) Stable during storage and sterilization.
(e) Well within its solubility, to avoid crystallization at low temperatures [8].

Unfortunately none of the preservatives has all the required qualities so as to be used universally for any ophthalmic preparation. Indeed preservatives that kill or damage growing microbial cells may also be toxic to growing cells of the ocular tissues. The following considerations must be given while selecting preservative for an ophthalmic formulation: (a) irritation potential, (b) pH range for maximal antimicrobial activity, (c) compatibility with other ingredients, (d) synergism or antagonism in antimicrobial activity, and (e) processing conditions such as heat or packaging. Because of potential toxic effects of the ocular preservatives, it is an urgent need for the healthcare providers to have knowledge on the chronic effects of the more frequently used preservatives. Consequently, pharmaceutical industry should try to develop newer preservatives that cause less-toxic manifestations during ocular therapy. This chapter deals with the description and limitations of some common ocular preservatives, their effectiveness, their ocular toxic effects on long-term use, factors affecting ocular tolerance of preservatives, and current alternatives in the ophthalmic medications.

13.2 Ocular Preservatives and Their Mechanism of Action

Preservatives can be divided into two types, depending on their origin. Class I preservatives refer to those preservatives which are naturally occurring, everyday substances. Examples include salt, honey, and woodsmoke. Class II preservatives refer to preservatives which are synthetically manufactured. According to chemical type, preservatives are classified as follows:

(i) *Quaternary ammoniums:* benzalkonium chloride, cetrimide or cetrimonium chloride or bromide, benzododecinium bromide, cetylpyridinium chloride, polidronium chloride or polyquaternium-1, polyquaternium-42 (also known as polixetonium), sepazonium, etc.
(ii) *Mercurial derivatives:* phenylmercuric acetate or nitrate, thimerosal, and mercurobutol
(iii) *Amidines:* chlorhexidine digluconate or polyhexamethylene biguanide (PHMB)
(iv) *Alcohols:* chlorobutanol or phenylethanol, phenoxyethanol, and benzyl alcohol
(v) *Parabens or esters:* parahydroxybenzoic acid, methylparaben, and propylparaben
(vi) *Carboxylic acids:* sorbic acid
(vii) *Miscellaneous:* antioxidant preservatives such as edetic acid (EDTA, disodium edetate), sodium silver chloride complex, and silver sulfate

Based on their mechanism of action, preservatives are classified into two categories: detergent and oxidizing.

Detergent Preservatives
Detergent preservatives such as BKC and benzododecinium bromide have surfactant effects that disrupt cell membrane permeability. These compounds cause lipid dispersion, which in turn leads to lysis of cytoplasmic contents in microorganisms and ultimately bacterial death. Mammalian cells cannot neutralize chemical preservatives, so the preservative is incorporated into the cell and causes cellular damage [9].

Oxidative Preservatives
Oxidative preservatives are usually small molecules that penetrate cell membranes and interfere with cellular function [10]. They can destabilize cell membranes, but to a lesser extent than detergent preservatives. At low levels, oxidative preservatives have an advantage over detergent preservatives because they can provide enough activity against microorganisms. Mammalian cells are equipped with antioxidants, oxidases, and catalases to neutralize the effect of a low-level oxidant.

13.3 Description and Limitations of Ocular Preservatives

13.3.1 Benzalkonium Chloride (BKC)

This is a white or yellow powder and one of the most useful preservatives in ophthalmic preparations. Its dissolution in water is slow; aqueous solutions are easier to handle and are preferred. Solutions foam profusely when shaken and have a bitter taste and a faint almond-like odor which is only detectable in concentrated solutions.

Due to its cationic nature, it should be avoided in ophthalmic preparations with anionic drugs and medicinal compounds. The use of salicylate and nitrate esters of medicaments (pilocarpine, physostigmine) has been replaced by their hydrochloride and sulfate salt, respectively, to permit the addition of this preservative in ophthalmic formulation. It is also found incompatible with methylcellulose and hypromellose at >5 % (w/v). Hence, for satisfactory use of this preservative, the concentration of hypromellose eye drops BPC is kept low. It also shows pH-dependent activity and is found less effective at pH <5.0. It shows rapid antibacterial activity however; some strains of *Pseudomonas aeruginosa* and other gram-negative organisms are relatively resistant. Chelating agent EDTA or its disodium salt can be used in combination with the bactericide to reduce this resistance. These additional agents increase the permeability of the certain bacteria to BKC, possibly by removing calcium and/or magnesium ions from the cell membrane.

BKC is effective at exceptionally low concentrations (0.01 %); contact lens solutions typically contain exceptionally low (0.002–0.01 %) concentrations for effective preservative action [11]. It has been found that repeated use of BKC at tenfold higher concentrations of 1:5000 (0.02 %) or stronger can denature corneal protein and cause damage to the eye. Concentrated solutions (10 % or more) are toxic to humans, causing irritation to the skin and mucosa under prolonged contact times and death if taken internally in sufficient volumes [12].

The bactericidal/microbicidal action is thought to be due to disruption of intermolecular interactions. This can cause dissociation of cellular membrane lipid bilayers, which compromises of cellular permeability controls and induces leakage of cellular contents. Enzymes, which finely control a wide range of respiratory and metabolic cellular activities, are particularly susceptible to deactivation. Critical intermolecular interactions and tertiary structures in such highly specific biochemical systems can be readily disrupted by cationic surfactants. BKC solutions are fast-acting biocidal agents with a moderately long duration of action. They are active against bacteria and some viruses, fungi, and protozoa. Bacterial spores are considered to be resistant. Solutions are bacteriostatic or bactericidal according to their concentration. Gram-positive bacteria are generally more susceptible than gram-negative. Its activity is not greatly affected by pH, but increases substantially at higher temperatures and prolonged exposure times.

Cetrimonium is also a detergent-type preservative. Its ophthalmic uses have included preservation of artificial tear preparations such as Civigel (Ciba Vision Ophthalmics, USA). Cetrimonium causes keratinization and inflammatory infiltrates at the limbus and within the conjunctival stroma and epithelium [13]. Its corneo–conjunctival cell toxicity has been deemed similar to BKC.

13.3.2 Organomercurials

Phenylmercuric nitrate or acetate is a white crystalline powder and is slightly soluble in water. Phenylmercuric nitrate and, to a lesser extent, phenylmercuric acetate are both extensively used in eye drops (0.002–0.004 %). Their solubility is adequate (acetate: 1 in 500; nitrate: 1 in 1,500) and they are nonirritant and stable. They are effective against a wide range of bacteria and fungi but are rather slow in action, particularly against heavy inoculums of *Pseudomonas aeruginosa*. Their main disadvantages are their incompatibility with halides, strong absorption by rubber, and capacity to cause mercurialentis.

Phenylmercuric acetate has promising antifungal activity, which is significantly superior to the amphotericin B and natamycin against a wide variety of ocular pathogenic fungi based on comparative minimum inhibitory concentration (MIC) values [14].

Phenylmercuric acetate may be used in preference to phenylmercuric nitrate owing to its greater solubility. Phenylmercuric acetate is a broad-spectrum antimicrobial preservative with slow bactericidal and fungicidal activity similar to that of phenylmercuric nitrate. They are incompatible to halides, emulsifying agents, and suspending agents. In a formulation, phenylmercuric nitrate may be preferred to phenylmercuric acetate as it does not precipitate. Phenylmercuric salts are broad-spectrum, growth-inhibiting agents at concentrations normally used for the preservation of pharmaceuticals. They have slow bactericidal and fungicidal activity. Antimicrobial activity tends to increase with increasing pH, although in solution of pH6.0 and below, activity against *P. aeruginosa* has been demonstrated. Its activity increases in the presence of phenylethyl alcohol and sodium metabisulfite at acidic pH1.0–3.0. Phenyl mercuric salts may leave mercury deposits on the lens, cornea, eyelids, and conjunctiva.

Gram-Positive (+) It shows good inhibitory activity and more bactericidal activity. MIC against *S. aureus* is found to be 0.5 μg/ml.

Gram-Negative (–) It shows inhibitory activity against most of the negative bacteria. Phenyl mercury salts are less active against *Pseudomonas* species (MIC 12 μg/ml).

Most fungi are inhibited by 0.3–1 μg/ml concentrations of organomercurials, and MIC for phenylmercuric acetate against *C. albicans* is 0.8 μg/ml and against *Aspergillus niger* is 10 μg/ml.

Thimerosal (0.01 % w/v) is the sodium salt of a weak organic acid. Because its stability is low in acid solutions, it is not widely used in eye drops, most of which are acid. Like phenylmercuric salts, its antibacterial activity is slow, but it differs from these compounds in being less stable in solution (protection from light and heavy metals is necessary) and does not cause mercurialentis.

Organomercurials bind to the cell membrane protein sulfhydryl groups, causing an increase in cellular permeability, and kill microorganisms by affecting internal cell respiration. Adverse ocular side effects due to these preservatives are rare. The most striking side effect is mercurial deposits in various ocular tissues. The optimal pH for the antimicrobial activity and chemical stability lies in the range of 7.0–8.0.

13.3.3 Chlorhexidine Acetate

Chlorhexidine is a cationic polybiguanide and primarily used as its acetate salts. Chlorhexidine acetate is a nonirritant bactericide having low toxicity. It is an antibacterial substance which is effective against a wide range of gram-positive and gram-negative bacteria. It inhibits potassium transmembrane transport. It is most effective against gram-positive bacteria but inactive against spores. The susceptibility of some *Proteus* and *Pseudomonas* species is low but increases in the presence of EDTA. It has shown highest activity in neutral or slightly alkaline pH.

Some organic matters such as serum and, particularly, phospholipids seriously reduce its efficiency, and like BKC, it is incompatible with anionic compounds and, due to binding, with methylcellulose and hypromellose.

Sparingly soluble salts are formed with bicarbonates, borates, carbonates, chlorides, citrates, phosphates, and sulfates of drugs, but precipitation (which is usually delayed for about 24 h) does not occur unless the concentration of chlorhexidine acetate is about 0.05 %. It is not true with sulfates, which cause difficulty even when the preservative concentration is as low as 0.005 %. The usual concentration is 0.01 % (w/v). Heat causes slight degradation to 1, 4 chloroaniline, an eye irritant, but no abnormalities were found in the eyes of guinea pigs after instilling solutions heated at 98–100 °C for 30 min. Only very minor effects were observed with solutions autoclaved at 115 °C for 30 min.

13.3.4 Hydroxybenzoates

Chemically, parabens are esters of p-hydroxybenzoic acid. The most common parabens used in ophthalmic products are methylparaben and propylparaben. They are solids that melt between 95 °C and 128 °C. Parabens are only slightly water-soluble and water solubility decreases from methylparaben (0.25 %) to propylparaben (0.05 %). Parabens are also easily soluble in other organic solvents. The concentration is usually between 0.1 and 1.0 % (w/v) which is considerably higher than for more recently developed preservatives. They are active against bacteria, yeast, and molds used to preserve drops. The relatively high price together with the low water solubility limits their use as preservatives. Eye drops containing methylparaben, even in small amounts, can damage the lining of the eyes, as well as the cornea. Maxitrol™ ointment contains anti-infective steroid with methylparaben (0.05 %) and propylparaben (0.01 %) preservatives. Prolonged use may result in ocular hypertension and/or glaucoma, with damage to the optic nerve, defects in the visual acuity and fields of vision, and posterior subcapsular cataract formation. It possesses slow antibacterial activity, particularly against *Pseudomonas aeruginosa*.

13.3.5 Chlorobutanol

In pure state it is a white, volatile solid and slightly soluble in water. Chlorobutanol is an alcohol-based preservative, with no surfactant actions. The antibacterial activity of chlorobutanol is slow, even at 0.5 % (w/v). As this concentration is near to saturation point at low temperatures, the solutions may deposit crystals in cold weather.

Since it is volatile and only slowly soluble, care is necessary to avoid loss into the atmosphere when solutions are being prepared. It hydrolyses in solution, producing hydrochloric acid. Acid solutions of pH less than six are most stable but they decompose appreciably when steamed or autoclaved. Alkaline or buffered neutral solutions are very unstable, even at room temperatures.

USP recommends it as an alternative to BKC and it is used in combination with other preservatives, in some commercial eye drops. Its optimal pH for antimicrobial activity is <5.0.

Although the antimicrobial activity of chlorobutanol is extensive (gram-positive and gram-negative bacteria and fungi), its use is limited due to its instability when stored at room temperature for an extended period. It causes cell lysis by way of disruption of microbial cell membrane lipid configuration [15]. Although chlorobutanol seems relatively safe, chlorobutanol (0.5 %) in artificial tears caused irritation in more than 50 % of the subjects in a double-blind crossover study [16]. To this end, research has shown that in human corneal cells, chlorobutanol causes cell retraction and cessation of normal cytokinesis, cell movement, and mitotic activity [17]. It also causes degeneration of human corneal epithelial cells and the formation of conspicuous membranous blebs. Chlorobutanol does not, however, affect the stability of the lipid component of the tear film. In human corneal epithelial cells, the cytotoxic effects of chlorobutanol occur less rapidly than BKC. Therefore, ophthalmic solutions preserved with chlorobutanol may be less damaging.

Chlorobutanol is mainly known to induce cytotoxicity in corneal epithelial cells. In previous studies, it was reported that occasional use (twice daily up to 12 days) of a chlorobutanol-preserved artificial tear resulted in only modest exfoliation of corneal epithelial cells (i.e., up to a maximum of 14 %) [18]. However, these changes were reversible and thus the eye could adapt to repeated use of the preserved artificial tears [19].

13.4 Factors Affecting the Ocular Tolerance of Preservatives

In usual concentrations and at low-frequency dosages, preservatives used selectively in eye medications do not notably damage the eye, as proven by the millions of daily applications all over the world [20]. However, the ocular adverse effects induced by preservatives may increase under certain circumstances.

13.4.1 Concentration of Preservative

The concentration of the preservative should be kept below a level; otherwise it may be toxic to human beings. The concentration of an added antimicrobial preservative can be kept at a minimum if the active ingredients of the formulation possess an intrinsic antimicrobial activity.

Increasing the concentration generally results in increased ocular injuries. The severity of ocular damage increases with increasing concentration of preservatives. More precisely, at concentrations below the effective antimicrobial concentration, no or little damage is detected, whereas at tenfold increased concentration, more severe lesions appeared.

Under electron microscope, the rabbit eyes treated with 1 % BKC (twice instillations/7 days) showed corneal cloudiness and conjunctival infection. However, no gross damage occurred at 0.1 % concentration. Conjunctival necrosis, corneal ulcers, haziness, massive iritis, and endothelial defects were observed at 2 % concentration following twice instillations for 1 day [21].

13.4.2 Frequency of Instillation

The number of times the eye drops are instilled can influence the exposure dose of ocular tissues and hence the extent of possible ocular damage. A fourfold increase in corneal damage was found with high instillation regimen (two drops of 0.02 % BKC solution every 3 min for 1 h) [22]. The mild regimen (two drops instilled every 30 min for 2 h) induced only partial loss of microvilli, whereas the exaggerated one caused cell peeling, retraction of cell membrane borders, and total loss of microvilli.

13.4.3 Duration of Treatment

In the treatment of chronic ocular diseases, topical ocular medications are applied several times a day, over years or for life. The long-term exposure of corneal epithelium to preservatives is susceptible to increasing the irritation potential of the preservatives [23]. In healthy volunteers, BKC in beta-blocker eye drops has been shown to break the tear film stability [24]. Schwab et al. [25] reported that antiglaucoma medication use for greater than 3 years was associated with significant foreshortening of conjunctiva. Longstaff et al. [26] conducted a study to identify potential risk factors for failure of pressure control following glaucoma triple procedure and found that the only significant adverse factor was cumulative years of preoperative topical therapy. The hazard ratio for this factor was 1.1 ? implying that the risk of failure doubled after 8 years of cumulative therapy.

13.4.4 Combination of Preservatives

The antimicrobial spectrum is extended by combining preservatives. EDTA helps in restricting metal-catalyzed oxidation of various drugs and promotes the action of antimicrobial preservatives [27]. EDTA is often combined with BKC (0.01–0.1 % w/v) or other preservatives to enhance the antimicrobial activity against strains of *Pseudomonas*. However, the preservative combination extends the antimicrobial spectrum but can also affect the tolerance of eye drops. The addition of EDTA (0.05 %) to thimerosal (0.02 %) or chlorhexidine (0.025 %) was found to induce morphological alterations on the corneal surface, whereas these alone were not found to cause any changes at the same strengths [20].

13.4.5 Physiological State of Cornea

After ophthalmic surgery, the eyes become more sensitive to the deleterious effect of some preservatives that delay the wound healing or influence the corneal permeability. The changes in corneal integrity due to preservative use like 0.01 % chlorhexidine may dramatically affect the corneal penetration of some inert excipients in ophthalmic formulations and lead to possible toxicity [28].

13.4.6 Use of Mucoadhesive Ophthalmic Product

The inert mucoadhesive polymers (carboxymethylcellulose, carbopol, polycarbophil, and sodium alginate) are used to enhance viscosity of ophthalmic preparations, which in turn prolong the corneal contact time of medications and increase ocular bioavailability of the drug. Such polymers have a property known as bioadhesion, meaning attachment of a drug carrier to a specific biological tissue. Sometimes, this may lead to ocular damages. For example, eye preparations containing hydroxyethylcellulose (HEC), a thickening agent, prolonged the contact time of BKC with the rabbit cornea and consequently resulted in epithelial cell damage that was not observed with either preservative or polymer alone [29]. The blurring of vision and esthetic disadvantages of caking and drying on eyelashes are drawbacks of highly viscous agents that patients with mild to moderate dry eye will not tolerate. Lower molecular-weight viscous agents help to minimize these problems. Hydroxypropyl-guar (HP-guar) has been used as a gelling agent in a solution containing glycol 400 and propylene glycol (Systane®). It has been suggested that HP-guar preferentially binds to the more hydrophobic, desiccated, or damaged areas of the surface epithelial cells, providing temporary protection for these cells [30]. Hyaluronic acid (0.2 %) has significantly longer ocular surface residence times than 0.3 % HPMC or 1.4 % polyvinyl alcohol. Some clinical studies reported improvement in dry eye in patients treated with sodium hyaluronate-containing solutions compared to other lubricant solutions, whereas others did not [31].

13.5 Newly Introduced Ocular Preservatives

Indeed, the effectiveness of microorganisms through cytotoxic effects cannot be achieved without a minimal amount of toxicity to tissues where preservatives are applied. The authorities are more concerned by such toxicological issues and nowadays they tend to favor the development of new preservatives by the industrials.

13.5.1 Polyquaternium-1

Polyquaternium-1 (polyquad or polidronium chloride) is a polymeric quaternary ammonium antimicrobial preservative. Polyquaternium-1 (0.001 %, w/v) is a detergent-type preservative derived from BKC. It was formulated in the mid of 1980s by Alcon as a preservative for contact lens storage solutions. Nowadays, it is being increasingly used as a preservative in ophthalmic drops for glaucoma and artificial tear solutions. Recent studies have assessed its suitability as a preservative for prostaglandin analogs used in antiglaucoma therapy [32]. The main disadvantage is its tendency to reduce the density of conjunctival goblet cells, thereby decreasing the aqueous tear film production [33]. Despite being less toxic to the corneal surface than BKC, it is known to induce damage to corneal epithelial cells as well [34]. Cationic polymer polyquad preservative was found less cytotoxic in vitro than the highest commercial BKC (0.02 %) concentration. The eye drops containing this preservative did induce even stronger inflammatory response than did BKC in human corneal epithelial cells [35]. Commercial eye drops containing this new preservative are Travatan and Systane Ultra. Recent findings reveal that this has detrimental effects on cell membrane integrity and induces cytotoxicity in ocular surface cells [36].

13.5.2 Sorbic Acid

It is a colorless solid that is slightly soluble in water and sublimes readily. Sorbic acid (pKa 4.76) and its salts, such as sodium sorbate, potassium sorbate, and calcium sorbate, are antimicrobial agents often used as food preservatives to prevent the growth of mold, yeast, and fungi. In general, the salts are preferred over the acid form because they are more soluble in water, but it is the acid form that is active. Sorbates are generally used at concentrations of 0.025–0.10 % in ocular formulations. Generally, it has limited antimicrobial activity and cannot eradicate many organisms on its own. Sorbic acid molecules pass through the plasma membrane, dissociate in the cytoplasm, release protons, and inhibit growth *via* acidification. This may activate energetically inefficient intracellular activities, such as energy-dependent ion pumps. Punctate keratitis may rarely result from the use of sorbate. Still, sorbic acid-preserved products are commonly promoted for sensitive eyes and for contact lens wearers [37]. It shows optimal antimicrobial activity below the pH6.5.

13.5.3 Sodium Perborate

Sodium perborate is a white, odorless, water-soluble chemical compound. It releases oxygen rapidly at temperatures over 60 °C. To make it active at lower temperatures (40–60 °C), it has to be mixed with a suitable activator, typically tetraacetylethylenediamine. It is used as a "disappearing" preservative in some brands of eye drops. Sodium perborate (GenAqua®) is an oxidative preservative and can destroy numerous types of bacteria and a fungus *Aspergillus niger*. Sodium perborate oxidizes cell walls or membranes, affects membrane-bound enzymes, and disrupts cellular function such as protein synthesis. When sodium perborate is combined with water, it is converted to hydrogen peroxide, an effective antimicrobial agent. In contact with the tear film, it is converted into oxygen and water molecules by catalase and other enzymes present in the conjunctival sac [38]. The hydrogen peroxide allows low levels of sodium perborate to be effective at destroying microbes, while hydrogen peroxide as a byproduct also permits ocular comfort. However, hydrogen peroxide levels above 100 ppm or 3 % can cause ocular stinging [39,40].

13.5.4 Stabilized Oxychloro Complex (SOC)

SOC is an oxidative ophthalmic preservative and is popularly known as Purite®. It is a microbicide with a broad spectrum of antimicrobial activity and very low toxicity to mammalian cells. SOC consists of an equilibrium mixture of oxychloro species: 99.5 % chlorite (ClO_2), 0.5 % chlorate (ClO_3), and trace amounts of chlorine dioxide (ClO_2) which have additional fungicidal (*Aspergillus niger*) and virucidal activity. Purite® preserves the solution in the bottle but ultimately, following exposure to light, dissociates into water, sodium, chloride ions, and oxygen. It is considered as a *disappearing preservative*. Because these substances are also found in natural tears, the risk of preservative-induced ocular irritation and corneal damage is minimized [40]. Most likely, the mechanism is due to the oxidation potential of chlorite and possibly from the generation of chlorine dioxide in the presence of microbial acidic environments, which leads to the disruption of protein synthesis in the bacteria and ultimately its antimicrobial activity. SOC is efficacious at low concentrations (0.005 % w/v). SOC is safe and well tolerated when administered frequently, as demonstrated in a 62-patient study in which Refresh Tears containing an SOC-based preservative was administered 4–8 times daily for a 4-week period [41].

Another SOC is also available under the trade name OcuPure™ (Advanced Medical Optics). It consists of a mixture of 80 % chlorite, 11 % sodium chloride, water, and trace electrolytes. OcuPure is also considered a *disappearing preservative* because when exposed to light it breaks down into water and sodium chloride, which are already present in natural tears [42]. "OcuPure" preservative reduces blurring and is clinically shown to improve tear film stability. Oxyd™ is another Chlorite-peroxy compound and is a unique new preservative which is safe for use with all contact lenses. When

Oxyal™ viscoelastic ocular product comes into contact with the enzymes and proteins naturally present in tear film, Oxyd™ is converted into physiological salt and oxygen. This results in users having no toxicity, allergy, and corneal irritation problems when using Oxyal™. Oxyd, the new antimicrobial preservative, is very effective as a broad-spectrum antimicrobial agent [43]. OxyChlorite™ is a chlorite-peroxy compound used in contact lens solution. Following application, it also breaks down into the components of natural tears (H_2O and NaCl) and releases)oxygen.

13.5.5 Ionic-Buffered Preservative

Ionic-buffering systems are the latest class of ophthalmic preservatives to be incorporated into topical medicines. This acts in a similar manner to oxidizing preservatives within multidose bottles. SofZia (Alcon, TX, USA), the most recent preservative of this kind, is a combination of boric acid, zinc, sorbitol, and propylene glycol. This ionic-buffered system has been shown to have both antibacterial and antifungal qualities. When exposed to cations, such as those that are normally encountered in the tear film of the eye, the substance is deemed inactive. This is thought to induce less cytotoxicity to the ocular surface compared with more conventional preservatives [44].

SofZia-preserved travoprost (Travatan Z, Alcon) formulation appears to have the same intraocular pressure (IOP)-lowering efficacy as BKC-preserved travoprost in controlled clinical trials [45]. Arguably, this formulation is the most potent non-BKC-containing glaucoma eye drop currently available. Preclinical studies comparing SofZia-preserved travoprost and latanoprost, which contain the highest amount of BKC, demonstrated the relative safety to the ocular surface of the non-BKC formulation [46].

13.5.6 Polixetonium Chloride (Polyquaternium-42)

Chemically, it is poly(oxyethylene (dimethyliminio) ethylene (dimethylimino) ethylene dichloride). Polixetonium is a highly effective antimicrobial that does not damage the ocular surface or tear film with long-term use. The product is known as polyquaternium-42 (Buckman Laboratories of Memphis, TN). Polixetonium does not contain benzene rings (aromatic) in the side chains, mostly straight polyethylene chain segments. Polixetonium actually facilitates wetting of surfaces (even contact lenses) and improves the ocular comfort and safety of patients wearing contact lenses. FreshKote (APL-105) is a new three-layered artificial tear with the highest oncotic pressure (65 mm Hg) to provide healing of the ocular surface. The normal human tear film is comprised of three layers (mucin-aqueous-lipid). This innovative product is preserved with the only totally nonirritating ophthalmic preservative (polixetonium chloride). Polixetonium (Polyquaternium-42)exhibits potent antimicrobial and antifungal activity.

13.5.7 Benzododecinium Bromide

Benzododecinium bromide (dimethyldodecylbenzylammonium bromide) is a quaternary ammonium compound. It is well soluble in water and has properties of cationic surfactant. Benzododecinium bromide is effective against gram-positive microbes. In lower concentrations, its activity against conditionally gram-negative microorganisms (such as *Proteus*, *Pseudomonas*, *Clostridium tetani*) is uncertain. Timoptol LA® gel-forming eye drops contain benzododecinium bromide (0.012 %). A study confirmed that quaternary ammoniums should be avoided as far as possible, especially in chronic ocular surface diseases such as glaucoma, dry eye, and allergy. A significant decrease of membrane integrity with chromatin condensation was observed with all the quaternary ammoniums tested at concentrations of 0.005 % and higher [47].

13.6 Toxicity of Ocular Preservatives

13.6.1 Conjunctival Toxicity

The conjunctiva is a delicate membrane that lines the posterior side of the eyelid and covers the anterior part of the eyeball. The conjunctival tissue reacts in a number of ways in response to topical ocular medication: allergic reactions, papillary or follicular irritative conjunctivitis (redness), dyschromia, and microbial imbalance accompanied by tearing (discharge). Thimerosal has been found to cause conjunctival hyperemia in rabbit and human eye [48]. Unlike monolayer cultures in which ocular medications containing BKC cause near-complete loss of cell viability, stratified, air-lifted cultures are not affected by the presence of BKC in ocular medications with up to 30 min exposures. Stratification shifts the dose–response curve to the right for BKC, thimerosal, chlorhexidine digluconate, potassium sorbate, and EDTA and thus demonstrates that stratification significantly affects cell viability of Chang conjunctival cells in response to preservatives [49]. No difference in the conjunctival toxicity exists between preservative-free gentamicin and preservative-free cefuroxime [50]. BKC can damage corneal and conjunctival epithelium by affecting cell-to-cell junctions and cell shape and microvilli and, eventually, lead to cell necrosis with sloughing of 1–2 layers of epithelial cells [51].

13.6.2 Tear Film Disruption

The tear film is tightly bounded to the corneal epithelium and protects the eyes from dehydration and other external harmful agents. It is a three-layer fluid of 7 mm thickness composed of lipid, aqueous, and mucin phases. In the classical three-layered tear film model, the mucin layer is usually thought of as a surfactant or wetting agent, acting to lower the surface tension of relatively hydrophobic ocular surface, rendering the corneal and conjunctival cells wettable. The stability of tear film depends on their chemical physical characteristics, interacting with the conjunctival and corneal epithelium via the membrane-spanning mucins (MUC-16 and MUC-4).

Any discontinuity and instability in the tear film over the cornea may lead to ocular non-wetting tears disorders called "dry eye" [52]. Besides polymers, artificial tears contain preservatives. In quaternary ammoniums, BKC is the most disruptive ophthalmic preservative and causes almost instantaneous disruption of the tear film by solubilizing the tear lipid layer. A single drop of 0.01 % BKC can break superficial lipid layers of tear film into numerous oil droplets which are made soluble due to its detergent action. Repeated blinking does not restore the lipid layer for some time [53]. It is estimated that dry eye may be present in approximately 40 % glaucoma patients. Research demonstrates that significant exposure to glaucoma medications, preserved with BKC, can produce alterations in tear film function and inflammatory conjunctival infiltrates [7].

The ocular surface inflammation associated with dry eye is exacerbated by the preserved lubricants; however, non-preserved solutions are themselves inadequate to improve the surface inflammation and epithelial pathology seen in dry eye disease.

13.6.3 Allergic Symptoms

Three predisposing factors are considered important to allergic reactions: individual susceptibility, prolonged use of medication, and pre-existing cutaneous disease of the eyelids. Evidence of allergic contact reactions includes follicular conjunctivitis, eczematoid blepharitis, periocular dermatitis preceded by conjunctival hyperemia, discharge, and edema. Other symptoms include itching and burning sensations of the eye, epiphora, and photophobia [54]. Topical medications act as haptens (partial antigens) and must combine with tissue protein to form complete antigens for sensitizing lymphocytes and producing inflammation. This sensitization takes weeks to years to develop, but once the patient has been sensitized, allergic reaction can appear in as little as 48 h. The commonly used preservatives are involved in allergic contact reactions. For example, thimerosal acts as hapten and is the most allergenic preservative [55]. BKC tends to substitute thimerosal because of a wider antibacterial spectrum and for environmental reasons (thimerosal contains mercury), but one must remember that BKC also has an allergenic potential and can cause severe allergic reactions.

According to their nature, the preservatives induce an allergic reaction but more frequently a cytotoxic reaction (90 %). The degree and type of allergy incurred depends upon patient factors and the type of preservative. Benzalkonium salts, which are considered moderately sensitizing, usually result in contact allergies and delayed hypersensitivity reactions. Furthermore, long-term use of these agents may result in conjunctival scarring known as drug-induced pemphigoid, in which chronic allergic reaction leads to a marked and self-sustaining inflammatory process [56,57].

13.6.4 Corneal Damages

The human cornea comprises of five layers, i.e., corneal epithelium, Bowman's membrane, stroma, Descemet's membrane, and endothelium. The corneal epithelium is particularly vulnerable to the toxic effects of preservatives due to its position. Corneal damages can be caused by preservatives contained in topical ocular medications. The preservatives with surfactant properties like BKC dissolve the membrane of corneal epithelial cells. At low concentration, they induce a loss of microvilli from the cell surface and at a higher concentration they cause lifting of cell borders and cellular desquamation [58]. Sometimes, cationic compounds also interact with cellular proteins containing glutamate and aspartate [59]. Another clinical evidence of preservative-induced ocular damages includes nonspecific irritation reactions (edema, eye redness) generated by chlorhexidine and alcoholic preservatives. The corneal endothelium is essential for maintaining dehydration of the cornea and its optical properties. Any damage to the endothelium is more serious than damage to the epithelium and is associated with marked corneal swelling and loss of corneal transparency. Prolonged use of topical medications containing BKC is assumed to induce toxic endothelial degeneration in human eyes requiring corneal transplantation [60]. Parabens are also found to cause morphologic changes in rabbit endothelium [61]. BKC (0.01 %), chlorobutanol (0.5 %), and thimerosal (0.001 %) cause immediate cell retraction, cessation of normal cytokinesis, cell movement, mitotic activity, and degeneration of the epithelial cells in 2, 8, and 9 h, respectively. Cultures treated with chlorobutanol develop conspicuous blebs on the cell surface after 3–5 h of exposure. Sorbic acid (0.1 % and 0.2 %) greatly reduces cell movement and suppresses mitotic activity, but causes no cell death. Polyquaternium ammonium chloride (0.001 %) and polyaminopropyl biguanide (0.00005 %) have no discernible effects on cytokinetic movement or mitotic activity of the epithelial cells [62].

Once-daily dosing of travoprost with SofZia produces fewer corneal changes and less conjunctival inflammation than latanoprost preserved with BKC in rabbit model. Corneal and conjunctival changes noted with travoprost with SofZia are similar to those induced by preservative-free artificial tears. Glaucoma medication with high levels of BKC may cause more deleterious effects on the ocular surface than non-BKC-preserved medications [63].

13.6.5 Toxicity to Contact Lens Users

A slight corneal abrasion may occur during normal wear, as well as anoxia and accidental trauma during insertion of the lens. These factors predispose the cornea to infection as the lens provides a medium for introducing pathogens into the eye. For this reason, the use of preservatives is sometimes required in some contact lens care solution. Preservatives are contained in cleaning solutions, soaking or disinfecting solutions, and wetting solutions. Hydrophilic polymers used for soft lenses can adsorb preservatives. Although this adsorption is not harmful per se, the later desorption presents a toxicity problem for the eye. Hard lenses have the same problems as soft lenses, but to a much smaller extent because they have a lower degree of adsorption.

The lens acts as a reservoir for the preservatives, thereby providing an opportunity for prolonged contact between these chemicals and the ocular surface. For this reason, substances that are not toxic in eye drops may become toxic when used with soft lenses. Allergic and toxic reactions to preservatives in soft contact lens solutions have been reviewed extensively. For instance, thimerosal (0.004–0.005 %, w/v) may induce delayed ocular hypersensitivity. The patients may develop superior limbic keratoconjunctivitis, conjunctival hyperemia, limbal follicles, papillary conjunctivitis, superficial punctuate keratitis, and others [64]. BKC, whose uptake by soft contact lenses is high, has been demonstrated to increase corneal epithelial desquamation and cause ocular inflammation [65].

13.6.6 Toxicity of Newer Preservatives

The toxicity of newer preservatives has been evaluated in animal models. In an investigation, the rabbits are treated four times a day with ocular products containing polyquaternium-1, sodium perborate (0.001 %), or SOC (0.005 %) and the corneas are examined after 1 week [66]. The untreated rabbit corneal epithelium has extensive microvilli and tight intercellular junctions. The rabbit eye treated with SOC product looked nearly identical to the untreated rabbit eye, and the eye treated with perborate-preserved product also shows a mostly normal epithelium with extensive microvilli and tight epithelial junctions. The same workers report that the rabbit eye treated with polyquaternium-1-preserved product has extensive superficial epithelial erosion and lack of protruding microvilli. Another study supports this finding in that polyquaternium-1 (0.001 %) causes a small but significant degree of epithelial damage [67]. However, Pham and Huff [68] have shown that polyquaternium-1 (0.001 %) has a low potential for ocular irritation. Compared to untreated eye, the damage to the corneal epithelium follows the order: polyquaternium-1> sodium perborate >SOC. Of the artificial tears compared in this study, the product with SOC causes the least damage to corneal epithelial cells.

Relatively few studies are available that compare the toxicity of preservatives used in ophthalmic preparations. Debbasch et al. [47] conducted a separate study to evaluate the toxicity of preservatives in a continuous human conjunctival cell line. The cell line is treated with different concentrations of various preservatives for 15 min and followed by 24 h of cell recovery: three BKCs with different hydrocarbon chain length, benzododecinium bromide, cetrimide, phenylmercuric nitrate, thimerosal, methyl parahydroxybenzoate, chlorobutanol, and EDTA. With all the quaternary ammoniums tested at concentrations of 0.005 % and higher, a significant decrease of membrane integrity with chromatin condensation is observed. The other preservatives tested do not decrease membrane integrity. They conclude that the quaternary ammoniums tested are the most cytotoxic preservatives.

Epstein and his coworkers [69] evaluated the potential toxicity of the common components of eye drops such as the preservatives in a tissue culture model of immortalized corneal and conjunctival epithelial cells (Fig. 13.1). These results suggest that even at low concentrations, all the preservatives tested cause some degree of damage to ocular tissue. Their toxicity increases with increasing concentration of all the preservatives. The order of decreasing toxicity of the tested agents is as follows: thimerosal (0.01 %)> BKC (0.01 %)> chlorobutanol (0.25 %) >methylparaben (0.01 %)> sodium perborate (0.0025 %) ≈ EDTA (0.01 %).

13.7 Newer Alternative Approaches

The research and development of new chemicals that are well tolerated and less toxic is a challenge. In addition to the search for a comparatively newer, less-toxic preservative, newer options must be sought and some of them are discussed herein.

	Formalin, 10%	Thiomersal, 0.01%	BKC, 0.01%	Chlorobutol, 0.25%	Methyl paraben, 0.01%	Perborate, 0.0025%	EDTA, 0.01%	Medium alone
HCE	100	96.49	78.32	75.94	57.09	21.77	24.74	0
CCC	100	92.62	74.24	70.35	53.11	31.4	31.05	0

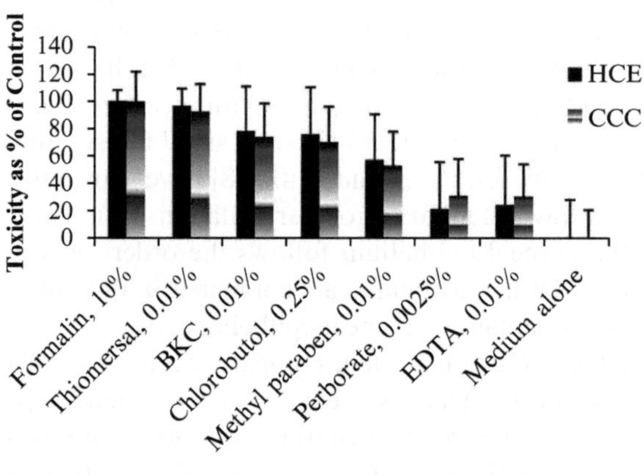

Fig. 13.1 Graphical representation of overall relative toxicities of preservatives to human corneal (HCE; *black bars*; 10.014 pRSV-T) and conjunctival epithelial cells (CCC, *gradient gray bars*; Wong–Kilbourne derivative of conjunctiva, clone 1-5c-4) after being put into contact with each of the experimental testing solutions for 1 h (Graph constructed from data reported by Epstein et al. [69]

13.7.1 Non-Preserved Unit-Dose Eye Drops

The package (0.1 and 1 ml) is often prepared in form-fill-seal equipment where the melted low-density polyethylene forms plastic containers, which are aseptically filled and sealed. The container has a pointed cone with a twist-off closure or a closure cap [70]. Because an opened package is discarded after application, a unit-dose package eliminates the risk of product contamination, making the addition of preservative unnecessary. However, they present some drawbacks: (i) relatively expensive (5–10 times) than multidose containers; (ii) waste of solution and plastic packaging on chronic use; (iii) clear-cut opening of the container that is difficult and, hence, dose variation depending on the dimensions of the orifice created; and (iv) possibility of remaining a barb of plastic around the orifice during twisting off the closure, causing irritation to the eye, if touched during application of drop [71]. Although the risk of contamination is theoretically nil, in actual practice, patients tend to keep the unit-dose container longer than for a single use, even for several days, with the ensuing risks of contamination.

13.7.2 Non-preserved Multidose Drops

To avoid preservative-induced ocular problems without compromising sterility of eye drops, the multidose bottles of non-preserved formulations that contain antibiotics or alkaloids have been suggested because the drugs show good inherent efficacy against microbes [72]. However, once opened by patients, the in-use storage life is limited to 7 days, provided the preparations are stored in a refrigerator. However, this approach has also some drawbacks: (i) drug specific (those having intrinsic antibacterial activity), (ii) uncomfortable for elderly patients to instill chilled eye drop, and (iii) refrigeration not an absolute protection against microbial contamination. Indeed, some mesophilic and many psychrophilic bacteria can multiply at temperature between 4 °C and 8 °C, especially since home refrigerators are hardly sterile.

13.7.3 Special Filter Devices

An interesting alternative to unit-dose containers consists of a packaging in multidose bottles with a special filter device. The device may contain a sterile preservative-free solution, protected against microbial contamination by a 0.2 μ membrane filter. Alternatively, the preservative is retained by a filter (adsorbing resin) upon instillation. In this way, preservative-free eye drop is delivered to the eye. An antiglaucoma medication is now commercially available (Timabak) that uses special filter devices. It is to remember that this new packaging system is costly compared to traditional multidose bottles, which is further lower than the unit-dose packages.

13.7.4 Collasomes

Collasomes are made up of collagen pieces suspended in a viscous vehicle; this formulation is similar to that of corneal collagen shields, but is more useful for chronic therapy because the collasomes can be instilled by the patient and cause less blurring of vision. Potential uses include delivery of hydrophobic antimetabolites such as cyclosporine to prevent graft rejection and delivery of lipids for the therapy of dry eyes. Ideal replacements for dysfunctional tear layers, especially the lipid layer, have not yet been developed [73]. For patients with conditions requiring chronic rather than acute therapy, the advantages of collagen shields in providing high and sustained levels of drugs and/or lubricants to the cornea are outweighed by the difficulty of insertion of the shield and the problem of blurred vision. In the normal eyes of volunteers, collasomes hydrated in a solution of sodium fluorescein and suspended in a methylcellulose vehicle as a model for the delivery of water-soluble drugs produced fluorescein concentrations 17–42 times higher in the cornea and 6–8 times higher in the aqueous humor, compared with fluorescein-containing vehicle alone [74].

13.7.5 Hypo-Osmotic Solution

Tears of patients with dry eye have a higher tear film osmolarity (crystalloid osmolarity) than do those of normal patients. Elevated tear film osmolarity causes morphological and biochemical changes to the corneal and conjunctival epithelium and is pro-inflammatory. This knowledge influenced the development of preservative-free hypo-osmotic artificial tears such as Hypotears® (230 mOsm/l) and subsequently Thera-Tears® (181 mOsm/l) [75].

13.8 Conclusion

In chronic eye infections or disorders, repetitive use of ophthalmic medications is necessary for the effective treatment of ocular diseases or disorders. Recently, a number of reports raise the possibility of imparting toxic effects to the ocular surface by preservatives added to the formulations. Patients frequently develop clinical signs of preservative toxicity. The telltale symptoms of toxicity include discomfort on instillation, irritation between instillations, foreign body sensation, stinging and burning, dry eye, tearing, and eyelid itching. More than half of patients experience such symptoms. Due to rising concerns of ocular toxicity of the commonly used preservatives, the manufacturers should reformulate the existing products without preservatives or with newer less-toxic preservatives. Furthermore, preservative-free products are now coming into the market and should be used instead of preserved formulations, where appropriate. Preservative-free products reduce some signs and symptoms of ocular irritation by more than half. In addition, the formulator may adopt some novel formulation (noisome, liposomes) or packaging approaches to minimize the risk of ocular damages. Physicians must select the treatments that maximize ocular health while minimizing damage. In current scenario, it has now become essential to make aware of the recent toxic ocular reactions/manifestations to the patients and the formulators as well as physicians. There is a need to take a much more holistic and long-term view of ophthalmic treatment in order to treat presenting ocular diseases without causing further damages.

References

1. Freeman PD, Kahook MY (2009) Preservatives in topical ophthalmic medications: historical and clinical perspectives. Expert Rev Ophthalmol 4:59–64
2. Colin C (2006) Are multidose over-the-counter artificial tears adequately preserved? Basic investigations. Cornea 25:432–437
3. Schein OD, Hibberd PL, Starck T, Baker AS, Kenyon KR (1992) Microbial contamination of in-use ocular medications. Arch Ophthalmol 110:82–85
4. Baudouin C, Labbé A, Liang H, Pauly A, Brignole-Baudouin F (2010) Preservatives in eyedrops: the good, the bad and the ugly. Prog Retin Eye Res 29:312–334
5. Burstein NL (1980) Preservative cytotoxic threshold for BKC and chlorhexidine digluconate in cat and rabbit corneas. Invest Ophthalmol Vis Sci 32:2259–2265
6. Kaur IP, Lal S, Rana C, Kakkar S, Singh H (2009) Ocular preservatives: associated risks and newer options. Cutan Ocul Toxicol 28:93–103
7. Baudouin C, Pisella PJ, Fillacier K et al (1999) Ocular surface inflammatory changes induced by topical antiglaucoma drugs: human and animal studies. Ophthalmology 106:556–563
8. Cooper JW, Gunn C (2000) Ophthalmic products. In: Carter SL (ed) Dispensing for pharmaceutical students, 12th edn. CBS Publisher and Distributors, New Delhi, pp 634–662
9. Shaw AJ, Balls M, Clothier RH, Bateman ND (1991) Predicting ocular irritancy and recovery from injury using Madin-Darby canine kidney cells. Toxicol In Vitro 5:169–174
10. Takahashi N (1982) Cytotoxicity of mercurial preservatives in cell culture. Ophthalmic Res 14:63–69
11. Martin SM, Tsao F-P (1998) Method of preserving ophthalmic solutions and compositions therefor. US patent 5725887
12. Block SS (2001) Disinfection, sterilization, and preservation, 5th edn. Lippincott Williams & Wilkins, Philadelphia, p 311
13. Becquet F, Goldschild M, Moldovan MS, Ettaiche M, Gastaud P, Baudouin C (1998) Histopathological effects of topical ophthalmic preservatives on rat corneoconjunctival surface. Curr Eye Res 17:419–425
14. Xu Y, Zhao D, Gao C, Zhou L, Pang G, Sun S (2012) In vitro activity of phenylmercuric acetate against ocular pathogenic fungi. J Antimicrob Chemother 67:1941–1944
15. Tomlinson A, Trees GR (1991) Effect of preservatives in artificial tear solutions on tear film evaporation. Ophthalmic Physiol Opt 11:48–52
16. Fassihi AR, Naidoo NT (1989) Irritation associated with tear-replacement ophthalmic drops. A pharmaceutical and subjective investigation. S Afr Med J 75:233–235
17. Tripathi BJ, Tripathi RC (1989) Cytotoxic effects of BKC and chlorobutanol on human corneal epithelial cells in vitro. Lens Eye Toxic Res 6:395–403
18. Nordt SP (1996) Chlorobutanol toxicity. Ann Pharmacother 30:1179–1180
19. Doughty MJ (1994) Acute effects of chlorobutanol- or BKC-containing artificial tears on the surface features of rabbit corneal epithelial cells. Optom Vis Sci 71:562–572

20. Kilp H, Brewitt H (1984) Cytotoxicity of preservatives: a scanning electron microscopic and biochemical investigation. Chibret Int J Ophthalmol 1:4–20
21. Li HF, Petroll WM, Mo¨ller-Perderson T, Maurer JK, Cavanagh HD, Jester JV (1997) Epithelial and corneal thickness measurements by in vivo confocal microscopy through focusing (CMTF). Curr Eye Res 16:214–221
22. Berdy GJ, Abelson MB, Smith LM, George MA (1992) Preservative-free artificial tear preparations. Assessment of corneal epithelial toxic effects. Arch Ophthalmol 110:528–532
23. de Saint JM, Brignole F, Bringuier A-F, Bauchet A, Feldman G, Baudouin C (1999) Effects of BKC on growth and survival of Chang conjunctival cells. Invest Ophthalmol Vis Sci 40:619–630
24. Herreras JM, Pastor C, Calonge M, Asensio VM (1992) Ocular surface alteration after long-term treatment with an antiglaucomatous drug. Ophthalmology 99:1082–1088
25. Schwab IR, Linberg JV, Gioia VM, Benson WH, Chao GM (1992) Foreshortening of the inferior conjunctival fornix associated with chronic glaucoma medications. Ophthalmology 99:197–202
26. Longstaff S, Wormald RP, Mazover A, Hitchings RA (1990) Glaucoma triple procedures: efficacy of intraocular pressure control and visual outcome. Ophthalmic Surg 21:786–793
27. Gangrade NK, Gaddipati NB, Ganesan MG, Reddy IK (1996) Topical ophthalmic formulations: basic considerations. In: Reddy IK (ed) Ocular therapeutics and drug delivery: a multidisciplinary approach. Technomic, Basel, pp 377–403
28. Ashton P, Diepold R, Pla¨tzer A, Lee VHL (1990) Effect of chlorhexidine acetate on the corneal penetration of sorbitol from an arnolol formulation in the pigmented rabbit. J Ocul Pharmacol 6:37–42
29. Durand-Cavagna G, Delort P, Duprat P, Bailly Y, Plazonnet B, Gordon LR (1989) Corneal toxicity studies in rabbits and dogs with hydroxyethylcellulose and BKC. Fundam Appl Toxicol 13:500–508
30. Christiansen M, Cohen S, Rinehart J et al (2004) Clinical evaluation of an HPguar gellable lubricant eye drop for the relief of dryness of the eye. Curr Eye Res 28:55–62
31. Nelson JD, Farris RL (1988) Sodium hyaluronate and polyvinyl alcohol artificial tear preparations a comparison in patients with keratoconjunctivitis sicca. Arch Ophthalmol 106:484–487
32. Brignole-Baudouin F, Riancho L, Liang H, Baudouin C (2011) Comparative in vitro toxicology study of travoprost polyquad-preserved, travoprost BAK-preserved, and latanoprost BAK-preserved ophthalmic solutions on human conjunctival epithelial cells. Curr Eye Res 36:979–988
33. Labbé A, Pauly A, Liang H, Brignole-Baudouin F, Martin C, Warnet JM, Baudouin C (2006) Comparison of toxicological profiles of BKC and polyquaternium-1: an experimental study. J Ocul Pharmacol Ther 22:267–278
34. Ammar DA, Noecker RJ, Kahook MY (2010) Effects of BKC-preserved, polyquad-preserved, and sofZia-preserved topical glaucoma medications on human ocular epithelial cells. Adv Ther 27:837–845
35. Paimela T, Ryhänen T, Kauppinen A, Marttila L, Salminen A, Kaarniranta K (2012) The preservative polyquaternium-1 increases cytoxicity and NF-kappaB linked inflammation in human corneal epithelial cells. Mol Vis 18:1189–1196
36. Choy CK, Cho P, Boost MV (2012) Cytotoxicity and effects on metabolism of contact lens care solutions on human corneal epithelium cells. Clin Exp Optom 95:198–206
37. Abelson MB, Udell IJ (2000) Conjunctiva, cornea, and sclera. In: Albert DM, Jakobiec FA (eds) Principles and practice of ophthalmology, 2nd edn. WB Saunders, Philadelphia, pp 1689–1692
38. Grant R, Ajello M, Vlass E (1996) Salt water or high tech? A look at two new rinsing solutions for contact lenses. Optician 212:38–41
39. Paugh JR, Brennan NA, Efron N (1988) Ocular response to hydrogen peroxide. Am J Optom Physiol Opt 65:91–98

40. Vaughan JS, Porter DA (1993) A new *in vitro* method for assessing the potential toxicity of soft contact lens care solutions. CLAO J 19:54–57
41. Rozen S, Abelson M, Giovanoni A, Welch D (1998) Assessment of the comfort and tolerance of 0.5% carboxymethylcellulose preserved with Purite (Refresh Tears) in dry eye sufferers. Invest Ophthalmol Vis Sci 39:451
42. Noecker R (2001) Effects of common ophthalmic preservatives on ocular health. Adv Ther 18:205–215
43. Karageszian VH, Karageozian HL, Climent A, Munoz S, Cuevas G, Gallegos VG (2003) Determine the in-vitro antimicrobial preservative effectiveness of a new broad spectrum antimicrobial preservative "Oxyd". Invest Ophthalmol Vis Sci 44:1469
44. Kahook MY (2007) Travoprost Z. Ophthalmic solution with sofZia: clinical safety and efficacy. Expert Rev Ophthalmol 2:363–368
45. Lewis RA, Katz G, Weiss MJ (2007) Travoprost 0.004% with and without BKC: a comparison of safety and efficacy. J Glaucoma 16:98–103
46. Yee RW, Norcom EG, Zhao XC (2006) Comparison of the relative toxicity of travoprost 0.004% without BAK and latanoprost 0.005% in an immortalized human cornea epithelial cell culture system. Adv Ther 23:511–518
47. Debbasch C, Brignole F, Pisella P-J, Warnet J-M, Rat P, Baudouin C (2001) Quaternary ammoniums and other preservatives' contribution in oxidative stress and apoptosis on Chang conjunctival cells. Invest Ophthalmol Vis Sci 42:642–652
48. Baines MG, Cai F, Backman HA (1991) Ocular hypersensitivity to thiomersal in rabbits. Invest Ophthalmol Vis Sci 32:2259–2265
49. Yanochko GM, Khoh-Reiter S, Evans MG, Jessen BA (2010) Comparison of preservative-induced toxicity on monolayer and stratified Chang conjunctival cells. Toxicol In Vitro 24:1324–1331
50. Brahma AK, Pande M (1994) Comparative conjunctival toxicity of subconjunctival preservative-free gentamicin and subconjunctival preservative-free cefuroxime. Eur J Implant Refract Surg 6:57–58
51. Smith L, George M, Berdy G, Abelson M (1991) Comparative effects of preservative free tear substitutes on the rabbit cornea: a scanning electron microscopic evaluation (ARVO abstract). Invest Ophthalmol Vis Sci 32 (Suppl):733
52. Tsubota K (1998) Tear dynamics and dry eye. Prog Retin Eye Res 17:565–596
53. Holy FJ (1978) Surface chemical evaluation of artificial tears and their ingredients II. Interaction with superficial lipid layer. Cont Intraoc LENS Med J 4:52–65
54. Arffa RC (1997) Toxic and allergic reactions to topical ophthalmic medications. In: Arffa RC (ed) Grayson's diseases of the cornea. Mosby, St. Louis, pp 669–683
55. Cai F, Backman HA, Baines MG (1988) Thiomersal: an ophthalmic preservative which acts as a hapten to elicit specific antibodies and cell mediated immunity. Curr Eye Res 7:341–351
56. Fuchs T, Meinert A, Aberer W et al (1993) BKC-a relevant contact allergen or irritant? Results of a multicentre study of the German contact allergy group. Houtzart 44:699–702
57. Pisella PJ, Pouliquen P, Baudouin C (2002) Prevalence of ocular symptoms and signs with preserved and preservative free glaucoma medication. Br J Ophthalmol 86:418–423
58. Green K (1992) The role of surfactants as bactericides in topical drug delivery. STP Pharm Sci 2:34–38
59. Grant RL, Acosta D (1996) Prolonged adverse effects of BKC and sodium dodecyl sulfate in a primary culture system of rabbit corneal epithelial cells. Fundam Appl Toxicol 33:71–82
60. Lemp MA, Zimmerman LE (1988) Toxic endothelial degeneration in ocular surface disease treated with topical medications containing BKC. Am J Ophthalmol 105:670–673
61. Weinreb RN, Wood I, Tomazzoli L, Alvarado J (1986) Subconjunctival injections. Preservatives-related changes in the corneal endothelium, Invest. Ophthalmol Vis Sci 27:525–531
62. Tripathi BJ, Tripathi RC, Kolli SP (1992) Cytotoxicity of ophthalmic preservatives on human corneal epithelium. Lens Eye Toxic Res 9:361–375

63. Kahook MY, Noecker RJ (2008) Comparison of corneal and conjunctival changes after dosing of travoprost preserved with sofZia, Latanoprost with 0.02% BKC, and preservative-free artificial tears. Cornea 27:339–343
64. Mondino BJ, Brawman-Mintzer O, Boothe WA (1987) Immunological complications of soft contact lenses. J Am Optom Assoc 58:832–835
65. Chapman JM, Cheeks L, Green K (1990) Interactions of BKC with soft and hard contact lenses. Arch Ophthalmol 108:244–246
66. Way WA, Matsumoto S, Apel LJ, Wiese A, Tarlo K, Vehige J (2001) Purite as a non-disruptive preservative for lubricating eye drop solutions in comparison to alternative preservatives. Invest Ophthalmol Vis Sci 42:S39
67. Lopez B, Ubel J (1991) Quantitative evaluation of the corneal epithelial barrier: effect of artificial tears and preservatives. Curr Eye Res 10:7645–7656
68. Pham XT, Huff JW (1999) Cytotoxicity evaluation of multipurpose contact lens solutions using an in vitro test battery. CLAO J 25:28–35
69. Epstein EP, Ahdoot M, Marcus E, Asbell PA (2009) Comparative toxicity of preservatives on immortalized corneal and conjunctival epithelial cells. J Ocul Pharmacol Ther 25:113–119
70. Gaddipati NB, Gangrade NK, Ganesan MG, Reddy IK (1996) Packaging aspects of ophthalmic products. In: Reddy IK (ed) Ocular therapeutics and drug delivery: a multi-disciplinary approach. Technomic, Basel, pp 529–541
71. Van Santvliet L, Sam T, Ludwig A (1996) Packaging of ophthalmic solutions – influence on stability, sterility, eye drops instillation and patient compliance. Eur J Pharm Biopharm 42:375–384
72. Oldham GB, Andrews V (1996) Control of microbial contamination in unpreserved eyedrops. Br J Ophthalmol 80:588–591
73. Palmer RM, Kaufman HE (1995) Tear film, pharmacology of eye drops, and toxicity. Curr Opin Ophthalmol 6:11–16
74. Kaufman HE, Steinemann TI, Lehman E, Thompson HW, Varnell ED, Jacob-labarre JT, Gebhardt BM (1994) Collagen-based drug delivery and artificial tears. J Ocul Pharmacol Ther 10:17–27
75. Luo L, Li D, Corrales R, Pflugfelder S (2005) Hyperosmolar saline is a proinflammatory stress on the mouse ocular surface. Eye Contact Lens 31:186–193

Chapter 14
Aging Disorders of the Eye: Challenges and Approaches for Their Treatment

Shruti Hazare, Rongbing Yang, Smita Chavan, Mala D. Menon, and Mahavir B. Chougule

Abstract The proportion of the global population aged 60 years and over is steadily increasing and projected to increase to almost 30 % in 2050. Among the various health problems, eye and vision problems are serious issues in the elderly. These may be manifested as basic functional disabilities or a decline in the receptive, storage, and analytical capacities of the central visual system. The major eye disorders of aging are cataract, age-related macular degeneration (AMD), glaucoma, and diabetic retinopathy. Although there are treatment methods (e.g., medications and surgical interventions) for these conditions, they are still very challenging areas due to the delicate and critical nature of the eye tissues. Compared with drug delivery to other parts of the body, drug delivery to the eye has met with significant challenges posed by various ocular barriers, which are inherent and unique to the ocular anatomy. In addition, in the case of the aging population, there are added difficulties due to multiple diseases and health problems, the several medications being taken together and the physical and psychological difficulties, including disabilities, dependence, fears, and apprehensions: opening packages, swallowing oral medication and/or reading leaflet information, fear of surgery, and device insertion and removal.

Contributions from *Mr. Rajesh Jain & Mr. Ravi Sonkamble*, Research Scholars, Department of Pharmaceutics, Bombay College of Pharmacy.

S. Hazare • S. Chavan • M.D. Menon (✉)
Department of Pharmaceutics, Bombay College of Pharmacy,
Kalina, Santacruz (E), 400098 Mumbai, India
e-mail: maladmbcp@yahoo.co.in

R. Yang • M.B. Chougule (✉)
The Daniel K. Inouye College of Pharmacy, University of Hawaii at Hilo,
34 Rainbow Drive, STE 300, Hilo, HI 96720, USA
e-mail: mahavir@hawaii.edu; mahavirchougule@gmail.com

© Springer International Publishing Switzerland 2016

Y. Pathak et al. (eds.), *Nano-Biomaterials For Ophthalmic Drug Delivery*,
DOI 10.1007/978-3-319-29346-2_14

This chapter discusses:

(a) Brief of anatomy of the eye and aging changes in the eye
(b) The major age-related eye problems, including cataract, AMD, diabetic retinopathy, and glaucoma
(c) The current treatment methods, novel drug delivery systems, and approaches being investigated for each of abovementioned eye problems

Maintaining good vision is an important part of "active aging," a concept promoted by the WHO. Active aging means continued health, security, and participation in society as people age, in order to ensure a good quality of life in later years. Nations and communities have to gear up to this challenge to ensure that good quality eye care and therapies are available to this group.

Keywords Aging disorders of the eye • Cataract • Diabetic retinopathy • Age-related macular degeneration • Glaucoma

14.1 Introduction

There is an overall change in the pattern of the world's population, mainly because of improved primary healthcare facilities and services; older people are making an ever greater proportion of the population. As this trend continues, it is projected that by 2050, 30 % of the population will be the elderly (more than 80 years). The process of aging involves a continuum of chemical, biological, functional, psychological, and social parameters which can vary depending on the genetic factors, age, vulnerability, and organ function and reserves. Among other health problems like hypertension, diabetes, arthritis, parkinsonism, and osteoporosis, eye and vision problems also emerge in the elderly; these may be due to decline in basic functions or in the receptive, storage, and analytical capacities. Various surveys and literature reports have highlighted the extent of the aging problems in vision and the challenges in this area [1–6].

14.1.1 Human Eye

The human eye is the organ that reacts to light and allows conscious light perception and vision including color differentiation and the perception of depth and can distinguish about 10 million colors. This important sense organ is highly complex and requires a great deal of care and specialized attention throughout life. Compared with drug delivery to other parts of the body, drug delivery to the eye has met with significant challenges posed by various ocular barriers. Many of these barriers are inherent and unique to ocular anatomy and physiology making it a challenging task for drug delivery scientists. These barriers are specific depending upon the route of administration, viz., topical, systemic, and injectable. These barriers are a part of the body's normal protective mechanisms to prevent external toxicants entering the eye.

14.1.2 Structure and Functions of the Eye

Each eyeball is contained in a bony cavity known as orbit, which is a pear-shaped structure formed by several bones; along with the eyeball, it also houses associated structures, muscles, nerves, and blood vessels, as well as the lachrymal glands, and is composed of several layers.

14.1.2.1 Sclera

Sclera is the outer covering, a relatively tough, opaque, fibrous, and protective white layer (white of the eye), and is composed of collagen and elastic fiber. It is continuous with the dura mater of the brain and the cornea and maintains the shape of the globe, offering resistance to internal and external forces; besides it provides an attachment for the extraocular muscle insertions. Near the front of the eye, in the area protected by the eyelids, a thin, transparent membrane (conjunctiva) running up to the edge of the cornea covers the sclera, as well as the moist back surface of the eyelids and eyeballs. Human eyes are somewhat distinctive in the animal kingdom in that the sclera is very plainly visible whenever the eye is open, due to its white color and the relatively small size of human iris, which comprises a significantly smaller portion of the exposed eye surface compared to other animals. This adaptation evolved because of our social nature, and thus the eye is considered to be a useful communication tool [7, 8].

14.1.2.2 Choroid

Choroid is the middle layer or tunic, which is the vascular layer of the eye, containing connective tissue and lying between the retina and the sclera. The human choroid is thickest at the far extreme rear of the eye (0.2 mm), while in the outlying areas, it narrows to 0.1 mm. This layer is comprised of blood vessels, melanocytes, fibroblasts, resident immunocompetent cells, and supporting collagenous and elastic connective tissue. As one of the most highly vascularized tissues of the body, its main function has been traditionally viewed as supplying oxygen and nutrients to the outer retina, and, in species with avascular retinas, to the inner retina as well. Other likely functions include light absorption (in species with pigmented choroids), thermoregulation via heat dissipation, and modulation of intraocular pressure (IOP) via vasomotor control of blood flow. The choroid also plays an important role in the drainage of the aqueous humor from the anterior chamber, via the uveoscleral pathway. This pathway is responsible for approximately 35 % of the drainage in humans; a higher percentage, between 40 % and 60 %, in nonhuman primates; and a much lower percentage in the cat (about 3 %) and rabbit (3–8 %) [9]. Along with the ciliary body and iris, the choroid forms the uveal tract. In addition some uveal regions have special functions of great importance, including secretion of the aqueous humor by the ciliary processes, control of accommodation (focus) by the ciliary body, and optimization of retinal illumination by the iris's control over the pupil. Many of these functions are under the control of the autonomic nervous system [10, 11].

14.1.2.3 Retina

Retina is the light-sensitive layer of tissue, lining the inner surface of the eye, and can be compared to the film of a camera. The optics of the eye creates an image of the visual world on the retina (through the cornea and lens). Light striking the retina initiates a cascade of chemical and electrical events that ultimately trigger nerve impulses, which are sent to various visual centers of the brain through the fibers of the optic nerve. The retina and the optic nerve originate as outgrowths of the developing brain; hence the retina is considered to be part of the central nervous system (CNS) and is actually a brain tissue [11]. It is the only part of the CNS that can be visualized noninvasively.

The retina contains several millions of photoreceptors, which sense light and the blood vessels that nourish them. There are basically two types of photoreceptors – rods and cones. Cones are responsible for sharp, detailed central vision and color vision and are clustered mainly in the macula, whereas the rods are responsible for night and peripheral (side) vision. Rods are more numerous than cones and much more sensitive to light and are grouped mainly in the peripheral areas of the retina. A high density of cones is present within a small area known as macula, which is the most sensitive part of the retina. This helps to create a detailed visual image. The nerve fibers linked to each photoreceptor are grouped and bundled together to form the optic nerve. The photoreceptors convert the image into electrical signals, which are carried to the brain by the optic nerve [10–12].

14.1.2.4 Process of Vision

Light enters the eye through the cornea, the clear, curved layer in front of the iris and pupil. The cornea serves as a protective covering for the front of the eye and also helps focus light on the retina at the back of the eye. After passing through the cornea, light travels through the pupil; the iris controls the amount of light that enters the eye, based on the intensity of light in surroundings. The size of the pupil is controlled by the action of the pupillary sphincter muscle and dilator muscle.

14.1.2.5 Lens of the Eye

The eye lens is situated behind the iris and is part of the anterior segment of the eye. The lens has an ellipsoid, biconvex shape. The anterior surface is less curved than the posterior. In the adult, the lens is typically circa 10 mm in diameter and has an axial length of about 4 mm, though it is important to note that the size and shape can change due to its power of accommodation and because the lens continues to grow throughout a person's lifetime. The lens is suspended in place by the suspensory ligament of the lens, a ring of fibrous tissue that attaches to the lens at its equator and connects it to the ciliary body. Posterior to the lens is the vitreous body, which, along with the aqueous humor on the anterior surface, bathes the lens. The power of accommodation of the lens is due to the action of small ciliary muscles. To focus on nearby objects, the lens becomes thicker, and to focus on faraway objects, it becomes thinner [13, 14].

14.1.2.6 Anterior and Posterior Segments of the Eye

The eyeball is divided into two sections – the anterior and posterior segments. The anterior segment extends from the inside of the cornea to the front surface of the lens and is filled with a fluid, the aqueous humor, which nourishes the internal structures. The posterior segment extends from the back surface of the lens to the retina and contains a jellylike fluid, the vitreous humor. The pressure generated by these fluids fills out the eyeball and helps maintain its shape. The anterior segment can be further divided into two chambers, the front (anterior) chamber extending from the cornea to the iris and the back (posterior) chamber, which extends from the iris to the lens. The aqueous humor is produced in the posterior chamber and flows slowly through the pupil into the anterior chamber and then drains out of the eyeball through outflow channels located where the iris meets the cornea [15, 16].

14.1.3 Aging Process in the Eye

An excellent review on aging eye is given by Salvi et al. [17]. With age, the functional abilities of the eye are affected; also the receptive, storage, and analytical capacities of the central visual system decline, resulting in the various vision problems seen in the elderly population. Two common theories to explain aging processes are the "biological clock theory" and the "wear-and-tear theory." The biological clock theory or programmed theory attributes aging to be an inherent characteristic of each individual governed by the unique genetic code in the DNA. However, lifestyle habits and other environmental factors can also contribute to the aging process. The wear-and-tear theory was first introduced in 1882, by Dr August Weismann, a German biologist. This theory explains aging to be a process wherein the body and its cells are damaged by overuse and abuse, and slowly over a period of time, all the organs, including the eye, are worn down by dietary and environmental toxins (e.g., excessive consumption of fat, sugar, caffeine, alcohol, and nicotine; the ultraviolet rays of the sun, and also the various physical and emotional stresses). At the cellular level, the wear-and-tear process is explained as being caused by production of "free radicals," commonly referred to as reactive oxygen species (ROS). The ROS possess free electrons and are highly reactive species, which are capable of attacking the structure of cell membranes and creating metabolic waste products, like lipofuscins. In this process, key cellular processes like DNA and RNA synthesis and protein synthesis are affected; energy levels are lowered, affecting the vital body building processes and enzyme functions. In younger healthy people, the effects of ROS are less pronounced, due to the extensive repair and replacement mechanisms. With age, however the accumulated effects of free radical damage begin to show the typical aging effects. The aging effects in the eye in each of the regions of the eye are summarized in the sections below.

14.1.3.1 Eyelids and Lacrimal System

Eyelids show the typical aged look evidenced by shrinkage, folds, and wrinkles. Loss of adnexal structural support of tarsus, canthal tendons, and orbicularis muscle with thinned skin leads to orbital fat prolapse, eyelid malposition, blepharoptosis, and tearing. In the lower eyelid, horizontal lid laxity is common. Reduction in the orbital fat with aging causes the eyes to "sink in" accentuating the lid laxity. Other aging effects include ectropion or eversion of the eyelid margin from the globe and subsequent symptoms of a watery eye, inversion (entropion) instead causing eyelashes to rub against the cornea and subsequent discomfort, involutional ptosis, age-related descent of the brow (brow ptosis), and dermatochalasis or pseudoptosis, also known as "baggy eyes." In all these conditions, if they interfere with vision, oculoplastic surgery is the remedy. These operations are often carried out before a cataract operation to avoid infection. In entropion, temporary relief may be achieved by simply taping the lid to pull it outwards.

Lachrymal Glands There are two extreme conditions seen:

(a) *Watery eye* in the elderly is often caused by eyelid malposition or sometimes from true lacrimal obstruction leading to distressful watering or recurrent infections; treatment is by dacryocystorhinostomy.
(b) *Dry eye syndrome* is due to reduction in the amount of tears produced by the lacrimal gland. This condition is treated with artificial tears or punctual plugs to retain tears in the conjunctival sac [18].

14.1.3.2 Changes in the Cornea

With age, changes in corneal toricity (curvature) can cause alteration in refraction, resulting in astigmatism; and regular refraction checkups are advised to detect these changes. Besides this, there may be a decrease in corneal luster and corneal sensitivity and increase in corneal fragility. Other observations include age-related dystrophic changes in the corneal epithelium, stroma and endothelium which involve deposition of iron, yellow white deposits of cholesterol esters and other lipids, deposition of uveal pigment, etc., and are referred to as Hudson-Stahli line, arcus senilis, Hassall-Henle bodies, Krukenberg spindle, and cornea guttata. These changes however do not interfere with vision and hence are not serious concerns [19].

Another observation is pleomorphism, which occurs as the endothelial cell density reduces with increasing age. As the cells cannot regenerate, the remaining endothelial cells enlarge to cover the gaps. Beyond a certain point, this affects corneal deturgescence, which can further cause corneal thickening, opacity, and decrease in quality of vision. If visual acuity is reduced, penetrating keratoplasty is recommended.

14.1.3.3 Changes in Trabecular Meshwork and Uvea

The trabecular meshwork is a spongy tissue in the eye located around the base of the cornea, near the ciliary body, and is responsible for draining the aqueous humor from the eye via the anterior chamber into a set of tubes called Schlemm's canal flowing into the blood system. On aging, there is often increased pigmentation of the trabecular meshwork and increase in the resistance to the outflow of aqueous humor; these changes can precipitate glaucoma. Besides this, the age-associated changes in the uvea are manifested as reduction in size of the pupil, pigment loss, and lower reactivity of the iris. Also, the shape and tone of ciliary body changes leading to decrease elasticity of the lens capsule and compactness of the lens fibers; this can affect the accommodative power of the lens, resulting in presbyopia.

14.1.3.4 Crystalline Lens Changes

On aging, the lens is known to selectively absorb more blue light (410 nm), which is a part of cataractogenic process, leading to a condition called "blue blindness." This has been attributed to the accumulation of yellow pigments in the lens. Other common and familiar sign of lens aging is the hardening (nuclear sclerosis) caused by various biochemical and photochemical changes, leading to presbyopia and further to cataract formation. Phacoemulsification and intraocular lens implantation are the methods to resolve these problems to ensure the performance of daily routine activities of the elderly [20]. Further details of cataract are given under Sect. 2.1.

14.1.3.5 Vitreous Aging

Aging processes in the vitreous humor are manifested as harmless floaters noticed by the elderly. The changes take place sequentially: condensation or liquefaction of the vitreous gel, followed by enhancement and increased mobility of the fibrillary structures of vitreous. As the liquefaction increases, there is formation of optically empty spaces called lacunae; further accentuation of the lacunae leads to larger cavities and finally to shrinkage of vitreous body from the retina. Complications like posterior vitreous detachment (PVD) may be evident at around 50 % liquefaction; a retinal tear during acute PVD can further lead to retinal detachment, evident as a curtain-like shadow in the field of vision. In such cases, laser treatment and surgical intervention is advocated to prevent retinal detachment [21].

14.1.3.6 Retinal Aging

Retinal aging is mainly manifested as decreased visual function in various forms – acuity, field, contrast sensitivity, and increased dark adaptation threshold. A combination of changes in neuronal elements of visual system, changes in the ocular media, and pupillary miosis are responsible. Neurosensory retinal changes include decrease in ganglion cells and photoreceptors, thickening of basement membrane, and increase in corpora amylacea bodies and lipofuscin content. Several changes observed in RPE are increased pleomorphism, decrease in cells in posterior pole, decreased melanin content, decreased volume of cytoplasm, and increased lipofuscin content. Retinal macular microcirculation reduces with age and slowly results into age-related macular degeneration (AMD). Other observations evident in the optic nerve include regional swollen axons, decrease in the number of axons, increase in the thickness of the connective tissue, and increased elastic fibers. Clinical manifestations are loss of fundus reflexes, gradual fading of fundus color, greater visibility of larger choroidal vessels (senile tigroid fundus), peripapillary atrophy, and peripheral retinal degenerations. Details of some of these changes and AMD are discussed under Sect. 2.3 [22, 23].

14.2 Age-Related Disorders of the Eye

14.2.1 Cataract

A cataract is clouding of the lens of the eye, causing an obstruction in the passage of light and thereby interference with clear vision. Most cataracts are related to aging, although occasionally children may be born with the condition, or cataract may develop after an injury, inflammation, or disease. Cataract is the leading cause of blindness and responsible for 51 % of world blindness [24]. The extent of the vision loss depends on the size and location of the cataract. Cataracts may be located in the center of the lens (nuclear), in the superficial cortex (cortical), or in the posterior subcapsular area. Cataracts are also classified according to their color, which is consistent with the location and density of the cataract. Pale yellow cataracts are typically slight opacities of the cortex, subcapsular region, or both; yellow or light brown cataracts are consistent with moderate to intense opacities of the cortex, nucleus, or both; and brown cataracts are associated with dense nuclear cataracts [25].

Symptoms of cataract include near-vision image blur, abnormal color perception, monocular diplopia, glare, and impaired visual activity and may vary depending on the location of the cataract. For example, if the opacity is located in the center of the lens (nuclear cataract), myopia is often a symptom, whereas posterior subcapsular cataracts tend to be most noticeable in bright light [26].

14.2.1.1 Risk Factors for Cataract

14.2.1.1.1 Diabetes

People with diabetes have an increased risk of cortical and posterior subcapsular cataract and are also more likely to have early cataract surgery. The enzyme aldose reductase (AR) catalyzes the reduction of glucose to sorbitol through the polyol pathway, which is linked to the development of diabetic cataract. Intracellular accumulation of sorbitol leads to osmotic changes and further to degeneration of hydropic lens fibers and formation of sugar cataracts.

14.2.1.1.2 Prolonged Exposure to UVB Radiation

Increase in ultraviolet radiation resulting from depletion of ozone layer is expected to increase the incidence of cataract. Many experimental studies have shown that an increased number of UVB rays lead to a profound decrease in corneal antioxidants, which results in oxidative injury of the cornea and damage to the inner parts of the eye [27].

14.2.1.1.3 Tobacco and Alcohol Addiction

A higher prevalence of nuclear and posterior subcapsular cataracts is reported in case of people who smoked and drank heavily [28].

14.2.1.1.4 High Body Mass Index (BMI)

Study findings suggest that elevated BMI may increase the risk of age-related cataract (ARC), especially posterior subcapsular cataracts, but further trials are needed to investigate the effect of weight reduction in obese populations on the risk of ARC.

14.2.1.2 Mechanisms in Pathophysiology of Cataract Include

14.2.1.2.1 Electrolyte Disturbances

Electrolyte disturbances can result in osmotic imbalances to cause derangements in the membrane functions. These imbalances may be due to increased membrane permeability or to a depression of the [Na.sup.+]/[K.sup.+] pump or because of interference with the enzyme [Na.sup.+]/[K.sup.+] ATPase [29].

14.2.1.2.2 Oxidative Damage

Oxidative damage plays a major role in the etiology and pathogenesis of cataract, wherein the lens proteins are subjected to extensive oxidative modifications. Oxidation of lens proteins occurs due to generation of reactive oxygen species (ROS) like superoxide, hydrogen peroxide, and hydroxyl radicals. In cataractous lenses, the proteins are found in an insoluble, oxidized form. In the aging process, several posttranslational changes like racemization, glycation, -COOH terminal degradation, deamidation, and noncovalent aggregation of the proteins in the inner region of lens take place, which may markedly modify the overall conformation of the proteins. Along with this, certain key metabolically active components which protect the lens from stress appear to decrease in activity. This results in the formation of high molecular weight protein aggregates, covalently linked by disulfides. Some of these aggregates are greater than 50×10^6 Da, large enough to scatter light and contribute to the loss of transparency. Studies have shown considerable *oxidation of membrane lipids* in cataract compared to normal tissue [30]. Some studies have shown increased levels of hydrogen peroxide and significantly lower levels of glutathione (GSH) in cataractous lenses, compared to normal controls [30, 31].

14.2.1.3 Prevention and Treatment of Cataract

Cataracts are a part of aging process and not completely preventable. However, their occurrence can be delayed, by controlling the risk factors – quitting smoking, avoiding overexposure to sunlight, and avoiding excessive amounts of alcohol, which can help to a great extent. Importance of proper diet has been stressed in the elderly, especially foods rich in antioxidants. Several studies have compared the nutritional status and the occurrence of senile cataract. These investigators have reiterated the usefulness of fresh fruits and vegetables containing dietary antioxidants like ascorbic acid, carotenoids, lutein, and zeaxanthin; these are reported to lower the incidence of cataracts, severe enough to require extraction [32]. The relationship between antioxidant nutrient status and senile cataract was examined in 77 subjects with cataracts and 35 control subjects with clear lenses [33, 34].

14.2.1.3.1 Surgery

The most common option for patients with advanced cataract is surgery. The procedure involves removal of the natural opaque lens followed by insertion/implantation of an artificial intraocular lens fabricated from silicone or acrylic polymeric materials. Enormous research and surveys have been carried out in surgical techniques and the development of the lenses [35, 36]. There are two basic surgical procedures – phacoemulsification and extracapsular cataract extraction (ECCE). Lenses of various types – foldable, accommodating, and providing protections from UV light or blue light – are some examples of the features designed [35, 36]. Although safe, cataract surgeries are not without complications – posterior vitreous detachment, posterior capsular opacification or tear, retinal detachment, toxic anterior segment syndrome, endophthalmitis, glaucoma, edema of the cornea, and dislocation of implanted lens. The cost of surgery and the postoperative care are prohibitive, especially for the elderly patients. Hence, developing alternative means to control and treat cataract is an important area which is slowly gaining attention.

14.2.1.3.2 Therapeutic Drugs

Natural therapies that have been investigated and may be successful in cataract treatment are lutein and zeaxanthin carotenoids, bilberry extract, and combinations of antioxidants with vitamins, zinc, bioflavonoids, and caffeine [33]. Several drugs have been developed and investigated, aiming to treat and reverse the lens opacification in cataract. Drugs likes orbenil, aspirin, and sodium monomethyl trisilanol orthohydroxybenzoate have been studied and have shown promise in slowing down and preventing the progression of experimental sugar cataracts or partial reversal of early morphological signs [37]. A lot of research efforts have been directed on two chemically very close non-steroid anti-inflammatory drugs (NSAID), bendazac and benzydamine. These two drugs are reported to reduce the biological liquid oxidant activity and have shown some promise in clinical studies; on oral administration (500 mg, three times daily), they were able to stabilize the progression of lens opacification in cataract patients. Preliminary studies evaluating bendazac lysine 0.5 % eyedrops have reported comparable results to those obtained with oral treatment. Although significant improvements in individual and mean visual acuities in treated patients have been reported by several studies, the parameters used have not been universally accepted as a reliable index of lens status; side effects observed include laxative effect and other gastrointestinal disturbances associated with oral therapy, and a transient burning sensation is the most commonly reported symptom occurring with eyedrop application [38]. The anti-cataract activity has been attributed to an active metabolite of the NSAID which has antioxidant activity. However, other possible mechanisms, such as protein and membrane stabilization and protective effect on photooxidative processes linked to free radicals, may also be responsible for effecting the reversal of cataractous opacity [39]. Further, these two drugs have also shown promise in the treatment of retinitis pigmentosa. as a drug capable of attenuating the biological effects of sun radiations on the retina [39, 40].

Based on above drugs, the potential of benzyl alcohol which is used as preservative in eyedrops has also been investigated and shown promise [41]. Pirenoxine (PRX) a pyridophenoxazine compound is another agent shown to protect the lens protein and crystallins against UVC-, selenite-, and calcium-induced lens protein turbidity; however, it was found detrimental under UVB exposure. A Japanese eye product, Catalin based on PRX, and also a Catalin-formulated vehicle are available and recommended for reducing various types of cataract development [42].

Clinical studies on Catalin eyedrops have been carried out and reports are variable. Kociecki et al. have observed that the eyedrops were well tolerated, and in patients with cortical cataract, the drug showed its effectiveness in inhibition of lens opacification and its progression, especially in group of patients with age up to 59 years [43]. Long-term studies spanning over 2 years, in 51 patients by Angra K et al., concluded that Catalin when instilled locally in the eye does not have any beneficial effect in checking the progress of senile and congenital cataract irrespective of the stage of its development. The natural course of progression of cataract was no way changed [38, 44, 45].

Another group of drugs investigated include the statins, and there are varied reports. Several groups have indicated that long-term use of statins conferred a protective effect against cataract surgery, whereas some of the reports conclude that statin therapy is associated with a modestly increased risk of cataract surgery [46].

N-Acetylcarnosine (NAC) is another important drug with potential anti-cataract efficacy. It is a derivative of carnosine, which is found in muscle tissue. These compounds have been reported to possess varying degrees of activity as free radical scavengers. NAC is particularly active against lipid peroxidation in the different parts of the lens in the eye [47]. Eyedrops containing *N*-acetylcarnosine have been studied in Russia for the treatment of cataracts and have been found to be successful. However, these drugs are not FDA approved and can be bought off the shelf as "lubricants" [48]. In China OTC products like Phacolysin, Catalin (a modified phenoxazine carboxylic acid) eyedrops, Tathion (glutathione eyedrops), and Zhangyanming tablets are used for the early and mid-stage cataracts. A Swiss product called Quinax (dihydroazapentacene or azapentacene polysulfonate) is used as a treatment for cataracts [49].

14.2.1.4 Drug Delivery Approaches for Treatment of Cataract

Topical drug application is the most widely accepted and convenient mode of ocular drug delivery to treat many eye disorders. But cornea is major barrier for drug delivery through this route for lens and retina, as large portion of drug is lost through precorneal losses. Considering the pathology of cataract, development of antioxidant formulation with optimum therapeutic efficiency and target specificity is necessary. To overcome the corneal barrier and to increase bioavailability of molecule, several novel approaches and delivery systems can be explored.

14.2.1.4.1 Liposomal Systems

Liposomal formulations are biodegradable and biocompatible in nature and have been studied and explored for ophthalmic drug delivery applications. The major focus has been to improve the corneal adhesion and permeation by incorporating various bioadhesive formulations based on hydrogels like Carbopol, polyacrylic acids, and chitosan. Another important approach involves the use of penetration enhancers, including different surfactants (Span 20, 40, and 85; Tween 20, 40, and 81; deoxycholic acid; taurocholic acid) and calcium chelators. The development of vitamin E-containing liposomes prepared with dipalmitoylphosphatidylcholine and dioleoylphosphatidylcholine indicated that instilled vitamin E-containing liposomes retard cataract progression in 12-month-old rats fed a 25 % galactose diet, mainly by the antioxidative and membrane-stabilizing actions. The development of disulfiram liposomes prevented cataract formation in rat pups with cataract induced by selenite [50]. Reports on anti-cataract effect of cationic freeze-dried liposomes containing cytochrome C were shown to be stable superior ophthalmic carriers and were able to markedly retard the onset of cataract development [51].

14.2.1.4.2 Nanoparticles (NP)

Biodegradable and biocompatible nanoparticles have been reported to enhance the pharmacokinetic and pharmacodynamic properties of encapsulated molecules. Studies on curcumin encapsulated PLGA NP for oral delivery showed better efficacy in delaying diabetic cataract in rat model [52]. NP of cerium oxide (CeO2) with antioxidant properties can also be applied in the prevention of cataract. The nano-encapsulation of quercitrin (potential anti-cataract agent) on PLA was found to enhance its antioxidant activity up to 40 % [53].

Encapsulation of biomolecules into nanoparticles may enhance protection against metabolic enzymes to decrease the dose and enhance uptake by corneal cells. This will aid in efficient treatment of cataract. Although surgery has been used for cataract treatment, it is a difficult procedure in the aged elderly population, who may also be having other disorders and complications. The aged often has difficulties in handling as well as fears and apprehensions in using devices; hence novel systems which are simple to administer should be developed as alternatives to surgery.

In a study by our group, mucoadhesive in situ gelling systems based on mucoadhesive polymers – poloxamer 407 and HPMC, incorporating catalase as active – were developed. Catalase is the endogenous enzyme responsible for eliminating the ROS. The feasibility for the prevention of cataract by these developed systems was assessed in vivo in oxidative stress-induced cataract by selenite induction, in Wistar rat pups [54]. Protective role of catalase was evident by the observation of transparency of excised lenses and also by the assessment of levels of malonaldehyde which is a marker lipid peroxidation and enzyme levels (Table 14.1 and Fig. 14.1) [55].

Table 14.1 Levels of catalase (CAT) and malonaldehyde (MDA) in the experimental selenite-induced Wistar rat pup lenses

Groups	CAT (units/mg protein)	nmol MDA/mg protein	Opacification score
(a) Normal control	1.685 ± 0.035	0.413 ± 0.015	0
(b) Cataract control	0.078 ± 0.054[a]	0.754 ± 0.172[b]	++ to ++++
(c) Cataract treated (plain enzyme solution)	1.18 ± 0.19[c]	0.474 ± 0.055[d]	0 to +
(d) Cataract treated (in situ gel)	0.93 ± 0.61[e]	0.479 ± 0.088[e]	0 to +

All values expressed as mean ± SD, (n = 6)
[a]$p < 0.01$ when compared with normal control
[b]$p < 0.05$ when compared with normal control
[c]$p < 0.001$ when compared with cataract control
[d]$p < 0.01$ when compared with cataract control
[e]$p < 0.05$
Opacification score: 0 no opacification, + slight opacification, ++ diffuse opacification involving periphery of lens, +++ diffuse opacification involving almost entire lens, ++++ extensively thick opacification involving almost entire lens

14.2.2 Diabetic Retinopathy

Diabetic retinopathy (DR) is one of the leading causes of blindness for diabetic patients due to the damaging effect on the capillaries in retina by hyperglycemia. In the count of 35 studies, 34.6 % of 20,000 diabetic patients were diagnosed to have DR which means about 90 million patients are suffering from this disease [56]. DR is defined as a microvascular disease that leads to capillary occlusion and damage of light-sensitive tissues in the back of the eye as a complication of diabetes [57]. The retinal precapillary arterioles, capillaries, and venules would be all affected during its development. In the early stage, thickening of the basement membrane, loss of pericytes, and the development of microaneurysms could be detected [57, 58]. In the cellular pathology, the vascular endothelium cells have been altered by the high level of blood glucose. The abnormal endothelium will cause the platelet and leucocyte activation and thus adhesion together to cause the occlusion. At the molecular level, the occlusion leads to an ischemic retina, and the level of cytokines get elevated to promote the growth of new blood vessels through vascular endothelial growth factor (VEGF) which increase the vascular permeability in the early stage [57–60].

The development of DR can be divided into different stages: mild non-proliferative DR, moderate non-proliferative DR, severe non-proliferative DR, and proliferative DR. In the first two early stages, the vascular blockage was found but not severe enough to trigger the signal for growth of new blood vessels [61]. In the severe non-proliferative stage, the growth of new blood vessels is not as significant as the proliferative stage. Therefore, it was also known as pre-proliferative stage. In the proliferative stage, two types of new vessels will develop depending on their growth location in the eye: forward new vessels developed into vitreous cavity and flat new vessels remained on the surface of retina. There is a possibility of growing new vessels on the surface of the iris

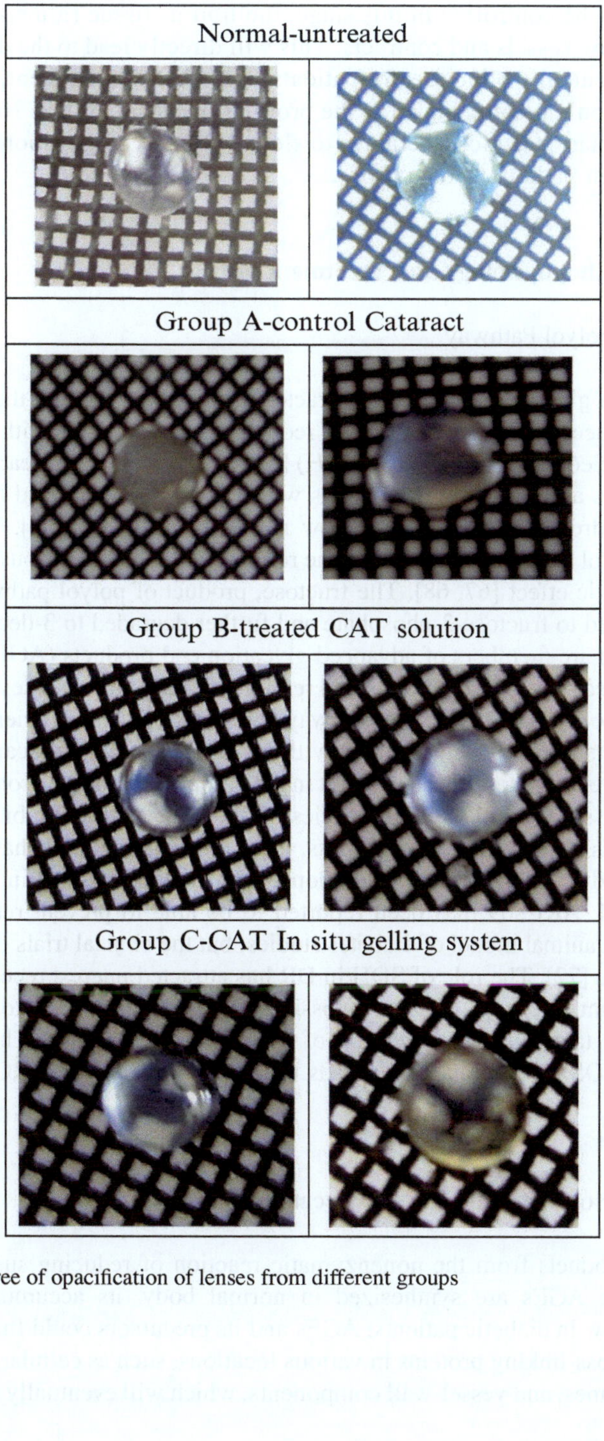

Fig. 14.1 Degree of opacification of lenses from different groups

which was known as rubeosis iridis [62]. These weak new vessels could lead to bleed, which will cause preretinal and vitreous hemorrhage. If the development of DR cannot be controlled in this stage, the fibrous tissue (gliosis) will appear around the new vessels and contracts. This will directly lead to the detachment of retina and eventually blindness for patients. The surgical treatment, such as laser treatment, is only recommended in the proliferative stage of DR. However, there are various pharmaceutical therapies to slow down the propagation of DR based on its different pathophysiology [63].

14.2.2.1 Pathophysiology and Treatment of DR

14.2.2.1.1 Polyol Pathway

The excessive glucose is the major characteristic for diabetes, and the polyol pathway is to reduce it to sorbitol by aldose reductase enzyme along with nicotinamide adenine dinucleotide phosphate (NADPH) [64, 65]. The poor permeability of sorbitol leads to its accumulation in the cells where it will be converted to fructose by sorbitol dehydrogenase (SDH) in a slow metabolism process [66]. The elevating level of sorbitol is reported to damage the retinal cells through various mechanisms, such as osmotic effect [67, 68]. The fructose, product of polyol pathway, could be phosphorylated to fructose-3-phosphate and further degraded to 3-deoxyglucosone, both of which are members of advanced glycation end products (AGEs), which are strongly related with DR conditions and retinal inflammatory diseases [68, 69]. The consumption of NADPH in polyol pathway could lead to its insufficiency in the formation of glutathione reductase which will weaken the antioxidant capacity of cells due to low level of glutathione and result in the compromise of the protection against oxidative stress [70]. Although the studies of aldose reductase inhibitors (ARIs) in animal models have shown some success, the human clinical trials have been a disappointment due to insufficient inhibition of the polyol pathway in human tissue [71]. An ARI, ARI-809, has been reported to be able to prevent retinopathy-like changes in an animal model of diabetes studies, but the clinical trials on humans are yet to be done [72]. The role of SDH in DR has attracted interest recently due to its more important role than ARIs and possibility to be a genetic factor in DR [73]. However, the targeted overexpression of SDH in retinal pericytes has shown the increase of ROS production which leads to toxicity via increased ROS production [68, 74].

14.2.2.1.2 Nonenzymatic Protein Glycation

AGEs are products from the nonenzymatic reaction of reducing sugars [64, 65]. Although the AGEs are synthesized in normal body, its accumulation rate is extremely slow in diabetic patients. AGEs and its precursors could further progress the DR by cross-linking proteins in various locations, such as cellular matrix, basement membranes, and vessel-wall components, which will eventually result into the alteration of their structures and functions [75–77]. Moreover, AGEs could combine with various receptors (receptor for advanced glycation end products (RAGEs), galectin-3, CD36, and the macrophage scavenger receptor [77]) which would lead to cellular activation and pre-oxidant, pre-inflammatory events [78]. During the development of DR, three main mechanisms of AGEs could be concluded: (1) as adducts occurring on modified serum proteins, (2) as endogenous adducts formed as a consequence of glucose metabolism, or (3) as extracellular matrix-immobilized modifications of long-lived structural proteins [77–79]. By decreasing the accumulation of AGEs through the treatment of AGE formation inhibitor, such as aminoguanidine (pimagedine) hydrochloride and vitamin B6 derivative, pyridoxamine, the development of DR could be slowed and prevented [79]. In the latest attempt, the progression of DR might be further prevented by breaking the cross-links caused by AGE, such as ALT-711, known as alagebrium [80].

14.2.2.1.3 Protein Kinase C Activation

Protein kinase C (PKC) is a serine/threonine kinase involved in signal transduction events responding to specific hormonal, neuronal, and growth factor stimuli [64, 65]. The studies have shown the β1/2 isoform of PKC possesses a close association with the development of DR [81]. An activator of PKC, diacylglycerol (DAG), is produced under the hyperglycemic conditions through the glycolysis pathway in diabetic patients, which leads to the activation of PKC [82]. Massive changes of functions were regulated by activation of PKC, such as increasing the endothelial permeability, altering the retinal hemodynamics, and expressing the vascular endothelial growth factor (VEGF) [82]. The inhibition of activation of PKC by a PKC-β1/2 inhibitor has been studied and reported to significantly reduce the progression of DR [83, 84]. The clinical trial is undergoing at Phase III.

14.2.2.1.4 Subclinical Inflammation and Leukostasis

Inflammation is a key characteristic for diabetic patients and its role in the development of DR has been reported [85, 86]. The previous discussed pathophysiologies of DR have a synergetic relationship with inflammation through multiple pathways, including cytokines, adhesion molecules, VEGF signaling, enhanced RAGE expression, changes in nitric oxide regulation, and NF-κB signaling. Leukostasis is another major property for DR and it could lead to capillary occlusion and reactive oxygen species (ROS)-mediated cell death. Its amplifying effect on inflammation of local retinal tissue was recognized [87]. A few of conditions related to DR (VEGF overexpression, irregular vascular permeability, high levels of cell death and leukostasis, and weak visual acuity) could be addressed by anti-inflammatory drugs such as the intravitreal triamcinolone acetonide (IVTA) and nonsteroidal anti-inflammatory drugs such as nepafenac [88, 89]. However, these drugs only have a significant impact during later stages of progression of DR and their side effects mainly due to the mode of delivery, intravitreal injections [89].

14.2.2.1.5 Oxidative Stress

Oxidative stress is defined as “a disturbance in the balance between the production of RPS (free radicals) and antioxidant defenses” [64, 65]. In the diabetes, its damaging effect on tissues was widely recognized and its increase contributes to the progression of DR [90]. A study has shown an important role of the ROS-mediated activation of metalloproteinase-2 (MMP-2) in the development of DR through mitochondrial dysfunction of retinal endothelial cells [91]. The contribution of oxidative stress to the development of retinopathy and the resistance of retinopathy to reverse after good glycemic control have been demonstrated on animal model [92]. The prevention of experimental DR has been reported based on animal studies, but there is no clinical trial to reach the same conclusion yet [93].

14.2.2.1.6 Growth Factors

A list of growth factors have been reported to be associated with the development of DR: fibroblast growth factor (bFGF) [94], insulin-like growth factor-1 (IGF-1) [95], angiopoietin-1 and angiopoietin-2 [96], stromal-derived factor-1 [97], epidermal growth factor (EGF) [98], transforming growth factor-beta 2 (TGF-β2) [99], platelet-derived growth factors (PDGFs) [100], and erythropoietin [101]. VEGF was the most intensively studied due to its capability of promoting angiogenesis, breaking down the blood-retinal barrier, stimulating endothelial cell growth, and increasing vascular permeability in the ischemic retina [102–105]. The treatment of DR with some of the current anti-VEGF agents like VEGF trap has shown great promise. The success of this strategy has shown in early clinical results for the treatment of diabetic macular edema (DME) [104]. However, there is a gap in our understanding of the variation inefficacy of anti-VEGF agents in individuals and also in the unknown long-term effect for this type of treatment.

14.2.2.1.7 Other Factors

Other pathophysiologies of DR include (a) carbonic anhydrases (CAs) which could cause rapid conversion of carbon dioxide to bicarbonate and protons and (b) neurodegeneration which occurs on retinal neurons and glial cells before the development of microaneurysms [64, 65, 106]. Blood pressure is another factor to play a significant role in the progression of DR through two mechanisms [107]: mechanical stretch and shear stress of blood [108] and misbehave of endocrine system [109]. The treatment of candesartan was able to reduce the incidence of retinopathy by 18 % [110]. However, the studies have shown no significant effect on the reduction of the DR progress except the regression of early stage of type II diabetic patients [110].

14.2.2.2 Drug Delivery Approaches for Treatment of DR

Although topical, systemic, intravitreal, and periocular routes have been used in the treatment of DR, the latter two have been more ideal to reach a high dosage of drug in the back of the eyes due to the bypass of the structural barriers of the eyes. Drug delivery to the posterior segment of the eye is challenging and a few of delivery systems were developed specifically for DR.

14.2.2.2.1 Polymeric Implants

Dexamethasone Polymeric Implants

Dexamethasone is a corticosteroid which has anti-inflammatory and immunosuppressant effects. Its intravitreal implant is a sustained release drug delivery system for DME. The study has found that dexamethasone implant could improve visual acuity significantly for 90 days after insertion [111]. The platform of this delivery system is based on a biodegradable poly(lactide-co-glycolide) (PLGA) polymer which has been applied into multiple FDA-approved implants [112]. Placid trial report in American Academy of Optometry (AAO), Oct. 2010, has reported their findings on the improvement of combining laser treatment with dexamethasone implant (Ozurdex) compared with laser treatment alone, besides the visual acuity improvement [113]. It was found that the slow-release implant was able to maintain 3-month drug release and achieve improvements in visual acuity and central macular thickness for DME patients from the third day [114].

Fluocinolone Acetonide (FA) Insert

Fluocinolone acetonide [115] is another corticosteroid and its intravitreal implant has been tested and approved for DME. It is reported to be able to improve visual acuity in DME. The clinical trial on an FA insert, Iluvien, has shown benefits in reducing DME. The visual acuity improvement was observed in 16.8 %, 16.4 %, and 31.8 % of implanted eyes at 6 months, 1 year, and 3 years, respectively. The retinal macula thickening is significantly improved in 2 years [116]. The low-dosage insert has shown a lower risk-benefit ratio comparing with high-dosage group; despite this both dosages have been reported to show a significant therapeutic effect in 3 weeks [117].

14.2.2.2.2 Celecoxib Microparticles

Celecoxib is a nonsteroid anti-inflammatory drug and its efficacy for DME management has been reported [118]. The sustained release could be obtained by using biodegradable PLGA (85:15) microparticles, which has shown a 49-day release profile in vitro. Animal study has revealed that no celecoxib was detected in the contralateral eye of rats 14 days after the injection. Improvement in the oxidative damage has been observed [119]. However, no clinical trial has been carried out with this type of treatment for DME or DR.

14.2.2.2.3 Budesonide Particles

Budesonide is a steroidal anti-inflammatory drug and an inhibitor for VEGF [120, 121]. Its PLA-based particles were studied for DME treatment. The subconjunctival administration of these particles in rats produced a sustained release profile to the posterior of the eyes, and the budesonide levels were shown to be maintained in the retina and other ocular tissues. The success of inhibition of VEGF was observed and no budesonide was detected in the contralateral eye [121]. Further preclinical and clinical studies are required for this delivery system.

14.2.2.3 Future Directions for Diabetic Retinopathy

Although multiple pathologies of DR are well recognized, the treatment of DR is still limited for their advanced stage and needs further investigations of newer therapies. The current therapies are merely targeting one or two pathways. Considering the pathogenesis of DR, there is need for therapies which will target multiple pathways. The utilization of the newer drug delivery approaches such as nano- and microparticles has a long way to clinical trial despite the effort on applying the FDA-approved material.

14.2.3 Age-Related Macular Degeneration (AMD)

AMD is an age-related, progressive degeneration of photoreceptors and their underlying retinal pigment epithelium (RPE) in the macular area of the retina, which eventually leads to vision loss. AMD is one of the leading causes of vision loss among elderly in western world, particularly in industrialized nations [122, 123]. More than 1.75 million people in the USA have AMD; also due to the rapid aging of the American population, this number is likely to increase to approximately three million by 2020 [124].

The macula, the most sensitive part of the eye, is essential for sharp, central vision and image resolution, which helps to see objects that are straight ahead. The retina consists of two layers, the inner neurosensory retina and the outer retinal pigment epithelium (RPE) cell layer. The Bruch's membrane (BM) separates RPE from the outer vascular choroid. The choroid present adjacent to BM consists of a network of fenestrated capillaries and an external large vessel layer. It plays an important role to supply oxygen and nutrients to and remove waste from the retina, especially at the macula [122].

14.2.3.1 AMD Types and Pathophysiology

AMD is mainly divided into two forms, dry and wet AMD. The dry type accounts for more than 85 % of the cases; in the dry type, the drusens, subretinal deposits start accumulating between the RPE and the underlying choroid due to aging and thinning of the macular tissue followed by atrophy of the RPE and adjacent cells in contiguous areas of the macula. The wet form of AMD, accounting for approximately 15 % of patients, is characterized by choroidal neovascularization (CNV). Although wet AMD is less prevalent than dry AMD, it is usually very destructive and can cause rapid and severe vision loss [125].

In the normal eye (Fig. 14.2a), the retinal pigment epithelial (RPE) layer sits on Bruch's membrane (BM), which is the relative diffusion barrier between RPE and the external choroid layers, and is composed of collagen and elastin. The RPE cells are large, the bipolar and ganglion cells at the macula are displaced, and there are no retinal blood vessels in this region; the fovea, forming the macular center, is concave in cross section and consists only of cone photoreceptors. Hence there is minimal interference with the incoming image. In dry AMD (Fig. 14.2b), drusen accumulate between RPE cells and BM. Reticular drusen accumulate between the RPE layer and photoreceptor cells. Pigmentary irregularities including atrophy and hypertrophy/hyperplasia of RPE cells appear, along with choriocapillaris atrophy and thickening of BM. Extensive atrophy leads to geographic atrophy, a form of late/advanced AMD. Wet AMD (Fig. 14.2c), a more severe form of AMD, is characterized by choroidal neovascularisation (CNV). In initial stages, within the choroid, new vessels break through a gap in BM to grow under/within the retina. The CNV leaks fluid and blood, disrupting the organized architecture of the retinal cells. This results in distortion of central vision, often the first symptom of CNV experienced by the patient. Wet AMD can also occur from intraretinal neovascularization (retinal angiomatous proliferation – RAP). Consequent thickening and elevation of the retina can be detected clinically. Eventually the CNV scars, with permanent disruption of the retinal architecture.

14.2.3.2 Risk Factors for AMD

14.2.3.2.1 Age

One of the major risk factors for the development of AMD is age. The prevalence of AMD is observed in subjects over the age of 50.

14.2.3.2.2 Smoking

Epidemiologic studies have shown cigarette smoke to be the single and greatest risk factor for development of both wet and dry AMD [126]. The smoker subjects aged more than 85 years possess a 6.6-fold increased risk of development of AMD than

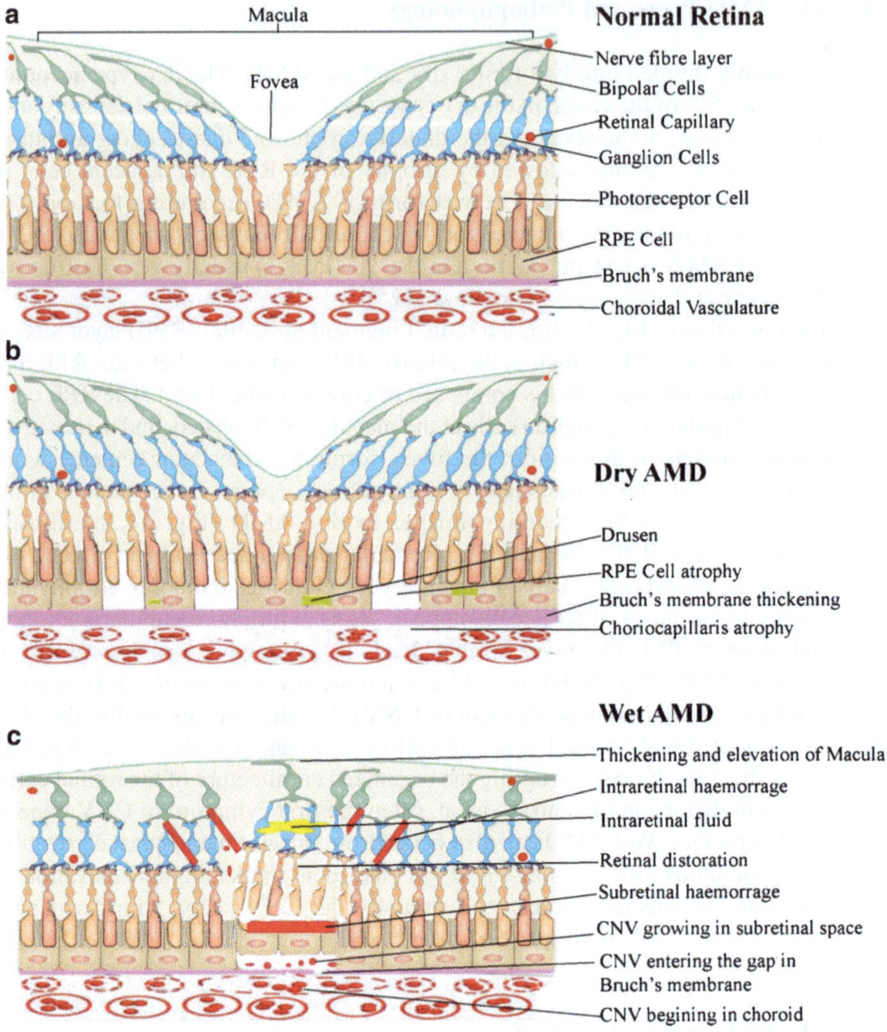

Fig. 14.2 Cross-sectional pathological changes occurring in macula in age-related macular degeneration (AMD). (**a**) Normal retina, (**b**) dry AMD, (**c**) wet AMD

nonsmokers (95 % CI, 2.8–15.9) [127, 128]. A study on the effect of smoking in male twins showed that both current and past smokers had risk of AMD development. Current smokers had about twofold, while past smokers had 1.7-fold risk of AMD development compared to nonsmokers [129, 130]. Studies have also shown a relationship between smoking and genetic polymorphisms. The risk for AMD is high in smokers bearing likely polymorphisms in the LOC387715 or CFH genes [130–132].

14.2.3.2.3 Light Exposure

Light exposure produces reactive oxygen species (ROS) that are harmful to the eye and responsible for retinal damage. Only blue light is able to reach retina as UV light is absorbed by the lens and the cornea [133]. The macular pigment protects macula against light damage by absorbing blue light [134–141]. Many studies have proven the effect of light exposure in the development of AMD [142].

14.2.3.2.4 Greater Body Mass Index (BMI)

Several studies reported the association of high BMI with AMD [143–145]. The prevalence of AMD in obese patients may be due to higher oxidative stress, changes in the lipid profile, and other physiological changes resulting in increased inflammation, leading to cellular destruction of the macular cells [145].

14.2.3.2.5 Cataract Surgery

There are several studies which have demonstrated that cataract surgery increases the incidence of AMD [145–152]. The data from the Andhra Pradesh Eye Disease Study in South India demonstrated the relationship between prior cataract surgery with increased occurrence of AMD [145].

14.2.3.2.6 Others

The other miscellaneous risk factors reported for the prevalence of AMD are systemic hypertension, genetic and family predisposition, cardiovascular (atherosclerotic) disease, and diet high in saturated fat.

14.2.3.3 Treatment of AMD

14.2.3.3.1 Dry AMD

The oxidative mechanisms play an important role in the visual system degeneration. Hence, the use of minerals, vitamins, and antioxidant supplements is recommended for treatment of AMD [153]. For dry AMD, daily dietary nutrient supplementation has been shown to reduce the risk of disease progression in individuals. The clinical trial of Age-Related Eye Disease Study (AREDS) showed reduction in AMD development with oral supplementation of a combination of vitamin E, vitamin C, beta-carotene, cupric oxide, and zinc oxide [122, 154].

14.2.3.3.2 Wet AMD

Laser Photocoagulation and Photodynamic Therapy (PDT)

The laser photocoagulation is a type of laser surgery involving the use of an intense beam of light, which is focused on the retinal region; this procedure helps to burn small areas of the retina and the abnormal blood vessels beneath the macula. The burns form scar tissue that seals the blood vessels, keeping them from leaking under the macula. Thus there is a slowing down of the buildup of fluid under the retina that distorts the shape and position of the macula and hence can arrest the progression of AMD by destroying choroidal neovascular membranes. One of the studies on laser photocoagulation demonstrated a 1.5 relative risk reduction of severe vision loss. However, recurrence of neovascularization was high [155].

Photodynamic therapy (PDT) was developed as an alternative to the laser photocoagulation. In this therapy pharmacologic photo-sensitizer (verteporfin) was administered intravenously followed by activation with a laser having drug absorption wavelength. Thus, PDT selectively damages the tissue that contains dye. When PDT was used in the treatment of wet AMD, there was an approximately 50 % reduction in vision loss, but improvement in vision was difficult [156, 157].

Anti-VEGF Therapy

Pegaptanib sodium (Macugen®) is the first intraocular Anti-VEGF therapy. Pegaptanib sodium is a pegylated aptamer that has high affinity to the heparin-binding domain of the EGF_{165} isoform and has demonstrated improved vision in patients with exudative AMD. Ranibizumab (Lucentis®), a humanized, recombinant, monoclonal anti-VEGF antibody fragment, demonstrated stable vision at 1 year in about 95 % of patients and improved vision in about 40 % of treated patients [158, 159]. Bevacizumab (Avastin®) is active against all VEGF-A isoforms. This drug has been approved for treatment of metastatic colon cancer [160]. Many studies demonstrated that there is no difference between ranibizumab and bevacizumab efficiencies [161, 162].

Surgical Treatments

The surgical treatment includes macular translocation, submacular surgery, and submacular hemorrhage displacement. However, these surgical procedures are not very useful in AMD, as the degenerative processes have already damaged the retina and macular tissues and have not resulted in significant improvement of vision [163].

Others

VEGF Trap-Eye (aflibercept), small interfering RNAs (siRNA), tyrosine kinase inhibitors, cytokine PEDF, epimacular brachytherapy, and combination treatment of anti-VEGF and PDT are under investigation for treatment of AMD [163].

14.2.3.4 Drug Delivery Approaches for Treatment of AMD

The treatment of posterior eye diseases such as AMD is difficult due to complexity in delivering effective doses of drugs to target tissues in the posterior eye. Various delivery systems are reported and developed to deliver drugs to the posterior segment of the eye for treatment of AMD. The standard treatment of AMD includes small amounts of intravitreal injections of anti-VEGF drugs. Repeated intravitreal injections, although effective in treatment of AMD, may lead to complications that include endophthalmitis, increased intraocular pressure, traumatic cataract, detached retina, and stroke. To overcome these problems, many sustained release formulations have been attempted using the novel particulate systems, which include the following.

14.2.3.4.1 Liposomes

Prolongation of effect on the administration of liposomal bevacizumab in the vitreous and aqueous humor has been investigated. A twofold and fivefold increase in concentrations of bevacizumab in the eye tissues was evident after the 28th day and 42nd day, respectively, after liposomal injection, compared to the levels after injection of soluble bevacizumab injection. The study proved the beneficial effects of prolonged release bevacizumab liposomes in the vitreous [164]. Prolonged liposomal formulations of SU5416 (a VEGF receptor tyrosine kinase inhibitor) [165, 166], siRNA along with PEGylation technology [167], were reported.

14.2.3.4.2 Micro-/Nanoparticles

In one of the studies on controlled release of bevacizumab, the sustained drug levels for over 90 days were obtained from nano- and microspheres fabricated from poly(ethylene glycol)-b-poly (D,L-lactic acid) and poly(DL-lactide-co-glycolide), respectively [168]. Folate-decorated polymeric nanoparticles of triamcinolone [169, 170], human serum albumin nanoparticles of Cu, Zn superoxide dismutase gene [171], and nanocarrier systems of signal peptide serine-threonine-tyrosine (ser-thr-tyr) with chitosan [172] are also reported for controlled ocular delivery.

14.2.3.4.3 Implants

Intravitreal administration of Fluocinolone acetonide conjugates with PAMAM dendrimers were selectively localized within activated outer retinal microglia in two rat models of retinal degeneration, and achieved sustained release for a period of over 90 days [173]. A novel intraocular implantable capsule drug device fabricated with polyvinyl alcohol (PVA) provided near-zero order release of Avastin® and has been reported for treatment of AMD [174]. Several animal experimentations and clinical trials have explored implantable systems of antiangiogenic steroids like triamcinolone acetonide, anercotave acetate , as well as α agonists like brimonidine; however no definite conclusions have been drawn from these studies [175, 176].

14.2.3.5 Animal Models for AMD

With the advancement of the research for treatment of the AMD, various animal models have been developed for better understanding of the underlying pathological mechanisms of AMD and replicate the features of human AMD. However it is difficult to develop the unique model which exactly resembles conditions of AMD, as it is a very complex and heterogeneous disease involving environmental and genetic factors. Animal models of pigs, rabbits, rats, mice, and nonhuman primates were reported in the literature.

14.2.3.5.1 Animal Models of Dry AMD

Complement Factor Pathway

Dry AMD is characterized by the formation of the drusens under RPE layer which leads to hypertrophy, hyperplasia, or atrophy. The complement system plays an important role in the body's defense mechanism against pathogens, apoptotic cells, immune responses, and elimination of immune complexes [177]. Several studies demonstrated the role of complement system in formation of drusen, including complement components C3a and C5a [178], C5 and C5b-9 terminal complement complex (TCC) [179, 180], as well as complement factor H-CFH, clusterin and vitronectin, and membrane-bound complement inhibitors (complement receptor 1-CR1, also called CD35, and membrane cofactor protein-MCP/CD46) [181, 182]. Dysregulation of the above complement components leads to damage and pathogenesis of AMD. Genetically engineered mice lacking complement factor H (Cfh$^{-/-}$ mice) [177, 183], transgenic CFH Y402H mice [184], and transgenic mice overexpressing C3 [185], C3a, and C5a receptor$^{-/-}$ mice [178] are the models used for the study of dry AMD.

Chemokine Models

Chemokines are the signaling proteins secreted by cells exerting their action by coupling with G protein-linked transmembrane receptors. The main functions of the chemokines are homeostasis and inflammation [186]. Ccl2$^{-/-}$ and Ccr2$^{-/-}$ mice and Cx3cr1$^{-/-}$ mice and Ccl2$^{-/-}$ Cx3cr1$^{-/-}$ double knockout mice are also reported as models used for the study of dry AMD.

Oxidative Damage Models

Oxidative stress from reactive oxygen species (ROS) is reported to be one of the key factors in the pathogenesis of aging-associated diseases, including age-related macular degeneration (AMD). Animal models with applied oxidative stress and those that lack antioxidant mechanisms express many features of AMD. Carboxyethyl pyrrole (CEP)-adducted proteins formed from docosahexaenoic acid

(DHA) in the retina are found in drusen. This could be the initiation of AMD. Hollyfield et al. demonstrated that immunization of mouse with CEP-adducted mouse serum albumin demonstrated many features of dry AMD [187, 188]. With aging iron plays an important role in the development of oxidative stress. Ceruloplasmin plays important role in cell iron export. A ceruloplasmin/hephaestin$^{-/-}$ mouse is another model of AMD in which lack of ceruloplasmin develops symptoms of AMD [189]. Sod1$^{-/-}$ mice [190, 191] and Sod2 knockdown mice [192] lacking function of superoxide dismutase which is a potent endogenous antioxidant develop AMD-like features. Exposure to cigarette smoke also plays an important role in the induction of AMD. Mice exposed to cigarette smoke demonstrated features of dry AMD. Other models such as OXYS rats, lipid/glucose metabolism, aging mice+/−high fat diet+/−light treatment, high glycemic index diet, ApoE−/−mice, APOEe2/e4 transgenic mice, APO/E3-Leiden transgenic mice, APOB100 transgenic mice + high fat diet, Ldl receptor−/−mice + high fat diet, Vldl receptor−/−mice, CD36−/−mice, and Mcd/mcd mice (transgenic mutant cathepsin D) are reported in the literature for dry AMD [193].

Other Models for Dry AMD

Senescence-accelerated mice [194] and macular dystrophies by single gene mutations, e.g., Timp3$^{-/-}$mice [195], the Abcr$^{-/-}$mice [196], the ELOVL4 transgenic mouse [197], and fibulin-3 transgenic mice [198], are reported in the literature.

Nonhuman Primate Models

Animal models of rhesus and cynomolgus macaques and Japanese and cynomolgus macaques are reported in the literature [193].

14.2.3.5.2 Animal Models of Wet AMD

Laser-Induced CNV

Rodent and nonhuman primate models (cynomolgus macaques, rhesus macaques, and African green monkeys) of laser-induced CNV are reported in the literature. In these models CNV is induced by the use of laser energy [199–202].

Subretinal Injection-Induced CNV

Subretinal matrigel injection causes CNV in mice and rabbits [201, 203]. Subretinal injections of adenovirus vectors expressing VEGF [204], macrophages [205], lipid hydroperoxide [206], and polyethylene glycol [207] are also reported to produce CNV in rats.

VEGF/bFGF Pellet-Induced CNV

Zhang et al. demonstrated that implantation of VEGF/bFGF-impregnated pellets in suprachoroidal space of rabbit eye developed CNV [51].

Surgical Model of CNV

Surgical rupture of Bruch's membrane leads to development of CNV [208]. In one of the studies, surgical model of CNV was compared with CNV induced by laser and xenon lamp [209].

14.2.4 Glaucoma

Glaucoma(GL)is a condition affecting over 66 million people worldwide and responsible for causing bilateral blindness in 6.8 million people, as per WHO statistics. GL results due to slow progressive degeneration of retinal ganglion cells (RGCs) and optic nerve axons. The most common form of the disease is primary open-angle glaucoma (POAG).

14.2.4.1 Types

GL is actually a group of diseases. The most common type is hereditary.

14.2.4.1.1 Primary Open-Angle Glaucoma

This is the most common type of GL caused due to clogging of eye's drainage canal over the time. This causes an increase in inner eye pressure (also called intraocular pressure or IOP), as the correct amount of fluid cannot drain out of the eye. It shows no symptoms and no early warning signs. If undiagnosed and untreated, it can cause a gradual loss of vision. The development is slow and sometimes without noticeable sight loss for many years. It usually responds well to medication, especially in the early stages.

14.2.4.1.2 Angle-Closure Glaucoma

This is a rare and very different condition compared to open-angle GL and is also called as narrow-angle GL, with quick rise in eye pressure. Symptoms of angle-closure glaucoma may include headaches, eye pain, nausea, rainbows around lights at night, and very blurred vision. Treatment of angle-closure glaucoma usually involves either laser or conventional surgery which helps unblock the drainage canals so that the extra fluid can drain. Success rate of surgery is high but still regular checkups are important to avoid chronic form of GL.

14.2.4.1.3 Normal-Tension Glaucoma

This condition is also called low-tension or normal-pressure glaucoma, in which the optic nerve is damaged even though the pressure in the eye is not very high. People with a family history of glaucoma, or with Japanese ancestry, and those with a history of systemic heart disease are at high risk of developing this condition. Treatment of normal tension glaucoma is by reducing the eye pressure as low as possible using medications, laser treatments, and conventional surgery.

14.2.4.1.4 Other Types of Glaucoma

There are several variations of open-angle or angle-closure types. These types can occur in one or both the eyes. It occurs as a result of an eye injury, inflammation, tumor, or in advanced cases of cataract or diabetes, or steroid drug medication. It includes secondary GL, pigmentary GL, pseudoexfoliative GL, traumatic GL, neovascular GL, iridocorneal endothelial syndrome (ICE), and congenital GL (childhood glaucoma). These forms of GL may be mild or severe. The type of treatment will depend on whether it is open-angle or angle-closure GL [210].

14.2.4.2 Pathophysiology and Treatment of GL

Underlying cause of open-angle glaucoma remains unclear and the pathophysiology of glaucoma is believed to be multifactorial. Multiple factors acting either on cell bodies or their axons are believed to lead to retinal ganglion cell (RGC) death.

14.2.4.2.1 Elevated Intraocular Pressure (IOP)

This form of GL is due to increased resistance to the draining of aqueous humor through the trabecular meshwork, which leads to an increase in IOP. This further leads to cell death from compression of the optic nerve axons [211].

14.2.4.2.2 Glutamate Excitotoxicity

A chronic increase in glutamate levels is seen in GL. This can result in activation of glutamate receptors, which further may lead to cellular apoptosis and neuronal cell death.

14.2.4.2.3 Nitric Oxide (NO) Neurotoxicity

Excessive levels of NO are associated with several degenerative aging disorders like Parkinson's disease, Alzheimer's disease, and multiple sclerosis. Studies have reported high levels of *nitric oxide synthase-2 (iNOS-2), in GL*; *iNOS-2* is an inducible form of NOS, capable of releasing an excessive quantity of NO.

14.2.4.2.4 Neuroinflammation

This is a sort of autoimmune phenomenon wherein secondary antigen-mediated neurotoxicity is seen, due to sustained neurodegeneration; this is primarily due to long-term presence of lymphocytes and antigens and subsequently increased T cells, followed by release of cytokines. Marked increase in cytokine receptors has been detected in normal-pressure glaucoma (NPG) patients (67 %) and POAG patients (77 %), indicating T-cell over-activation in GL.

14.2.4.3 Drug Delivery Approaches for Treatment of GL

The major aim in GL treatment is to prevent or delay the loss of visual field. Since neuronal cell death is irreversible, no cure is available once the visual field is lost. However, since IOP is the primary risk factor, the strategies for treatment attempt to lower the raised IOP, for which several drugs are available that are effective in reducing IOP. These drugs are typically applied as eyedrops.

Different delivery systems like oral tablets (acetazolamide), topical eyedrops or gels (timolol), inserts (pilocarpine, timolol), and surgical implants (dexamethasone) are clinically available for the treatment of IOP. These have short-lived effect and need to be administered/instilled several times a day, which would be cumbersome from patient point of view. An ideal drug delivery system for GL should offer sustained release of the drug for several hours/days or months from a single application without any surgery. Using one or more of the existing IOP-lowering medications, such slow-release ocular delivery systems will circumvent patient adherence factors and may offer an attractive alternative to traditional topical eyedrops for many elderly patients. Some such novel delivery systems have been developed in the form of injectable systems, medicated contact lenses, ocuserts, and implantable devices [212].

14.2.4.3.1 Liposomes and Nanospheres

Pilocarpine-loaded liposomal dispersions, which can be instilled as eyedrops, are reported to increase the residence time of the drug to twice that obtained with conventional eyedrops. Effects of charge on the liposomes have also been studied, and the neutral uncharged systems were recommended [213].

Colloidal dispersions of neutral nanocapsules instilled as eyedrops, incorporating rhodamine as model drug, indicated that nanocapsules take an intracellular route through the corneal epithelium. Another study on polymeric nanocapsules based on diblock caprolactone copolymers for ocular delivery has compared PEG and chitosan coatings for surface modification; PEG coatings provided fast release, whereas chitosan-coated nanocapsules provided greater retention in corneal layers [214].

14.2.4.3.2 Contact Lenses as Delivery Vehicles

Contact lenses can be an attractive system to provide prolonged levels of drugs in eye. Soft contact lenses, fabricated from polymers of *N*, *N*-diethylacrylamide and methacrylic acid, have shown extended release of timolol for up to 24 h. However, with reference to elderly patients, contact lenses may be unwieldy and not suitable.

14.2.4.3.3 Sophisticated Surgical Implants

These represent advanced technological reservoir devices to be implanted in the subconjuctival space. One such example is a microelectromechanical system (MEMS) which employs electrolysis to create bubbles that push the drug out of the reservoir of the device. Surgical steps required would be similar to currently available glaucoma drainage devices. Reloading of the device and controlling the delivery rate afford flexibility to the clinician.

14.2.4.3.4 Injectable Systems

Existing drugs may be injected into the subconjunctival space to achieve localization and prolonged delivery compared with simple topical application. Formulations based on degradable and nondegradable polymers have been studied for drugs like antibiotics after cataract surgery, carboplatin for murine retinoblastoma, and celecoxib to reduce oxidative stress in the rat. Timolol encapsulated in polyester microspheres was found to afford sustained release for greater than 90 days in vitro [212].

14.2.4.3.5 Other Systems Suitable for Elderly

In general, considering the aging population, the sophisticated systems may be not very suitable. Simple eyedrops, gels, and in situ gelling systems may be more user-friendly. Studies in our group have led to the development of simple gels, in situ gelling systems, and disks of timolol maleate, based on various mucoadhesive polymers – polycarbophil and sodium alginate. In vivo studies in rabbits

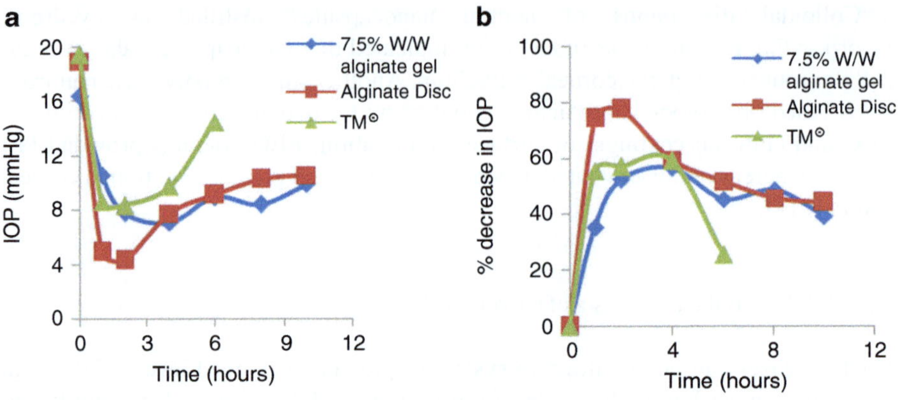

Fig. 14.3 (**a**) Effect of alginate formulation of timolol maleate and marketed eyedrops on intraocular pressure in rabbits. (**b**) Effect of alginate formulation of timolol maleate and marketed eyedrops on % reduction of intraocular pressure in rabbits

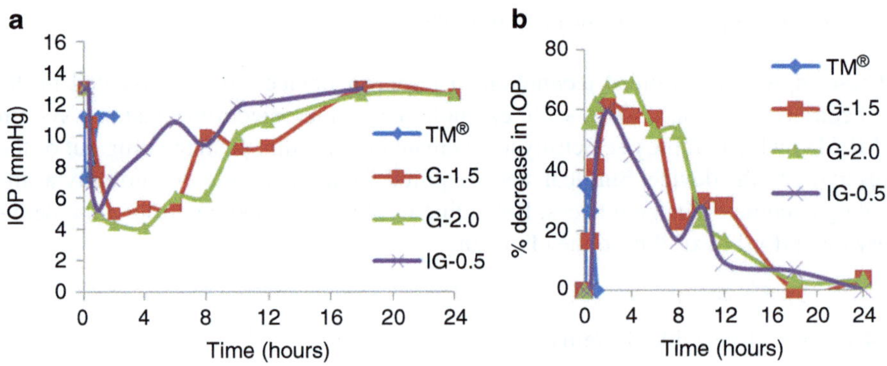

Fig. 14.4 (**a**) Effect of polycarbophil gels (G-1.5 % and G-2 %), polycarbophil 0.5 %, and methocel K4M 1 % gel (IG-0.5 %) of timolol maleate and marketed eyedrops on intraocular pressure in rabbits. (**b**) Effect of polycarbophil gels (G-1.5 % and G-2 %), polycarbophil 0.5 %, and methocel K4M 1 % gel (IG-0.5 %) of timolol maleate and marketed eyedrops on % reduction of intraocular pressure in rabbits

have revealed marked reduction in intraocular pressure (40–70 %) which was sustained for a period of up to 5 h, in the case of gels and in situ gelling systems, and up to 10 h in case of disks, as against conventional eyedrops (Figs. 14.3 and 14.4) [215, 216].

14.3 Conclusion and Future Directions

Overall, it can be summarized that the global scenario is indicative of a significant increase in the elderly population. This can have an impact on the healthcare resources and systems, which will have to gear toward the aging population.

Along with other aging disorders, visual impairment and eye problems like cataract, diabetic retinopathy, AMD, and glaucoma will be on the rise. Visual impairment dramatically reduces the ability of older people to contribute to their full capacity, which has a negative impact on society as a whole. Maintaining good vision is an important part of "active aging," a concept promoted by the WHO. Active aging means continued health, security, and participation in society as people age, in order to ensure a good quality of life in later years. Nations and communities have to gear up to this challenge to ensure that good quality eye care and therapies are available to this group. Currently the surgical cure for cataract is well established. However, surgery in the elderly is often risky and an inherent fear factor may deter people from surgical interventions, especially when they are coupled with many other concomitant diseases. In the case of other conditions, viz., glaucoma, AMD, and diabetic retinopathy, therapies and drugs are not adequate; they are simply used to provide symptomatic relief, and a great deal of research still needs to be focused in developing cures for them. Also, newer approaches based on the novel drug delivery systems (NDDS) which can overcome the formidable barriers of the eye for effective therapy need to be explored. These NDDS should be developed as simple noninvasive systems which can be easily handled, inserted, and used by the aged persons, which can ensure good patient compliance and confidence to them.

Acknowledgments The authors acknowledge the support given by the following funding agencies:

(a) National Institute of General Medical Science of the National Institutes of Health (award number SC3 SC3GM109873, 2011); (b) Hawaii Community Foundation, Honolulu, HI, USA, for research support on asthma and mesothelioma research (Leahi Fund) 2013; (c) The 2013 George F. Straub Trust and Robert C. Perry Fund of the Hawaii Community Foundation, Honolulu, HI, USA, for research support on lung cancer; (d) Seed grant from the Research Corporation of the University of Hawaii at Hilo, HI, USA, and University of Hawaii at Hilo, the Daniel K Inouye College of Pharmacy for providing start-up financial support to our research group; (e) University Grants Commission, India – Major Research Project funding titled "Studies in the Development of Mucoadhesive Systems for Ocular and Nasal Delivery of Drugs" (Sanction No. F-7-11/99-SR, Aug 2000–2003); (f) Amrut Mody research fund (AMRF) for project titled "Studies in the development of novel enzyme delivery systems" (Year 2007–2010).

References

1. Abdullah KN, Abdullah MT (2002) Management and planning for primary eye care of the elderly: the need to create public awareness of age-related cataract in Pakistan. Community Eye Health 15(43):45
2. Control Centers for Disease (2006) Improving the nation's vision health: a coordinated public health approach. Centre for disease control, Atlanta
3. Evans J (2008) Eye care for older people. Community Eye Health J 21(66):21–23
4. Nobili A, Garattini S, Mannucci PM (2011) Multiple diseases and polypharmacy in the elderly: challenges for the internist of the third millennium. J Comorbidity 1(1):28–44
5. Team Ve-r (2010) Vision 2020 e-resource for eye care management worldwide. Vision 2020 e-resource team. http://www.v2020eresource.org/newsitenews.aspx?tpath=news012010

6. Wong TY, Loon SC, Saw SM (2006) The epidemiology of age related eye diseases in Asia. Br J Ophthalmol 90(4):506–511
7. Gentry LR (1998) Anatomy of the orbit. Neuroimaging Clin N Am 8(1):171–194
8. Hughes MS (1991) Dictionary of eye terminology. Arch Ophthalmol 109(9):1208
9. Alm A, Nilsson SFE (2009) Uveoscleral outflow – A review. Exp Eye Res 88(4):760–768
10. Hayreh SS (1975) Segmental nature of the choroidal vasculature. Br J Ophthalmol 59(11):631–648
11. Oyster CW (1999) The human eye: structure and function. Sinauer Associates, Inc., Sunderland
12. Hubel DH (1995) Eye, brain, and vision. Scientific American Library series (Issue 22), Henry Holt and Company, New York
13. Forrester J, Dick A, McMenamin P, Lee W (1996) The eye: basic sciences in practice. WB Saunders Company Ltd, London
14. Venes D (2013) Taber's cyclopedic medical dictionary. FA Davis, Philadelphia
15. Baerveldt G (2000) Method and apparatus for inserting a glaucoma implant in an anterior and posterior segment of the eye. Google patents
16. Kronfeld P (1962) Gross anatomy and embryology of the eye. Eye 1:1–66
17. Salvi SM, Akhtar S, Currie Z (2006) Ageing changes in the eye. Postgrad Med J 82(971):581–587
18. Van Haeringen NJ (1997) Aging and the lacrimal system. Br J Ophthalmol 81(10):824–826
19. Faragher RGA, Mulholland B, Tuft SJ, Sandeman S, Khaw PT (1997) Aging and the cornea. Br J Ophthalmol 81(10):814–817
20. Duncan G, Wormstone IM, Davies P (1997) The aging human lens: structure, growth, and physiological behaviour. Br J Ophthalmol 81(10):818–823
21. Le Goff MM, Bishop PN (2008) Adult vitreous structure and postnatal changes. Eye 22(10):1214–1222
22. Grunwald JE, Metelitsina TI, DuPont JC, Ying G-S, Maguire MG (2005) Reduced foveolar choroidal blood flow in eyes with increasing AMD severity. Invest Ophthalmol Vis Sci 46(3):1033–1038
23. Grunwald JE, Piltz J, Patel N, Bose S, Riva CE (1993) Effect of aging on retinal macular microcirculation: a blue field simulation study. Invest Ophthalmol Vis Sci 34(13):3609–3613
24. Brian G, Taylor H (2001) Cataract blindness: challenges for the 21st century. Bull World Health Organ 79(3):249–256
25. Rao GN, Sadasivudu B, Cotlier E (1983) Studies on glutathione S-transferase, glutathione peroxidase and glutathione reductase in human normal and cataractous lenses. Ophthalmic Res 15(4):173–179
26. Beers MH, Berkow R (1999) The Merck manual of diagnosis and therapy. Merck and Co. Inc., Whitehouse Station
27. Cejková J, Stípek S, Crkovska J, Ardan T, Platenik J, Cejka C, Midelfart A (2004) UV rays, the prooxidant/antioxidant imbalance in the cornea and oxidative eye damage. Physiol Res 53:1–10
28. Cumming RG, Mitchell P (1997) Alcohol, smoking, and cataracts: the Blue Mountains eye study. Arch Ophthalmol 115(10):1296–1303
29. Auricchio G, Libondi T (1982) The physiologic and pharmacologic factors protecting the lens transparency and the update approach to the prevention of experimental cataracts: a review. Metab Pediatr Syst Ophthalmol 7(2):115–124
30. Spector A, Garner WH (1981) Hydrogen peroxide and human cataract. Exp Eye Res 33(6):673–681
31. Wakamatsu TH, Dogru M, Tsubota K (2008) Tearful relations: oxidative stress, inflammation and eye diseases. Arq Bras Oftalmol 71(6):72–79
32. Brown L, Rimm EB, Seddon JM, Giovannucci EL, Chasan-Taber L, Spiegelman D, Willett WC, Hankinson SE (1999) A prospective study of carotenoid intake and risk of cataract extraction in US men. Am J Clin Nutr 70(4):517–524

33. Gritz DC, Srinivasan M, Smith SD, Kim U, Lietman TM, Wilkins JH, Priyadharshini B, John RK, Aravind S, Prajna NV (2006) The antioxidants in prevention of cataracts study: effects of antioxidant supplements on cataract progression in South India. Br J Ophthalmol 90(7):847–851
34. Jacques PF, Chylack LT (1991) Epidemiologic evidence of a role for the antioxidant vitamins and carotenoids in cataract prevention. Am J Clin Nutr 53(1):352S–355S
35. Calladine D, Evans JR, Shah S, Leyland M (2012) Multifocal versus monofocal intraocular lenses after cataract extraction. Cochrane Database Syst Rev 9
36. Leung TG, Lindsley K, Kuo IC (2014) Types of intraocular lenses for cataract surgery in eyes with uveitis. Cochrane Database Syst Rev 3:Cd007284
37. Chanalet L, Lapalus P (1994) Drugs designed to maintain the transparence of the ocular lens. Fundam Clin Pharmacol 8(4):322–341
38. Testa M, Iuliano G, Marino E, Buongiovanni C, Paolercio F, Trapanese A, Mortow P (1986) Bendazac and benzydamine for treatment of cataract: individualized therapy by the "BLOA test". J Ocul Pharmacol Ther 2(3):251–266
39. Toh TY, Morton J, Coxon J, Elder MJ (2007) Medical treatment of cataract. Clin Experiment Ophthalmol 35(7):664–671
40. Balfour JA, Clissold SP (1990) Bendazac lysine. Drugs 39(4):575–596
41. Testa M, Iuliano G, Morton P, Longoni A (1987) Topical benzyl alcohol reduces cataract surgery need: two long-term double blind studies. J Ocul Pharmacol Ther 3(3):211–225
42. Hu C-C, Liao J-H, Hsu K-Y, Lin IL, Tsai M-H, Wu W-H, Wei T-T, Huang Y-S, Chiu S-J, Chen H-Y (2011) Role of pirenoxine in the effects of catalin on in vitro ultraviolet-induced lens protein turbidity and selenite-induced cataractogenesis in vivo. Mol Vis 17: 1862
43. Kociecki J, Załecki K, Wasiewicz-Rager J, Pecold K (2003) Evaluation of effectiveness of Catalin eyedrops in patients with presenile and senile cataract. Klinika oczna 106(6): 778–782
44. Angra SK, Mohan M, Saini JS (1983) Medical therapy of cataract (evaluation of Catalin). Indian J Ophthalmol 31(1):5
45. Hockwin O, Laser H, De Gregorio M, Carrieri MP (1989) Bendazac lysine in selected types of human senile cataract. Ophthalmic Res 21(3):141–154
46. Leuschen J, Mortensen EM, Frei CR, Mansi EA, Panday V, Mansi I (2013) Association of statin use with cataracts: a propensity score-matched analysis. JAMA Ophthalmol 131(11):1427–1434
47. Bonnefont-Rousselot D (2000) Antioxidant and anti-AGE therapeutics: evaluation and perspectives. J Soc Biol 195(4):391–398
48. Babizhayev MA, Deyev AI, Yermakova VN, Semiletov YA, Davydova NG, Doroshenko VS, Zhukotskii AV, Goldman IM (2002) Efficacy of *N*-acetylcarnosine in the treatment of cataracts. Drugs R&D 3(2):87–103
49. Stankiewicz A, Poppe E, Stasiewicz B, Gołebiowska-Hrycukowa A (1990) Evaluation of the effectiveness of Quinax in the prevention of the development of senile cataract. Klinika oczna 92(3–4):52–54
50. Ito Y et al (1999) Correlation between prevention of cataract development by disulfiram and fates of selenium in selenium-treated rats. Curr Eye Res 18(4):292–299
51. Zhang J, Guan P, Wang T, Chang D, Jiang T, Wang S (2009) Freeze-dried liposomes as potential carriers for ocular administration of cytochrome c against selenite cataract formation. J Pharm Pharmacol 61(9):1171–1178
52. Grama CN, Suryanarayana P, Patil MA, Raghu G, Balakrishna N, Kumar MNVR, Reddy GB (2013) Efficacy of biodegradable curcumin nanoparticles in delaying cataract in diabetic rat model. PLoS One 8(10), e78217
53. Sunkireddy P, Jha SN, Kanwar JR, Yadav SC (2013) Natural antioxidant biomolecules promises future nanomedicine based therapy for cataract. Colloids Surf B Biointerfaces 112:554–562

54. Ito Y, Nagai N, Cai H, Takeda M, Koizumi Y (2006) Preventive effect of eye drops of liposomes containing disulfiram and cefmetazole on selenite-induced cataract in rat pups. J Oleo Sci 55(1):15–22
55. Hazare SA (2010) Studies in the development of novel carrier systems for enzymes. University of Mumbai, Mumbai
56. Ramos D, Carretero A, Navarro M, Mendes-Jorge L, Rodriguez-Baeza A, Nacher V, Ruberte J (2014) Mouse models of diabetic retinopathy. Drug Discov Today Dis Models. doi:10.1016/j.ddmod.2014.02.002
57. Macleod S, Forrester JV (2002) Diabetic retinopathy. Medicine 30(2):41–44. doi:10.1383/medc.30.2.41.28272
58. Scanlon PH (2010) Diabetic retinopathy. Medicine 38(12):656–660. doi:10.1016/j.mpmed.2010.08.010
59. Alghadyan AA (2011) Diabetic retinopathy – an update. Saudi J Ophthalmol 25(2):99–111. doi:10.1016/j.sjopt.2011.01.009
60. Fante RJ, Durairaj VD, Oliver SC (2010) Diabetic retinopathy: an update on treatment. Am J Med 123(3):213–216. doi:10.1016/j.amjmed.2009.09.020
61. National Eye Institute (2003) Diabetic retinopathy: what you should know. National Eye Institute, Bethesda
62. Laatikainen L (1977) Preliminary report on effect of retinal panphotocoagulation on rubeosis iridis and neovascular glaucoma. Br J Ophthalmol 61(4):278–284
63. Turner R, Holman R (1995) Lessons from UK prospective diabetes study. Diabetes Res Clin Pract 28:S151–S157
64. Cunha-Vaz J (1978) Pathophysiology of diabetic retinopathy. Br J Ophthalmol 62(6):351–355
65. Tarr JM, Kaul K, Chopra M, Kohner EM, Chibber R (2013) Pathophysiology of diabetic retinopathy. ISRN Ophthalmol. http://dx.doi.org/10.1155/2013/343560
66. Dagher Z, Park YS, Asnaghi V, Hoehn T, Gerhardinger C, Lorenzi M (2004) Studies of rat and human retinas predict a role for the polyol pathway in human diabetic retinopathy. Diabetes 53(9):2404–2411
67. Lorenzi M (2007) The polyol pathway as a mechanism for diabetic retinopathy: attractive, elusive, and resilient. Exp Diabetes Res 2007:61038. doi: 10.1155/2007/61038
68. Van den Enden MK, Nyengaard JR, Ostrow E, Burgan JH, Williamson JR (1995) Elevated glucose levels increase retinal glycolysis and sorbitol pathway metabolism. Implications for diabetic retinopathy. Invest Ophthalmol Vis Sci 36(8):1675–1685
69. Hammes H-P, Du X, Edelstein D, Taguchi T, Matsumura T, Ju Q, Lin J, Bierhaus A, Nawroth P, Hannak D (2003) Benfotiamine blocks three major pathways of hyperglycemic damage and prevents experimental diabetic retinopathy. Nat Med 9(3):294–299
70. Kern TS, Kowluru RA, Engerman RL (1994) Abnormalities of retinal metabolism in diabetes or galactosemia: ATPases and glutathione. Invest Ophthalmol Vis Sci 35(7):2962–2967
71. Robison W, Nagata M, Laver N, Hohman T, Kinoshita J (1989) Diabetic-like retinopathy in rats prevented with an aldose reductase inhibitor. Invest Ophthalmol Vis Sci 30(11):2285–2292
72. Hotta N, Akanuma Y, Kawamori R, Matsuoka K, Oka Y, Shichiri M, Toyota T, Nakashima M, Yoshimura I, Sakamoto N (2006) Long-term clinical effects of epalrestat, an aldose reductase inhibitor, on diabetic peripheral neuropathy the 3-year, multicenter, comparative aldose reductase inhibitor-diabetes complications trial. Diabetes Care 29(7):1538–1544
73. Sun W, Oates PJ, Coutcher JB, Gerhardinger C, Lorenzi M (2006) A selective aldose reductase inhibitor of a new structural class prevents or reverses early retinal abnormalities in experimental diabetic retinopathy. Diabetes 55(10):2757–2762
74. Amano S, S-i Y, Kato N, Inagaki Y, Okamoto T, Makino M, Taniko K, Hirooka H, Jomori T, Takeuchi M (2002) Sorbitol dehydrogenase overexpression potentiates glucose toxicity to cultured retinal pericytes. Biochem Biophys Res Commun 299(2):183–188
75. Stitt AW (2003) The role of advanced glycation in the pathogenesis of diabetic retinopathy. Exp Mol Pathol 75(1):95–108

76. Stitt AW, Li YM, Gardiner TA, Bucala R, Archer DB, Vlassara H (1997) Advanced glycation end products (AGEs) co-localize with AGE receptors in the retinal vasculature of diabetic and of AGE-infused rats. Am J Pathol 150(2):523
77. Grossin N, Wautier MP, Mes T, Guillausseau PJ et al (2008) Severity of diabetic microvascular complications is associated with low soluble RAGE level. Diabetes Metab 34:392–395
78. Zong H, Ward M, Stitt AW (2011) AGEs, RAGE, and diabetic retinopathy. Curr Diab Rep 11(4):244–252
79. Ahmed N, Thornalley P (2007) Advanced glycation end products: what is their relevance to diabetic complications? Diabetes Obes Metab 9(3):233–245
80. Thallas-Bonke V, Lindschau C, Rizkalla B, Bach LA, Boner G, Meier M, Haller H, Cooper ME, Forbes JM (2004) Attenuation of extracellular matrix accumulation in diabetic nephropathy by the advanced glycation end product cross-link breaker ALT-711 via a protein kinase C-α-dependent pathway. Diabetes 53(11):2921–2930
81. Koya D, King GL (1998) Protein kinase C activation and the development of diabetic complications. Diabetes 47(6):859–866
82. Aiello LP, Bursell S-E, Clermont A, Duh E, Ishii H, Takagi C, Mori F, Ciulla TA, Ways K, Jirousek M (1997) Vascular endothelial growth factor–induced retinal permeability is mediated by protein kinase C in vivo and suppressed by an orally effective β-isoform–selective inhibitor. Diabetes 46(9):1473–1480
83. Aiello LP, Clermont A, Arora V, Davis MD, Sheetz MJ, Bursell S-E (2006) Inhibition of PKC β by oral administration of ruboxistaurin is well tolerated and ameliorates diabetes-induced retinal hemodynamic abnormalities in patients. Invest Ophthalmol Vis Sci 47(1):86–92
84. Strøm C, Sander B, Klemp K, Aiello LP, Lund-Andersen H, Larsen M (2005) Effect of ruboxistaurin on blood–retinal barrier permeability in relation to severity of leakage in diabetic macular edema. Invest Ophthalmol Vis Sci 46(10):3855–3858
85. Klein BE, Knudtson MD, Tsai MY, Klein R (2009) The relation of markers of inflammation and endothelial dysfunction to the prevalence and progression of diabetic retinopathy: Wisconsin epidemiologic study of diabetic retinopathy. Arch Ophthalmol 127(9): 1175–1182
86. Tang J, Kern TS (2011) Inflammation in diabetic retinopathy. Prog Retin Eye Res 30(5):343–358. doi:10.1016/j.preteyeres.2011.05.002
87. Chibber R, Ben-Mahmud BM, Mann GE, Zhang JJ, Kohner EM (2003) Protein kinase C β2-dependent phosphorylation of core 2 GlcNAc-T promotes leukocyte-endothelial cell adhesion a mechanism underlying capillary occlusion in diabetic retinopathy. Diabetes 52(6):1519–1527
88. Gillies MC, Sutter FK, Simpson JM, Larsson J, Ali H, Zhu M (2006) Intravitreal triamcinolone for refractory diabetic macular edema: two-year results of a double-masked, placebo-controlled, randomized clinical trial. Ophthalmology 113(9):1533–1538
89. Kuppermann BD, Blumenkranz MS, Haller JA, Williams GA, Weinberg DV, Chou C, Whitcup SM (2007) Randomized controlled study of an intravitreous dexamethasone drug delivery system in patients with persistent macular edema. Arch Ophthalmol 125(3):309–317
90. Baynes JW, Thorpe SR (1999) Role of oxidative stress in diabetic complications: a new perspective on an old paradigm. Diabetes 48(1):1–9
91. Mohammad G, Kowluru RA (2011) Novel role of mitochondrial matrix metalloproteinase-2 in the development of diabetic retinopathy. Invest Ophthalmol Vis Sci 52(6):3832–3841
92. Kowluru RA (2003) Effect of reinstitution of good glycemic control on retinal oxidative stress and nitrative stress in diabetic rats. Diabetes 52(3):818–823
93. Haskins K, Bradley B, Powers K, Fadok V, Flores S, Ling X, Pugazhenthi S, Reusch J, Kench J (2003) Oxidative stress in type 1 diabetes. Ann N Y Acad Sci 1005(1):43–54
94. Hueber A, Wiedemann P, Esser P, Heimann K (1997) Basic fibroblast growth factor mRNA, bFGF peptide and FGF receptor in epiretinal membranes of intraocular proliferative disorders (PVR and PDR). Int Ophthalmol 20(6):345–350
95. Haurigot V, Villacampa P, Ribera A, Llombart C, Bosch A, Nacher V, Ramos D, Ayuso E, Segovia JC, Bueren JA (2009) Increased intraocular insulin-like growth factor-I triggers blood-retinal barrier breakdown. J Biol Chem 284(34):22961–22969

96. Rangasamy S, Srinivasan R, Maestas J, McGuire PG, Das A (2011) A potential role for angiopoietin 2 in the regulation of the blood–retinal barrier in diabetic retinopathy. Invest Ophthalmol Vis Sci 52(6):3784–3791
97. Brooks HL, Caballero S, Newell CK, Steinmetz RL, Watson D, Segal MS, Harrison JK, Scott EW, Grant MB (2004) Vitreous levels of vascular endothelial growth factor and stromal-derived factor 1 in patients with diabetic retinopathy and cystoid macular edema before and after intraocular injection of triamcinolone. Arch Ophthalmol 122(12):1801–1807
98. Lev-Ran A, Hwang DL, Miller JD, Josefsberg Z (1990) Excretion of epidermal growth factor (EGF) in diabetes. Clin Chim Acta 192(3):201–206
99. Ie D, Gordon LW, Glaser BM, Pena RA (1994) Transforming growth factor-beta 2 levels increase following retinal laser photocoagulation. Curr Eye Res 13(10):743–746
100. Praidou A, Klangas I, Papakonstantinou E, Androudi S, Georgiadis N, Karakiulakis G, Dimitrakos S (2009) Vitreous and serum levels of platelet-derived growth factor and their correlation in patients with proliferative diabetic retinopathy. Curr Eye Res 34(2):152–161
101. Eckardt K-U (2009) Erythropoietin and microvascular diabetic complications. Nephrol Dial Transplant 24(2):388–390
102. Awata T, Inoue K, Kurihara S, Ohkubo T, Watanabe M, Inukai K, Inoue I, Katayama S (2002) A common polymorphism in the 5′-untranslated region of the VEGF gene is associated with diabetic retinopathy in type 2 diabetes. Diabetes 51(5):1635–1639
103. Boulton M, Foreman D, Williams G, McLeod D (1998) VEGF localisation in diabetic retinopathy. Br J Ophthalmol 82(5):561–568
104. Joussen AM, Poulaki V, Qin W, Kirchhof B, Mitsiades N, Wiegand SJ, Rudge J, Yancopoulos GD, Adamis AP (2002) Retinal vascular endothelial growth factor induces intercellular adhesion molecule-1 and endothelial nitric oxide synthase expression and initiates early diabetic retinal leukocyte adhesion in vivo. Am J Pathol 160(2):501–509
105. Ray D, Mishra M, Ralph S, Read I, Davies R, Brenchley P (2004) Association of the VEGF gene with proliferative diabetic retinopathy but not proteinuria in diabetes. Diabetes 53(3):861–864
106. Park Y, Freedman B, Lee E, Park S, Jameson J (2003) A dominant negative PPARγ mutant shows altered cofactor recruitment and inhibits adipogenesis in 3T3-L1 cells. Diabetologia 46(3):365–377
107. Estacio R, Jeffers B, Gifford N, Schrier R (2000) Effect of blood pressure control on diabetic microvascular complications in patients with hypertension and type 2 diabetes. Diabetes Care 23:B54–B64
108. Rassam S, Patel V, Kohner E (1995) The effect of experimental hypertension on retinal vascular autoregulation in humans: a mechanism for the progression of diabetic retinopathy. Exp Physiol 80(1):53–68
109. Suzuma I, Hata Y, Clermont A, Pokras F, Rook SL, Suzuma K, Feener EP, Aiello LP (2001) Cyclic stretch and hypertension induce retinal expression of vascular endothelial growth factor and vascular endothelial growth factor receptor-2 potential mechanisms for exacerbation of diabetic retinopathy by hypertension. Diabetes 50(2):444–454
110. DPS Group (2005) The DIabetic REtinopathy Candesartan Trials (DIRECT) programme: baseline characteristics. J Renin Angiotensin Aldosterone Syst 6(1):25–32
111. Corneli HM, Zorc JJ, Mahajan P, Shaw KN, Holubkov R, Reeves SD, Ruddy RM, Malik B, Nelson KA, Bregstein JS (2007) A multicenter, randomized, controlled trial of dexamethasone for bronchiolitis. New Engl J Med 357(4):331–339
112. Makadia HK, Siegel SJ (2011) Poly lactic-co-glycolic acid (PLGA) as biodegradable controlled drug delivery carrier. Polymers 3(3):1377–1397
113. Boyer DS, Faber D, Gupta S, Patel SS, Tabandeh H, Li X-Y, Liu CC, Lou J, Whitcup SM, OCS Group (2011) Dexamethasone intravitreal implant for treatment of diabetic macular edema in vitrectomized patients. Retina 31(5):915–923
114. Pacella E, Vestri AR, Muscella R, Carbotti MR, Castellucci M, Coi L, Turchetti P, Pacella F (2013) Preliminary results of an intravitreal dexamethasone implant (Ozurdex®) in patients with persistent diabetic macular edema. Clin Ophthalmol 7:1423

115. Pearson P, Levy B, Comstock T, Group FAIS (2006) Fluocinolone acetonide intravitreal implant to treat diabetic macular edema: 3-year results of a multi-center clinical trial. Invest Ophthalmol Vis Sci 47(5):5442
116. Pearson PA, Comstock TL, Ip M, Callanan D, Morse LS, Ashton P, Levy B, Mann ES, Eliott D (2011) Fluocinolone acetonide intravitreal implant for diabetic macular edema: a 3-year multicenter, randomized, controlled clinical trial. Ophthalmology 118(8):1580–1587
117. Campochiaro PA, Brown DM, Pearson A, Ciulla T, Boyer D, Holz FG, Tolentino M, Gupta A, Duarte L, Madreperla S (2011) Long-term benefit of sustained-delivery fluocinolone acetonide vitreous inserts for diabetic macular edema. Ophthalmology 118(4):626–635. e622
118. Chew EY, Kim J, Coleman HR, Aiello LP, Fish G, Ip M, Haller JA, Figueroa M, Martin D, Callanan D (2010) Preliminary assessment of celecoxib and microdiode pulse laser treatment of diabetic macular edema. Retina 30(3):459
119. Ayalasomayajula SP, Kompella UB (2005) Subconjunctivally administered celecoxib-PLGA microparticles sustain retinal drug levels and alleviate diabetes-induced oxidative stress in a rat model. Eur J Pharmacol 511(2):191–198
120. Felinski EA, Antonetti DA (2005) Glucocorticoid regulation of endothelial cell tight junction gene expression: novel treatments for diabetic retinopathy. Curr Eye Res 30(11):949–957
121. Kompella UB, Bandi N, Ayalasomayajula SP (2003) Subconjunctival nano-and microparticles sustain retinal delivery of budesonide, a corticosteroid capable of inhibiting VEGF expression. Invest Ophthalmol Vis Sci 44(3):1192–1201
122. Prasad PS, Schwartz SD, Hubschman J-P (2010) Age-related macular degeneration: current and novel therapies. Maturitas 66(1):46–50
123. Swaroop A, Branham KEH, Chen W, Abecasis G (2007) Genetic susceptibility to age-related macular degeneration: a paradigm for dissecting complex disease traits. Hum Mol Genet 16(R2):R174–R182
124. Friedman DS, O'Colmain BJ, Munoz B, Tomany SC, McCarty C, De Jong PT, Nemesure B, Mitchell P, Kempen J (2004) Prevalence of age-related macular degeneration in the United States. Arch Ophthalmol 122(4):564–572
125. Kowluru RA, Zhong Q (2011) Beyond AREDS: is there a place for antioxidant therapy in the prevention/treatment of eye disease? Invest Ophthalmol Vis Sci 52(12):8665–8671
126. Evans JR (2001) Risk factors for age-related macular degeneration. Prog Retin Eye Res 20(2):227–253
127. Seddon JM, Reynolds R, Rosner B (2010) Associations of smoking, body mass index, dietary lutein, and the LIPC gene variant rs10468017 with advanced age-related macular degeneration. Mol Vis 16:2412
128. Vingerling JR, Hofman A, Grobbee DE, De Jong PTVM (1996) Age-related macular degeneration and smoking: the Rotterdam study. Arch Ophthalmol 114(10):1193–1196
129. Chakravarthy U, Augood C, Bentham GC, de Jong P, Rahu M, Seland J, Soubrane G, Tomazzoli L, Topouzis F, Vingerling JR (2007) Cigarette smoking and age-related macular degeneration in the EUREYE Study. Ophthalmology 114(6):1157–1163
130. Seddon JM, George S, Rosner B, Klein ML (2006) CFH gene variant, Y402H, and smoking, body mass index, environmental associations with advanced age-related macular degeneration. Hum Hered 61(3):157–165
131. Francis PJ, George S, Schultz DW, Rosner B, Hamon S, Ott J, Weleber RG, Klein ML, Seddon JM (2007) The LOC387715 gene, smoking, body mass index, environmental associations with advanced age-related macular degeneration. Hum Hered 63(3–4):212–218
132. Schmidt S, Hauser MA, Scott WK, Postel EA, Agarwal A, Gallins P, Wong F, Chen YS, Spencer K, Schnetz-Boutaud N (2006) Cigarette smoking strongly modifies the association of LOC387715 and age-related macular degeneration. Am J Hum Genet 78(5):852–864
133. Algvere PV, Marshall J, Seregard S (2006) Age related maculopathy and the impact of blue light hazard. Acta Ophthalmol Scand 84(1):4–15
134. Cruickshanks KJ, Klein R, Klein BEK (1993) Sunlight and age-related macular degeneration: the Beaver Dam eye study. Arch Ophthalmol 111(4):514–518

135. Darzins P, Mitchell P, Heller RF (1997) Sun exposure and age-related macular degeneration: an Australian case-control study. Ophthalmology 104(5):770–776
136. Delcourt C, Carriere I, Ponton-Sanchez A, Fourrey S, Lacroux A, Papoz L (2001) Light exposure and the risk of age-related macular degeneration: the Pathologies Oculaires Liees a l'Age (POLA) study. Arch Ophthalmol 119(10):1463–1468
137. McCarty CA, Mukesh BN, Fu CL, Mitchell P, Wang JJ, Taylor HR (2001) Risk factors for age-related maculopathy: the visual impairment project. Arch Ophthalmol 119(10):1455–1462
138. Pham TQ, Rochtchina E, Mitchell P, Smith W, Wang JJ (2009) Sunlight-related factors and the 10-year incidence of age-related maculopathy. Ophthalmic Epidemiol 16(2):136–141
139. Taylor HR, West S, Muñoz B, Rosenthal FS, Bressler SB, Bressler NM (1992) The long-term effects of visible light on the eye. Archives Ophthalmol 110(1):99–04
140. Tomany SC, Cruickshanks KJ, Klein R, Klein BEK, Knudtson MD (2004) Sunlight and the 10-year incidence of age-related maculopathy: the Beaver Dam eye study. Arch Ophthalmol 122(5):750–757
141. Whitehead AJ, Mares JA, Danis RP (2006) Macular pigment: a review of current knowledge. Arch Ophthalmol 124(7):1038–1045
142. Fletcher AE, Bentham GC, Agnew M, Young IS, Augood C, Chakravarthy U, de Jong PTVM, Rahu M, Seland J, Soubrane G (2008) Sunlight exposure, antioxidants, and age-related macular degeneration. Arch Ophthalmol 126(10):1396–1403
143. Johnson EJ (2005) Obesity, lutein metabolism, and age-related macular degeneration: a web of connections. Nutr Rev 63(1):9–15
144. Klein R, Klein BEK, Tomany SC, Cruickshanks KJ (2003) The association of cardiovascular disease with the long-term incidence of age-related maculopathy: the Beaver Dam eye study. Ophthalmology 110(4):636–643
145. Krishnaiah S, Das T, Nirmalan PK, Nutheti R, Shamanna BR, Rao GN, Thomas R (2005) Risk factors for age-related macular degeneration: findings from the Andhra Pradesh eye disease study in South India. Invest Ophthalmol Vis Sci 46(12):4442–4449
146. Cugati S, Mitchell P, Rochtchina E, Tan AG, Smith W, Wang JJ (2006) Cataract surgery and the 10-year incidence of age-related maculopathy: the Blue Mountains eye study. Ophthalmology 113(11):2020–2025
147. Ho L, Boekhoorn SS, van Duijn CM, Uitterlinden AG, Hofman A, de Jong PTVM, Stijnen T, Vingerling JR (2008) Cataract surgery and the risk of aging macula disorder: the Rotterdam study. Invest Ophthalmol Vis Sci 49(11):4795–4800
148. Kaiserman I, Kaiserman N, Elhayany A, Vinker S (2007) Cataract surgery is associated with a higher rate of photodynamic therapy for age-related macular degeneration. Ophthalmology 114(2):278–282
149. Klein R, Klein BEK, Wong TY, Tomany SC, Cruickshanks KJ (2002) The association of cataract and cataract surgery with the long-term incidence of age-related maculopathy: the Beaver Dam eye study. Arch Ophthalmol 120(11):1551–1558
150. Pham TQ, Cugati S, Rochtchina E, Mitchell P, Maloof A, Wang JJ (2006) Early age-related maculopathy in eyes after cataract surgery. Eye 21(4):512–517
151. Pollack A, Marcovich A, Bukelman A, Oliver M (1996) Age-related macular degeneration after extracapsular cataract extraction with intraocular lens implantation. Ophthalmology 103(10):1546–1554
152. Wang JJ, Klein R, Smith W, Klein BEK, Tomany S, Mitchell P (2003) Cataract surgery and the 5-year incidence of late-stage age-related maculopathy: pooled findings from the Beaver Dam and Blue Mountains eye studies. Ophthalmology 110(10):1960–1967
153. Jarrett SG, Boulton ME (2012) Consequences of oxidative stress in age-related macular degeneration. Mol Aspects Med 33(4):399
154. Group A-REDSR (2001) A randomized, placebo-controlled, clinical trial of high-dose supplementation with vitamins C and E, beta carotene, and zinc for age-related macular degeneration and vision loss: AREDS report no 8. Arch Ophthalmol 119(10):1417

155. Macular Photocoagulation Study Group (1991) Argon laser photocoagulation for neovascular maculopathy: five-year results from randomized clinical trials. Arch Ophthalmol 109(8):1109
156. Bressler NM (2002) Verteporfin therapy of subfoveal choroidal neovascularization in age-related macular degeneration: two-year results of a randomized clinical trial including lesions with occult with no classic choroidal neovascularization-verteporfin in photodynamic therapy report 2. Am J Ophthalmol 133(1):168–169
157. Treatment of Age-Related Macular Degeneration with Photodynamic Therapy Study Group (1999) Photodynamic therapy of subfoveal choroidal neovascularization in age-related macular degeneration with verteporfin: one-year results of 2 randomized clinical trials – TAP report 1. Arch Ophthalmol 117(10):1329
158. Brown DM, Kaiser PK, Michels M, Soubrane G, Heier JS, Kim RY, Sy JP, Schneider S (2006) Ranibizumab versus verteporfin for neovascular age-related macular degeneration. New Engl J Med 355(14):1432–1444
159. Rosenfeld PJ, Brown DM, Heier JS, Boyer DS, Kaiser PK, Chung CY, Kim RY (2006) Ranibizumab for neovascular age-related macular degeneration. New Engl J Med 355(14):1419–1431
160. Kabbinavar F, Hurwitz HI, Fehrenbacher L, Meropol NJ, Novotny WF, Lieberman G, Griffing S, Bergsland E (2003) Phase II, randomized trial comparing bevacizumab plus fluorouracil (FU)/leucovorin (LV) with FU/LV alone in patients with metastatic colorectal cancer. J Clin Oncol 21(1):60–65
161. Fong DS, Custis P, Howes J, Hsu J-W (2010) Intravitreal bevacizumab and ranibizumab for age-related macular degeneration: a multicenter, retrospective study. Ophthalmology 117(2):298–302
162. Landa G, Amde W, Doshi V, Ali A, McGevna L, Gentile RC, Muldoon TO, Walsh JB, Rosen RB (2009) Comparative study of intravitreal bevacizumab (Avastin) versus ranibizumab (Lucentis) in the treatment of neovascular age-related macular degeneration. Ophthalmologica 223(6):370–375
163. Thomas L, Berenberg MD, Ying GS et al (2012) The association between Drusen extent and foveolar choroidal blood flow in AMD. Retina 32(1):25–31
164. Abrishami M, Ghanavati SZ, Soroush D, Rouhbakhsh M, Jaafari MR, Malaekeh-Nikouei B (2009) Preparation, characterization, and in vivo evaluation of nanoliposomes-encapsulated bevacizumab (avastin) for intravitreal administration. Retina 29(5):699–703
165. Honda M, Asai T, Umemoto T, Araki Y, Oku N, Tanaka M (2011) Suppression of choroidal neovascularization by intravitreal injection of liposomal SU5416. Arch Ophthalmol 129(3):317–321
166. Katanasaka Y, Ida T, Asai T, Shimizu K, Koizumi F, Maeda N, Baba K, Oku N (2008) Antiangiogenic cancer therapy using tumor vasculature-targeted liposomes encapsulating 3-(3, 5-dimethyl-1H-pyrrol-2-ylmethylene)-1, 3-dihydro-indol-2-one, SU5416. Cancer Lett 270(2):260–268
167. Wang C-H, Lu D-W, Chiang C-H (2010) Gene therapy using SiRNA for treatment of ocular neovascularization. J Med Sci 30(3):79–84
168. Li F, Hurley B, Liu Y, Leonard B, Griffith M (2012) Controlled release of bevacizumab through nanospheres for extended treatment of age-related macular degeneration. Open Ophthalmol J 6:54
169. Kadam RS, Tyagi P, Edelhauser HF, Kompella UB (2012) Influence of choroidal neovascularization and biodegradable polymeric particle size on transscleral sustained delivery of triamcinolone acetonide. Int J Pharm 434(1):140–147
170. Suen W-LL, Chau Y (2013) Specific uptake of folate-decorated triamcinolone-encapsulating nanoparticles by retinal pigment epithelium cells enhances and prolongs antiangiogenic activity. J Control Release 167(1):21–28
171. Mo Y, Barnett ME, Takemoto D, Davidson H, Kompella UB (2007) Human serum albumin nanoparticles for efficient delivery of Cu, Zn superoxide dismutase gene. Mol Vis 13:746

172. Jayaraman MS, Bharali DJ, Sudha T, Mousa SA (2012) Nano chitosan peptide as a potential therapeutic carrier for retinal delivery to treat age-related macular degeneration. Mol Vis 18:2300
173. Iezzi R, Guru BR, Glybina IV, Mishra MK, Kennedy A, Kannan RM (2012) Dendrimer-based targeted intravitreal therapy for sustained attenuation of neuroinflammation in retinal degeneration. Biomaterials 33(3):979–988
174. Molokhia SA, Sant H, Simonis J, Bishop CJ, Burr RM, Gale BK, Ambati BK (2010) The capsule drug device: novel approach for drug delivery to the eye. Vision Res 50(7):680–685
175. Geltzer A, Turalba A, Vedula SS (2007) Surgical implantation of steroids with antiangiogenic characteristics for treating neovascular age-related macular degeneration. Cochrane Database Syst Rev 4, 2013 Jan 31;1:CD005022. doi: 10.1002/14651858.CD005022.pub3
176. Cantor LB (2006) Brimonidine in the treatment of glaucoma and ocular hypertension. Ther Clin Risk Manag 2(4):337
177. Pickering MC, Cook HT, Warren J, Bygrave AE, Moss J, Walport MJ, Botto M (2002) Uncontrolled C3 activation causes membranoproliferative glomerulonephritis in mice deficient in complement factor H. Nat Genet 31(4):424–428
178. Nozaki M, Raisler BJ, Sakurai E, Sarma JV, Barnum SR, Lambris JD, Chen Y, Zhang K, Ambati BK, Baffi JZ (2006) Drusen complement components C3a and C5a promote choroidal neovascularization. Proc Natl Acad Sci 103(7):2328–2333
179. Johnson LV, Ozaki S, Staples MK, Erickson PA, Anderson DH (2000) A potential role for immune complex pathogenesis in drusen formation. Exp Eye Res 70(4):441–449
180. Mullins RF, Russell SR, Anderson DH, Hageman GS (2000) Drusen associated with aging and age-related macular degeneration contain proteins common to extracellular deposits associated with atherosclerosis, elastosis, amyloidosis, and dense deposit disease. FASEB J 14(7):835–846
181. Hageman GS, Anderson DH, Johnson LV, Hancox LS, Taiber AJ, Hardisty LI, Hageman JL, Stockman HA, Borchardt JD, Gehrs KM (2005) A common haplotype in the complement regulatory gene factor H (HF1/CFH) predisposes individuals to age-related macular degeneration. Proc Natl Acad Sci 102(20):7227–7232
182. Johnson LV, Leitner WP, Staples MK, Anderson DH (2001) Complement activation and inflammatory processes in drusen formation and age related macular degeneration. Exp Eye Res 73(6):887–896
183. Coffey PJ, Gias C, McDermott CJ, Lundh P, Pickering MC, Sethi C, Bird A, Fitzke FW, Maass A, Chen LL (2007) Complement factor H deficiency in aged mice causes retinal abnormalities and visual dysfunction. Proc Natl Acad Sci 104(42):16651–16656
184. Ufret-Vincenty RL, Aredo B, Liu X, McMahon A, Chen PW, Sun H, Niederkorn JY, Kedzierski W (2010) Transgenic mice expressing variants of complement factor H develop AMD-like retinal findings. Invest Ophthalmol Vis Sci 51(11):5878–5887
185. Cashman SM, Desai A, Ramo K, Kumar-Singh R (2011) Expression of complement component 3 (C3) from an adenovirus leads to pathology in the murine retina. Invest Ophthalmol Vis Sci 52(6):3436–3445
186. Mélik-Parsadaniantz S, Rostène W (2008) Chemokines and neuromodulation. J Neuroimmunol 198(1):62–68
187. Hollyfield JG, Bonilha VL, Rayborn ME, Yang X, Shadrach KG, Lu L, Ufret RL, Salomon RG, Perez VL (2008) Oxidative damage-induced inflammation initiates age-related macular degeneration. Nat Med 14(2):194–198
188. Hollyfield JG, Perez VL, Salomon RG (2010) A hapten generated from an oxidation fragment of docosahexaenoic acid is sufficient to initiate age-related macular degeneration. Mol Neurobiol 41(2–3):290–298
189. Hahn P, Qian Y, Dentchev T, Chen L, Beard J, Harris ZL, Dunaief JL (2004) Disruption of ceruloplasmin and hephaestin in mice causes retinal iron overload and retinal degeneration with features of age-related macular degeneration. Proc Natl Acad Sci 101(38):13850–13855

190. Crabb JW, Miyagi M, Gu X, Shadrach K, West KA, Sakaguchi H, Kamei M, Hasan A, Yan L, Rayborn ME (2002) Drusen proteome analysis: an approach to the etiology of age-related macular degeneration. Proc Natl Acad Sci 99(23):14682–14687
191. Imamura Y, Noda S, Hashizume K, Shinoda K, Yamaguchi M, Uchiyama S, Shimizu T, Mizushima Y, Shirasawa T, Tsubota K (2006) Drusen, choroidal neovascularization, and retinal pigment epithelium dysfunction in SOD1-deficient mice: a model of age-related macular degeneration. Proc Natl Acad Sci 103(30):11282–11287
192. Justilien V, Pang J-J, Renganathan K, Zhan X, Crabb JW, Kim SR, Sparrow JR, Hauswirth WW, Lewin AS (2007) SOD2 knockdown mouse model of early AMD. Invest Ophthalmol Vis Sci 48(10):4407–4420
193. Pennesi ME, Neuringer M, Courtney RJ (2012) Animal models of age related macular degeneration. Mol Aspects Med 33(4):487–509
194. Majji AB, Cao J, Chang KY, Hayashi A, Aggarwal S, Grebe RR, de Juan E (2000) Age-related retinal pigment epithelium and Bruch's membrane degeneration in senescence-accelerated mouse. Invest Ophthalmol Vis Sci 41(12):3936–3942
195. Weber BHF, Lin B, White K, Kohler K, Soboleva G, Herterich S, Seeliger MW, Jaissle GB, Grimm C, Reme C (2002) A mouse model for Sorsby fundus dystrophy. Invest Ophthalmol Vis Sci 43(8):2732–2740
196. Mata NL, Weng J, Travis GH (2000) Biosynthesis of a major lipofuscin fluorophore in mice and humans with ABCR-mediated retinal and macular degeneration. Proc Natl Acad Sci 97(13):7154–7159
197. Karan G, Lillo C, Yang Z, Cameron DJ, Locke KG, Zhao Y, Thirumalaichary S, Li C, Birch DG, Vollmer-Snarr HR (2005) Lipofuscin accumulation, abnormal electrophysiology, and photoreceptor degeneration in mutant ELOVL4 transgenic mice: a model for macular degeneration. Proc Natl Acad Sci 102(11):4164–4169
198. Marmorstein LY, McLaughlin PJ, Peachey NS, Sasaki T, Marmorstein AD (2007) Formation and progression of sub-retinal pigment epithelium deposits in Efemp1 mutation knock-in mice: a model for the early pathogenic course of macular degeneration. Hum Mol Genet 16(20):2423–2432
199. Dobi ET, Puliafito CA, Destro M (1989) A new model of experimental choroidal neovascularization in the rat. Arch Ophthalmol 107(2):264–269
200. Ryan S (1979) The development of an experimental model of subretinal neovascularization in disciform macular degeneration. Trans Am Ophthalmol Soc 77:707
201. Shen D, Wen R, Tuo J, Bojanowski CM, Chan CC (2006) Exacerbation of retinal degeneration and choroidal neovascularization induced by subretinal injection of Matrigel in CCL2/MCP-1-deficient mice. Ophthalmic Res 38(2):71–73
202. Tobe T, Ortega S, Luna JD, Ozaki H, Okamoto N, Derevjanik NL, Vinores SA, Basilico C, Campochiaro PA (1998) Targeted disruption of the FGF2 gene does not prevent choroidal neovascularization in a murine model. Am J Pathol 153(5):1641–1646
203. Qiu G, Stewart JM, Sadda S, Freda R, Lee S, Guven D (2006) A new model of experimental subretinal neovascularization in the rabbit. Exp Eye Res 83(1):141–152
204. Baffi J, Byrnes G, Chan CC, Csaky KG (2000) Choroidal neovascularization in the rat induced by adenovirus mediated expression of vascular endothelial growth factor. Invest Ophthalmol Vis Sci 41(11):3582–3589
205. Grossniklaus HE, Ling JX, Wallace TM, Dithmar S, Lawson DH, Cohen C, Elner VM, Elner SG, Sternberg P Jr (2002) Macrophage and retinal pigment epithelium expression of angiogenic cytokines in choroidal neovascularization. Mol Vis 8(8):119–126
206. Tamai K, Spaide RF, Ellis E, Iwabuchi S, Ogura Y, Armstrong D (2002) Lipid hydroperoxide stimulates subretinal choroidal neovascularization in the rabbit. Exp Eye Res 74(2):301–308
207. Lyzogubov VV, Tytarenko RG, Liu J, Bora NS, Bora PS (2011) Polyethylene glycol (PEG)-induced mouse model of choroidal neovascularization. J Biol Chem 286(18):16229–16237
208. Goldberg MF (1976) Bruch's membrane and vascular growth. Invest Ophthalmol Vis Sci 15(6):443–446

209. Kiilgaard JF, Andersen MVN, Wiencke AK, Scherfig E, La Cour M, Tezel TH, Prause JU (2005) A new animal model of choroidal neovascularization. Acta Ophthalmol Scand 83(6):697–704
210. Glaucoma Research Foundation (2014) Types of glaucoma. Glaucoma Research Foundation. Accessed Sept 2014
211. Kumarasamy NA, Lam FS, Wang AL, Theoharides TC (2006) Glaucoma: current and developing concepts for inflammation, pathogenesis and treatment. Eur J Inflamm 4(3):129
212. Lavik E, Kuehn MH, Kwon YH (2011) Novel drug delivery systems for glaucoma. Eye 25(5):578–586
213. Monem AS, Ali FM, Ismail MW (2000) Prolonged effect of liposomes encapsulating pilocarpine HCl in normal and glaucomatous rabbits. Int J Pharm 198(1):29–38
214. De Campos AM, Sánchez A, Gref R, Calvo P, Alonso MJ (2003) The effect of a PEG versus a chitosan coating on the interaction of drug colloidal carriers with the ocular mucosa. Eur J Pharm Sci 20(1):73–81
215. Date RD (1999) Studies in the development of new drug delivery systems- mucoadhesive systems. University of Mumbai, Mumbai
216. Gaikwad DD (2000) New drug delivery systems: mucoadhesive opthalmic formulations. University of Mumbai, Mumbai

Chapter 15
Colloidal Carriers in Ophthalmic Drug Delivery

Deepak Gupta

Abstract Ocular diseases like cataract, glaucoma, retinoblastoma, diabetic retinopathy, retinal degeneration, retinitis pigmentosa, and choroidal neovascularization are the leading causes of blindness all over the world. Drugs for treating eye diseases are limited in their efficacy due to different ocular barriers. Furthermore, most of the drugs show either poor ocular permeation or rapid clearance from the eyes. The use of colloidal carriers such as nanoparticles, dendrimers, and liposomes has shown the potential to overcome these barriers to the usefulness of a drug. Preclinical and a few clinical studies illustrate that these carrier systems are very useful for delivering wide variety of drugs and genes. These carriers have found applications not only in drug delivery but also in imaging and diagnostics. This chapter focuses on some recent developments in these colloidal systems and discusses how drug delivery to both anterior and posterior eye can be optimized by understanding route of delivery and properties of colloidal systems and their suitability based on drug and disease targeted.

Keywords Colloidal carrier • Ophthalmic formulation • Ophthalmic drug delivery • Nanoparticles • Dendrimers • Liposomes • Niosomes • Cubosomes • Micelles • Intraocular delivery • Suprachoroidal delivery • Periocular delivery • Topical delivery • Subconjunctival delivery • Anterior eye • Posterior eye

15.1 Introduction

Ocular drug delivery is an extensive research area for finding therapies to treat various ocular diseases. Some of these diseases are preventable by use of drugs suitable for the disease. Traditional ocular formulations like solutions, suspensions, and ointments are commonly used for treating some of these disorders. However, most

D. Gupta (✉)
Associate Professor of Pharmaceutical Sciences, L|E|C|O|M Bradenton School of Pharmacy, 5000 Lakewood Ranch Blvd, Office#281, Bradenton 34211, FL, USA
e-mail: dgupta@lecom.edu

© Springer International Publishing Switzerland 2016

Y. Pathak et al. (eds.), *Nano-Biomaterials For Ophthalmic Drug Delivery*,
DOI 10.1007/978-3-319-29346-2_15

of these formulations suffer major drawbacks like faster clearance or poor bioavailability due to ocular barriers [1]. This problem becomes more critical for drug delivery to the posterior eye. Development of an ideal colloidal carrier system is an important research area that offers both challenges and opportunities in the field of ocular research. Colloidal carriers are currently being investigated as drug delivery approaches with main focus on the posterior of the eye. Various colloidal systems have shown significant advantages over traditional formulations and will more likely shift eye therapeutics to these innovative and advanced drug delivery systems [2]. Nanoparticles (nanospheres and nanocapsules), dendrimers, liposomes, micelles, niosomes, cubosomes, nanoemulsions, microemulsions, and microparticles are examples of significant ocular colloidal carriers reported in the literature. Figure 15.1 represents some of these colloidal carriers [3].

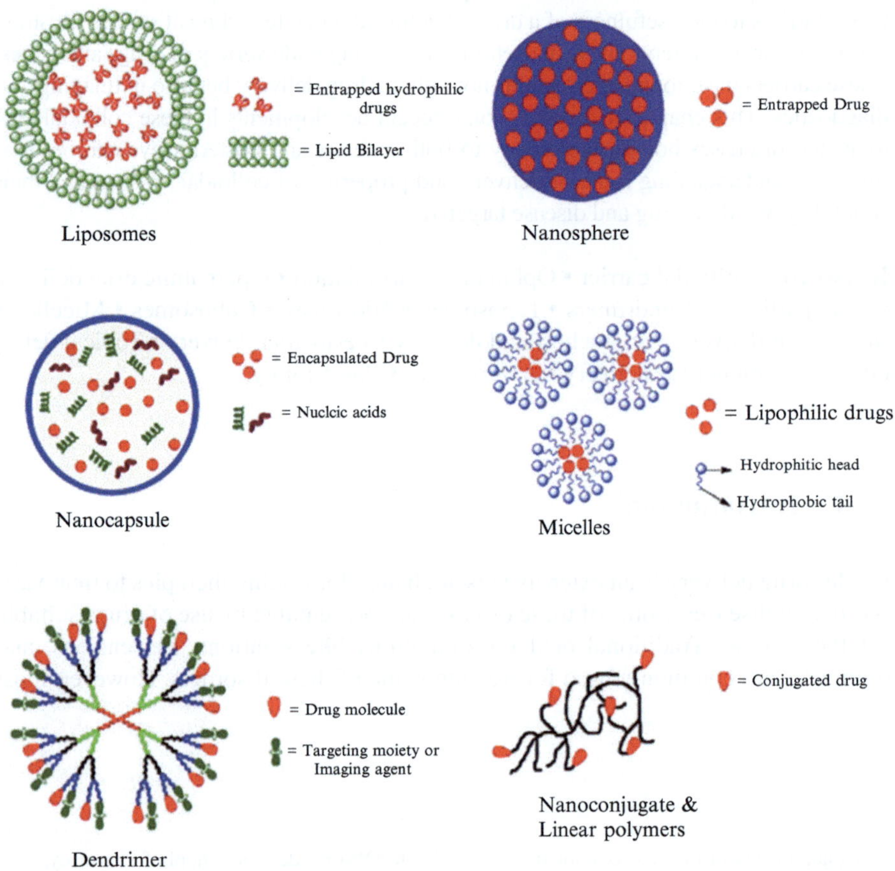

Fig. 15.1 Schematic representation of different colloidal carriers for ophthalmic drug delivery (Reproduced from Xu et al. [3])_

15.2 Rationale for Colloidal Carriers

15.2.1 Factors Limiting Usefulness of Ophthalmic Therapeutics

Fate of a drug is determined by ocular physiological and anatomical factors that can reduce the amount of time therapeutic drug concentration is present at the site of action. These factors can be divided into dynamic and static barriers. It has been found that these barriers pose significant treatment failure risks. Some of these factors are listed below:

- Tear formation
- Tear dynamics
- Tear turnover
- Tear dilution
- Nasolacrimal Drainage
- Conjunctival absorption
- Low permeability of the cornea
- Transient residence in the cul-de-sac
- Poor ocular drug bioavailability
- High administration frequency
- Systemic exposure
- Poor targeting to the posterior segment due to lens–iris diaphragm

15.2.2 Novel Drug Delivery Technologies Addressing Factors to the Usefulness of Ophthalmic Therapeutics

Innovative technologies like use of delivery systems have been explored in detail and have met some success in bypassing these ocular barriers. These newer drug delivery technologies have been designed by considering these barriers along with target diseases and possible routes of delivery. Table 15.1 shows examples of a few delivery methods targeting static and/or dynamic barriers [4].

15.2.3 Advantages and Disadvantages of Colloidal Drug Carriers

As mentioned earlier, use of advanced drug delivery systems like colloidal carriers offers both challenges and opportunities. These drug delivery systems offer multiple advantages, mainly focusing on improving pharmacokinetics like increase in

Table 15.1 Examples of delivery methods and strategies for bypassing ocular drug delivery barriers (Reproduced from Rowe-Rendleman et al. [4])

Delivery method	Example	Static* barriers	Dynamic† barriers	Innovation
Intraocular	Intravitreal injection	Vitreous	Anterior aqueous humor flow, posterior trans-retinal flow, efflux transporters	Self-assembling DNA NPs
				PEGylated-FAB fragments
				Bioerodible microspheres
				Adenoviral transfection
Suprachoroidal	Suprachoroidal injection	Choroid, Bruch's membrane, Retinal Pigment Epithelium (RPE) tight junctions, retina	Choroidal circulation	Hollow microneedle
			Retinal circulation	
Periocular	Sub-Tenon's injection	Sclera, choroid, Bruch's membrane, RPE tight junctions, retina	Subconjunctival–episcleral lymph and blood vessels, choroidal circulation, retinal circulation	Subconjunctival hydrogel injection
Topical	Eye drop	Cornea, conjunctiva, sclera, choroid, Bruch's membrane, RPE tight junctions, retina	Tears and lacrimal drainage, subconjunctival–episcleral lymph and blood vessels, choroidal circulation, retinal circulation	Dendrimeric-polyguanidylated translocators, Hybrid Dendrimer Hydrogel/Nanoparticle Platform (HDNP) formulations

Innovative drug delivery systems that have been useful in animal studies and clinical trials include nanoparticles, hydrogels, dendrimers, and adenoviruses

*Static barriers are generally regarded as tissues that provide a passive physical barrier to drug and gene penetration

†Dynamic barriers are physiologic clearance mechanisms and barriers that include aqueous turnover, vitreous flow, ocular drug metabolism, and metabolizing enzymes in the eye

half-life of the drug at the site of action. However, some limitations warrants thorough understanding of drug delivery systems, and benefits should outweigh risks associated with these therapeutics before they can be approved by Food and Drug Administration (FDA). Table 15.2 lists some of the potential advantages and disadvantages with some important drug delivery systems.

Table 15.2 Potential advantages and disadvantages of some drug delivery systems

Colloidal carrier	Advantages	Disadvantages
Dendrimers	• Nano-size range (1–110 nm) • Lower polydispersity index • Modifiable surface groups to tailor according to the needs • Ease of synthesis • Ease of modification/functionalization • Can be useful to enhance solubility and/or permeability • Charge modifications: Cationic dendrimers interact with lipid bilayers for enhancing permeability but can also increase cytotoxicity • Can be designed to be biocompatible with low immunogenic potential • Can enhance corneal residence time • Can deliver drugs to the retina systemically • Can be useful in retinal degeneration	• Potential cytotoxicity • Release of nucleic acid from the dendrimer complex • Complexity can sometimes be a limiting factor and can add to the cost of therapy
Nanoparticles and liposomes	• Stable carriers for drugs, peptides, vaccines, and oligonucleotides • Sustained release can decrease clearance, toxicity, and frequency of administration • Potential applications in gene therapy • Can be delivered to multiple targets and for multiple disease states • Offers localized delivery • Can be useful for treating genetic defects	• Side effects due to injections • Patient acceptance remains to be seen • Vitreous clouding
Cell encapsulation	• Offers localized delivery thus limiting toxicity • Increased half-life of the drug • Good patient compliant with the treatment • Reversible treatment as implants can be removed as needed	• Surgical intervention • Invasive method that can lead to ocular complications • Poor patient acceptance

15.3 Type of Colloidal Carriers

15.3.1 Nanoparticles

15.3.1.1 Introduction

Nanoparticles are submicron-sized colloidal systems that have been projected to hold enormous potential in ocular drug delivery. These nanoparticles can either be based on natural or synthetic polymers and are susceptible to modification to offer unique properties. Most commonly used polymers useful for synthesizing nanoparticles include polylactides (PLAs), polylactide-polyglycolide (PGA), polyacrylates, polyalkylcyanoacrylates, polycaprolactones, chitosan, hyaluronic acid, collagen, alginate, gelatin, albumin, and dextran. Active ingredient can either be dissolved, entrapped, or encapsulated based on the requirements, and these nanoparticles can be synthesized with different sizes, shapes, charges, and physicochemical properties. Nanoparticles can be broadly divided into two types: nanospheres and nanocapsules. Nanospheres are nanostructures where drug is uniformly and physically distributed either on the surface or entrapped within the sphere. On the other hand, nanocapsules are reservoir-type nanostructures having oil core with dissolved drug and surrounded by a polymeric membrane. These nanostructures are composed of various biodegradable polymers including PLA, PLGA, polycaprolactones, or natural polymers like chitosan, alginate, and dextran. Further, these can be coated with additional polymers or functionalized with antibodies for targeting purposes. Tremendous research in the field of nanoparticles has resulted in unique delivery vehicles offering significant advantages. For example, opsonization of nanoparticles can help them internalize to the target cells due to *biological fingerprinting* role of opsonins [5]. On the other hand, if nanoparticles with long circulating half-lives are desired, coating with PEG can help them escape internalization and allow them to stay longer in the blood [6]. Nanoparticles for ocular application offer many exclusive advantages. These are generally easy to synthesize and characterize and hold the potential for the treatment of a wide variety of ocular diseases. They can be fabricated to be nonirritating and noncytotoxic and can serve as delivery vehicles for hydrophobic/hydrophilic drugs, peptides, vaccines, oligonucleotides, and genes. Additionally, nanoparticles can be designed to have desirable properties like increased permeability to the cornea and conjunctiva, prolonged residence time, and sustained release and ease of administration. These nanoparticle formulations can be given by different routes of administration for eliciting therapeutic effects in different parts of the eye as shown in Fig. 15.2 [7].

15.3.1.2 Applications in the Anterior Eye

Topical ocular application of the drugs for diseases like keratitis, conjunctivitis, and topical and corneal infections is sometimes limited in their use. This may be due to poor precorneal residence time and other related issues like frequent administration leading to decrease patient compliance. Use of nanoparticles to increase precorneal residence time can offer many advantages for anterior applications, and it can also lead to enhanced absorption to the posterior of the eye. Figure 15.3 illustrates some applications of the nanoparticles for the treatment of different ocular diseases [8].

Different mucoadhesive polymers or sustained release devices have been extensively explored for anterior applications. Chitosan is a well-known natural polymer with excellent ocular tissue compatibility. It is noncytotoxic and has mucoadhesive properties due to its ability to interact with negatively charged cornea and conjunctiva. Nagarwal et al. reported chitosan-tripolyphosphate nanoparticles of 5-fluorouracil (5-FU) for ocular application [9]. In vivo studies showed that bioavailability of 5-FU incorporated into these nanoparticles was significantly improved as compared to the control. Similarly, other hydrophilic drugs like proteins, peptides, and genes can be incorporated in these nanoparticles. Rajendran et al. reported in vitro sustained release (over a period of 24 h) of acyclovir by using chitosan nanoparticles after ionic gelation with tripolyphosphate anions [10].

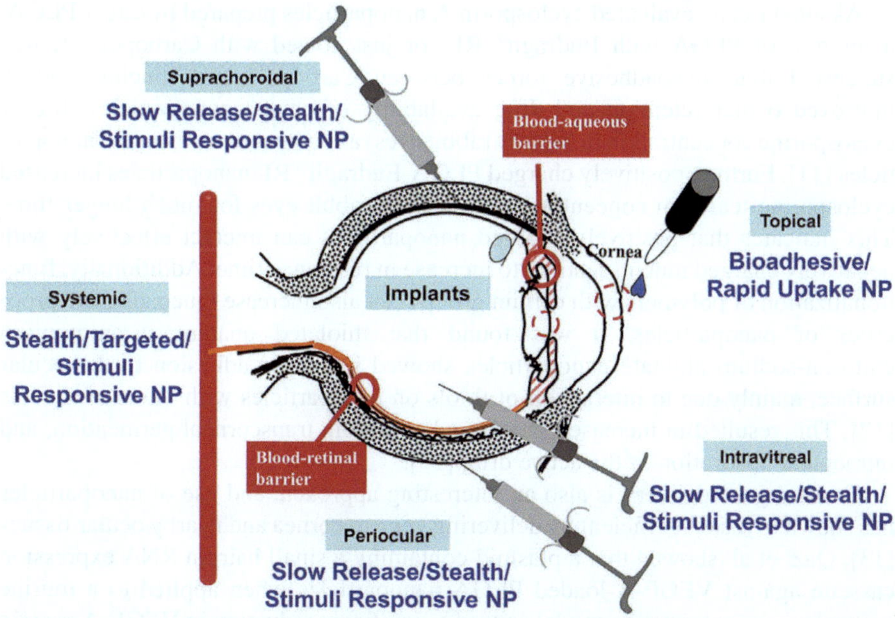

Fig. 15.2 Different routes of ocular administration for delivering different types of nanoparticles to the posterior eye (Reproduced from Kompella et al. [7])

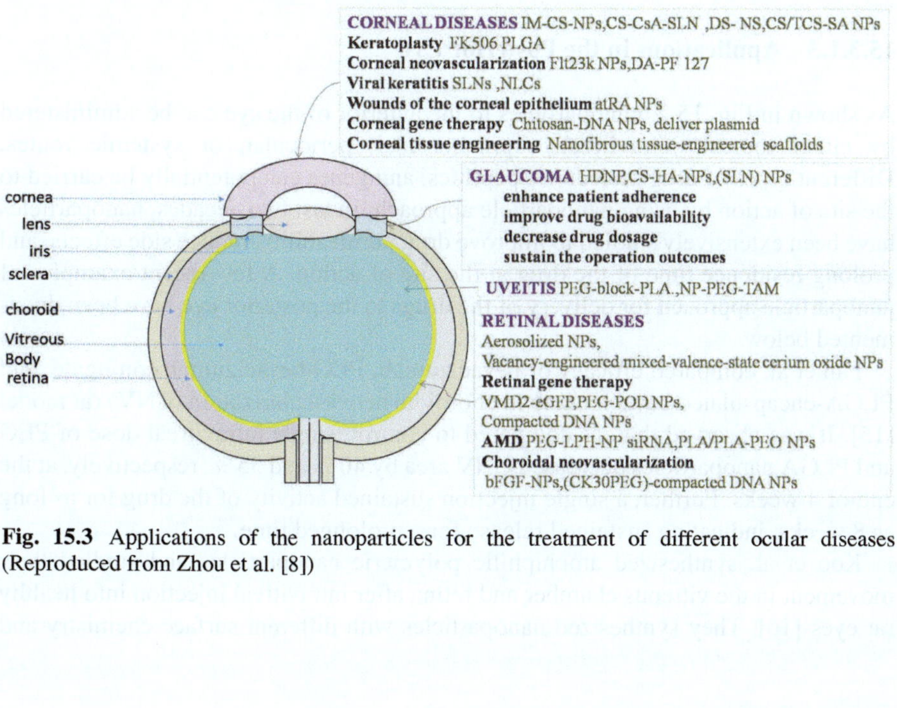

Fig. 15.3 Applications of the nanoparticles for the treatment of different ocular diseases (Reproduced from Zhou et al. [8])

Aksungur et al. evaluated cyclosporin A nanoparticles prepared by either PLGA, a mixture of PLGA with Eudragit® RL, or just coated with Carbopol®. It was suggested that mucoadhesive forces between Carbopol® and mucins lead to improved ocular retention and drug availability of coated nanoparticles (higher cyclosporine concentration in healthy rabbit eyes) as compared to uncoated nanoparticles [11]. Further, positively charged PLGA-Eudragit® RL nanoparticles increased cyclosporine tear film concentration in healthy rabbit eyes for much longer time. This indicates that positively charged nanoparticles can interact effectively with negatively charged mucins leading to increase in residence time. Additionally, functionalization of polymers with certain groups can also increase mucoadhesive properties of nanoparticles. It was found that thiolated quaternary ammonium chitosan-sodium alginate nanoparticles showed increased adhesion to the ocular surface, mainly due to interaction of thiols on nanoparticles with thiols of mucins [12]. This resulted in increased precorneal retention, transcorneal permeation, and intraocular absorption of the active drug [12].

Corneal gene delivery is also an interesting approach, and use of nanoparticles has shown to make it efficient by delivering gene to cornea and nearby ocular tissues [13]. Qazi et al. showed that a plasmid containing a small hairpin RNA expression cassette against VEGF-A-loaded PLGA nanoparticles when applied to a murine corneal neovascularization model caused a significant reduction in VEGF-A protein expression, when compared to control. It demonstrated that PLGA nanoparticles can work as effective, nontoxic, and sustainable form of gene therapy [14]. Thus, gene therapy based on nanoparticles has the potential for sustained delivery leading to better transfection efficiencies.

15.3.1.3 Applications in the Posterior Eye

As shown in Fig. 15.2, nanoparticles to the anterior of the eye can be administered by either topical, intravitreal, suprachoroidal, periocular, or systemic routes. Different types of drug molecules, peptides, and genes can potentially be carried to the site of action by using nanoparticle approach. In last two decades, nanoparticles have been extensively studied to improve drug permeability, reduce side effects, and prolong residence time of the drug at the site of action. A few recent examples of nanoparticle approach for delivery of the drugs to the posterior eye have been documented below.

Pan et al. compared efficacy of bevacizumab, PEG-bevacizumab conjugate, and PLGA-encapsulated bevacizumab in choroidal neovascularization (CNV) rat model [15]. It was observed that, as compared to control, single intravitreal dose of PEG and PLGA nanoparticles decreased CNV area by 40 % and 35 %, respectively, at the end of 4 weeks. Further, a single injection sustained activity of the drug for as long as 8 weeks, indicating sustained release for a prolonged time.

Koo et al. synthesized amphiphilic polymeric nanoparticles and studied their movement in the vitreous chamber and retina after intravitreal injection into healthy rat eyes [16]. They synthesized nanoparticles with different surface chemistry and studied their distribution in the eye. It was observed that cationic glycol chitosan nanoparticles and glycol chitosan-polyethylenimine blended nanoparticles were able to penetrate vitreal barrier. This showed that cationic nanoparticles with glycol on the surface can serve as efficient carrier of drugs and genes for various retinal disorders like glaucoma, age-related macular degeneration, and diabetic retinopathy.

Singh et al. developed transferrin, arginine–glycine–aspartic acid (RGD) peptide, or dual-functionalized poly-(lactide-co-glycolide) nanoparticles to target the delivery of anti-VEGF interceptor plasmid Flt23K to choroidal neovascularization (CNV) lesions [17]. The plasmid was injected IV and it was found that plasmid-loaded functionalized nanoparticles reduced CNV significantly as compared to both plasmid-loaded non-functionalized nanoparticles as well as naked plasmid treated groups. Additionally, it was observed that functionalization of nanoparticles helped their localization to the CNV site. This suggests that functionalized nanoparticles can lead to localized drug delivery for smaller drug molecules and plasmids. These results were further supported by similar studies in different CNV models [18].

Cai et al. tested compacted-DNA nanoparticle-mediated gene delivery by subretinal injection to a juvenile mouse model of retinitis pigmentosa [19]. A block polymer of 30-mer lysine (with terminal cysteine group) was conjugated with methoxy-PEG-maleimide. This conjugate was then condensed with plasmid DNA to form neutral nanoparticles. Nanoparticles were injected subretinally and were taken up by photoreceptors. It was observed that these nanoparticles can efficiently drive gene expression for structural and functional rescue in retinitis pigmentosa and retard degeneration in the mouse model used. These observations illustrate the potential of nanoparticle-based gene delivery for the treatment of human retinal degenerations.

Some of these examples imply that nanoparticles varying in size, shape, and charge can be synthesized and tailored based on properties/features desired for efficient drug delivery. Nanoparticles hold potential for enhanced in vivo performance of a drug. However, opportunity comes with challenges and potential risks. Pharmacokinetics, metabolism, disposition, elimination, and other properties of nanoparticles need to be studied in detail before they can be approved for therapeutic uses. Biocompatibility of these foreign molecules is also an important criterion and needs to be studied in detail along with efficacy studies.

15.3.2 Dendrimers

15.3.2.1 Introduction

Dendrimers are treelike nanostructured monodisperse macromolecules consisting of a) central core and b) highly branched repeating units in different layers. Multivalent functional groups in the dendrimers allow them to be useful with different types of drug molecules. This multibranching property and flexibility of the cargo make dendrimers quite remarkable for drug delivery approaches. Their synthesis generally starts with central core molecule, and at least two reactive groups and layers are strategically added with repeated branches allowing them to grow into different layers/generations depending upon the requirement. This type of synthesis is generally referred to as *divergent* as compared to *convergent*, where distinct dendrons are combined with reactive core to produce different generation dendrimers [20, 21]. Literature reports more than 100 different dendrimer families with synthetic variations in both *divergent* and *convergent* methods. Some of the recent advances include *hypercore and branched monomers' growth* where preassembled oligomeric branches were fused, thus reducing number of synthetic steps [22]. In *lego* synthetic approach, phosphorus dendrimers are synthesized allowing up to fivefold increase in peripheral functional groups for potential attachments [23]. In *click chemistry* approach, triazoles are synthesized from azides and alkynes using Cu(I) as a catalyst. Reduced separation and purification steps in *click chemistry* allow efficient and cost-effective synthesis of dendrimers [24]. Based on the polymers used, different kinds of dendrimers can be synthesized and a detailed explanation is beyond the scope of this chapter.

Due to flexibility in the types of cargo that can be attached to dendrimers, these delivery agents are extensively studied in ocular applications. Dendrimers have applications in delivering small drug molecules, oligonucleotides, peptides, and genes as well as applications in genetic testing.

15.3.2.2 Applications in the Anterior Eye

Dendrimers have been reported to offer improved pharmacokinetics of drug molecules in different ocular applications. Multifunctional groups in dendrimers offer mucoadhesive properties. Thus, they can be useful as surface-coating agents alongside their applications as drug delivery vehicles to improve PK properties. Mucoadhesive properties lead to reduced clearance and improved precorneal residence. In vitro studies have shown stronger interactions of cationic as well as neutral dendrimers with mucins. These interactions indicate that dendrimers can be useful in the anterior of the eye due to their bioadhesive and drug delivery properties [25].

PAMAM G1.5-4 dendrimers of pilocarpine nitrate and tropicamide were synthesized by Vandamme et al., and these were found to have increased corneal residence time and lead to increased bioavailability with prolonged duration of action when tested in New Zealand albino rabbits [26]. Similarly, Lys_xCys_y dendritic polymers in situ gel incorporating brimonidine and timolol maleate were found to be effective when topically for securing cataract incisions [27]. Durairaj et al. reported that dendrimeric polyguanidilyated translocators (DPTs) when complexed with fluoroquinolone gatifloxacin lead to enhanced corneal transport in human corneal epithelial cells (HCECs) and across isolated bovine sclera-choroid-RPE (SCRPE). Further in vitro studies showed that these DPTs increased antimicrobial (including anti-MRSA) activity for the treatment of conjunctivitis and ocular infections [28]. In vivo delivery in New Zealand white rabbits showed approximately 13- and 2-fold higher AUCs for tissue concentrations in the conjunctiva and cornea, respectively. Grinstaff et al. designed and evaluated corneal adhesives in ex vivo and White

Leghorn chicken models. These formulations were prepared from modified G1, G2, and G3 dendrimers both by photo-crosslinking reaction and by using peptides to ligate individual dendrimers. These adhesives were found to seal the wounds and, thus, repair corneal perforations along with closing the flap produced in LASIK procedures [29]. Examples of some of the dendrimers useful in the anterior of the eye are given in Table 15.3 [30].

15.3.2.3 Applications in the Posterior Eye

Advanced drug delivery systems like dendrimers find applications in targeting posterior eye and that too with much reduced toxicity and selective distribution. Iezzi et al. explored strategies for delivering fluocinolone acetonide using hydroxyl-terminated polyamidoamine (PAMAM) dendrimers [44]. Intravitreal administration of these dendrimers showed pathology-dependent distribution and selective localization in retinal microglial cells in rats. The localization was observed for up to 30 days and this approach was able to arrest retinal degeneration. The same dendrimers were able to selectively localize in activated microglia upon systemic administration to newborn rabbits [47]. These studies propose that dendrimers can serve as important tools for selective distribution and sustained delivery of the drugs.

Marano et al. used modified dendrimers (having lipophilic tail of lipoamino acid, LAA, and polylysine amino acids) bound with anti-VEGF oligodeoxynucleotide (ODN-1) [34]. LAA was supposed to help in the transport across cell membranes and tissues and protect DNA from nucleases. These dendrimers were administered in a rat model by intravitreal injection for treating laser-induced CNV. These were well tolerated in vivo and showed a significant inhibition of CNV for up to 6 months [34].

Overall, dendrimers offers great control over size, shape, charge, as well as flexibility of modifiable surface functional groups. They have proven to offer increased corneal residence, sustained delivery, and selective distribution allowing their uses and potential applications in delivering diverse cargo as well as applications in biomedical imaging and in genetic testing. Some of the recent ocular applications of dendrimers and dendrimeric delivery systems are given also in Table 15.3 [30].

15.3.3 Liposomes

15.3.3.1 Introduction

Liposomes are spherical lipid vesicles which can either be unilamellar, multilamellar, or multivesicular. Liposomes can be prepared by different techniques like thin-film hydration, sonication, French press, dialysis, extrusion, or reverse-phase evaporation. Different methods of preparation dictate different sizes, circulation stability, and altered pharmacokinetics [48, 49]. Addition of other components is sometimes desirable to further increase stability. For example, incorporating PEG

Table 15.3 Ocular applications of dendrimers and dendrimeric delivery systems (Reproduced from Yavuz et al. [30])

Drug	Dendrimer type	Administration	Treatment	Outcomes	References
Pilocarpine nitrate and tropicamide	PAMAM G1.5–4	Topical	Myosis and mydriasis	Increased corneal residence and prolonged reduction of IOP	[26]
Carteolol	Phosphorus-containing dendrimers	Topical	Glaucoma	Increased corneal residence and reduced toxicity and IOP	[31]
Gatifloxacin	Dendrimeric polyguanidilyated translocators	Topical	Conjunctivitis and intraocular infections	Enhanced corneal transport and increased antimicrobial activity	[28]
Glucosamine and glucosamine 6-sulfate	PAMAM G3.5-COOH	Subconjunctival injection	Antiangiogenic in glaucoma surgery	Reduced inflammation and no scar formation	[32]
Carboplatin	PAMAM G3.5-COOH (dendrimeric nanoparticles)	Subconjunctival injection	Retinoblastoma	Increased half-life and bioavailability reduced drug toxicity and tumor mass	[33]
VEGF-ODN	Lipophilic amino acid dendrimer	Intravitreal injection	CNV	Prolonged suppression of VEGF and neovascularization	[34]
–	Polypropylenimine octaamine G2	Corneal scaffold	Corneal tissue engineering	Enhanced human corneal epithelial cell growth	[35]
–	Surface modified-COOH ending dendrimers	Corneal scaffold	Corneal tissue engineering	Promoted adhesion and proliferation of human corneal epithelial cells	[36]
–	Modified G1, G2, and G3 dendrimers	Topical (corneal hydrogel adhesive)	Corneal wounds	Wound sealing and no scar formation	[29]
Photosensitizer	G3 aryl ether dendrimer zinc porphyrin	Intravenous injection-photodynamic therapy	CNV	Accumulation in neovascularized area	[37]

Drug	Dendrimer type	Administration	Treatment	Outcomes	References
Concanavalin A	Porphyrin glycodendrimers	Topical photodynamic therapy	Intraocular tumors and retinoblastoma	Enhanced targeting and reduced toxicity	[38]
–	Porphyrin dendrimers	Topical-photodynamic therapy	AMD and CNV	Selective accumulation in inflammatory cells and prolonged retention time	[39, 40]
DNA	Phthalocyanine dendrimers			Accumulation in photo-irradiated areas and increased transgene expression	[41]
–	Anionic and cationic carbosilane dendrimers	Topical	Tolerance	Hydrogen bonding between mucin and PAMAM-enhanced retention time	[42]
Puerarin	PAMAM	Topical	Ocular hypertension and cataract	Increased bioavailability	[43]
Fluocinolone acetonide	PAMAM G4-OH	Intravitreal injection	Retinal neuroinflammation	Reduced inflammation	[44]
Brimonidine and timolol maleate	PAMAM hydrogel (G3)	Topical	Glaucoma	Increased uptake	[45]
Brimonidine and timolol maleate	Hybrid PAMAM dendrimer hydrogel/PLGA nanoparticle	Topical	Glaucoma	Increased uptake	[46]
–	Lys_xCys_y dendritic polymers in situ gel	Topical	Cataract incisions	Wound sealing	[27]

or gangliosides results in better stability and, thus, increases the circulation half-life of liposomes. Generally, both hydrophilic and hydrophobic drug molecules can be incorporated into the liposomes.

Liposomes offer advantages for both anterior and posterior segment of the eye. While enhanced corneal penetration is the main goal in the anterior segment, the research for liposomes in the posterior segment has focused on developing formulations with less toxicity, longer clearance times, and site-specific delivery to the retinal pigment epithelium (RPE).

15.3.3.2 Applications in the Anterior Eye

Anterior segment application of liposomes mainly focuses on advantages like increasing tissue specificity, decreasing systemic toxicity, and bypassing the blood–ocular barrier [50–52]. Liposomes have been studied as vehicles for genetic transfection using DNA or other similar genetic materials. In one study, liposomes were found to be promising gene delivery vehicles to mammalian intraocular tissue. It was concluded that efficient transfer of liposomes carrying functional gene is feasible in the cornea, iris, ciliary body, and retina of the rats [53]. In a similar study, Matsuo et al. found that liposomes can serve as delivery vehicles to the retinal ganglion cells of rats for a plasmid vector for β-galactosidase [54].

Liposomes have been hypothesized to offer advantage in preventing graft rejection as they are preferentially taken up by the phagocytes. It has been shown that subconjunctival injections of liposomes containing dichloromethylene diphosphate (clodronate) prevented corneal allograft rejections in treated rats [55]. Same group later showed that these injections also prevented cytotoxic T lymphocyte activity, thus helping in preventing graft rejections [56].

Liposomes have also been evaluated in inflammatory conditions like keratitis and uveitis. Different antibiotics like gentamicin, norfloxacin, and tobramycin have been tested as liposomal preparations delivered to the anterior segment of the eye. One in vitro study by Lin et al. illustrated increased accumulation of liposomal norfloxacin formulation in the cornea when compared to the non-liposomal formulation. This increased accumulation was due to slow drainage of these liposomal formulations from corneal surface [57]. Immunosuppressive drugs like cyclosporine and tacrolimus have also found to be quite useful in decreasing inflammation when used in liposomal preparations. Pleyer et al. showed that liposomes containing tacrolimus demonstrate increased ocular penetration when compared with non-liposomal formulations [58].

15.3.3.3 Applications in the Posterior Eye

Drug delivery to the posterior segment of the eye is one of the major challenges in ophthalmic drug delivery. Although liposomes offer significant advantages for delivering drugs to the anterior eye, delivery to the posterior eye is much more complex, and main obstacles include faster clearance, poor penetration (due to either tighter junctions in the corneal cells), and unfavorable physicochemical properties of drug molecules [50, 59–61]. Blood–ocular barrier is found to be one of the main obstacles leading to less than adequate intraretinal and/or intravitreal drug concentrations [62–64]. Although repeated intravitreal injections can be useful in achieving therapeutic concentration, it suffers major downside of increased chances of infection and vitreous hemorrhage. Liposomal formulations can potentially decrease drug clearance and limit frequency of injections, thus, minimizing these side effects (Fig. 15.4).

Liposomal formulations have shown decreased clearance and toxicity of some drugs. As shown in Fig. 15.4, Alghadyan et al. showed that half-life of cyclosporine was increased 12-fold when using intravitreal injection of liposomal cyclosporine [65]. It was also found that these liposomal doses of cyclosporine showed no histopathological or electroretinographic changes when compared with the free form [65, 66]. Thus, it was concluded that liposome-bound cyclosporine prolonged drug availability while reducing its toxicity.

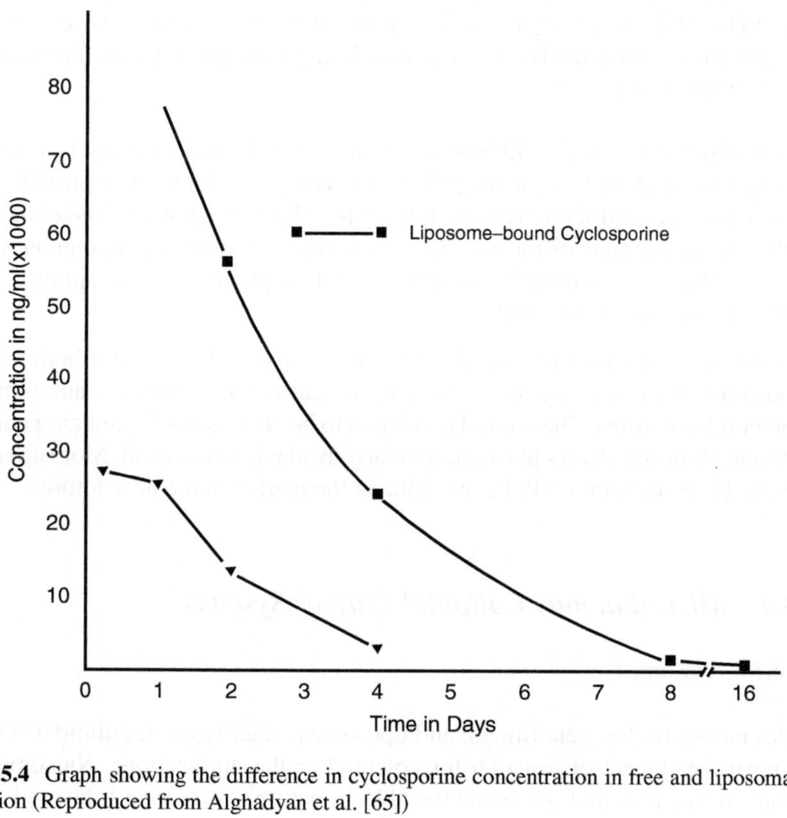

Fig. 15.4 Graph showing the difference in cyclosporine concentration in free and liposomal formulation (Reproduced from Alghadyan et al. [65])

Similarly, Liu et al. found that intravitreal injection of liposomal trifluorothymidine showed much longer half-life and minimal retinal toxicity in albino rabbits when compared with the free drug [67].

These examples illustrate that liposomal application in the posterior segment of the eye has the potential to achieve therapeutic concentrations for prolonged time and with much lesser toxicity. Besides that, liposomal formulations have been shown to be useful in hemodynamic measurements. Khoobehi et al. used a liposomal dye (calcein) to measure blood velocity and volumetric flow rates in the retina of rhesus monkey. This experiment helped them quantify these important hemodynamic parameters in retinal arteries and veins [68].

15.3.3.4 Commercially Available Ophthalmic Liposomal Preparations

Verteporfin (Visudyne®) Verteporfin is a light-activated drug available in the form of liposomes. It is injected by intravenous infusion followed by activation using nonthermal diode laser. It is transported by lipoproteins to the site of action and once activated by light, reactive oxygen species cause local damage to neovascular endothelium. Verteporfin has been approved as a liposomal photosensitizer by the FDA. It is indicated for the treatment of predominantly classic subfoveal CNV due to age-related macular degeneration, pathologic myopia, or presumed ocular histoplasmosis [52].

Tears Again Advanced Eyelid Spray® Tears Again liposomal spray is a cooling, soothing mist marketed by OCuSOFT™. The company claims it to provide relief from dry, irritated, and scratchy eyes. It is further claimed to provide longer duration of action with minimal distortion and a very useful remedy for computer vision syndrome. It acts as sealant for patient's own tear production and sprayed on to closed eyes for immediate relief.

Literature survey demonstrates that liposomes have applications for both anterior and posterior eye diseases and can be a useful carrier for antibiotics, autoimmune, and other related drugs. They have been found to be very useful for imaging contrast agents and photosensitizers like commercially available verteporfin. More approved liposome-based therapies will be available in the market in the near future.

15.3.4 Miscellaneous Colloidal Carrier Systems

15.3.4.1 Niosomes

Besides nanoparticles, dendrimers, and liposomes, other types of colloidal systems have been the subject of research for potential ocular applications. Niosomes use nonionic surfactants and are biodegradable in nature. They have bilayered structures like liposomes and can act as a carrier for both hydrophilic and hydrophobic drugs. Generally, they are more stable and less toxic than liposomes. Timolol maleate niosomes coated with chitosan and Carbopol® were prepared by Aggarwal et al. and were found to have superior therapeutic effect in albino rabbits (decrease in intraocular pressure, IOP) as compared to the solution form [69]. Same group also studied the extent of ocular absorption of acetazolamide niosomal formulation in combination with mucoadhesive polymer Carbopol® in albino rabbits [70]. As compared to aqueous suspension of acetazolamide, niosomal formulation exhibited prolonged effect for 6 h and lowered IOP much more effectively.

One of the recent studies mentioned proniosome-derived niosomes for tacrolimus [71]. The group developed tacrolimus-loaded proniosomes containing poloxamer 188 and lecithin as stabilizer. These proniosomes were reconstituted to niosomes prior to use and applied topically in an in vitro freshly excised rabbit corneal model. As compared to tacrolimus 0.1 % ointment, these proniosomes exhibited significant permeation and drug retention. In vivo anti-allograft rejection studies in Sprague–Dawley (SD) rat models showed that niosomes delayed the occurrence of corneal allograft rejection when compared with different controls. These studies suggest that niosomes can be potential carriers for some drugs and research in this field can further be extended to drugs targeting other diseases.

15.3.4.2 Cubosomes

Cubosomes are colloidal carrier systems mainly characterized by a cubic crystallographic symmetry. They are similar to nanoparticles and are formed by self-assembly of surfactant-like molecules. Monoglyceride glycerol monoolein is a well-known surfactant useful to make cubosomes. Surfactant-like nature allows them to incorporate both hydrophilic and hydrophobic drug molecules. Generally, cubosomes have biodegradable and bioadhesive constituents and can provide sustained release of a drug and also work as a penetration enhancer. Han et al. developed flurbiprofen cubosomes with main emphasis on reducing ocular irritation and increasing drug bioavailability [72]. In vitro corneal permeation using modified Franz-type cells revealed up to twofold increase in permeation with these cubosomes. These were further evaluated for ocular irritation using Draize method and histological examination. Cubosomes exhibited excellent ocular tolerance with improved AUC as measured by microdialysis. These studies suggested cubosomes as promising delivery vehicles. In the same year, Gan et al. reported cubosomes as delivery vehicles for dexamethasone [73]. In vitro studies showed up to 3.5-fold increase in permeability with dexamethasone cubosomes, when compared to that of dexamethasone eye drops. Additionally, these cubosomes showed longer retention and up to 1.8-fold increase in AUC as measured by microdialysis. These studies further emphasize that these highly biocompatible, biodegradable delivery vehicles can be an effective approach for certain ocular therapies.

15.3.4.3 Other Colloidal Carriers

Literature studies advocates that micro- and nanoemulsions can sometimes be useful as colloidal carriers. Klans et al. evaluated piroxicam microemulsion on a rabbit cornea model with alkali burn [74]. 0.1 % positively charged microemulsion showed significant reduction in corneal deterioration rate as compared to different types of controls. There are a few other examples in literature supporting the use of micro-/nanoemulsions for specific drugs [75, 76]. The use of microparticles for ocular drug delivery has also been documented in the literature. Kompella group has shown that celecoxib-PLGA microparticles when given subconjuctivally prolonged retinal drug levels and showed significant therapeutic effects in a rat model [77]. Same group also showed that budesonide microparticles sustained drug levels in the retina for much longer time as compared to the control [78].

Besides smaller drug molecules, colloidal delivery systems have been found to be very useful in ocular gene therapy. These nonviral nucleic acid delivery systems have been studied by different groups and some of the examples are given in Table 15.4 [79].

Table 15.4 Miscellaneous colloidal carrier systems for nucleic acids (Reproduced from Chaplot et al. [79])

Delivery system	Nucleic acid	Target disease/cell line/tissue	References
PLGA microspheres	PS-AsODN (TGFβ2) (with PEI)	Bleb survival after trabeculectomy	[80]
PLGA microspheres	RNA aptamer (VEGF)	Human umbilical vein endothelial cells (HUVECs)	[81]
PEI	PS-AsODN (anti- TGFβ2)	Rat retinal Muller glial cells	[82]
Cell-penetrating peptide	pDNA (GPF_{His})	Human embryonic retinoblast (HER) cells	[83]
Cationic nanoemulsion	AsODN (VEGF-R2)	Corneal neovascularization	[84]
Cationic core–shell liponanoparticles	pDNA (EGFP)	Human conjunctival epithelial cells and rabbit eye	[85]
Electrotransfer to ciliary muscle	pDNA (hTNFR-Is)	Inflammatory, degenerative, or angiogenic diseases	[86]
Ultrasound with DOTAP/DOPE/ DSPE–PEG liposomes	pDNA (luciferase)	Bovine neural retina	[87]
In situ gelling systems	AsODN (Cx43)	Persistent epithelial defect wounds Corneal wound healing after PRK Bleb survival after trabeculectomy Inflammatory posterior eye diseases	[88–92]
Solution	AsODN (VEGF)	Iris neovascularization	[93]

DOPE 1,2-dioleoyl-3-phosphatidylethanolamine, *DOTAP* 1,2-Dioleoyl-3-trimethyl ammonium propane, *EGFP* enhanced green fluorescent protein, *hTNFR-Is* chimeric tumor necroses factor alpha soluble receptor, *pDNA* plamid DNA, *DSPE–PEG* 1,2-distearoyl-sn-glycero-3--phosphatidylethanolamine-N-[poly(ethyleneglycol)-2000], *PEI* polyethylenimine, *PS-AsODN* phosphorothionate antisense oligodeoxynucleotide, *TGFβ* transforming growth factor beta, *VEGF* vascular endothelial growth factor

15.4 Routes of Application

Nanotechnology has been utilized in delivering drugs to the anterior and posterior segments of the eye. Topical applications targeted for anterior segment mainly focus on increasing drug's residence time by the use of mucoadhesive polymers. Topical application to deliver drug to the back of the eye mainly uses conjunctival or corneal routes. Drugs following conjunctival pathway generally cross conjunctiva followed by the sclera, choroid, retinal pigment epithelium (RPE), and retina. Drug entering corneal pathway generally traverses the iris, aqueous humor, lens, and ciliary body. Colloidal carriers tagged with certain ligands can help in drug delivery to the site of action after topical application. Other significant routes of drug delivery to the posterior eye include intravenous (IV), intravitreal, subconjunctival, and suprachoroidal routes. Liposomes and nanoparticles of anticancer agents are routinely given by IV routes. Anatomy and physiology of tumors permit enhanced retention of these drug formulations at the tumor site; thus, allowing increased drug at the site of action leading to both reduced frequency and drug dose. Intravitreal drug delivery offers unique advantages due to proximity of the drug to the choroid, RPE, and retina. However, major disadvantages like retinal detachment, increased IOP, and endophthalmitis associated with intravitreal formulations are major limiting factors. Intravitreal route has also been studied for gene delivery and other biologicals like bevacizumab, plasminogen kringle-5, and human serum albumin (HSA) [15, 94, 95]. Due to certain disadvantages of intravitreal route as stated above, subconjunctival route for colloidal carriers is a good alternative and has also been exploited for gene delivery and small molecule therapeutics. It has been shown that altering the particle size of the formulation can alter residence time in subconjunctival space and larger particles generally provide sustained delivery for longer periods of time [96]. Suprachoroidal route of drug delivery is another potential route and offers unique advantages like close proximity of the drug to the choroid, macula, and retina. Drug in the suprachoroidal space can easily diffuse due to less kinetic barriers, and the use of nanoparticles, dendrimers, or similar formulations can provide sustained release for longer periods. Some examples of the use of colloidal carriers via different routes of administration are shown in Table 15.5.

15.5 Factors Affecting Transport of Colloidal Dosage Forms

All the factors that affect transport of traditional formulations like ointments and solutions can pose as a barrier to the transport of colloidal dosage forms as well. Some significant problems are rapid clearance of the drugs and their low bioavailability especially to the posterior part of the eye. However, one of the main reasons to develop nanotechnology-based medicines is to overcome these barriers to the usefulness of the drug. Dosage forms are designed with these issues in mind and can address problems specific to the disease and the site of action. For example, some

Table 15.5 Colloidal carriers administered via different routes of administrations (Reproduced from Kompella et al. [7])

Route of administration	Target disease	Experimental model	Colloidal carrier	Observation	Mechanism or principle	References
Topical (drug delivery to the anterior part)	N/A	Rabbit	Poly(lactic-co-glycolic acid) (PLGA) nanoparticles blended with cationic Eudragit® RL polymer surfactant	Increase in the tear film concentration and the area under the concentration–time curve between 0 and 24 h. Increased residence time and cellular uptake	Negatively charged mucins on the preocular surface interact with positively charged Eudragit® RL	[25]
Topical (drug delivery to the posterior part)	Bacterial infections of the eye	Human corneal epithelial cells (HCECs), New Zealand White rabbits	An optimized dendritic polyguanidilyated translocator (DPT)-gatifloxacin (GFX) formulation	Rapid uptake of dendrimers in HCECs within 5 min, following topical application, with 13-fold and twofold higher areas under the curve, with increased residence times, in rabbit conjunctiva and cornea, respectively	Mucoadhesion and bioadhesion, rapid internalization, prolonged residence time in tissues	[28]
	Glaucoma	Dutch-belted rabbits	Polyamidoamine dendrimer G3.0 tethered with three PEG-acrylate chains and integrated with PLGA NPs, for co-delivery of brimonidine and timolol maleate	Single topical application of the hybrid dendrimer NP formulation maintained an effective IOP reduction for 4 days and better sustained drug levels for one week compared to controls	Mucoadhesion and bioadhesion, sustained release, sustained residence time	[46]

Route of administration	Target disease	Experimental model	Colloidal carrier	Observation	Mechanism or principle	References
Intravenous	Choroidal neovascularization (CNV)	Brown Norway rats with CNV	Transferrin, RGD peptide, or dual-functionalized PLGA NPs loaded with anti-VEGF intraceptor plasmid Flt23K	Surface functionalization of the NPs increased delivery to CNV lesions due to receptor-mediated targeting	Target recognition and receptor-mediated endocytosis	[17]
	Choroidal neovascularization (CNV)	Primate and murine age-related macular degeneration (AMD) model	Biodegradable nanoparticles delivering a recombinant Flt23k intraceptor plasmid	Regressed CNV with suppressed subretinal fibrosis	Extended-release and long-term inhibition of angiogenesis and fibrosis	[18]
Intravitreal	Choroidal neovascularization	Brown Norway rats with CNV	Bevacizumab-loaded PEG and PLGA NPs	Persistent effects of bevacizumab with both PEG and PLGA NPs for up to 6 weeks, following laser induction of CNV	Sustained release	[15]
	Diabetic retinopathy	Brown Norway rats with diabetic complications	Plasminogen kringle 5 (K5) plasmid-loaded PLGA:chitosan NPs	Significant attenuation of VEGF and ICAM-1 overexpression and reduction of leukostasis, vascular leakage, and neovascularization for up to 4 weeks	Enhanced and sustained gene transfection	[95]
	Choroidal neovascularization and retinal inflammation	Brown Norway rats with CNV	Plasminogen kringle 5 plasmid-loaded PLGA:chitosan NPs	Significant expression of K5 and reduction of CNV area (40 %) at the end of 2 weeks following laser induction of CNV	Enhanced and sustained gene transfection	[97]

(continued)

Table 15.5 (continued)

Route of administration	Target disease	Experimental model	Colloidal carrier	Observation	Mechanism or principle	References
	Retinal degenerative diseases	ARPE-19 cells and SOD 1-deficient mice	Cu, Zn superoxide dismutase (SOD1) gene encapsulated human serum albumin NPs	80 % transfection efficiency in ARPE-19 cells with receptor-mediated uptake and rapid endo-lysosomal escape. Significant levels of protein expression were detected in vivo at 48 h post injection	Enhanced and sustained gene transfection	[94]
Posterior subcon junctival	Inflammatory diseases and diseases with VEGF elevation	Sprague–Dawley rats	Budesonide-loaded PLA NPs	Polymeric nanoparticles sustained drug delivery better than plain drug	Sustained release	[78]
	Retinoblastoma	LHβ-Tag transgenic mice	Carboplatin-loaded polyamidoamine dendrimer G3.5 NPs	Significant reduction in mean tumor burden at day 22, following a single injection of the drug containing nanoparticles	Sustained release	[33]

Route of administration	Target disease	Experimental model	Colloidal carrier	Observation	Mechanism or principle	References
Suprachoroidal	Choroidal neovascularization	Ex vivo rabbit eye	Bevacizumab entrapped in a hydrogel formed by UV-induced crosslinking of polycaprolactone dimethacrylate and hydroxyethyl methacrylate polymers sustained in vitro drug release in phosphate buffered saline for up to 4 months	UV (365 nm) exposure for a duration of 10 min following suprachoroidal injection of polymers better confined the hydrogel spread in the choroid, the site of injection, when compared to polymers without stimuli or lower durations of light exposure	Stimuli responsive, sustained release	[98]

nanoparticles are designed to be highly permeable with sustained release properties. Although innovative drug delivery technologies can address important limiting factors, the complexity of these colloidal carriers makes them prone to additional issues. When designing a colloidal delivery system, these additional factors should be taken into account. Table 15.6 summarizes some of the important factors that dictate in vivo fate of a drug.

Table 15.6 Factors affecting transport of drug molecule or its colloidal dosage form (Reproduced from Mitra et al. [99])

Factors affecting in vivo fate of drug	Important features and conclusions of the study	References
Size of NPs	20 nm NPs could not retain in the periocular site due to rapid elimination, while 200 nm and 2 μm particles were detected after 2 months at the site of periocular injection in rats	[96]
Melanin pigment	Transscleral transport of celecoxib was significantly higher in the Sprague–Dawley rats (nonpigmented) than Brown Norway rats (pigmented)	[100]
Blood and lymphatic circulation	Transport of 20 nm NPs across the sclera was minimal due to rapid disposition in the presence of both blood and lymphatic circulations. Higher size (200 nm) is required to avoid clearance via both circulations	[101]
Disease condition (diabetes)	In diabetic rats, transscleral retinal and vitreal levels of celecoxib were significantly higher than nondiabetic rats. Transscleral delivery in pigmented rats was higher than nonpigmented rats	[102]
Functionalization of NPs	Transport across cornea was enhanced due to functionalization of polystyrene NPs (20 nm) with deslorelin, a luteinizing hormone-releasing hormone (LHRH) agonist and transferring	[103]
Charge on the NPs	Positively and negatively charged NPs were shown to distribute to the inner ocular tissues such as retina and vitreous humor following application of iontophoresis technique. Positively charged NPs have shown higher penetration than negatively charged NPs	[104]
Charge on NPs	Following intravitreal injection, anionic NPs were shown to distribute to the subretinal space and RPE in higher extent than cationic NPs. Endocytosis of NPs in Muller cells was also shown in the study	[105]
Charge and PEGylation	After intravitreal injection of NPs, neural retina restricts the movement of NPs into RPE, which can be overcome by application of ultrasound	[87]

15.6 Summary

In summary, colloidal carriers offer significant opportunities for different ocular applications. Literature studies suggest that these delivery systems can be designed in a way that they permeate well through ocular barriers and reach to the site of action in therapeutic concentrations. Functionalized nanoparticles coated with certain proteins or peptide ligands can help them take the cargo inside the cells. Using receptors that can be internalized can help in efficient cell entry for therapeutic purposes. Use of deslorelin or transferrin on nanoparticle surfaces showed increase in cellular uptake and transport. These can potentially be useful as drug delivery vehicles. Similarly, sustained release properties can be achieved by smartly engineered colloidal systems. Mucoadhesive polymers like chitosan have shown significant improvement in prolonged retention and slower clearance of drugs. Use of polymers like PEG or cationic systems has shown increased interaction with mucins. This ultimately resulted in enhanced corneal and conjunctival adhesion for longer retention. These approaches were able to address important ocular issues like rapid clearance and increased frequency of administration. Sustained release property of colloidal carrier was also useful in delivery of the drug to the retina by intravitreal and periocular routes. In addition to incorporating smaller drug molecules into different colloidal systems, various macromolecules like oligonucleotides, siRNA, DNA, peptides, and genes have also been tested for efficient delivery. Macromolecules like plasmids need to enter cells to alter protein expression. Same is also true for RNA-based therapeutics. These issues can be addressed by carefully selecting appropriate delivery systems. Use of different delivery systems has indicated that these macromolecules can be protected from undesired metabolism and efflux transporters. Additionally, improved and sustained delivery at the site of action with biocompatible polymers can offer advantages in enhancing therapeutic drug concentrations for longer period of time. Besides that, stimuli-based therapeutics offers numerous advantages of having much tighter control on the drug release that can be initiated based on the requirements of the disease. Literature studies show light, heat, pH, ultrasound, magnetic, and electric current based colloidal carriers, where drug release can be triggered in a localized and controlled manner.

Different polymers, peptides, carbohydrates, and lipid chains can be used in making different delivery vehicles based on properties of drug (like molecular weight, charge), targeted disease, site of action, and other desired outcomes. Safety and efficacy with improved pharmacokinetics are the main driving force behind these carriers, and a variety of ocular routes (like topical, intravenous, intravitreal) can be explored to optimize ocular drug therapy.

15.7 Conclusion

The last two decades have witnessed exponential growth in number of publications and research based on colloidal systems for ocular applications. Various worldwide and US patents point to potentially useful colloidal system-based therapeutics and hold the promise to make some of these innovative technologies to the market. Recently, paclitaxel albumin-stabilized nanoparticle formulation for patients with metastatic melanoma of the eye completed phase 2 trial. Liposomal vincristine (Marqibo®) is undergoing phase 2 clinical trial for its efficacy to control metastatic uveal melanoma. Similarly, dexamethasone–cyclodextrin microparticle eye drops are currently undergoing phase 2 and 3 clinical trials for the treatment of diabetic macular edema (DME). Although only a few nanotechnology-based therapeutics made it to the market, preclinical ocular studies for drug and gene delivery, imaging, and diagnostic devices especially based on stimuli holds real potential, and one will see an increased approval of these safer and effective systems for diagnosing ocular problems and targeting different ocular diseases in a much safer and efficient manner.

References

1. Urtti A (2006) Adv Drug Deliv Rev 58:1131–1135
2. Rawas-Qalaji M, Williams CA (2012) Curr Eye Res 37:345–356
3. Xu Q, Kambhampati SP, Kannan RM (2013) Middle East Afr J Ophthalmol 20:26–37
4. Rowe-Rendleman CL, Durazo SA, Kompella UB, Rittenhouse KD, Di Polo A, Weiner AL, Grossniklaus HE, Naash MI, Lewin AS, Horsager A, Edelhauser HF (2014) Invest Ophthalmol Vis Sci 55:2714–2730
5. Aggarwal P, Hall JB, McLeland CB, Dobrovolskaia MA, McNeil SE (2009) Adv Drug Deliv Rev 61:428–437
6. Gref R, Minamitake Y, Peracchia MT, Trubetskoy V, Torchilin V, Langer R (1994) Science 263:1600–1603
7. Kompella UB, Amrite AC, Pacha Ravi R, Durazo SA (2013) Prog Retin Eye Res 36:172–198
8. Zhou HY, Hao JL, Wang S, Zheng Y, Zhang WS (2013) Int J Ophthalmol 6:390–396
9. Nagarwal RC, Singh PN, Kant S, Maiti P, Pandit JK (2011) Chem Pharm Bull (Tokyo) 59:272–278
10. Rajendran N, Natrajan R, Kumar R, Selvaraj S (2010) Acyclovir-loaded chitosan nanoparticles for ocular delivery. Asian J Pharm 4:220–226
11. Aksungur P, Demirbilek M, Denkbas EB, Vandervoort J, Ludwig A, Unlu N (2011) J Control Release 151:286–294
12. Zambito Y, Di Colo G (2010) Eur J Pharm Biopharm 75:194–199
13. Mohan RR, Tovey JC, Sharma A, Tandon A (2012) Prog Retin Eye Res 31:43–64
14. Qazi Y, Stagg B, Singh N, Singh S, Zhang X, Luo L, Simonis J, Kompella UB, Ambati BK (2012) Invest Ophthalmol Vis Sci 53:2837–2844
15. Pan CK, Durairaj C, Kompella UB, Agwu O, Oliver SC, Quiroz-Mercado H, Mandava N, Olson JL (2011) J Ocul Pharmacol Ther 27:219–224
16. Koo H, Moon H, Han H, Na JH, Huh MS, Park JH, Woo SJ, Park KH, Kwon IC, Kim K, Kim H (2012) Biomaterials 33:3485–3493
17. Singh SR, Grossniklaus HE, Kang SJ, Edelhauser HF, Ambati BK, Kompella UB (2009) Gene Ther 16:645–659

18. Luo L, Zhang X, Hirano Y, Tyagi P, Barabas P, Uehara H, Miya TR, Singh N, Archer B, Qazi Y, Jackman K, Das SK, Olsen T, Chennamaneni SR, Stagg BC, Ahmed F, Emerson L, Zygmunt K, Whitaker R, Mamalis C, Huang W, Gao G, Srinivas SP, Krizaj D, Baffi J, Ambati J, Kompella UB, Ambati BK (2013) ACS Nano 7:3264–3275
19. Cai X, Conley SM, Nash Z, Fliesler SJ, Cooper MJ, Naash MI (2010) FASEB J 24:1178–1191
20. Liu M, Frechet JM (1999) Pharm Sci Technol Today 2:393–401
21. Tomalia DA (2005) Prog Polym Sci 30:294–324
22. Nanjwade BK, Bechra HM, Derkar GK, Manvi FV, Nanjwade VK (2009) Eur J Pharm Sci 38:185–196
23. Maraval V, Pyzowski J, Caminade AM, Majoral JP (2003) J Org Chem 68:6043–6046
24. Wu P, Feldman AK, Nugent AK, Hawker CJ, Scheel A, Voit B, Pyun J, Frechet JM, Sharpless KB, Fokin VV (2004) Angew Chem Int Ed Engl 43:3928–3932
25. Bravo-Osuna I, Noiray M, Briand E, Woodward AM, Argueso P, Molina Martinez IT, Herrero-Vanrell R, Ponchel G (2012) Pharm Res 29:2329–2340
26. Vandamme TF, Brobeck L (2005) J Control Release 102:23–38
27. Wathier M, Jung PJ, Carnahan MA, Kim T, Grinstaff MW (2004) J Am Chem Soc 126:12744–12745
28. Durairaj C, Kadam RS, Chandler JW, Hutcherson SL, Kompella UB (2010) Invest Ophthalmol Vis Sci 51:5804–5816
29. Grinstaff MW (2007) Biomaterials 28:5205–5214
30. Yavuz B, xe; in; Bozda; x11f; Pehlivan, S.; xdc; nl; xfc; Nur; x15f; en (2013) Sci World J 13
31. Spataro G, Malecaze F, Turrin CO, Soler V, Duhayon C, Elena PP, Majoral JP, Caminade AM (2010) Eur J Med Chem 45:326–334
32. Shaunak S, Thomas S, Gianasi E, Godwin A, Jones E, Teo I, Mireskandari K, Luthert P, Duncan R, Patterson S, Khaw P, Brocchini S (2004) Nat Biotechnol 22:977–984
33. Kang SJ, Durairaj C, Kompella UB, O'Brien JM, Grossniklaus HE (2009) Arch Ophthalmol 127:1043–1047
34. Marano RJ, Toth I, Wimmer N, Brankov M, Rakoczy PE (2005) Gene Ther 12:1544–1550
35. Duan X, Sheardown H (2006) Biomaterials 27:4608–4617
36. Duan X, McLaughlin C, Griffith M, Sheardown H (2007) Biomaterials 28:78–88
37. Sugisaki K, Usui T, Nishiyama N, Jang WD, Yanagi Y, Yamagami S, Amano S, Kataoka K (2008) Invest Ophthalmol Vis Sci 49:894–899
38. Makky A, Michel JP, Maillard P, Rosilio V (2011) Biochim Biophys Acta 1808:656–666
39. Nishiyama N, Morimoto Y, Jang WD, Kataoka K (2009) Adv Drug Deliv Rev 61:327–338
40. Nishiyama N, Stapert HR, Zhang GD, Takasu D, Jiang DL, Nagano T, Aida T, Kataoka K (2003) Bioconjug Chem 14:58–66
41. Nishiyama N, Iriyama A, Jang WD, Miyata K, Itaka K, Inoue Y, Takahashi H, Yanagi Y, Tamaki Y, Koyama H, Kataoka K (2005) Nat Mater 4:934–941
42. Bravo-Osuna I, Herrero-Vanrell R, Molina Martinez IT, Giron N, Gomez R, de la Mata F, Guzman Navarro M (2010) Invest Ophthalmol Vis Sci 51:437
43. Yao W, Sun K, Mu H, Liang N, Liu Y, Yao C, Liang R, Wang A (2010) Drug Dev Ind Pharm 36:1027–1035
44. Iezzi R, Guru BR, Glybina IV, Mishra MK, Kennedy A, Kannan RM (2012) Biomaterials 33:979–988
45. Holden CA, Tyagi P, Thakur A, Kadam R, Jadhav G, Kompella UB, Yang H (2012) Nanomedicine 8:776–783
46. Yang H, Tyagi P, Kadam RS, Holden CA, Kompella UB (2012) ACS Nano 6:7595–7606
47. Kannan S, Dai H, Navath RS, Balakrishnan B, Jyoti A, Janisse J, Romero R, Kannan RM (2012). Sci Transl Med 4:130ra46
48. Szoka F Jr, Papahadjopoulos D (1980) Annu Rev Biophys Bioeng 9:467–508
49. Szoka F Jr, Papahadjopoulos D (1978) Proc Natl Acad Sci U S A 75:4194–4198
50. Damji KF, Noel LP, Peterson RG, Ma PE, Clarke WN, Gilberg S, Crouch ER Jr, Williams PB (1998) Can J Ophthalmol 33:308–313

51. Frucht-Perry J, Assil KK, Ziegler E, Douglas H, Brown SI, Schanzlin DJ, Weinreb RN (1992) Cornea 11:393–397
52. Scott LJ, Goa KL (2000) Drugs Aging 16:139–46; discussion 147–8
53. Masuda I, Matsuo T, Yasuda T, Matsuo N (1996) Invest Ophthalmol Vis Sci 37:1914–1920
54. Matsuo T, Masuda I, Yasuda T, Matsuo N (1996) Biochem Biophys Res Commun 219:947–950
55. Van der Veen G, Broersma L, Dijkstra CD, Van Rooijen N, Van Rij G, Van der Gaag R (1994) Invest Ophthalmol Vis Sci 35:3505–3515
56. Van der Veen G, Broersma L, Van Rooijen N, Van Rij G, Van der Gaag R (1998) Curr Eye Res 17:1018–1026
57. Lin HH, Ko SM, Hsu LR, Tsai YH (1996) J Pharm Pharmacol 48:801–805
58. Pleyer U, Lutz S, Jusko WJ, Nguyen KD, Narawane M, Ruckert D, Mondino BJ, Lee VH, Nguyen K (1993) Invest Ophthalmol Vis Sci 34:2737–2742
59. McCalden TA, Levy M (1990) Experientia 46:713–715
60. Kawakami S, Yamamura K, Mukai T, Nishida K, Nakamura J, Sakaeda T, Nakashima M, Sasaki H (2001) J Pharm Pharmacol 53:1157–1161
61. Frangie JP (1995) Clin Pharmacokinet 29:130–138
62. BenEzra D, Maftzir G (1990) Arch Ophthalmol 108:584–587
63. Fiscella RG, Gieser J, Phillpotts B, Gilmartin C, Labib S, Cwik MJ, Solomon MJ, Shapiro MJ (1998) Retina 18:339–342
64. Schulman J, Peyman GA, Fiscella R, Greenberg D, Horton MB, de Miranda P (1986) Br J Ophthalmol 70:138–140
65. Alghadyan AA, Peyman GA, Khoobehi B, Milner S, Liu KR (1988) Int Ophthalmol 12:109–112
66. Alghadyan AA, Peyman GA, Khoobehi B, Liu KR (1988) Int Ophthalmol 12:105–107
67. Liu KR, Peyman GA, Khoobehi B, Alkan H, Fiscella R (1987) Ophthalmology 94:1155–1159
68. Khoobehi B, Peyman GA, Niesman MR, Oncel M (1989) Ophthalmology 96:905–912
69. Aggarwal D, Kaur IP (2005) Int J Pharm 290:155–159
70. Aggarwal D, Pal D, Mitra AK, Kaur IP (2007) Int J Pharm 338:21–26
71. Li Q, Li Z, Zeng W, Ge S, Lu H, Wu C, Ge L, Liang D, Xu Y (2014) Eur J Pharm Sci 62:115–123
72. Han S, Shen JQ, Gan Y, Geng HM, Zhang XX, Zhu CL, Gan L (2010) Acta Pharmacol Sin 31:990–998
73. Gan L, Han S, Shen J, Zhu J, Zhu C, Zhang X, Gan Y (2010) Int J Pharm 396:179–187
74. Klang SH, Siganos CS, Benita S, Frucht-Pery J (1999) J Control Release 57:19–27
75. Gan L, Gan Y, Zhu C, Zhang X, Zhu J (2009) Int J Pharm 365:143–149
76. Klang S, Abdulrazik M, Benita S (2000) Pharm Dev Technol 5:521–532
77. Ayalasomayajula SP, Kompella UB (2005) Eur J Pharmacol 511:191–198
78. Kompella UB, Bandi N, Ayalasomayajula SP (2003) Invest Ophthalmol Vis Sci 44:1192–1201
79. Chaplot SP, Rupenthal ID (2013) J Pharm Pharmacol 66:542–556
80. Gomes dos Santos AL, Bochot A l, Doyle A, Tsapis N, Siepmann J, Siepmann F, Schmaler J, Besnard M, Behar-Cohen F, Fattal E (2006) J Control Release 112:369–381
81. Carrasquillo KG, Ricker JA, Rigas IK, Miller JW, Gragoudas ES, Adamis AP (2003) Invest Ophthalmol Vis Sci 44:290–299
82. Gomes dos Santos A, Bochot A l, Tsapis N, Artzner F, Bejjani R, Thillaye-Goldenberg B, de Kozak Y, Fattal E, Behar-Cohen F (2006) Pharm Res 23:770–781
83. Johnson LN, Cashman SM, Read SP, Kumar-Singh R (2010) Vision Res 50:686–697
84. Hagigit T, Abdulrazik M, Valamanesh F, Behar-Cohen F, Benita S (2012) J Control Release 160:225–231
85. Jiang M, Gan L, Zhu C, Dong Y, Liu J, Gan Y (2012) Biomaterials 33:7621–7630
86. Bloquel C, Bejjani R, Bigey P, Bedioui F, Doat M, BenEzra D, Scherman D, Behar-Cohen F (2006) FASEB J 20:389–391

87. Peeters L, Lentacker I, Vandenbroucke R, Lucas B, Demeester J, Sanders N, De Smedt S (2008) Pharm Res 25:2657–2665
88. Ormonde S, Chou C-Y, Goold L, Petsoglou C, Al-Taie R, Sherwin T, McGhee CJ, Green C (2012) J Membr Biol 245:381–388
89. Deva N, Zhang J, Green C, Danesh-Meyer H (2012) Inflammation 35:1276–1286
90. Grupcheva CN, Laux WT, Rupenthal ID, McGhee J, McGhee CNJ, Green CR (2012) Invest Ophthalmol Vis Sci 53:1130–1138
91. Rupenthal ID, Alany RG, Green CR (2011) Mol Pharm 8:2282–2290
92. Thrimawithana TR, Young SA, Bunt CR, Green CR, Alany RG (2011) Eur J Pharm Sci 44:399–409
93. Bhisitkul RB, Robinson GS, Moulton RS, Claffey KP, Gragoudas ES, Miller JW (2005) Arch Ophthalmol 123:214–219
94. Mo Y, Barnett ME, Takemoto D, Davidson H, Kompella UB (2007) Mol Vis 13:746–757
95. Park K, Chen Y, Hu Y, Mayo AS, Kompella UB, Longeras R, Ma JX (2009) Diabetes 58:1902–1913
96. Amrite AC, Kompella UB (2005) J Pharm Pharmacol 57:1555–1563
97. Jin J, Zhou KK, Park K, Hu Y, Xu X, Zheng Z, Tyagi P, Kompella UB, Ma JX (2011) Invest Ophthalmol Vis Sci 52:6230–6237
98. Tyagi P, Kadam RS, Kompella UB (2012) PLoS One 7, e48188
99. Gaudana R, Ananthula HK, Parenky A, Mitra AK. AAPS J 12:348–360
100. Cheruvu NP, Amrite AC, Kompella UB (2008) Invest Ophthalmol Vis Sci 49:333–341
101. Amrite AC, Edelhauser HF, Singh SR, Kompella UB (2008) Mol Vis 14:150–160
102. Cheruvu NP, Amrite AC, Kompella UB (2009) Pharm Res 26:404–414
103. Kompella UB, Sundaram S, Raghava S, Escobar ER (2006) Mol Vis 12:1185–1198
104. Eljarrat-Binstock E, Orucov F, Aldouby Y, Frucht-Pery J, Domb AJ (2008) J Control Release 126:156–161
105. Kim H, Robinson SB, Csaky KG (2009) Pharm Res 26:329–337

Chapter 16
Polymeric Nanoparticulate Systems: A Potential Approach for Ocular Drug Delivery

Rutika Godse, Kavita Singh, Agnivesh Shrivastava, and Ujwala Shinde

Abstract Ocular disorders are a major health concern worldwide that has a powerful impact on afflicted individuals and their families and are associated with enormous socioeconomical consequences. The prevalence of ocular disease conditions is remarkably increasing, and the pharmaceutical industry has taken a note of this significant unmet need and is putting enormous efforts in the development of safe and effective drug candidates for the treatment of ocular diseases. In addition to developing safe and effective drug candidates, their delivery to the target tissues is also of critical importance. Drugs can be delivered to the eye following local or systemic administration. The major goal in ocular drug delivery is to obtain therapeutic drug concentrations at the intended site of action for a reasonable period of time by overcoming biological barriers. Among the delivery systems designed so far for these purposes, those of a nanoscale size are particularly attractive from the easier administration and patient acceptability point of view. Multifaceted features of polymers can be efficiently exploited by engineering them into polymeric nanoparticles for ocular purpose to overcome the drawbacks of existing ocular therapy. This chapter aims to describe the actual potential of polymer-based nanoparticles for anterior and posterior segment ocular diseases. The chapter explores the suitable polymers and fabrication technology for ocular nanoparticles. It further summarizes the ability of polymeric nanoparticles by citing studies done in the field of treatment of anterior and posterior ocular conditions.

Keywords Ocular delivery • Nanoparticles • Polymers • Anterior segment • Posterior segment

R. Godse • K. Singh • A. Shrivastava • U. Shinde (✉)
Bombay College of Pharmacy, Kalina, Santacruz (E), Mumbai 400098, India
e-mail: rutika113@gmail.com; singhkavi@gmail.com; agnishri@gmail.com; ujwala29@gmail.com

© Springer International Publishing Switzerland 2016

Y. Pathak et al. (eds.), *Nano-Biomaterials For Ophthalmic Drug Delivery*,
DOI 10.1007/978-3-319-29346-2_16

16.1 Introduction

The eye, a precious, unique, and anatomically isolated specialized sensory organ, shows distinctive pharmacodynamic and pharmacokinetic properties. This is a consequence of its desolation from systemic access by the blood–retinal, blood–aqueous humor, and blood–vitreous humor barriers. Targeting the drug to the appropriate site of action in the eye is usually one of the greatest challenges in drug delivery because of its anatomical and physiological defense mechanisms. Ocular drug delivery systems thus compel specified criteria according to the physiological structure of the eye.

The conventional eyedrop solution suffers from three main impediments, i.e., low bioavailability, non-specificity of drug action toward a specific ocular target, and enzyme inactivation. Ocular disorders like glaucoma, diabetic retinopathy, macular degeneration, and squamous cell carcinoma require continuous treatment. However, conventional eyedrops cannot maintain desired therapeutic concentration for a prolonged period of time.

In addition to the three main impediments stated above, the other modified ocular delivery systems like ocular inserts and in situ gelling systems, though providing some advantages in terms of extended drug delivery, could not overcome the problems of blurred vision, sticking of the eyelids, undesirable systemic absorption, and low patient acceptance.

The presence of several types of barriers in the eye impedes the effective passage of many drugs (Fig. 16.1), leading to a minimal dose absorption. These barriers consist of:

(i) Muco-aqueous layer of the tear film that protects the anterior surface of the eye.
(ii) Tight junctions of epithelial cells and desmosomes in the cornea impede any substance larger than 500 Da.
(iii) Iris blood vessels that lack fenestrations.
(iv) Blood–aqueous barrier (BAB) is formed by tight junctions of iris vascular endothelium and nonpigmented layer of ciliary epithelial cells.
(v) Blood–retinal barrier (BRB) contains two types of cells, i.e., (1) the retinal capillary endothelial (RCE) cells and (2) the retinal pigment epithelial (RPE) cells. These cells form the inner and outer BRB, respectively.
(vi) For transscleral drug delivery, there are a few static and dynamic barriers. Scleral permeability depends on the molecular radius rather than the molecular lipophilicity, and molecules of up to 70,000 Da can readily penetrate the sclera [2–4].

Additionally, certain physiological processes contribute to the poor efficacy of conventional drug formulations. A number of factors, namely, rapid tear turnover and the resulting precorneal loss, induction of tear flow due to irritation caused by the drug preparation, as well as the relatively large volume of the administered eyedrop, lead to a high rate of lacrimal drainage [5]. The conjunctival cul-de-sac normally contains about 7–9 μL of tears. The maximum quantity of fluid held in the cul-de-sac without overflow is about 30 μL. Instillation of eyedrop stimulates the nerve in the

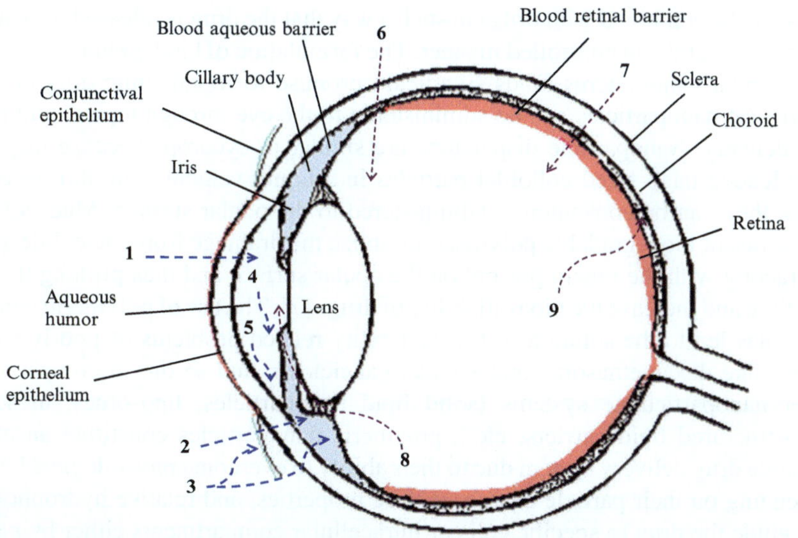

Fig. 16.1 *Schematic presentation of the ocular structure including the barriers along with the probable pathways for nanoparticles following topical, periocular, and intravitreal routes of administration.* The numbers indicate following processes: (*1*) corneal route of absorption(main route), (*2*) non-corneal route of absorption(conjunctiva–sclera), (*3*) systemic absorption (absorption through nasal mucosa, pharynx, gastrointestinal tract(main route), conjunctival, and other ocular blood vessels), (*4*) elimination from the aqueous humor into the systemic circulation across the blood–aqueous barrier, (*5*) removal from the anterior chamber by the aqueous humor turnover to the trabecular meshwork and Schlemm's canal, (*6*) intravitreal drug administration, (*7*) periocular drug administration, (*8*) elimination from the vitreous via posterior route across the blood–retinal barrier, and (*9*) elimination from the vitreous via anterior route to the posterior chamber (Adapted with permission from [1])

cornea which evokes reflex tears, and the volume of same varies from 3 to 400 μL/min according to the irritating power of the solution. The blink movement of the eyelid quickly brings the excess fluid to the lacrimal drainage system which leads to elimination of instilled drug from the precorneal area within 90 s [6]. Altogether, as a consequence, only a very small amount of about 1–3 % of the dosage actually penetrates through the cornea and is able to reach intraocular tissues [5].

Hence, the existence of several ocular tissue and cell barriers, along with the physiological processes, restricts the effective passage of many drugs, leading to a minimum dose absorption into the eye. As a result, intensive investigations for ameliorating these constraints and colloidal carriers like biodegradable polymeric nanoparticles have been broadly studied in the last two decades [7].

Nanoparticles, as the very name implies, are particles varying in size from 10 to 1000 nm, and the drug may be adsorbed, dissolved, encapsulated, or entrapped, giving rise to different terminologies such as nanoparticles, nanospheres, or nanocapsules [7]. Since the early study of [8] showing the intrinsic capacity of nanoparticles to adhere to the ocular surface and interact with the epithelium, applications of nanotechnologies to solve eye problems have been sought [2]. Biodegradable polymers can be combined with drugs in such a way that the drug is released into the eye in a very careful and controlled manner. The formulation of biodegradable polymers as colloidal nanosystems holds significant promise for ophthalmic drug delivery. Polymeric nanoparticles can be administered to the eye through topical and posterior delivery. Nanoparticle dispersions are similar to eyedrops because they have drug-loaded nano-sized colloidal particles in a liquid vehicle with akin viscosity. Thus, they can be conveniently administered to the ocular surface. Mucoadhesive properties of biodegradable polymers minimize the drainage from the cul-de-sac by interacting with the mucin present on the ocular surface and thus prolong the contact time and increase the bioavailability of drugs [7]. The use of polymeric nanoparticles has led to the solution of bioavailability-related problems of poorly soluble drugs, like dexamethasone, budesonide, ganciclovir, and so on. In comparison to other nanoparticulate systems (solid lipid nanoparticles, liposomes, niosomes, nanostructured lipid carriers, etc.), polymeric nanoparticles constitute an almost versatile drug delivery system due to their ability to overcome physiological barriers depending on their particle charge, surface properties, and relative hydrophobicity and guide the drug to specific cells or intracellular compartments either by passive or ligand-mediated targeting mechanisms [7]. In addition to these points, encapsulation of drugs in polymeric nanoparticles can also provide protection to the loaded drug and prolong exposure of the drug by controlled release. The release of loaded drug from polymeric nanoparticles can be sustained through various mechanisms, i.e., diffusion, chemical reaction, polymer degradation, or ion exchange, and by strategically selecting the polymers with requisite properties. Those properties include bioadhesion, swelling, hydration time, molecular weight, and degree of cross-linking. Such an intelligently tailored nanoparticle system may prove to be one of the best drug delivery tools for the treatment of some chronic ocular diseases, in which frequent drug administration is necessary, for example, in ophthalmic diseases like chronic cytomegalovirus retinitis [9].

Polymeric nanoparticles are also suitable in the case of the retinal disorders, as it has no lymph system; hence, retinal neovascularization and choroidal neovascularization (CNV) have similar environments to that of solid tumors, in which enhanced permeability and retention effect may be available for drug targeting by nanoparticles. Moreover, active targeting can be achieved by conjugating small molecules like galactose, mannose, and folic acid to polymeric nanoparticles as receptors on these molecules are expressed on retinal cells [10]. Thus, polymeric nanoparticles hold promise to be an important part of the new therapeutic armamentarium in ophthalmology [11]. This is further consolidated due to various benefits exhibited by polymeric nanoparticles in comparison to solid lipid nanoparticles which show low drug loading and drug expulsion during storage due to the transition of highly ordered lipid particles [12, 13].In addition, large quantities of surfactants and cosurfactants are required for their production which may cause damage to the ocular tissues; also these systems are not suitable for hydrophilic drugs [9, 14]. Other than these, various nanocarriers are also getting into trend for ocular drug delivery, i.e., dendrimer, niosomes, micelles, etc.

However, their stability, clinical suitability, safety, and tolerability are still to be proven with extensive in vitro/in vivo studies.

16.2 Polymers Used for Formulating Nanoparticles in Ophthalmic Drug Delivery

Polymers are large-molecular-weight compounds consisting of systematic or random repeat units. Polymers can be classified based on their structure (e.g., polycaprolactone, polyanhydrides), stability (e.g., biodegradable, nondegradable), charge (e.g., cationic, anionic), nature (e.g., hydrophobic, hydrophilic, amphiphilic), origin (e.g., synthetic, natural, semisynthetic), architecture (e.g., linear, branched, cross-linked), and nature of repeating units (e.g., homopolymers, copolymers, block copolymers, random copolymers). Further, polymers can be formed in various supramolecular architectures (e.g., interpenetrating and non-interpenetrating networks, micelles). Also, polymers can be designed for sensitivity to various stimuli including pH and temperature. Given the versatility of polymer design, polymer-based nanosystems hold an important position in drug delivery systems. Depending on their chemical composition and architecture, polymeric delivery systems can be designed to have certain properties such as biodegradation, sustained release, increased gene transfection efficiency, and controlled release by actuated or stimuli-sensitive physicochemical changes [15].

Eyes being the most sensitive body organ responsible for vision, it is important to carefully select the polymers for delivering the drugs through this route. As potential carriers in ophthalmic drug delivery, polymers need to have favorable properties such as lack of immunogenicity; inertness; being nontoxic, biodegradable, biocompatible, bioadhesive, easily available, and inexpensive; and ability to be modified to obtain the desired effect. The summarized information about suitable polymers with respect to their properties and nanoparticle formulation in ocular drug delivery is presented in Table 16.1.

16.3 Fabrication Techniques of Polymeric Nanoparticles

Polymeric nanoparticles for pharmaceutical use are prepared from a variety of methods mainly classified into two categories (Fig. 16.2):

(i) Polymerization of monomers
(ii) Dispersion of the preformed polymers

These nanoparticles generally comprised of synthetic and natural polymers. Synthetic polymers such as polyacrylates, PCLs, PLAs, and its copolymers with PLGs [7, 87, 88] and natural polymers like albumin, alginate [89], gelatin, and chitosan [7, 90] are widely used.

Table 16.1 Summary of characteristics of polymers and their polymeric nanoparticle formulations as ocular drug delivery

Sr.no.	Polymer	Description	Solubility	Nanoparticles
Natural polymers				
1.	Chitosan	Chitosan is a natural, cationic amino polysaccharide copolymer of glucosamine and N-acetylglucosamine. It is obtained by the alkaline, partial deacetylation of chitin, which originates from shells of crustaceans such as crabs and prawns. Chitosan is a biodegradable, biocompatible, less toxic, and mucoadhesive biopolymer [16]	Dissolves readily in dilute and concentrated solutions of most organic acids and to some extent in mineral inorganic acids (except phosphoric and sulfuric acids) Sparingly soluble in water, practically insoluble in ethanol (95 %), other organic solvents [17]	[16, 18–31]
2.	Sodium hyaluronate	Hyaluronic acid is negatively charged and is ubiquitously found in the connective tissue, umbilical cord, vitreous humor, synovial fluid of the human body, and rooster comb, and microorganisms from *Streptococcus* species can form sodium hyaluronate by the process of fermentation. Chemically, it is a high-molecular-weight polyanionic linear mucopolysaccharide composed of alternating units of N-acetylglucosamine and D-glucuronic acid [32]	Soluble in water, slightly soluble in mixtures of organic acids and water[17]	[17, 19, 32–35]
3.	Sodium alginate	It is a negatively charged hydrophilic colloidal polysaccharide of sodium salt of alginic acid. Alginic acid is present in the cell wall of the brown seaweeds *Laminaria*, *Macrocystis*, and *Ascophyllum* (class Phaeophyceae). Chemically, it is polyuronic acid formed of D-mannuronic acid and L-guluronic acid [17]	Soluble in alkali hydroxides, very slightly soluble in ethanol (95 %) and other organic solvents [17]	[17, 36–38]
4.	Gelatin	Gelatin is the partially hydrolyzed product of collagen and is composed of a unique sequence of amino acids with high content of glycine, proline, and hydroxyproline. The gelatin molecules contain repeating sequences of glycine-X-Y triplets, where X and Y are frequently proline and hydroxyproline amino acids. It is a mucoadhesive polymer and possess amphoteric charge[17]	Soluble in hot water, glycerin, weak acids, and weak alkalis. Practically insoluble in acetone, chloroform, ethanol (95 %), ether, and methanol[17]	[17, 39–43]

5.	Albumin	Albumin is a protein naturally found in blood and also found in many food products like milk and egg white. Human serum albumin is a single polypeptide chain of 585 amino acids and contains seven disulfide bridges. It is negatively charged, biodegradable, highly compatible, and nonantigenic and has no toxic effects [17]	Freely soluble in water and salt solutions [17]	[5, 17, 44–48]
6.	Tamarind seed polysaccharide (xyloglucan)	It is a nonionic linear plant polysaccharide with side chains and is present as a storage polysaccharide in the cell walls of higher plants. The backbone structure is (β1→4)-linked D-glucan. A part of glucose unit is substituted with α-D-xylose at the O6 position [49]	Due to the balancing of hydrophilic and hydrophobic character, xyloglucan shows solubility in water, but the individual macromolecules do not hydrate fully leading to the presence of aggregated species in water [49]	[49, 51]
7.	Gum cordia	It is a mucilaginous substance extracted from the raw fruits of *Cordia obliqua* Willd and is a negatively charged polymer [50]	Soluble in water on addition of sodium hydroxide [52]	[52, 53]
Synthetic polymers				
8.	Acrylates	They are bioadhesive, anionic polymers having acrylic acid as the monomer. The approximate pK^a for the poly(acrylic acid) family of polymers is between 4 and 5. The maximum adhesive strength of these polymers is observed around pH 4–5 and decreases gradually above a pH of 6 [17, 54]	Polymers belonging to the acrylate family vary in their solubilities [17]	[17, 54–63]
9.	Poly(D,L-lactic acid) (PLA) and poly(D,L-lactide-co-glycolide) (PLGA)	Polyester PLGA is a copolymer of PLA and polyglycolic acid (PGA). It is the best defined polymer available for drug delivery with respect to design and performance. The enantiomeric forms of the polymer PLA are poly D-lactic acid and poly L-lactic acid [64]	PLGAs are soluble in organic solvents such as chloroform and dichloromethane but insoluble in common alcohols like ethanol [65]	[58, 64–77]
10.	Poly-ε-caprolactone (PCL)	Semicrystalline polymer having a low glass transition temperature (260°C) and low melting point (55–60°C). It has the ability to form compatible blends with other polymers, which provides opportunities to manipulate the drug release rate from matrix. Its slow degradation, high permeability to many drugs, and nontoxicity lead to study for a long-term drug/vaccine delivery vehicle [78]	PCL is soluble in chloroform, dichloromethane, carbon tetrachloride, benzene, toluene, cyclohexanone, and 2-nitropropane at room temperature and is insoluble in alcohol, petroleum ether, and diethyl ether [79]	[78–86]

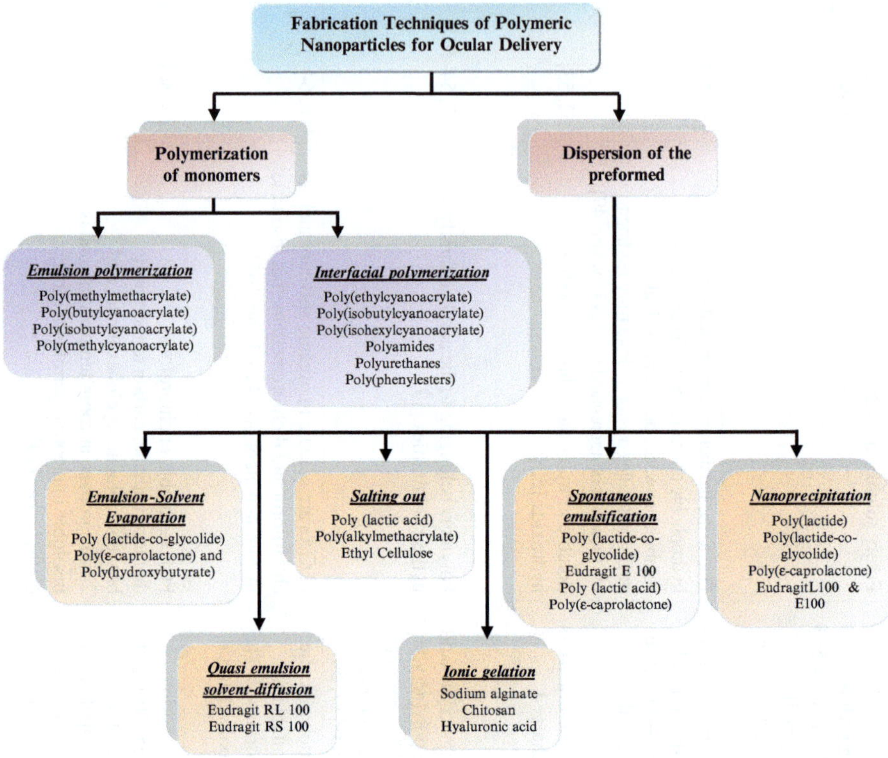

Fig. 16.2 Flow chart for fabrication techniques of polymeric nanoparticles for ocular delivery

16.3.1 Polymerization Method

Polymerization methods are classified into emulsion and interfacial polymerization technique. Emulsion polymerization could be either organic or aqueous depending on the continuous phase.

The continuous organic phase technique involves the dispersion of a monomer into an emulsion or a non-solvent. This procedure requires toxic organic solvents, surfactants, monomers, and an initiator, which are subsequently difficult to eliminate from the formed nanoparticles. However, in the aqueous continuous phase method, the monomer is dissolved in a continuous phase that is usually an aqueous solution, and the surfactants or emulsifiers are not essential. The polymerization process can be initiated by different mechanisms, i.e., the use of an initiator like an ion or a free radical, high-energy radiation like γ-radiation or ultraviolet, or strong visible light [91]. Chain growth initiates when monomer ions or monomer radicals collide with other monomer molecules according to an anionic polymerization mechanism. This technique is suitable for water-soluble drugs [92–94]. Drug is incorporated either by dissolving in the polymerization medium or by adsorption onto the nanoparticles after polymerization is completed. This technique has been reported for making polybutylcyanoacrylate (PBC) or poly(alkylcyanoacrylate) (PACA) nanoparticles [92, 94]. Nanocapsule formation and their particle size depend on the concentration of the surfactants and stabilizers used [93, 95]. Poly(methylmethacrylate) (PMMA), poly(ethylcyanoacrylate) (PECA), and PBC nanoparticles were produced by dispersion via surfactants into solvents such as cyclohexane, n-pentane, and toluene as the organic phase. For ophthalmic use, the drugs like triamcinolone [96], pilocarpine, and timolol [97] were encapsulated in these systems. However, the toxicity of the organic solvent used in the preparation of nanoparticles limits the feasibility of this method for ocular purpose.

Interfacial polymerization technique reported for cyanoacrylates; in this method, the monomer of cyanoacrylate and drug were dissolved in a mixture of an oil and absolute ethanol [98, 99]. It was then followed by slowly injecting the mixture into an aqueous solution containing surfactant and initiating ions under stirring to form nanoparticles by polymerization. This technique gives high entrapment efficiency of drugs [99, 100].

16.3.2 Dispersion of Preformed Polymers

Several methods have been reported to prepare biodegradable nanoparticles from PLA, PLG, PLGA, and PCL by dispersing the preformed polymers [75, 85].

Synthetic polymers are typically dissolved in a convenient solvent followed by precipitation in a liquid environment leading to nanoparticle formation. The drug intended to be encapsulated in the particles is usually incorporated during the polymer solvation and precipitation. Emulsification/solvent diffusion, emulsification/solvent evaporation, nanoprecipitation, salting out, and quasi-emulsion methods are widely applied techniques and are discussed here.

16.3.2.1 Emulsification/Solvent Evaporation Method

In this method, the polymer is dissolved in an organic solvent like dichloromethane, chloroform, or ethyl acetate. The drug is dissolved or dispersed into the preformed polymer solution, and this mixture is then emulsified into an aqueous solution to make an O/W emulsion by using a surfactant/emulsifying agent like polysorbate 80, poloxamer 188, gelatin, polyvinyl alcohol, etc. After the formation of a stable emulsion, the organic solvent is evaporated either by increasing the temperature and/or under reduced pressure or by continuous stirring. The W/O/W method has also been used to prepare the water-soluble drug-loaded nanoparticles [101]. Both the above methods involve the use of a high-energy technique, i.e., high-speed homogenization or sonication. However, these procedures are good for a laboratory-scale operation, but in the case of large-scale production, low-energy emulsification is a more appropriate and viable technique. Dillen et al. have reported development of PLGA ocular nanoparticles loaded with ciprofloxacin HCl by W/O/W emulsification/solvent evaporation method using polyvinyl alcohol as stabilizer [102].

16.3.2.2 Spontaneous Emulsification/Solvent Diffusion Method

In a modified version of the solvent evaporation method [88, 99, 103], the water-soluble solvent like acetone or methanol along with the water-insoluble organic solvent like dichloromethane or chloroform was used as an oil phase. Due to the spontaneous diffusion of the water-soluble solvent (acetone or methanol), an interfacial turbulence is created between two phases leading to the formation of smaller particles. As the concentration of water-soluble solvent increases, a considerable decrease in particle size can be observed. Literature reports various drug-containing polymeric nanoparticles by solvent diffusion method. Triptolide-loaded PLA nanoparticles were prepared by the modified spontaneous emulsification/solvent diffusion method by Liu et al. This technique had presented some advantages such as avoiding the use of higher-energy apparatus and chlorinated solvent, reproducibility, and the possibility of scale-up [104].

16.3.2.3 Salting Out/Emulsification–Diffusion Method

Salting out is based on the separation of a water-miscible solvent from aqueous solution via a salting out effect. The salting out procedure can be considered as a modification of the emulsification/solvent diffusion method. Polymer and drug are initially dissolved in a solvent such as acetone, which is subsequently emulsified into an aqueous gel containing the salting out agent (electrolytes such as magnesium chloride, calcium chloride, and magnesium acetate or non-electrolytes such as sucrose) and a colloidal stabilizer such as polyvinylpyrrolidone or hydroxyethylcellulose. This oil/water emulsion is diluted with a sufficient volume of water or aqueous solution to enhance the diffusion of acetone into the aqueous phase, thus inducing the formation of nanospheres. The selection of the salting out agent is important, as it is a critical factor which determines the encapsulation efficiency of the drug. Since this method does not require increase in temperature, it is a recommended method for heat-labile materials. This method can only be used for lipophilic drugs and also requires thorough washing of nanoparticles [99, 105]. A modified emulsification–diffusion–evaporation technique was reported for preparation of gatifloxacin-loaded PLGA nanoparticles stabilized by polyvinyl alcohol and chitosan for ocular purpose [36]. The verteporfin-loaded nanoparticles were prepared using the salting out technique for effective treatment of certain types of CNV associated with age-related macular degeneration (AMD) [106].

16.3.2.4 Nanoprecipitation Method

Nanoprecipitation differs from the emulsification–diffusion and salting out methods in that formally no precursor emulsion is formed during nanoparticle preparation. In fact, this process can be achieved with or without mechanical work. Basically, nanoparticle formation is explained in terms of the interfacial turbulence and the "diffusion-stranding" processes between two unequilibrated liquid phases [107, 108]. In the standard procedure, the organic phase containing the polymer is poured into the aqueous phase under slight magnetic stirring to form nanoparticles. When both phases are in contact, it is assumed that the solvent diffuses from the organic phase into the water and carries with it some polymer chains which are still in the solution. Then, as the solvent diffuses further into the water, the associated polymer chains aggregate forming nanoparticles. Sparfloxacin-loaded PLGA nanoparticles using nanoprecipitation were prepared by Gupta et al. for sustained ocular drug delivery in treatment of conjunctivitis [109].

16.3.2.5 Quasi-emulsion Solvent Diffusion (QESD)

The method was first described by Kawashima et al. for the preparation of microspheres of ibuprofen with acrylic polymers. In the typical QESD approach, a solution of the drug and polymer in solvents exhibiting high solubility is poured into an anti-solvent (poor solvent) with rapid mixing to achieve quasi-emulsion (short-lived) droplets. These quasi-emulsion droplets (translucent) are stirred and mixed with additional anti-solvent to remove the solvent from the emulsion droplets (via diffusion) and to precipitate the drug in the polymer matrix. After solvent diffusion, emulsion droplets solidify and turn into nanoparticles with the drug entrapped in the polymer matrix (solid dispersions). To control the size of droplets and achieve homogeneous distribution of the drug and polymer within the emulsion, high-speed mixing is a requisite [110]. Ibuprofen-loaded polymeric ophthalmic nanoparticle suspensions were made from inert polymer resins (Eudragit RS100®) by a similar quasi-emulsion solvent diffusion technique using variable formulation parameters [62].

16.3.2.6 Production of Nanoparticles Using Supercritical Fluid Technology

Literature on the production of drug-loaded microparticles using supercritical fluids is enormous. However, comparatively much less have been investigated to produce ocular nanoparticles. Two most commonly used methods of producing micro- or nanoparticles using supercritical fluids include rapid expansion of supercritical solution (RESS) method and supercritical anti-solvent (SAS) method [88, 111].

16.3.2.7 Coacervation or Ionic Gelation Method

Remarkable research has been focused on the preparation of nanoparticles using biodegradable hydrophilic polymers such as chitosan, sodium alginate, hyaluronic acid, albumin, gelatin, etc. These methods involve ionic interaction between oppositely charged polymer and cross-linking agent. These are simple processes with mild preparation conditions. Polymers containing positively charged groups interact with negatively charged cross-linking agent to form coacervates with a size in the range of nanometer and vice versa. Coacervates are formed as a result of electrostatic interaction between two aqueous phases, whereas ionic gelation involves ionic interaction in which transition of polymeric solutions from sol to gel takes place at room temperature. Calvo et al. developed a method for preparing polymeric nanoparticles by ionic gelation. The method involves a mixture of two aqueous phases, of which one is the polymer chitosan or a di-block copolymer ethylene oxide or propylene oxide (PEO–PPO) and the other is a polyanion sodium tripolyphosphate [81]. Chitosan nanoparticles prepared by ionic gelation are extensively studied for ocular delivery of drugs like acyclovir [27], betaxolol hydrochloride [112],brimonidine tartrate [29], cyclosporine A [16], etc. Nanoparticles of sodium alginate loaded with brimonidine tartrate are reported wherein sodium alginate nanoparticles were prepared by the cation-induced controlled gelation of negatively charged sodium alginate and positively charged calcium chloride [29].

16.4 Processing of Nanoparticles After Preparation

Several types of treatments can be applied to nanoparticle suspensions after preparation. They include purification, sterilization, drying, and characterization.

16.4.1 Purification of Nanoparticle Suspension

Before in vivo administration of nanoparticles, purification is mandatory to achieve therapeutically safe and acceptable product. To remove the excess of reagents and by-products formed during nanoparticle preparation, the nanoparticle suspension is subjected to purification. Organic solvents, oil, surfactants, residual monomers, polymerization initiators, salts or stabilizing agents, and excess polymer aggregates need to be separated from nanoparticles after preparation. These purification methods are specific to methods employed to manufacture the nanoparticles [113].

Most common methods include evaporation under reduced pressure, centrifugation, ultracentrifugation techniques, filtration, dialysis, gel filtration, dia-filtration, and cross-flow microfiltration. For methods like nanoprecipitation, salting out, emulsification/solvent diffusion, and interfacial polycondensation, evaporation under reduced pressure is the most appropriate option since the mixture contains a large quantity of solvents[113]. Filtration processes are utilized in the case of intravenous and ophthalmic products to remove large particles and polymer aggregates. Similarly, the centrifugation method is also used to remove large particles and aggregates but cannot separate defined particle size like filtration process and may not purify high-density nanoparticles due to sedimentation. Ultracentrifugation is used to sediment and concentrate nanoparticle to obtain a pellet (at 100,000–110,000 × g for 30–45 min), which is resuspended into fresh biocompatible medium. However, the method suffers from a disadvantage, which includes breaking of fragile nanoparticles upon repeated cycles of ultracentrifugation, and it is not always possible to redisperse the obtained pellet [113].

Dialysis is another method involving the use of different cellulose membranes of various molecular cutoffs which permit diffusion of lower-molecular-weight component in the medium outside the dialysis bag [113]. For efficient purification by dialysis method, large excess of counter-dialysis medium is required so that the concentration gradient is maintained throughout the purification process. Also it is not suitable for large-scale purification because it requires long time and supports microbial growth, rendering cross-flow filtration, dia-filtration, ultrafiltration, and microfiltration as the methods of choice. Cross-flow filtration utilizes ultrafiltration membrane and has advantage of being a faster method with negligible effect on nanoparticle properties. Although various filtration methods offer many advantages, problems like caking, precipitation of particles, and decreased membrane permeability are of concern. Gel filtration can also be used to separate high-molecular-weight compounds since it is a faster method compared to dialysis [113]. Gel filtration has disadvantages like irreversible adsorption of actives in column, poor resolution due to large size of impurity, and that only a small volume sample can be purified. Among these methods, cross-flow filtration and dia-filtration are most efficient; they do not cause aggregation of nanoparticles and can be used in large-scale production.

16.5 Sterilization of Purified Nanoparticles

Sterility is an essential requisite for clinical use of an ophthalmic drug delivery system. The most convenient and widely accepted methods for sterilization of ocular nanoparticulate systems include moist heat sterilization, gamma radiation, and sterile filtration. Although these sterilization techniques are popular, they also have various limitations which may adversely affect the key features of nanoparticles. Autoclaving is one of the most effective techniques but may induce decomposition or degradation of polymers having a glass transition and/or a melting point below 120 °C and can also catalyze some reactions with additives such as surfactants, which can modify the polymers constituting the nanoparticles [113, 114]. Gamma radiation can be applied on heat-sensitive materials and has the advantage of a homogeneous sterilization. However, the exposure to radiation energy may induce fragmentation of covalent bonds and produce free radicals which, in turn, can damage the polymer forming the nanoparticles or induce production of compounds with toxicological hazard [113, 115]. Sterile filtration may be used for chemically or thermally sensitive materials. It involves filtration of the nanoparticles through membrane filters with pore diameters of 0.22 μm. Indeed, this technique of sterilization can only be applied on nanoparticle suspensions of low viscosity and containing particles with size having a diameter below the pore size [113, 116]. Therefore, considering these limitations, it is necessary to carefully choose a suitable technique as per the physical properties of the nanoparticle and its components.

16.6 Drying of Nanoparticles

As in the case of many pharmaceutical preparations, the storage of nanoparticles as suspensions presents many disadvantages such as risk of microbiological contamination, hydrolysis, and degradation of polymer, physicochemical instability due to particle aggregation and sedimentation, and loss of the biological activity of the drug [117]. To circumvent such problems, pharmaceutical preparations are stored in dry form. In general, the transformation of a liquid preparation into a dry product can be achieved using freeze-drying or spray-drying processes [113]. Both processes can be used at an industrial scale, and their applications were tested on nanoparticle suspensions.

16.7 Characterization of Nanoparticles

Key parameters for characterizing a nanoparticle suspension are particle size, size distribution, surface properties of nanoparticles, drug loading, in vitro drug release, and ocular tolerability studies. Particle size and size distribution are the most important characteristics of a nanoparticle system. They determine the in vivo distribution, biological fate, toxicity, and targeting ability of nanoparticle systems. In addition, they can also influence the drug loading, drug release, and stability of nanoparticles. Surface properties of nanoparticles play a vital role since the immune system easily identifies the nanoparticles as foreign bodies and then clears them by phagocytes [118]. The in vivo fate of nanoparticles is closely dependent on the surface hydrophobicity because it leads to the adsorption of blood components, mainly proteins (opsonins) [119].

An acceptable nanoparticulate system should have a high drug-loading capacity thereby reducing the quantity of polymeric matrix during ocular administration. In vitro release profiles of nanoparticles are determined by the following methods [93]:

1. Side-by-side diffusion cells with artificial or biological membranes
2. Dialysis bag diffusion technique
3. Reverse dialysis bag technique
4. Agitation followed by ultracentrifugation/centrifugation
5. Using flow through cell (USP type IV dissolution apparatus) [120]

Draize test and HET-CAM studies give idea about the ocular tolerability.

16.8 Polymeric Nanoparticles for Ocular Drug Delivery Systems

The anterior segment and the posterior segment are two entirely different ocular regions, and the challenges faced in delivering therapeutic drugs to each of these areas are unique; hence, they are dealt separately in the following section.

16.8.1 Polymeric Nanoparticles for Anterior Segment Drug Delivery

The anterior segment of the eye consists of the cornea, conjunctiva, iris, ciliary body, and lens with its zonules. The major routes for anterior segment drug delivery are topical delivery and subconjunctival injection [121] as shown in Fig. 16.1. Nano-sizing has offered a whole new array of scientific development toward ocular drug administration.

One goal of topical delivery systems is to enhance the precorneal residence time of the drug which can be catered by nanoparticles prepared using natural and synthetic polymers. Nanoparticulate drug delivery to anterior segment of the eye has been rigorously researched by scientists, and polymeric nanoparticles have been proved to be an advantageous alternate over other nanocarriers. The most preferred natural polymers for ocular purpose are chitosan, sodium alginate, albumin, gelatin, and sodium hyaluronate. Synthetic polymers like PLGA, PLA, PCL, Eudragit® L100, and Eudragit® E100 have also been extensively studied in the ophthalmic arena.

"One of the first clinical studies addressing ocular disease treatment have been reported by Ticho et al. using Piloplex systems consisting of pilocarpine-loaded nanospheres produced with PMMA–acrylic acid copolymer" [122]. Piloplex was found to lower the IOP in clinical trials; however, due to their nonbiodegradability, local toxicity, and difficulty in the large-scale manufacturing, it was proved commercially nonviable.

In 1997, Calvo et al. reported ionic gelation as a method for development of chitosan nanoparticles; since then several drug-loaded nanoparticles of chitosan were explored for ocular delivery [82]. In this method, the particle size of nanoparticles is affected by chitosan/sodium tripolyphosphate ratio, whereas it's not affected by the type of drug. Today chitosan nanoparticles are established as a promising approach for management of glaucoma and external inflammatory and autoimmune ocular diseases.

Papadimitriou et al. developed dorzolamide-loaded chitosan nanoparticles by ionic gelation method. The formulated nanoparticles showed higher mucoadhesion with sustained in vitro release profile. The study suggested potential for efficacious glaucoma management and warranted in vivo evaluation into a suitable animal model [123].

Nanoparticles of chitosan loaded with brimonidine tartrate have been developed. Brimonidine tartrate is an α-agonist used in treatment for open-angle glaucoma. Patients with this form of glaucoma need to frequently instill brimonidine eyedrops, which lowers intraocular pressure (IOP) only for 3–4 h. When the eyedrop formulation of brimonidine was evaluated in comparison with the developed brimonidine-loaded chitosan nanoparticles in healthy albino rabbits, a prolonged lowering in IOP was observed up to 8 h [29]. The in vitro release profile studies showed an initial burst release followed by a sustained release over the period of 24 h. These results were promising, and further evaluation in a glaucoma animal model is recommended.

Recently, another chitosan-based nanosystem has been reported by our group, wherein chitosan is modified by o-carboxymethylation. The developed o-carboxymethylated chitosan nanoparticles (OCM-CSNP) were found to be a potential reservoir for topical delivery of dorzolamide [124]. The foremost advantage of this system is that the o-carboxymethyl-chitosan derivative has high solubility at neutral pH. This is a desired property for the ocular route which otherwise faces major drawback of ocular irritation and tear flow causing immediate drug washout prepared with chitosan nanoparticles in which the vehicle is acidic. The nanoparticles prepared with modified chitosan possessed higher mucoadhesive property than a native polymer. The developed OCM-CSNPs were found to be tolerable in ocular cul-de-sac when tested by HET-CAM eye irritation test. In vivo studies of dorzolamide-loaded OCM-CSNPs on normotensive New Zealand rabbits exhibited a promising prolonged antiglaucoma effect as compared to chitosan nanoparticles.

In a study by Motawani et al., the potential of chitosan–sodium alginate nanoparticles as drug carriers for ocular delivery was investigated, where gatifloxacin was used as a model drug. In vitro release studies demonstrated sustained release of gatifloxacin from polymeric nanoparticles over a period of 24 h, primarily by non-Fickian diffusion. Authors have accounted this new formulation as a viable alternative to conventional eyedrops by virtue of its ability to sustain the drug release and its ease of administration because of reduced dosing frequency resulting in better patient compliance [24].

Chitosan has also been used to coat the nanoparticles, since the surface characteristics of the nanocarriers also have an influence on the interaction with the ocular surface structures. Bioadhesion and in vivo fate of nanoparticles can be improved by surface characteristics of polymeric nanoparticles. Literature reports development of cyclosporine A-loaded PLGA nanoparticles coated with chitosan, prepared using the O/W emulsification/solvent evaporation method. It was observed that PLGA nanoparticles coated with 0.3 % (w/v) chitosan solution showed higher muco-/bioadhesion due to the positive charge of chitosan-coated nanoparticles. There may be electrostatic interaction between the positively charged chitosan and the negatively charged mucus layer of the eye. In vitro tests revealed low to absent cytotoxic effects along with significant suppression of IL-2 secretion in concanavalin A-stimulated Jurkat T cells by the nanoparticles [125]. For instance, chitosan-coated PCL nanocapsules enter the corneal epithelium in vivo more efficiently than uncoated PCL or polyethylene glycol-coated PCL nanocapsules [83].

Nanoparticles composed of hyaluronic acid and chitosan (HA–CS nanoparticles) are reported by de la Fuente et al. which are capable of encapsulating hydrophilic as well as hydrophobic drugs such as the polypeptide cyclosporine A [20]. HA–CS nanoparticle matrix may be suitable for gene delivery to carry plasmid DNA as shown with ocular surface-derived cell lines [21] with good in vivo tolerance [18].

Sodium alginate is another potential natural polymer offering desirable mucoadhesive and in situ gelling properties hence explored for ocular nanoparticulate delivery. Brimonidine tartrate-loaded sodium alginate nanoparticles were investigated for improved ocular retention and to achieve better efficacy of brimonidine tartrate in lowering IOP generated in patients suffering from open-angle glaucoma [126].

Nanoparticles were evaluated in albino rabbits for ocular efficacy of brimonidine tartrate and exhibited sustained lowering in IOP. Therefore, it was proved to be a potential delivery platform for ocular delivery and can be evaluated for incorporation of other suitable drugs in market for ophthalmic use.

Albumin is a widely available natural polymer well accepted for its safety; hence, it is utilized to prepare nanoparticles for ocular route. Drugs like pilocarpine and hydrocortisone are entrapped in the polymer matrix of albumin which facilitates anterior delivery and may prolong its residence in the eye, minimizing the opacification observed for macroscopic implants [48, 127]. Likewise, gelatin nanoparticles of pilocarpine and hydrocortisone are also reported in literature. Gelatin offers advantages like biocompatibility and biodegradability, along with the presence of collagen, the native protein from which gelatin is derived; is found in the eye, specifically in the stroma, the middle cell layer of the cornea; and has been extensively employed in ocular applications [128, 129]. The drug encapsulation was found to be lower in the case of pilocarpine hydrochloride-loaded gelatin nanoparticles. The release mechanism for hydrocortisone was observed to be anomalous and demonstrated close to zero-order release kinetics. However, compared to the aqueous drug solutions, a sustained release for both drugs was observed when delivered in nanoparticulate form [130].

Eudragit® RS and RL polymers are copolymers of partial esters of acrylic and methacrylic acids with alcohols containing a low amount of quaternary ammonium groups (approximately 5 % and 10 % for RS and RL, respectively). The resulting matrix shows water insolubility having permeability profile independent of pH variations and thus represents a valid material for controlled drug delivery by anterior route.

Ibuprofen-loaded Eudragit®RS 100 and RL 100 nanoparticles demonstrated improved drug levels in the aqueous humor after application, and it did not show toxicity in ocular tissues [56, 62]. Similarly, flurbiprofen-loaded nanoparticle suspensions of Eudragit® RS 100 and RL 100 were found to exhibit improved drug concentration in the aqueous humor and an antimycotic activity higher than a reference commercial eyedrop formulation during in vivo test in rabbits. It was observed that flurbiprofen-loaded nanosuspensions are devoid of ocular toxicity [63]. Cloricromene-loaded Eudragit® RS100 and RL 100 nanoparticles have been investigated for in vivo performance. The study revealed that the inflammation can be inhibited by the nanoparticles more effectively compared to the microsuspension of drug alone in the rabbits with endotoxin-induced uveitis [131, 132]. Diclofenac diethyl ammonium, piroxicam [133], and methylprednisolone [134] have also been encapsulated in Eudragit® nanoparticles with excellent encapsulation efficiency.

PLGA is one of the most successfully used biodegradable nanosystems for the development of nanomedicines because it undergoes hydrolysis in the body to produce the biodegradable metabolite monomers, lactic acid and glycolic acid. Since the body effectively deals with these two monomers, there is very minimal systemic toxicity associated with PLGA in drug delivery or biomaterial applications [135]. Flurbiprofen-loaded nanoparticles of PLGA, coated with poloxamer 188, were examined in vivo in rabbits and demonstrated devoid of toxicity or irritation. After induction of inflammation by instillation of sodium arachidonate into the animal model, nanoparticle formulations were evaluated in comparison with the commercial eyedrops (Ocoflur® Allergan, Belgium). The results indicated a very good anti-inflammatory efficacy for flurbiprofen-loaded PLGA nanoparticles providing controlled and continuous drug delivery [136].

Polyisobutylcyanoacrylate/PBC polymers have been examined for ocular delivery of metipranolol, amikacin sulfate, betaxolol-chlorohydrate, and progesterone. In the case of metipranolol-loaded polyisobutylcyanoacrylate, although the polymer exhibited reduction in systemic toxicity, it could not control the release of metipranolol from dosage form over the period of time [137]. Achieving therapeutically efficacious dose of the drug for prolonged duration in the target and surrounding tissues of the eye has always been a challenge in ocular delivery. Losa et al. tested PBC nanoparticles with different stabilizing agents to improve the binding of the antibiotic amikacin sulfate to the nanoparticles [138]. For the formulation, containing dextran 70,000 as a stabilizer, significant increase in amikacin concentration was observed. The increase of the amikacin concentration in the cornea and aqueous humor was statistically significant for this nanoparticle formulation compared to other formulations and the control solution. Similarly, nanoparticles of polyisobutylcyanoacrylate containing betaxolol-chlorohydrate were studied by Marchal-Heussler et al. The surface charge of the particles and the binding type of the drug with the nanoparticles were found to be the factors of greater significance in comparison to the drug adsorption percentage onto the nanoparticles [139].

The biodegradable as well as biocompatible PACAs are degraded by esterases in biological fluids and produce some toxic products that will stimulate or damage the central nervous system. Thus, this polymer is not recommended for application in humans [140]. However, PACA nanoparticles are reported for drug delivery in ocular anti-inflammatory therapy, for the delivery of dexamethasone [141] and triamcinolone [9], revealing improved encapsulation efficiencies for the anti-inflammatory drugs and acceptable physicochemical characteristics for ocular administration. PACA nanoparticles and nanocapsules have been shown to improve and prolong the corneal penetration of hydrophilic [57] and lipophilic drugs [137]. Despite these positive results, the potential of the PACA nanoparticles is limited because of the disruption they cause to the corneal epithelium cell membrane [5]. Hence, literature doesn't show sufficient work done using these polymers for research and development of polymeric nanoparticles suitable for drug delivery to the anterior segment of the eye.

PLA is a biocompatible and biodegradable material which undergoes scission in the body to monomeric units of lactic acid as a natural intermediate in carbohydrate metabolism. Acyclovir-loaded nanoparticle of PLA developed by Giannavola et al. has shown potential for ophthalmic delivery of acyclovir for the treatment of ocular viral infections [69].

PCL is degraded by hydrolysis of its ester linkages in physiological conditions and has therefore received a great attention for use in drug delivery. In particular, it is especially interesting for the preparation of long-term implantable devices, owing to its degradation slower than that of polylactide [135]. Carteolol-loaded nanoparticles of

PCL are reported to show improvement in ocular penetration. The therapeutic results, i.e., decrease in IOP with these nanoparticles, were much more pronounced than with the commercial eyedrops [137, 142]. Literature also shows existence of colloidal polymeric suspension of PCL containing metipranolol. The pharmacologic response was not affected by the encapsulated metipranolol compared with the commercial eyedrops, but a drastic reduction of the drug's systemic side effects was observed. Hence PCL nanoparticles offer a good opportunity for ophthalmic delivery of drugs.

16.8.1.1 Disposition of Nanoparticles in Anterior Segment of Eye

The transport pathways by which nanoparticles penetrate the ocular surface tissues are of great interest. Elimination of topically instilled nanoparticle suspensions may not be very different from the conventional ophthalmic solutions. Ocularly instilled nanoparticles may enter through following three passages: (i) the cornea, (ii) the conjunctiva, or (iii) drain via the nasolacrimal duct. The drug in the nanocarriers passes through cornea and conjunctiva to reach various ocular tissues primarily to the anterior segment. The drug in the nanoparticulate form enters the nose, from where it gets eliminated via the gastrointestinal tract. If the drug carried as nanoparticle is sufficiently stable in nasal and gastrointestinal tissues, it may result in systemic absorption of the drug. Following topical administration of nanoparticles, a significant portion of it is lost due to drainage through the nasolacrimal duct resulting in drug levels in the circulation. It is possible that very small nanoparticles might escape the nasal and gastrointestinal epithelial barriers to enter the systemic and portal circulations, respectively [143].

A precorneal clearance study of 100–300 nm dextran particles in a rabbit model using gamma scintigraphy indicated biphasic drainage of particles from the cornea with an initial rapid decline in 15 s and a later slower terminal half-life of 2.15 ± 0.09 min [144]. Following topical administration in rabbits, radiolabeled poly(hexyl-cyanoacrylate) particles exhibited high tears which declined over a period of 6 h, whereas the levels in the cornea and the conjunctiva did not change significantly throughout the 6 h. The authors speculated that the nanoparticles might have adhered to the cornea and conjunctiva, with nasolacrimal drainage being the major route of precorneal elimination. Another study showed that PCL nanocapsules coated with chitosan has potential to elevate the precorneal residence time of nanoparticles [145]. A second objective of topical nanoparticles is to enhance the cellular uptake of the drug. Zimmer et al. studied the ocular transport pathway of fluorescein-labeled PBCA nanoparticles in rabbits [146]. They found that the fluorescence signal localized inside conjunctival and corneal epithelial cells and observed differences in depth of tissue penetration. They proposed that the superficial cellular layers of the cornea revealed higher penetration, whereas there were no particles observed in the corneal stroma and endothelium. In vitro studies with conjunctival cells have demonstrated uptake of the nanoparticles by an endocytic process not mediated by clathrin- or caveolin-containing vesicles [147, 148]. The transcellular route was also proposed for coated PCL nanocapsules [83].

In summary, a significant proportion of the topical dose instilled in the form of drug-loaded polymeric nanoparticles remains in the tears and is lost through nasolacrimal drainage; however, drug clearance kinetics is changed by nanoparticles in comparison to a solution form. Conjunctival uptake of topical nanoparticles is higher than the corneal uptake, and inflammatory conditions modify the uptake mechanism with elevated tissue levels. The nanoparticles entering the tissue are cleared slowly and hence may offer prolonged drug levels.

16.8.2 Polymeric Nanoparticulates as Posterior Segment Ocular Drug Delivery Systems

The most prevalent posterior eye diseases which cause visual impairment include AMD, macular edema secondary to retinal vein occlusion, uveitis, proliferative vitreoretinopathy (PVR), CNV, diabetic retinopathy, cytomegalovirus (CMV) retinitis, retinitis pigmentosa, and intraocular solid tumors [4, 149].

16.8.2.1 Routes of Drug Delivery to the Posterior Eye

Topical, systemic, intravitreal, and periocular routes can be used to deliver pharmaceuticals to the posterior segment of the eye. The topical route is inefficient in delivering therapeutic concentration of a drug to the posterior segment, owing to rapid drainage through the nasolacrimal ducts, low permeability of the corneal epithelium, systemic absorption, and the blood–aqueous barrier. Conversely, the BRB hinders the diffusion of systemically administered drugs to the posterior segment of the eye. Frequent systemic administration of high doses is also likely to exacerbate drug-related toxicities owing to nonspecific absorption. Consequently, the ideal routes of drug delivery to the posterior segment are the intravitreal and the periocular routes [4].

16.8.2.1.1 Intravitreal Route

Intravitreal injection of drugs to the eye involves direct injection of the formulation, in the form of solution, particles, suspension, depot, or implants, into the vitreous through the pars plana [4]. The vitreous is a gelatinous, cell-free structure that is capable of retaining molecules and also delivering them to nearby structures, such as the ciliary body or the RPE, a vital component of the retina [2]. Many pharmaceuticals aimed to treat posterior eye diseases are delivered via this route. Intravitreal injection provides increased drug concentration at the neural retina and minimizes systemic side effects. Nonetheless, frequent administration of drugs via this route can lead to retinal detachment, retinal hemorrhage, endophthalmitis, increased IOP, higher risk of infections, and poor acceptance by the patient [2, 150].

The requirement of frequent intravitreal injections to treat serious intraocular disorders affecting the choroid and retina has compelled researchers to look for better solutions derived from the use of nanoparticles. Polymeric nanoparticles have the ability to achieve long-term drug delivery in the retina and nearby tissues hence reducing the number of intraocular injections. Several kinds of polymeric nanoparticles carrying active molecules, including genetic material, are currently in preclinical studies. The advantage of the vitreous capacity to act as a biological reservoir and its ability to retain and deliver molecules to surrounding tissues can be exploited for delivery of the polymeric nanoparticles via intravitreal route. Patient compliance can be enormously improved if drug levels sustained for months after a single intravitreal injection. There are promising studies reported in recent literature on the use of intravitreally injected polymeric nanoparticles [2]. Ganciclovir-loaded albumin nanoparticles are an interesting example. In vitro experiments demonstrated that albumin nanoparticles released ganciclovir in a sustained way, with a significant improvement of drug uptake by CMV-infected human cells. For single intravitreal injections in rats, these nanoparticles showed prolonged residence in the eye and were found to be well tolerated, and the histological evaluation of the retina and adjacent tissues revealed the absence of inflammatory reactions or alterations in the tissue architecture (i.e., cellular infiltrations or vascular inflammation). These polymeric nanoparticles were safe, well-tolerated carriers, not only for ganciclovir but also for the anti-cytomegaloviral oligonucleotide analog fomivirsen. They were present in the vitreous and ciliary body for at least two weeks. deKozak et al. presented data demonstrating that tamoxifen-loaded nanoparticles prepared using the polymer poly[methoxy poly(ethylene glycol) cyanoacrylate-co-hexadecyl cyanoacrylate] may be of great benefit in the treatment of experimental autoimmune uveoretinitis, on a rat model of ocular inflammation. Indeed, no influence on EAU was observed after intravitreal treatment with free tamoxifen. This could be related to the encapsulation of molecules within biodegradable polymeric nanoparticles, which enables their progressive release and also prevents their in vivo degradation and rapid metabolization. The study demonstrated that the use of polymeric nanoparticles for the delivery of immunoregulatory molecules may open new perspectives for the treatment of uveitis [151]. El-Samaligy et al. utilized PECA nanoparticles of acyclovir and ganciclovir for sustained delivery to the retina after intravitreal administration in rabbits. The drug tissue concentrations were significantly higher in the retina with the nanoparticles when compared to the drug solutions for a period of over 10 days. Compared to the solution-dosage form, the polymeric nanoparticle formulations resulted in lower plasma concentrations of ganciclovir. Sustained delivery using intravitreal polymeric nanoparticles is an area of intense preclinical and clinical research [152].

RPE cells have the capacity to take up different kinds of nanoparticles opening the possibility of using them to treat retinal disorders. The purpose will be to target these cells with specific molecules or genetic material capable of reversing or stopping the processes leading to these diseases [2]. Merodio et al. demonstrated on rats that intravitreal PLA nanoparticles loaded with fluorochromes were able to retain in the RPE cells and continuously deliver the fluorochrome for months after the single injection. Fluorescence diffusing from the nanoparticles was observed in distant parts of retinal tissue, ganglion cells, and rod outer segments for up to four months after the injection as opposed to the free fluorochrome injection which lasted for barely one week [46]. Similar result was observed in a study carried out by Bejjani et al. wherein in vitro and in vivo evaluation of PLA and PLGA nanoparticles loaded with fluorochromes and model plasmids were studied. These nanoparticles encapsulating a plasmid-encoding red nuclear fluorescent protein were found to be localized in the RPE cells 24 h after intravitreal injection in rats. Effective plasmid expression was achieved after four days of injection, and expression-associated red fluorescence remained detectable in RPE cells during the following three weeks, with no apparent tissue damage or toxicity. The ability of PLGA nanoparticles to sequester plasmids and their nontoxic characteristics and rapid internalization enable gene transfer and expression in RPE cells [153]. Intravitreal administration of nanoparticles containing anti-transforming growth factor (TGF)-β2 phosphorothioate oligonucleotides (PS-ODNs) has also been evaluated for retinal drug delivery. Results demonstrated that in vivo particles were homogeneously distributed in the retina within 24 h of intravitreal administration. It was also found that anti-TGF-β2 PS-ODN was accumulated in retinal Muller glial cells after 72 h of intravitreal administration, demonstrating targeted gene delivery. These results showed nanoparticle prepared using cationic polymers such as polyethylenimine can be used to improve retinal gene delivery [154]. Folic acid is actively transported by RPE cells via folate receptors, which are only present in this cell layer in the retina, and when nano-drug carriers are injected into the vitreous, RPE cells represent an important gatekeeper before they are cleared dynamically into the blood circulation. Suen and Chau formulated targeted and sustained nanoparticles to take advantage of receptor on RPE cells for delivering therapeutics. In this study, it was found that triamcinolone acetonide-loaded folate-anchored poly(ethylene glycol)-b-polycaprolactone nanoparticles enhance drug uptake via folate receptor-mediated endocytosis and extend drug effects by slow release inside the cells. They exhibited low cytotoxicity against cultured ARPE-19 cells and promoted the antiangiogenesis gene expression for at least 3 weeks in vitro [155]. Thus, these folate-decorated polymeric nanoparticles were found to be efficient carriers of drugs to target and prolong delivery inside RPE cells.

Several in vivo studies have been conducted, for small molecules, peptides, and plasmid-loaded nanoparticles in treating neovascularization and retinal degeneration. Efficacy of intravitreally injected controlled release bevacizumab-loaded PEG (b-PEG) and PLGA (b-PLGA) nanoparticles was assessed in laser-induced CNV rat model. In addition to successful retention of bevacizumab antiangiogenic activity, the investigators also noticed that a single dose of b-PEG sustained bevacizumab release for as long as eight weeks [156]. In another study conducted on a laser-induced CNV rat model, PLGA nanoparticles loaded with plasminogen kringle-5 (K5-NP), a natural angiogenic inhibitor, were injected intravitreally. A single dose of K5-NP resulted in significant expression of K5 in the retina for up to 2 weeks, leading to reduction of VEGF levels, suppression of angiogenesis and retinal vascular leakage, and a 40 % decrease in CNV area compared to the untreated group. Thus, K5-NP mediated a sustained inhibitory effect on CNV and thus has therapeutic potential for AMD[71]. Park et al. also reported that K5-NP mediated efficient and sustained K5 expression in the retina and thus has the therapeutic potential for nanoparticle-mediated gene delivery [157]. Similarly, nanoparticles prepared from human serum albumin (HSA) were also evaluated for encapsulating plasmid-encoding Cu and Zn superoxide dismutase. These polymeric nanoparticles presented high encapsulation efficiency (84 %) and sustain the expression of the fusion protein in vivo for up to 48 h following a single intravitreal dose of 130 ng. Thus, HSA nanoparticles showed the potential to overcome several barriers associated with conventional nonviral gene delivery and were found to be promising for gene therapy applications [158].

Boddu et al. prepared biodegradable doxorubicin (DOX)-loaded PLGA–poly(ethylene glycol)-folate micelles (DOXM). Uptake of DOX was four times higher with DOXM than DOX in Y-79 cells overexpressing folate receptors. DOXM exhibited higher cytotoxicity in Y-79 cells when compared with pure DOX. These nanoparticles may provide sustained and targeted delivery of anticancer agents to retinoblastoma cells following intravitreal administration [159].

16.8.2.1.2 Periocular Routes

The periocular route of drug delivery enables the deposition of molecules against the external surface of the sclera, thereby minimizing the risk of endophthalmitis and retinal damage associated with the intravitreal route of administration [4]. The sclera allows diffusion of low-molecular-weight drugs as well as macromolecules. Periocularly administered drugs can gain entry into the retina and vitreous by diffusing across the sclera, choroid, and RPE layers. Periocular injections are less invasive to the globe and are potentially safer than intravitreal injections [160]. Periocular pathways used for the delivery of drugs to the posterior tissues of the eye include the retrobulbar, peribulbar, sub-Tenon, and subconjunctival routes (Fig. 16.1). Both the subconjunctival and the sub-Tenon routes are widely used in research into transscleral drug delivery owing to their proximity to the sclera. With subconjunctival injection, the formulation is placed beneath the conjunctival membrane that covers the sclera. This enables the drugs to bypass the conjunctiva–cornea barrier, giving direct access to the transscleral route. Sub-Tenon injection involves the placement of a formulation between the sclera and Tenon's capsule, an avascular membrane. As such, the contact time between the administered drug and the sclera is prolonged. Consequently, the sub-Tenon route is considered to be one of the most promising routes for targeting the posterior segment of the eye. The direct penetration pathway is the main route in achieving high concentration of a drug in the vitreous following sub-Tenon injection [4].

There are some studies reported in the literature on polymeric nanoparticles administered by periocular route. Sustained-release nanoparticles of budesonide were successfully prepared with PLA which sustained retinal budesonide levels in the retina and other ocular tissues when compared with a solution form in a rat model when administered subconjunctivally [161].

After subconjunctival administration in normotensive rabbit eyes, latanoprost-loaded poly(lactide)/monomethoxy-poly(ethylene glycol) (PLA–PEG) nanoparticles provided sustained latanoprost delivery and exhibited superior IOP-lowering effect compared to free latanoprost. IOP-lowering effect by PLA–PEG nanoparticles was sustained for eight consecutive days in contrast to the free drug group in which the IOP returned to baseline levels 84 h post-administration. The nanoparticles appeared to be well tolerated by rabbit eyes, and no signs of adverse tissue reactions were observed. Thus, these polymeric nanoparticles appeared to be a promising system for the controlled subconjunctival delivery of latanoprost [162].

Shome et al. carried out a study to compare intravitreal concentration of commercially available carboplatin (CAC) and the novel nanomolecule carboplatin, after periocular injection. CAC was bound with a nanoparticulate carrier by coacervation with a biocompatible and biodegradable protein bovine serum albumin. Nanoparticulate-bound carboplatin has greater transscleral transport than commercially available carboplatin, especially in the first week after injection, and may help enhance the proven adjuvant efficacy of periocular carboplatin in treating human retinoblastoma, especially those with vitreal seeds [163].

Mitra et al. prepared etoposide-loaded PLGA nanoparticles as a novel drug delivery system for retinoblastoma. Greater antiproliferative activity ~100 times was observed with nanoparticles than that of native etoposide in retinoblastoma cell line (Y-79). Nanoparticles demonstrated greater (G1/S) blocking and decreased mitochondrial membrane potential as measured by flow cytometry. There was upregulation of apoptotic gene activity in nanoparticles than native etoposide, as revealed through microarray analysis (Mitra et al., 2011).

16.8.2.2 Disposition of Nanoparticles Administered Through Posterior Route

16.8.2.2.1 Intravitreal Disposition of Nanoparticles

Following intravitreal injection, nanoparticles settle onto the inner limiting membrane of the retina within a few hours. The settling is governed by Stokes' law. As Stokes' law is based on the settling of a particle in a stagnant liquid layer, and because the vitreous in vivo will have some mobility because of the eye movements and the fluid clearance, the nanoparticle sedimentation times in vivo could be shorter than those estimated using Stokes' law. Thus, besides the particle properties, the properties of the vitreous including viscosity, density, convection currents, and fluid clearance determine particle settling. The site of injection within the vitreous will also determine when and where the particles settle within the vitreous cavity. Following deposition on the inner limiting membrane of the retina, nanoparticle penetration into the various retinal layers has been observed. Once the particles gain access to the retina, they might gain partial access to the systemic circulation via the retinal or choroid vasculature [15].

Influence of the particle's size on disposition is best understood using nondegradable particles. Sakurai et al. investigated the effect of particle size on the intravitreal disposition of nanoparticles, by selecting nonbiodegradable polystyrene particles of three sizes 50, 200, and 2000 nm. The investigators observed a decrease in intravitreal half-life with an increase in particle size. It was suggested that the larger particles (2000 nm) are mainly eliminated through the anterior chamber angle and smaller particles are cleared via the retina and anterior chamber angle. However, inflammatory conditions within the vitreous can enhance particle clearance [15]. Following intravitreal administration, 300 nm albumin nanoparticles were retained in the vitreous at the end of two weeks in rats. The nanoparticles were mainly seen overlaying the retina on the inner limiting membrane. Polylactide–rhodamine nanoparticles of 140 ± 20 nm were shown to be retained to a small extent within the vitreous, for at least one month after administration in rats. The presence of particles in the inner layers of the retina and the RPE has also been demonstrated by Merodio et al. [46]. The RPE has a natural mechanism of phagocytosing the rod outer segments, and investigators believe that it could be one of the reasons for increased uptake of nanoparticles by the RPE. The uptake of nanoparticles of different sizes has been investigated. Fluorescent carboxylate-modified polystyrene particles with sizes ranging from 20 to 2000 nm were incubated with the ARPE-19 cells for 3 h. The authors observed that the percentage uptake of the nanoparticles increased with a decrease in particle size [15].

Thus, intravitreally administered nanoparticles are removed by the retinal as well as the anterior segment pathways, with the nanoparticles better persisting in the vitreous compared to microparticles based on a few limited studies. The half-lives of nanoparticles are much greater than those for a solution-dosage form. Nanoparticles are more permeable through the various layers of the retina compared to microparticles. Particulate systems have a tendency to accumulate in the RPE consistent with the phagocytotic nature of this cell layer. The more prolonged retention of nanoparticles in the vitreous compared to microparticles might be due to their extremely low settling velocities as opposed to reduced clearance by cells. Most of the studies performed in the vitreal disposition of particles to date are qualitatively based on tissue images. In the future, more quantitative studies should be undertaken [15].

16.8.2.2.2 Periocular Disposition of Nanoparticles

A probable route for the disposition of nanoparticles from the periocular space is schematically presented in Fig. 16.1. From periocularly administered particulate systems, the drug can be released to enter the conjunctiva, tear film, cornea, and aqueous humor, or the drug might cross the sclera and choroid to reach the retina and the vitreous. Among the different periocular routes, viz., the subconjunctival, sub-Tenon, and retrobulbar administrations, the disposition can differ, and it is not yet investigated. There have been a few studies with periocular microparticles but no published studies with periocular nanoparticles. Indirect evidence of disposition can be obtained from studies that have investigated the use of micro- or nanoparticulate systems for sustained drug delivery. In one such study, Kompella et al. investigated the use of nano- and microparticles of budesonide formulated using polylactide as the encapsulating polymer [161]. There was a higher burst and lower subsequent release rate with the nanoparticles as compared to the microparticles. Also, the ocular tissue levels of budesonide were several folds higher with microparticles at the end of seven days as compared to the nanoparticles of equivalent dose. However, when compared to the solution of budesonide, the nanoparticles provided significantly higher budesonide levels in the retina and other ocular tissues at the end of seven days post-administration. Though the authors primarily consider the differences in drug release rates to be the reason for the higher tissue levels of budesonide following subconjunctival administration, the possibility of differential disposition of the particles from the subconjunctival space could not be ruled out from their study. In trying to investigate the disposition of particulate systems from the subconjunctival space, Amrite and Kompella, using nonbiodegradable fluorescent particles of 20 nm and 2 μm sizes, demonstrated that the larger particles were almost completely retained in the periocular space up to 60 days post-administration [160]. However, they observed that the 20 nm particles disappeared rapidly from the site of injection with 25 % of the dose remaining at the end of day one and <15 % at the end of seven days post-administration. The authors did not observe the presence of 20 nm nanoparticles at the site of administration at 60 days post-administration. There was no significant penetration of either the nanoparticles or the microparticles into the ocular tissues except the sclera up to 60 days. Further investigations by the authors suggested that particles 200 nm in diameter show a similar retention behavior in the subconjunctival space as the larger 2 μm particles. The authors have also shown that the ocular penetration of the different particles is <0.1 % of the administered dose for 20 nm particles with no particles detected in any of the ocular tissues for the 200 and 2 μm particles at the 1-, 7-, and 60-day time points they studied. In summary, periocularly administered particulate systems exhibit unique disposition behavior. Micro- and nanoparticles above 200 nm do not gain access to the intraocular tissues including choroid, retina, and vitreous. Particles of this size range can be completely retained at the site of injection for at least up to two months, indicating their potential usefulness as sustained drug delivery systems. The comparison of various periocular routes with respect to the differences in particle disposition has yet to be undertaken [15].

16.9 Patented Technologies on Polymeric Ocular Nanoparticles

Various patents reveal that significant research has been carried out on various nanoparticle technologies for ocular drug delivery. Few of the recent patented technologies on polymeric ocular nanoparticle are discussed here.

Banerjee et al. patented a system which includes composition of nanoparticulate gelling vitreous substitute, comprising of alpha-tocopheryl polyethylene glycol succinate (TPGS), a polymeric blend of gellan and polyvinyl alcohol (PVA), and a cross-linker, to form a gel having the properties of the vitreous humor. The system is liquid at room temperature to facilitate easy administration through a small needle incision and forms a gel within the eye. The cross-linking may occur in situ upon administration to the posterior segment of the eye. The biodegradable nanoparticles within the gel allow sustained release of hydrophilic drugs or actives in posterior segment of the eye [164, 165].

Ravi invented a hydrogel system made of copolymer solution containing nanoparticles (<150 nm) having the ability to form a reversible in situ hydrogel when triggered. The triggering mechanisms include (a) change in oxidation state, i.e., a solution of the copolymer can be oxidized to form a hydrogel, and upon reduction, it can convert again in solution; (b)using polymers sensitive to mechanical stress, i.e., they behave like a solution when sheared and as a gel when the shearing force is reduced/removed; or (c) using light frequency and intensity, i.e., formation of hydrogel upon exposure of light at a first wavelength and formation of a solution when exposed to light at a second wavelength [165–167].

Irache and Yancheva patented a pegylated nanoparticle composite formed by a polyvinyl methyl ether and maleic anhydride and polyethylene glycol copolymer which provide excellent bioadhesion, size, and zeta potential characteristics and can be used as a drug carrier; this method is suitable to produce nanoparticles in the range of 200–500 nm [168].

A method to prepare PLGA nanospheres or microspheres is patented by Raicheet al. which can produce nanoparticles ranging between 200 and 400 nm in size. The method involves the use of water, acetonitrile, and sodium chloride system, which is derived by a solvent and non-solvent system and depends on immiscibility induced by ternary agent (i.e., ammonium azide, cadmium chloride, lithium sulfate, etc.) concentration and temperature alteration [169].

Kararli et al. patented a treatment or prevention of a cyclooxygenase-2-mediated ophthalmic disease. They reported two different processes to prepare these nanoparticles including wet milling and nanoprecipitation. Various mucoadhesive polymers, like carboxymethyl cellulose, carbomer (acrylic acid polymer), PMMA, etc., can be employed for preparation of nanoparticles to produce effective residence time in the eye of about 2–24 h. This method is suitable for poorly water-soluble drugs, and the nanoparticles with average particle size ranging from 500 to 900 nm can be produced [170].

Xu et al. in 2011 patented a multifunctional biodegradable nanoparticle for selectable targeting, imaging, and delivery of therapeutic agents for the treatment of ocular disorders. These particles comprise of (a) a biodegradable shell like PLA, PCL, PGA, PLGA, chitosan, Eudragit, hyaluronic acid, alginate, carboxymethyl cellulose, Carbopol, polyethylene glycol, PECA, polystyrene, etc., (b) at least one ocular targeting agent like antibody or protein coupled to the exterior surface of the shell, and (c) at least one imaging agent, like fluorescence imaging agents, radionuclide-labeled imaging agents, etc., encapsulated within the shell. These nanoparticles can migrate to the ocular target after administration and deliver at least one therapeutic agent [171].

A drug delivery system comprising soft contact lenses dispersed with drug-loaded nanoparticles (50–200 nm) is patented by Chauhanet al. This delivery system is capable to address limitations of ocular delivery such as drug loss, provide sustained delivery, improve efficiency, and eliminate side effects. These soft lenses are made of poly-2-hydroxyethyl methacrylate (HEMA) polymer [168, 172].

16.10 Clinical Development of Ocular Nanoparticles

Ocular nanoparticulate systems have been extensively exploited to treat various ophthalmic disorders with proven efficacy and safety by in vitro and preclinical in vivo studies. Despite of remarkable development in ocular nanotechnology, these systems could reach very limited clinical trials. Therefore, additional research focus is required to establish correlations between human and animal studies for successful clinical use of ocular nanoparticles. A phase II clinical trial to assess safety and efficacy of paclitaxel albumin-stabilized nanoparticle formulation in treating patients with metastatic melanoma of the eye that cannot be removed by surgery has been completed successfully (Clinical trials registry and [173]). Another clinical study of cyclodextrin nanoparticles to treat diabetic macular edema is reported. Cyclodextrin nanoparticle eyedrop suspension can deliver drugs to the posterior part of the eye, thus solving one of the biggest obstacles in ocular drug delivery (Clinical trials registry and [174]). Clinical studies are essential to prove the efficacy and safety of nanoparticles to the ocular tissue. Therefore, it is recommended to carry out clinical assessment of novel technologies speedily for patient welfare.

16.11 Conclusion

Polymeric nanoparticles, since their first appearance in the 80s, have been keenly explored as delivery systems for ocular disease conditions. Nanoparticles represent promising drug carriers for ophthalmic applications. In general, the polymeric drug delivery systems offer advantages such as the reduction in the dosing frequency, reduced side effects, delivery with enhanced efficiency, and, most importantly, improved patient compliance. When designing a polymer-based nanoparticulate drug delivery system, the choice of synthetic or biopolymer is an important consideration. Equally important is the choice of technique for preparation of polymeric nanoparticles which is governed by physicochemical properties of drug and polymer characteristics. Advancement in modification of polymer properties opened new avenues for designing effective polymeric nanoparticles which could overcome drawbacks associated with existing polymeric nanoparticles. Furthermore, significant efforts are currently underway to develop highly sophisticated nanoparticles functionalized with specific targeting ligands (i.e., lectins, sugars, folate, and antibodies). However, it is daunting to recognize that while research on polymeric nanoparticles as ophthalmic drug delivery has been ongoing for the past decades, very few systems have reached the clinical development stage. This striking deficiency points out the complexity of meeting all criteria necessary to achieve a commercially acceptable product. That being said, polymeric nanoparticles can form a very powerful drug delivery technology for the pharmaceutical industry since the patent protection to a number of blockbuster drugs is expiring in this decade. Such technology-based products can be used for the extension of the patent life of the drug or to prevent/delay the entry of generic versions into the specialized markets.

References

1. Del Amo EM, Urtti A (2008) Current and future ophthalmic drug delivery systems. A shift to the posterior segment. Drug Discov Today 13:135–143
2. Diebold Y, Calonge M (2010) Applications of nanoparticles in ophthalmology. Prog Retin Eye Res 29:596–609
3. Nakhlband A, Barar J (2011) Impacts of nanomedicines in ocular pharmacotherapy. Bioimpacts 1:7
4. Thrimawithana TR, Young S, Bunt CR, Green C, Alany RG (2011) Drug delivery to the posterior segment of the eye. Drug Discov Today 16:270–277
5. Zimmer A, Kreuter J (1995) Microspheres and nanoparticles used in ocular delivery systems. Adv Drug Deliv Rev 16:61–73
6. Sasaki H, Yamamura K, Nishida K, Nakamura J, Ichikawa M (1996) Delivery of drugs to the eye by topical application. Prog Retin Eye Res 15:583–620
7. Nagarwal RC, Kant S, Singh P, Maiti P, Pandit J (2009) Polymeric nanoparticulate system: a potential approach for ocular drug delivery. J Control Release 136:2–13
8. Wood RW, Lee HK, Kreuter J, Robinson JR (1985) Ocular disposition of polyhexyl-2-cyano(3-14C) acrylate nanoparticles in the albino rabbit. Int J Pharm 23:175–183.
9. Araújo J, Gonzalez E, Egea MA, Garcia ML, Souto EB (2009) Nanomedicines for ocular NSAIDs: safety on drug delivery. Nanomed: Nanotechnol Biol Med 5:394–401
10. Jwala J, Vadlapatla RK, Vadlapudi AD, Boddu SHS, Pal D, Mitra AK (2012) Differential expression of folate receptor-alpha, sodium-dependent multivitamin transporter, and amino acid transporter (B (0,+)) in human retinoblastoma (Y-79) and retinal pigment epithelial (ARPE-19) cell lines. J Ocul Pharmacol Ther 28:237–244
11. Sahoo SK, Dilnawaz F, Krishnakumar S (2008) Nanotechnology in ocular drug delivery. Drug Discov Today 13:144–151
12. Souto E, Mehnert W, Müller R (2006) Polymorphic behaviour of Compritol® 888 ATO as bulk lipid and as SLN and NLC. J Microencapsul 23:417–433
13. Souto E, Müller R (2006) Investigation of the factors influencing the incorporation of clotrimazole in SLN and NLC prepared by hot high-pressure homogenization. J Microencapsul 23:377–388
14. Chan J, EL Maghraby GM, Craig JP, Alany RG (2007) Phase transition water-in-oil microemulsions as ocular drug delivery systems: in vitro and in vivo evaluation. Int J Pharm 328:65–71
15. Gupta RB, Kompella UB (2006) Nanoparticle technology for drug delivery. Taylor & Francis, New York
16. De Campos AM, Sánchez A, Alonso MAJ (2001) Chitosan nanoparticles: a new vehicle for the improvement of the delivery of drugs to the ocular surface. Application to cyclosporin A. Int J Pharm 224:159–168

17. Rowe RC, Sheskey PJ, Owen SC, American Pharmacists Association (2006) Handbook of pharmaceutical excipients. Pharmaceutical Press, London
18. Contreras-Ruiz L, De La Fuente M, García-Vázquez C, Sáez V, Seijo B, Alonso MJ, Calonge M, Diebold Y (2010) Ocular tolerance to a topical formulation of hyaluronic acid and chitosan-based nanoparticles. Cornea 29:550–558
19. De La Fuente M, Seijo B, Alonso M (2008) Bioadhesive hyaluronan–chitosan nanoparticles can transport genes across the ocular mucosa and transfect ocular tissue. Gene Ther 15:668–676
20. De La Fuente M, Seijo B, Alonso MJ (2008) Novel hyaluronan-based nanocarriers for transmucosal delivery of macromolecules. Macromol Biosci 8:441–450
21. De La Fuente M, Seijo B, Alonso MJ (2008) Novel hyaluronic acid-chitosan nanoparticles for ocular gene therapy. Invest Ophthalmol Vis Sci 49:2016–2024
22. De Salamanca AE, Diebold Y, Calonge M, García-Vazquez C, Callejo S, Vila A, Alonso MJ (2006) Chitosan nanoparticles as a potential drug delivery system for the ocular surface: toxicity, uptake mechanism and in vivo tolerance. Invest Ophthalmol Vis Sci 47:1416–1425
23. Diebold Y, Jarrín M, Saez V, Carvalho EL, Orea M, Calonge M, Seijo B, Alonso MJ (2007) Ocular drug delivery by liposome–chitosan nanoparticle complexes (LCS-NP). Biomaterials 28:1553–1564
24. Motwani SK, Chopra S, Talegaonkar S, Kohli K, Ahmad FJ, Khar RK (2008) Chitosan–sodium alginate nanoparticles as submicroscopic reservoirs for ocular delivery: formulation, optimisation and in vitro characterisation. Eur J Pharm Biopharm 68:513–525
25. Nagarwal RC, Nath Singh P, Kant S, Maiti P, Pandit JK (2011) Chitosan nanoparticles of 5-fluorouracil for ophthalmic delivery: characterization, in-vitro and in-vivo study. Chem Pharm Bull 59:272–278
26. Qu X, Khutoryanskiy VV, Stewart A, Rahman S, Papahadjopoulos-Sternberg B, Dufes C, Mccarthy D, Wilson CG, Lyons R, Carter KC (2006) Carbohydrate-based micelle clusters which enhance hydrophobic drug bioavailability by up to 1 order of magnitude. Biomacromolecules 7:3452–3459
27. Rajendran N, Natrajan R, Kumar R, Selvaraj S (2010) Acyclovir-loaded chitosan nanoparticles for ocular delivery. Asian J Pharm 4:220
28. Shao T, Li X, Ge J (2011) Target drug delivery system as a new scarring modulation after glaucoma filtration surgery. Diagn Pathol 6:64
29. Singh K, Shinde U (2011) Chitosan nanoparticles for controlled delivery of brimonidine tartrate to the ocular membrane. Die Pharmazie 66:594–599
30. Sonvico F, Cagnani A, Rossi A, Motta S, DI Bari M, Cavatorta F, Alonso M, Deriu A, Colombo P (2006) Formation of self-organized nanoparticles by lecithin/chitosan ionic interaction. Int J Pharm 324:67–73
31. Yuan X-B, Li H, Yuan Y-B (2006) Preparation of cholesterol-modified chitosan self-aggregated nanoparticles for delivery of drugs to ocular surface. Carbohydr Polym 65:337–345
32. Kogan G, Šoltés L, Stern R, Gemeiner P (2007) Hyaluronic acid: a natural biopolymer with a broad range of biomedical and industrial applications. Biotechnol Lett 29:17–25
33. Contreras-Ruiz L, De La Fuente M, Párraga JE, López-García A, Fernández I, Seijo B, Sánchez A, Calonge M, Diebold Y (2011) Intracellular trafficking of hyaluronic acid-chitosan oligomer-based nanoparticles in cultured human ocular surface cells. Mol Vis 17:279
34. Ibrahim HK, El-Leithy IS, Makky AA (2010) Mucoadhesive nanoparticles as carrier systems for prolonged ocular delivery of gatifloxacin/prednisolone bitherapy. Mol Pharm 7:576–585
35. Wadhwa S, Paliwal R, Paliwal SR, Vyas S (2010) Hyaluronic acid modified chitosan nanoparticles for effective management of glaucoma: development, characterization, and evaluation. J Drug Target 18:292–302
36. Motwani SK, Ahmad F, Iqbal Z, Talegaonkar S, Khar R (2007) Gatifloxacin nanoparticles for ophthalmic delivery. Nanotechnology 2:310–312

37. Nagarwal RC, Kumar R, Pandit J (2012) Chitosan coated sodium alginate–chitosan nanoparticles loaded with 5-FU for ocular delivery: in vitro characterization and in vivo study in rabbit eye. Eur J Pharm Sci 47:678–685
38. Zhu X, Su M, Tang S, Wang L, Liang X, Meng F, Hong Y, Xu Z (2012) Synthesis of thiolated chitosan and preparation nanoparticles with sodium alginate for ocular drug delivery. Mol Vis 18:1973
39. Kaul G, Amiji M (2005) Cellular interactions and in vitro DNA transfection studies with poly (ethylene glycol)-modified gelatin nanoparticles. J Pharm Sci 94:184–198
40. Konat Zorzi G, Contreras-Ruiz L, Párraga JE, López-García A, Romero Bello R, Diebold Y, Seijo BA, Sánchez A (2011) Expression of MUC5AC in ocular surface epithelial cells using cationized gelatin nanoparticles. Mol Pharm 8:1783–1788
41. Sakai T, Kuno N, Takamatsu F, Kimura E, Kohno H, Okano K, Kitahara K (2007) Prolonged protective effect of basic fibroblast growth factor–impregnated nanoparticles in royal college of surgeons rats. Invest Ophthalmol Vis Sci 48:3381–3387
42. Tseng C-L, Chen K-H, Su W-Y, Lee Y-H, Wu C-C, Lin F-H (2013) Cationic gelatin nanoparticles for drug delivery to the ocular surface: in vitro and in vivo evaluation. J Nanomater 2013:7
43. Zorzi GK, Párraga JE, Seijo B, Sánchez A (2011) Hybrid nanoparticle design based on cationized gelatin and the polyanions dextran sulfate and chondroitin sulfate for ocular gene therapy. Macromol Biosci 11:905–913
44. Das S, Banerjee R, Bellare J (2005) Aspirin loaded albumin nanoparticles by coacervation: implications in drug delivery. Trends Biomater Artif Organs 18:203–212
45. Merodio M, Arnedo A, Renedo MJ, Irache JM (2001) Ganciclovir-loaded albumin nanoparticles: characterization and in vitro release properties. Eur J Pharm Sci 12:251–259
46. Merodio M, Irache JM, Valamanesh F, Mirshahi M (2002) Ocular disposition and tolerance of ganciclovir-loaded albumin nanoparticles after intravitreal injection in rats. Biomaterials 23:1587–1594
47. Zimmer AK, Maincent P, Thouvenot P, Kreuter J (1994) Hydrocortisone delivery to healthy and inflamed eyes using a micellar polysorbate 80 solution or albumin nanoparticles. Int J Pharm 110:211–222
48. Zimmer AK, Zerbe H, Kreuter J (1994) Evaluation of pilocarpine-loaded albumin particles as drug delivery systems for controlled delivery in the eye I. In vitro and in vivo characterisation. J Control Release 32:57–70
49. Dilbaghi N, Kaur H, Ahuja M, Kumar S (2013) Evaluation of tropicamide-loaded tamarind seed xyloglucan nanoaggregates for ophthalmic delivery. Carbohydr Polym 94:286–291
50. Parmar C, Maughal MK (1982) Cordia oblique. In: Wild fruits. Kalyani publishers, New Delhi, pp 19–22
51. Nishinari K, Takemasa M, Zhang H, Takahashi R (2007) Storage plant polysaccharides: xyloglucans, galactomannans, glucomannans. Comprehensive glycoscience: from chemistry to systems biology, 1st edn. Elsevier, Oxford
52. Yadav M, Ahuja M (2010) Preparation and evaluation of nanoparticles of gum cordia, an anionic polysaccharide for ophthalmic delivery. Carbohydr Polym 81:871–877
53. Parmar C, Kaushal MK (1982) Cordia obliqua wild fruits. Kalyani Publishers, New Delhi
54. Ahuja A, Khar RK, Ali J (1997) Mucoadhesive drug delivery systems. Drug Dev Ind Pharm 23:489–515
55. Bhagav P, Upadhyay H, Chandran S (2011) Brimonidine tartrate–eudragit long-acting nanoparticles: formulation, optimization, in vitro and in vivo evaluation. AAPS PharmSciTech 12:1087–1101
56. Bucolo C, Maltese A, Puglisi G, Pignatello R (2002) Enhanced ocular anti-inflammatory activity of ibuprofen carried by an Eudragit RS100® nanoparticle suspension. Ophthalmic Res 34:319–323
57. Desai SD, Blanchard J (2000) Pluronic® F127-based ocular delivery system containing biodegradable polyisobutylcyanoacrylate nanocapsules of pilocarpine. Drug Deliv 7:201–207

58. Fresta M, Fontana G, Bucolo C, Cavallaro G, Giammona G, Puglisi G (2001) Ocular tolerability and in vivo bioavailability of poly (ethylene glycol)(PEG)-coated polyethyl-2-cyanoacrylate nanosphere-encapsulated acyclovir. J Pharm Sci 90:288–297
59. Hsiue G-H, Hsu S-H, Yang C-C, Lee S-H, Yang I (2002) Preparation of controlled release ophthalmic drops, for glaucoma therapy using thermosensitive poly- *N*-isopropylacrylamide. Biomaterials 23:457–462
60. Inoue T, Chen G, Nakamae K, Hoffmana AS (1998) An AB block copolymer of oligo (methyl methacrylate) and poly (acrylic acid) for micellar delivery of hydrophobic drugs. J Control Release 51:221–229
61. Langer K, Mutschler E, Lambrecht G, Mayer D, Troschau G, Stieneker F, Kreuter J (1997) Methylmethacrylate sulfopropylmethacrylate copolymer nanoparticles for drug delivery: part III: evaluation as drug delivery system for ophthalmic applications. Int J Pharm 158:219–231
62. Pignatello R, Bucolo C, Ferrara P, Maltese A, Puleo A, Puglisi G (2002) Eudragit RS100® nanosuspensions for the ophthalmic controlled delivery of ibuprofen. Eur J Pharm Sci 16:53–61
63. Pignatello R, Bucolo C, Spedalieri G, Maltese A, Puglisi G (2002) Flurbiprofen-loaded acrylate polymer nanosuspensions for ophthalmic application. Biomaterials 23:3247–3255
64. Makadia HK, Siegel SJ (2011) Poly lactic-co-glycolic acid (PLGA) as biodegradable controlled drug delivery carrier. Polymers 3:1377–1397
65. Södergård A, Stolt M (2002) Properties of lactic acid based polymers and their correlation with composition. Prog Polym Sci 27:1123–1163
66. Agnihotri SM, Vavia PR (2009) Diclofenac-loaded biopolymeric nanosuspensions for ophthalmic application. Nanomed: Nanotechnol Biol Med 5:90–95
67. Chen Y-S, Alany RG, Young SA, Green CR, Rupenthal ID (2011) In vitro release characteristics and cellular uptake of poly (D, L-lactic-co-glycolic acid) nanoparticles for topical delivery of antisense oligodeoxynucleotides. Drug Deliv 18:493–501
68. Dillen K, Vandervoort J, Van Den Mooter G, Ludwig A (2006) Evaluation of ciprofloxacin-loaded Eudragit® RS100 or RL100/PLGA nanoparticles. Int J Pharm 314:72–82
69. Giannavola C, Bucolo C, Maltese A, Paolino D, Vandelli MA, Puglisi G, Lee VH, Fresta M (2003) Influence of preparation conditions on acyclovir-loaded poly-d, l-lactic acid nanospheres and effect of PEG coating on ocular drug bioavailability. Pharm Res 20:584–590
70. Gupta H, Aqil M, Khar R, Ali A, Bhatnagar A, Mittal G (2011) Biodegradable levofloxacin nanoparticles for sustained ocular drug delivery. J Drug Target 19:409–417
71. Jin J, Zhou KK, Park K, Hu Y, Xu X, Zheng Z, Tyagi P, Kompella UB, Ma J-X (2011) Anti-inflammatory and antiangiogenic effects of nanoparticle-mediated delivery of a natural angiogenic inhibitor. Invest Ophthalmol Vis Sci 52:6230–6237
72. Nagarwal RC, Singh P, Kant S, Maiti P, Pandit J (2010) Chitosan coated PLA nanoparticles for ophthalmic delivery: characterization, in-vitro and in-vivo study in rabbit eye. J Biomed Nanotechnol 6:648–657
73. Sakai T, Ishihara T, Higaki M, Akiyama G, Tsuneoka H (2011) Therapeutic effect of stealth-type polymeric nanoparticles with encapsulated betamethasone phosphate on experimental autoimmune uveoretinitis. Invest Ophthalmol Vis Sci 52:1516–1521
74. Sakai T, Kohno H, Ishihara T, Higaki M, Saito S, Matsushima M, Mizushima Y, Kitahara K (2006) Treatment of experimental autoimmune uveoretinitis with poly (lactic acid) nanoparticles encapsulating betamethasone phosphate. Exp Eye Res 82:657–663
75. Vega E, Egea MA, Calpena AC, Espina M, García ML (2012) Role of hydroxypropyl-β-cyclodextrin on freeze-dried and gamma-irradiated PLGA and PLGA–PEG diblock copolymer nanospheres for ophthalmic flurbiprofen delivery. Int J Nanomedicine 7:1357
76. Vega E, Gamisans F, Garcia M, Chauvet A, Lacoulonche F, Egea M (2008) PLGA nanospheres for the ocular delivery of flurbiprofen: drug release and interactions. J Pharm Sci 97:5306–5317
77. Zhang L, Li Y, Zhang C, Wang Y, Song C (2009) Pharmacokinetics and tolerance study of intravitreal injection of dexamethasone-loaded nanoparticles in rabbits. Int J Nanomedicine 4:175

78. Jain JP, Yenet Ayen W, Domb AJ, Kumar N (2011) Biodegradable polymers in drug delivery. In: Domb AJ, Kumar N, Ezra A (eds) Clinical use and clinical development. John Wiley & Sons, Inc., Hoboken, pp 1–58
79. Coulembier O, Degée P, Hedrick JL, Dubois P (2006) From controlled ring-opening polymerization to biodegradable aliphatic polyester: especially poly (β-malic acid) derivatives. Prog Polym Sci 31:723–747
80. Barbault-Foucher S, Gref R, Russo P, Guechot J, Bochot A (2002) Design of poly-ε-caprolactone nanospheres coated with bioadhesive hyaluronic acid for ocular delivery. J Control Release 83:365–375
81. Calvo P, Remuñan-López C, Vila-Jato JL, Alonso MJ (1997) Chitosan and chitosan/ethylene oxide-propylene oxide block copolymer nanoparticles as novel carriers for proteins and vaccines. Pharm Res 14:1431–1436
82. Calvo P, Remunan-Lopez C, Vila-Jato J, Alonso M (1997) Novel hydrophilic chitosan-polyethylene oxide nanoparticles as protein carriers. J Appl Polym Sci 63:125–132
83. De Campos AM, Sánchez A, Gref R, Calvo P, Alonso MAJ (2003) The effect of a PEG versus a chitosan coating on the interaction of drug colloidal carriers with the ocular mucosa. Eur J Pharm Sci 20:73–81
84. Guha R, Chowdhury S, Palui H, Mishra A, Basak S, Mandal TK, Hazra S, Konar A (2013) Doxorubicin-loaded MePEG-PCL nanoparticles for prevention of posterior capsular opacification. Nanomedicine 8:1415–1428
85. Tamboli V, Mishra GP, Mitra AK (2013) Novel pentablock copolymer (PLA–PCL–PEG–PCL–PLA)-based nanoparticles for controlled drug delivery: effect of copolymer compositions on the crystallinity of copolymers and in vitro drug release profile from nanoparticles. Colloid Polym Sci 291:1235–1245
86. Yenice İ, Mocan MC, Palaska E, Bochot A, Bilensoy E, Vural İ, Irkeç M, Atilla Hincal A (2008) Hyaluronic acid coated poly-ε-caprolactone nanospheres deliver high concentrations of cyclosporine A into the cornea. Exp Eye Res 87:162–167
87. Hans M, Lowman A (2002) Biodegradable nanoparticles for drug delivery and targeting. Curr Opin Solid State Mater Sci 6:319–327
88. Soppimath KS, Aminabhavi TM, Kulkarni AR, Rudzinski WE (2001) Biodegradable polymeric nanoparticles as drug delivery devices. J Control Release 70:1–20
89. Zahoor A, Sharma S, Khuller G (2005) Inhalable alginate nanoparticles as antitubercular drug carriers against experimental tuberculosis. Int J Antimicrob Agents 26:298–303
90. Agnihotri SA, Mallikarjuna NN, Aminabhavi TM (2004) Recent advances on chitosan-based micro-and nanoparticles in drug delivery. J Control Release 100:5–28
91. Andonova VY, Georgiev GS, Georgieva VT, Petrova NL, Kasarova M (2013) Indomethacin nanoparticles for applications in liquid ocular formulations. Folia Med (Plovdiv) 55:76–82
92. Boudad H, Legrand P, Lebas G, Cheron M, Duchene D, Ponchel G (2001) Combined hydroxypropyl-β-cyclodextrin and poly (alkylcyanoacrylate) nanoparticles intended for oral administration of saquinavir. Int J Pharm 218:113–124
93. Mohanraj V, Chen Y (2007) Nanoparticles-a review. Trop J Pharm Res 5:561–573
94. Zhang Q, Shen Z, Nagai T (2001) Prolonged hypoglycemic effect of insulin-loaded polybutylcyanoacrylate nanoparticles after pulmonary administration to normal rats. Int J Pharm 218:75–80
95. Puglisi G, Fresta M, Giammona G, Ventura C (1995) Influence of the preparation conditions on poly (ethylcyanoacrylate) nanocapsule formation. Int J Pharm 125:283–287
96. Krause H-J, Schwarz A, Rohdewald P (1986) Interfacial polymerization, a useful method for the preparation of polymethylcyanoacrylate nanoparticles. Drug Dev Ind Pharm 12:527–552
97. Harmia-Pulkkinen T, Tuomi A, Kristoffersson E (1989) Manufacture of polyalkylcyanoacrylate nanoparticles with pilocarpine and timolol by micelle polymerization: factors influencing particle formation. J Microencapsul 6:87–93

98. Al Khouri Fallouh N, Roblot-Treupel L, Fessi H, Devissaguet JP, Puisieux F (1986) Development of a new process for the manufacture of polyisobutylcyanoacrylate nanocapsules. Int J Pharm 28:125–132
99. Pinto Reis C, Neufeld RJ, Ribeiro AJ, Veiga F (2006) Nanoencapsulation I. Methods for preparation of drug-loaded polymeric nanoparticles. Nanomed: Nanotechnol Biol Med 2:8–21
100. Couvreur P, Barratt G, Fattal E, Legrand P, Vauthier C (2002) Nanocapsule technology: a review. Crit Rev Ther Drug Carrier Syst 19:99–134
101. Zambaux M, Bonneaux F, Gref R, Maincent P, Dellacherie E, Alonso M, Labrude P, Vigneron C (1998) Influence of experimental parameters on the characteristics of poly (lactic acid) nanoparticles prepared by a double emulsion method. J Control Release 50:31–40
102. Dillen K, Vandervoort J, Van Den Mooter G, Verheyden L, Ludwig A (2004) Factorial design, physicochemical characterisation and activity of ciprofloxacin-PLGA nanoparticles. Int J Pharm 275:171–187
103. Niwa T, Takeuchi H, Hino T, Kunou N, Kawashima Y (1993) Preparations of biodegradable nanospheres of water-soluble and insoluble drugs with D, L-lactide/glycolide copolymer by a novel spontaneous emulsification solvent diffusion method, and the drug release behavior. J Control Release 25:89–98
104. Liu M, Dong J, Yang Y, Yang X, Xu H (2005) Characterization and release of triptolide-loaded poly (D, L-lactic acid) nanoparticles. Eur Polym J 41:375–382
105. Lambert G, Fattal E, Couvreur P (2001) Nanoparticulate systems for the delivery of antisense oligonucleotides. Adv Drug Deliv Rev 47:99–112
106. Konan-Kouakou Y, Boch R, Gurny R, Allemann E (2005) In vitro and in vivo activities of verteporfin-loaded nanoparticles. J Control Release 103:83–91
107. Galindo-Rodriguez S, Allemann E, Fessi H, Doelker E (2004) Physicochemical parameters associated with nanoparticle formation in the salting-out, emulsification-diffusion, and nano-precipitation methods. Pharm Res 21:1428–1439
108. Miller CA (1988) Spontaneous emulsification produced by diffusion—a review. Colloids Surf 29:89–102
109. Gupta H, Aqil M, Khar RK, Ali A, Bhatnagar A, Mittal G (2010) Sparfloxacin-loaded PLGA nanoparticles for sustained ocular drug delivery. Nanomed: Nanotechnol Biol Med 6:324–333
110. Kumar S, Linehan B, Tseng Y-C (2014) A new combination approach of CI jet and QESD to formulate pH-susceptible amorphous solid dispersions. Int J Pharm 466:368–374
111. Tom JW, Debenedetti PG (1991) Formation of bioerodible polymeric microspheres and microparticles by rapid expansion of supercritical solutions. Biotechnol Prog 7:403–411
112. Jain K, Suresh Kumar R, Sood S, Dhyanandhan G (2013) Betaxolol hydrochloride loaded chitosan nanoparticles for ocular delivery and their anti-glaucoma efficacy. Curr Drug Deliv 10:493–499
113. Vauthier C, Bouchemal K (2009) Methods for the preparation and manufacture of polymeric nanoparticles. Pharm Res 26:1025–1058
114. Masson V, Maurin F, Fessi H, Devissaguet J (1997) Influence of sterilization processes on poly (ε-caprolactone) nanospheres. Biomaterials 18:327–335
115. Sintzel MB, Merkli A, Tabatabay C, Gurny R (1997) Influence of irradiation sterilization on polymers used as drug carriers-a review. Drug Dev Ind Pharm 23:857–878
116. Bos GW, Trullas-Jimeno A, Jiskoot W, Crommelin DJ, Hennink WE (2000) Sterilization of poly (dimethylamino) ethyl methacrylate-based gene transfer complexes. Int J Pharm 211:79–88
117. Magenheim B, Benita S (1991) Nanoparticle characterization: a comprehensive physico-chemical approach. STP Pharma Sci 1:221–241
118. Müller R, Wallis KH (1993) Surface modification of iv injectable biodegradable nanoparticles with poloxamer polymers and poloxamine 908. Int J Pharm 89:25–31
119. Brigger I, Dubernet C, Couvreur P (2002) Nanoparticles in cancer therapy and diagnosis. Adv Drug Deliv Rev 54:631–651

120. Shrivastava A, Shimpi H, Roy S, Galdhar C, Shirole R, Gaikwad R, Rajan M, Iyer K, Menon M (2012) Nanoparticle based DPI formulation for active targeting of etoposide in lung cancer. In: Dalby R, Byron P, Peart J, Farr S, Suman J (eds) Respiratory drug delivery; Phoenix, USA. DHI Publishing, River Grove, pp 427–432
121. Kim YC, Chiang B, Wu X, Prausnitz MR (2014) Ocular delivery of macromolecules. J Control Release 190:172–181
122. Ticho U, Blumenthal M, Zonis S, Gal A, Blank I, Mazor Z (1979) A clinical trial with Piloplex–a new long-acting pilocarpine compound: preliminary report. Ann Ophthalmol 11:555–561
123. Papadimitriou S, Bikiaris D, Avgoustakis K, Karavas E, Georgarakis M (2008) Chitosan nanoparticles loaded with dorzolamide and pramipexole. Carbohydr Polym 73:44–54
124. Shinde U, Ahmed MH, Singh K (2013) Development of dorzolamide loaded 6-o-carboxymethyl chitosan nanoparticles for open angle glaucoma. J Drug Delivery 2013:15
125. Hermans K, Van Den Plas D, Everaert A, Weyenberg W, Ludwig A (2012) Full factorial design, physicochemical characterisation and biological assessment of cyclosporine A loaded cationic nanoparticles. Eur J Pharm Biopharm 82:27–35
126. Singh K, Shinde U (2010) Development and evaluation of novel polymeric nanoparticles of brimonidine tartrate. Curr Drug Deliv 7:244–251
127. Leucuta SE (1989) The kinetics of in vitro release and the pharmacokinetics of miotic response in rabbits of gelatin and albumin microspheres with pilocarpine. Int J Pharm 54:71–78
128. Friess W (1998) Collagen–biomaterial for drug delivery. Eur J Pharm Biopharm 45:113–136
129. Sintzel MB, Bernatchez SF, Tabatabay C, Gurny R (1996) Biomaterials in ophthalmic drug delivery. Eur J Pharm Biopharm 42:358–374
130. Vandervoort J, Ludwig A (2004) Preparation and evaluation of drug-loaded gelatin nanoparticles for topical ophthalmic use. Eur J Pharm Biopharm 57:251–261
131. Bucolo C, Maltese A, Maugeri F, Busà B, Puglisi G, Pignatello R (2004) Eudragit RL100 nanoparticle system for the ophthalmic delivery of cloricromene. J Pharm Pharmacol 56:841–846
132. Pignatello R, Ricupero N, Bucolo C, Maugeri F, Maltese A, Puglisi G (2006) Preparation and characterization of eudragit retard nanosuspensions for the ocular delivery of cloricromene. AAPS PharmSciTech 7:E192–E198
133. Adibkia K, Shadbad MRS, Nokhodchi A, Javadzedeh A, Barzegar-Jalali M, Barar J, Mohammadi G, Omidi Y (2007) Piroxicam nanoparticles for ocular delivery: physicochemical characterization and implementation in endotoxin-induced uveitis. J Drug Target 15:407–416
134. Adibkia K, Omidi Y, Siahi MR, Javadzadeh AR, Barzegar-Jalali M, Barar J, Maleki N, Mohammadi G, Nokhodchi A (2007) Inhibition of endotoxin-induced uveitis by methylprednisolone acetate nanosuspension in rabbits. J Ocul Pharmacol Ther 23:421–432
135. Kumari A, Yadav SK, Yadav SC (2010) Biodegradable polymeric nanoparticles based drug delivery systems. Colloids Surf B Biointerfaces 75:1–18
136. Vega E, Egea M, Valls O, Espina M, Garcia M (2006) Flurbiprofen loaded biodegradable nanoparticles for ophthalmic administration. J Pharm Sci 95:2393–2405
137. Losa C, Marchal-Heussler L, Orallo F, Jato JLV, Alonso MJ (1993) Design of new formulations for topical ocular administration: polymeric nanocapsules containing metipranolol. Pharm Res 10:80–87
138. Losa C, Calvo P, Castro E, Vila-Jato J, Alonso M (1991) Improvement of ocular penetration of amikacin sulphate by association to poly (butylcyanoacrylate) nanoparticles. J Pharm Pharmacol 43:548–552
139. Marchal-Heussler L, Maincent P, Hoffman M, Spittler J, Couvreur P (1990) Antiglaucomatous activity of betaxolol chlorhydrate sorbed onto different isobutylcyanoacrylate nanoparticle preparations. Int J Pharm 58:115–122
140. Davis S (2000) Drug delivery systems. Interdiscip Sci Rev 25:175–183

141. Gómez-Gaete C, Tsapis N, Besnard M, Bochot A, Fattal E (2007) Encapsulation of dexamethasone into biodegradable polymeric nanoparticles. Int J Pharm 331:153–159
142. Marchal-Heussler L, Sirbat D, Hoffman M, Maincent P (1993) Poly (ε-caprolactone) nanocapsules in carteolol ophthalmic delivery. Pharm Res 10:386–390
143. Amrite AC, Kompella UB (2006) Nanoparticles for ocular drug delivery. In: Gupta RB, Kompella UB (eds) Nanoparticle technology for drug delivery. Taylor & Francis Group, LLC, New York, pp 319–360
144. Fitzgerald P, Hadgraft J, Kreuter J, Wilson C (1987) A γ-scintigraphic evaluation of microparticulate ophthalmic delivery systems: liposomes and nanoparticles. Int J Pharm 40:81–84
145. Calvo P, Vila-Jato JL, Alonso MJ (1996) Comparative in vitro evaluation of several colloidal systems, nanoparticles, nanocapsules, and nanoemulsions, as ocular drug carriers. J Pharm Sci 85:530–536
146. Zimmer A, Kreuter J, Robinson J (1991) Studies on the transport pathway of PBCA nanoparticles in ocular tissues. J Microencapsul 8:497–504
147. Qaddoumi MG, Gukasyan HJ, Davda J, Labhasetwar V, Kim K-J, Lee V (2003) Clathrin and caveolin-1 expression in primary pigmented rabbit conjunctival epithelial cells: role in PLGA nanoparticle endocytosis. Mol Vis 9:559–568
148. Qaddoumi MG, Ueda H, Yang J, Davda J, Labhasetwar V, Lee VH (2004) The characteristics and mechanisms of uptake of PLGA nanoparticles in rabbit conjunctival epithelial cell layers. Pharm Res 21:641–648
149. Yasukawa T, Ogura Y, Tabata Y, Kimura H, Wiedemann P, Honda Y (2004) Drug delivery systems for vitreoretinal diseases. Prog Retin Eye Res 23:253–281
150. Hsu J (2007) Drug delivery methods for posterior segment disease. Curr Opin Ophthalmol 18:235–239
151. De Kozak Y, Andrieux K, Villarroya H, Klein C, Thillaye-Goldenberg B, Naud MC, Garcia E, Couvreur P (2004) Intraocular injection of tamoxifen-loaded nanoparticles: a new treatment of experimental autoimmune uveoretinitis. Eur J Immunol 34:3702–3712
152. El-Samaligy MS, Rojanasakul Y, Charlton JF, Weinstein GW, Lim JK (1996) Ocular disposition of nanoencapsulated acyclovir and ganciclovir via intravitreal injection in rabbit's eye. Drug Deliv 3:93–97
153. Bejjani RA, Benezra D, Cohen H, Rieger J, Andrieu C, Jeanny J-C, Gollomb G, Behar-Cohen FF (2005) Nanoparticles for gene delivery to retinal pigment epithelial cells. Mol Vis 11:124–132
154. Dos Santos ALG, Bochot A, Tsapis N, Artzner F, Bejjani RA, Thillaye-Goldenberg B, De Kozak Y, Fattal E, Behar-Cohen F (2006) Oligonucleotide-polyethylenimine complexes targeting retinal cells: structural analysis and application to anti-TGFβ-2 therapy. Pharm Res 23:770–781
155. Suen W-LL, Chau Y (2013) Specific uptake of folate-decorated triamcinolone-encapsulating nanoparticles by retinal pigment epithelium cells enhances and prolongs antiangiogenic activity. J Control Release 167:21–28
156. Pan CK, Durairaj C, Kompella UB, Agwu O, Oliver SC, Quiroz-Mercado H, Mandava N, Olson JL (2011) Comparison of long-acting bevacizumab formulations in the treatment of choroidal neovascularization in a rat model. J Ocul Pharmacol Ther 27:219–224
157. Park K, Chen Y, Hu Y, Mayo AS, Kompella UB, Longeras R, Ma J-X (2009) Nanoparticle-mediated expression of an angiogenic inhibitor ameliorates ischemia-induced retinal neovascularization and diabetes-induced retinal vascular leakage. Diabetes 58:1902–1913
158. Mo Y, Barnett ME, Takemoto D, Davidson H, Kompella UB (2007) Human serum albumin nanoparticles for efficient delivery of Cu, Zn superoxide dismutase gene. Mol Vis 13:746
159. Boddu SH, Jwala J, Chowdhury MR, Mitra AK (2010) In vitro evaluation of a targeted and sustained release system for retinoblastoma cells using Doxorubicin as a model drug. J Ocul Pharmacol Ther 26:459–468
160. Amrite AC, Kompella UB (2005) Size-dependent disposition of nanoparticles and microparticles following subconjunctival administration. J Pharm Pharmacol 57:1555–1563

161. Kompella UB, Bandi N, Ayalasomayajula SP (2003) Subconjunctival nano-and microparticles sustain retinal delivery of budesonide, a corticosteroid capable of inhibiting VEGF expression. Invest Ophthalmol Vis Sci 44:1192–1201
162. Giarmoukakis A, Labiris G, Sideroudi H, Tsimali Z, Koutsospyrou N, Avgoustakis K, Kozobolis V (2013) Biodegradable nanoparticles for controlled subconjunctival delivery of latanoprost acid: in vitro and in vivo evaluation. Preliminary results. Exp Eye Res 112:29–36
163. Shome D, Poddar N, Sharma V, Sheorey U, Maru GB, Ingle A, Sarin R, Banavali S, Dikshit R, Jain V (2009) Does a nanomolecule of carboplatin injected periocularly help in attaining higher intravitreal concentrations? Invest Ophthalmol Vis Sci 50:5896–5900
164. Banerjee R, Carvalho E (2013) Nanoparticulate in-situ gels of TPGS, gellan and PVA as vitreous humor substitutes. US patent 8343471
165. Gupta N, Goel S, Gupta H (2013) Patent review on nanotechnology in ocular drug delivery. Recent Pat Nanomed 3:37–46
166. Gupta H, Aqil M, Khar RK, Ali A, Bhatnagar A, Mittal G (2013) Nanoparticles laden in situ gel for sustained ocular drug delivery. J Pharm Bioallied Sci 5:162
167. Ravi N (2012) Hydrogel nanocomposites for ophthalmic applications. US patent 8153156
168. DasPaul S, Mazumder R, Bhattacharya S, Kumar Jha A (2012) A review on patented nanotechnology used for ocular drug delivery. Recent Pat Nanomed 2:156–163
169. Raiche A, Salamone J (2004) Method for the production of nanoparticles and microparticles by ternary agent concentration and temperature alteration induced immiscibility. US patent 20060057215
170. Kararli TT, Bandyopadhyay R, Singh SK, Hawley LC (2002) Ophthalmic formulation of a selective cyclooxygenase-2 inhibitory drug. US patent 20,020,035,264
171. Xu R, Roberts C, Zhang L (2010) Multi-functional biodegradable particles for selectable targeting, imaging, and therapeutic delivery and use thereof for treating ocular disorders. US patent 20110104069
172. Chauhan A, Gulsen D (2004) Ophthalmic drug delivery system. US patent 8273366
173. Clinical Trials Registry and Results Database (2014a) https://clinicaltrials.gov/ct2/show/NCT00738361?term=eye+nanoparticle&rank=1 [Online]. Accessed 08 Aug 2014
174. Clinical Trials Registry and Results Database (2014b) https://clinicaltrials.gov/ct2/show/NCT01523314?term=eye+nanoparticle&rank=2 [Online]. Accessed 08 Aug 2014

Chapter 17
Barriers to Glaucoma Drug Delivery and Resolving the Challenges Using Nanotechnology

Morgan V. Fedorchak

Abstract As with other diseases of the eye, glaucoma patients face a number of challenges to efficient drug delivery such as low bioavailability due to transport barriers in the eye. It is important to note, however, that the pathophysiology of glaucoma, though not well understood, makes it a particularly challenging disease to address using traditional drug delivery techniques. Researchers have therefore begun to investigate approaches using nanotechnology, and in particular nanoscale biomaterials, to improve upon the delivery of approved and pipeline therapeutic agents. In addition to well-characterized vehicles like liposomes and polymer formulations, a wide variety of other devices like drug-loaded contact lenses and intraocular implants are in development. The primary goal of these drug delivery systems is to improve bioavailability, which may lead to increased adherence to treatment and decreased systemic side effects. Secondary goals like imaging and anti-scarring applications are also relevant to this widespread, vision-threatening disease.

Keywords Glaucoma • Drug delivery • Bioavailability • Biomaterial • Mucoadhesion • Emulsion • Nanoparticle

Drug delivery to the eye presents a vast array of challenges that researchers are trying to address with modern nanotechnology. Much interest has been generated recently in this area, particularly for the treatment of glaucoma [1–3]. Emerging drugs [4–6], new therapeutic targets [7, 8], and an evolving understanding of the progression of the disease [9] make glaucoma an attractive application for

M.V. Fedorchak (✉)
Departments of Ophthalmology and Chemical Engineering, University of Pittsburgh and the Louis J. Fox Center for Vision Restoration, Pittsburgh, PA, USA
e-mail: fedorchak@pitt.edu

© Springer International Publishing Switzerland 2016

Y. Pathak et al. (eds.), *Nano-Biomaterials For Ophthalmic Drug Delivery*,
DOI 10.1007/978-3-319-29346-2_17

translation of nanobiomaterials to the clinic. However, the eye is a complex organ, and each potential application possesses a unique set of issues to address. This chapter will focus on the additional barriers to drug delivery presented by glaucoma.

17.1 Drug Delivery Barriers Specific to Glaucoma

One must appreciate the complexities of glaucoma and its progression to understand why treating it is different from other ocular conditions. First, there is still much to learn about the pathogenesis of glaucoma [10], which makes treatment a much more formidable task. Much of the burden is placed on the patient for self-administration of eyedrop medication to treat the associated increase in intraocular pressure (IOP) [11]. Doctors treating patients with glaucoma must determine which, if any, of the many drug formulations are suitable given a patient's age, disease severity, contraindications, and other factors [12]. Often, patients must also undergo surgery, which can introduce even more barriers to successful drug delivery. These and other challenges are described in detail below.

17.1.1 Patient Adherence

Perhaps the greatest barrier to effective glaucoma drug delivery is patient adherence to the prescribed treatment regimen. One study reports less than 30 % compliance with eyedrops for glaucoma [13], with another study determining that over a quarter of newly diagnosed patients discontinue use of their medication within 3 months [14]. Furthermore, even when patients do administer the drops, it is estimated that nearly 90 % do not instill drops correctly, leading to unintentional noncompliance [15]. The issue of patient adherence in glaucoma is of particular importance because of the implications of not treating the disease. As a progressive and chronic neurological disease, patients who are not properly treated are at a greatly increased risk of vision loss and blindness [16]. In fact, glaucoma accounts for up to 12 % of all cases of blindness in the US alone and is the second leading cause of blindness worldwide [17].

Most glaucoma medications aim to lower IOP in order to slow or stop the progression of the disease [11]. While these modern medications, the most common of which is the prostaglandin analog latanoprost (with over two million prescriptions in Europe alone in 2012 [18]), are quite effective when used properly, many patients struggle with doing so. One of the primary reasons for this lack of compliance is the frequency with which the drops must be self-administered. Latanoprost is administered once daily, which in itself can be challenging, while others such as brimonidine and timolol are delivered two to three times as often [19, 20].

Another significant obstacle is the difficulty in self-administration that many patients experience. Approximately one in ten individuals over 75 are diagnosed with glaucoma [21], which contributes to the high rates of patients who experience difficulty properly instilling the topical drops [15]. Many elderly patients lack the dexterity to self-administer their daily drops. Mobile phone applications, such as "EyeDROPS" (HarPas International), and educational programs [22] have been developed to try to improve compliance with regular reminders, yet these still cannot address the inherent challenges of drop self-administration.

One additional reason why patients may not adhere strictly to their topical drop regimen is the apparent lack of symptoms in those whose disease has not yet progressed [23]. Glaucoma is a largely painless disease and can go untreated for months or years before patients begin to see a decrease in vision [22]. The sometimes slow and often unnoticeable progression of glaucoma can result in patients placing less priority on their medication [15]. Doing so can have devastating effects, however, as vision loss due to glaucoma is irreversible [24].

17.1.2 Side Effects

One 2010, study showed a significant positive correlation between concerns about the side effects of eyedrop medication and intentional noncompliance with the prescribed treatment regimen [25]. Side effects such as blurred vision, burning or itching of the eyes, dry eye, foreign body sensation, and tearing are frequently reported with topical glaucoma medication use [26]. These inconvenient and sometimes painful side effects may prevent patients from appropriately administering drops [23, 25]. Additionally, any reformulation of existing glaucoma medication must take into account the potential side effects associated with the drug. As will be described further below, controlled release of drugs may be able to circumvent this issue by administering lower doses of drug while maintaining effectiveness [3].

17.1.3 Site of Action

When determining the best way to deliver glaucoma drugs, one must consider the site of action. Some drugs act on multiple sites [27], while for others, the primary site of action remains somewhat unclear, particularly for emerging neuroprotective agents [28, 29]. Brimonidine, for instance, is an alpha-adrenergic agonist used to lower IOP primarily through suppressing aqueous humor production and increasing uveoscleral outflow [30] that has also demonstrated potent systemic effects [31] and neuroprotective qualities [32, 33]. Successful drug delivery for glaucoma must factor in the targeted site of action and take steps to deliver the drug most efficiently to that site. For example, drugs acting on the posterior of the eye, like neurotropic factors, would be ideally delivered as an intravitreal injection to bypass the diffusion hindrances of the anterior tissues [24].

17.1.4 Limited Uptake

Depending on the site of action, as described above, the medication delivered can have variable uptake to the affected tissue. Topical drops in particular are unable to reach intraocular tissues in appreciable amounts [24]. The amount of drug that reaches various parts of the eye is heavily influenced by the structure of the drug. For example, small, lipophilic drugs are more readily absorbed through the cornea to the aqueous humor [34], while the permeability of the conjunctiva to larger, hydrophilic molecules is higher than both the sclera and cornea [35, 36]. A portion of drug is also taken up systemically, which can result in unwanted adverse effects [37, 38]. These factors cause the overall uptake to the aqueous humor to remain low, at less than 10 % [24]. This amount is difficult to increase through drop volume alone, due to the constant tear film turnover and size limitations of the conjunctival cul-de-sac [39].

17.1.5 Contraindications

While understanding the site of action and desired drug uptake are critical to designing a successful drug delivery system for glaucoma, some patients are contraindicated for certain methods. One such method is in intraocular injection, which can cause a temporary spike in IOP as well as a risk of retinal detachment, inflammation, and hemorrhage [40]. Patients with a history of acute inflammation in reaction to previous injections, active external ocular infection, or a recent history of thromboembolic events are not recommended for such treatments [41]. The design of intraocular implants or injections for glaucoma patients who are able to receive them would still need to ensure that there was no prolonged increase in IOP due to the implant. For example, one study demonstrates an increase over time in IOP for blank particles injected in the subconjunctival space [42], suggesting that the safety of this potential nanomaterial delivery route should be investigated further.

17.1.6 Pharmacokinetic Considerations

Current approved glaucoma medications have the intended effect of IOP reduction as a method of treating glaucoma. The hypotensive effect of these drugs can be achieved with discrete doses throughout the day, but as with many drugs, the pharmacokinetic profile is not ideal. More specifically, IOP can fluctuate greatly throughout the day based on a number of factors including activity level, posture, and time of day [43–45]. These fluctuations can lead to high IOP levels in certain patients and have been identified as a risk factor for progression of the disease [46, 47]. Peak IOP in patients has been shown to reach dangerous levels (greater than 21 mmHg) due to the peak/trough concentration dynamics of drug throughout the day as drops are administered, which may contribute to the percentage of patients whose disease progresses despite proper treatment [48]. An ideal drug release profile for many glaucoma drugs may be constant, linear ("zero order") release kinetics, which can be difficult to achieve with common controlled-release biomaterials [49].

17.1.7 Other Challenges

Glaucoma presents a number of other challenges that could potentially be addressed using nanotechnology. One such challenge is that of IOP monitoring, which is of primary importance because of the aforementioned fluctuations. These fluctuations vary widely from patient to patient and can be unpredictable [50]. Clinicians should monitor IOP closely in high-risk patients; however, the current standard of care provides only discrete measurements during office visits. One retrospective analysis determined that 24-h IOP monitoring led to increased early detection of glaucoma and changes in treatment for 79 % of patients in the studies included [51].

Uncontrolled IOP or other complications in glaucoma patients can require that surgery be performed to provide filtration of the aqueous humor. Trabeculectomy or shunting surgeries, as with any ocular procedure, introduce a risk of infection or inflammation. Patients are prescribed drops to decrease this risk, which can often lead to further complications. Drug delivery in these situations for glaucoma is made even more challenging by the need to incorporate postoperative care.

Another significant challenge to glaucoma drug delivery is the lack of information about the pathogenesis of the disease. Researchers are working to elucidate the mechanisms of the disease using imaging techniques that explore the outflow pathways and pathology associated with the disease [52–54]. Though not drug delivery in the traditional sense, nanotechnology could also aid in these studies to further improve the treatment options available to glaucoma patients.

17.2 Potential Advantages of Nanotechnology

Recently, there has been rapid progress in the field of nanoscale materials and devices for a number of drug delivery applications. In particular, this field has produced promising advances for a number of diseases that, similar to glaucoma, have unique challenges to successful drug delivery. One obvious example is cancer, where chemotherapeutic agents are being reformulated with nanocarriers for targeted delivery and diagnostics [55]. Nanomedicine has also been used to develop tuberculosis treatments aimed at increasing patient compliance rates, which are thought to be low due to the daily dosing requirements and significant side effects [56].

These examples highlight the versatility of nanotechnology for drug delivery. Such nanoscale systems can theoretically be used to deliver any ocular therapeutic, from small molecule drugs to viruses [2]. They can be administered through a variety of methods, including but not limited to drops, injections, and implants [57]. This adaptability is a prime advantage for a disease like glaucoma, where the individual needs of the patient must be considered when designing a drug delivery system. For instance, an elderly patient may not have the ability to instill topical drops, while a younger patient may prefer the freedom of self-administration.

The incredibly small size afforded by nanotechnology also offers the advantage of high payloads in a small dose [58], which is important in parts of the eye with limited volume like the anterior chamber. Additionally, the relatively small size of most nanocarriers and nanoscale excipients is ideal when conjugated with larger molecules like proteins and antibodies because they less likely affect their function in vivo [57]. Many of the materials that can be used in these drug delivery systems are already being used in FDA-approved formulations or devices, such as poly(lactic-co-glycolic acid) (PLGA). The proven track record of FDA approval for such nanomaterials can facilitate a faster and easier translation to the clinic [59].

17.3 Nanomaterials for Glaucoma Treatment

Perhaps the greatest research effort to use nanotechnology in treating glaucoma has been the field of nanoscale biomaterials for drug delivery. These materials aim to improve upon current drug formulations in a number of ways, which include lowering dosages, localized delivery, sustained release, and improved retention time. Yet other classes of nanomaterials for glaucoma are investigating experimental treatments for glaucoma, often taking advantage of the protection and targeted administration that is offered by some nanocarriers. Here we review various therapeutic biomaterials that aim to address the aforementioned challenges in treating glaucoma. As with current drug formulations, these materials primarily seek to treat glaucoma through IOP reduction. Often, these materials involve a combination of approaches, such as sustained-release formulations incorporating a mucoadhesive carrier.

17.3.1 Vehicle Additives

Although topical drops for glaucoma are associated with low patient adherence rates, they are still a convenient and familiar method for administering antiglaucoma medication. Thus, many researchers have persisted with this route of administration using vehicle additives to improve the retention time of the drug in the precorneal surface and thereby increase the bioavailability of drug. The estimated retention time in the tear film for drugs is approximately 5–7 min due to the constant turnover of tear film, at a rate of about 13–20 % per minute [60].

One method to increase bioavailability is to add hydrogels to the eyedrop formulation. Some examples include hydroxypropyl methylcellulose (HPMC), carboxymethylcellulose (CMC), and propylene glycol [61, 62]. These additives and their polymeric variations act to absorb water and expand the tear film (which also makes them ideal for lubricants in dry eye therapy), thereby increasing retention time [63]. Increased retention time can result in fewer daily doses, leading to fewer systemic side effects and a reduced burden on the patient. Such was the case in a clinical investigation of 0.5 % aqueous timolol maleate (TM) drops versus 0.1 % hydrogel TM drops [64]. The hydrogel TM drops were administered once daily, resulting in a peak drug concentration of 1/6 that of the twice-daily aqueous drops and fewer cardiovascular side effects. These excipients can also serve to stabilize suspensions of nanocrystalline preparations of poorly soluble drugs, such as brinzolamide [65], to increase absorption [66].

Other additives focus on increasing mucoadhesion, the adhesion of two surfaces including a mucosal layer. Chitosan, a biodegradable polysaccharide, and hyaluronic acid (HA), a glycosaminoglycan abundantly found in extracellular matrix, are frequently added to aqueous eyedrop suspensions to increase bioavailability via mucoadhesion [67]. One study demonstrated significantly decreased IOP and improved mucoadhesion using a HA-modified chitosan carrier for the glaucoma drugs timolol and dorzolamide [68]. Surface modification of drug-loaded micellar structures demonstrated similarly positive results in vitro and in vivo [69].

The addition of bioadhesive polymers like CMC or HA can also help protect the ocular surface and reduce toxicity effects. In vitro tolerance of human corneal-limbal epithelial and conjunctival cells to bioadhesive formulations of TM was shown to be significantly higher in one study [70]. Similar results were seen in a separate in vivo study of the melatonin receptor agonist 5-methoxy-carbonylamino-N-acetyltryptamine, with the added benefit of over 30 % decrease in IOP for up to 7 h [71].

17.3.2 Nanoemulsions and Liposomes

Submicron emulsions offer a potential solution for delivering poorly water-soluble drugs. Also called nanoemulsions, these systems comprise a surfactant molecule surrounding a core of the hydrophobic drug [72]. The drug is dissolved in an oil phase which is then encapsulated in a surfactant usually labeled "generally recognized as safe" (GRAS) by the Food and Drug Administration (FDA), forming droplets tens to hundreds of nanometers in diameter. These droplets can be formed acoustically [73] or mechanically [74]. While simpler autoemulsification processes are possible, the high concentration of surfactant required generally makes these formulations unsafe for use in the eye [72].

Often, these formulations will also contain a mucoadhesive additive to further improve bioavailability, as with the formulation described by Ying et al. [75]. A nanoscale lipid emulsion of fluorescently labeled drug-containing poloxamer and chitosan surface modifiers was administered as a topical drop and resulted in significantly greater uptake to the posterior segment versus relevant control groups [75].

The glaucoma drug dorzolamide was also tested as a nanoemulsion using a number of different surfactants [76]. The emulsified forms of the drug showed no signs of irritation in a rabbit model and demonstrated significantly prolonged drug release behavior in vitro. These and similar results have the potential to reduce toxicity effects and dosing frequency for common hydrophobic glaucoma medications, notably prostaglandin analogs [77].

Liposomes are similar to nanoemulsions in that the drug material is contained within an outer layer, in this case made up of a lipid bilayer [78]. The advantages offered by liposomes are the ability to functionalize the surface for targeted delivery, increased solubility (as with submicron emulsions), enhanced biocompatibility, and protection of the drug from degradation [79, 80]. One such study demonstrated that liposomal acetazolamide (ACZ) resulted in less irritation (as determined by increased tear production), increased stability, and an extended hypotensive effect compared to ACZ solution [81].

One of the main drawbacks of both nanoemulsions and liposomes is the rapid release of drug from the core due to membrane diffusion across a very short path length (5–10 nm), which typically results in drug delivery times that cannot be sustained beyond several hours. The recent publication by Natarajan et al. [82], however, describes a liposomal nanocarrier with the latanoprost embedded in the bilayers that can sustain IOP-lowering effect for 120 days [82]. The authors hypothesize that drug release is sustained because transport is controlled by partitioning rather than by diffusion, as with traditional nanovesicles. The liposomes are administered as a subconjunctival injection, which has been reported in separate studies to be a preferred treatment method for glaucoma patients when the frequency of injections is low enough [83].

17.3.3 Polymeric Nanoparticles

Any particle with a diameter on the order of one to hundreds of nanometers can technically be considered a nanoparticle, which would include several of the aforementioned formulations such as nanocrystals, nanoemulsions, and liposomes. Also included within this definition are nanostructures consisting of degradable polymer or dendrimer matrices or combinations of materials. These nanoparticles are distinct because drug is embedded within a solid polymer matrix. The primary advantage offered by these solid nanospheres is the ability to sustain drug release for long periods of time while protecting the unreleased drug from the surrounding environment [78].

Drug release from solid nanospheres is controlled primarily by degradation of the polymer matrix, allowing for diffusion of the drug. One common material is poly(lactic-co-glycolic) acid (PLGA), which is commonly used to make slightly larger microparticles for glaucoma drug delivery lasting up to 1 month [42, 84]. Smaller, nano-sized particles typically cannot sustain drug release for as long and may undergo faster clearance when injected transsclerally, making them better suited for topical delivery to the anterior chamber [85]. The incorporation of drug-loaded PLGA nanoparticles in a polyamidoamine (PAMAM) dendrimer hydrogel has been shown to sustain drug release for 1 week in vivo, with ocular hypotensive effects lasting nearly as long [86]. These particles have the added advantage of low cytotoxicity and high versatility, with single and dual drug-releasing systems in development [87].

Polymer and dendrimer particles are also particularly well suited for delivery of biological therapeutic agents because they are protected from denaturation for the duration of drug release, which can last up to 1 month or more. One such formulation seeks to deliver matrix metalloproteinase-3 (MMP-3) to the trabecular meshwork to prevent buildup of extracellular matrix materials and subsequent IOP increase [88]. Biodegradable nano- or microspheres are also commonly used to deliver neurotrophic agents to the retina for neuroprotection in glaucoma models [89–91].

17.3.4 Hydrogels

A simple definition of a hydrogel is a water-soluble polymer whose properties allow it to be formed into particles, films, coatings, or formed solids [92]. These materials are attractive for drug delivery applications because their physical properties are highly tunable, like porosity, swelling ratio, and degradability. They are also highly biocompatible, primarily because of their high water content and mechanical properties resembling that of extracellular matrix (ECM) [93].

Drug release from the gel matrix is typically controlled by diffusion through the cross-linked polymer network, which often results in faster drug release than from water-insoluble polymer formulations [92, 94]. The highly porous structure also leads to low tensile strength and instability upon injection [92]. Many hydrogel formulations take advantage of copolymer additives to increase the cross-linking density and therefore alter the physical properties. Some examples include hydrogels that triggerably form a solid matrix after a change in temperature, pH, ionic strength, shear stress, and more [95, 96].

Hydrogel-based formulations have been widely investigated for ocular drug delivery via subconjunctival injection [97], nanogel eyedrops [98], and combination systems such as hydrogel-embedded liposomes. One system containing colloidal nanocarriers in a chitosan-based gel was able to sustain IOP reduction for 40 days in a rabbit glaucoma model [99]. Hydrogel-based intravitreal injections are also used for administering anti-VEGF (bevacizumab) in neovascular age-related macular degeneration [100, 101] and may also be used as a carrier for sustained delivery of a neuroprotective payload to the retina.

17.3.5 Contact Lenses

Since the widespread adoption of soft, gas permeable contact lenses in the 1980s, contact lenses have become a familiar and convenient option for vision correction, with a 2010 FDA report estimating 30 million users in the USA. The same materials used to make vision-correcting lenses are now being investigated for their potential as a drug delivery system. One major consideration for contact lens-based drug delivery is maintaining oxygen permeability and optical transparency.

Adsorption of drug onto traditional contact lenses is a simple method for increasing residence time of the drug on the cornea and offers a potential replacement for eyedrop administration [102]. However, these systems cannot sustain drug release beyond 1 day and require high levels of drug to ensure adequate loading, which can lead to an unwanted burst release [103]. To remedy this, other groups have investigated novel materials for contact lens-based drug delivery [104, 105]. One such system uses timolol-loaded nanoparticles within a silicone hydrogel contact lens [106]. Another system uses lenses molecularly imprinted by timolol at the nanoscale that can sustain drug delivery for up to twice as long as lenses without pre-imprinting [107].

The largest drawback to contact lenses for drug delivery is the potential for low patient compliance rates, estimated to be as low as 53 % for replacement of lenses and 45 % for proper handling [108]. This tendency could be problematic for drug delivery applications, as lenses would need to be changed at the appropriate intervals to ensure that therapeutic drug levels are being delivered. Additionally, improper handling and poor compliance could lead to additional complications such as contact lens-related dry eye and ocular surface infections.

17.3.6 Implants or Inserts

Although eyedrops may be more practical for some patients, such as those in developing nations, inserts, implants, and refillable devices have been explored for other patient populations that may benefit from a clinician-controlled drug delivery system. These systems are more frequently investigated for their use in treating posterior segment diseases, as with dexamethasone (Ozurdex®, Allergan) and fluocinolone acetonide (Retisert®, Bausch + Lomb) intravitreal implants for treating macular edema and uveitis, respectively [109]. This type of treatment would especially hold potential for treatment of neovascular glaucoma, using anti-VEGF [110], or neurotrophic factors for neuroprotection of retinal ganglion cells [111]. Care must be taken, however, as secondary glaucoma is a potential side effect for intravitreal implants and injection [40]. Similar rod-shaped implants have been tested for use in the subconjunctival space, providing months of release [112] from an administration method that one study suggests over 62 % of patients would prefer as a replacement for frequent eyedrops [113].

Nanoporous ocular inserts have also been described for use in treating glaucoma by releasing antiglaucoma drugs from within the conjunctival cul-de-sac for up to 28 days [114]. This device uses modern controlled-release polymer technology to improve upon passive diffusion-based systems that are no longer used [115, 116]. Building on the ocular insert concept is the latanoprost-loaded nanosheet, which can be applied directly to the cornea and provides up to 9 days of IOP reduction [117]. Similar results were also seen using an electrospun nanofiber patch placed in the conjunctival cul-de-sac [118].

Another unique device utilizes a microelectromechanical drug pump implanted similarly to a glaucoma drainage device [119]. This electrolysis-based pump is designed to provide 4–6 weeks of drug delivery with each transconjunctival drug refill. One distinct advantage of this type of nanodevice is that it could theoretically be loaded with any drug without having to redesign the device.

17.4 Other Uses of Nanotechnology for Glaucoma

Beyond antiglaucoma drug delivery, nanotechnology can serve other roles in the treatment of glaucoma patients. In patients receiving glaucoma filtration surgery, concomitant placement of drug-loaded nanoparticles can modulate the wound-healing response, thereby reducing scarring and improving the function of the bleb [120–122]. Nanocarriers loaded with corticosteroids that would traditionally be administered topically following trabeculectomy can also be used to provide autonomous postoperative care [123, 124].

Nanotechnology has also led to recent advances in the overall understanding of outflow pathways in the eye. Nano-sized tracers are injected into the anterior chamber and monitored over time [125] and can even be used in conjunction with antiglaucoma medication to determine the effect on drainage [54]. These techniques may aid in elucidating the pathophysiology of glaucoma and help identify potential new treatment methods for the disease.

As more is understood about glaucoma, the need for more reliable frequent IOP monitoring arises [126]. However, most patients only have IOP measured during visits to the clinic, which may miss peak IOP levels occurring at other times throughout the day [127]. Thus, some groups are using nanoscale devices, either implanted [128, 129] or embedded in a soft contact lens [130, 131], as a way of tracking IOP continuously and wirelessly reporting measurements back to the clinician. One recent publication describes a sensitive and precise microfluidic device whose output can be read using a smart phone camera, enabling at-home monitoring of IOP [132].

These and other applications of nanotechnology for glaucoma could potentially be combined with sustained-release drug delivery systems described above. Despite the challenges inherent to glaucoma drug delivery, nanoscale materials and devices offer myriad solutions that may one day improve the diagnosis, treatment, and outcomes for patients with glaucoma.

References

1. Egan P, Harris A, Siesky B, Abrams-Tobe L, Gerber AL, Park J, Holland S, Kim NJ, Januleviciene I (2014) Comparison of intraocular pressure in glaucoma subjects treated with Xalatan((R)) versus generic latanoprost. Acta Ophthalmol 92(5):e415–e416. doi:10.1111/aos.12321

2. Kim NJ, Harris A, Gerber A, Tobe LA, Amireskandari A, Huck A, Siesky B (2014) Nanotechnology and glaucoma: a review of the potential implications of glaucoma nanomedicine. Br J Ophthalmol 98(4):427–431. doi:10.1136/bjophthalmol-2013-304028
3. Pita-Thomas DW, Goldberg JL (2013) Nanotechnology and glaucoma: little particles for a big disease. Curr Opin Ophthalmol 24(2):130–135. doi:10.1097/ICU.0b013e32835cfe92
4. Bagnis A, Papadia M, Scotto R, Traverso CE (2011) Current and emerging medical therapies in the treatment of glaucoma. Expert Opin Emerg Drugs 16(2):293–307. doi:10.1517/14728214.2011.563733
5. Wang SK, Chang RT (2014) An emerging treatment option for glaucoma: Rho kinase inhibitors. Clin Ophthalmol 8:883–890. doi:10.2147/OPTH.S41000
6. Wu JH, Zhang SH, Gao FJ, Lei Y, Chen XY, Gao F, Zhang SJ, Sun XH (2013) RNAi screening identifies GSK3beta as a regulator of DRP1 and the neuroprotection of lithium chloride against elevated pressure involved in downregulation of DRP1. Neurosci Lett 554:99–104. doi:10.1016/j.neulet.2013.08.057
7. Dey S, Mitra AK (2005) Transporters and receptors in ocular drug delivery: opportunities and challenges. Expert Opin Drug Deliv 2(2):201–204. doi:10.1517/17425247.2.2.201
8. Crooke A, Colligris B, Pintor J (2012) Update in glaucoma medicinal chemistry: emerging evidence for the importance of melatonin analogues. Curr Med Chem 19(21):3508–3522
9. Yucel YH, Johnston MG, Ly T, Patel M, Drake B, Gumus E, Fraenkl SA, Moore S, Tobbia D, Armstrong D, Horvath E, Gupta N (2009) Identification of lymphatics in the ciliary body of the human eye: a novel "uveolymphatic" outflow pathway. Exp Eye Res 89(5):810–819. doi:10.1016/j.exer.2009.08.010
10. Gemenetzi M, Yang Y, Lotery AJ (2012) Current concepts on primary open-angle glaucoma genetics: a contribution to disease pathophysiology and future treatment. Eye (Lond) 26(3):355–369. doi:10.1038/eye.2011.309
11. Quigley HA (2011) Glaucoma. Lancet 377(9774):1367–1377. doi:10.1016/S0140-6736(10)61423-7, Epub 2011/04/02. S0140-6736(10)61423-7 [pii]
12. Morrison JC, Pollack IP (2011) Glaucoma: science and practice. Thieme, New York
13. Campbell JH, Schwartz GF, LaBounty B, Kowalski JW, Patel VD (2014) Patient adherence and persistence with topical ocular hypotensive therapy in real-world practice: a comparison of bimatoprost 0.01% and travoprost Z 0.004% ophthalmic solutions. Clin Ophthalmol 8:927–935. doi:10.2147/OPTH.S49467
14. Kashiwagi K, Furuya T (2014) Persistence with topical glaucoma therapy among newly diagnosed Japanese patients. Jpn J Ophthalmol 58(1):68–74. doi:10.1007/s10384-013-0284-2
15. Gupta R, Patil B, Shah BM, Bali SJ, Mishra SK, Dada T (2012) Evaluating eye drop instillation technique in glaucoma patients. J Glaucoma 21(3):189–192. doi:10.1097/IJG.0b013e31820bd2e1
16. Musch DC, Gillespie BW, Niziol LM, Lichter PR, Varma R, Group CS (2011) Intraocular pressure control and long-term visual field loss in the collaborative initial glaucoma treatment study. Ophthalmology 118(9):1766–1773. doi:10.1016/j.ophtha.2011.01.047
17. Quigley HA, Vitale S (1997) Models of open-angle glaucoma prevalence and incidence in the United States. Invest Ophthalmol Vis Sci 38(1):83–91
18. Painter SL, Mead AL (2014) Patient experience of the transition from Xalatan to generic latanoprost. Eye (Lond) 28(7):911. doi:10.1038/eye.2014.71
19. Cantor LB (2006) Brimonidine in the treatment of glaucoma and ocular hypertension. Ther Clin Risk Manag 2(4):337–346
20. Moisseiev E, Kurtz S, Lazar M, Shemesh G (2013) Intraocular pressure reduction using a fixed combination of timolol maleate 0.5% and brimonidine tartrate 0.2% administered three times daily. Clin Ophthalmol 7:1269–1273. doi:10.2147/OPTH.S47760
21. Klein R, Klein BE (2013) The prevalence of age-related eye diseases and visual impairment in aging: current estimates. Invest Ophthalmol Vis Sci 54(14):ORSF5–ORSF13. doi:10.1167/iovs.13-12789
22. Waterman H, Evans JR, Gray TA, Henson D, Harper R (2013) Interventions for improving adherence to ocular hypotensive therapy. Cochrane Database Syst Rev 4:CD006132. doi:10.1002/14651858.CD006132.pub3

23. Lacey J, Cate H, Broadway DC (2009) Barriers to adherence with glaucoma medications: a qualitative research study. Eye (Lond) 23(4):924–932. doi:10.1038/eye.2008.103
24. Ghate D, Edelhauser HF (2008) Barriers to glaucoma drug delivery. J Glaucoma 17(2):147–156. doi:10.1097/IJG.0b013e31814b990d
25. Rees G, Leong O, Crowston JG, Lamoureux EL (2010) Intentional and unintentional nonadherence to ocular hypotensive treatment in patients with glaucoma. Ophthalmology 117(5):903–908. doi:10.1016/j.ophtha.2009.10.038
26. Pisella PJ, Pouliquen P, Baudouin C (2002) Prevalence of ocular symptoms and signs with preserved and preservative free glaucoma medication. Br J Ophthalmol 86(4):418–423
27. Robin AL, Burnstein Y (1998) Selectivity of site of action and systemic effects of topical alpha agonists. Curr Opin Ophthalmol 9(2):30–33
28. Levkovitch-Verbin H, Waserzoog Y, Vander S, Makarovsky D, Ilia P (2014) Minocycline mechanism of neuroprotection involves the Bcl-2 gene family in optic nerve transection. Int J Neurosci. doi:10.3109/00207454.2013.878340
29. Roos JC, Haridas AS (2014) Prolonged mydriasis after inadvertent topical administration of the calcium channel antagonist amlodipine: implications for glaucoma drug development. Cutan Ocul Toxicol. doi:10.3109/15569527.2014.896016
30. Greenfield DS, Liebmann JM, Ritch R (1997) Brimonidine: a new alpha2-adrenoreceptor agonist for glaucoma treatment. J Glaucoma 6(4):250–258
31. Sy GY, Bruban V, Bousquet P, Feldman J (2002) Nitric oxide discriminates the sites and mechanisms of action of centrally acting anti-hypertensive drugs in rabbits. Neuropharmacology 43(8):1330–1338
32. Dong CJ, Guo Y, Agey P, Wheeler L, Hare WA (2008) Alpha2 adrenergic modulation of NMDA receptor function as a major mechanism of RGC protection in experimental glaucoma and retinal excitotoxicity. Invest Ophthalmol Vis Sci 49(10):4515–4522. doi:10.1167/iovs.08-2078
33. Fujita Y, Sato A, Yamashita T (2013) Brimonidine promotes axon growth after optic nerve injury through Erk phosphorylation. Cell Death Disease 4:e763. doi:10.1038/cddis.2013.298
34. Schoenwald RW (1997) Ocular pharmacokinetics. In: Z TJ (ed) Textbook of ocular pharmacology., pp 119–138
35. Ahmed I, Gokhale RD, Shah MV, Patton TF (1987) Physicochemical determinants of drug diffusion across the conjunctiva, sclera, and cornea. J Pharm Sci 76(8):583–586
36. Sasaki H, Igarashi Y, Nagano T, Nishida K, Nakamura J (1995) Different effects of absorption promoters on corneal and conjunctival penetration of ophthalmic beta-blockers. Pharm Res 12(8):1146–1150
37. Korte JM, Kaila T, Saari KM (2002) Systemic bioavailability and cardiopulmonary effects of 0.5% timolol eyedrops. Graefes Arch Clin Exp Ophthalmol 240(6):430–435. doi:10.1007/s00417-002-0462-2
38. Lama PJ (2002) Systemic adverse effects of beta-adrenergic blockers: an evidence-based assessment. Am J Ophthalmol 134(5):749–760
39. Mishima S, Gasset A, Klyce SD Jr, Baum JL (1966) Determination of tear volume and tear flow. Invest Ophthalmol 5(3):264–276
40. Frenkel MP, Haji SA, Frenkel RE (2010) Effect of prophylactic intraocular pressure-lowering medication on intraocular pressure spikes after intravitreal injections. Arch Ophthalmol 128(12):1523–1527. doi:10.1001/archophthalmol.2010.297
41. Semeraro F, Morescalchi F, Duse S, Gambicorti E, Romano MR, Costagliola C (2014) Systemic thromboembolic adverse events in patients treated with intravitreal anti-VEGF drugs for neovascular age-related macular degeneration: an overview. Expert Opin Drug Saf 13(6):785–802. doi:10.1517/14740338.2014.911284
42. Fedorchak MV, Conner IP, Medina CA, Wingard JB, Schuman JS, Little SR (2014) 28-day intraocular pressure reduction with a single dose of brimonidine tartrate-loaded microspheres. Exp Eye Res 125:210–216. doi:10.1016/j.exer.2014.06.013
43. Liu JH, Zhang X, Kripke DF, Weinreb RN (2003) Twenty-four-hour intraocular pressure pattern associated with early glaucomatous changes. Invest Ophthalmol Vis Sci 44(4):1586–1590

44. Mansouri K, Weinreb RN, Liu JH (2012) Effects of aging on 24-hour intraocular pressure measurements in sitting and supine body positions. Invest Ophthalmol Vis Sci 53(1):112–116. doi:10.1167/iovs.11-8763
45. Wilensky JT, Gieser DK, Dietsche ML, Mori MT, Zeimer R (1993) Individual variability in the diurnal intraocular pressure curve. Ophthalmology 100(6):940–944
46. Lee YR, Kook MS, Joe SG, Na JH, Han S, Kim S, Shin CJ (2012) Circadian (24-hour) pattern of intraocular pressure and visual field damage in eyes with normal-tension glaucoma. Invest Ophthalmol Vis Sci 53(2):881–887. doi:10.1167/iovs.11-7846
47. Renard E, Palombi K, Gronfier C, Pepin JL, Noel C, Chiquet C, Romanet JP (2010) Twenty-four hour (Nyctohemeral) rhythm of intraocular pressure and ocular perfusion pressure in normal-tension glaucoma. Invest Ophthalmol Vis Sci 51(2):882–889. doi:10.1167/iovs.09-3668
48. Kass MA, Heuer DK, Higginbotham EJ, Johnson CA, Keltner JL, Miller JP, Parrish RK 2nd, Wilson MR, Gordon MO (2002) The ocular hypertension treatment study: a randomized trial determines that topical ocular hypotensive medication delays or prevents the onset of primary open-angle glaucoma. Arch Ophthalmol 120(6):701–713, discussion 829–30
49. Mohammadi-Samani S, Taghipour B (2014) PLGA micro and nanoparticles in delivery of peptides and proteins; problems and approaches. Pharm Dev Technol. doi:10.3109/10837450.2014.882940
50. Tanaka S, Watanabe M, Inatomi S, Umeda K, Yoshida K, Ohguro I, Ohguro H (2014) Effects of several anti-glaucoma medications on the circadian intraocular pressure fluctuations in patients with primary open-angle glaucoma. J Ocul Pharmacol Ther 30(1):12–20. doi:10.1089/jop.2013.0121
51. Hughes E, Spry P, Diamond J (2003) 24-hour monitoring of intraocular pressure in glaucoma management: a retrospective review. J Glaucoma 12(3):232–236
52. Ho LC, Conner IP, Do CW, Kim SG, Wu EX, Wollstein G, Schuman JS, Chan KC (2014) In vivo assessment of aqueous humor dynamics upon chronic ocular hypertension and hypotensive drug treatment using gadolinium-enhanced MRI. Invest Ophthalmol Vis Sci 55(6):3747–3757. doi:10.1167/iovs.14-14263
53. Tam AL, Gupta N, Zhang Z, Yucel YH (2011) Quantum dots trace lymphatic drainage from the mouse eye. Nanotechnology 22(42):425101. doi:10.1088/0957-4484/22/42/425101
54. Tam AL, Gupta N, Zhang Z, Yucel YH (2013) Latanoprost stimulates ocular lymphatic drainage: an in vivo nanotracer study. Transl Vis Sci Technol 2(5):3. doi:10.1167/tvst.2.5.3
55. Aslan B, Ozpolat B, Sood AK, Lopez-Berestein G (2013) Nanotechnology in cancer therapy. J Drug Target 21(10):904–913. doi:10.3109/1061186X.2013.837469
56. Dube A, Lemmer Y, Hayeshi R, Balogun M, Labuschagne P, Swai H, Kalombo L (2013) State of the art and future directions in nanomedicine for tuberculosis. Expert Opin Drug Deliv 10(12):1725–1734. doi:10.1517/17425247.2014.846905
57. Kompella UB, Amrite AC, Pacha Ravi R, Durazo SA (2013) Nanomedicines for back of the eye drug delivery, gene delivery, and imaging. Prog Retin Eye Res 36:172–198. doi:10.1016/j.preteyeres.2013.04.001
58. Zarbin MA, Montemagno C, Leary JF, Ritch R (2010) Nanomedicine in ophthalmology: the new frontier. Am J Ophthalmol 150(2):144–162.e2. doi:10.1016/j.ajo.2010.03.019
59. Danhier F, Ansorena E, Silva JM, Coco R, Le Breton A, Preat V (2012) PLGA-based nanoparticles: an overview of biomedical applications. J Control Release 161(2):505–522. doi:10.1016/j.jconrel.2012.01.043, Epub 2012/02/23
60. Greaves JL, Wilson CG (1993) Treatment of diseases of the eye with mucoadhesive delivery systems. Adv Drug Deliver Rev 11(3):349–383. doi:10.1016/0169-409x(93)90016-W
61. Cohen S, Martin A, Sall K (2014) Evaluation of clinical outcomes in patients with dry eye disease using lubricant eye drops containing polyethylene glycol or carboxymethylcellulose. Clin Ophthalmol 8:157–164. doi:10.2147/OPTH.S53822
62. Jansook P, Stefansson E, Thorsteinsdottir M, Sigurdsson BB, Kristjansdottir SS, Bas JF, Sigurdsson HH, Loftsson T (2010) Cyclodextrin solubilization of carbonic anhydrase inhibitor drugs: formulation of dorzolamide eye drop microparticle suspension. Eur J Pharm Biopharm 76(2):208–214. doi:10.1016/j.ejpb.2010.07.005

63. Mayol L, Quaglia F, Borzacchiello A, Ambrosio L, La Rotonda MI (2008) A novel poloxamers/hyaluronic acid in situ forming hydrogel for drug delivery: rheological, mucoadhesive and in vitro release properties. Eur J Pharm Biopharm 70(1):199–206. doi:10.1016/j.ejpb.2008.04.025
64. Uusitalo H, Kahonen M, Ropo A, Maenpaa J, Bjarnhall G, Hedenstrom H, Turjanmaa V (2006) Improved systemic safety and risk-benefit ratio of topical 0.1% timolol hydrogel compared with 0.5% timolol aqueous solution in the treatment of glaucoma. Graefes Arch Clin Exp Ophthalmol 244(11):1491–1496. doi:10.1007/s00417-006-0328-0
65. Wu W, Li J, Wu L, Wang B, Wang Z, Xu Q, Xin H (2013) Ophthalmic delivery of brinzolamide by liquid crystalline nanoparticles: in vitro and in vivo evaluation. AAPS PharmSciTech 14(3):1063–1071. doi:10.1208/s12249-013-9997-2
66. Tuomela A, Liu P, Puranen J, Ronkko S, Laaksonen T, Kalesnykas G, Oksala O, Ilkka J, Laru J, Jarvinen K, Hirvonen J, Peltonen L (2014) Brinzolamide nanocrystal formulations for ophthalmic delivery: reduction of elevated intraocular pressure in vivo. Int J Pharm 467(1–2):34–41. doi:10.1016/j.ijpharm.2014.03.048
67. Kaur IP, Smitha R (2002) Penetration enhancers and ocular bioadhesives: two new avenues for ophthalmic drug delivery. Drug Dev Ind Pharm 28(4):353–369. doi:10.1081/DDC-120002997
68. Wadhwa S, Paliwal R, Paliwal SR, Vyas SP (2010) Hyaluronic acid modified chitosan nanoparticles for effective management of glaucoma: development, characterization, and evaluation. J Drug Target 18(4):292–302. doi:10.3109/10611860903450023, Epub 2009/12/01
69. Lin HR, Chang PC (2013) Novel pluronic-chitosan micelle as an ocular delivery system. J Biomed Mater Res B Appl Biomater 101(5):689–699. doi:10.1002/jbm.b.32871
70. Andres-Guerrero V, Vicario-de-la-Torre M, Molina-Martinez IT, Benitez-del-Castillo JM, Garcia-Feijoo J, Herrero-Vanrell R (2011) Comparison of the in vitro tolerance and in vivo efficacy of traditional timolol maleate eye drops versus new formulations with bioadhesive polymers. Invest Ophthalmol Vis Sci 52(6):3548–3556. doi:10.1167/iovs.10-6338
71. Andres-Guerrero V, Molina-Martinez IT, Peral A, de las Heras B, Pintor J, Herrero-Vanrell R (2011) The use of mucoadhesive polymers to enhance the hypotensive effect of a melatonin analogue, 5-MCA-NAT, in rabbit eyes. Invest Ophthalmol Vis Sci 52(3):1507–1515. doi:10.1167/iovs.10-6099
72. Vandamme TF (2002) Microemulsions as ocular drug delivery systems: recent developments and future challenges. Prog Retin Eye Res 21(1):15–34
73. Sivakumar M, Tang SY, Tan KW (2014) Cavitation technology – a greener processing technique for the generation of pharmaceutical nanoemulsions. Ultrason Sonochem 21(6):2069–2083. doi:10.1016/j.ultsonch.2014.03.025
74. Benita S, Levy MY (1993) Submicron emulsions as colloidal drug carriers for intravenous administration: comprehensive physicochemical characterization. J Pharm Sci 82(11):1069–1079
75. Ying L, Tahara K, Takeuchi H (2013) Drug delivery to the ocular posterior segment using lipid emulsion via eye drop administration: effect of emulsion formulations and surface modification. Int J Pharm 453(2):329–335. doi:10.1016/j.ijpharm.2013.06.024
76. Ammar HO, Salama HA, Ghorab M, Mahmoud AA (2009) Nanoemulsion as a potential ophthalmic delivery system for dorzolamide hydrochloride. AAPS PharmSciTech 10(3):808–819. doi:10.1208/s12249-009-9268-4
77. Linden C (2001) Therapeutic potential of prostaglandin analogues in glaucoma. Expert Opin Investig Drugs 10(4):679–694. doi:10.1517/13543784.10.4.679
78. Honda M, Asai T, Oku N, Araki Y, Tanaka M, Ebihara N (2013) Liposomes and nanotechnology in drug development: focus on ocular targets. Int J Nanomedicine 8:495–503. doi:10.2147/IJN.S30725
79. Gan L, Wang J, Jiang M, Bartlett H, Ouyang D, Eperjesi F, Liu J, Gan Y (2013) Recent advances in topical ophthalmic drug delivery with lipid-based nanocarriers. Drug Discov Today 18(5–6):290–297. doi:10.1016/j.drudis.2012.10.005

80. Swaminathan J, Ehrhardt C (2012) Liposomal delivery of proteins and peptides. Expert Opin Drug Deliv 9(12):1489–1503. doi:10.1517/17425247.2012.735658
81. Pisal PB, Joshi MA, Padamwar MN, Patil SS, Pokharkar VB (2014) Probing influence of methodological variation on active loading of acetazolamide into nanoliposomes: biophysical, in vitro, ex vivo, in vivo and rheological investigation. Int J Pharm 461(1–2):82–88. doi:10.1016/j.ijpharm.2013.11.034
82. Natarajan JV, Darwitan A, Barathi VA, Ang M, Htoon HM, Boey F, Tam KC, Wong TT, Venkatraman SS (2014) Sustained drug release in nanomedicine: a long-acting nanocarrier-based formulation for glaucoma. ACS Nano 8(1):419–429. doi:10.1021/nn4046024
83. Chong RS, Su DH, Tsai A, Jiang Y, Htoon HM, Lamoureux EL, Aung T, Wong TT (2013) Patient acceptance and attitude toward an alternative method of subconjunctival injection for the medical treatment of glaucoma. J Glaucoma 22(3):190–194. doi:10.1097/IJG.0b013e318237c6c4
84. Bertram JP, Saluja SS, McKain J, Lavik EB (2009) Sustained delivery of timolol maleate from poly(lactic-co-glycolic acid)/poly(lactic acid) microspheres for over 3 months. J Microencapsul 26(1):18–26. doi:10.1080/02652040802095250
85. Amrite AC, Edelhauser HF, Singh SR, Kompella UB (2008) Effect of circulation on the disposition and ocular tissue distribution of 20 nm nanoparticles after periocular administration. Mol Vis 14:150–160
86. Yang H, Tyagi P, Kadam RS, Holden CA, Kompella UB (2012) Hybrid dendrimer hydrogel/PLGA nanoparticle platform sustains drug delivery for one week and antiglaucoma effects for four days following one-time topical administration. ACS Nano 6(9):7595–7606. doi:10.1021/nn301873v, Epub 2012/08/11
87. Zhou HY, Hao JL, Wang S, Zheng Y, Zhang WS (2013) Nanoparticles in the ocular drug delivery. Int J Ophthalmol 6(3):390–396. doi:10.3980/j.issn.2222-3959.2013.03.25
88. Turturro S, Sunoqrot S, Ying H, Hong S, Yue BY (2013) Sustained release of matrix metalloproteinase-3 to trabecular meshwork cells using biodegradable PLGA microparticles. Mol Pharm 10(8):3023–3032. doi:10.1021/mp4001052
89. Checa-Casalengua P, Jiang C, Bravo-Osuna I, Tucker BA, Molina-Martinez IT, Young MJ, Herrero-Vanrell R (2011) Retinal ganglion cells survival in a glaucoma model by GDNF/Vit E PLGA microspheres prepared according to a novel microencapsulation procedure. J Control Release 156(1):92–100. doi:10.1016/j.jconrel.2011.06.023
90. Ward MS, Khoobehi A, Lavik EB, Langer R, Young MJ (2007) Neuroprotection of retinal ganglion cells in DBA/2J mice with GDNF-loaded biodegradable microspheres. J Pharm Sci 96(3):558–568. doi:10.1002/jps.20629
91. Xiao JH, Zhang MN (2010) Neuroprotection of retinal ganglion cells with GDNF-Loaded biodegradable microspheres in experimental glaucoma. Int J Ophthalmol 3(3):189–191. doi:10.3980/j.issn.2222-3959.2010.03.01
92. Hoare TR, Kohane DS (2008) Hydrogels in drug delivery: progress and challenges. Polymer 49(8):1993–2007. doi:10.1016/J.Polymer.2008.01.027
93. Gauvin R, Parenteau-Bareil R, Dokmeci MR, Merryman WD, Khademhosseini A (2012) Hydrogels and microtechnologies for engineering the cellular microenvironment. Wiley Interdiscip Rev Nanomed Nanobiotechnol 4(3):235–246. doi:10.1002/wnan.171
94. Yang H, Leffler CT (2013) Hybrid dendrimer hydrogel/poly(lactic-co-glycolic acid) nanoparticle platform: an advanced vehicle for topical delivery of antiglaucoma drugs and a likely solution to improving compliance and adherence in glaucoma management. J Ocul Pharmacol Ther 29(2):166–172. doi:10.1089/jop.2012.0197
95. Joglekar M, Trewyn BG (2013) Polymer-based stimuli-responsive nanosystems for biomedical applications. Biotechnol J 8(8):931–945. doi:10.1002/biot.201300073
96. Klouda L, Mikos AG (2008) Thermoresponsive hydrogels in biomedical applications. Eur J Pharm Biopharm 68(1):34–45. doi:10.1016/j.ejpb.2007.02.025
97. Cheng YH, Hung KH, Tsai TH, Lee CJ, Ku RY, Chiu AW, Chiou SH, Liu CJ (2014) Sustained delivery of latanoprost by thermosensitive chitosan-gelatin-based hydrogel for controlling ocular hypertension. Acta Biomater. doi:10.1016/j.actbio.2014.05.031.PubMed

98. Abd El-Rehim HA, Swilem AE, Klingner A, el Hegazy SA, Hamed AA (2013) Developing the potential ophthalmic applications of pilocarpine entrapped into polyvinylpyrrolidone-poly(acrylic acid) nanogel dispersions prepared by gamma radiation. Biomacromolecules 14(3):688–698. doi:10.1021/bm301742m
99. Hsiao MH, Chiou SH, Larsson M, Hung KH, Wang YL, Liu CJ, Liu DM (2014) A temperature-induced and shear-reversible assembly of latanoprost-loaded amphiphilic chitosan colloids: characterization and in vivo glaucoma treatment. Acta Biomater 10(7):3188–3196. doi:10.1016/j.actbio.2014.03.016
100. Rauck BM, Friberg TR, Medina Mendez CA, Park D, Shah V, Bilonick RA, Wang Y (2014) Biocompatible reverse thermal gel sustains the release of intravitreal bevacizumab in vivo. Invest Ophthalmol Vis Sci 55(1):469–476. doi:10.1167/iovs.13-13120
101. Wang CH, Hwang YS, Chiang PR, Shen CR, Hong WH, Hsiue GH (2012) Extended release of bevacizumab by thermosensitive biodegradable and biocompatible hydrogel. Biomacromolecules 13(1):40–48. doi:10.1021/bm2009558
102. Peng CC, Ben-Shlomo A, Mackay EO, Plummer CE, Chauhan A (2012) Drug delivery by contact lens in spontaneously glaucomatous dogs. Curr Eye Res 37(3):204–211. doi:10.3109/02713683.2011.630154
103. Karlgard CC, Wong NS, Jones LW, Moresoli C (2003) In vitro uptake and release studies of ocular pharmaceutical agents by silicon-containing and p-HEMA hydrogel contact lens materials. Int J Pharm 257(1–2):141–151
104. Ciolino JB, Stefanescu CF, Ross AE, Salvador-Culla B, Cortez P, Ford EM, Wymbs KA, Sprague SL, Mascoop DR, Rudina SS, Trauger SA, Cade F, Kohane DS (2014) In vivo performance of a drug-eluting contact lens to treat glaucoma for a month. Biomaterials 35(1):432–439. doi:10.1016/j.biomaterials.2013.09.032
105. Kim HJ, Zhang K, Moore L, Ho D (2014) Diamond nanogel-embedded contact lenses mediate lysozyme-dependent therapeutic release. ACS Nano 8(3):2998–3005. doi:10.1021/nn5002968
106. Jung HJ, Abou-Jaoude M, Carbia BE, Plummer C, Chauhan A (2013) Glaucoma therapy by extended release of timolol from nanoparticle loaded silicone-hydrogel contact lenses. J Control Release 165(1):82–89. doi:10.1016/j.jconrel.2012.10.010
107. Hiratani H, Fujiwara A, Tamiya Y, Mizutani Y, Alvarez-Lorenzo C (2005) Ocular release of timolol from molecularly imprinted soft contact lenses. Biomaterials 26(11):1293–1298. doi:10.1016/j.biomaterials.2004.04.030
108. Ramamoorthy P, Nichols JJ (2014) Compliance factors associated with contact lens-related dry eye. Eye Contact Lens 40(1):17–22. doi:10.1097/ICL.0000000000000009
109. Choonara YE, Pillay V, Danckwerts MP, Carmichael TR, du Toit LC (2010) A review of implantable intravitreal drug delivery technologies for the treatment of posterior segment eye diseases. J Pharm Sci 99(5):2219–2239. doi:10.1002/jps.21987
110. Simha A, Braganza A, Abraham L, Samuel P, Lindsley K (2013) Anti-vascular endothelial growth factor for neovascular glaucoma. Cochrane Database Syst Rev 10:CD007920. doi:10.1002/14651858.CD007920.pub2
111. Payne AJ, Kaja S, Sabates NR, Koulen P (2013) A case for neuroprotection in ophthalmology: developments in translational research. Mo Med 110(5):429–436
112. Jessen BA, Shiue MH, Kaur H, Miller P, Leedle R, Guo H, Evans M (2013) Safety assessment of subconjunctivally implanted devices containing latanoprost in Dutch-belted rabbits. J Ocul Pharmacol Ther 29(6):574–585. doi:10.1089/jop.2012.0190
113. Foo RC, Lamoureux EL, Wong RC, Ho SW, Chiang PP, Rees G, Aung T, Wong TT (2012) Acceptance, attitudes, and beliefs of Singaporean Chinese toward an ocular implant for glaucoma drug delivery. Invest Ophthalmol Vis Sci 53(13):8240–8245. doi:10.1167/iovs.12-10393
114. Mealy JE, Fedorchak MV, Little SR (2014) In vitro characterization of a controlled-release ocular insert for delivery of brimonidine tartrate. Acta Biomater 10(1):87–93. doi:10.1016/j.actbio.2013.09.024

115. Macoul KL, Pavan-Langston D (1975) Pilocarpine ocusert system for sustained control of ocular hypertension. Arch Ophthalmol 93(8):587–590
116. Strohmaier K, Snyder E, Adamsons I (1998) A multicenter study comparing dorzolamide and pilocarpine as adjunctive therapy to timolol: patient preference and impact on daily life. J Am Optom Assoc 69(7):441–451
117. Kashiwagi K, Ito K, Haniuda H, Ohtsubo S, Takeoka S (2013) Development of latanoprost-loaded biodegradable nanosheet as a new drug delivery system for glaucoma. Invest Ophthalmol Vis Sci 54(8):5629–5637. doi:10.1167/iovs.12-9513
118. Gagandeep, Garg T, Malik B, Rath G, Goyal AK (2014) Development and characterization of nano-fiber patch for the treatment of glaucoma. Eur J Pharm Sci 53:10–16. doi:10.1016/j.ejps.2013.11.016
119. Saati S, Lo R, Li PY, Meng E, Varma R, Humayun MS (2010) Mini drug pump for ophthalmic use. Curr Eye Res 35(3):192–201. doi:10.3109/02713680903521936
120. Butler MR, Prospero Ponce CM, Weinstock YE, Orengo-Nania S, Chevez-Barrios P, Frankfort BJ (2013) Topical silver nanoparticles result in improved bleb function by increasing filtration and reducing fibrosis in a rabbit model of filtration surgery. Invest Ophthalmol Vis Sci 54(7):4982–4990. doi:10.1167/iovs.13-12047
121. Chen X, Zhu X, Li L, Xian G, Wang W, Ma D, Xie L (2013) Investigation on novel chitosan nanoparticle-aptamer complexes targeting TGF-beta receptor II. Int J Pharm 456(2):499–507. doi:10.1016/j.ijpharm.2013.08.028
122. Shao T, Li X, Ge J (2011) Target drug delivery system as a new scarring modulation after glaucoma filtration surgery. Diagn Pathol 6:64. doi:10.1186/1746-1596-6-64
123. Sabzevari A, Adibkia K, Hashemi H, De Geest BG, Mohsenzadeh N, Atyabi F, Ghahremani MH, Khoshayand MR, Dinarvand R (2013) Improved anti-inflammatory effects in rabbit eye model using biodegradable poly beta-amino ester nanoparticles of triamcinolone acetonide. Invest Ophthalmol Vis Sci 54(8):5520–5526. doi:10.1167/iovs.13-12296
124. Swaminathan S, Vavia PR, Trotta F, Cavalli R (2013) Nanosponges encapsulating dexamethasone for ocular delivery: formulation design, physicochemical characterization, safety and corneal permeability assessment. J Biomed Nanotechnol 9(6):998–1007
125. Guignier B, Bourahla K, Bekaert V, Brasse D, Gaucher D, Speeg-Schatz C, Bourcier T (2013) Scintigraphic study of the lymphatic drainage of the anterior chamber of the mouse eye and its pathophysiological implications. J Fr Ophtalmol 36(10):836–842. doi:10.1016/j.jfo.2012.11.021
126. Mansouri K, Weinreb RN (2012) Meeting an unmet need in glaucoma: continuous 24-h monitoring of intraocular pressure. Expert Rev Med Devices 9(3):225–231. doi:10.1586/erd.12.14
127. Agnifili L, Mastropasqua R, Frezzotti P, Fasanella V, Motolese I, Pedrotti E, Iorio AD, Mattei PA, Motolese E, Mastropasqua L (2014) Circadian intraocular pressure patterns in healthy subjects, primary open angle and normal tension glaucoma patients with a contact lens sensor. Acta Ophthalmol. doi:10.1111/aos.12408
128. Chitnis G, Maleki T, Samuels B, Cantor LB, Ziaie B (2013) A minimally invasive implantable wireless pressure sensor for continuous IOP monitoring. IEEE Trans Biomed Eng 60(1):250–256. doi:10.1109/TBME.2012.2205248
129. Ha D, de Vries WN, John SW, Irazoqui PP, Chappell WJ (2012) Polymer-based miniature flexible capacitive pressure sensor for intraocular pressure (IOP) monitoring inside a mouse eye. Biomed Microdevices 14(1):207–215. doi:10.1007/s10544-011-9598-3
130. Chen GZ, Chan IS, Leung LK, Lam DC (2014) Soft wearable contact lens sensor for continuous intraocular pressure monitoring. Med Eng Phys 36(9):1134–1139. doi:10.1016/j.medengphy.2014.06.005
131. Sanchez I, Laukhin V, Moya A, Martin R, Ussa F, Laukhina E, Guimera A, Villa R, Rovira C, Aguilo J, Veciana J, Pastor JC (2011) Prototype of a nanostructured sensing contact lens for noninvasive intraocular pressure monitoring. Invest Ophthalmol Vis Sci 52(11):8310–8315. doi:10.1167/iovs.10-7064
132. Araci IE, Su B, Quake SR, Mandel Y (2014) An implantable microfluidic device for self-monitoring of intraocular pressure. Nat Med. doi:10.1038/nm.3621

Chapter 18
Intraocular Drug Delivery Technologies: Advancing Treatment of Posterior Segment Disorders of the Eye

Viness Pillay, Yahya E. Choonara, and Lisa C. du Toit

Abstract Posterior segment ocular (vitreoretinal) disorders are the foremost contributors to visual impairment and, ultimately, blindness. This is placing an increasing demand on the pharmaceutical scientist to formulate effective ophthalmic drug delivery systems. Biophysiological blood–ocular barriers present a major challenge in achieving adequate bioactive bioavailability to the posterior segment, necessitating the search for pathways that can achieve access to the posterior ocular tissues for successful management of these disorders. Despite drug advances, the pharmacological management of these severe ocular pathologies is still a major hurdle, but a surmountable one. Research has been implicit in conveying that innovative polymeric drug delivery systems are essential for realizing a superlative pharmaceutical intervention, where effective bioactives are available for intraocular disease treatment. Increasingly innovative approaches are being investigated to address intraocular drug delivery challenges, each striving to achieve enhanced targeted bioactive delivery to the posterior ocular tissues. This chapter provides an overview of pertinent developments in the design of intraocular delivery systems; all of which are attempts at improving the treatment and/or management of posterior segment/vitreoretinal pathologies of the eye.

Keywords Vitreoretinal disorders • Polymeric intraocular delivery systems • Biodegradable and nonbiodegradable intraocular systems • Intraocular microsystems and nanosystems • Hybrid intraocular delivery systems • Microneedles • Biologics

V. Pillay (✉) • Y.E. Choonara • L.C. du Toit
Wits Advanced Drug Delivery Platform Research Unit, Department of Pharmacy and Pharmacology, School of Therapeutic Sciences, Faculty of Health Sciences, University of the Witwatersrand, Johannesburg, Parktown 2193, South Africa
e-mail: viness.pillay@wits.ac.za

© Springer International Publishing Switzerland 2016

Y. Pathak et al. (eds.), *Nano-Biomaterials For Ophthalmic Drug Delivery*,
DOI 10.1007/978-3-319-29346-2_18

18.1 The Intricacies of the Eye: Rationalizing Intraocular Drug Delivery

The eye – an anatomically isolated specialized sensory organ – exhibits unique pharmacodynamic and pharmacokinetic properties and is a prime site for practically investigating drug delivery system implantation. This is a consequence of its systemic seclusion by the blood–ocular barriers (retinal, aqueous, and vitreous). The eye is divided into two ocular fluid-filled cavities (the anterior and posterior segment), creating an essentially large implantation site. Furthermore, the surface of tissues to be targeted is large despite their associated small volume. The ocular space also possesses the advantage of "immune privilege," mainly evident in the anterior chamber and in the subretinal space, reducing the risk of foreign antigen-induced inflammatory reactions and rejection of cell grafts. The visual functions of the eye are, however, exceptionally delicate, and more complex means of drug targeting are required. Significantly, the eye can be viewed as an apt "laboratory" for the investigation of drug delivery in inflammatory and infectious afflictions. According to Robinson [1] "no other organ is so readily accessible or as visible for observation," undoubtedly a distinct test for the pharmaceutical scientist in drug delivery system design. Successful drug delivery is thus considered the limiting factor for an optimal therapeutic strategy [2].

In terms of research spurring the pharmaceutical scientist, ophthalmic drug delivery is an increasingly demanding sector [3, 4]. In their investigations, authors have pointed to inflammatory posterior segment ocular (vitreoretinal) disorders as the foremost contributors to visual impairment and, ultimately, blindness [5, 6]. The posterior segment disorders of prevalence and importance culminating in visual impairment include age-related macular degeneration (AMD), macular edema (a result of retinal vein occlusion), uveitis, diabetic retinopathy, proliferative vitreoretinopathy (PVR), choroidal neovascularization (CNV), retinitis pigmentosa, and cytomegalovirus (CMV) retinitis [7] – these largely affect the retina. Ensuring delivery of the indicated bioactive to the posterior segment of the eye (which includes the vitreous humor, retina, choroid, and optic nerve) is fundamental for the effectual treatment of internal eye structure disorders [5]. Nonetheless anterior segment drug delivery systems, most notably eye drops, still lead trends in the ocular market. When it comes to the disorders perpetuating visual impairment and even blindness, inflammatory disorders of the posterior segment are the ultimate perpetrators [5].

As emphasized by Yasukawa and co-workers [8], advancement of the ocular drug delivery field is delayed when problems of drug availability to the posterior segment are encountered; furthermore indirect bioactive pathways (topical, systemic, or periocular) to the vitreous have in the past suffered from the disadvantage of poor penetration of the biophysiological blood–ocular barriers, necessitating considerations such as direct intravitreal drug delivery to access the posterior ocular tissues for successful management of posterior segment disorders [9]. Intravitreal injection of drug and sustained drug delivery systems fabricated from polymers (from the nano- to the macroscale, biodegradable or nonbiodegradable) for delivery via injection or implantation target the posterior segment [6, 9]. Implementation of a sustained drug delivery strategy is preferred to repeated injections [6].

The field of ocular therapeutics is at the stage where appropriate active compounds for the management and treatment of posterior segment disorders are available; but in order to render the pharmaceutical intervention a success, innovative intraocular controlled drug delivery systems are essential. Controlled release of the drug can be fulfilled through the use of polymeric drug delivery systems, for bioavailability optimization and potential side effect minimization. Simultaneously, advances in nanotechnology could elaborate nano-sized delivery vehicles for enhanced drug permeation of blood–ocular barriers [10].

Despite drug advances, the pharmacological management of these severe ocular pathologies is still a major hurdle, but a surmountable one. There is widespread procedural occurrence of elevated intraocular pressure and cataract progression [11]. Other risks, particularly associated with intravitreal injection of corticosteroids, include endophthalmitis [12], retinal detachment [13], hemicentral vein occlusion [14], preretinal hemorrhage [15], pseudohypopyon [15], and vitreous hemorrhage [12].

Research has been implicit in conveying that controlled polymeric drug delivery systems are essential for realizing a superlative pharmaceutical intervention, where effective bioactives are available for intraocular disease treatment. Increasingly innovative approaches are being investigated to address intraocular drug delivery challenges, each striving to achieve enhanced targeted bioactive delivery to the posterior ocular tissues (choroid, retina, and vitreous) [16].

This chapter provides an all-encompassing overview of pertinent developments in the design of intraocular delivery systems; all of which are attempts at improving the treatment and/or management of posterior segment/vitreoretinal pathologies of the eye. Diverse intraocular implants (on the market or still under investigation), nano- and microsystems, hybrid systems, and future trends are discussed. Figure 18.1 depicts a representative overview of some of the diverse approaches that have been innovated to date for enhancing intraocular drug delivery.

18.2 A Synopsis of Drug Delivery Approaches to the Eye

A brief step back in time, prior to the mid-1970s, divulges that somewhat archaic approaches were implemented for the treatment of endophthalmitis, including the intensive application of topical antibiotics used in conjunction with intravenous and subconjunctival therapy [17, 18]. Unsurprisingly, suboptimal outcomes were attained and the prognosis for visual recovery was poor. Treatment failure, with reference to extraocular treatment modalities for intraocular disorders such as endophthalmitis and uveitis, has, as emphasized, been largely ascribed to the poor penetration of the drug of interest into the posterior segment [19].

Previously, where the institution of intraocular agents was constrained to endophthalmitis treatment, intraocular approaches are now considered for the entire panoply of posterior segment disorders. Four routes of administration can be employed

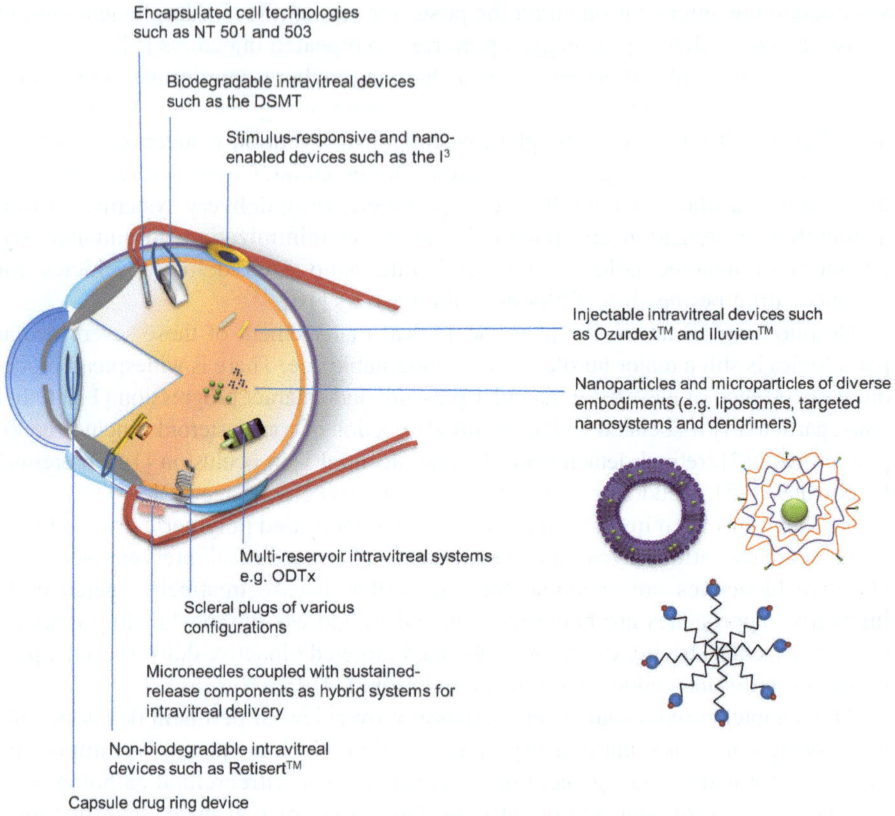

Fig. 18.1 Representative intraocular delivery systems for posterior segment disorders (I^3 = intelligent intraocular implant, described in Sect. 18.5.5)

for drug delivery to the posterior ocular tissues: topical, systemic, intraocular, and periocular (i.e., subconjunctival, retrobulbar, and sub-Tenon). The diverse routes of ocular administration and associated kinetic pathways are divulged in Fig. 18.2.

Static barriers, which comprise the corneal layers, sclera, and retina and encompass the blood–aqueous barrier (BAB) and blood–retinal barrier (BRB), and dynamic barriers, including blood flow in the choroid and conjunctiva, lymphatic clearance, and dilution by tears, as well as efflux pumps, present a substantial challenge for ocular drug delivery, most specifically to the posterior segment. These barriers to ocular drug delivery are intrinsic and distinctive to the anatomical and physiological functioning of the eye and are encountered via most administration routes, i.e., topical, systemic, and periocular. Approaches specifically focused on overcoming these barriers include colloidal embodiments of drug delivery systems (e.g., nanoparticles, liposomes, dendrimers, and nanoemulsions). Furthermore, other innovative drug delivery strategies have been designed to achieve targeted and sustained release such as bioadhesive gels. Designing fairly noninvasive sustained drug delivery systems for availing drugs to the posterior segment seems the most

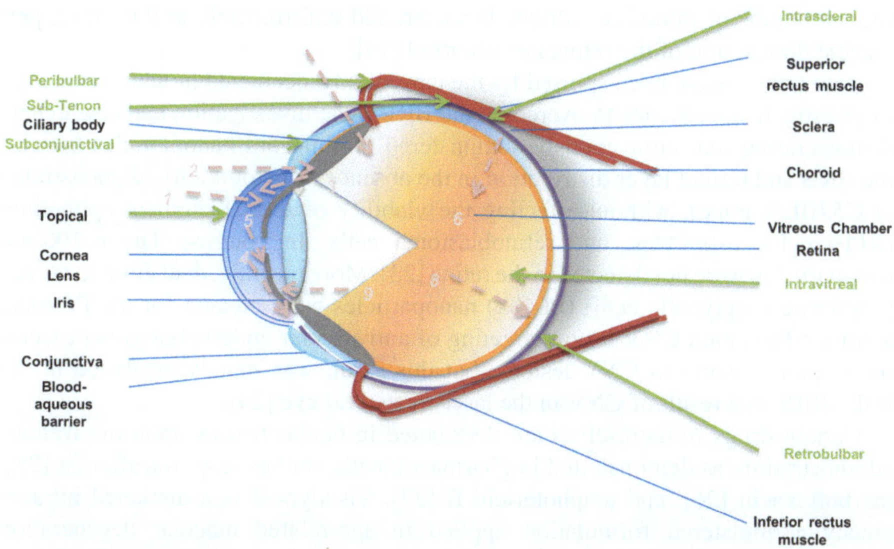

Fig. 18.2 Schematic representing the ocular structures, routes of drug administration to the eye, and drug kinetic pathways. Note that the iris, ciliary body and choroid form the uvea. The following processes are referred to: (1) trans-corneal permeation from the lachrymal fluid into the anterior chamber, (2) non-corneal drug permeation across the conjunctiva and sclera into the anterior uvea, (3) drug distribution from the blood stream via blood–aqueous barrier into the anterior chamber, (4) elimination of drug from the anterior chamber by the aqueous humor turnover to the trabecular meshwork and Sclemm's canal, (5) drug elimination from the aqueous humor into the systemic uveoscleral circulation, (6) drug distribution from the blood into the posterior eye across the BRB, (7) intravitreal drug administration, (8) drug elimination from the vitreous via posterior route across the blood retina barrier, and (9) drug elimination from the vitreous via anterior route to the posterior chamber (Adapted from Urtti [21])

feasible option prospectively for providing acceptable means for clinicians and patients to address posterior segment disorders [20]. Consideration of diverse delivery routes, preformulation, and formulation factors is thus paramount in the design of an effective ocular delivery system for vitreoretinal disorders [20].

18.2.1 Systemic Administration

The major barriers encountered following systemic administration in ocular drug delivery are the BAB (anterior segment) and BRB (posterior segment). There is thus a need for specific targeting systems to enable transport of molecules through the choroid and into the underlying confines of the retina [20, 21]. Reports have highlighted poor concentrations in the vitreous following intravenous administration, with bactericidal antibiotic concentrations in the posterior chamber being problematic to attain [22]. Ultimately, the passage of materials, including antibiotics, into the vitreous cavity is inhibited by the tight junctions of these barriers. These junctions may be disrupted in cases of infection or inflammation, to allow increased concentrations of intravitreal drugs; however, and unfortunately at this stage, permanent destruction of the retina has occurred [19].

Researchers have been spurred by the current advancements in nanotechnology to investigate approaches to overcome the BRB. One investigation had researchers demonstrating that intravenously administered 20-nm gold nanoparticles traverse the BRB and retinal layer distribution in the absence of cytotoxicity (demonstrated in C57BL/6 mice), without affecting the viability of retinal pigment epithelium (RPE) cells, astrocytes, and retinoblastoma cells. In contrast, larger 100-nm nanoparticles were not detected in the retina [23]. More recently, dual-functionalized poly(lactic-co-glycolic acid) (PLGA) nanoparticles were created for the IV route and found to attain CNV lesion targeting of antivascular endothelial growth factor intraceptor plasmid to CNV lesions, but this result was mainly attributed to the leaky BRB as a result of CNV in the laser-treated rat eye [24].

Certain drugs in themselves are distributed in ocular tissues upon intravenous administration, as demonstrated in pharmacokinetic studies, e.g., micafungin [25], marbofloxacin [26], and amphotericin B [27]. Visudyne® is a marketed intravenously administered formulation applied in age-related macular degeneration (AMD) as photodynamic therapy. However, intravenous administration is not favored for treating ocular disorders because of associated toxicity and delivery issues [20].

18.2.2 Topical Administration

Eye drops are the most common topical preparation and are generally employed to treat anterior segment diseases, possessing their site of action in the various layers of the cornea, conjunctiva, sclera, and associated tissues of the anterior segment (i.e., iris and ciliary body). Upon topical drug administration, the resistance encountered to drug penetration when crossing the cornea, aqueous humor, and lens generally emanates in nontherapeutic levels of drug in the posterior vitreous, retina, and choroid [6]. Precorneal factors (e.g., implicating lachrymation events) and anatomical barriers significantly reduce the bioavailability of topical formulations following instillation [28]. The first defense encountered is the tear film which has a high turnover rate. Mucin, as a component of the tear film, acts protectively through formation of a hydrophilic layer that is distributed over the ocular surface glycocalyx, clearing debris and pathogens [29]. Consider this: the volume of human tears is ~7 μL, and the cul-de-sac can transiently hold ~30 μL of the administered eye drop, but because tear film is rapidly restored within 2–3 min, most of the topically administered solutions are thus washed away within ~15–30 s following instillation. All these factors emanate in a short contact time with the absorptive membranes, the rationale for not even 5 % of the applied dose gaining access to the intraocular tissues [30]. Furthermore, the drug has yet to encounter and permeate the various layers of the cornea, conjunctiva, and sclera. Ultimately the corneal layers such as the epithelium and stroma impose themselves as major ocular drug delivery barriers. A pertinent consideration is thus that the permeant should possess amphipathic properties to facilitate permeation through these layers [31].

Investigations have highlighted that bioactive permeability across the sclera is inversely proportional to the molecular radius, indicating that compounds with linear structures are less likely to permeate than globular compounds [32]. Permeability across the sclera is also affected by the charge of the drug molecule, with positively charged molecules exhibiting poor permeability as a result of attachment to the negatively charged proteoglycan matrix [20, 33], although this method has been exploited to enhance corneal retention time of nanosystems.

Although investigators continue to develop new topical drug delivery systems, past efforts (in the 1990s) to realize reasonable therapeutic drug levels in the posterior segment have fallen short [19].

18.2.3 Oral Administration

There have been diverse reasons for the investigation of oral delivery [34–36], often in combination with topical delivery [37], one being that isolated use of topical delivery did not achieve therapeutic concentrations in the posterior segment. Secondly, the lower degree of invasiveness and higher patient acceptance of the oral route spurred consideration of this approach (compared to IV route). The restrictions, however, are apparent: there is inadequate access to a number of ocular tissues of interest which then requires high dosage for therapeutic efficacy, especially in the vitreous–retina, resulting in systemic side effects [6, 9]. This route is, therefore, not the primary choice and only select compounds have been investigated for ocular drug delivery, e.g., analgesics [38], antibiotics [39, 40], antivirals [41], antineoplastic agents [42], and omega-6 fatty acids [43]. Should a bioactive be considered for the oral route, it must possess a high oral bioavailability and have the propensity to cross the BAB and BRB [20].

18.2.4 Periocular and Intravitreal Administration

The periocular route encompasses subconjunctival, sub-Tenon, retrobulbar, and peribulbar administration and is advantageously less invasive than the intravitreal route (see Fig. 18.2). Following administration of a drug via periocular injection, three routes enable access to the posterior segment: (1) transscleral, (2) systemic circulation via the choroid, and (3) the anterior pathway traversing the tear film, cornea, aqueous humor, and vitreous humor [20].

Administering drug directly to the subconjunctival space could be more advantageous than topical delivery for anterior segment disorders as it avoids the conjunctival epithelial barrier [44]. However, investigators have reported in the past that this route was not entirely successful in achieving therapeutically efficacious drug concentrations posteriorly [19, 45, 46]. A transscleral route bypasses the cornea–conjunctiva barrier, but there are still various barriers to overcome before the posterior segment is reached. Dynamic barriers attain rapid drug elimination following subconjunctival administration, and the formulation is essentially drained into the systemic circulation, thus reducing ocular bioavailability [47, 48]. The pertinent factor impacting vitreous drug levels following conjunctival administration is therefore drug elimination from the subconjunctival space and molecules evading the conjunctival vasculature permeating through the sclera (which is not a significant barrier, more permeable than the cornea, and independent of lipophilicity but dependent on the molecular radius). It would then permeate the choroid, gaining access to the neural retina and photoreceptor cells [21]. The choroid nonetheless poses a formidable obstacle as increased choroidal blood flow may cause elimination of an extensive amount of drug prior to it arriving at the neural retina [20]. The BRB is subsequently encountered which limits drug availability to the photoreceptor cells; however, the periocular route necessitates smaller amounts of drugs compared to systemically and in essence achieves therapeutic drug levels in the vitreous–retina with lower systemic side effects [6]. This route is thus still a contender for drug delivery considerations.

Overall, where the indirect route attains inefficient access of bioactives to the vitreous; the direct intravitreal delivery of drugs is necessitated for the effective treatment of a number of posterior segment diseases. Compared to periocular injections, intravitreal injections avail direct insertion of drug molecules into the vitreous, which furnishes an advantage. A challenge, however, is that there is irregular drug distribution in the vitreous with small molecules rapidly distributing throughout the vitreous and restricted larger molecule diffusion. Factors such as the disease state and molecular weight of the administered drug also affect distribution [49]. HA, a negatively charged glycosaminoglycan occupying the vitreous, has the propensity to undergo interaction with polymeric DNA complexes and cationic lipoidal leading to extensive aggregation and immobilization of the complexes [50, 51]. The nanoparticle structure and surface charge further have pertinent effects on mobility in the vitreous [50]. This is where surface modification of nanosystems comes into play, and hydrophilic PEG chains are often included. Studies have also demonstrated better diffusion of anionic nanoparticles throughout the vitreous than cationic systems [52].

Separation of the vitreous and retina is via a cell layer known as the inner limiting membrane (ILM), serving as a barrier, following intravitreal administration, to retinal delivery. In addition, the presence of the RPE presents an inherent complexity in drug transport from the vitreous to the outer retinal and choroidal segments. Therapeutic efficacy is also affected by the half-life in the vitreous. Elimination pathways play a role as drug is cleared either anteriorly or posteriorly following intravitreal injection. Elimination via the anterior route implicates passive drug diffusion via the zonular spaces from the vitreous into the aqueous humor. This is ensued with further clearance via aqueous turnover and uveal blood flow. Elimination via the posterior pathway implicates BRB permeation by the drug; therefore, optimum passive permeability or active transport mechanisms are necessitated. The half-life of compounds is thus increased by hydrophilicity and large molecular weight [20, 21].

Intraocular drug delivery can be achieved through application of intravitreal drug injection and diverse bioerodible or non-bioerodible polymeric controlled-release systems that are injected (i.e., colloidal systems) or implanted (i.e., naillike scleral plugs, disks, rods, pumps of diverse embodiments, matrices, rings, etc.) in the posterior segment of the eye and maintain therapeutic levels over an extended period of time. Sites at which these systems can be introduced include the vitreous cavity, the subretinal space, the sclera, the sub-Tenon capsule, and the suprachoroidal and the peribulbar space. Other alternatives include transdermal patches and iontophoretic devices. Biodegradable devices are inherently advantageous when compared to non-erodible systems in that as they are degrading, they are gradually disappearing from the site of implantation and eliminating the risk of fibrous encapsulation and the necessity of surgical removal [5, 9, 19].

Before implants made an appearance on the ocular drug delivery front, intravitreal injection into the posterior segment was deemed adequate [19]. Intravitreal injection can achieve targeted therapeutic concentrations, thus diminishing systemic side effects. The concept of an intravitreal injection was presented in 1944 by Von Sallmann and co-workers but was not widely implemented at that stage because toxicity studies on the directly injected intraocular drugs were lacking [53, 54]. From their investigations emanated reports that there could be maintenance of therapeutic levels of penicillin in the posterior segment by intravitreal injection. In a series of experiments, Peyman and associates (as reported in [19]) positively ascertained the level of toxicity of a number of intravitreal antibiotics, determined clearance times, and undertook a comparison of intravitreal injection with combined intravenous and subconjunctival administration to obtain clarity on this treatment modality [19]. Currently treatment of various acute and chronic posterior segment diseases, including PVR, recurrent uveitis, and CMV retinitis, is by repeat intravitreal injections for maintenance of drug concentrations within the therapeutic window. However, the side effects of these multiple intraocular injections required to reach therapeutic efficacy are manifold, being not only disagreeable to patients, but also instigating the risk of cataract formation, retinal detachment, endophthalmitis, and vitreous hemorrhage. Because the majority of drugs employed for posterior segment disorders possess a low therapeutic index, this can necessitate intravitreous injections at drug concentrations fast approaching the toxic level for the retina [6].

18.3 Intraocular Drug Delivery: Nonbiodegradable and Biodegradable Embodiments, Diverse Biomaterials, and Mechanisms of Release

Sustained drug delivery devices are proffered as a superlative alternative to multiple intravitreous injections [6]. The fabrication of an intraocular controlled-release system that can maintain therapeutic drug levels of drug at the posterior ocular tissues would proposedly embody significant progress for the treatment of ocular disease.

The platform on which implantable intraocular technologies are based may be nonbiodegradable or biodegradable, reservoir, or matrix, most often with injection or implantation into the vitreous or sclera. Reservoir systems are often non-eroding devices that require surgical placement and replacement but purportedly provide more "predictable" control of drug release furnished by an inbuilt rate control mechanism connected to an internal drug reservoir, such as a device opening, a porous membrane or a diffusion-controlling coating. Zero-order drug release kinetics is more readily achieved via a reservoir system due to the fact that a single mechanism controls delivery. Nonbiodegradable implants generally provide longevity and simplicity. Non-eroding intraocular devices are engineered employing biocompatible biomaterials including polyimides, polysulfones, polyvinyl alcohols, polyvinylidene fluorides, ethylene vinyl acetates, siloxane polymers, and various methacrylate and ethylacrylate polymers.

Comparatively, in a matrix system, drug is blended with the rate-controlling polymer. If the matrix is biodegradable implant, it negates the need for posttreatment removal, which enhances patient acceptance and compliance [55, 56]. Mechanisms of attaining zero-order drug release kinetics in a matrix device include use of a non-eroding polymer enabling control of drug release via polymeric chain relaxation and drug dissolution rate. A high drug concentration is required such that erosion of the matrix creates adequate pathways, enabling dissolution media to navigate to the drug. The increasing presence of voids created from dissolved drug could compromise the integrity of a non-eroding matrix system. In a bioerodible matrix system, more complex behavior ensues due to either bulk or surface polymeric erosion. Biocompatible polymers commonly used for erodible intraocular matrices are poly-lactides, poly-glycolides, and polycaprolactone. Polyanhydrides and polyorthoesters are exploited for their surface-eroding characteristics. Biopolymers such as alginate, chitosan, and cross-linked hyaluronic acid (HA) may also be employed to achieve unique bioresponsive erosion. Polymeric system selection for intraocular drug delivery design is reliant on the projected length of treatment, with identification of drugs possessing high potencies (μg/day range), owing to the small intraocular space to accommodate the limited space of the ocular cavity [57–68].

The release kinetics of both bioerodible and non-bioerodible implants depends on the drug solubility and diffusion coefficient, the drug-loading capacity, the configuration of the device, and the rate of degradation of the polymer in vivo. Pertinent considerations for the design of intraocular delivery systems are biocompatibility with a requirement for all components to be chemically inert, non-carcinogenic, hypoallergenic, and mechanically stable at the implantation site – no immune or inflammatory response should be instigated at the implantation/delivery site. Materials employed must possess stability and should be resistant to chemical or physical modification by local tissue [69, 70], unless this is an inbuilt bioresponsive mechanism. It is clear that the design of a suitable intraocular system requires a number of considerations, rendering the process inherently complex and lengthy. Table 18.1 is a summary of intravitreal implants (nonbiodegradable and biodegradable embodiments) that have run this gauntlet to produce a generally successful device and are available on the market.

Table 18.1 Market-available nonbiodegradable and biodegradable intravitreal implants

Implant	Description	Pros	Cons
Nonbiodegradable/reservoir implants			
Vitrasert® by pSivida	Ethylene vinyl acetate and poly(vinyl) alcohol reservoir device loaded with ganciclovir	Limited ocular complications on initial implantation and efficacy exceeding standard of care, being intravenous ganciclovir	Multiple implantations elicit complications – vitreous hemorrhage, rhegmatogenous retinal detachment, endophthalmitis, and cystoid macular edema with epiretinal membrane [67]
	Approved in 1996 for the treatment of AIDS-related CMV retinitis	Simple and good operation [66]	
	Surgically implanted through a 5.5-mm pars plana incision. Lasts 5–8 months [66]		
Retisert® by pSivida marketed by Bausch and Lomb	Implant generation following Vitrasert®	Attained enhanced targeted delivery and duration compared to Vitrasert® [66]	Increased intraocular pressure and cataract progression [67]
	FDA approved in 2005 for noninfectious uveitis	Significant decrease in disease recurrence [67]	Long duration elicits ocular side effects [66]
	Scleral suture following surgical implantation through a 3.5-mm pars plana incision		Toxicities too severe for patients with diabetic macular edema
	Releases fluocinolone acetonide ~ 30 months [66]		Steroid-induced glaucoma [66]
Biodegradable implants			
Ozurdex™ (previously Posurdex) by Allergan	Inserted surgically or with a special injector	Incorporates more water-soluble steroid than triamcinolone or fluocinolone acetonide – can provide improved disease control and minimize ocular complications	Preliminary data indicated a lower rate of glaucoma vs. Retisert® – but a head-to-head comparison is still required
	FDA approved in June 2009 for macular edema due to vein occlusion	Could be more economical than injections if glaucoma and cataract risks reduced [66]	
	Pelletized biodegradable copolymer that hydrolyzes to lactic and glycolic acids		
	Releases either 350 or 700 μg dexamethasone over 6 months [66]		

18.4 Intraocular Delivery Systems at the Clinical Stage of Testing

In addition to intraocular systems on the market, a number are in the clinical phase of testing and approaching market readiness. Table 18.2 divulges the latest stage of development (referring to updated 2012/2013 reports) of related ocular drug delivery systems for posterior segment disorders.

18.5 Review of Specialized Investigational Intraocular Implantable Technologies for the Treatment of Vitreoretinal Disorders

The pertinent advantages of sustained-release intraocular device technology has being over other delivery modes for the management of posterior segment disorders of the eye have been elaborated [8, 85]. There has been a surge in the design of diverse implantable systems for controlled intraocular drug delivery to the posterior ocular tissues, including the design of osmotic mini-pumps, biodegradable and non-biodegradable scleral plugs, scleral disks, capsular systems, bioadhesive polymeric gels, micromachined devices, minirobots, and polymeric matrices and scaffolds of various geometries, each enabling a distinctive modus operandi for enhancing intraocular drug delivery. Cutting-edge delivery approaches can be evaluated for intraocular potential due to the confinement and isolation of the eye, in addition to its "immune privilege," restricting the foreign body reaction [86].

Systems investigated in the recent past such as biodegradable microsystems, emulsions, and liposomes offer the advantage of being injected into the posterior segment, but ocular media clarity may be impaired if they ultimately become suspended in the vitreous. Transscleral iontophoresis is a noninvasive technique that has been described as holding some promise for intravitreal delivery, the disadvantages being the complicated procedural requirements and limited application to only ionizable drugs [9].

Scientists have pinpointed biodegradable polymeric nanoparticles as demonstrating significant innovative potential as drug delivery devices for ocular application. Biodegradable polymers can be combined with drugs at the nano-level in such a way to enable drug release from the nanosystem into the eye in a predesigned fashion, which may be constant or cyclic over an extensive period or stimulus responsive. Following delivery of the drug load, the polymeric carrier is naturally degraded in the body [87]. According to Tsibouklis in Barbu et al. [88], "the formulation of biodegradable or bioerodible polymers as colloidal [nanosystems] holds significant promise for ophthalmic drug delivery." However, "formulation stability, particle size uniformity, control of drug release rate, and the large-scale manufacture of sterile preparations are challenges that still require directed investigation" [87]. These systems are further elaborated on in this chapter.

Table 18.2 Ocular drug delivery systems at the preclinical or clinical stage of development

Ocular delivery system	Description	Clinical trial status	Digital image
Nonbiodegradable polymeric drug delivery systems			
I-vation (SurModics Inc.)	Sustained vitreal triamcinolone acetonide (TA) delivery for DME	Patients demonstrating improved visual profile were 64 % in the slow-release group and 72 % in the fast-release group	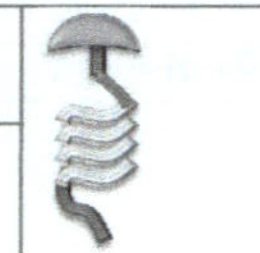
	Titanium helical coil coated with TA (925 μg) and nonbiodegradable polymers poly(methyl methacrylate) and ethylene vinyl acetate. Predicted in vivo sustained delivery of a minimum of 2 years	No further clinical trial has commenced [71]	
Medidur (pSivida)	Incorporates fluocinolone	Phase III clinical trial has commenced with continued enrollment	
	Significantly smaller device than Retisert® and Vitrasert®	pSivida plans to seek US approval for Medidur for posterior uveitis – a result form data generated from one Phase III trial. Availability of supplemental clinical data if Iluvien for chronic DME is approved by the FDA	
	Easier to surgically implant – through a 3.5-mm incision with a 25G needle	pSivida will be able to reference the Iluvien New Drug Application (NDA) in a Medidur NDA – could potentially accelerate US commercial availability and reduce overall development costs [72]	
	Reservoir type, not sutured, free floating intravitreally		
	Medidur employs same injectable sustained-release micro-insert delivering the same dosage of the same drug as Iluvien		
Iluvien by pSivida	Non-erodible intravitreal implant injected through a 25G needle during outpatient procedure – self-healing wound	A trial of 37 patients on Iluvien demonstrated better visual acuity in high-dose group than low-dose group – but also suffered higher rates of glaucoma [66]	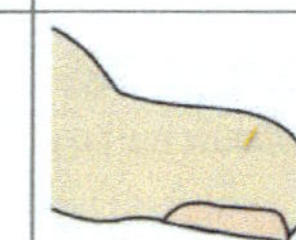
	Composed of polyimide – 3.7 mm in length		
	Delivery of fluocinolone acetonide		
	Suspended in vitreous cavity for DME treatment		
	Two release rates – 0.5 μg/day for 18 months or 0.2 μg/day for 36 months [66]		

(continued)

Table 18.2 (continued)

Ocular delivery system	Description	Clinical trial status	Digital image
Renexus, formerly NT-501 (Neurotech)	Encapsulated cell technology (ECT) providing extracellular delivery of ciliary neurotrophic factor (CNTF)	Phase I and Phase II clinical studies for various indications (retinitis pigmentosa and age-related macular degeneration) have demonstrated stabilized visual acuity together with dose-dependent structural changes	
	Protracted and stable release of constant intraocular doses		
	ECT – implicates the genetic engineering via plasmid transfection of human RPE cells. A polymer membrane capsule holds the engineered cells [66]		
	Renexus contains human cell line of RPE, NTC-200 – genetically modified to secrete recombinant human CNTF		
	Sealed semipermeable hollow fiber membrane capsule enclosing a scaffold comprising six strands of polyethylene terephthalate yarn – can be loaded with cells		
	Small scleral incision with vitreal implantation, secured by a single suture through a titanium loop at device end		
	Enables outward diffusion of CNTF+cellular metabolites and the inward diffusion of nutrients for cell survival in the vitreous and simultaneously preventing host cellular immunologic attack		
NT-503 (Neurotech)	NT-503 encapsulates VEGF receptor Fc-fusion protein (VEGFR-Fc)-releasing cells	A Phase I clinical trial of NT-503 for neovascular AMD is currently underway	As for Renexus
	VEGFR-Fc – reportedly 20x enhanced capability in neutralizing VEGF compared to ranibizumab		
	NT-503 – releases VEGFR-Fc constantly intravitreally in the rabbit eye model for up to 1 year [73]		

ODTx (On Demand Therapeutics)	An injectable, biocompatible, nonresorbable device containing drug release reservoirs for noninvasive activation via laser (OnDemandTx.com) Reservoirs – store small- or large-molecule drugs Multiple reservoir system – enables ophthalmologists to control drug delivery through selective reservoir activation	Clinical trials of ODTx have not commenced	
Biodegradable polymeric drug delivery systems			
Brimonidine intravitreal implant (Allergan)	Biodegradable, rod-shaped poly(lactide-co-glycolide) intravitreal implant Releases brimonidine tartrate (an alpha-2 adrenergic agonist enabling neurotrophin release)	A Phase II/III clinical trial has been conducted in patients with retinitis pigmentosa Phase II clinical study in atrophic AMD is currently underway [74]	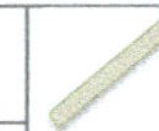
Tethadur (pSivida)	Honeycomb-like nanostructured porous biodegradable silicon (BioSilicon) Can purportedly deliver various drugs, e.g., small chemical entities, peptides, and proteins, and can provide sustained drug release [75]	Further details regarding Tethadur still to be disclosed by the company	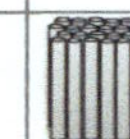
DE-102 (Santen Pharmaceutical Co. Ltd)	Injectable biodegradable microspheres loaded with betamethasone for sub-Tenon injection Provide sustained release for the treatment of DME [76] and for branch retinal vein occlusion [77]	Have been subjected to a randomized, multicenter, controlled, Phase II/III clinical trial in Japan	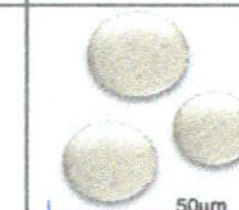
Ranibizumab-loaded microspheres (SurModics Inc. and Genentech)	SurModics Inc. and Genentech are in the process of developing biodegradable microparticles incorporating ranibizumab Anticipated to deliver the agent over a period of ~4 to 6 months	Reportedly in preclinical stages, but the details of this cannot be readily accessed online	Microparticles as above

(continued)

Table 18.2 (continued)

Ocular delivery system	Description	Clinical trial status	Digital image
Oil-/lipid-based drug delivery systems			
IBI-20089 (Icon Bioscience Inc.)	Verisome drug delivery platform technology – has been employed by Icon Bioscience Inc. in developing IBI-20089 containing TA	An open-label, Phase I study for cystoid macular edema associated with branch retinal vein occlusion or central retinal vein occlusion has been completed	
	Verisome is a translucent liquid	Published results – decrease in mean central subfield OCT thickness	
	The combination of IBI-20089 with saline – creates milky, slightly opaque color and forms a gel	Reported side effects – IOP elevation with neovascular glaucoma [78]	
	Patent of Icon states that IBI-20089 might be a solution of TA in benzyl benzoate	Phase II study of IBI-20089 adjunctively with 0.5 mg of ranibizumab – commenced for neovascular AMD	
	Single intravitreal injection – lasts up to a year [79]	Results highlight that IBI-20089 could reduce frequency of as-needed ranibizumab treatment [80]	
Cortiject (NOVA63035, Novagali Pharma SA, acquired by Santen Pharmaceutical Co. Ltd.)	Emulsion composed of oily carrier and phospholipid as surfactant	Open-label, Phase I study (dexamethasone) is currently underway [81]. Interim report – foveal thickness decreased following intravitreal injection [82]	
	Encapsulates a target tissue-activated dexamethasone prodrug and provides sustained release over 6–9 months following a single intravitreal injection [83]	Further results of this trial – not reported since 2009	

Adapted from Kuno and Fujii [84]

Systems that permit intraocular perfusion or microdialysis offer the advantages of allowing continuous drug administration into the vitreous cavity and possess the ability to change or stop drug administration at any time. Nevertheless, because there is continual access to the intraocular space, there is an increased risk of intraocular infection as well as patient discomfort [9].

To follow are in-depth discussions on specific implantable intraocular developments some have which have been critically analyzed in our past reviews and investigations [3, 89–91]. In addition, we have extensively focused on systems extending into the nano-realm as well as reviewing hybrid systems. The ensuing discussion is related to all systems designed for drug delivery to the *posterior ocular tissues.* Table 18.3 is a representative summary with selected associated schematics (Fig. 18.3) of intraocular implants reviewed by Choonara et al. [89].

18.5.1 Nonbiodegradable Scleral Devices

There is an increase in popularity of implants for transscleral administration for delivery of drug to the retina–choroid. Transscleral systems may hold an advantage for the treatment of certain ocular pathologies as drugs delivered into the vitreous can possess the disadvantage of poor access to the outer retina and choroid, which are affected by diseases such as AMD. Okabe and co-workers [102] developed an intrascleral implant formulated from ethylene vinyl acetate (EVA) copolymer incorporating a betamethasone disk (4 mg in mass, 1 mm thick, and 4 mm in diameter). In vivo studies involved implantation in the scleral pocket of rabbits. Therapeutic betamethasone levels were maintained for 4 weeks in the absence of burst release. Kato et al. [103] designed nonbiodegradable scleral implants for retino-choroidal delivery of betamethasone. The implants achieved zero-order release, and after 4 weeks of testing following implantation in the episcleral space of rabbits, therapeutic levels of betamethasone were still detected. Pontes de Carvalho et al. [104] also developed nonbiodegradable exoplants, incorporating fluorescein, which achieved more effective delivery in rabbits compared to periocular injection, specifically to the posterior ocular tissue. The exoplant facilitated diffusion of fluorescein through the sclera to the retina and posterior vitreous. Systemic levels of fluorescein were minimal. There is no perforation of the ocular wall during intrascleral insertion; therefore these devices may reduce complications after implantation. Further human studies on these intrascleral devices are necessary to prove clinical efficacy.

18.5.2 Episcleral Implant for Drug Delivery to the Retina

Kawashima et al. [105] undertook the design and evaluation of an episclerally implantable system for the zero-order controlled release of brain-derived neurotrophic factor (BDNF) and related protein-type drugs. Formulation of the controlled-release membrane was via photopolymerized polyethylene glycol dimethacrylate

Table 18.3 Summary of intraocular drug delivery devices under investigation, as reviewed by Choonara et al. [89]

Device and investigators	Description	In vitro, ex vivo, and in vivo studies	Conclusive remarks and investigative progress
A cyclosporine-loaded discoid device [92]	Discoid intravitreal device	Tested for inflammatory episodes of uveitis in horses	CsA device – effective for the treatment of immune-mediated intraocular inflammation
	Constant cyclosporine (CsA) release	Inflammatory cell counts, infiltrates, and biomarkers were notably lower in CsA device-implanted eyes vs. the placebo	To date, no further studies, with progression to human studies, have been reported for this device
		Device failed to prevent a recurrent experimental inflammatory episode (horse model). However – the inflammation was short-lived and less severe where the CsA device was implanted	
A microelectromechanical device (MEM) [93, 94] See Fig. 18.3a	Treatment of incurable ocular diseases	Ex vivo testing – highlighted MEM device feasibility	Flexible delivery system for long-term therapy and minimized surgeries
	MEM device is refillable, features electronic control of drug delivery – enables targeted intraocular drug delivery	Saait et al. [94] implanted elliptical ($9.9 \times 7.7 \times 1.8$ mm) non-electrically active pumps into rabbits	Further testing required – the long-term biocompatibility of an electrically controlled implanted pump
	Long-term drug therapy	Benchtop testing – 2.0 μL/min delivery of trypan blue employing 0.4 mW for electrolysis	Potential to delivery diverse pharmacologic agents is required
	Repetitive surgeries avoided	Implanted devices refilled at 4–6-week intervals for 4–6 months	Production of device is difficult
	Electronic control of dosing – electrolysis-actuated pumping for posterior chamber drug delivery	No infection, device extrusion, or filtering bleb was observed	Necessitates patient intervention for attaining therapeutic efficacy
	Flexible transscleral cannula – tissue-specific delivery to anterior or posterior segment.		
	Continuous and bolus drug delivery modes		

An osmotic minipump device [95]	An implantable osmotic minipump device	Subcutaneous implantation in the ear region of a rabbit eye model of endophthalmitis	Similar pump models attempted by other investigators [96, 97] – clinical acceptance not achieved
	Pars plana incision for connective tubing entry into vitreous cavity	Via tubing – maintained a calculated dose of the antibiotic gentamicin (0.01 mg/h) over 4 days	
A hyaluronic acid plug device [98] See Fig. 18.3b	HA intravitreal plug device composed of three different HA esters (100 % ethyl ester, 100 % benzyl ester, and 75 % benzyl ester – Hyaff7 and Hyaff11, and partial Hyaff11p75)	Plugs implanted through a sclerotomy at 3.5 mm from the limbus of rabbit eyes	Safety and efficacy of esters of HA as ophthalmic implants demonstrated
		Shaft diameter measured by ultrasound biomicroscopy as a measure of in vivo biodegradation of the device	Potential of this system to deliver ophthalmic drugs yet to be ascertained
		Biocompatibility of esters demonstrated	
		100 % HA esters – slow erosion kinetics. 75 % ester – complete resorption by 15 days	
		Correlation demonstrated between the biodegradation rate and resorption time, which was related to chemical structure of HA	
		Potential application in posterior segment disorders	
A micromachined drug delivery device (the Ocular Drug Delivery Group at the University of California, Irvine, USA)	Potential application of micromachining technology to engineer a passive, programmable, pulsatile drug delivery device	–	No further published reports on progress
	Small size for ocular implantation		
	Possesses polymeric null zones and drug-loaded zones		
	Drug levels cycle up and down in a preprogrammed manner		
	Enables native expression of endogenous cytokines and growth factors [99]		

Table 18.3 (continued)

Device and investigators	Description	In vitro, ex vivo, and in vivo studies	Conclusive remarks and investigative progress
A doughnut-shaped minitablet device (DSMT) [100, 101]. See Fig. 18.3c	Biodegradable intraocular drug delivery system employing poly(lactic-*co*-glycolic) acid polymer combinations	DSMT retained geometry and zero-order release kinetics over 24 weeks of erosion in vitro	Provides geometrically controlled rate-modulated release of antiviral drugs to the posterior segment of the eye
	Rate-modulated delivery of antiviral drugs	In vivo studies – twelve New Zealand White rabbits. The DSMT implanted through the pars plana/peripheral retina of the right eye and sutured with 9–0 nylon	Application in treatment of CMV retinitis
	Manufactured – specialized punch set with a central rod in a Manesty tableting press	DSMT was well tolerated and remained visible in the supero-temporal quadrant of the rabbit eye	Human studies are to be conducted
	Manufacturing is simple and reproducible – produced on a conventional tableting press via novel tooling	No retinal detachments, significant vitreous hemorrhages, or clinical changes in the vitreous, retina, and choroid for duration of study	
		DSMT device elaborated constant GCV release within the ED_{50} for human CMV-R (0.1–2.75 μg/mL)	

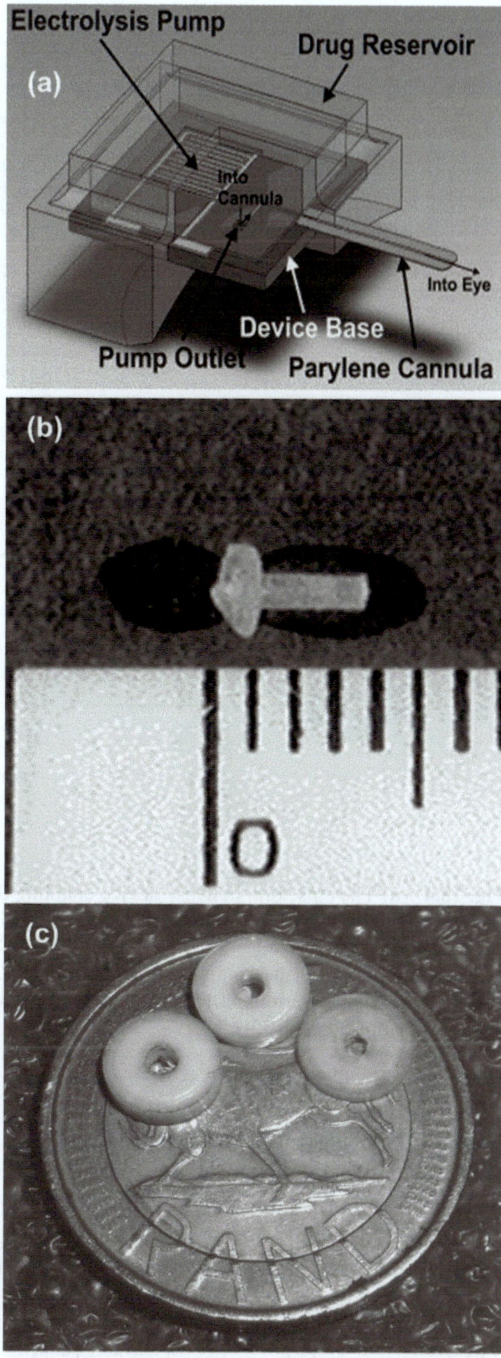

Fig. 18.3 (**a**) Schematic depicting the components of the micro-machined drug delivery device. The arrows indicate the flow path from the reservoir through the Parylene cannula (Reproduced from Li et al. [93]). (**b**) Architecture of the hyaluronic acid scleral plug delivery device (Reproduced from Avitabile et al. [98]). (**c**) A digital image depicting the DSMT device in relation to a typical South African five rand coin (Reproduced from Choonara et al. [100])

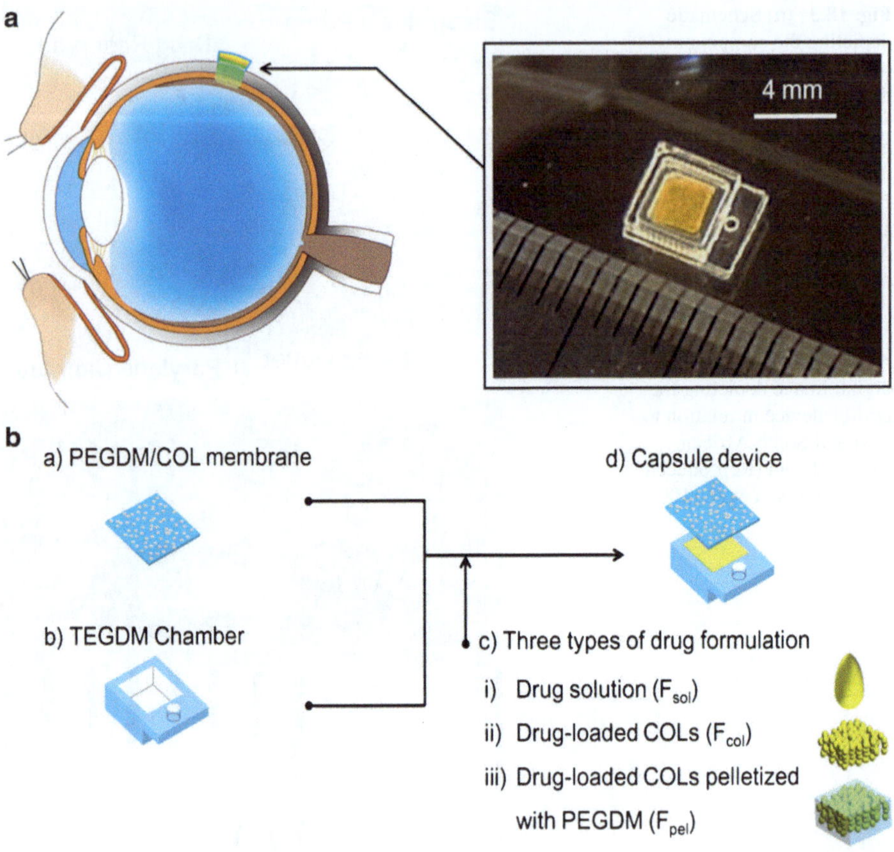

Fig. 18.4 (**a**) The episcleral delivery device enabling delivery of protein-type drugs. Photograph: capsule holding FD40-loaded COLs pelletized with PEGDM with a hole allowing for suturing on the sclera. (**b**) The capsule comprised of the TEGDM drug reservoir and photopolymerized PEGDM controlled-release membrane composed of COLs as drug permeation routes. Capsule design could enable various drug formulations to be incorporated in the reservoir (Reproduced from Kawashima et al. [105])

(PEGDM) incorporating an interconnected collagen microparticulate network (COLs) through which drug permeated (Fig. 18.4). Bioactive release from the device was dependent on the membrane COL concentration; thus, modification in its concentration creates diverse rates for protein-type drugs. They also report ease of implantation and removal of the device episclerally with minor surgery. The device thus exhibits good potential as a conduit for continuous, controlled drug release. Protein-type drugs can be effective for the treatment of vitreoretinal disorders, but few systems are available for enabling their controlled release to the retina, which has currently been described for this device. Preclinical animal studies reporting on the safety and efficacy of this device have yet to be conducted by Kawashima and co-workers.

18.5.3 Polymeric Dispersion for Transscleral Delivery of Macromolecules

Thrimawithana and co-workers [106] prepared polymeric dispersions combining methylcellulose (MC) and either kappa carrageenan (KC) or iota carrageenan (IC) as a transscleral delivery platform for macromolecules. In-depth investigation of additive effects of the two polymers was performed via physicochemical and physicomechanical analyses. Favorable bioadhesion was observed for MC–IC polymeric system, attributed to the high negative charge; this, however, did not impact the in vitro scleral permeability of sodium fluorescein and 10 kDa FITC-dextran. Interestingly, a notable effect was observed in vivo, where efficacy of transscleral drug delivery was assessed in a rat model with altered choroidal connexin 43 (Cx43) levels, a gap junction protein. The MC–IC system incorporated Cx43 antisense oligonucleotides (AsODN), which was injected periocularly (Fig. 18.5). The Cx43 levels in the choroid of rats were notably reduced at 24 h of treatment, possibly a result of control of drug reflux from the site as an exerted effect of the bioadhesive polymeric dispersion. A reduction in Cx43 levels was also noted for the AsODN in PBS; however results were variable and less significant. Data highlighted that choroidal delivery could be successfully attained via the transscleral route for the management of posterior segment disorders as well as demonstrated the potential of a novel polymeric system enabling aqueous loading of bioactives with prolonged retention at the administration site. Studies in a larger animal model should be undertaken to validate preclinical efficacy prior to progression to human studies.

18.5.4 A Capsule Drug Ring as an Intraocular Implant

The costly and risky treatment of AMD via monthly intravitreal injections was emphasized by Molokhia et al. [107]. They therefore set out to design an innovative intraocular device as a capsule drug ring (CDR), serving as a reservoir, for insertion within the

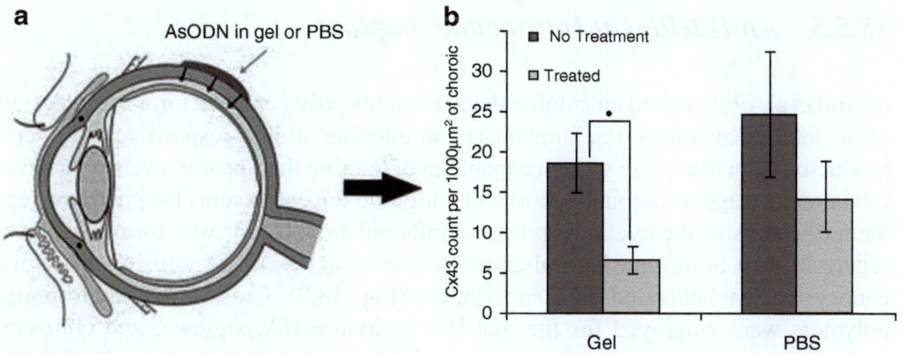

Fig. 18.5 Schematic highlighting (**a**) placement of the AsODN-loaded gel system enabling choroidal bioactive delivery and (**b**) the ensuing reduction in Cx43 count in the choroid (Reproduced from Thrimawithama et al. [106])

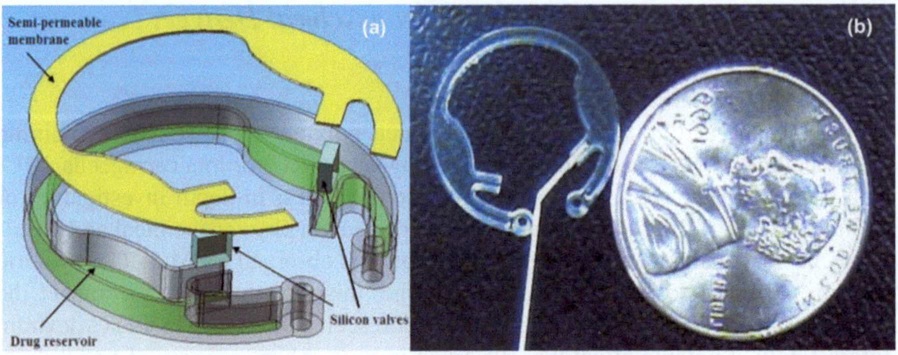

Fig. 18.6 (**a**) Schematic demonstrating the CDR architecture highlighting the drug reservoir, semi-permeable membrane and valves. (**b**) Photographic image of the prototype device prior to membrane and valve assembly (27G cannula inserted in valve access port) (Reproduced from Molokhia et al. [107])

lens capsule at the point of cataract surgery. The implant is refillable and has the potential for multiple drug delivery. A circular prototype was developed incorporating the antibody, bevacizumab (Avastin®), with the reservoir material composed of polymethyl methacrylate sheets, controlled delivery achieved with a semipermeable membrane, and refilling of the device via silicon check valves (Fig. 18.6). In order to enhance release control and drug stability, polyvinyl alcohol (PVA) was incorporated as a polymeric carrier. The circular shape maximized the capsular bag volume. Near-zero-order release kinetics was attained from the implant. Molokhia and co-workers [107] suggest a versatile device due to the multiple valves and associated multiple drug chambers for diverse ocular disease management. Preclinical studies of the CDR in a rabbit model are proposed in the future in order to establish whether zero-order release can be achieved for 6 months and to identify whether prolonged implantation with ensuing fibrosis may instigate occlusion of pores.

18.5.5 An Intelligent Intraocular Implant

du Toit et al. [90] created an autofeedback complex polymeric platform in the design of an intelligent intraocular implant (I^3) employing stimuli-responsive polymers, producing a smart release system capable of delivering therapeutic levels of an anti-inflammatory agent (indomethacin) and antibiotic (ciprofloxacin) for posterior segment disorders of the eye in response to inflammation. The I^3 was formulated as a secure-fit dual bioresponsive polymeric matrix system with an anti-inflammatory nanosystem enclatherated as a superlattice (Fig. 18.7). Cross-linked hydrophilic polymers were employed for the matrices including HA, alginate, and chitosan, with the nanosystem possessing a composite lipoidal polymeric composition for inflammatory tissue targeting. In silico modeling simulating a pathological

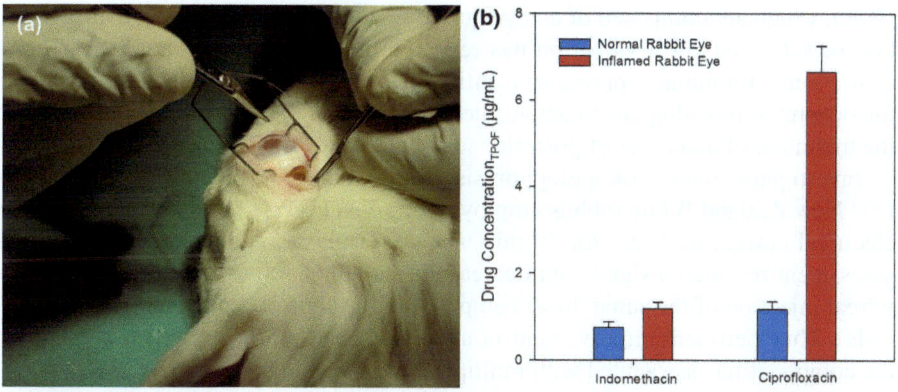

Fig. 18.7 (**a**) Photographic image highlighting removal of the I3 from the rabbit eye at day 7 prior to enucleation for vitreous and tissue sampling. (**b**) Drug levels attained at 7 days in the posterior ocular fluids in the normal and inflamed rabbit eye (Reproduced from du Toit et al. [90])

inflammatory intraocular state highlighted the interaction potential of hydroxyl radicals with the selected polysaccharides comprising the I^3. This highlights that the cross-linked polymeric system forming the I^3 thus responded at an innate level predicted by its molecular makeup to inflammatory conditions, which was confirmed in vitro. In vivo drug release analysis demonstrated notably enhanced indomethacin and ciprofloxacin levels in the inflamed rabbit eye in which uveitis was induced via lipopolysaccharide (LPS) injection, compared to the normal rabbit eye. du Toit and co-workers [90] propose that the I^3 is a step forward from the traditional sustained-release ocular system owing to its combined bioresponsive, nano-enabled architecture. Further validation in humans is required.

18.5.6 Ocular Iontophoresis Systems for Transscleral Drug Delivery

Iontophoresis is a noninvasive process which exploits an electrical field to achieve ionized drug permeation across biological barriers. The inception of ocular iontophoresis was by Wurtz in 1908, using saturated cotton sponges on the eye globe to conduct an electric current. The basis of the technique is eletrorepulsion and electroosmosis in the presence of a low voltage (~10 V) providing a continuous direct current (~0.5 mA/cm2). The composition of the iontophoretic device includes a power source, milliampere meter, a rheostat, a drug delivery electrode, and a dispersive electrode. These devices have application in the management of anterior and posterior segment disorders, via transcorneal or transscleral application, respectively. Systems under current investigation include OcuPhor® (Iomed Inc., USA), EyeGate II Delivery System® (EyeGate Pharma, USA), and Visulex® (Aciont Inc.,

USA), which are composed of an eye cup drug delivery probe holding a drug solution or saturated gel. Evaluation has revealed the attainment of therapeutic concentrations of immunosuppressants, antibiotics, and corticosteroids via transscleral iontophoresis, providing an attractive alternative to multiple intravitreal injections for the treatment of a number of posterior segment disorders [108–110].

Investigators undertook iontophoresis of dexamethasone disodium phosphate in EIU New Zealand White rabbits employing the Visulex® [111]. They employed an electrical current of 4 mA for 20 min using a Visulex®. Histological and clinical assessment revealed a significant decrease in inflammation (as conferred via intravitreal injection of the same drug) compared to the control group (saline iontophoresis). Thus pertinent improvement in uveitis scores could be attained, precluding the complications associated with multiple intravitreal injections.

There are a number of innovative iontophoretic devices with patents pending with US Patent and Trade Mark office (USPTO) and European Patent Office. Phase II clinical trials were recently completed for the EyeGate II Delivery System® for anterior uveitis treatment [110].

18.5.7 Additional Diverse Intravitreal Delivery Systems

Further developments in intraocular delivery systems, highlighting the diversity of approaches, have seen various modifications of photo-cross-linked poly(propylene fumarate)/poly(*N*-vinyl pyrrolidone) (PPF/PNVP) matrices by Haesslein et al. [112] to investigate their effect on the release kinetics of acetazolamide and timolol maleate. The hydrophilicity of solid PPF/PNVP matrices was increased following the addition of different amounts of poly(ethylene glycol) or employing increasing levels of *N*-vinyl pyrrolidone in the polymer mixture prior to cross-linking. Prior to this, Haesslein and co-workers [9] fabricated these same rod-shaped matrices for the delivery of fluocinolone acetate for the prevention of PVR and treatment of intraocular and inflammatory proliferative diseases. Saliba et al. [113] synthesized and characterized a nanostructured biodegradable intraocular implant based on PLGA (75:25) with cyclosporine A (CyA) and evaluated its in vitro drug delivery profile. The results confirmed that the device achieved prolonged delivery of cyclosporine. No in vivo studies have been undertaken on this system. Allergan Inc. [114] has developed biocompatible intraocular implants including a steroid and a polymer associated with each other to facilitate release of the steroid into an eye for a period of time greater than about 2 months (specified in European Patent 1750688). Two embodiments are described: in one, the steroid was associated with a biodegradable polymer matrix. Alternatively, the steroid was associated with a polymeric coating having one or more openings permitting release of the steroid into the external environment. Developed and characterized biodegradable poly(ε-caprolactone) implants, for intraocular prolonged and controlled release of dexamethasone.

18.6 Nanosystems and Microsystems for Intraocular Drug Delivery

18.6.1 The Benefits of Nanosystem Components in Ocular Drug Delivery

Advances in micro- and nanosystems have accelerated the development of innovative drug delivery technologies, enabling transformation of biological potential into medical reality [115]. Nanotechnology is an all-encompassing term that describes a fast progressing interdisciplinary technological field, which implicates engineering or manipulating matter at a submicron level. It encircles systems, assemblages, or mechanistic approaches in the size range of smaller than the micrometer and larger than 1 nm or about 10 atomic diameters [116].

It is evident from our discussions that the goal in ocular therapeutics is to achieve and maintain an effective drug concentration at the site of action of adequate duration, in order to elicit the appropriate pharmacological response [117]. We have highlighted that "drug delivery systems undoubtedly form a crucial component of the *therapeutic armamentarium* in ophthalmology" [3]. Accordingly, investigators have stated that "ophthalmic drug delivery, probably more than any other route of administration, may benefit from the characteristics of nanotechnology-based drug delivery systems" [3, 10]. Promising results obtained from drug delivery systems employing non-ocular routes of administration have stimulated researchers to discover applications for nanotechnology in ophthalmology.

The inherent advantages of nanopharmaceutical systems (nanosystems, NS) are intrinsic to their colloidal nature and have been specified in our previous works [3, 91]. In du Toit et al. [3] we state that:

> NS are poised, and have thus been conceptualized and fabricated, to circumvent the problems associated with conventional ocular systems to essentially increase the ocular bioavailability of drugs (especially of poorly water-soluble or poorly permeable drugs) and maintain activity at the site of action thus enhancing the therapeutic effect (due to reduced cellular and tissue clearance of drugs, sustained drug delivery, and uptake of drugs by ocular epithelia). NS have the ability to provide protection of the entrapped bioactive while conveying it to the desired compartments of the eye, as well as providing targeted delivery of bioactives [118–124]. This is a result of the capability to overcome blood–ocular barriers and efflux of the parent drug. Additionally, various stability-related concerns of drug molecules are bypassed or overcome through creation of a nanocarrier, as evidenced for proteins and peptides [20]. Furthermore, NS can provide controlled drug delivery for extended periods of time: an attractive benefit for the treatment of certain chronic ocular diseases [122, 125]. Furthermore, NS purportedly greatly reduce toxicity compared to the free drug and they overcome the poor acceptability of viscous or tacky preparations, including ointments, which cause extensive blurring of vision on correct administration [126].

Compared to normal eye drop formulations, drug-loaded nanoparticles, as a colloidal dispersion, possess the added advantage of enhancing patient compliance, due to less frequent application and prolonged retention in the extraocular portion due to the extended intraocular half-life of the drug (which is important where the use of the drug preparation by the patient is lifelong, e.g., glaucoma) [126, 127]. Significantly, with future developments firmly in mind, NS are paramount to exploitation of the emerging field of novel gene therapies for ocular disorders management [10, 128].

In order to visualize how NS "slot in" to the ocular delivery world, one must consider the main colloidal carriers under scrutiny for ocular delivery due to their favorably small size, which include liposomes and nanoparticles (NPs). These NS can be coated with a hydrophilic polymer or further functionalized with antibodies for the enhancement of the targeting potential. Diverse methods exist for the preparation of NS and loading with therapeutic molecules. Ultimately NS are a highly adaptable drug delivery system, having the ability to surmount physiological barriers and transporting the drug to specific cells or intracellular compartments via passive or ligand-mediated targeting mechanisms [126, 129–132]. The last two decades have thus seen directed efforts in the rational design of intraocular drug delivery NS. Thus, falling under ocular NS, one can find an extensive array of polymeric formulations comprised of nondegradable and biodegradable polymers possessing either hydrophilicity or hydrophobicity. Examples of NS that vary in composition and application for ocular drug delivery are highlighted in Table 18.4.

18.6.2 Considerations in Nanosystem Design for Management of Posterior Segment Disorders

Important considerations are necessitated in designing a nanosystem for drug delivery to posterior ocular tissues. The propensity of a compound to attain transscleral permeation is dependent on bioactive and carrier molecular properties, such as overall size, charge, and hydophilicity. Additionally, with regard to factors impacting the in vivo success of the nanosystem, dynamic barriers (i.e., blood and lymphatic flow), entrapped bioactive affinity toward melanin in the choroid–RPE, and surface conjugation of NS are all pertinent. NS size thus has a considerable effect subsequent to periocular administration as clearance of 20 nm particles by conjunctival, choroidal, and lymphatic circulations has been demonstrated to be significantly higher, whereas particles of 200 nm diameter were detected at the injection site for over 2 months. Furthermore, 20-nm particles have been shown to minimally cross the scleral tissue, while 200-nm particles did not permeate [138]. A size range between 20 and 200 nm could thus be seen as most preferable. Bioactive melanin affinity influences ocular disposition by delaying delivery to the inner retinal layers and the vitreous humor [139]. Drug targeting by functionalization for tissue-specific uptake has demonstrated some success. Drug availability into the inner retinal tissues is also impacted on by disease states [140]. The fate of NS depends on numerous critical parameters following intravitreal delivery of a dosage form. The composition of the vitreous humor is colloidal in nature, incorporating solutes, ions, and low-molecular-weight compounds, containing mainly collagen (40–120 μg/mL) and HA (100–400 μg/mL). Positively charged molecules display a propensity for aggregation in the vitreous humor due to the negative charge of vitreal

Table 18.4 Description of potential nanosystems for ocular drug delivery

Nanosystem	Definition	Composition	Potential/manipulation for ODD[a]
Nanoparticles (NPs)	Nanoparticles for drug delivery are mostly polymerically based and vary in size from 1 to 1,000 nm. Varying drug-loading configurations are evident, either the drug is attached to a nanoparticle matrix, dissolved, encapsulated, or entrapped, leading to different terminologies, all of which embody the nanosize characteristic – i.e., nanospheres or nanocapsules	Natural or synthetic polymers, metals, lipids, phospholipids	Yes
Liposomes and niosomes	Liposomes can be defined as biocompatible and biodegradable microscopic vesicles consisting of membrane-like lipid bilayers, composed mostly of phospholipids, surrounding aqueous compartments. These phospholipids are amphiphilic, having a hydrophilic head and a lipophilic tail, which spontaneously arrange in aqueous solution as bilayers (the fatty acid tails are located in the membrane's interior, and the polar heads point outward) to form closed vesicles	Phospholipids	Yes
	Niosomes, a variant of liposomes, are nonionic surfactant vesicles. They are also bilayered structures which can entrap both hydrophilic and lipophilic drugs (aqueous layer or lipid vesicular membrane)	Nonionic surfactants (niosomes)	
Emulsions and microemulsions	Consist of interfacially stabilized (with surfactant and cosurfactant) oil-in-water or water-in-oil mixtures	Oil-in-water and water-in-oil mixtures that require surfactants	Yes
Nanosuspensions	–	Inert polymer resins	Yes
Dendrimers	Dendrimers differ from conventional polymers as they possess a multi-branched, three-dimensional architecture with minimal polydispersity and high functionality. Dendrimers are formed from three chemically pertinent components: (1) a polyfunctional core, (2) interior layers, and (3) a multivalent surface. They are novel and intriguing scaffolds for ocular drug delivery	Synthetic polymers	Yes

(continued)

Table 18.4 (continued)

Nanosystem	Definition	Composition	Potential/manipulation for ODD[a]
Nanoparticle-loaded contact lenses	–	Various hydrogel-based lenses incorporating different drug-loaded nanoparticulate formulations into the lens matrix	Yes
Nanotubes and fullerenes	A fullerene is a form of carbon comprising a large spheroidal molecule incorporating a hollow cage of atoms. Carbon nanotubes are a fullerene having a cylindrical or toroidal configuration. They are large molecules, approximately 1–3 nm in diameter and 100–1,000 of nanometers long	Carbon-based materials. There is a shift to producing nanotubes from polymeric materials to enhance safety	Not tested yet
Quantum dots	A quantum dot is a semiconductor whose excitons are confined in all three spatial dimensions	Semiconductor materials possessing fluorescent properties coated with various materials	Yes (diagnostic)
Cyclodextrins	Cyclodextrins are a group of cyclic oligosaccharides. They have the propensity to form inclusion complexes with numerous drugs		Yes

[a]ODD = ocular drug delivery

Adapted from Carafa et al. [133], Svenson and Tomalia [134], Barbu et al. [88], Sahoo et al. [130], Gaudana et al. [135], Diebold and Calonge [136], Kaur et al. [137], and du Toit et al. [91]

HA. Hence, it is crucial to ascertain the zeta potential of an NS under physiological conditions [51]. PEGylation or alternative hydrophilic coatings on particles can significantly minimize this problem [141]. The ILM (separating the retina and vitreous) also plays a role in restricting the movement of particles from the vitreous humor to inner retinal layers [20].

NS of various sizes for the improvement of ocular drug delivery have been based on polymers and biomaterials such as PLGA, poly(lactide) (PLA), poly (ε-caprolactone), albumin, and chitosan [65, 88, 142–147]. The versatility of NS in terms of their size, shape, surface charge, and other physicochemical features, in addition to drug loading, enhances their attractiveness for drug delivery; however, these features also influence their cellular uptake and intracellular trafficking as well as interaction with vitreal proteins, other intraocular components [148], and with immune cells [149], all of which are relevant to their ultimate distribution. Opsonization of NS (emanating in the covering of the NS by opsonins) furnishes a "molecular signature" [150] predicting the route of NS internalization and subsequent fate in phagocytic cells. Covering the NS surface with a hydrophilic coat prevents opsonization via the "stealth effect" [151] and thus decreases the cellular uptake rate, enhancing NS presence [136].

The location of bioactive therapeutic targets may be extra- or intracellular. On consideration of classical drugs exceeding 1 kDa, stability conditions or hydrophilicity renders them unsuitable for traversing the cell membrane. NS thus provide a suitable vehicle for enabling the bioactive molecule to pass through cell membranes and epithelial barriers via exploitation of different internalization pathways – which highlights a very significant feature of NS and is one of the most interesting aspects of this technology. As early as the 1970s, Couvreur and co-workers [152] demonstrated that submicrometer delivery systems such as NS can "enter cells and become concentrated into nondiffusible molecules" [152]. As emphasized, it is the physicochemical traits of nanocarriers, i.e., size, shape, surface charge, coating, and functionalization with targeting ligands, as well as the target cell type, that predict the internalization pathway and the intracellular fate [153–155]. The established internalization pathways are phagocytosis, endocytosis, and macropinocytosis. It is of importance to note that several endocytic mechanisms can take place simultaneously. Remarkably, there is a scarcity of studies regarding internalization pathways and intracellular trafficking of NS intended for ocular drug delivery. Internalization pathways are a significant consideration as this will impact on the intracellular metabolism of delivered drugs and imply that NS physicochemical makeup will determine the degradation ratio, which ultimately impacts on the pharmacological activity of the encapsulated molecule.

The release profiles of drugs from NS are controlled by various kinetic processes, namely: (i) drug desorption from the surface (bound or adsorbed), (ii) diffusion through the NS matrix or the polymeric wall, (iii) erosion of the NS wall, and (iv) a combination of erosion and diffusion processes [156, 157]. Uniformly distributed or dissolved drug in the NS matrix displays kinetic release attributes dependent on diffusion and biodegradation of the matrix. Should polymeric degradation be slower than drug diffusion, drug release will ensue mainly through diffusion; alternatively, drug release follows the degradation of the polymeric carrier.

Lastly, consideration of the mode of administration (e.g., injection, implantation) is required during the design of the ocular NS. These complexities all tie together to furnish the requirements of minimal potential toxicity and improved efficiency. A number of investigations have been undertaken in the search for alternative NS in ophthalmic drug delivery [136].

NS and microsystems can attain sustained release of the entrapped bioactive for weeks to months. Their clinical potential for treatment posterior segment disorder of the eye is still limited; however a number of preclinical studies have highlighted the efficacy of these systems in enabling drug delivery via periocular and intravitreal pathways to the posterior ocular tissues [67].

18.6.3 Microsystem Technologies for Targeted Delivery to the Posterior Ocular Tissues

Microspheres loaded with the kinase inhibitor, PKC412, were developed by Saishin et al. [158] for the treatment of choroidal neovascularization (CNV) in a porcine model. CNV sites were notably reduced in areas to which a periocular injection of microspheres encapsulating 25 % of PKC412 was administered, compared to control microspheres. There was detection of PKC412 after 10 days in the choroid, with no detection in the vitreous or retina; however after 20 days, PKC412 was detected in all three ocular sites. Barcia et al. [159] evaluated the short- and long-term abilities of dexamethasone-loaded PLGA microspheres for reducing ocular inflammation in rabbits provoked by intravitreal LPS injection. The formulated microspheres effectively reduced short- and long-term intraocular inflammations caused by LPS.

18.6.4 Nanosystem Technologies for Targeted Delivery to the Posterior Ocular Tissues

As highlighted by du Toit and co-workers [3], "few reports emphasize the intravitreal delivery of drugs through ocular barriers via NS, however some recent studies have shown commendable results on the use of intravitreally injected NPs." For example, ganciclovir-loaded albumin NPs have been developed as ganciclovir is one of the standard treatments for CMV retinitis (Fig. 18.8). These carriers were prepared by a coacervation method and chemical cross-linking with glutaraldehyde. Sustained drug release was evident following in vitro experiments [160] as well as a notable enhancement of drug uptake by CMV-infected human cells [161, 162]. Following single intravitreal injections in rats, safety and tolerance was demonstrated, and drug levels were detected in the vitreous and ciliary body for at least 2 weeks [163]. The risks associated with injectable systems, however, must be borne in mind.

Because RPE cells have the potential to internalize various kinds of NPs [119, 164, 165], NPs can be designed for targeted treatment of age-related ocular degeneration or photoreceptor dystrophies, following loading with the appropriate drug or genetic material for reversing or discontinuing the disease process. Bourges et al.

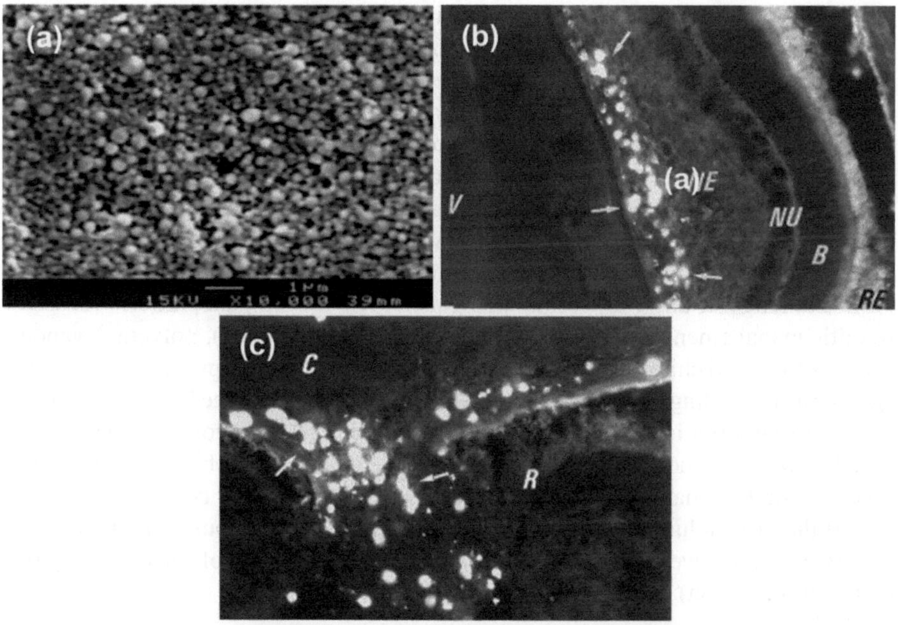

Fig. 18.8 (**a**) Scanning electron micrograph of albumin nanospheres developed for the treatment of cytomegalovirus retinitis possessing an average diameter of ~250nm (Reproduced from Merodio et al. [160]); and biotin-labeled NP immunochemical detection by streptavidin-FITC staining 2 weeks after (**b**) their injection in the vitreous cavity and (**c**) in the BAB adjacent area and ciliary muscle (200x magnification). B: Bacillary layer; C: ciliary muscle; NE: neuronal interplay area; NU: outer and inner nuclear layers; R: retina; RE: retinal epithelium; and V: vitreous cavity (Reproduced from Merodio et al. [162])

[119] developed intravitreal PLA NPs loaded with fluorochromes and tested them in rats; they demonstrated a preferential localization at the RPE cells after 24 h. Notably, the RPE cells retained the NPs, allowing continuous delivery of the fluorochrome for months following a single injection.

Bejjani et al. [164] undertook in vitro and in vivo investigations on PLA and PLGA NPs incorporating fluorochromes and model plasmids. Post-intravitreal injection in rats (24 h), NPs entrapping a plasmid encoding red nuclear fluorescent protein were localized in the RPE, and plasmid expression was achieved 4 days subsequent to injection. There was no apparent tissue damage or toxicity, and red fluorescence associated with expression was detectable in RPE cells for the ensuing 3 weeks.

Other intravitreal systems include injectable NPs for gene therapy as extensively reviewed by Conley and Naash [166]. Gene therapy is establishing itself as a firm contender in the ocular drug delivery field. Of late, the development of nonviral gene therapies has been an area of intense focus, and one technology, polymer-compacted DNA NPs, is becoming an excellent candidate for a retinal gene delivery vehicle. Safety, tolerability, and extensive and persistent expression throughout the retina have been observed for these NPs. The ability to target NPs to different cell types and/or regions of the cell furnishes them customizable for diverse applications. The outstanding obstacles tackling the application of these vectors is a means to efficiently target the extensive array of disease genes and actually curing genetic diseases diagnosed following the inception of degeneration.

Experimental agents have also been incorporated within nanoparticles for intravitreal delivery, such as tamoxifen-loaded nanoparticles prepared by de Kozak et al. [167] for the treatment of experimental uveitis. The polyethylene glycol (PEG)-coated nanoparticles (NP-PEG-TAM) demonstrated extracellular distribution throughout the ocular tissues as well as localization in resident ocular cells and in infiltrating macrophages. The NP-PEG-TAM system significantly halted the course of EAU, compared to injection of free tamoxifen

El-Samaligy et al. [168] formulated ganciclovir-loaded nanoparticles for CMV retinitis management employing bovine serum albumin (BSA), polyethylcyanoacrylate (PEC), and chitosan as biodegradable carriers. The three products were evaluated for drug-loading capacity, particle size, and drug release. The PEC nanoparticles (being the smallest in size comparatively) containing [^{3}H]acyclovir and ganciclovir together were deemed the most suitable and evaluated in a rabbit eye model. Over 10 days, the PEC nanoparticle product attained notably higher concentrations than that of the drug solution, with particular reference to the vitreous and retina. These results, however, were observed in the presence of a degree of lens opacification (cataract formation) and vitreous turbidity [169].

18.6.4.1 Polymeric Triamcinolone-Loaded Microparticles and Nanoparticles

The study of Kadam and co-workers [170] set out to ascertain the potential of nanoparticles and/or microparticles to provide sustained delivery transscleral, choroidal, and retinal delivery of triamcinolone acetonide (TA) over 2 months following periocular administration. The particulate systems were formed from poly-L-lactide (PLA) originated via *o/w* emulsification-solvent evaporation. In vivo evaluation was undertaken in brown Norway rats in which CNV was induced. Nanoparticles with an average size of 551 nm and microparticles of 2,090 nm were obtained, providing sustained release for 45 and 120 days, respectively, following in vitro assessment (Fig. 18.9). In vivo assessment of the microparticulate system revealed sustained TA delivery to all intraocular tissues for 2 months; furthermore, TA levels were higher in CNV-induced rats, notably reducing VEGF secretion. However, TA was not detected for the nanoparticulate system or plain TA suspension [170].

18.6.4.2 Nanotube-Based Intravitreal Implant

As an alternative to prolonged release intravitreal implants injected via large 22 or 25 gauge needles, such as Medidur™, Panda et al. [171] devised innovative dipeptide-based (phenylalanine-α,β-dehydrophenylalanine, Phe-ΔPhe) nanotubes (~15–30 × 1,500 nm). The nanotubes were loaded with a multi-targeted pyrosine kinase inhibitor, pazopanib, at the point of nanotube assembly or following nanotube formation. Advantageously, these could be injected via a 33

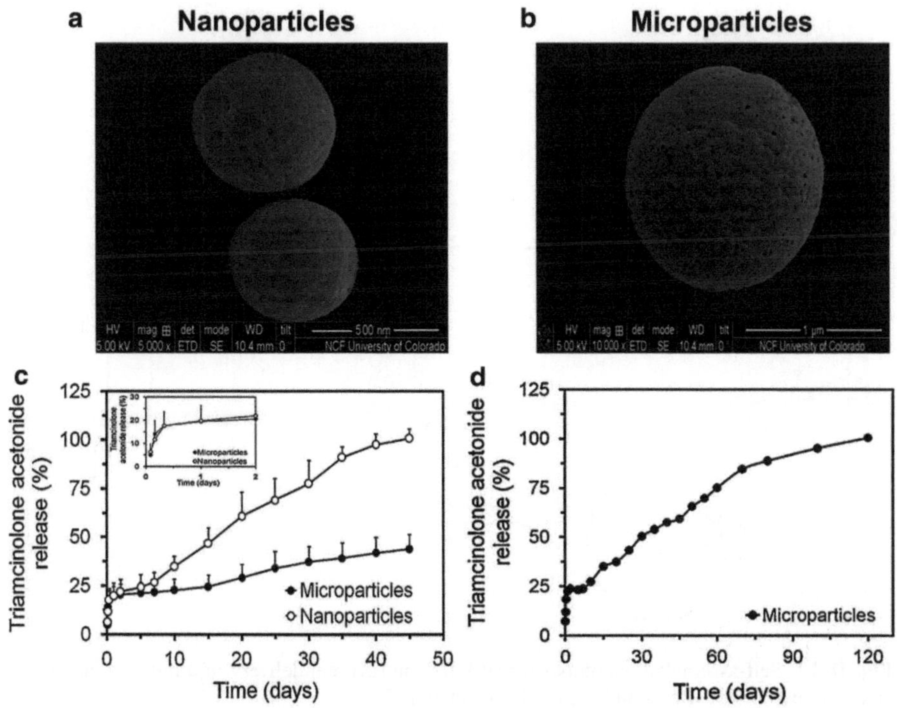

Fig. 18.9 Morphological architecture of (**a**) nanoparticles and (b) microparticles; (**c**) in vitro release TA release from nanoparticles and microparticles. (**d**) In vitro TA release from microparticles over 120 days (Reproduced from Kadam et al. [170])

gauge needle (Fig. 18.10). The resultant nanotubes proved non-cytotoxic to RPE cells and could be detected in the vitreous humor of a rat model over the 15 day study period, with pazopanib levels being 4.5, 5, and 2.5 times greater in the vitreous humor, retina, and choroid–RPE, in relation to unincorporated drug. This places the nanotubular system as a potentially promising intravitreal delivery system [171]. Studies in a larger animal model and humans would be necessary for definitive establishment of safety and efficacy.

18.6.4.3 Hyaluronan-Modified Core-Shell Intravitreal Liponanoparticles

In an attempt to address the issues associated with intravitreal injection, which include a limited intraocular retention time and non-specific biodistribution to the retina, Gan and team [172] designed HA-modified core-shell liponanoparticles (LCS-NPs) as intravitreal delivery vehicles. Formation of the LCS-NPs was via hydration of a dried lipid film with chitosan nanoparticles solution followed by conjugation with hyaluronan to produce HA-LCS-NPs via covalent linkage of the glucuronic acid moiety of HA with the amine of the lipid component. This achieved particles of 450 nm and a stable negative surface charge. An increased HA density,

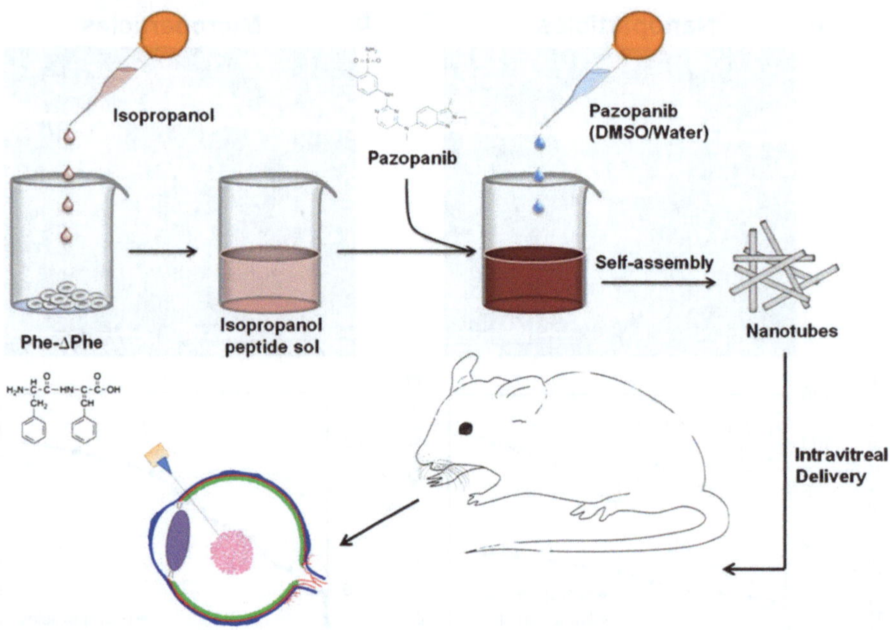

Fig. 18.10 Self-assembled nanotubes for sustained intravitreal delivery of a multi-targeted tyrosine kinase inhibitor (Reproduced from Panda et al. [171])

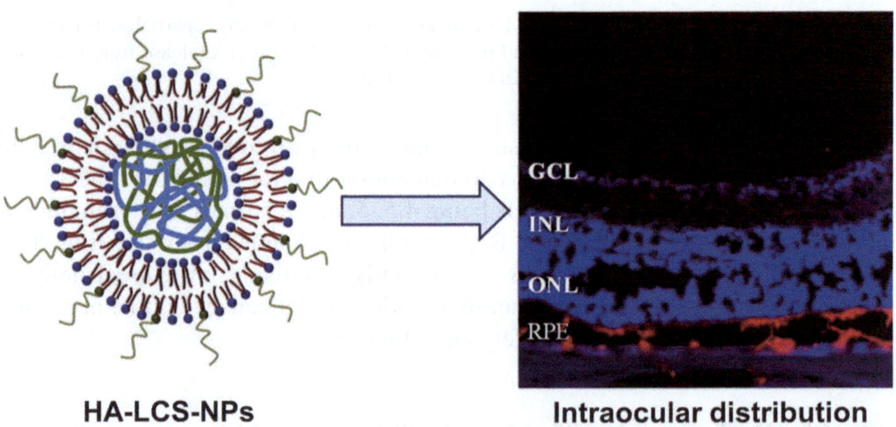

Fig. 18.11 (**a**) Schematic representation of the HA-LCS-NPs (**b**) CLSM image depicting penetrating ability of the HA-LCS-NPs across the whole retina to RPE (Reproduced from Gan et al. [172])

as well as molecular weight, enhanced cellular uptake of the NPs by ARPE-19 cells. CLSM image analysis revealed that the majority of HA-LCS-NPs overcame the vitreous barriers, passing through the inner limiting membrane, to a greater degree that CS-NPs or LCS-NPs (Fig. 18.11). This composite system could prove to be a viable option for intravitreal drug delivery but further in vivo analysis is necessitated.

18.6.4.4 Nanosystems for Inflammatory Ocular Tissue Targeting

Certain NS have been conceptualized, having specific applications in uveitis and associated intraocular inflammatory disorders. Drug-loaded liposomes have been employed for intraocular injections to reduce dosing frequency. The distribution of fluorescent PEG-coated liposomes following intravitreal injections in healthy rats was investigated by Camelo et al. [173]. Liposomes were largely identified in the vitreous humor, the inner retina, and to a lesser extent in the iris root and ciliary body. Lajavardi et al. [174] developed sterically stabilized PEG-coated liposomes incorporating vasoactive intestinal peptide (VIP) for the purposes of stability enhancement in biological media and controlled release. The VIP-loaded liposomes were injected intravitreally in rats in which EIU was induced. Their findings demonstrated that VIP only demonstrated efficacy in treating EIU when encapsulated in liposomes [175]. In an extension to this study, Lajavardi et al. [176] subsequently determined the influence of HA addition to the VIP liposomal delivery system. The system instigated a significant reduction in inflammation due to the improved efficacy and duration afforded by HA to the effect of liposomal VIP [177].

Ryu et al. [178] investigated the success of NPs composed of poly(γ-glutamic acid) conjugated with L-phenylalanine (γ-PGA-Phe NPs) for management of retinal diseases. The γ-PGA-Phe NPs (200 nm) were evaluated with macrophages and microglia in vitro and in vivo via intravitreal delivery into normal or pathological rat eyes. The NPs were efficiently taken up by cultured macrophages or microglia and still detectable in the cytoplasm of these cells at day 7. No accumulation of NPs was noted in the retinas of normal rat eyes with no inflammatory cell recruitment. Pathological conditions saw the converse outcome, with NPs localized in activated CD11b-positive cells in the retina. Suppression of the expression of TNF-α and MCP-1 in cultured macrophages or microglia was thus achieved by dexamethasone NPs. Data generated suggested that γ-PGA-Phe NPs could be an effective DDS for lowering the inflammatory cell burden in pathological retinal afflictions and also demonstrated the concept of a more bioresponsive approach toward ocular drug delivery.

Groups have focused on the intravenous administration of NS for inflammatory tissue targeting. The efficacy and distribution of poly(lactic acid) (PLA) NPs encapsulating betamethasone phosphate on rats with experimental autoimmune uveoretinitis (EAU) receiving a single IV injection was evaluated by Sakai et al. [179]. Confocal images highlighted no NP accumulation in the normal retina of healthy rats, whereas several NPs accumulated in the retina and a few NPs were detected in the outer nuclear layer, observed 3 h after injection, in EAU rats. These NPs remained in the retina of EAU rats up to 7 days after the administration. The anti-inflammatory effects observed were rapid and prolonged, being equal to a fivefold dose of free betamethasone phosphate. Further investigations are necessary to identify whether an enhanced inflammatory effect is due to either repeated injections or higher doses of betamethasone phosphate. However, even though intravenously administered NPs achieved sustained ocular targeting of ocular inflammation, originating endothelially, >95 % of the NPs were caught in the liver and spleen. Hashida et al. [180] brought an alternative approach to endothelial delivery to the inflamed eye to the fore in their work in which an active targeting strategy was exploited. Novel dexamethasone-loaded sialyl-Lewis X (sLe^X) conjugated liposomes (sLe^XL) were injected into EAU mice. The interaction of PMNs and endothelial cells exerts a pertinent role at inflammatory sites and is characterized by E-selectin- and P-selectin-mediated "rolling" of PMNs along the endothelium [181]. E-selectin and P-selectin display recognition of the tetrasaccharides sLe^X and A – these are either protein or lipid bound on the PMN surface [182, 183]. Dexamethasone-sLe^XL achieved a twofold higher dexamethasone concentration and selective targeting to the inflamed eye, relative to free dexamethasone (1 mg). A binding assay was performed to confirm whether sLe^XL liposomes target E-selectin, P-selectin, or both – this highlighted that sLe^XL liposomes had the propensity for site-directed delivery to both selectin types on the activated endothelial cells in vivo [175]. Inclusion of sLe^XL could thus be considered an effective inflammation-targeted delivery system in vivo for posterior segment inflammatory diseases, emanating in enhanced pharmacologic effects with and reduced side effects.

18.6.4.5 Stimulus-Responsive Ocular Nanosystems

Light-induced systems encompass light-activated drugs (photodynamic therapy, PDT) and light-activated drug delivery systems and employ a delivery mechanism for target tissue localization of the photosensitizing compound, ensued by activation via a nonthermal laser light. The light-activated drug verteporfin (Visudyne®) is one such example – activated 15 min after IV administration via a nonthermal red laser light – and is available clinically for the treatment of AMD. Verteporfin provides only adjunctive therapy; following activation, the drug instigates neovascular vessel occlusion; however, these vessels reappear. The institution of polymeric micelles for delivery of photosensitizers has been recently considered for reducing the potential for skin hypersensitivity and enhancing the selectivity and efficacy of PDT [67, 184]. The institution of micelles incorporating dendritic porphyrins was explored by Ideta and co-workers [185] who created this "supramolecular nanomedical device" comprising a dendritic photosensitizer (DP) encapsulated by a polymeric micellar system for the management of CNV induced in rats (Fig. 18.12). The micelles were injected into the tail vein of the rats and micelles selectively accumulated at CNS sites and enabled a 60–78 % CNV lesion occlusion following laser application at varied time points of 0.25 and 4 h post-micelle administration. Occlusion was attained at a comparatively low light power, a possible result of the unique capabilities of the porphyrins, where there is steric inhibition of aggregation of the core sensitizer, enabling an effective photochemical reaction. Following treatment, there was maintenance of 80 % of vessel occlusion for 7 days, with limited phototoxicity.

Other innovative research enabling targeted ocular gene delivery have focused on vectosomes: vesicles composed of a structural protein derived from herpes simplex virus (VP22). This fact was exploited by Normand et al. [165] as the VP22 was found to be capable of forming spherical particles of 300–1,000 nm in diameter, and the C-reactive terminal amino acid of VP22 enables binding to antisense oligonucleotides. VP22 was thus expressed in *Escherichia coli*, purified and combined with 20 mer phosphorothioate oligonucleotides originating the light-sensitive complex

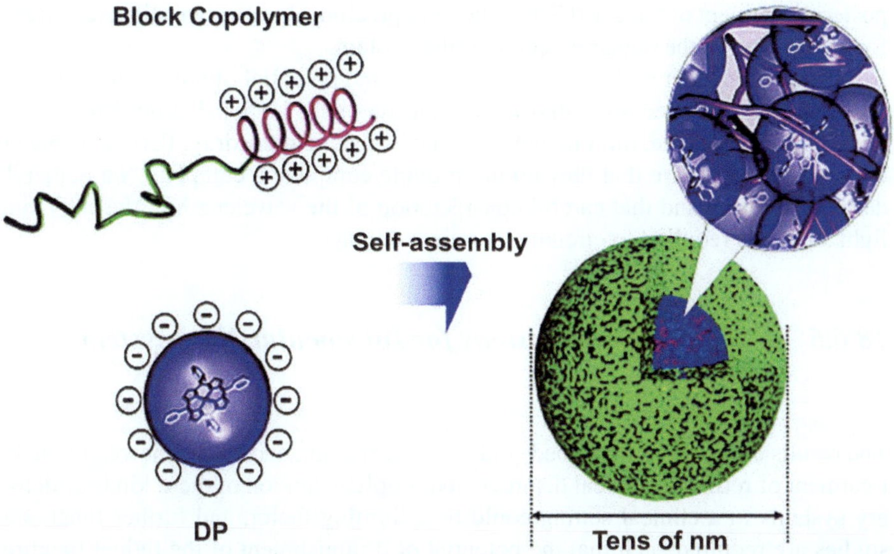

Fig. 18.12 Schematic highlighting PDT achieved via a supramolecular nanomedical device comprised of a dendritic photosensitizer encapsulated by a polymeric micelle formulation, DP = dendritic photosensitizer (Reproduced from Ideta et al. [185])

particles (vectosomes). Confocal and electron microscopy was employed to evaluate vectosome uptake and light-induced redistribution of the oligonucleotides in human choroid melanoma cells (OCM-1) and in human retinal pigment epithelial cells (ARPE-19). Their stability in the cytoplasm was maintained for weeks until exposed to cold white light or laser beam transscleral illumination which triggered the release of the oligonucleotides. In vivo evaluation in rats highlighted that vectosomes underwent intracellular internalization and could be detected in all retinal layers and were mainly concentrated at the external limiting membrane and RPE cytoplasm; thereafter, activation promoted vectosome distribution to the nuclei of the ganglion cell layer, inner nuclei layer, and RPE nuclei. In vivo studies would be imperative for this system for indication of clinical efficacy.

Surface plasmon resonance demonstrated by noble metal NPs enables them to absorb energy at a distinctive wavelength in UV–vis region. When metal nanoparticles are subjected to light-induced heat, they can cause damage to NP-labeled cells. This observation could be exploited for therapeutic applications; thus in efforts to develop a light-responsive DDS, Paasonen et al. [186] developed liposomes loaded with lipid-functionalized gold NPs (3–5 nm) on the inner and outer liposomal surfaces, together with calcein as a marker substance. This system was compared with liposomes in which hydrophobic NPs were embedded in the lipid bilayer and core-encapsulated negatively charged hydrophilic NPs. Exposure to light irradiation caused calcein release from the otherwise intact liposomes as a result of the gold NPs absorbing heat; the transferred heat therefrom instigates the ensuing liposome phase transition and calcein release. Gold NP-loaded liposomes remained intact at 37°C but released a marker substance, calcein, by light irradiation, and the authors have proposed the application of this system for localized treatment of afflictions of the posterior segment of the eye. Cell studies and preclinical testing are still necessitated for explication of the ocular potential of this system.

Thrimawithana et al. [67] have emphasized the potential of these light-sensitive DDS in terms of their potential to provide targeted, responsive delivery, to the posterior ocular tissue, minimizing systemic exposure to the drug. Pertinent considerations, however, are that they do not provide complete a complete "on-demand" delivery solution and that careful consideration of the wavelength of the activating light source is required to circumvent retinal damage.

18.6.5 Further Considerations for Intraocular Nanosystem Delivery

The results of the afore-described studies are promising and shine a new light on the treatment of retinal and uveal disorders, but implementation of these kinds of delivery systems in a clinical setting could be a limiting factor, and further functional studies are required such that the potential of diminishment of the retinal function and vision and the occurrence of accompanying chronic inflammatory processes are dismissed [136]. The total cost implications of implementing active targeting approaches are a further pertinent consideration.

It is apparent that design of an ocular NS could be the key to enhancing overall bioactive bioavailability for the treatment of posterior segment disorders. To add to this, instituting the afore-described approaches, several kinds of NS encapsulating diverse active molecules have reached the stage of preclinical analysis [119, 163–165, 186, 187]. A further consideration, however, is that implantable drug delivery systems are also moving to the fore as alternatives to intravitreal or periocular injections. Ocular implants provide the capacity to achieve long-term drug delivery in the retina and associated posterior ocular tissues. Realization of the advantages of combining these two drug delivery approaches would appear to be an intuitive step. The concept here is to exploit the implant capacity to retain and deliver molecules to tissues that it has close or direct contact with and employ it as a biological reservoir for the NS. The key is to attain therapeutic levels of a drug for extended periods following implantation, which could see a vast improvement in the quality of life of many patients.

18.7 Hybrid Delivery Systems for Posterior Segment Eye Disorders

18.7.1 Microneedles as a Hybrid Intraocular Delivery System

Hollow and solid microneedles possessing minimal invasiveness are being investigated for avoiding the described safety and bioavailability limitations associated with intravitreal and periocular injections. These microneedles possess application as sustained-release drug depots for intrascleral and suprachoroidal delivery, enabling

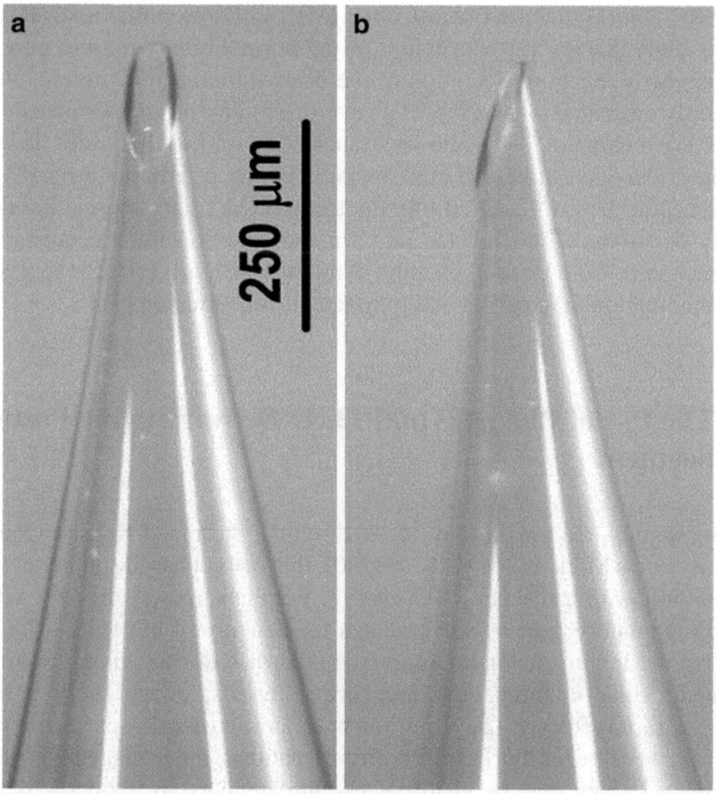

Fig. 18.13 Representative hollow microneedle: (**a**) front view and (**b**) side view. The microneedle shown has an approximately elliptical tip opening with a long axis of ~100 μm and a short axis of ~40 μm with a bevel tip angle of 25° (Reproduced from Jiang et al. [189])

reduced drug clearance. Retinochoroidal targeting is also enhanced compared to periocular delivery. The major reported challenge of this route is the short to medium duration of action; thus, it requires hybridization with a sustained-release delivery system, which may include a suprachoroidal microcannulation technique, alteration in drug formulations to maximize drug depot retention, enhancing drug binding in the delivered tissues, and intrascleral implants or devices [188].

Jiang et al. [189] endeavored to put the hypothesis of minimally invasive infusive delivery of soluble molecules, nanoparticles, and microparticles into sclera via hollow microneedles (Fig. 18.13). They inserted individual hollow microneedles into human cadaver sclera (ensuring the needles did not actually cross the sclera). In this case, hybridization with a sustained-release component was enabled through aqueous solutions containing sulforhodamine or fluorescently tagged nanoparticles or microparticles. These were infused via the microneedles into the sclera. There was successful, minimally invasive delivery of nanoparticle suspensions into the sclera, with microparticles only being delivered in the presence of hyaluronidase and collagenase spreading enzymes, which disrupted the microstructure of the scleral tissue.

At SERC 2009, Francine Behar-Cohen (MD, PhD) proposed a novel technique instituting gene therapy for conversion of the normal ciliary muscle into an intraocular bioreactor, enabling secretion of the desired therapeutic protein, which was subsequently evaluated in a rat model of uveitis employing "a microneedle-enabled gene delivery technique." This allowed transfection of the ciliary muscle with plasmids. These encoded three different variants of the p55 tumor necrosis factor-α soluble receptor. It was observed that the transfected ciliary muscle possessed the capability to produce and release its own monomeric tumor necrosis factor-α-soluble receptors for a period reaching 8 months and successfully achieved local immunomodulation of experimental intraocular inflammation [188].

18.8 The Use of Biologics and Their Delivery to the Posterior Segment

According to press release on May 15, 2014 on the Sustained-Release Ocular Drug Delivery Systems Market (2014–2024) [190], this market has the potential to exceed USD 1 billion within the next 10 years. A pertinent reason for this will be the increasing regulatory approval of implants for delivery of biologics. The first-line therapy for the treatment for immune-mediated diseases, including noninfectious uveitis and ocular inflammation, are corticosteroids, owing to their low cost, potency, and rapid efficacy. The requirement for corticosteroid-sparing immunosuppressive agents is generally in a minority of patients for management of autoimmune disease (which may be vision or life threatening), for reducing the toxicity associated with prolonged corticosteroid use. The application of biologic therapy occurs in situations patients exhibit a poor response or tolerance to conventional immunotherapy. The initial development of biologics was for the treatment of systemic inflammatory diseases or preventing organ transplant rejection, but they have been employed off-label for uveitis or ocular inflammation treatment [191].

Ocular disorders associated with rheumatic conditions can present in numerous ways, including dry eye syndrome, acute and chronic anterior uveitis, panuveitis or vitritis, scleritis, keratitis, retinal vasculitis, and ischemic optic neuropathy. Where corticosteroid therapy becomes inadequate, biologic therapies have been instituted with varying efficacy in the control of a number of the primary manifestations of rheumatic disease, with their use ultimately extending to ocular manifestations, as highlighted in Table 18.5 [192].

In terms of appropriate delivery systems for ocular delivery of biologics, some have already been divulged in this chapter. As discussed in Table 18.2, ECT such as Renexus (Neurotech) provides a novel therapy for biologics delivery. The NT-503 ECT technology, also by Neurotech, was also designed for constant VEGFR-Fc release intravitreally. The capsule drug ring developed by Molokhia et al. [107] incorporates the anti-VEGF-A antibody, bevacizumab (Avastin®).

Bushell–Embling [193] posted in "Medical Devices" in May 2014 on the preclinical data that pSivida (ASX:PVA) presented at the 14th Annual Meeting of the Association for Research in Vision and Ophthalmology (ARVO). pSivida pinpointed

Table 18.5 Rheumatic conditions and their ocular manifestations for potential treatment with biologics

Rheumatic condition	Associated posterior segment disorder	Biologic treatment
Adamantiades–Behçet's disease (ABD)	Posterior or diffuse uveitis with accompanying retinal vasculitis	Anti-TNF-α – first line – reduces symptoms, relapse, and reduction of corticosteroid dose
Aggressive, sight-threatening disease with males more commonly affected	Necessitates immunosuppressive therapy – counteracts vision loss	Infliximab (TNF-α blocker) – foremost agent demonstrating remission of anterior and posterior segment inflammation, resolution of macular edema, and refractory uveitis control. Favored anti-TNF-α agents in controlling ocular symptoms
Ocular manifestations in 43–72 % of patients		Etanercept and adalimumab (TNF-α blockers) have shown efficacy in ameliorating ABD disease severity. Rituximab (lymphocyte inhibitor) – application in retinal vasculitis associated with ABD
Juvenile idiopathic arthritis (JIA)	Uveitis in 10–15 % of patients	Of the anti-TNF-α agents, infliximab proved effective – but had high side effect profile
Primary cause of childhood uveitis	Inflammation observed in children commonly asymptomatic and bilateral – chronic course	Adalimumab – preferred anti-TNF-α agent – effective against uveitis and better tolerated by children
	Severe vision loss with ultimate blindness	Non-anti-TNF-α agents = rituximab (lymphocyte inhibitor); high-dose IV administration of daclizumab (IL-2 antagonist); abatacept (lymphocyte inhibitor)
		Head-to-head comparison of these newer agents is required
Relapsing polychondritis (RP)	Uveitis observed in 25 % of patients, retinal vasculitis is less frequent	Immunosuppressive chemotherapy is generally necessary
59 % of patients possess ocular manifestations		Infliximab – decreased ocular manifestations
Giant cell arteritis (GCA)	Rapid or near or total vision loss	Proposed – maintenance therapy with biologic agents – avoids chronic steroids and associated adverse effects
In the absence of rapid implementation of high-dose IV steroids, bilateral vision loss can occur		Anti-TNF-α – varied outcomes
		Potential agents = adalimumab and rituximab

Adapted from Kraus and Culican [192]

the potential for application of the Tethadur implant (based on their BioSilicon technology) for providing sustained release of antibodies such as bevacizumab (Avastin). The long-term release of the antibodies is modifiable through adjustment of the pore size and surface area of the ocular implant. This could present a significant advanced compared to the frequent injections that would otherwise be required (as often as once a month).

18.9 Innovative Frontiers in Intraocular Drug Delivery

The future of implant technology at the biotechnology industry level sees the incorporation of a nano- or microstructured component coming to the fore. These are some examples of devices at the preclinical stage of investigation [66]:

- SurModics furnished Genentech, a member of the Roche group, an exclusive license to employ its proprietary biodegradable microparticulate drug delivery system to develop the biologic, ranibizumab, as a sustained-release formulation (Lucentis).
- The design of a bioerodible nanostructured porous silicon (BioSilicon Technology) has been the current focus of pSivida. It is proposed to surmount challenges associated with long-term delivery of proteins and peptides through accommodation of various molecular sizes as well as enabling drug stabilization within the device.

In the search for additional innovative sustained delivery systems, the employment of drug-loaded intraocular lenses (IOLs) is an attractive option. A pertinent obstacle here would be ensuring that therapeutic levels of drug actually reach the posterior ocular tissues [66].

Other highly novel approaches include the design of microrobots for minimally invasive intraocular surgery. Ullrich et al. [194] illuminated the fact of the movement of ophthalmology toward more minimally invasive approaches. The advantages of this are reduced inflammation, shorter operating times, and minimized conjunctival damage. In this vein, researchers have shifted their focus to conceptualization of miniaturized robotic-assisted devices for precise assistance with ocular surgery (overcoming human error). The microrobots described and designed by Ullrich and his team [194] enable minimally invasive surgery in the posterior segment of the eye, following injection into the vitreous via a small pars plana incision. The robot is subsequently wirelessly guided to the pathological site, and this could enable guidance of intravitreal implants to the site of interest, such as the retina, to provide enhanced bioavailability over an extended period. Following ex vivo and in vivo evaluations in the rabbit eye, the untethered mobile microrobots were controlled by an electromagnetic control system and could be manipulated in the eye and allowed suture-less and targeted ophthalmic drug delivery, with potential application in AMD, intravenous deployment of anti-coagulation agents for retinal vein occlusion (RVO), in addition to mechanical applications, such as manipulation of epiretinal membrane (ERM) peeling.

18.10 Potential Pitfalls of Sustained-Release Intraocular Implants

Despite the extensive potential and advantages of innovative intraocular systems, several risk factors could be associated with sustained-release implants. In certain disease states, a simpler, safe, and more effective option may be available, an example being the suspension of an I-vation trial [195]. Price can also be a concern; however, as has been highlighted by Dr. Nguyen, Retisert® can be a cost-effective therapy when appropriately assigned, as it provides drug release for 30 months. The requirement for additional agents and evaluations, such as monthly cyclosporine and associated tests every 3–4 months, which is costly, is thus eliminated. Depending on the device, implantation could be complicated and require surgery in an operating theater. Implants which allow a simple outpatient procedure to be performed by the ophthalmologist are at an advantage here (e.g., Ozurdex or Iluvien).

18.11 Considerations in the Design of Intraocular Implants

The burgeoning design of intraocular systems has revealed a few guidelines for designing an effective device. The positioning of an implant within the globe is of importance – movement of an implant 4 mm farther into the vitreous provides macular drug levels which are four times greater than if they are positioned at the ocular wall. The effect of position of an implant on side effects is still unclear. Most implantable systems have aimed for achieving a consistent drug concentration over an extended period – this is ultimately impacted on by the physicochemical characteristics of the drug. However, in some instances, pulsatile release may be preferable, giving the ocular environment a "drug holiday", and reducing the incidence of an elevated IOP. This release profile is attained with a number of the stimulus-responsive systems discussed (e.g., inflammation and light-responsive devices) [66].

18.12 Conclusions

Effective treatment of ocular diseases presents a formidable task, dually attributed to the nature of intraocular diseases and the presence of ocular barriers. Challenges in ocular drug delivery have been partially addressed through the design of innovative drug delivery systems which all strive to enhance bioactive bioavailability. Development of compliance-promoting and disease-responsive delivery techniques will revolutionize ocular drug delivery. Evolution of sustained drug delivery systems integrating advanced polymers is infinite, and conceptualization of novel polymeric systems with intelligent (e.g., stimulus-responsive) mechanisms and advances in nanotechnology are at the forefront of achieving directed and controlled delivery for treating vision-threatening diseases.

References

1. Robinson JC (1993) Ocular anatomy and physiology relevant to ocular drug delivery. In: Mitra AK (ed) Ophthalmic drug delivery systems. Marcel Dekker, New York, pp 29–58
2. Henderer JD, Rapuano CJ (2011) Chapter 64. Ocular pharmacology. In: Chabner BA, Brunton LL, Knollman BC (eds). Goodman & Gilman's the pharmacological basis of therapeutics, 12th edn. McGraw-Hill, New York. Retrieved from URL: http://www.accesspharmacy.com/content.aspx?aID=16681771.
3. du Toit LC, Govender T, Carmichael T et al (2013) Design of an anti-inflammatory composite nanosystem and evaluation of its potential for ocular drug delivery. J Pharm Sci 102:2780–2805
4. Nanjawade BK, Manvi FV, Manjappa AS (2007) In situ-forming hydrogels for sustained ophthalmic drug delivery. J Control Release 122:119–134
5. Del Amo EM, Urtti A (2008) Current and future ophthalmic drug delivery systems: a shift to the posterior segment. Drug Discov Today 13:135–143
6. Herrero-Vanrell R, Refojo MF (2001) Biodegradable microspheres for vitreoretinal drug delivery. Adv Drug Deliv Rev 52:5–16
7. Janoria KG, Gunda S, Boddu SH et al (2007) Novel approaches to retinal drug delivery. Expert Opin Drug Deliv 4:371–388
8. Yasukawa T, Ogura Y, Sakurai E et al (2005) Intraocular sustained drug delivery using implantable polymeric devices. Adv Drug Deliv Rev 57:2033–2046
9. Haesslein A, Ueda H, Hacker MC et al (2006) Long-term release of fluocinolone acetonide using biodegradable fumarate-based polymers. J Control Release 114:251–260
10. de la Fuente M, Raviña M, Paolicelli P et al (2010) Chitosan-based nanostructures: a delivery platform for ocular therapeutics. Adv Drug Deliv Rev 62:100–117
11. Roth DB, Chieh J, Spirn MJ et al (2003) Noninfectious endophthalmitis associated with intravitreal triamcinolone injection. Arch Ophthalmol 121:1279–1282
12. Moshfeghi DM, Kaiser PK, Scott IU et al (2003) Acute endophthalmitis following intravitreal triamcinolone acetonide injection. Am J Ophthalmol 136:791–796
13. Jonas JB, Kreissig I, Degenring RF (2003) Intravitreal triamcinolone acetonide for pseudophakic cystoid macular edema. Am J Ophthalmol 136:384–386
14. Gillies MC, Simpson JM, Billson FA et al (2004) Safety of an intravitreal injection of triamcinolone. Arch Ophthalmol 122:336–340
15. Jonas JB, Hayler J, Sofker A et al (2001) Intravitreal injection of crystalline cortisone as adjunctive treatment of proliferative diabetic retinopathy. Am J Ophthalmol 131:468–471
16. Ashton P (2000) Sustained-release inserts help deliver intraocular drugs. (Brief article) (Column) *Ophthalmol Times*, May.
17. Allen HF, Mangiaracine AB (1964) Bacterial endophthalmitis after cataract extraction: a study of 22 infections in 20.000 operations. Arch Ophthalmol 72:454–462
18. Okada AA, Johnson RP, Liles CW et al (1994) Endogenous bacterial endophthalmitis: report of a ten-year retrospective study. Ophthalmology 101:832–838
19. Peyman GA, Ganiban GJ (1995) Delivery systems for intraocular routes. Adv Drug Deliv Rev 16:107–123
20. Gaudana R, Krishna Ananthula H, Parenky A et al (2010) Ocular drug delivery. AAPS J 12:348–360
21. Urtti A (2006) Challenges and obstacles of ocular pharmacokinetics and drug delivery. Adv Drug Deliv Rev 58:1131–1135
22. Espana E, Chipont E, Sanchez S et al (1993) Collagen shields enhanced penetration of topical cyclosporine. Invest Ophthalmol Vis Sci 34(Suppl):1488
23. Kim JH, Kim KW, Kim MH et al (2009) Intravenously administered gold nanoparticles pass through the blood-retinal barrier depending on the particle size, and induce no retinal toxicity. Nanotechnology 20:505101
24. Singh SR, Grossniklaus HE, Kang SJ et al (2009) Intravenous transferrin, RGD peptide and dual-targeted nanoparticles enhance anti-VEGF intraceptor gene delivery to laser-induced CNV. Gene Ther 16:645–659

25. Suzuki T, Uno T, Chen G et al (2008) Ocular distribution of intravenously administered micafungin in rabbits. J Infect Chemother 14:204–207
26. Regnier A, Schneider M, Concordet D et al (2008) Intraocular pharmacokinetics of intravenously administered marbofloxacin in rabbits with experimentally induced acute endophthalmitis. Am J Vet Res 69:410–515
27. Goldblum D, Rohrer K, Frueh BE et al (2002) Ocular distribution of intravenously administered lipid formulations of amphotericin B in a rabbit model. Antimicrob Agents Chemother 46:3719–3723
28. Ananthula HK, Vaishya RD, Barot M et al (2009) Duane's ophthalmology. In: Tasman W, Jaeger EA (eds) Bioavailability. Lippincott Williams & Wilkins, Philadelphia
29. Gipson IK, Argueso P (2003) Role of mucins in the function of the corneal and conjunctival epithelia. Int Rev Cytol 231:1–49
30. Ahmed I (2003) The noncorneal route in ocular drug delivery. In: Mitra AK (ed) Ophthalmic drug delivery systems. Marcel Dekker, New York, pp 335–363
31. Barar J, Javadzadeh AR, Omidi Y (2008) Ocular novel drug delivery: impacts of membranes and barriers. Expert Opin Drug Deliv 5:567–581
32. Geroski DH, Edelhauser HF (2001) Transscleral drug delivery for posterior segment disease. Adv Drug Deliv Rev 52(1):37–48
33. Kim SH, Lutz RJ, Wang NS et al (2007) Transport barriers in transscleral drug delivery for retinal diseases. Ophthalmic Res 39:244–254
34. Kampougeris G, Antoniadou A, Kavouklis E et al (2005) Penetration of moxifloxacin into the human aqueous humour after oral administration. Br J Ophthalmol 89:628–631
35. Santulli RJ, Kinney WA, Ghosh S et al (2008) Studies with an orally bioavailable alpha V integrin antagonist in animal models of ocular vasculopathy: retinal neovascularization in mice and retinal vascular permeability in diabetic rats. J Pharmacol Exp Ther 324:894–901
36. Shirasaki Y, Miyashita H, Yamaguchi M (2006) Exploration of orally available calpain inhibitors. Part 3: Dipeptidyl alpha-ketoamide derivatives containing pyridine moiety. Bioorg Med Chem 14:5691–5698
37. Sakamoto H, Sakamoto M, Hata Y et al (2007) Aqueous and vitreous penetration of levofloxacin after topical and/or oral administration. Eur J Ophthalmol 17:372–376
38. Coppens M, Versichelen L, Mortier E (2002) Treatment of postoperative pain after ophthalmic surgery. Bull Soc Belge Ophtalmol 285:27–32
39. Rajpal T, Srinivas A, Azad RV et al (2009) Evaluation of vitreous levels of gatifloxacin after systemic administration in inflamed and non-inflamed eyes. Acta Ophthalmol 87:648–652
40. Samtani S, Amaral J, Campos MM et al (2009) Doxycycline-mediated inhibition of choroidal neovascularization. Invest Ophthalmol Vis Sci 50:5098–5106
41. Chong DY, Johnson MW, Huynh TH et al (2009) Vitreous penetration of orally administered famciclovir. Am J Ophthalmol 148:38–42
42. Takahashi K, Saishin Y, King AG et al (2009) Suppression and regression of choroidal neovascularization by the multitargeted kinase inhibitor pazopanib. Arch Ophthalmol 127:494–499
43. Kokke KH, Morris JA, Lawrenson JG (2008) Oral omega-6 essential fatty acid treatment in contact lens associated dry eye. Cont Lens Anterior Eye 31:141–146
44. Clement DB, Tailor V (1987) A study of aqueous and serum levels of ceftazidime following subconjunctival administration. Br J Ophthalmol 71:433–435
45. Axelrod J-L, Newton JC, Klein RM et al (1987) Penetration of imipenem into human aqueous and vitreous humor. Am J Ophthalmol 104:649–653
46. Sharir M, Triester G, Kneer J et al (1989) The intravitreal penetration of ceftriaxone in man following systemic administration. Invest Ophthalmol Vis Sci 30:2179–2183
47. Hosseini K, Matsushima D, Johnson J et al (2008) Pharmacokinetic study of dexamethasone disodium phosphate using intravitreal, subconjunctival, and intravenous delivery routes in rabbits. J Ocul Pharmacol Ther 24:301–308
48. Kim SH, Csaky KG, Wang NS et al (2008) Drug elimination kinetics following subconjunctival injection using dynamic contrast-enhanced magnetic resonance imaging. Pharm Res 25:512–520

49. Mitra AK, Anand BS, Duvvuri S (2006) Drug delivery to the eye. In: Fischbarg J (ed) The biology of the eye. Academic, New York, pp 307–351
50. Peeters L, Sanders NN, Braeckmans K et al (2005) Vitreous: a barrier to nonviral ocular gene therapy. Invest Ophthalmol Vis Sci 46:3553–3561
51. Pitkanen L, Ruponen M, Nieminen J et al (2003) Vitreous is a barrier in nonviral gene transfer by cationic lipids and polymers. Pharm Res 20:576–583
52. Kim H, Robinson SB, Csaky KG (2009) Investigating the movement of intravitreal human serum albumin nanoparticles in the vitreous and retina. Pharm Res 26:329–337
53. Von Sallmann L (1948) Controversial points in penicillin therapy for ocular diseases. Arch Ophthalmol 39:752–804
54. Von Sallmann L, Meyer K, Di Grandi J (1944) Experimental study on penicillin treatment of ectogenous infection of vitreous. Arch Ophthalmol 32:179–189
55. Lewis DH (1990) Controlled release of bioactive agents from lactide/glycolide polymers. In: Chasin M, Langer R (eds) Biodegradable polymers as drug delivery systems. Marcel Dekker, New York, pp 1–41
56. Wood DA (1980) Biodegradable drug delivery systems. Int J Pharm 7:1–18
57. Chowhan M, Weiner AL, Bhagat H (2002) Drug delivery-ophthalmic route. In: Swarbrick J, Boylan JC (eds) Encyclopedia of pharmaceutical technology. Marcel Dekker, New York, pp 863–870
58. Davis JL, Gilger BC, Robinson MR (2004) Novel approaches to ocular drug delivery. Curr Opin Mol Ther 6:195–205
59. Duvvuri S, Majumdar S, Mitra AK (2003) Drug delivery to the retina: challenges and opportunities. Expert Opin Biol Ther 3:45–56
60. Edlund U, Albertsson AC (2002) Degradable polymer microspheres for controlled drug delivery. In: Albertsson AC (ed) Advances in polymer science, vol 157. Springer, Berlin, pp 68–112
61. Ghate D, Edelhauser HF (2006) Ocular drug delivery. Expert Opin Drug Deliv 3:275–287
62. Heller J (2005) Ocular delivery using poly(orto esters). Adv Drug Deliv Rev 57:2053–2062
63. Hughes PM, Olejnik O, Chang TS et al (2005) Topical and systemic drug delivery to the posterior segments. Adv Drug Deliv Rev 57:2010–2032
64. Kumar MNVR (2000) Nano and microparticles as controlled drug delivery devices. J Pharm Pharm Sci 3:234–258
65. Ludwig A (2005) The use of mucoadhesive polymers in ocular drug delivery. Adv Drug Deliv Rev 57:1595–1639
66. Stuart A (2010) The promise of implantable drug delivery systems. Eyenet 3:33–37
67. Thrimawithana TR, Young SA, Bunt CR et al (2011) Drug delivery to the posterior segment of the eye. Drug Discov Today 16:270–277
68. Weiner AL (1994) Polymeric drug delivery systems for the eye. In: Domb AJ (ed) Polymeric site specific pharmacotherapy. Wiley, New York, pp 316–346
69. Park H, Park K (1996) Biocompatibility issues of implantable drug delivery systems. Pharm Res 13:1770–1776
70. Zaheer S, Lehman J, Stevenson G (1982) Capsular contracture around silicone implants: the role of intraluminal antibiotics. Plast Reconstr Surg 69:809–812
71. ClinicalTrials.gov(a). A study of MK0140 in diabetic patients with macular edema. Retrieved from URL: http://clinicaltrials.gov/ct2/show/NCT00692614?term=I-vation&rank=1. ClinicalTrials.gov updated this record on 29 January 2013.
72. Ashton P (President and CEO of pSivida) (2014) http://www.theflyonthewall.com/permalinks/entry.php/PSDVid2010764/PSDV-pSivida-plans-to-seek-US-approval-for-Medidur-for-posterior-uveitis
73. Kauper K, Ling V, Elliot S et al (2012) Long-term, sustained intraocular delivery of escalating doses of VEGF antagonist using encapsulated cell technology implant for the treatment of choroidal neovascular diseases. Presented at the Association for Research in Vision and Ophthalmology Annual Meeting, May 2012, Fort Lauderdaule, FL

74. ClinicalTrials.gov(b). Safety and efficacy of brimonidine intravitreal implant in patients with geographic atrophy due to age-related macular degeneration (AMD). Retrieved from URL: http://clinicaltrials.gov/ct2/show/NCT00658619. ClinicalTrials.gov processed this record on 13 March 2013. Accessed 14 Feb 2013.
75. pSivida.com. Products/Thethadur. Available at: http://psivida.com/products-biosilicon.html
76. ClinicalTrials.gov(c). Efficacy and safety of betamethasone micropsheres in patients with diabetic macular edema (Tsubasa). Retrieved from URL: http://clinicaltrials.gov/ct2/show/NCT01411254?term=DE-102&rank=1. ClinicalTrials.gov processed this record on 13 March 2013.
77. ClinicalTrials.gov(d). Efficacy and safety of betamethasone microspheres in patients with macular edema following branch retinal vein occlusion (HIKARI). Retrieved from URL: http://clinicaltrials.gov/ct2/show/NCT01512901?term=DE-102&rank=2. ClinicalTrials.gov processed this record on 29 July 2012.
78. Lim JI, Fung AE, Wieland M, Hung D, Wong V (2011) Sustained-release intravitreal liquid drug delivery using triamcinolone acetonide for cystoid macular edema in retinal vein occlusion. Ophthalmology 118:1416–1422
79. Wong V, Wood, L (2006) Conveniently Implantable Sustained Release Drug Compositions. US Patent WO2006/039336, 6 April.
80. ClinicalTrials.gov(e). IBI 20089+Lucentis combo intravitreal injections for treatment of neovascular age-related macular degeneration (AMD) (Icon Combo). Retrieved from URL: http://clinicaltrials.gov/ct2/show/NCT01175395?term=IBI20089&rank=1. ClinicalTrials.gov processed this record on 28 June 2012.
81. ClinicalTrials.gov(f). Safety and tolerability of NOVA63035 "Corticosteroid" in patients with diabetic macular edema secondary to diabetic retinopathy. Available at: http://clinicaltrials.gov/ct2/show/NCT00665106?term=NOVA63035&rank=1. ClinicalTrials.gov processed this record on 15 July 2010.
82. Kuppermann B (2009) Cortiject dexamethasone lipid emulsion for the treatment of diabetic macular edema: a pilot study. The 8th international symposium on ocular pharmacology and therapeutics: 33–31.
83. Novagali.com. Novagli Pharma. Eyeject. Retrieved from URL: http://www.novagali.com/en/our-research/eyeject/
84. Kuno N, Fujii S (2012) Ocular drug delivery systems for the posterior segment: a review. Retina Today May/June:54–59.
85. Sultana Y, Jain R, Aqil M et al (2006) Review of ocular drug delivery. Curr Drug Deliv 3:207–217
86. Danckwerts MP, Fassihi A (1991) Implantable controlled release drug delivery systems: a review. Drug Dev Ind Pharm 17:1465–1502
87. Corcoran S (2006) Drug delivery to the eye. Highlights in Chem Sci 9. Retrieved from URL: http://www.rsc.org/Publishing/ChemScience/Volume/2006/09/nanoparticles_in_the_eye.asp. Accessed 12 Mar 2008.
88. Barbu E, Verestiuc L, Nevell TG et al (2006) Polymeric materials for ophthalmic drug delivery: trends and perspectives. J Mater Chem 16:3439–3443
89. Choonara YE, Pillay V, Danckwerts MP et al (2010) A review of implantable intravitreal drug delivery technologies for the treatment of posterior segment eye diseases. J Pharm Sci 99:2219–2239
90. du Toit LC, Carmichael T, Govender T et al (2014) In vitro, in vivo, and in silico evaluation of the bioresponsive behavior of an intelligent intraocular implant. Pharm Res 31:607–634
91. du Toit LC, Pillay V, Choonara YE et al (2011) Ocular drug delivery – a look towards nanobioadhesives. Expert Opin Drug Deliv 8:71–94
92. Gilger BC, Malok GE, Stewart T et al (2000) Effect of an intravitreal cyclosporine implant on experimental uveitis in horses. Vet Immunol Immunopathol 76:239–255
93. Li PY, Shih J, Lo R et al (2008) An electrochemical intraocular drug delivery device. Sens Actuat A: Phys 143:41–48 (based on contributions revised from the Technical Digest of the IEEE 20th international conference on Micro Electro Mechanical Systems (MEMS 2007) — MEMS 2007)

94. Saait S, Lo R, Li P-Y et al (2009) Mini drug pump for ophthalmic use. Trans Am Ophthalmol Soc 107:60–70
95. Michelson JB, Nozik RA (1979) Experimental endophthalmitis treated with an implantable osmotic minipump. Arch Ophthalmol 97:1345–1346
96. Eliason JA, Maurice DM (1980) An ocular perfusion system. Invest Ophthalmol Vis Sci 19:102–105
97. Miki K, Ohkuma H, Ryan SJ (1984) A method for chronic drug infusion into the eye. Jpn J Ophthalmol 28:140–146
98. Avitabile T, Marano F, Castiglione F et al (2001) Biocompatibility and biodegradation of intravitreal hyaluronan implants in rabbits. Biomaterials 22:195–200
99. Kuppermann BD (2006) Implant delivery of corticosteroids and other pharmacologic agents. Presented at Retina 2006: Emerging new concepts. Held in conjunction with the American Academy of Ophthalmology 2006 Annual Meeting, 10–11 November, Las Vegas
100. Choonara YE, Pillay V, Carmichael T et al (2006) An in vitro study of the design and development of a novel doughnut-shaped minitablet for intraocular implantation. Int J Pharm 310:15–24
101. Choonara YE, Pillay V, Carmichael T et al (2007) Studies on a novel doughnut-shaped Minitablet for intraocular drug delivery. AAPS PharmSciTech 8, Article 118. DOI: 10.1208/pt0804118
102. Okabe K, Kimura H, Okabe J et al (2003) Intraocular tissue distribution of betamethasone after intrascleral administration using a non-biodegradable sustained drug delivery device. Invest Ophthalmol Vis Sci 44:2702–2707
103. Kato A, Kimura H, Okabe K et al (2004) Feasibility of drug delivery to the posterior pole of the rabbit eye with an episcleral implant. Invest Ophthalm Vis Sci 45:238–244
104. Pontes de Carvalho RA, Krausse ML, Murphree AL et al (2006) Delivery from episcleral exoplants. Invest Ophthalmol Vis Sci 47:4532–4539
105. Kawashima T, Nagal N, Kaji H et al (2011) A scalable controlled-release device for transscleral drug delivery to the retina. Biomaterials 32:1950–1956
106. Thrimawithana TR, Young SA, Bunt CR et al (2011) *In-vitro* and *in-vivo* evaluation of carrageenan/methylcellulose polymeric systems for transscleral delivery of macromolecules. Eur J Pharm Sci 44:399–409
107. Molokhia SA, Sant H, Simonis J et al (2010) The capsule drug device: novel approach for drug delivery to the eye. Vis Res 50:680–685
108. Csaky KG (2007) New developments In the transscleral delivery of ophthalmic agents. The profile of the drug being delivered is as important as the delivery method. Retina Today 3:32–33
109. Eljarrat-Binstock E, Domb AJ (2006) Iontophoresis: a non-invasive ocular drug delivery. J Control Release 110:479–489
110. Karla PK, Ako-Adounvo A-M (2012) Advances in ocular iontophoresis research. Recent Pat Nanomed 2:126–132
111. Hastings MS, Li SK, Miller DJ et al (2004) Visulex: advancing iontophoresis for effective non-invasive back-to-the-eye therapeutics. J Drug Deliv Technol 4:53–57
112. Haesslein A, Hacker MC, Ueda H et al (2009) Matrix modifications modulate ophthalmic drug delivery from photo-cross-linked poly(propylene fumarate)-based networks. J Biomat Sci 20:49–69
113. Saliba JB, Gomes Faraco AA, Yoshida MI et al (2008) Development and characterization of an intraocular biodegradable polymer system containing cyclosporine A for the treatment of posterior uveitis. Mater Res 11:207–211
114. Allergan Inc. EP1750688 (2007) Steroid intraocular implants having an extended sustained release for a period of greater than 2 months (patent filed). Retrieved from URL: https://register.epo.org/espacenet/application?number=EP05744945. Accessed 5 Mar 2008.
115. Hilt JZ, Peppas NA (2005) Microfabricated drug delivery devices. Int J Pharm 306:15–23
116. Bawa R (2004) Nanotechnology patents and challenges, ipFrontline.com.
117. Bucolo C, Maltese A, Drago F (2008) When nanotechnology meets the ocular surface. Expert Rev Ophthalmol 3:325–332

118. Badawi AA, El-Laithy HM, El Qidra RK et al (2008) Chitosan based nanocarriers for indomethacin ocular delivery. Arch Pharm Res 31:1040–1049
119. Bourges J-L, Gautier SE, Delie F et al (2003) Ocular drug delivery targeting the retina and retinal pigment epithelium using polylactide nanoparticles. Invest Ophthalmol Vis Sci 44:3562–3569
120. De Campos AM, Diebold Y, Carvalho EL et al (2004) Chitosan nanoparticles as new ocular drug delivery systems: in vitro stability, in vivo fate, and cellular toxicity. Pharm Res 21:803–810
121. De Campos AM, Sanchez A, Alonso MJ (2001) Chitosan nanoparticles: a new vehicle for the improvement of the delivery of drugs to the ocular surface. Application to cyclosporin A. Int J Pharm 224:159–168
122. de la Fuente M, Csaba N, Garcia-Fuentes M et al (2008) Nanoparticles as protein and gene carriers to mucosal surfaces. Nanomedicine 3:845–857
123. Kayser O, Lemke A, Hernandez-Trejo N (2005) The impact of nanobiotechnology on the development of new drug delivery systems. Curr Pharm Biotechnol 6:3–5
124. Sanchez A, Alonso MJ (2006) Nanoparticular carriers for ocular drug delivery. In: Torchilin VP (ed) Nanoparticulates as drug carriers. Imperial College Press, London, pp 649–673
125. Mainardes RM, Urban MC, Cinto PO et al (2005) Colloidal carriers for ophthalmic drug delivery. Curr Drug Targets 6:363–371
126. Nagarwal RC, Kant S, Singh PN et al (2009) Polymeric nanoparticulate system: a potential approach for ocular drug delivery. J Control Release 136:2–13
127. Araújo J, Vega E, Lopes C et al (2009) Effect of polymer viscosity on physicochemical properties and ocular tolerance of FB-loaded PLGA nanospheres. Colloids Surf B Biointerfaces 72:48–56
128. Bloquel C, Bourges JL, Touchard E et al (2006) Non-viral ocular gene therapy: potential ocular therapeutic avenues. Adv Drug Deliv Rev 58:1224–1242
129. Hamidi M, Azadi A, Rafiei P (2008) Hydrogel nanoparticles in drug delivery. Adv Drug Deliv Rev 60:1638–1649
130. Sahoo SK, Dilnawaz F, Krishnakumar S (2008) Nanotechnology in ocular drug delivery. Drug Discov Today 13(3–4):144–151
131. Sahoo SK, Labhasetwar V (2003) Nanotech approaches to drug delivery and imaging. Drug Discov Today 8:1112–1120
132. Vasir JK, Reddy MK, Labhasetwar V (2005) Nanosystems in drug targeting: opportunities and challenges. Curr Nanosci 1:47–64
133. Carafa M, Santucci E, Alhaique F et al (1998) Preparation and properties of new unilamellar non-ionic surfactant vesicles. Int J Pharm 160:51–59
134. Svenson S, Tomalia D (2005) Dendrimers in biomedical applications – reflections on the field. Adv Drug Deliv Rev 57:2106–2129
135. Gaudana R, Jwala J, Boddu SHS et al (2009) Recent perspectives in ocular drug delivery. Pharm Res 26(5):1197–1216
136. Diebold Y, Calonge M (2010) Applications of nanoparticles in ophthalmology. Prog Retin Eye Res 29:596–609
137. Kaur IP, Aggarwal D, Singh H et al (2010) Improved ocular absorption kinetics of timolol maleate loaded into a bioadhesive niosomal delivery system. Graefes Arch Clin Exp Ophthalmol 248:1467–1472
138. Amrite AC, Edelhauser HF, Singh SR et al (2008) Effect of circulation on the disposition and ocular tissue distribution of 20 nm nanoparticles after periocular administration. Mol Vis 14:150–160
139. Cheruvu NP, Amrite AC, Kompella UB (2008) Effect of eye pigmentation on transscleral drug delivery. Invest Ophthalmol Vis Sci 49:333–341
140. Cheruvu NP, Amrite AC, Kompella UB (2009) Effect of diabetes on transscleral delivery of celecoxib. Pharm Res 26:404–414
141. Peeters L, Lentacker I, Vandenbroucke RE et al (2008) Can ultrasound solve the transport barrier of the neural retina? Pharm Res 25:2657–2665

142. Calvo P, Remuñán-López C, Vila-Jato JL et al (1997) Novel hydrophilic chitosan-polyethylene oxide nanoparticles as protein carriers. J Appl Polym Sci 63:125–132
143. Calvo P, Vila-Jato JL, Alonso MJ (1997) Evaluation of cationic polymer-coated nanocapsules as ocular drug carriers. Int J Pharm 153:41–50
144. Ding S (1998) Recent developments in ophthalmic drug delivery. Pharm Sci Technol Today 1(8):328–335
145. Lang JC (1995) Ocular drug delivery conventional ocular formulations. Adv Drug Deliv Rev 16:39–43
146. Meadows DL, Paugh JR, Joshi A et al (2002) A novel method to evaluate residence time in humans using a nonpenetrating fluorescent tracer. Invest Ophthalmol Vis Sci 43:1032–1039
147. Zimmer A, Kreuter J (1995) Microspheres and nanoparticles used in ocular delivery systems. Adv Drug Deliv Rev 16:61–73
148. Dobrovolskaia MA, Aggarwal P, Hall JB et al (2008) Preclinical studies to understand nanoparticles interaction with the immune system and its potential effects on nanoparticles biodistribution. Mol Pharmacol 5:487–495
149. Dobrovolskaia MA, McNeil SE (2007) Immunological properties of engineered nanomaterials. Nat Nanotechnol 2:469–478
150. Aggarwal P, Hall JB, McLeland CB et al (2009) Nanoparticle interaction with plasma proteins as it relates to particle biodistribution, biocompatibility and therapeutic efficacy. Adv Drug Deliv Rev 61:428–437
151. Gref R, Minamitake Y, Peracchia MT, Trubetskoy V, Torchilin V, Langer R (1994) Biodegradable long-circulating polymeric nanospheres. Science 263:1600–1603
152. Couvreur P, Tulkenst P, Roland M et al (1977) Nanocapsules: a new type of lysosomotropic carrier. FEBS Lett 84:323–326
153. Bareford LM, Swaan PW (2007) Endocytic mechanisms for targeted drug delivery. Adv Drug Deliv Rev 59:748–758
154. Frenkel V (2008) Ultrasound mediated delivery of drugs and genes to solid tumors. Adv Drug Deliv Rev 30:1193–1208
155. Hillaireau H, Couvreur C (2009) Nanocarriers' entry into the cell: relevance to drug delivery. Cell Mol Life Sci 66:2873–2896
156. Harush-Frenkel O, Altschuler Y, Benita S (2008) Nanoparticle-cell interactions: drug delivery implications. Crit Rev Ther Drug Carrier Syst 25:485–544
157. Soppimath KS, Aminabhavi TM, Kulkarni AR et al (2001) Biodegradable polymeric nanoparticles as drug delivery devices. J Control Release 70:1–20
158. Saishin Y, Silva RL, Saishin Y et al (2003) Periocular injection of microspheres containing PKC412 inhibits choroidal neovascularization in a porcine model. Invest Ophthalmol Vis Sci 44:4989–4993
159. Barcia E, Herrero-Vanrell R, Díez A et al (2009) Downregulation of endotoxin-induced uveitis by intravitreal injection of polylactic-glycolic acid (PLGA) microspheres loaded with dexamethasone. Exp Eye Res 89:238–245
160. Merodio M, Arnedo A, Renedo MJ et al (2001) Ganciclovir-loaded nanoparticles: characterization and in vitro release properties. Eur J Pharm Sci 12:251–259
161. Merodio M, Espuelas MS, Mirshahi M et al (2002) Efficacy of ganciclovir-loaded nanoparticles in human cytomegalovirus (HCMV)-infected cells. J Drug Target 10:231–238
162. Merodio M, Irache JM, Valamanesh F et al (2002) Ocular disposition and tolerance of ganciclovir-loaded albumin nanoparticles after intravitreal injection in rats. Biomaterials 23:1587–1594
163. Irache JM, Merodio M, Arnedo A et al (2005) Albumin nanoparticles for the intravitreal delivery of anticytomegaloviral drugs. Mini Rev Med Chem 5:293–305
164. Bejjani RA, BenEzra D, Cohen H et al (2005) Nanoparticles for gene delivery to retinal pigment epithelial cells. Mol Vis 11:124–132
165. Normand N, Valamanesh F, Savoldelli M et al (2005) VP22 light controlled delivery of oligonucleotides to ocular cells *in vitro* and *in vivo*. Mol Vis 11:184–191

166. Conley SM, Naash MI (2010) Nanoparticles for retinal gene therapy. Prog Retin Eye Res 29(5):376–397
167. de Kozak Y, Andrieux K, Villarroya H et al (2004) Intraocular injection of tamoxifen-loaded nanoparticles: a new treatment of experimental autoimmune uveoretinitis. Eur J Immunol 34:3702–3712
168. El-Samaligy MS, Rojanasakul Y, Charlton JF et al (1996) Ocular disposition of nanoencapsulated acyclovir and ganciclovir via intravitreal injection in rabbit's eye. Drug Deliv 3:93–97
169. Christoforidis JB, Chang S, Jiang A et al (2012) Intravitreal devices for the treatment of vitreous inflammation. Mediators of Inflammation 2012 (ID 126463):8 pages.
170. Kadam RS, Tyagi P, Edelhauser HF et al (2012) Influence of choroidal neovascularization and biodegradable polymeric particle size on transscleral sustained delivery of triamcinolone acetonide. Int J Pharm 434:140–147
171. Panda JJ, Yandrapu S, Kadam RS et al (2013) Self-assembled phenylalanine-α, β-dehydrophenylalanine nanotubes for sustained intravitreal delivery of a multi-targeted tyrosine kinase inhibitor. J Control Release 172:1151–1160
172. Gan L, Wang J, Liu J et al (2013) HA-modified core–shell liponanoparticles for efficient intravitreal drug delivery. J Control Release 172:e48
173. Camelo S, Lajavardi L, Bochot A et al (2009) Protective effect of intravitreal injection of vasoactive intestinal peptide-loaded liposomes on experimental autoimmune uveoretinitis. J Ocul Pharmacol Ther 25:9–21
174. Lajavardi L, Bochot A, Camelo S et al (2007) Downregulation of endotoxin-induced uveitis by intravitreal injection of vasoactive intestinal peptide encapsulated in liposomes. Invest Ophthalmol Vis Sci 48:3230–3238
175. Ulbrich W, Lamprecht A (2010) Targeted drug-delivery approaches by nanoparticulate carriers in the therapy of inflammatory diseases. J R Soc Interface 7(Suppl 1):S55–S66
176. Lajavardi L, Camelo S, Agnely F et al (2009) New formulation of vasoactive intestinal peptide using liposomes in hyaluronic acid gel for uveitis. J Control Release 139:22–30
177. Haghjou N, Soheilian M, Abdekhodaie MJ (2011) Sustained release intraocular drug delivery devices for treatment of uveitis. J Ophthalm Vis Res 6:317–329
178. Ryu M, Nakazawa T, Akagi T et al (2011) Suppression of phagocytic cells in retinal disorders using amphiphilic poly(γ-glutamic acid) nanoparticles containing dexamethasone. J Control Release 151:65–73
179. Sakai T, Kohno H, Ishihara T et al (2006) Treatment of experimental autoimmune uveoretinitis with poly(lactic acid) nanoparticles encapsulating betamethasone phosphate. Exp Eye Res 82:657–663
180. Hashida N, Ohguro N, Yamazaki N et al (2008) High-efficacy site-directed drug delivery system using sialyl-Lewis X conjugated liposome. Exp Eye Res 86:138–149
181. McEver RP (1992) Leukocyte-endothelial cell interactions. Curr Opin Cell Biol 4:840–849
182. Foxall C, Watson SR, Dowbenko D et al (1992) The three members of the selectin receptor family recognize a common carbohydrate epitope, the sialyl Lewis (x) oligosaccharide. J Cell Biol 117:895–902
183. Polley MJ, Phillips ML, Wayner E et al (1991) CD62 and endothelial cell-leukocyte adhesion molecule 1 (ELAM-1) recognize the same carbohydrate ligand, sialyl-Lewis X. Proc Natl Acad Sci U S A 88:6224–6228
184. Zeimer R, Goldberg MF (2001) Novel ophthalmic therapeutic modalities based on noninvasive light-targeted drug delivery to the posterior pole of the eye. Adv Drug Deliv Rev 52:49–61
185. Ideta R, Tasaka F, Jang W-D et al (2005) Nanotechnology-based photodynamic therapy for neovascular disease using a supramolecular nanocarrier loaded with a dendritic photosensitizer. Nano Lett 5:2426–2431
186. Paasonen L, Laaksonen T, Johans C et al (2007) Gold nanoparticles enable selective light-induced contents release from liposomes. J Control Release 11:86–93

187. Farjo R, Skaggs J, Quiambao AB et al (2006) Efficient nonviral ocular gene transfer with compacted DNA nanoparticles. PLoS One 1(1):e38.
188. Edelhauser HF, Rowe-Rendleman CL, Robinson MR et al (2010) Ophthalmic drug delivery systems for the treatment of retinal diseases: basic research to clinical applications. Invest Ophthalmol Vis Sci 51(11):5403–5420
189. Jiang J, Moore JS, Edelhauser HF (2009) Intrascleral drug delivery to the eye using hollow microneedles. Pharm Res 26:395–403
190. Sustained release ocular drug delivery systems market, 2014–2024. Press Release, 15 May 2014. http://www.prnewswire.com/news-releases/sustained-release-ocular-drug-delivery-systems-market-2014-2024-259368521.html
191. Pasadhika S, Suhler EB, Cunningham ET (2010) Use of biologic agents in the treatment of uveitis: these potent new agents may help some patients who are refractory to more traditional immunosuppressive therapies. Rev Ophthalmol 10(1). http://www.reviewofophthalmology.com/content/d/retinal_insider/i/1208/c/22770/
192. Kraus CL, Culican SM (2012) Use of biologic agents in ocular manifestations of rheumatic disease. Int J Rheumatol 2012(Article ID 203819):6 pages.
193. Bushell-Embling (2014) pSivida implant provides sustained antibody release. Medical devices, 08 May 2014. http://lifescientist.com.au/content/biotechnology/news/psivida-implant-provides-sustained-antibody-release-28600494#sthash.Rv8YjuQq.dpuf
194. Ullrich F, Bergeles C, Pokki J et al (2013) Mobility experiments with microrobots for minimally invasive intraocular surgery. Invest Ophthalmol Vis Sci 54:2853–2863
195. Helzner J (ed) (2010) Sustained-release drugs: heralds of the future. 3 January 2010. http://www.ophthalmologymanagement.com/articleviewer.aspx?articleid=104070

Chapter 19
Animal Models to Evaluate Nanoparticulate Ocular Drug Delivery Systems

Amy Broadwater, Anjali Hirani, and Yashwant Pathak

Abstract Ophthalmic injuries and diseases can be devastating, especially when eyesight is at risk. Steps have been taken to enhance treatment options for patients with conditions such as glaucoma or fungal eye infections, to name a few. Nanoparticle ophthalmologic drug delivery offers a chance for a more effective treatment against many common eye malignancies by increasing the efficacy of many current eye medications. Additionally, nanoparticles open a door to new treatment approaches, such as sustained release for chronic conditions or for drugs with low patient compliance, by discovering new drug delivery methods. Without the use of animal models in medical research, we would not be able to take great strides in the direction of optimum patient health. Evaluation of these modern drug delivery systems in animal models shows great promise for the future of ophthalmologic health.

Keywords Ocular drug • Animal studies • Nanoparticles • Glaucoma • Ocular infection • Ocular cancer

19.1 Nanoparticle Drug Delivery in Animal Models

19.1.1 Ocular Surface

Typically, eye drops are used to topically administer medications to the ocular surface. However, the protective barriers of the eye usually reduce the bioavailability of the drug significantly [1]. The corneal and conjunctival epithelium, sealed with

A. Broadwater
College of Pharmacy, University of South Florida, Tampa, FL, USA

Y. Pathak (✉)
Department of Pharmaceutical Sciences, USF College of Pharmacy, University of South Florida, Tampa, FL 33612, USA
e-mail: ypathak1@health.usf.edu

A. Hirani
Department of Pharmaceutical Sciences, USF College of Pharmacy, University of South, Tampa, FL 33612, USA

School of Biomedical Engineering and Sciences, Virginia Tech-Wake Forest University, Blacksburg, VA 24061, USA

© Springer International Publishing Switzerland 2016

Y. Pathak et al. (eds.), *Nano-Biomaterials For Ophthalmic Drug Delivery*,
DOI 10.1007/978-3-319-29346-2_19

tight junctions, and tear drainage through the nasolacrimal duct function as barriers of the eye [2, 3]. These mechanisms make it difficult for both drug adhesion to the ocular surface as well as surface penetration.

Nanoparticles can be used to drastically increase the bioavailability of medications applied to the eye surface. One type of nanomaterial, chitosan, is preferred for topical drug delivery. Chitosan is a positively charged molecule that can interact with the negative charge of the eye surface, increasing its adhesive properties. Additionally, it is biodegradable and nontoxic, making it safe for human use [3].

The usage of chitosan may vary greatly depending on the intended effects. For example, one study developed a pluronic-chitosan micelle which carried metipranolol, a lipophilic nonselective β adrenergic blocking agent which was easily carried in the micelles' hydrophobic core [1]. This drug is typically utilized to decrease intraocular pressure (IOP), a common cause of glaucoma. The nanoparticle was tested on New Zealand white rabbits; the results show that the nanoparticle treatment was therapeutically active even 6 h after application, while commercially available eye drops were no longer effective after 2 h. Overall, the micelles showed sustained-release properties as well as an improved therapeutic response, making them a potential candidate in treating glaucoma and increasing patient compliance [1].

Another study used liposome-chitosan nanoparticle complexes to test for ocular bioavailability and ability to cross the barriers of the eye. The nanoparticles were applied to albino New Zealand rabbits' eyes. After testing, it was found that there was no damage to the eye or eyelid tissue of the rabbits. In addition, the rabbits showed no signs of discomfort during the application period [2].

Fluorescent microscopy of the ocular tissue revealed that the liposome-chitosan nanoparticles were able to cross the physical barrier of the rabbit eye and enter into the corneal and conjunctival cell layers according to their composition [2]. The results of the study point to these nanoparticle complexes as strong candidates for topical ocular drug delivery by combining the potency of the chitosan with the surface-permeating properties of the liposome.

19.1.2 Intraocular

19.1.2.1 Polylactide Nanoparticles

Intraocular drug delivery is one of the greater challenges in ocular disease treatment, with the retina and retinal pigment epithelium being one of the harder areas for drugs to reach. One of the solutions to this problem is the possibility of using nanoparticles as drug delivery vehicles to successfully reach the retina. A benefit of nanoparticles is their ability to create sustained release and increase intracorneal and transcorneal penetration of medications.

In treating the retina, one particular type of nanoparticle is of interest: polylactides. These molecules have the ability to carry a variety of drugs and can be changed to achieve a certain rate of degradation [4]. A study in rat eyes showed the effective approach of using polylactides to reach the retina.

In the study, the right eye of 88 male rats was injected with one of the following: loaded polylactide nanoparticles, blank nanoparticles, free drug, or saline solution. The results showed that after 1 day, the loaded nanoparticles were accumulating in the retinal pigment epithelium. After approximately 4 months, the results were even more drastic: the nanoparticles were not found in significant concentrations anywhere besides the retinal pigment epithelium [4].

Targeting of the nanoparticles was successful in that they were able to accumulate in the retinal layer. Usage of polylactides as an intraocular drug delivery system targeted at the retina may create a new method of retinal disease treatment. Additionally, the nanoparticles were found to still be in the retinal cell layer 4 months after injection, creating a possibility for treatment of chronic retinal diseases [4].

19.1.2.2 Compacted DNA Nanoparticles

Treatment of many ocular diseases revolves around surgery and/or topical drugs. However, recovery from diseases would vastly improve if the underlying cause of the disorder could be fixed alternatively. Gene therapy offers a route to, not only treat the symptoms or slow the progression of ocular diseases, but to cure it altogether. Currently, viral routes are being utilized to deliver DNA or RNA to diseased cells. Unfortunately, this method comes with a lot of baggage, such as inflammation or infection with disease, and limitations such as size of the gene being transferred or cellular protection mechanisms [5].

Many ocular diseases' genetic cause is known in the medical field so the ability to provide gene treatment is possible. However, instead of utilizing viruses as vehicles for gene treatment, there is a new focus on nanoparticle drug delivery instead. A proven method of gene transfer is with compacted DNA nanoparticles: a neutral complex composed of a single strand of plasmid DNA and polyethylene glycol-substituted lysine particles [5]. Compacted DNA nanoparticles have been studied in mouse eyes to test the ability for gene transfer to various eye tissues.

Balb/cJ mice were studied after injection with the aforementioned nanoparticles. Two days after injection, the eyes were enucleated and preserved for tissue study. It was shown that the nanoparticles were in high concentrations in the lens, and in significant concentrations in other eye tissues, including the retina [5]. Additionally, no inflammation was found in the eye tissue 2 days after injection.

The study showed that the compacted DNA nanoparticles were very effective in gene delivery to various tissues of the eye. The retina, retinal pigment epithelium, lens, and other types of tissue showed increased gene expression postinjection. The nanoparticles could be potentially used to treat many types of ocular diseases in which the genetic mutation is known.

19.1.2.3 Poly(Caprolactone) Films

Polycaprolactone (PCL) is a commonly used polymer in medicine because of its unique degradation. PCL can remain structurally sound throughout its elimination primarily because it can only be degraded by a randomized hydrolytic chain cut, and secondly, only monomers and/or dimers of the polymer can be eliminated [6]. Because of these properties, the film may be ideal for sustained drug delivery for chronic conditions. A study was done to test the film devices in vivo in rabbit eyes.

New Zealand white female rabbits were obtained to study the PCL films. The films were surgically inserted either in the aqueous humor or the vitreous humor of the rabbits, and the conditions were observed over a period of 6 months. It was shown that there was inflammation throughout the first month postoperation, but this inflammation was absent from months 1 to 6. Additionally, there were no toxic effects, signs of degeneration, or morphological changes within the rabbit eye [6].

SEM studies on the PCL nanosheet showed that the sheet was still structurally sound up to 9 months within the eye [6]. Because this new technology is both structurally sound and carries little to no known long-term side effects, it is a real possibility that the nanosheet could be utilized in the future to create a new type of technology that retains its structural integrity over time.

19.2 Ocular Disease Models: Nanoparticle Treatments

19.2.1 Conjunctival/Corneal Squamous Cell Carcinoma in Rabbit Eye

Conjunctival/corneal squamous cell carcinoma (CCSC) is the most common eye cancer in the United States; one to three in every 100,000 suffer from the malignancy annually [7]. The cancer has been linked to UV radiation which deactivates the tumor suppressor gene p53; therefore, regions that are sunnier tend to have a higher occurrence of the disease.

Tumors tend to be in the interpalpebral area of the conjunctiva and show up commonly as nodular or gelatinous, but can show up as other tumor growth patterns [7]. Current treatment options for the disease include removal of the tumor, topical chemotherapy, cryotherapy, and/or radiotherapy. Typically, simple surgical excision is not sufficient to eliminate the disease. If the invasion is too severe, the eye may have to be completely removed.

In the study by Nagarwal et al., the rabbit eye was treated with a drug called 5-fluorouracil [8]. 5-fluorouracil's labeled use is management of superficial basal cell carcinomas [9]. The drug works by interfering with cells in S-phase of the cell cycle, or the time when the DNA is being actively synthesized to prepare for mitosis. 5-fluorouracil is an antimetabolite that blocks the methylation of deoxyuridylic acid, inhibits thymidylate synthesis, or is integrated into RNA [9]. The properties of this drug make it a good candidate for the treatment of CCSC; additionally it has little to no side effects with normal eye tissue.

The drug delivery method chosen was nanoparticles because of their excellent targeting properties. Typically, with a normal topical solution, the drug could reach systemic circulation by absorbing through mucous membranes after traveling through the nasolacrimal duct [8]. By choosing a nanoparticle system, the likelihood of absorption into circulation is decreased. Another positive quality of using nanoparticles would be increasing the bioavailability of the drug, enhancing cellular uptake of the drug thereby reducing clearance and sustaining drug delivery [8].

The nanoparticles were also modified to form a gel in situ. A gel was selected because of its ability to stay on the surface of the eye for extended periods of time which could lower the frequency of dose needed and increase the penetration of the drug. The nanoparticles were suspended in a polymeric solution and, once put onto the surface of the eye, the new chemical environment caused the solution to form a gel with nanoparticles suspended throughout.

In the experiment, three New Zealand albino rabbits of varying sex were selected for study. There were four installations of the nanoparticle 5-fluorouracil system into their right eyes with saline solution used as a control into their left eyes. Aqueous humor was drawn from their eyes after certain increments of time to keep track of the concentration of the drug within their eye. The concentration of the drug within the aqueous humor was measured with liquid chromatography, and the physical state of the rabbit eyes, including damage to the iris, lens, and cornea and congestion, swelling, and discharge of the conjunctiva, was observed and recorded [8].

The concentration of drug over time was tested using pharmacokinetic parameters. The peak plasma concentration of the 5-fluorouracil nanoparticles was shown to be almost five times greater than the same measurement of a typically used 5-fluorouracil solution. The results also showed very high ocular bioavailability and a higher half-life, suggesting greater retention of the drug.

The results of the study of 5-fluorouracil in the rabbit eye shows that a nanoparticle in situ gel-forming system is a potent tool for the delivery of drugs into the eye. It allows a sustained release of medication and a longer period of time for elimination. Finally, the specific area of the eye can be targeted to avoid widespread effects.

19.2.2 Treatment of Fungal Keratitis in Rabbit Eye

Fungal keratitis is a serious disease caused by fungi invading corneal tissue damaged by trauma or disease, including species *Curvularia*, *Aspergillus*, *Candida*, and *Fusarium* [10]. These species of fungi usually exist in tropical climates. Additionally, the disease can be caused by: wearing contact lenses, prolonged chemotherapy, or immunosuppression [11]. The disease, if left untreated, can cause a loss of vision.

Typical treatments at this time vary. Oral treatment usually requires high doses in order to reach therapeutic concentrations, but this can also cause unwanted side effects. Additionally, injections into the eye are possible, but are very uncomfortable for the patient [11]. The worst case scenario for treatment of the disease would be a corneal transplant or complete eye removal [10]. For this reason, aggressive and prolonged treatment of the disease is necessary to provide a chance of recovery.

In a study by Tayel et al., terbinafine hydrochloride (THCl) was used to treat fungal keratitis in rabbits. THCl works by inhibiting squalene epoxidase, an enzyme involved in sterol biosynthesis. Inhibition of this enzyme decreases the amount of ergosterol in the fungal cell wall, which results in lysis and cell death [9].

The nanoparticles in which the drug was loaded, were specially created for delivery into the cornea. The cornea and conjunctiva have overall negative charges, so a positive nanoparticle from chitosan and Eudragit® RS100 was made to interact with the eye surfaces and maintain effective eye concentrations. Chitosan nanoparticles have many desirable properties, including stability, low toxicity, and ability to facilitate intraocular drug delivery [10].

The animals selected for the experiment were 20 male albino New Zealand rabbits. The rabbits were split into two groups: one group was given the nanoparticle suspension in the eye and the other group was given the drug in an oily vehicle in the eye [10]. After certain increments of time, aqueous humor was drawn from each of the rabbits and an ophthalmologist studied the eyes of the rabbits for signs of damage or inflammation. Over the course of the study, ten of the twenty rabbits were eliminated because of severe damage to avoid negative influence on the end results. The aqueous humor was run through high-performance liquid chromatography to graph the concentration of the drug over time [10].

The peak plasma concentration for the oil-vehicle drug was higher than that of the nanoparticle drug; however, the nanoparticles displayed much more of a controlled-release character. The ionic interactions between the nanoparticles and the eye surfaces could have been an explanation for this. Additionally, the bioavailability of the nanoparticles was found to be higher than that of the oil because of the colloidal carrier quality of nanoparticles [10].

Based on the results of the study in the rabbits, the developed nanoparticles could be used as an excellent vehicle for ophthalmic drug delivery; however, more studies should be conducted to decrease the toxicity and increase the chemical stability of the drug in vivo [10].

19.2.3 Treatment of Glaucoma in Rats and Rabbits

Glaucoma is the second leading cause of vision loss in the United States [12]. There are two different types of glaucoma: primary open-angle glaucoma (POAG) and angle-closure glaucoma (ACG); with POAG being the more dangerous of the two because of its seemingly sudden onset [12].

Glaucoma causes irreversible eye damage and POAG cannot be cured by neither surgery nor transplant. The disease's causes are unknown, but there has been a correlation between high intraocular pressure (IOP) and high chances of glaucoma [13]. Therefore, the primary treatment for glaucoma is a multidrug regimen, usually eye drops, which work to reduce the IOP, thus slowing the progression of the disease. The problem with this method of treatment is that many patients do not adhere to the medication regimen and their glaucoma progresses [14].

As a way to enhance patient compliance as well as patient care, a study was created in which a latanoprost-loaded nanosheet (LLNS) was made to treat glaucoma [14]. Latanoprost is a common antiglaucoma drug, which works to decrease the IOP by acting as a prostaglandin F2-alpha analog. The IOP is decreased by increasing the outflow of aqueous humor from the eye [9].

The LLNS that was created was tested on both rat cornea and rabbit eyes. The LLNS was applied to rat eye to study the local adverse effects of the drug. It was shown that the rat eyes were subject to hyperemia after only 6 h, but this condition quickly disappeared. The conditions were then monitored for erosion, edema, inflammation, and lens opacity. Overall, there were no severe adverse effects of the medication on the rat eye throughout the trial period [14].

Japanese white rabbits were selected to test the release of latanoprost from the LLNS into the aqueous humor of the eye. Samples of aqueous humor were taken in increments and the concentrations of drug were recorded. The findings show that the LLNS had longer linger time in the aqueous humor than 0.005 % latanoprost drops did [14].

One application of LLNS to the eye was sufficient to find humoral concentrations a week after application; however, the concentrations after a day fell below therapeutic levels, and decreased daily thereafter. Nevertheless, the LLNS was found to decrease IOP for a week after administration and offer a new way to increase patient adherence and, subsequently, patient outcome.

19.2.4 Inhibition of Corneal Angiogenesis

Angiogenesis is the process of growing new blood vessels from existing ones [15]. This process is a normal and necessary one; it improves blood flow to damaged or new tissue. However, if the process grows out of control, it can lead to serious conditions such as diabetes and cancer [16].

Changes in metabolic activity lead to regression of angiogenesis, naturally, and scientists use gene therapy to prevent it therapeutically. Typically, gene therapy is delivered to the patient through a virus, because it is capable of inserting the DNA it carries into live cells while bypassing membrane transport proteins. However, this type of treatment may have some adverse reactions as well as risks, such as inflammation and possible infection by the virus.

Instead, albumin can be used as a better vehicle for gene therapy. Albumin is a naturally-occurring serum protein already used for the delivery of many types of chemotherapy. Additionally, it can be used for the trapping and sustained release of DNA in the inhibition of corneal angiogenesis.

In the eye, angiogenesis can be caused by physical or chemical injury. In a study on the prevention of angiogenesis in damaged tissue, (before an injury occurs) albumin nanoparticles encapsulating pCMV.Flt23K were injected into a mouse cornea [16]. pCMV.Flt23K is a gene that interacts with cell receptors, preventing angiogenesis.

The eyes were injected with 1 μg/mL nanoparticulate solution 3 weeks prior to the injury to the eye. The eyes were monitored for toxicity the 3 weeks before the injury, and none was found. At 3 weeks, the eyes underwent alkali mechanical trauma and were monitored for subsequent angiogenesis [16].

The corneal angiogenesis assessment after injury showed that the eyes with the nanoparticle gene therapy exhibited almost half as much vessel growth as eyes treated with free genes, empty nanoparticles, and simple saline solution [16]. Additionally, the nanoparticles showed sustained release and prevented angiogenesis long after the time of injury.

These findings could be applied to avoid possible negative effects from an eye surgery or transplant. Presurgical treatment with this drug could prevent outcomes like cancer, and long-term treatment with the medication could slow the progression of long-term injuring diseases [16].

19.3 Conclusion

In conclusion, the use of nanoparticle ocular drug delivery systems is an important new way to treat ocular diseases. Nanoparticle treatments are advantageous because of their targeting abilities and their qualities based on physical shape. They have the ability to penetrate into cells and work directly from within. Additionally, their special properties such as adhesion, degradation, and bonding make them excellent for harder-to-treat diseases or harder-to-transport medications.

Nanoparticles show much promise in the studies using animal models. The toxicity studies of the nanoparticles and the medications have shown that adverse reactions are rarely linked to the usage of nanoparticle systems. Additionally, the therapeutic responses of the nanoparticle ODD systems typically show that the modern nanoparticle vehicle is conducive to a greater response than the typical commercially available drug. Overall, animal studies have shown that nanoparticle ODD is an excellent method of treating ocular diseases and hopefully will be integrated into health care in the near future.

References

1. Lin HR, Chan PC (2013) Novel pluronic-chitosan micelle as an ocular delivery system. J Biomed Mater Res B Appl Biomater 101B(5):689–699
2. Diebold Y, Jarrín M, Sáez V, Carvalho EL, Orea M, Calonge M et al (2007) Ocular drug delivery by liposome-chitosan nanoparticle complexes (LCS-NP). Biomaterials 28:1553–1564
3. Salamanca AE et al (2006) Chitosan nanoparticles as a potential drug delivery system for the ocular surface: toxicity, uptake mechanism and in vivo tolerance. Investig Ophthalmol Vis Sci 47(4):1416–1425
4. Bourges JL et al (2003) Ocular drug delivery targeting the retina and retinal pigment epithelium using polylactide nanoparticles. Investig Ophthalmol Vis Sci 4(8):3562–3569

5. Farjo R et al (2006) Efficient non-viral ocular gene transfer with compacted DNA nanoparticles. PLoS One 1(1):e38
6. Bernards DA et al (2013) Ocular biocompatibility and structural integrity of micro- and nano-structured poly(caprolactone) films. J Ocul Pharmacol Ther 29(2):249–257
7. Tunc M et al (1999) Intraepithelial and invasive squamous cell carcinoma of the conjunctiva: analysis of 60 cases. Br J Ophthalmol 83:98–103
8. Nagarwal RC et al (2011) Modified PLA nano in situ gel: a potential ophthalmic drug delivery system. Colloids Surf B Biointerfaces 86(1):28–34
9. Lexi-Comp, Inc. (Lexi-DrugsTM) (2014) Lexi-Comp, Inc.
10. Tayel SA, El-Nabarawi MA, Tadros MI, Adb-Elsalam WH (2013) Positively charged polymeric nanoparticles of terbinafine hydrochloride: preclinical implications for controlled drug delivery in the aqueous humor of rabbits. AAPS PharmSciTeh 14(2):782–793
11. Gratieri T et al (2011) Enhancing and Sustaining the topical ocular delivery of fluconazole using chitosan solution and poloxamer/chitosan in situ forming gel. Eur J Pharm Biopharm 79(2):320–327
12. http://www.glaucoma.org/glaucoma/. Accessed 25 Oct 2014
13. http://www.mayoclinic.org/diseases-conditions/glaucoma/basics/causes/con-20024042. Accessed 25 Oct 2014
14. Kashiwagi K et al (2013) Development of latanoprost-loaded nanosheet as a new drug delivery system. Investig Ophthalmol Vis Sci 54(8):5629–5637
15. Adair TH, Montani JP (2010) Angiogenesis. Morgan and Claypool Life Sciences, San Rafael
16. Jani PD, Singh N, Jenkins C, Raghava S, Mo Y, Amin S et al (2007) Nanoparticles sustain expression of Flt interceptors in the cornea and inhibit injury-induced corneal angiogenesis. Invest Ophthalmol Vis Sci 48:2030–2036

Chapter 20
Ophthalmic Nanosuspensions: Toxicity and Formulation

Preya Patel, Aditya Grover, Anjali Hirani, Yong W. Lee, and Yashwant Pathak

Abstract Treatment of posterior regions of the eye is challenging due to the difficulties faced with overcoming the blood-retinal barrier and the blood-aqueous barrier. In efforts to find a nontoxic and less invasive approach, nanoparticles have shown significant potential based on their size and composition. Many of the vectors discussed, such as PLGA, are biocompatible. Eudragit nanoparticles show low levels of toxicity in ocular models. Chitosan used in nanoparticle formulation is derived from chitin, an organic source. Additionally, human serum albumin-based nanoparticles show effective drug delivery through in vivo studies. The antibacterial properties of silver and its accessibility have led to studies investigating its potential with nanoparticles. Liposomes and nanomicelles interact similarly as they are able to entrap hydrophobic drugs for delivery. With proper use of these vectors in terms of formulation of nanosuspensions, there is potential for treating ocular diseases. Furthermore, the biological breakdown of these

P. Patel
College of Arts and Sciences, University of South Florida, Tampa, FL 33612, USA

Department of Pharmaceutical Sciences, College of Pharmacy, University of South Florida, Tampa, FL 33612, USA

A. Grover
Department of Pharmaceutical Sciences, College of Pharmacy, University of South Florida, Tampa, FL 33612, USA
e-mail: ypathak1@health.usf.edu

A. Hirani
Department of Pharmaceutical Sciences, USF College of Pharmacy, University of South, Florida, Tampa, FL 33612, USA

School of Biomedical Engineering and Sciences, Virginia Tech-Wake Forest University, Blacksburg, VA 24061, USA

Y.W. Lee
School of Biomedical Engineering and Sciences, Virginia Tech-Wake Forest University, Blacksburg, VA 24061, USA

Y. Pathak (✉)
Department of Pharmaceutical Sciences, USF College of Pharmacy, University of South Florida, Tampa, FL 33612, USA
e-mail: ypathak1@health.usf.edu

© Springer International Publishing Switzerland 2016
Y. Pathak et al. (eds.), *Nano-Biomaterials For Ophthalmic Drug Delivery*,
DOI 10.1007/978-3-319-29346-2_20

nanosuspensions emphasizes their biocompatibility. Incorporating other vectors and biomolecules will allow for new avenues with ocular drug delivery and treatment.

Keywords Eye • Nanoparticle • Blood-aqueous barrier • Blood-retinal barrier • Eudragit • Chitosan • Albumin • Silver • PLGA • Liposomes • Nanomicelles

20.1 Introduction

The anatomy of the eye complicates the delivery of drugs through the many ocular barriers protecting the posterior segment. Because of this, diseases pertaining to the posterior segment of the eyes result in the greatest incidences of blindness in the industrial world [1]. Two of the more prominent ocular barriers that act as obstacles to drug delivery are the (1) blood-retinal barrier (BRB) between systemic blood and the retina and (2) blood-aqueous barrier (BAB) between systemic blood and the anterior chamber of the eye. The physiology of these barriers results from the interactions of retinal pigment epithelium (RPE) with retinal capillary walls in the BRB and uveal capillary endothelia with ciliary epithelia in the BAB [1].

Current drug delivery methods differ for the anterior eye and posterior eye. Eye drops are used as the primary route of drug administration for the anterior eye; however, only around 5 % of the drug is absorbed by the eye and the rest enters systemic circulation through the nasal and conjunctival route [1, 2]. The corneal barrier limits the absorption of drug delivered via eye drops to the anterior chamber of the eye and the maximum amount of drug absorbed by the anterior eye via the topical route is around 10 % [2]. Little to no drug is able to target the posterior segment of the eye via the topical route, which leads to the application of alternate methods of drug delivery. Intravitreal injection is the method of choice for targeting the posterior segment of the eye. However, the invasive nature of intravitreal injections creates a number of unwanted side effects, including corneal detachment and ocular inflammation, associated with repeated injections to achieve therapeutic concentrations of the drug in the eye. In addition, it is common for drugs injected intravitreally to be absorbed into systemic circulation via retinal capillaries and RPE [3]. Furthermore, high concentrations of drugs are required to be administered systemically to target the posterior eye, thereby exposing healthy cells and tissues to potential toxicity and side effects [3].

Nanoparticles have the potential to solve the problems faced by drug delivery to posterior segments of the eye. The size range of nanoparticles (1–1000 nm) is suitable for injection. They can be composed of a number of different types of materials which impart various surface chemistries and determine the nanoparticles' interactions with drugs and tissues. Herein, we will describe popular nanoparticle vectors and their toxicities in ophthalmic drug delivery, as well as common formulation methods.

20.2 Eudragit

Acrylate polymers such as Eudragit have been used for decades in a number of forms including nanoparticles and tablets to maximize the effects of drugs, offer drug stability, and exhibit controlled drug release profiles in tissues. Eudragit nanoparticles are commonly used topically in ophthalmic drug administration. Physical characteristics of the nanoparticles can be altered by using different copolymers; this influences surface chemistry and pH, charge, solubility, and permeability [4].

A number of studies in rabbit models have found that Eudragit nanoparticles are nontoxic for ocular drug delivery. Pignatello et al. used the Druze test and methylene blue for ocular irritability to investigate the effect of blank Eudragit nanoparticles in the rabbit model. The nanoparticles were formulated by utilizing a quasi-emulsion solvent diffusion (QESD) method. The resultant nanoparticles were treated to the eyes by topical application. There were no reported incidences of inflammation or discomfort in the rabbit conjunctiva for up to 24 h after treatment as determined by slit lamp [5].

The same group tested similar Eudragit nanoparticles loaded with flurbiprofen, a nonsteroidal anti-inflammatory drug, in the rabbit eye model. Flurbiprofen (0.03 % w/v) was loaded in a number of nanoparticle formulations ranging in size from 50.4 to 96.4 nm, and were made according to the QESD technique described above. The nanoparticles were instilled onto rabbit eyes every half hour for 6 h. Similarly, no toxicity was observed in the rabbit eyes for up to 24 h after drug administration [6].

Low toxicity was also observed in rabbits treated with ibuprofen-loaded Eudragit nanoparticles as treatment for an ocular trauma model of paracentesis. The QESD technique was also utilized in the study for the formulation of nanoparticles. Rabbits treated with drug Eudragit nanoparticles showed drug permeation into aqueous humor and inhibition of inflammatory mitotic responses apart from no toxic response [7]. Numerous other studies have reported similar low-toxic effects of Eudragit in ocular models, establishing this vector as a suitable ophthalmic drug delivery carrier.

QESD is a popular formulation method for the precipitation of Eudragit nanoparticles. The miscibility of both phases is essential for the proper formulation of nanoparticles from this method, as the first phase is injected into the second. Cold temperatures, preferably by use of ice bath, are maintained when the first phase is injected into the water phase so as to not allow for the rapid evaporation of ethanol out of water phase before the proper formation of nanoparticles. As the solution is agitated to yield a pseudoemulsion and ethanol is subsequently evaporated, the nanoparticles precipitate out of solution and are suspended in the water phase. The stabilization of the precipitated nanoparticles is facilitated by the supplementation of the water phase by surfactants such as Tween 80 and charged species such as benzalkonium chloride.

20.3 Chitosan

Chitosan nanoparticles have been widely studied for ophthalmic drug delivery. The nano size of chitosan nanoparticles make this model suitable for cellular uptake. In addition, the vector exhibits low toxicity and controlled drug release [8].

Chitosan vector is derived from the modification of chitin found in crustacean exoskeletons, giving chitin the distinction of being considered an organic nanoparticle vector. Hydrogen bonds formed through the interaction of protein functional groups present on the chitosan vector with glycoprotein in mucous lead to adhesion and increased bioavailability of the nanoparticle. The adhesive effect can be strengthened by altering molecular weight and deacetylation during the formation of chitosan [8]. Chitosan is biodegradable and is cleared through the renal route; the rate of degradation in vivo is directly dependent on the rate of chitosan deacetylation [9].

Chitosan nanoparticles are reported to be well tolerated in ocular models. Histological analysis of rabbit ocular tissue treated topically with albumin-conjugated chitosan nanoparticles revealed the lack of necrosis, through normal conjunctival- and corneal tissue-exhibited morphology and proliferation, without any evidence of edema 24 h after treatment [10]. However, another study revealed that intravitreal injection of chitosan nanoparticles led to the induction of acute inflammation in a rabbit model [11]. Histological analysis revealed the presence of immune cells at the injection site. In addition, inflammation and polymorphonuclear cell infiltrates were visible in the eye cup 7 days after injection. The infection could have resulted from the body's removal of the chitosan invasion in the immunocompromised vitreous humor rather than from the effect of chitosan alone [4].

The method by which to formulate chitosan nanoparticles requires the mixture and sonication of different solutions. Chitosan nanoparticles form spontaneously from this procedure. An ionic buffer is used to dissolve and solubilize the chitosan vector. Another solution is simultaneously prepared in which the drug or product of interest is dissolved; the solution can be comprised of any buffer that ensures the stabilization of the drug or compound of interest. The two solutions are rapidly mixed together and vortexed or spun to allow for the spontaneous formation of chitosan nanoparticles in solution from agitation. It is essential that both solutions of choice are miscible for the proper formation of chitosan nanoparticles. The rapid and spontaneous formation of chitosan nanoparticles upon agitation imparts simplicity in the formulation technique.

20.4 Albumin

Albumin nanoparticles are a common vector of choice for ophthalmic drug delivery. The physical characteristics of human serum albumin (HSA) allow it to bind to a number of biological compounds, including fatty acids, proteins, and vitamins, in either covalent or transient bonds [12]. Because the HSA vector is derived from a protein found abundantly in blood, nanoparticles made from HSA exhibit low toxicity and increased bioavailability [13]. HSA nanoparticles are an interesting drug delivery system for ophthalmic purposes due to the presence of HSA in the aqueous humor and the ability to penetrate the BAB [14]; In addition, the transcytotic route allows for the passage of HSA nanoparticles into lens cells through the aqueous humor [15].

HSA nanoparticles have shown success in drug delivery to the eye. The surface charge of the compound is an important factor in ocular drug delivery; anionic HSA nanoparticles are reported to target the retina through intravitreal injection while cationic HSA nanoparticles injected similarly aggregate in the vitreous humor, possibly due to interactions with vitreal proteins [16]. Although particle size is an important determinant of ophthalmic drug delivery, HSA nanoparticles up to 300 nm have been shown to be dispersed in vitreous humor; however, surface charge of the HSA nanoparticles, whether cationic or anionic, seems to be a more important factor in drug delivery [16–18].

Studies carried out by Kompella et al. revealed that HSA nanoparticles did not cause significant cytotoxicity in human retinal pigment epithelium (ARPE-19) cells over a 96-h study period. Furthermore, the same study revealed that HSA nanoparticles can serve as efficient gene delivery molecules through the nuclear uptake of HSA nanoparticles via caveolae- and clathrin-mediated endocytosis pathways [19].

Merodio et al. carried out in vivo studies of intravitreally injected HSA nanoparticles for the delivery of ganciclovir in Wistar rats [20]. The nanoparticles were localized overlying the retina in the vitreous cavity 2 weeks postinjection, suggesting potential use as an effective sustained release delivery system for retinal disorders. In addition, the injected nanoparticles did not cause any inflammatory reactions nor induce the presence of infiltrates at the site of injection, vitreous, or retinal cells. There was no evidence of degeneration of photoreceptor cells postinjection by the HSA nanoparticles or byproducts of biodegradation processes in vivo [20].

The physical nature of albumin facilitates the formulation technique for HAS nanoparticles. Two phases are utilized in the formulation of HSA nanoparticles: (1) in which the albumin is solubilized and (2) into which or from which the albumin solution is added or removed. Water and ethanol are used as separate phases in the formulation technique. Some protocols call for the solvation of albumin into water which is added to ethanol solution, while others call for the solvation of albumin into ethanol which is then added to water. Albumin forms coacervates when the two phases are mixed together. The final step in formation of HSA nanoparticles involves the cross-linking of the albumin coacervates by glutaraldehyde. Glutaraldehyde can be added in various concentrations depending on the amount of cross-linking desired. Functional group side chains of glutaraldehyde induce the chemical linkage of multiple albumin protein functional groups to form HSA nanoparticles.

20.5 Silver

Silver nanoparticles are an interesting therapeutic option for ocular diseases. Silver has been used for millennia for its antibacterial properties and is still a common component of antibacterial ointments. Furthermore, silver nanoparticles have been reported to have antiviral, antifungal, and anti-inflammatory properties [21]. Of primary interest to ocular disorders are the potential angiogenic effects of silver nanoparticles [22].

Gurunathan et al. synthesized silver nanoparticles through the activity of the bacteria, *Bacillus licheniformis* to investigate the effect on vascular endothelial growth factor (VEGF) expression in bovine retinal endothelial cells [23]. The study found that silver nanoparticles are cytotoxic at concentrations above 500 nM/mL in endothelial cells when treated for 24 h. The nanoparticles were also shown to inhibit cell migration and proliferation characteristic of angiogenesis in bovine retinal endothelial cells over 24 h of treatment. The relative low cost and easy formulation technique behind the synthesis of silver nanoparticles makes it a vector of interest for inhibition of VEGF-induced angiogenic effects in disease models.

Interestingly, silver nanoparticles can be easily formed by the use of microorganisms such as *Pseudomonas stutzeri* and *Bacillus licheniformis*, as a number of organisms are resistant to silver and intake the metal into the cell which can then be retrieved for a number of uses. The bacteria are grown in nutrient broth and silver nitrate solution is added to the growth media when critical bacterial growth level is achieved. The silver and nitrate ions are separated in aqueous solution. The bacteria are able to uptake and internalize the silver ions and form up to 200 nm complexes in the bacteria cell. The bacterial cells can later be harvested and lysed for the acquisition of silver nanoparticles [24].

20.6 PLGA

Poly(lactide-co-glycolide) (PLGA) nanoparticles are made from the conjugation of polylactic acid and poly glycolic acid into chains and are formulated with drugs encapsulated into the hollow sphere inside the nanoparticles. Benefits for the use of PLGA nanoparticles include biocompatibility, biodegradability, sustained drug release, and FDA approval. Furthermore, the PLGA vector is commonly used in many research and clinical settings for drug delivery to many different tissues and diseases, making this a common vector of choice for ocular drug delivery.

Relative concentrations of each component can be modified during conjugation to alter the physical chemistries of the vector by influencing the resultant molecular weight of the nanoparticle vector. Respective D/L chirality when selecting lipid chains for conjugation is also a common method by which to influence the physical chemistry of the vector. In addition, various end groups and functionalities can be conjugated to the chains to induce further property modifications to induce greater hydrophilicity or stability in biological systems. Addition of carboxyl groups (-COOH) and poly(ethylene glycol) (PEG) are common surface modifications to PLGA vectors during chemical synthesis. Addition of these end groups induce greater hydrophilicity, with PEG end groups having an added benefit of shielding the nanoparticles from uptake by macrophages and the reticuloendothelial (RES) system in the blood. PEG end groups also assist in the nanoparticles' self-assembly into micellar systems during the formulation process [25].

The uptake of nanoparticles is essential for drug delivery purposes. Lee et al. investigated the uptake of coumarin 6 (fluorescent dye)-loaded PLGA nanoparticles in rabbit conjunctival epithelial cell layers [26]. The group found through confocal microscopy that the uptake of nanoparticles in these cells was dependent on the size of the nanoparticles, with smaller nanoparticles (100 nm) being uptaken more easily than larger particles (800–10,000 nm). Furthermore, the group reported that the nanoparticles were internalized within the cells rather than absorbing to the cell surface. Furthermore, the group found no evidence of cytotoxicity or stress response in the eye tissue.

PLGA nanoparticles are commonly used to administer steroids in therapy aimed toward age-related macular degeneration. The benefit of using nanoparticles is the sustained drug release properties as opposed to regular intravitreal injections and associated complications. PLGA nanoparticles loaded with dexamethasone acetate were investigated as a potential therapeutic option against choroidal neovascularization (CNV) in a laser-induced rat model by Song et al. [27]. Reverse-phase high-performance liquid chromatography (RP HPLC) analysis of ocular tissues showed that vitreal concentrations of drug following intravitreal injection remained at a therapeutically effective range for over 50 days postinjection, suggesting adequate time for the treatment of inflammation in CNV. Furthermore, 2 weeks after photocoagulation, all treatment groups had significantly less incidence of CNV as opposed to control groups, with a visible dose-dependent effect. The group reported no signs of toxicity during the treatment regimen.

The nanoprecipitation method is commonly utilized in the formulation of PLGA nanoparticles. The nanoprecipitation method is also known as the solvent evaporation method, which, as the name implies, involves the evaporation of an oil solvent phase from a water phase to result in the formation of nanoparticles. The PLGA vector and drug of interest are dissolved into the oil phase which is added to the water phase; it is important that both phases are miscible. Because the PLGA vector is made from the chemical linkage of lipid chains, it readily dissolves into the oil phase but self-assembles into spheres in water phase upon the evaporation of oil phase. Drugs can be encapsulated into the hollow core of the nanoparticle during the self-assembly of the PLGA vector sphere. PLGA nanoparticles are used as a drug delivery system for a number of drug compounds; hydrophobic drugs are easily encapsulated by the PLGA vector due to hydrophobic interactions between PLGA and drug molecule. However, hydrophilic drugs commonly escape into the water phase prior to proper encapsulation by PLGA nanoparticles.

20.7 Liposomes

Liposomes are a drug delivery system comprised of hollow spheres made from the self-assembly of an amphipathic phospholipid bilayer [28]. Due to the existence of charged and nonpolar parts of the liposomal vector, both hydrophobic and hydrophilic drug molecules can be incorporated into and delivered by liposomes [28, 29]. Furthermore, liposomes can be conjugated with a number of different surface functional groups to directly target tissues [30], are biocompatible and biodegradable due to their lipid nature, and are not shown to induce cytotoxicity in cellular studies [31].

Liposomes loaded with ganciclovir were investigated as a drug delivery method against cytomegalovirus (CMV)-induced blindness in AIDS patients by Romero et al. [32]. Treatments were injected intravitreally into rabbit eyes and no signs of ocular inflammation were visible immediately after the injection. In addition, rabbit eyes injected with the liposome-entrapped treatment showed no signs of toxicity over the two-week study period. The injection of the liposome treatment caused turbidity in the vitreous due to the physical nature of the liposomes, but no toxicity was visible to the lens or retina after injection of the treatment by biomicroscopic investigation.

Amphotericin B liposomes were injected intravitreally into the eyes of a rabbit model to test the toxicity of the therapeutic route by Danziger et al. at predetermined intervals over 21 days until animal death [33]. The injection technique resulted in the formation of cataracts in the rabbit eyes. Mild vitreal inflammation was evidenced by histological examination of ocular tissues which increased to moderate inflammation with increased dosage of drug formulation. However, the incidence of inflammation was not thought to be attributed to the liposomal drug delivery system but to the drug concentration in vitreous and injection technique. Furthermore, inflammation did not worsen nor was there any evidence of necrosis in tissue samples.

The formulation of liposomes involves two main steps: (1) the formation of the liposomes and (2) the hydration of liposomes. First, the liposome vector is dissolved in some type of volatile, organic solvent. Once the vector is properly solubilized, the organic solvent is evaporated from solution by a number of different methods including use of vacuuming pump or argon steam. The resultant lipid "cake" is lyophilized to properly dry the liposomes. An aqueous solution is added to the lyophilized liposomes and agitated for hydration. The liposomes maintain their spherical shape when hydrated due to the self-assembling property of lipid compounds in aqueous solutions. When loading hydrophobic drugs, the drug molecule can be added to the organic phase prior to evaporation to allow for the encapsulation of the molecule by the liposomes. The encapsulation of the drug is maintained during hydration. However, when drugs that cannot dissolve in organic phase are used, remote loading of the drug is utilized to load the drug into the liposomes after hydration [34].

20.8 Nanomicelles

Nanomicelles are amphiphilic compounds similar to liposomes that self-assemble into micellar, spherical structures in aqueous solution based on hydrophobic interactions. The sizes and shapes of the nanomicelles can vary depending on the molecular structure of the vector as well as formulation technique, ranging in size up to 1000 nm and in shape from spherical to cylindrical and even star-shaped. Hydrophobic drugs can be incorporated into the nanomicelle complex through hydrophobic and van der Waals' interactions [35].

Möller et al. investigated the cytotoxicity of cyclosporine A-loaded poly(ethylene glycol)-hexyl-substituted poly(lactide) nanomicelles through in vitro and in vivo tests [36]. MTT assay of nanomicelles in human corneal epithelial (HCE) cell after 1 h of incubation showed no cell death (100 % cell survival) in all treatment groups with no significant difference in cell survival. Even at higher copolymer concentrations (3 mg/mL), no significant cell death was noted. However, lower copolymer concentrations

(0.003 mg/mL) resulted in cell death but not at a statistically significant level. Furthermore, immunohistochemical analysis of unloaded nanomicelles in HCE cells did not result in activation of caspase-independent apoptosis pathways. Ocular tolerance of nanomicelle formulations in rabbit eyes resulted in insignificant changes in fluorescent injured ocular surfaces as compared to control groups.

Nanomicelles can be formed according to the solvent evaporation method similar to that of PLGA nanoparticles in which nanomicelles are formed from the evaporation of organic phase solvent from the water phase. The hydration method similar to that of liposomes formation can also be used in which nanomicelles are formed from the evaporation of the organic phase and subsequent hydration of the lyophilized nanomicelles by aqueous media. The direct dissolution method is another simple method by which nanomicelles can be formed; this method does not require the use of organic solvents. In this method the nanomicelle vector is directly dissolved in an aqueous medium of choice and thermally activated for the suspension of the nanomicelle vector. The thermal solution is cooled to room temperature and agitated by ultrasonication to result in the formation of nanomicelles, which can be subsequently collected by ultracentrifugation [35].

20.9 Conclusion and Future Perspectives

The majority of nanosuspensions described herein are of biological origin and are therefore well tolerated by the eyes. The surfaces of the nanosuspensions can be conjugated with a number of different molecules, including proteins, carbohydrates, or other similar molecules to induce greater hydrophilicity in biological systems or tissue targeting functions. The breakdown of these nanosuspensions occurs through biological processes in cells with catalytic products shuttled into biological pathways or excreted through renal or other waste pathways. The method of nanosuspension administration, such as intravitreal injection, causes inflammation and damage at the site of injection but the drug delivery systems themselves seem to be well tolerated in ocular tissues.

There are a number of ways to synthesize nanomedicines for ocular drug delivery. Organic or lipid-based nanoparticle vectors are commonly formulated by the evaporation of an organic solvent from an aqueous one, resulting in the precipitation of spherical nanoparticles. The spherical shape of formulated nanoparticles is maintained by the aqueous solution, as lipid-based vectors self-assemble into spherical compounds to reduce the amount of hydrophobic surface area exposed to hydrophilic solvent. On the other hand, silver nanoparticles are created through exploiting biological processes present in microorganisms.

Nanosuspensions and nanomedicines are becoming widely incorporated into clinical settings due to biocompatibility, biodegradability, and sustained release of drug products. As the future of nanomedicines progresses, such nanosuspensions will become commonplace in therapeutic intervention of ocular diseases. The sustained drug release profiles of nanomedicines will act as a boon against posterior eye segment diseases that cause blindness in millions of patients worldwide.

References

1. Del Amo EM, Urtti A (2008) Current and future ophthalmic drug delivery systems: a shift to the posterior segment. Drug Discov Today 13(3):135–143
2. Urtti A et al (1990) Controlled drug delivery devices for experimental ocular studies with timolol 2. Ocular and systemic absorption in rabbits. Int J Pharm 61(3):241–249
3. Maurice D, Mishima S (1984) Ocular pharmacokinetics. In: Pharmacology of the eye. Springer, New York, pp 19–116
4. Prow TW (2010) Toxicity of nanomaterials to the eye. Wiley Interdiscip Rev Nanomed Nanobiotechnol 2(4):317–333
5. Pignatello R, Bucolo C, Puglisi G (2002) Ocular tolerability of Eudragit RS100® and RL100® nanosuspensions as carriers for ophthalmic controlled drug delivery. J Pharm Sci 91(12):2636–2641
6. Pignatello R et al (2002) Flurbiprofen-loaded acrylate polymer nanosuspensions for ophthalmic application. Biomaterials 23(15):3247–3255
7. Pignatello R et al (2002) Eudragit RS100® nanosuspensions for the ophthalmic controlled delivery of ibuprofen. Eur J Pharm Sci 16(1):53–61
8. Wang JJ et al (2011) Recent advances of chitosan nanoparticles as drug carriers. Int J Nanomedicine 6:765–774
9. Kean T, Thanou M (2010) Biodegradation, biodistribution and toxicity of chitosan. Adv Drug Deliv Rev 62(1):3–11
10. Dornish M et al (1997) Safety of protasan: ultrapure chitosan salts for biomedical and pharmaceutical use. In: 7th international conference on chitin and chitosan, Lyon
11. Prow TW et al (2008) Ocular nanoparticle toxicity and transfection of the retina and retinal pigment epithelium. Nanomedicine 4(4):340–349
12. Kragh-Hansen U (1990) Structure and ligand binding properties of human serum albumin. Dan Med Bull 37(1):57–84
13. Kragh-Hansen U, Chuang VTG, Otagiri M (2002) Practical aspects of the ligand-binding and enzymatic properties of human serum albumin. Biol Pharm Bull 25(6):695–704
14. Tripathi RC, Millard CB, Tripathi BJ (1989) Protein composition of human aqueous humor: SDS-PAGE analysis of surgical and post-mortem samples. Exp Eye Res 48(1):117–130
15. Sabah JR et al (2004) In vivo passage of albumin from the aqueous humor into the lens. Mol Vis 10:254–259
16. Kim H, Robinson SB, Csaky KG (2009) Investigating the movement of intravitreal human serum albumin nanoparticles in the vitreous and retina. Pharm Res 26(2):329–337
17. Irache J et al (2005) Albumin nanoparticles for the intravitreal delivery of anticytomegaloviral drugs. Mini Rev Med Chem 5(3):293–305
18. Pitkänen L et al (2003) Vitreous is a barrier in nonviral gene transfer by cationic lipids and polymers. Pharm Res 20(4):576–583
19. Mo Y et al (2007) Human serum albumin nanoparticles for efficient delivery of Cu, Zn superoxide dismutase gene. Mol Vis 13:746
20. Merodio M et al (2002) Ocular disposition and tolerance of ganciclovir-loaded albumin nanoparticles after intravitreal injection in rats. Biomaterials 23(7):1587–1594
21. Kalishwaralal K et al (2010) Silver nano—a trove for retinal therapies. J Control Release 145(2):76–90
22. Gurunathan S et al (2009) Antiangiogenic properties of silver nanoparticles. Biomaterials 30(31):6341–6350
23. Kalishwaralal K et al (2009) Silver nanoparticles inhibit VEGF induced cell proliferation and migration in bovine retinal endothelial cells. Colloids Surf B Biointerfaces 73(1):51–57
24. Kalimuthu K et al (2008) Biosynthesis of silver nanocrystals by *Bacillus licheniformis*. Colloids Surf B Biointerfaces 65(1):150–153
25. Sah H et al (2013) Concepts and practices used to develop functional PLGA-based nanoparticulate systems. Int J Nanomedicine 8:747

26. Qaddoumi MG et al (2004) The characteristics and mechanisms of uptake of PLGA nanoparticles in rabbit conjunctival epithelial cell layers. Pharm Res 21(4):641–648
27. Xu J et al (2007) Inhibitory efficacy of intravitreal dexamethasone acetate-loaded PLGA nanoparticles on choroidal neovascularization in a laser-induced rat model. J Ocul Pharmacol Ther 23(6):527–540
28. Honda M et al (2013) Liposomes and nanotechnology in drug development: focus on ocular targets. Int J Nanomedicine 8:495
29. Allen TM, Cullis PR (2004) Drug delivery systems: entering the mainstream. Science 303(5665):1818–1822
30. Sapra P, Tyagi P, Allen TM (2005) Ligand-targeted liposomes for cancer treatment. Curr Drug Deliv 2(4):369–381
31. Van Rooijen N, van Nieuwmegen R (1980) Liposomes in immunology: multilamellar phosphatidylcholine liposomes as a simple, biodegradable and harmless adjuvant without any immunogenic activity of its own. Immunol Invest 9(3):243–256
32. Díaz-Llopis M et al (1992) Liposomally-entrapped ganciclovir for the treatment of cytomegalovirus retinitis in AIDS patients. Doc Ophthalmol 82(4):297–305
33. Cannon JP et al (2003) Comparative toxicity and concentrations of intravitreal amphotericin B formulations in a rabbit model. Invest Ophthalmol Vis Sci 44(5):2112–2117
34. Dua J, Rana A, Bhandari A (2012) Liposome: methods of preparation and applications. Int J Pharm Stud Res 3:14–20
35. Vaishya RD et al (2014) Controlled ocular drug delivery with nanomicelles. Wiley Interdiscip Rev Nanomed Nanobiotechnol 6:422–437
36. Di Tommaso C et al (2011) Ocular biocompatibility of novel cyclosporin a formulations based on methoxy poly(ethylene glycol)-hexylsubstituted poly(lactide) micelle carriers. Int J Pharm 416(2):515–524

Chapter 21
Ocular Drug Delivery: Impact of In Vitro Cell Culture Models

Aum Solanki, Sumir Desai, Aditya Grover, Anjali Hirani, Yashwant Pathak, and Vijaykumar Sutariya

Abstract The physiological characteristics make effective drug delivery very challenging. Various structures such as the cornea, conjunctival epithelium, blood-retinal barrier (BRB), and the retinal epithelium block or severely reduce the concentration of drugs that can reach particular locations in the eye. Because of this, in vitro models that can effectively model each of these barriers are highly desirable. Such in vitro modeling allows researchers to minimize the use of animal studies, as drug delivery experiments require frequent euthanization. As a substitute for animal experiments, various types of in vitro models have been developed that are made up of primary cell cultures or immortalized cell lines. These cell lines allow for the detailed study of the individual processes that determine the ability of drugs and drug delivery systems to reach their desired locations in therapeutic concentrations. This chapter discusses various examples of such cell lines and their applicability in various drug delivery studies.

A. Solanki
Honors College, University of South Florida, Tampa, FL, USA

S. Desai
College of Pharmacy, University of South Florida, Tampa, FL, USA
e-mail: ypathak1@health.usf.edu

A. Grover
College of Medicine, University of South Florida, Tampa, FL, USA

A. Hirani
Department of Pharmaceutical Sciences, USF College of Pharmacy, University of South Florida, Tampa, FL 33612, USA

School of Biomedical Engineering and Sciences, Virginia Tech-Wake Forest University, Blacksburg, VA 24061, USA

Y. Pathak (✉)
Department of Pharmaceutical Sciences, USF College of Pharmacy, University of South Florida, Tampa, FL 33612, USA
e-mail: ypathak1@health.usf.edu

V. Sutariya (✉)
Department of Pharmaceutical Sciences, College of Pharmacy, University of South Florida, Tampa, FL 33612, USA
e-mail: vsutariy@health.usf.edu

© Springer International Publishing Switzerland 2016

Y. Pathak et al. (eds.), *Nano-Biomaterials For Ophthalmic Drug Delivery*,
DOI 10.1007/978-3-319-29346-2_21

Keywords In vitro modeling • Corneal epithelium • Corneal stroma • Conjunctival epithelium • Blood-retinal barrier (BRB) • Retinal pigment epithelium (RPE) • Tight junctions • Primary cell culture • Immortalized cell culture

21.1 Introduction

In vitro models, when compared to their in vivo counterparts, are generally the more desirable option for researchers because of many reasons. Firstly, it is very difficult for researchers to study individual mechanisms in vivo because of the multitude of mechanisms that are working simultaneously, while in vitro models give them the ability to isolate the individual processes that they wish to study. Furthermore, in vitro research is inherently a more ethical form of research that is generally more cost-effective.

One particular mechanism where in vitro models are significant is the delivery of drugs through ocular barriers. These barriers are extremely important in maintaining proper ocular function because of their ability to restrict and regulate the uptake of fluids and solutes, but they also present a huge challenge to pharmaceutical scientists as they prevent many different therapies from reaching the desired destinations [1]. Animal experiments, usually rabbit, are the alternative to using in vitro models and these can be extremely inefficient and costly as rabbits must be sacrificed at each experimental time point to collect samples that must be tested. Furthermore, there are some significant differences between the human eye and the rabbit eye that must be accounted for when testing with rabbits. Some of these include the extremely small blinking rate and the large corneal surface area in comparison to the human eye [1].

Because of these issues associated with using animals for ocular testing, it becomes clear that in vitro models allow researchers to study ocular drug delivery more effectively. The three cellular barriers that are present in the eye can be found in the cornea, conjunctiva, and in the retina. The corneal barrier is primarily formed by the layer of epithelial cells present in the cornea, which protects the eye from any toxic penetration but makes it difficult for topically delivered drugs to penetrate into the eye [2]. There is an epithelium present in the conjunctiva that causes significant portions of any delivered drug to be lost to the systemic circulation [3]. The blood-retinal barrier (BRB) is formed by a combination of two different barriers: the retinal pigment epithelium (RPE) [1] and the inner BRB, which is composed of the tight junctions between the endothelial cells of the retinal vascular tree [4].

For each of these ocular barriers, immortalized cell lines have been developed that allow for effective in vitro modeling. This chapter will discuss the different cell culture models that can and have been used for developing new methods of ocular drug delivery.

21.2 Cornea

21.2.1 Pharmacokinetic Barriers of the Cornea

The cornea is vital to proper ocular function as it is a clear, mechanical, and chemical barrier that limits the invasion of exogenous substances into the eye while allowing for light to pass through the lens and to the retina [1, 5]. It is made up of five layers: corneal epithelium, basement membrane, Bowman's layer, stroma, Descemet's membrane, and endothelium [1]. The corneal epithelium is made up of highly regenerative cells that allow for quick healing of superficial injuries [6], and it also acts as the primary pharmacokinetic barrier as removal of this layer greatly increased permeability in comparison to the removal of any of the other four layers [7]. The epithelium's barrier properties constitute a very high impermeability toward hydrophilic drugs, while the stroma acts as a diffusional barrier to highly lipophilic drugs [1]. Additionally, the tight junctions in the epithelium allow for an extremely small average pore size (2.0 nm), which generally prevents the passive diffusion of large drugs (molecular weight >500) [8].

21.2.2 Cell Culture Models of Cornea

21.2.2.1 Corneal Epithelium

Having an immortalized cell line is essential for corneal research because of the scarcity of available tissue for primary cell culture models [5]. These immortalized cells must behave similarly to the primary cells in that they must form the cobblestone morphology that replicates the multilayered nature of the epithelium, which results in the formation of tight junctions [1, 5]. Kahn et al. were able to immortalize such cells using the Ad12-SV40 hybrid virus while maintaining the desired morphology [5]. By using culture techniques such as use of serum-free media and raft systems, they provided an air-liquid interface that replicates the in vivo environment of these cells [5].

Kruszewski et al. tested the ability of this model (referred to as "HCE-T") to replace the Draize test for ocular irritancy [9]. Alternatives to the Draize tests have been investigated constantly because of the high number of rabbit sacrifices necessary. The effect of different chemicals on the structural integrity of the cell layer was tested by measuring transepithelial permeability to sodium fluorescein (TEP) and transepithelial electrical resistance (TER) [9]. They found, by comparing their data to previously collected Draize test values, that using TEP and TER measurements in the HCE-T cells is a suitable substitute to the Draize test when studying ocular irritancy of chemicals [9].

An issue with the HCE-T cell line was that it produced infectious virus that shedded into supernatant, which limited its applications [1, 10]. Araki-Sasaki et al. were able to solve this issue and establish an immortalized SV40 cell line (HCE) that does not have any viral shedding [10]. This is a very useful cell culture model due to its effectiveness in both toxicity studies and drug permeability studies [11, 12].

Gelatinous drop-like dystrophy (GDLD) is a genetic disorder that results in the reduction of the barrier functions of the corneal epithelium, which leads to pathogen invasions into the corneal stroma [13]. It is an autosomal disease resulting from the loss of expression of the tumor-associated calcium signal transducer 2 (TACSTD2) gene [14]. Symptoms such as severe photophobia, lacrimation, and gradual vision loss can arise in the first 20 years of life with the severity of the symptoms gradually increasing with age [15]. Current treatments for the disorder are supportive treatments for temporary system relief, leading Kitazawa et al. to develop an immortalized human corneal epithelial cell line that can be used for in vitro studies of GDLD [13]. This was done through immortalizing corneal epithelial cells from a GDLD patient by the transfection of SV40 large T antigen and hTERT genes [13]. This cell line reflects the in vivo behavior of corneal epithelial cells well, making it a good in vitro model for future studies.

A cell culture model similar to this one was used to study the efficacy of nonthermal plasmas in the treatment of Herpes keratitis infections [16]. The corneal epithelial cells were infected with a strand of the virus and then exposed to culture medium that was treated with a nonthermal plasma [16]. The results of this experiment showed the potential of plasmas in the treatment of viral infections due to their versatile and adaptive nature [16]. Cell culture models of the corneal epithelium also allow for researchers to study different methods of using nanotechnology as a means of permeating drugs through the corneal epithelium [17, 18].

21.2.2.2 Three-Dimensional Models of Cornea

Three-dimensional models of the cornea have been created to contain multiple layers of the cornea instead of the epithelium alone. Griffith et al. created one such model consisting of three layers (epithelium, stroma, and endothelium) of immortalized human cells [19]. Their model successfully replicated human corneas in morphology, biochemical marker expression, transparency, ion and fluid transport, and gene expression [19]. These physiological and functional similarities made this model an incredibly useful one, especially for toxicity studies [19].

Recently, Engelke et al. created a model consisting of the corneal epithelium and stroma to potentially be a substitute for the Draize test for eye irritancy [20]. Firstly, their project showed that their model can be reproduced effectively in other independent laboratories [20]. The tissues were then treated with 20 chemicals with different eye-irritating potentials to find the best prediction model [20]. The best model showed an in vitro-in vivo concordance of 80 %, which reflects very good potential for this model, but some physiological changes must be made to make it a reliable substitute to the Draize test [20].

There is also some potential in using such models for drug permeation studies. Tegtmeyer et al. used a model similar to the one discussed above to study its permeability in comparison to that of excised bovine cornea [21]. The permeation of pilocarpine hydrochloride was tested in both the in vitro model and the excised bovine cornea, showing that the permeation constant for the model is two- to fourfold higher than that of the bovine cornea [21]. Even though this difference in the constant is not optimal, it is possible to reconstruct such models and use them for ocular drug delivery studies [21].

21.3 Conjunctiva

21.3.1 Pharmacokinetics of Conjunctiva

The conjunctiva is a very important pathway for topically applied drugs because hydrophilic drugs cannot effectively be absorbed through the cornea [1]. The conjunctiva is a mucous membrane consisting of an epithelium covering the anterior surface of the sclera (bulbar conjunctiva) and folding at the fornix [1]. A major issue that the conjunctiva presents is that much of the drug is lost to the systemic circulation while crossing the conjunctiva, but the drug that successfully reaches the sclera is able to diffuse rather easily [1]. In vitro models that can possibly help solve this issue are desirable because this is the only pathway for topically applied, hydrophilic drugs to enter the eye.

21.3.2 Cell Culture Models of the Conjunctiva

Similar to the corneal epithelium, having an immortalized cell line of the conjunctival epithelium is very important, as the primary cell cultures obtained through conjunctival biopsy are not very useful in terms of controlled research. Diebold et al. were able to spontaneously create one such cell line from a healthy donor who was undergoing cataract surgery [22]. This cell line was characterized to test its efficacy as an in vitro model, yielding results that show standard in vitro behavior [22]. The similarity with the in vivo conjunctival epithelium was also tested through karyotype analysis and Western blot analysis, and results were very promising, making this cell line an effective tool for research regarding ocular surface disorders [22].

Gao et al. used this cell culture model to study the role of mitochondrial protein transfer pores (MPTP) on inflammation-induced apoptosis and to test the mechanisms that cyclosporin A has in regulating the disorder [23]. Their findings firstly showed that opening of the MPTP corresponded with apoptosis in the IOBA-NHC cells, while cyclosporin A was able to stabilize the MPTP and prevent apoptosis [23]. These findings show the potential that cyclosporin A has in the treatment of ocular disorders such as dry eye disease [23].

Prior to the IOBA-NHC cell line, the Wong-Kilbourne derivative of the Chang cell line was widely used for in vitro toxicology studies, but its contamination with HeLa cells gave the IOBA-NHC the distinction of being a better model [22]. Brasnu et al. evaluated the relevance of the IOBA-NHC cell line in toxicological studies by comparing the effects of benzalkonium chloride (BAC), which is known to have toxic effects on conjunctival cells on both the IOBA-NHC and the Chang cells [24]. Findings showed that the toxic effects of BAC were consistently observed in both cell lines, with the Chang cell line exhibiting a slightly higher sensitivity to the toxic effects of BAC [24]. Despite this slight difference, the results validate both cell lines as effective models for toxicological studies and also show that the reported HeLa contamination of the Chang cell line does not affect its performance in such studies [24].

Kinoshita et al. established a conjunctival epithelial cell line from a GDLD patient [14]. This cell line lacks expression of the TACSTD2 gene, resulting in decreased expression of tight junction-related proteins [14]. This model shows great promise for GDLD research, as previous GDLD treatments have been relatively ineffective and this model could promote novel treatments for GDLD patients [14].

21.4 Blood-Ocular Barriers

21.4.1 Pharmacokinetics of the Blood-Ocular Barriers

The blood-ocular barrier is comprised of the blood-aqueous barrier (BAB) and the blood-retinal barrier (BRB); both limit the permeability of toxins while maintaining homeostasis of the eye [25]. Systemic oral and intravenous administrations of therapeutic drugs are ineffective due to the limited quantity that is able to permeate the ocular barrier to reach the ocular tissues [1]. Additionally, the remainder of the drug passes through the body causing unwanted side effects [1]. This barrier can be circumvented via intravitreal drug administration, although there is a risk for endophthalmitis, damage to the lens, retinal damage, and a low patient compliance [26].

21.4.1.1 Blood-Aqueous Barrier

The BAB is located in the anterior of the eye and is formed by endothelial cells lining the blood vessels in the iris and a ciliary epithelium cell layer [1]. These layers form a tight junction that prevents nonspecific solutes from passing to the intraocular milieu and maintain homeostasis [25]. However, the BAB does not function as a complete barrier. Due to low levels of plasma protein in the aqueous humor, molecules are able to pass through the capillaries of the ciliary processes and reach the aqueous humor [1, 26]. Movement from the aqueous humor to the iris is less controlled [27]. Iris tissue is more porous, so drugs dissolved in the aqueous humor cross into the iris tissue and are either absorbed by iris pigments or eliminated from the iris via blood vessels [1, 28]. Small and lipophilic drugs are eliminated more easily than larger, hydrophilic drugs [29].

21.4.1.2 Blood-Retinal Barrier

The BRB is located in the posterior of the eye and formed by endothelial cells of retinal blood vessels (inner BRB) and retinal pigment epithelial (outer BRB) cells. The retina is attached to the retinal pigment epithelium (RPE), and the RPE is responsible for maintaining retina viability [30]. The RPE has tight junctions that form a strong barrier preventing the permeation of many solutes. The inner BRB is very similar to the blood-brain barrier (BBB) in the regards that glial cells in proximity to endothelial cells induce formation of tight junctions [4].

While this barrier prevents the passage of toxic molecules from reaching the retina, it also prevents systemically administered drugs from entering [25]. The concentration of the drug found in the vitreous is typically only 1–2 % of the concentration of the drug found in plasma [31]. The RPE and retinal vessel walls prevent many proteins and small hydrophilic compounds from permeating while lipophilic material is more permeable in the retinal endothelial cells [32].

The BRB eliminates many drugs via active and passive transport; drugs with high lipophilicity are removed rapidly [1]. This is important in regards to intravitreal drug administration because drugs with a lower diffusion rate through the BRB will remain in the vitreous for longer periods of time [28].

21.4.2 Cell Culture Models of Blood-Ocular Barriers

21.4.2.1 Blood-Aqueous Barrier

Currently there is no literature regarding functional models of the BAB.

21.4.2.2 Blood-Retinal Barrier

There are many cell culture models created for both the outer and inner blood-retinal barriers. These models are vital for drug delivery systems targeting the posterior of the eye and understanding the interaction between RCE and RPE cells [1]. For instance, deterioration of the BRB occurs in diseases such as diabetic retinopathy or age-related macular degeneration (AMD) [33]. The breakdown of this barrier causes the development of retinal edema, which may lead to vision loss [25].

21.4.2.2.1 Retinal Pigment Epithelium

The barrier function of the RPE is challenging to emulate in vitro due to the special functions such as nutrient, metabolite, ion, and fluid transport between the neural retina and choroid [1]. Cell culture growth and functionality is impacted by the RPE species, donor age, isolation process time, and culture conditions [33].

Primary Cell Cultures

Primary cell cultures models of the RPE have been established from frog, rat, chick, bovine, and human RPE cells. Cell isolation from animal tissue is more readily available compared to human donor eyes, although many researchers choose to work with human RPE cells to eliminate species-related problems [1]. One example of differences is in TER. TER is used as a parameter to determine the tightness of the epithelium, and values in animal cell culture models vary from human RPE cells. For this reason, human RPE cells were used for primary cell culturing in recent experiments involving uptake and transport, protein expression, and cytotoxicity because animal cell culture models would not have [1].

Immortalized Cell Lines

As mentioned previously, immortalized cell lines are important due to the scarcity of tissue available for primary cell culture models [2]. Three immortalized cell lines exist for human retinal epithelial cells.

Davis et al. cloned D407, a spontaneously arising cell line originating from human RPE [34]. The cells shared morphological characteristics of in vivo RPE cells such as cobblestone morphology and metabolic characteristics such as keratin expression and production of retinal-specific CRALBP protein [1]. One issue with the cell line was the observed loss of enzymatic activity and cytoskeletal polarization. For that reason, the cell line is not used as a model for barriers but primarily for cytotoxicity studies, retinal metabolism, and retinal gene studies [1, 35, 36].

Dunn et al. created the second human RPE cell line, ARPE-19, arising spontaneously [11]. These cells were characterized by morphology, expression of two retina-specific markers known as CRAPL and RPE65, and barrier properties [1]. Morphological features of the cell characterized by the authors included a cobblestone pigmentation and appearance along with defined borders. CRAPL and RPE65 were both identified in the cell. Barrier properties were determined via TER tests with the permeation of H-inulin [1]. Cells grown in a specialized RPE medium as opposed to DMEM-F12 with 1 % FBS resulted in higher TER readings, signifying tighter complex junctions. However a different method creating ARPE-19 cells through viral gene transfer to express exogenous proteins in cell monolayers improved morphological polarization [1, 37]. This information determined this cell line was more appropriate for RPE polarity studies and monolayers are now prevalent in in vitro models of the outer BRB [1]. Monolayers of ARPE-19 are used frequently in vitro models of the outer BRB seen in transport studies, barrier function, influences on age-related macular degeneration, and many other retinal diseases.

Bodnar et al. created the third immortalized cell line of human RPE, hTERT-RPE, by encoding RPE cells with human telomerase vectors [38]. This line prevents telomerase loss, increasing cellular life span while being able to maintain normal growth and normal gene expression [38, 39]. In the absence of serum, this cell line is able to differentiate and undergo arrested cell growth [40]. Reintroduction of serum after 4 or 8 weeks allowed the cells to reenter their cell growth cycle, which is essential for long-term culturing in vitro.

Although working with human RPE has many benefits over animal tissue, there are some issues. Firstly, long culture periods greater than 2 months occasionally lead to loss of tissue-specific features and functionality as seen by TER values [37, 41]. Another barrier is the limited donor base for human eyes because it hinders the progress of the in vitro BRB model; experiments in vivo are essential in validating the in vitro model [1].

21.4.2.2.2 Retinal Capillary Endothelium

As mentioned previously, the deterioration of the inner BRB occurs in diseases including AMD and diabetic retinopathy [33]. Treatments of these diseases are dependent on factors that influence the permeability of the RCE. Bovine retinal capillary endothelial cell (BRCEC) is the currently accepted in vitro model and allows for a viable model- although it is not completely similar to the barrier offered by human RCE in vivo [42–49].

Immortalized Cell Lines

Hosoya et al. created immortalized rat retinal capillary endothelial cells (TR-iBRB) from transgenic mice expressing antigen gene SV40 T [50]. This gene is affected by temperature and therefore the growth of the cell line is temperature dependent [1]. Shen et al. evaluated the barrier functions of the TR-iBRB cell line by plating the cells on permeable supports and evaluating the drug permeability [51]. The cells were similar to that of endothelial cells in primary culture but did not form tight barriers determined by low TER reading values [46]. Additional permeation studies were completed with other model substances including rhodamine and beta-blockers with different lipophilicity for cellular uptake [1]. These results continued to show a discrepancy from primary culture cells [1].

Ultimately, an RCE model with in vivo barrier function has not been established. Blood-brain barrier (BBB) models share similar characteristics to the BRB and have been studied in much more detail, but are not a feasible replacement because gene expression in capillary endothelial cells is inhibited when cultured with endothelial cells. The results from the BBB model would not give accurate results. The barrier functions of current models need to be studied more in depth before progress of inner BRB models can occur [1].

21.5 Conclusion

There has been rapid growth in ocular barrier development in the past years and these models have been vital in different divisions of research including drug uptake and transport, toxicity tests, cell physiology, and many others [1]. The corneal and the conjunctival barriers are major barriers for topically applied drugs. In vitro models of these barriers have been extremely effective and continue to show potential for developing novel methods for drug delivery.

Ocular barriers prevent permeation of toxic substances but also prevent therapeutic drugs from entering. The blood-aqueous barrier and the blood-retinal barrier are the two ocular barriers that limit solute permeability. Oral and intravenous administrations are ineffective because limited drug is able to permeate the ocular barriers. Intravitreous administration circumvents this barrier.

The blood-aqueous barrier is porous and therefore does not function as a complete barrier so focus is on the blood-retinal barrier. Many ocular diseases cause the BRB to deteriorate, causing retinal edema, and in turn leading to vision loss. Current research focuses on creating a viable BRB in vitro model similar to that of an in vivo model. An appropriate model will give insight to barrier functions and determine drug permeation.

These models are essential for prevention of ocular diseases and establishing treatment strategies. Regardless, in vitro ocular barrier models are not currently equivalent to in vivo environments. Further development on current models is necessary before testing treatments provide reliable results.

References

1. Hornof M, Toropainen E, Urtti A (2005) Cell culture models of the ocular barriers. Eur J Pharm Biopharm 60(2):207–225
2. Offord EA et al (1999) Immortalized human corneal epithelial cells for ocular toxicity and inflammation studies. Invest Ophthalmol Vis Sci 40(6):1091–1101
3. Urtti A, Salminen L (1993) Minimizing systemic absorption of topically administered ophthalmic drugs. Surv Ophthalmol 37(6):435–456
4. Gardner TW et al (1999) The molecular structure and function of the inner blood-retinal barrier. Doc Ophthalmol 97(3):229–237
5. Kahn C et al (1993) Human corneal epithelial primary cultures and cell lines with extended life span: in vitro model for ocular studies. Invest Ophthalmol Vis Sci 34(12):3429–3441
6. Grolik M et al (2011) Regeneration of corneal epithelium using keratin modified chitosan membranes. Przegl Lek 69(10):992–997
7. Huang A, Tseng S, Kenyon K (1989) Paracellular permeability of corneal and conjunctival epithelia. Invest Ophthalmol Vis Sci 30(4):684–689
8. Hämäläinen K et al (1997) Characterization of paracellular and aqueous penetration routes in cornea, conjunctiva, and sclera. Invest Ophthalmol Vis Sci 38(3):627–634
9. Kruszewski F, Walker T, DiPasquale L (1997) Evaluation of a human corneal epithelial cell line as an in vitro model for assessing ocular irritation. Toxicol Sci 36(2):130–140
10. Araki-Sasaki K et al (1995) An SV40-immortalized human corneal epithelial cell line and its characterization. Invest Ophthalmol Vis Sci 36(3):614–621
11. Huhtala A et al (2002) Comparison of an immortalized human corneal epithelial cell line and rabbit corneal epithelial cell culture in cytotoxicity testing. J Ocul Pharmacol Ther 18(2):163–175
12. Toropainen E et al (2003) Paracellular and passive transcellular permeability in immortalized human corneal epithelial cell culture model. Eur J Pharm Sci 20(1):99–106
13. Kitazawa K et al (2013) Establishment of a human corneal epithelial cell line lacking the functional TACSTD2 gene as an in vitro model for gelatinous drop-like dystrophy. Invest Ophthalmol Vis Sci 54(8):5701–5711
14. Kinoshita S et al (2012) Establishment of a human conjunctival epithelial cell line lacking the functional Tacstd2 gene (an American ophthalmological society thesis). Trans Am Ophthalmol Soc 110:166
15. Movahedan H, Anvari-Ardekani HR, Nowroozzadeh MH (2013) Limbal stem cell transplantation for gelatinous drop-like corneal dystrophy. J Ophthalmic Vis Res 8(2):107
16. Alekseev O et al (2014) Nonthermal dielectric barrier discharge (DBD) plasma suppresses herpes simplex virus type 1 (HSV-1) replication in corneal epithelium. Translat Vis Sci Technol 3(2)
17. Zhu X et al (2012) Synthesis of thiolated chitosan and preparation nanoparticles with sodium alginate for ocular drug delivery. Mol Vis 18:1973

18. Yang J et al (2014) Dithiol-PEG-PDLLA micelles: preparation and evaluation as potential topical ocular delivery vehicle. Biomacromolecules 15(4):1346–1354
19. Griffith M et al (1999) Functional human corneal equivalents constructed from cell lines. Science 286(5447):2169–2172
20. Engelke M et al (2013) A human hemi-cornea model for eye irritation testing: quality control of production, reliability and predictive capacity. Toxicol In Vitro 27(1):458–468
21. Tegtmeyer S, Papantoniou I, Müller-Goymann CC (2001) Reconstruction of an in vitro cornea and its use for drug permeation studies from different formulations containing pilocarpine hydrochloride. Eur J Pharm Biopharm 51(2):119–125
22. Diebold Y et al (2003) Characterization of a spontaneously immortalized cell line (IOBA-NHC) from normal human conjunctiva. Invest Ophthalmol Vis Sci 44(10):4263–4274
23. Gao J et al (2013) Mitochondrial permeability transition pore in inflammatory apoptosis of human conjunctival epithelial cells and T cells: effect of cyclosporin a. Invest Ophthalmol Vis Sci 54(7):4717–4733
24. Brasnu E et al (2008) Comparative study on the cytotoxic effects of benzalkonium chloride on the Wong-Kilbourne derivative of Chang conjunctival and IOBA-NHC cell lines. Mol Vis 14:394–402
25. Cunha-Vaz JG (1997) The blood-ocular barriers: past, present, and future. Doc Ophthalmol 93(1–2):149–157
26. Bochot A, Couvreur P, Fattal E (2000) Intravitreal administration of antisense oligonucleotides: potential of liposomal delivery. Prog Retin Eye Res 19(2):131–147
27. Jumbe NL, Miller MH (2003) Ocular drug transfer following systemic drug administration. Drugs Pharm Sci 130:109–134
28. Maurice D, Mishima S (1984) Ocular pharmacokinetics. In: Pharmacology of the eye. Springer, New York, pp 19–116
29. Mitra AK (2003) Ophthalmic drug delivery systems. Marcel Dekker, New York
30. Marmor M (1998) Structure, function, and disease of the retinal pigment epithelium. In Marmor MF, Wolfensberger TJ (eds) The retinal pigment epithelium, vol 3. Oxford University Press, New York, p 12
31. Duvvuri S, Majumdar S, Mitra AK (2003) Drug delivery to the retina: challenges and opportunities. Expert Opin Biol Ther 3(1):45–56
32. Sunkara G, Kompella UB (2003) Membrane transport processes in the eye. Drugs Pharm Sci 130:13–58
33. Eldem T et al (2002) Cell cultures of the retinal pigment epithelium to model the blood–retinal barrier for retinal drug and gene delivery. Cell culture models of biological barriers: in vitro test systems for drug absorption and delivery, 2002, p 271
34. Davis AA et al (1995) A human retinal pigment epithelial cell line that retains epithelial characteristics after prolonged culture. Invest Ophthalmol Vis Sci 36(5):955–964
35. Mannerström M et al (2001) The phagocytosis of rod outer segments is inhibited by selected drugs in retinal pigment epithelial cell cultures. Pharmacol Toxicol 88(1):27–33
36. Hyvönen Z et al (2000) Novel cationic amphiphilic 1, 4-dihydropyridine derivatives for DNA delivery. Biochim Biophys Acta 1509(1):451–466
37. Dunn KC et al (1998) Use of the ARPE-19 cell line as a model of RPE polarity: basolateral secretion of FGF5. Invest Ophthalmol Vis Sci 39(13):2744–2749
38. Bodnar AG et al (1998) Extension of life-span by introduction of telomerase into normal human cells. Science 279(5349):349–352
39. Jiang X-R et al (1999) Telomerase expression in human somatic cells does not induce changes associated with a transformed phenotype. Nat Genet 21(1):111–114
40. Rambhatla L et al (2002) In vitro differentiation capacity of telomerase immortalized human RPE cells. Invest Ophthalmol Vis Sci 43(5):1622–1630
41. Rossem V, Vos D (1998) Polarized secretion of IL-6 and IL-8 by human retinal pigment epithelial cells. Clin Exp Immunol 112(1):34–43
42. Gillies MC, Su T (1995) Interferon-α 2b enhances barrier function of bovine retinal microvascular endothelium in vitro. Microvasc Res 49(3):277–288

43. Gillies MC et al (1997) Effect of high glucose on permeability of retinal capillary endothelium in vitro. Invest Ophthalmol Vis Sci 38(3):635–642
44. Gardner TW (1995) Histamine, ZO-1 and increased blood-retinal barrier permeability in diabetic retinopathy. Trans Am Ophthalmol Soc 93:583
45. Gardner T et al (1996) Histamine reduces ZO-1 tight-junction protein expression in cultured retinal microvascular endothelial cells. Biochem J 320:717–721
46. Yaccino JAM et al (1997) Physiological transport properties of cultured retinal microvascular endothelial cell monolayers. Curr Eye Res 16(8):761–768
47. Chang YS et al (2000) Effect of vascular endothelial growth factor on cultured endothelial cell monolayer transport properties. Microvasc Res 59(2):265–277
48. Lakshminarayanan S et al (2000) Effect of VEGF on retinal microvascular endothelial hydraulic conductivity: the role of NO. Invest Ophthalmol Vis Sci 41(13):4256–4261
49. Antonetti DA et al (2002) Hydrocortisone decreases retinal endothelial cell water and solute flux coincident with increased content and decreased phosphorylation of occludin. J Neurochem 80(4):667–677
50. Hosoya K-I et al (2001) Conditionally immortalized retinal capillary endothelial cell lines (TR-iBRB) expressing differentiated endothelial cell functions derived from a transgenic rat. Exp Eye Res 72(2):163–172
51. Shen J et al (2003) Evaluation of an immortalized retinal endothelial cell line as an in vitro model for drug transport studies across the blood-retinal barrier. Pharm Res 20(9):1357–1363

Chapter 22
Nanotoxicity of Nanobiomaterials in Ocular System and Its Evaluation

Bibhuti B. Kakoti, Manjir Sarma Kataki, and Yashwant Pathak

Abstract Nanotoxicology is an emerging science under the tremendously growing field of nanotechnology. The unique advantageous properties of nanoparticles or nanobiomaterials have become a challenging issue of primary concern in terms of possible unpredictable toxic interactions of these nanoparticles with the biomolecules in the physiological system. The scanty in vivo studies are insufficient to advocate the therapeutic success stories of nanobiomaterials. A plethora of reports demonstrated only in vitro studies of nanotoxicity assessment. This chapter discusses toxicity issues related to nanobiomaterials targeted for ocular delivery and current methodologies used to assess nanomaterial toxicity along with the challenges and limitations.

Keywords Nanoparticles • Nanostructures • Nanobiomaterial • Nanotoxicity • Ocular system

22.1 Introduction

The tantalizing potential of nanotechnology is to fabricate and combine nanoscale approaches (1–100 nm) to make useful tools with novel properties. In nanotechnology-derived products, at least one dimension of the smallest fundamental unit of organization is on the nanometer scale. Nanotechnology is now considered as the next big revolution. Nanotechnology has tremendous potential for bringing benefits in areas as diverse as drug delivery system development, water decontamination, information and communication technologies, and the production of lighter but stronger materials. Nanomedicine, the application of nanotechnology

B.B. Kakoti • M.S. Kataki
Department of Pharmaceutical Sciences, Dibrugarh University,
Dibrugarh, Assam 786004, India

Y. Pathak (✉)
Department of Pharmaceutical Sciences, USF College of Pharmacy, University of South, Florida, Tampa, FL 33612, USA
e-mail: ypathak1@health.usf.edu

© Springer International Publishing Switzerland 2016
Y. Pathak et al. (eds.), *Nano-Biomaterials For Ophthalmic Drug Delivery*,
DOI 10.1007/978-3-319-29346-2_22

to healthcare, also holds great promise for revolutionizing medical treatments and therapies in areas such as imaging, faster diagnosis, drug delivery, and tissue regeneration as well as development of new medical products. Nanomaterials have different physiochemical properties, such as ultrasmall size, large surface area-to-mass ratio, and high reactivity, which are different from bulk materials of the same composition. These unique properties of nanomaterial find its way to applications in diverse fields including healthcare. But these advantageous unique features of nanomaterial make them unique in terms of toxicity also. Miniaturization is what makes nanotechnology so useful in medicine and industry; it is also one of the important factors that might make them potentially dangerous to human health. Research also shows that smaller size has a direct correlation to higher toxicity. With increase in surface area per unit mass, intrinsic properties of particles will likely be stressed, which ultimately leads to nanotoxicity.

The interactions and possible toxicity of nanomaterial with the biological system components are an essential focus in assessing and understanding the compatibility of nanomaterial versus toxicity (Fig. 22.1). With ample opportunities for nanotechnology in pharmaceuticals and healthcare industries, a careful and comprehensive understanding and assessment of possible toxicity in the ocular system is crucial.

22.2 Nanotechnology and Ocular Systems (Ophthalmology)

Designing an ophthalmic drug delivery system is one of the challenging tasks for the administration of drugs and therapeutic agents to the human body. The eye is considered to be a small but complex organ. There are multiple biological barriers which separate the eye from the rest of the body. The eye is a protected closed system, and therefore, the internal ocular structures and tissues are safe from the external environment because of the tight corneal epithelium junctions and mucosal layer.

22.2.1 Eye as a Target Organ

The anatomy and physiology of the eye makes it a highly protected organ. There are various vision-threatening ocular disorders as suggested by a growing body of published scientific research. It is also evident that there is a shortage of really curative treatments for most of the eye diseases, particularly with those targeting drugs to the posterior segment of the eye. The reason behind the scarcity of novel treatment strategies includes the highly isolated and restricted characteristic of the eye leading to low bioavailability of drugs. The various types of barriers in the eye negatively influence the effective passage of drugs to the eye [156]. The various barriers in the eye include:

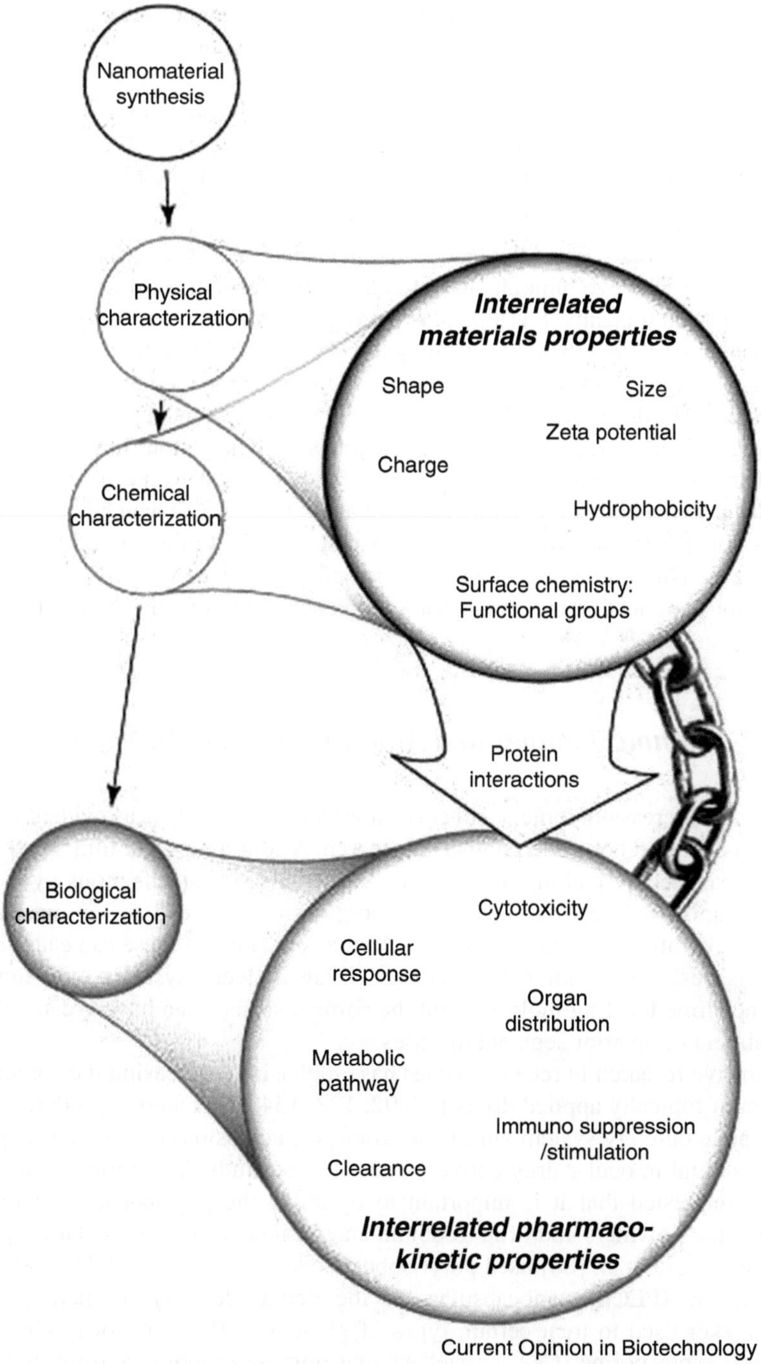

Fig. 22.1 The nanostructure physicochemical property relationship to in vivo responses (Adapted with permission from Hans C Fischer, Warren CW Chan, Nanotoxicity: the growing need for in vivo study, Current Opinion in Biotechnology, Volume 18, Issue 6, December 2007, Pages 565–571)

- Tear film mucous layer which protects anterior segment of the eye
- Tight junctions in and along with the corneal epithelium
- Iris blood vessels
- Blood–aqueous barrier (nonpigmented layer of the ciliary epithelium)
- Retinal pigment epithelium (RPE)
- Blood–retinal barrier (endothelium of the retinal blood vessels)

These barriers restrict the bioavailability of drugs, especially to the posterior segment of the eye. Physiological phenomena like tear drainage from lachrymal system along with eye blinking also reduce the residence time of drug molecules in the eye applied topically. For topical eye drops, it is evident that the drug molecules are washed out in less than 30 s, leading to very low residence time [72]. Due to the presence of tight junctions and desmosomes in the corneal epithelium, it is highly impermeable to drugs larger than 500 Da size [52]. Keister et al. [75] also suggested less than 5 % topical drug reaches intraocular tissues. The various limitations as mentioned above have stimulated research in the field of drug delivery in ophthalmology. And the growing body of literature has built a knowledge base which has initiated the application of new and novel techniques to overcome the limitations of drug delivery to the eye. Nanotechnology and nanomedicine is the new concept which finds tremendous application in ophthalmology for successful drug delivery to the eye.

22.2.2 Nanotechnology and Anterior Segment Diseases

There is an increase in topical and corneal infections nowadays as evident from the plethora of case reports and published research. A continuous tear film only protects the cornea, thereby making the cornea a vulnerable site of infections by invading microorganisms. The invasion by microorganisms can lead to various diseases including keratitis or conjunctivitis. Without any treatment, these can cause serious health issues, even leading to blindness. Drug delivery systems with improved residence time for drug molecules on the corneal surface can have greater efficacy in treatment of anterior segment diseases.

Intensive research in recent decades has emphasized increasing the corneal penetration of topically applied drugs [37, 92, 128, 134]. NPs were considered to be a more facile delivery system with better transport across tissues, and consequently, their potential in ocular drug delivery started to be studied. Experiences from past studies suggested that it is important to optimize the physicochemical relations between the polymers and drug to obtain an efficient carrier for ocular drug delivery. Later, Losa et al. [88] used poly(isobutyl-cyanoacrylate) (PIBCA) and poly-e-caprolactone (PECL) nanocapsules for the ocular delivery of metipranolol, a beta-blocker used to treat certain types of glaucoma. PECL nanocapsules loaded with 1 % cyclosporine (CsA), a well-known immunosuppressive drug, have been used in a rat model of corneal transplantation rejection [67]. This formulation had been developed to improve the corneal penetration of CsA applied topically [20].

By using PECL nanocapsules, toxic effects of systemic administration of CsA were avoided along with improved ocular penetration. Improved formulations using polymers with known biocompatibility and biodegradability, such as poly-D-lactic acid (PLA) and its copolymer glycolic acid (PLGA), were also developed. Flurbiprofen-loaded PLGA NPs were found to be successful in reducing inflammation in an in vivo rabbit model of ocular inflammation [160]. In comparison to commercial flurbiprofen formulations, flurbiprofen-loaded NPs were more effective in reducing inflammation as evaluated by direct observation of clinical signs.

More recently, scientists have developed applications using chitosan (CS)-based nanosystems. CS is a natural polysaccharide with interesting features such as biocompatibility and biodegradability, mucoadhesiveness, and the ability to transiently enhance the permeability of mucosal barriers. These features have made it quite useful in the development of drug delivery systems for transmucosal administration [7, 109]. CS/carbopol NPs can be loaded with pilocarpine, a parasympathomimetic drug used as treatment for open-angle glaucoma. CS/Carbopol NPs were found to last up to 24 h, much longer than the other formulations. Another CS-based nanosystem, CS/sodium alginate NPs, has been reported as a potential reservoir for topical delivery of gatifloxacin, a potent fourth-generation fluoroquinolone with low intrastromal penetration that is mainly used for microbial keratitis [98].

22.2.3 Nanotechnology and Posterior Segment Diseases

Neovascular and neurodegenerative retinal diseases are the common cause of vision impairment, as suggested by the growing research in ophthalmology. Neuroinflammation and ocular neovascularization are the main features in many of the retinal diseases including diabetic retinopathy, wet AMD, and retinitis pigmentosa. Currently, the best way to treat intraocular inflammations is by injecting drugs into the vitreous chamber. The vitreous is a cell-free structure and gelatinous in nature, which is capable of retaining drug molecules and also delivering them to nearby structures, such as the RPE or ciliary body. Frequent intraocular injections are required to treat retinal disorders effectively. Due to these frequent injections, potential undesired side effects are likely due to a higher risk of infections. Other frequent side effects associated with the repeated intravitreal injections to the eye include increased risk of cataract development, vitreous hemorrhage, retinal detachment, and endophthalmitis. These negative issues have moved researchers to look for better solutions derived from the use of NPs as drug carriers for effective drug delivery to the posterior segment of the eye. In recent years, much effort has been made to optimize drug delivery systems for intraocular treatment. Various other alternative ways have been explored for efficacious ophthalmic drug delivery including scleral implants and devices, transdermal patches, and different iontophoretic devices including hydrogel reservoirs [8]. Ganciclovir is one of the standard treatments for cytomegalovirus retinitis, a prevalent infectious retinal disease in immunosuppressed patients, such as those with AIDS. In vitro experiments demonstrated that albumin NPs released ganciclovir in a sustained way [95], with a significant improvement of drug uptake by cytomegalovirus-infected human cells [39]. RPE cells in the eye also have the ability to absorb different kinds of NPs [12, 16, 106], which has paved a new way for nanoparticle ophthalmic drug delivery. Bourges et al. [16] investigated in rats intravitreal PLA NPs loaded with fluorochromes and demonstrated a preferential localization at the RPE cells after 24 h. The most interesting achievement was that RPE cells retained the NPs, which continuously delivered the fluorochrome for months after the single injection. Later, Bejjani et al. [12] investigated in vitro and in vivo PLA and PLGA NPs loaded with fluorochromes and also with model plasmids. The results indicated that NPs encapsulating a plasmid encoding red nuclear fluorescent protein were localized in the RPE cells 24 h after intravitreal injection in rats. Effective plasmid expression was achieved after four days of injection, and expression-associated red fluorescence remained detectable in RPE cells during the following three weeks, with no significant tissue damage or toxicity. Pignatello et al. [115] developed ibuprofen-loaded NPs made of inert polymeric resins (Eudragit RS100). The outcome of this study demonstrated a controlled ibuprofen release profile from the NPs in vitro, and they had a high efficacy in reducing miosis with an in vivo model of ocular trauma. The therapeutic effect was found to be achieved with lower drug concentration as compared to eyedrop formulation, and the outcome was without any toxic effect on ocular tissues. Recently, Kassem et al. [70] studied different nanosuspensions prepared by high-pressure homogenization of three insoluble glucocorticoids, hydrocortisone, prednisolone, and dexamethasone, and revealed successful higher bioavailability and therapeutic effect with a significant extension of the glucocorticoid action. Interestingly, another study suggested that intravenously injected PLA NPs encapsulating betamethasone phosphate effectively controlled inflammation in a rat model of experimental autoimmune uveoretinitis [127]. The studies mentioned above have opened a new perspective for the treatment of retinal and uveal disorders. And we are expecting more success stories in the coming years.

22.3 Nanotechnology-Based Formulations in Ocular Drug Delivery

As discussed, designing an effective therapy for ocular diseases is a formidable task. Nanoparticulate drug delivery has generated a great deal of interest because of its potential to overcome many barriers associated with the current therapy. Some of the nanotechnology-based ocular formulations are listed below:

22.3.1 Liposomes

Liposomes are biocompatible, biodegradable lipid vesicles of about 25–10,000 nm in diameter made up of natural lipids. They can be prepared of different sizes, stability, and pharmacokinetics by different methods. Liposomes are like membranes made up of lipids similar to plasma membranes which surround an aqueous core [34]. Therefore, both hydrophilic and lipophilic drugs can be encapsulated to the lipid walls or the aqueous interior of the liposomes, respectively [8].

The properties of liposomes vary with lipid composition, size, surface charge, and the method of preparation. Surface charge property of liposomes plays an important role for ocular drug delivery. It is assumed that positively charged liposomes preferentially bind with the negatively charged corneal surface, compared with neutral or negatively charged liposomes. Felt et al. have explained that elimination of cationic vesicles along with the drug is slowed down from lachrymal flow by increasing the viscosity of the solution and by interacting with the negative charges of the mucus. Binding affinity of liposomes to cornea is as follows: positively charged liposomes > negatively charged liposomes > neutral liposomes. Initial interaction between corneal surface and liposomes is electrostatic in nature. Transcorneal flux of penicillin G across isolated rabbit cornea is fourfold more for positively charged unilamellar liposomes. Hence, liposomes can enhance corneal penetration of drug by being adsorbed onto the corneal surface, with direct transfer of drug from liposomal to epithelial cell membranes. Immunoliposomes are also under investigation and have been reported that site-specific and sustained release immunoliposomes could act as improved vehicles for drug delivery in treatment of ocular herpes simplex virus infection [126]. Intracellular drug delivery of liposomes is possible as they are taken up by phagocytic cells, like the RPEs (retinal pigment epithelial cells). Surface modification of liposomes is also possible, enabling preferential binding to endothelia of proliferating neovascular vessels [8].

Antigenicity of liposomes is low with almost no toxicity. Membrane structures of liposomes are highly stable, which can be deformed during injection through a 27- or 30-gauge needle. Studies on intravitreal injection of drug-bearing liposomes have demonstrated that toxicity of the drug can be reduced, the half-life of the drug inside the vitreous body can be prolonged, and the release of the drug can be controlled. Liposomes can also be used to treat cytomegalovirus retinitis in acquired immune deficiency syndrome (AIDS) patients encapsulating ganciclovir. Studies administrating free ganciclovir (GCV) and liposome-encapsulated GCV into the vitreous body of rabbits have shown no retinopathy in the group that received 1 mg of liposome encapsulated with GCV. Moreover, the plasma concentration of the drug demonstrated therapeutic levels up to 14 days after injection [4]. Cytomegalovirus retinitis (CMV) can also be treated with antisense oligonucleotides targeted to the retina, encapsulated in the liposomes. Thirty-seven percent of the administered oligonucleotides retain in the vitreous humor even after 15 days as shown in studies conducted by Bochot et al. As compared with the release from the solution, liposome-encapsulated phosphodiester (16-mer oligothymidylate) (pdT16) oligonucleotides resulted in sustained release into the vitreous and retinachoroid as well as a reduced distribution to the nontargeted tissues (sclera, lens) [126]. Apart from GCV, other drugs in studies encapsulated in liposomal formulation includes amikacin, amphotericin B35 (a model antisense oligonucleotide), bevacizumab, cyclosporine, 5-fluorouridine, 50 monophosphate, fluconazole, tacrolimus, tobramycin, vasoactive intestinal peptide, an angiogenesis inhibitor tilisolol and ofloxacin [4].

Despite all these advantages, limitations include potential intraocular clouding after the intravitreal injections [8]; liposomes are less preferred as they have short shelf life, limited drug-loading capacity, usage of aggressive conditions for preparation, and problems in sterilization [126]. Therefore, liposome technology has also been used to develop light-induced systems for treating retinal diseases (e.g., Visudyne1, Novartis Pharmaceuticals, USA). A brief description of light-induced system is given below:

Light-induced systems include light-activated drugs (i.e., photodynamic therapy) and light-activated drug delivery systems. Verteporfin (Visudyne1) is the only light-activated (i.e., photodynamic therapy) ocular liposomal drug currently in the clinical use. It is used to treat choroidal neovascularization and age-related macular degeneration. Visudyne1 is insufficient in some cases, and patients need repeated treatments. This is so because after intravenous infusion of Visudyne1, a nonthermal red laser is applied to the retina to activate verteporfin that causes local damage to neovascular endothelium, thus resulting in occlusion of the targeted vessels. Moreover, photodynamic therapy induces an increased local production of VEGF and potential reappearance of the choroidal neovessels. Another liposomal photosensitizing agent that aims to treat age-related macular degeneration is rostaporfin (Photrex1, Miravant Medical Technologies, USA) whose FDA approval is still pending. The frequency of the required treatments is significantly lower than Visudyne1. Paasonen et al. describes another light-triggered drug delivery system that immobilizes gold nanoparticles into the liposomes. When exposed to light, the liposomal contents were released, presumably owing to the heat transfer from the gold nanoparticles into the lipid bilayers and subsequent increased leakiness of the lipid layer. This light-controlled drug delivery to the intraocular structures is preferable and popular as there is wide use of lasers in ophthalmology. However, precaution and attention should be paid to the choice of the wavelength and energy of the laser to avoid retinal adverse reactions [8].

22.3.2 Niosomes

These are modified forms of liposomes. Niosomes are nonionic bilayered surfactant vesicles that can encapsulate both hydrophilic and lipophilic drugs. Niosomes are preferred over liposomes due to their enhanced chemical stability. They also have low toxicity because of their nonionic nature. Since surfactants are used, special precautions and conditions are not required like phospholipids. Therefore, niosomes are flexible in their structural composition, fluidity, and size. Niosomes are biodegradable, biocompatible, and non-immunogenic and increase therapeutic effect of the drug by better bioavailability and controlled delivery to the particular site. Successful niosomes for ocular delivery include cyclopentolate-loaded delivery systems, which improve the bioavailability of cyclopentolate as compared to the reference buffer solution, as niosomes behave independent of their pH. Some modified forms of niosomes like discomes are also utilized in ocular drug delivery. Although disk-shaped discomes have larger size (12–16 mm) than niosomes, they have added advantages in ocular drug delivery. Discomes are obtained from niosomes by incorporating nonionic surfactant Solulan C24. Owing to the larger shape of discomes, drainage into the systemic pool is prevented. The disk shape of niosomes also fits better in the cul-de-sac of the eyes. When Solulan C24 is progressively added into vesicular dispersion, it leads to partitioning of the soluble surfactant into a lipid bilayer at a particular critical level. As a result, large flattened disklike structures are formed. Increased entrapment efficiency, as well as increased ocular bioavailability of timolol maleate discomes, is observed and reported by Vyas et al. as compared to timolol maleate solution [126].

Bioavailability of water-soluble drugs entrapped in niosomes increases due to the fact that surfactants (main ingredient of niosomes) act as penetration enhancers and hence remove the mucus layer and break junctional complexes. Moreover, the irritation power of surfactants decreases in the following order: cationic > anionic > ampholytic > nonionic; therefore, nonionic surfactants are least irritating [71].

22.3.3 Nanoemulsions/Microemulsions

Micro- and nanoemulsions comprise of surfactant and co-surfactant, thereby increasing membrane permeability and drug uptake. They form micelles of oil and water that are of approximately 100 nm or less in diameter. Micro- and nanoemulsions are highly preferred in ophthalmic drug delivery because of their intrinsic properties and specific structures. These formulations could be easily prepared through emulsification, can be sterilized, and have high efficiency for dissolving drugs. As the surfactant and cosurfactant increases membrane permeability and drug uptake, the delivery system can act as a penetration enhancer to facilitate drug delivery. Emulsion-based delivery systems achieve sustained release of a drug when applied to the cornea and hence enhance higher penetration into the deeper layers of the ocular structure and the aqueous humor than the native drug. The prolonged release potential makes these delivery systems very attractive for ocular administration and significantly decreases the frequency of application of eye drops. For example, studies have proved that some pilocarpine-based microemulsions delay the therapeutic effect of the drug in such a way that two daily instillations of these systems were equivalent to four instillations of conventional eye drops. Instillation of dexamethasone-containing microemulsions in the eyes of rabbits has resulted in increased intraocular permeability. However, as compared to nanospheres and liposomes, nanoemulsions are unsuitable for long-term sustained drug release [4].

22.3.4 Nanoparticles (Nanospheres, Nanocapsules, and Nanosuspensions)

Nanospheres are metrical-type nanostructures that entrap or adsorb the biologically active molecules onto the surface [34]. They are of uniform size ranging from 50 to 1000 nm. Active molecules are encapsulated either in synthetic or natural polymers that permit sustained local release and tissue targeting of the drugs. Commonly used biomaterials include polyacrylates, polyalkylcyanoacrylates, polycaprolactones, polylactic acid (PLA), polyglycolic acid (PGA), and their copolymers poly(lactic–co-glycolic acid) PLGA dextran, albumin, gelatin, alginate, collagen, hyaluronic acid, and chitosan. Nanospheres of PLA and PLGA do not show any electrophysiological or histological toxicity in the retina. When fluorescent 2000-, 200-, and 50-nm nanospheres were injected into the vitreous body of rabbits, the 2000-nm particles were found in the intravitreal cavity and the trabecular, whereas the 200- and 50-nm particles were found even inside the retina [4].

Nanocapsules are reservoir-type nanostructures within a surrounding polymeric wall containing an oil core where the active molecule is dissolved. Both nanospheres and nanocapsules could be modified by coating with a hydrophilic polymer or even functionalized with antibodies attached to the coating. Numerous nanoparticles for ocular drug delivery are developed, and some are still under study. For example, flurbiprofen-loaded PLGA nanoparticles successfully reduced inflammation in an in vivo rabbit model of ocular inflammation. As compared to commercial flurbiprofen formulation, nanoparticles were more effective in reducing inflammation as evaluated by direct observation of clinical signs. Glaucoma and cytomegalovirus (CMV) retinitis, a prevalent infectious retinal disease in immunosuppressed patients in AIDS, could be treated with ganciclovir-loaded albumin nanoparticles [34]. Nanoparticles can be targeted in RPE cells. Studies by Bouarges et al. have shown that nanoparticles get retained in the RPE cells even after four months after a single intravitreal injection. In vivo rat studies of albumin nanoparticles have demonstrated their presence in the vitreous cavity and ciliary body for minimum 2 weeks after a single invitreal injection [8].

Nanosuspensions have the advantage in that poor water-soluble drugs could be suspended in a suitable dispersion medium. This nanosuspension technology is appropriate for drug molecules that form crystals with high energy content and makes them insoluble in either organic (lipophilic) or hydrophilic media. Nanosuspensions prepared from polymeric resins serve as important drug delivery vehicles that prolong drug release and enhance bioavailability. They can be successfully used for delivery of ophthalmic drugs, as they are inert and do not irritate the cornea, iris, or conjunctiva. Nanosuspensions form an attractive delivery system for drug delivery to the eye as it outshines the inherent difficulties associated with conventional ocular drug delivery. For example, flurbiprofen (FLU)-loaded nanosuspensions of Eudragit RS 100 and RL 100 polymer resins are used to prevent miosis induced during extracapsular cataract surgery. Nanosuspensions are mostly prepared by techniques that avoid toxic chemicals like QESD (quasi-emulsion solvent diffusion). Sometimes positive charge imparted to nanosuspensions aids their interaction to the corneal surface [126].

22.3.5 Cyclodextrin Nanosuspensions

Delivery of drugs to the posterior segment of the eyes could be made effective by cyclodextrin nanoparticle suspension in the form of eyedrops. Therefore, cyclodextrin nanosuspensions are promising for ocular drug delivery. Cyclodextrins, comprising a group of cyclic oligosaccharides, form inclusion complexes with many drugs. They have the ability to permeate biological membranes, increasing the aqueous solubility of hydrophobic drugs without changing their molecular structure by forming cyclodextrin complexes. Cyclodextrin complexes for ocular delivery increase corneal penetration, ocular absorption, and the efficacy of poor water-soluble drugs like dexamethasone, cyclosporine, acetazolamide, and so on. In conventional ocular drug delivery, penetration enhancers such as benzalkonium chloride are used to disrupt biological barriers. Cyclodextrins containing ocular delivery portion on the surface of corneal epithelium increase the bioavailability of dexamethasone and dexamethasone acetate (HP-b-CD) as reported by Usayapant et al. Irritation caused to the eyes by drugs could be masked by forming complexes with cyclodextrins. For example, pilocarpine prodrug/sulfobutyl ether b-cyclodextrin (SBE-7-b-CD) complexes act as a reservoir that controls free prodrug concentration at the precorneal area to a nonirritating level as reported. High concentration of 45 % or more hydrophobic HP-b-CD has been found to be nonirritating when applied to rabbit eye in the form of aqueous eyedrop solutions. Derivatives of cyclodextrins like hydroxypropyl derivatives of band c-cyclodextrin, randomly methylated b-cyclodextrin, and sulfobutyl ether b-cyclodextrin are utilized in nanotechnology-based ophthalmic drug delivery [4, 126].

22.3.6 Dendrimers

Dendrimers possess attractive properties that make them an effective vehicle for ocular drug delivery. These are macromolecule compounds encompassing a number of branches around an inner core. Biological recognition of dendrimers is easier because of their nanometer size range, ease of preparation, surface functionalization, and ability to display multiple copies of surface groups. Polymers like poly(amidoamine) (PAMAM) are used for the preparation of dendrimers rather than other bioadhesive polymers, as they are associated with difficulties like blurred vision caused by the formation of layer in the corneal area causing loss of eyesight. PAMAM are liquid or semiliquid having several amine, carboxylic, and hydroxyl surface groups which keep on increasing with the generation number (G0,G1,G2, and so on). There are greater possibilities of developing better ocular drug delivery using dendrimers due to this unique inherent architecture. This unique potential architecture facilitates solubilization of strongly and poor water-soluble drugs into their inner zones, which contains cascading tiers of branch cells with radial connectivity to the initiator core and an exterior or surface region of terminal moieties [126]. PMAM dendrimers containing carboplatin administered in subconjuctival space reduce retinoblastoma tumor growth to a much greater extent than drug in solution on day 22 after dosage in a mouse model. It sustains drug delivery to the tumor when its particle size ranges from approximately 260 nm rather than 20 nm, as particle size $\geq$200 nm is retained for longer periods as measured by dynamic light scattering [78].

22.3.7 Nanoparticle-Loaded Contact Lenses

Contact lenses containing drugs are effective in ocular drug delivery. Contact lenses reside for a longer time as compared to about 2–5 min for conventional eyedrops. This long residence time increases drug permeation through the cornea and hence the therapeutic effect. Drug-loaded contact lenses no doubt provide continuous drug release by slow diffusion of the drug molecules through the lens matrix. However, these conventional soaked lenses have certain limitations like low drug loading. Low drug loading in contact lenses is due to the fact that the amount of drug loaded in the lens matrix depends on the equilibrium solubility of the drug in the matrix, which is very less for most of the drugs. Moreover, it cannot be used for long-term delivery as the loaded drugs diffuse easily within few hours. Therefore, these limitations could be overcome by encapsulating the drug in nanoparticles like liposomes, microemulsions, etc., and then dispersing it into the contact lens material or polymer matrix. Since drugs are already encapsulated into the nanoparticles, further interaction with polymer is prevented. Therefore, these delivery systems could be used for long-term therapy as the drug first diffuses through the nanoparticles and then penetrates into the particle surface to reach the hydrogel matrix. Hydrogels of poly-2-hydroxyethyl methacrylate (p-HEMA) loaded with oil-in-water (o/w) microemulsions and drug were prepared and investigated by Gulsen et al. They found that drug residence time was increased significantly, and drug was released in two distinct separate phases. Initially 50 % of the drug was released as burst in the first few hours and then slower release for a few days. Initially drug absorbed on the surface of the nanoparticles and the gel is released, and later on drug entrapped inside the oil drops is released [126]. This type of nanoparticle-loaded contact lenses can be used for retinal damage, including glaucoma and proliferative vitreoretinopathy [103] (Tables 22.1 and 22.2).

22.4 Nanomaterials and the Cause of Their Toxicity

Nanomaterials as unique structure have been demonstrated to possess electronic, optical, and magnetic properties, which are due to their physical nanoscale dimensions, and the possible lysis of these nanomaterials could lead to unique type of toxic effects in vivo. These types of possible toxic effect are difficult to predict. A growing body of literature also suggested reactive oxygen species (ROS) production by nanostructures as the surface of nanostructures is involved in many catalytic and oxidative reactions. As a result, cytotoxicity may be induced which could be greater than a similar bulk material's due to the increase in the surface area-to-volume ratio for nanomaterials. In nanomedicine, metals and compounds used have their own toxic effect, and breakdown of these materials could elicit similar toxic effects as the components themselves. Sometimes, the toxic response can be more pronounced due to scaling the materials in nanolevel. Some of the mechanisms of toxicity by nanoparticles are depicted in Fig. 22.2.

Table 22.1 List of some nanomaterials in ocular drug delivery system with clinical status

Nanomaterial	Drug	Formulation	Indication	Clinical status	References
Micelles	Dexamethasone	NIPAAM-VP-MAA (eyedrop)	Inflammation	Preclinical	[119]
	Dexamethasone	Pluronic F127/chitosan (eyedrop)	Ocular hypertension	Preclinical	[112]
	Cyclosporin A	MPEG-hexPLA (eyedrop)	Dry eye, autoimmune uveitis	Preclinical	[33]
	Dexamethasone	PHEA-PEG (eyedrop)	Ocular hypertension	Preclinical	[27]
Liposomes	Flurbiprofen	Stearic acid + castor oil (eyedrop)	Inflammation	Preclinical	[46]
	Ciprofloxacin HCl	Chitosan-coated liposomes (eyedrop)	Conjunctivitis	Preclinical	[1]
	Verteporfin	Liposomes (Visudyne) (intravitreal injection)	Classic subfoveal choroidal neovascularization	Launched	[42]
	Fluconazole	Liposomes (eyedrop)	Fungal infection	Preclinical	[51]
	Bevacizumab	Liposomes (intravitreal injection)	Ocular neovascular activity	Preclinical	[2]
	Ciprofloxacin HCl	Chitosan-coated liposomes (eyedrop)	Bacterial growth of *P. aeruginosa*	Preclinical	[94]
Microspheres	Triamcinolone acetonide	PLGA (RETAAC) (intravitreal injection)	Diabetic macular edema	Launched	[21]
Cyclodextrins	Methazolamide	HPbCD and HPMC (eyedrop)	Ocular hypertension	Randomized human	[49]
Dendrimers	Pilocarpine nitrate and tropicamide	PAMAM (eyedrop)	Miosis and mydriasis	Preclinical	[158]
	Carteolol	Phosphorus-containing dendrimers (eyedrop)	Glaucoma	Preclinical	[144]
Hydrogels	Timolol	Topical	Ocular hypertension	Randomized human	[161]
		Transceral iontophoresis		Randomized human	[110]
Nanosuspension (colloidal system)	Sparfloxacin	PLGA (nanosuspension)	Bacterial growth of *P. aeruginosa*	Preclinical	[50]

Table 22.2 List of some nanobiomaterials/polymers used in ocular drug delivery system

Drug	Nanomaterial/polymer	Formulation	Indication	References
Cyclosporine A	PLGA	Nanoparticle	Dry eye syndrome	[3]
	A mixture of PLGA with Eudragit®R	Nanoparticle	Dry eye syndrome	
	Coated with Carbopol®	Nanoparticle	Dry eye syndrome	
Betamethasone sodium phosphate	Chitosan–sodium alginate	Nanoparticles topical nanoreservoir system	Macular edema	[135]
Aceclidin	Polymethyl methacrylate and sulfopropyl methacrylate copolymer	Nanoparticles	Cause miosis	[82]
Amikacin	Polybutyl cyanoacrylate	Nanoparticle	Bacterial infection	[82]
Ketotifen fumarate	Silicone hydrogel	Contact lenses	Inflammation	[169]
Puerarin	Complex of puerarin and poly(amidoamine) (PAMAM)	Dendrimers	Inflammation	[173]
None	Poly(amidoamine) (PAMAM)	Dendrimers solution	None	[158]
Transgene delivery	Polyethylenimine conjugated to gold nanoparticles (PEI2-GNP)	Nanoparticle for gene delivery	Gene therapy to cornea	[136]

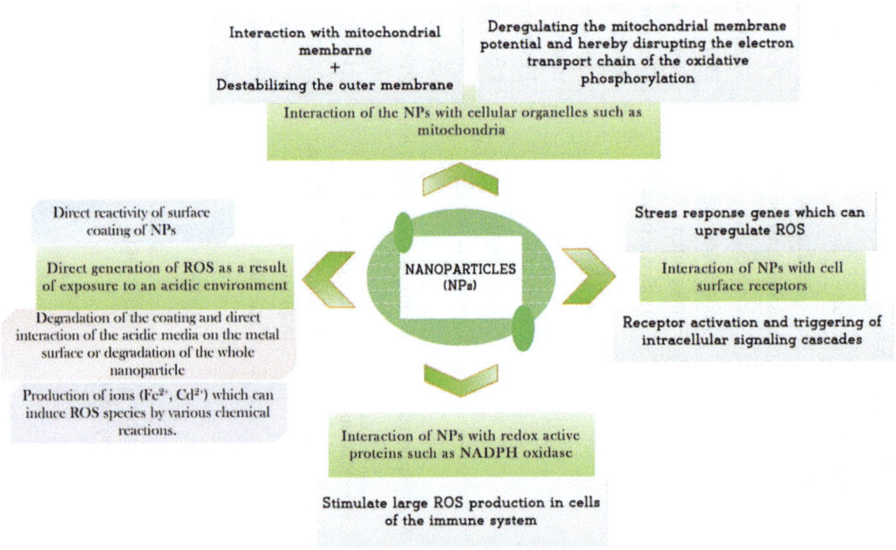

Fig. 22.2 Some predicted mechanisms of nanotoxicity

22.5 Ocular Nanobiomaterials or Nanoformulations: Nanotoxicity Issues

22.5.1 Liposomes

Liposomes are phospholipid vesicles (50–100 nm) with a bilayer membrane structure which is similar to the structure of biological membranes. As liposomes are amphiphilic in nature, they can entrap hydrophilic drugs in the aqueous interior and hydrophobic drugs into the membrane itself. This particular property enables liposomes to be excellent carriers for drugs. Moreover, liposomes show excellent circulation, penetration, and diffusion properties due to their specific physiochemical properties. Based upon the size and number of layers, liposomes are mainly classified as multi-, oligo-, or unilamellar. For specific drug delivery, the surface of liposomes can be modified with ligands and/or polymers, which is proved to be another advantage supporting the scaled-up research in the field of novel drug delivery [153].

Although liposomes have been promoted as a revolutionary concept for the encapsulation and delivery of drugs and genes, the toxicity concerns associated with the use of liposomes are also becoming very much evident as suggested by the growing body of literatures. The advantageous and unique physiochemical properties associated with the liposomes have challenged the safety and efficacy of the same. Literature suggests that there is an urgent need for the assessment of risk and toxicity aspects of liposomes. Studies suggested cationic liposomes as inappropriate to deliver nucleic acids to phagocytic cells. Cationic liposomes have been shown to be toxic toward phagocytic macrophages in vitro irrespective of the formation of complex with DNA. Pro-inflammatory modulators are found to be downregulated in vivo by the administration of cationic liposomes. Cationic liposomes are found to be extremely toxic to CD1 mice following the administration of a single dose, leading to lethal hypothermia. Studies have identified various adverse effects associated with the use of cationic lipids or cationic liposome used for DNA (or drug) delivery [40].

22.5.2 Dendrimers

Dendrimers are three-dimensional, highly branched, reactive macromolecules, with all bonds radiating from a central core. Addition of successive layers is called generations (G), and it gradually increases the molecular weight of the dendrimers. The surface groups of dendrimers can be neutral or positively/negatively charged. Poly(amidoamine) (PAMAM) dendrimers are the first and widely studied family of dendrimers for biomedical use. Their amphiphilic nature and easily modifiable surface make them attractive vehicles for targeted novel drug delivery [114]. Due to the unique structure and properties, dendrimers have emerged as a potential polymeric carrier for novel drug delivery since the mid-1980s. Dendrimers have a globular shape with much more accurately controlled structures. Dendrimers also have a single molecular weight rather than a distribution of molecular weights and a large number of controllable "peripheral" functionalities as compared to traditional linear polymers [87].

Recently, studies have demonstrated poly(amidoamine) (PAMAM) dendrimers as toxic to bacteria and animal cells. PAMAM dendrimers were found to be potentially toxic in *C. reinhardtii* (a green microalga). The median inhibitory concentration (IC_{50}) for cell viability was found as 2, 3, and 5 mg/L for G2, G4, and G5 dendrimers, respectively. PAMAM dendrimers of the tested generations are classified in the "toxic" group (range of 1–10 mg/L) toward *C. reinhardtii* as per the EU Directive 93/67/EEC classification scheme [114]. Another study evaluated the immunotoxicity of three generations of poly(amidoamine) (PAMAM) dendrimers (G4, G5, and G6) in mouse macrophage cells in vitro by utilizing alamarBlue and MTT assays. The results revealed a generation-dependent cytotoxicity of the PAMAM dendrimers in the order of G6 > G5 > G4. The study also correlated well with the observed toxicity to the number of surface primary amino groups, with increasing number resulting in an increase in toxic response. Reactive oxygen species (ROS) generation by dendrimers is another potential toxicity issue of primary concern. PAMAM dendrimers were found to be significantly cytotoxic in mouse macrophage cells (J774A.1) in vitro at a concentration of 0.013–6 μM. PAMAM dendrimers caused an increase in intracellular ROS generation and generation of inflammatory mediators in the order of G6 > G5 > G4 [104]. The results of the previous cytotoxicity studies revealed the toxicity of the PAMAM dendrimers in HaCaT and also in SW480 cells [99, 100]. The growing body of published literature confirmed the toxic pathway of the cationic PAMAM dendrimers through localization in the mitochondria leading to ROS production and resulting in oxidative stress, apoptosis, and DNA damage [100]. Moreover, toxicity studies involving zebra fish as alternative model demonstrated toxicity of G4 dendrimers with amino functional groups revealing attenuation of growth and development of zebra fish embryos at sublethal concentrations [55, 76].

22.5.3 Fullerene/Carbon Nanotube

Carbon nanotubes (CNTs) belong to the family of fullerenes and are composed of graphite sheets (˂100 nm) rolled up into cylinders. They are a distinct molecular form of carbon atoms and are solely composed of carbon atoms. Single-walled carbon nanotubes (SWCNTs) and multiwalled carbon nanotubes (MWCNTs) are the two classes of CNTs. Single-walled carbon nanotubes (SWCNTs) are long wrapped graphene sheets that have a length-to-diameter ratio of about 10^3; thus, their structure can be conceived as nearly one dimensional. Multiwalled carbon nanotubes (MWCNTs) are comparatively larger and consist of many single-walled tubes stacked one inside the other [130]. With superior mechanical, chemical, and electrical properties, carbon nanotubes (CNTs), in both the single-walled and multiwalled (SWCNT and MWCNT) forms, are widely considered to be the wonder materials of the twenty-first century. Despite of having such attractive features, the toxicity of CNTs is a prime concern, with several evidences indicating their similarity to asbestos fibers [116]. Results of many studies involving CNT toxicity suggested mostly DNA damage as the clear mechanism of toxicity. In a study, mesothelial cells exposed to SWCNTs at concentrations ~25 μg/cm^2 activated DNA recovery along with changes in the cell cycle and generation of apoptotic signals [41, 89]. Another study showed that most cells incubated with CNTs get arrested at the G1 phase of the cell cycle [41, 107]. It was also observed that DNA damage was the preferred mechanism of toxicity in a 3-h incubation study with 96 μg SWCNT/cm^2, which indicated DNA damage through micronucleus generation in lung fibroblasts [155].

The respiratory tract system is regarded as the main route for dust or nanoparticle entering the human body, followed by ingestion. Nanoparticulate entities can enter living organisms through inhalation (respiratory tract), ingestion (gastrointestinal tract), dermal absorption (skin), and injection (blood circulation) [73]. Deposited, nanosized particles in the respiratory tract appear to translocate readily to extrapulmonary sites and reach other target organs by different transfer routes and mechanisms [73]. Studies on the toxicity of CNTs have mainly focused on the pulmonary effects of CNTs administered through the intratracheal or pharyngeal routes [93]. Both SWCNTs and MWCNTs might induce cytotoxic effects leading to apoptosis in different cell types, as suggested by several studies [15, 29]. Nanotoxicity reports enumerate the effect of high-purity carbon black (CB; 14 nm and 16 nm) on pulmonary tissue, in which the high-purity CB was found to cause increased oxidative stress in human type II alveolar epithelial cells in vitro [143]. On the other hand, CB (16 nm) was found to cause pulmonary tumor development in rats through subchronic inhalations [143].

CNTs were also found to cause pulmonary inflammation, pulmonary fibrosis, accumulation of neutrophils and eosinophils, mechanical blockage, and increase in various cytotoxicity/inflammatory markers in the lungs [101]. A significant increase in total bronchoalveolar lavage cells, polymorphonuclear leukocytes, protein, lactate dehydrogenase (LDH), tumor necrosis factor-α (TNF-α), interleukin-1β (IL-1β), and mucin levels is also observed [53]. In a study by Jacobsen et al. involving an ApoE$^{-/-}$ mice, significant increases in IL-6, Mip-2, and Mcp-1 messenger RNA were detected in lung tissue, 3 and 24 h following instillation of SWCNTs and CB [62].

Carbon nanotubes caused an inflammatory response, which is more pronounced as compared to ground CNTs. A dose-dependent increase in LDH release (a marker of cell toxicity during the inflammation of lungs) was observed after administration of CNTs as well as ground CNTs. Lung hydroxyproline (OH-proline) and soluble collagen type I contents are important hallmark markers in pulmonary fibrosis. The OH-proline levels were found to be increased significantly and dose dependently after administration of CNTs. CNTs as well as ground CNTs induced a significant increase of the type I collagen lung levels which indicated the dose-dependent fibrotic response to carbon nanotubes [101]. Chou et al. reported severe pulmonary granuloma formation after intratracheal instillation of 0.5 mg of SWCNTs into male imprinting control region (ICR) mice [26]. SWCNTs were found to induce interstitial granulomas and pulmonary injuries in a dose-dependent manner [81].

It was observed that SWCNTs caused profound dose-dependent cytotoxicity in alveolar macrophages isolated from guinea pigs in vitro for 6 h. Both SWCNTs and MWCNTs showed characteristic features of apoptosis at different dosages in alveolar macrophages isolated from guinea pigs. As compared to MWCNTs, quartz, or fullerene used in this study, SWCNTs were found to be more toxic [65]. SWCNTs were also found to possess toxicity in studies involving monocytes/macrophage (THP-1) cells [68]. Stoker et al. also reported inflammatory and cytotoxic effect of SWCNTs in a study assessing the health risk of CNTs on the human respiratory system by using coculture of normal bronchial epithelial cells and normal human fibroblasts [147]. SWCNTs (about 1 nm in diameter and between 100 and 1000 nm in length) were related to the activation of a gene that produces lung cancer [138].

A large proportion of metal catalyst (iron and nickel) is the other cause of toxicity related to CNTs, as they are involved in the formation of free radicals and accumulation of peroxidative products, depletion of total antioxidant reserve, and a loss of cell viability, leading to significant oxidative stress [139]. Ultrafine nature or scaling at nanolevel was found to be responsible for the enhanced toxicity responses by nanoparticles compared with larger-sized particles of similar chemical composition. Magrez [91] examined the cellular toxicity of MWCNTs and other carbon-based nanomaterials in terms of their aspect ratio and surface chemistry using lung tumor cells in vitro and reported the toxicity of these materials as a function of their size [91].

22.5.4 Aluminum Nanoparticles

Wagner et al. [162] examined the effect of aluminum oxide and aluminum nanomaterials on cell viability and cell phagocytosis, in relation to their particle size and chemical composition. Rat alveolar macrophages (NR8383) were exposed to aluminum oxide nanoparticles (Al_2O_3 NP at 30 and 40 nm) and aluminum metal NPs containing a 2–3-nm oxide coat (Al NP at 50, 80, and 120 nm) for screening of initial in vitro cellular effects. Characterization of the nanomaterials, both as received and in situ, was performed using transmission electron microscopy (TEM), dynamic light scattering (DLS), laser Doppler velocimetry, and/or CytoViva150 Ultra Resolution Imaging (URI). DLS and the URI studies showed significant particle agglomeration in vitro. Cell phagocytotic ability was found to be significantly hindered by exposure to 50-, 80-, or 120-nm Al NPs at 25 μg/mL for 24 h. At doses from 100 to 250 μg/mL, significant reduction of cell viability was observed after 24 h of continuous exposure of Al NP. These results indicate that Al NPs exhibit greater toxicity and significantly diminish the phagocytotic ability of macrophages after 24 h of exposure when compared to Al_2O_3 NP. Significant increases in lactate dehydrogenase (LDH) leakage and the induction of apoptosis were also observed by Al NPs in mammalian germ line stem cells at concentrations from 1 to 100 μg/mL after 48 h has been shown [17, 162].

22.5.5 Gold Nanoparticles

Most of the studies involving the assessment of gold nanoparticles are related to their synthesis methods and resulting Au NP size, shape, surface chemistry, and surface charge. Recently, a study in different cell lines demonstrated that toxicity was size dependent; the smallest NPs tested (1.4 nm) had the greatest toxicity compared to other Au NPs with sizes up to 15 nm [108]. Smaller Au nanoparticles (10–50 nm) caused more toxicity in vivo compared to the larger particles (100–200 nm) when intravenously injected into mice [66]. Spherical Au NPs produced by the grinding and vaporizing of bulk gold (2.8, 5.5, and 38 nm in size) were also found to be toxic [174]. Smaller Au NPs produced immunological responses by upregulation of the expression of pro-inflammatory genes interleukin-1 (IL-1), interleukin-6 (IL-6), and tumor necrosis factor (TNF-α) [174]. On the other hand, CTAB-coated Au nanorods were also found to be more toxic than spherical Au NPs (~30 nm) to human HaCaT keratinocytes as observed in a study [163].

Pernodet et al. examined that 14-nm Au NPs cause expression of abnormal actin and extracellular matrix in dermal fibroblasts which lead to major decrease in cell proliferation, adhesion, and motility [113]. Average 20-nm-sized Au NPs induced significant amounts of oxidative DNA damage and downregulated the expression of DNA damage and cell-cycle genes [85]. Au NPs can cross the small intestine and further distribute into the blood, brain, lung, heart, kidney, spleen, liver, intestine, and stomach as suggested by in vivo studies [57]. Embryos when incubated with 0.025–1.2-nm Au NPs for 120 h indicated a 74 % development into normal zebra fish, 24 % dying, and 2 % displaying deformities [18]. Surface charge on Au NP is also considered to be an important factor in assessing nanotoxicity. Cationic (amine) spherical Au NPs (~2 nm in size) were found to be more toxic compared to the anionic or negatively charged Au NPs of the same size on Cos-1 cells, red blood cells, and *E. coli* bacteria [47].

Despite the growing body of literature for Au NP toxicity, there are other studies that report the nontoxic and nonreactive nature of Au NPs toward cells of the body. Many studies indicated a relationship between synthesis conditions (resultant surface chemistry) of Au NPs and their ability to modify the biological response.

22.5.6 Quantum Dots

Quantum dots (QDs) also known as fluorescence nanocrystals are engineered nanoparticles (NPs) that show potential for many biological and biomedical applications, especially in drug delivery/activation and cellular imaging. QDs are also known as semiconductor nanocrystals and are often assembled from atoms from groups II–VI or III–V elements in the periodic table. These particles when compared to their bulk have smaller exciton Bohr radius (EBR) that describes their definition. This property signifies what is described as quantum confinement (QC) when special optical and electronic properties are produced compared to bulk. Major development in the past has focused on the photophysics of nanostructures mainly in microelectronics and optoelectronics. Gradually, there has been great interest in developing nanoparticles for biological and medical applications. The development of cadmium-derived QDs shows great potential for treatment and diagnosis of cancer and site-directed delivery by virtue of their size-tunable fluorescence and with highly customizable surface for directing their bioactivity and targeting [44].

Interestingly, pharmaceutical sciences are exploiting NPs to minimize toxicity and undesirable side effects of drugs. However, unforeseen hazardous properties of the carrier NPs themselves have given rise to some concern in a clinical setting. The kind of hazards encountered with these new nanotechnology materials is complex compared to conventional limitations created by traditional delivery systems. There is no consistent data on QD toxicity as there are many varieties of QD synthesized. Each QD type possesses its own inherent physicochemical properties, which in turn dictates its potential toxicity or lack of it. There is no systematic body of data currently in literature regarding the toxicity of QD, and this can be attributed to a number of factors: absence of proper toxicology-based studies, the many types of QD dosage/exposure concentrations reported in scientific literature, and the diverse physicochemical properties of individual QD. Data from studies primarily designed for toxicological assessment (e.g., dose, duration, frequency of exposure, and mechanism of action) are far and few.

Understanding the relative toxicities of different modes of nanoparticle exposure as compared with their dissolved metal ions is an emerging area in ecotoxicology. Bioavailability, toxicity, and bioaccumulation of carboxyl-functionalized CdSe/ZnS quantum dots (QDs) to the amphipod *Leptocheirus plumulosus* exposed to equivalent Cd concentrations via dissolved Cd, QDs in water, or QDs in algal food are reported in a study. Exposure to QDs via algae resulted in high amphipod mortality. Cadmium and Se in amphipods exposed to QDs in water were highly correlated and spatially localized within the amphipod. In contrast, when exposed to QDs via algae, the metals were more disperse and not highly correlated, suggesting QD dissolution and resultant metal ion toxicity. The study suggests that QDs are accumulated to a greater extent than the dissolved ion and could lead to trophic transfer. QDs ingested with algae are bioavailable and result in toxicity, which is not observed in the absence of algae [61].

Another study reported a comparative study on the toxicities of quantum dots (QDs) and silica-coated CdTe nanoparticles (NPs) to mice after intravenous injection. The blood cells' analysis showed significant increased level of white blood cells (WBCs) in groups treated with CdTe QDs as compared to the control, while red blood cells (RBCs) and platelet counts were normal in treated as well as control groups. The concentration of biochemical markers of hepatic damage, alanine aminotransferase (ALT) and aspartate aminotransferase (AST), was in the normal range in all the groups. However, renal function analyses of mice showed significant increase in the concentration of blood urea nitrogen (BUN) and creatinine (CREA) in mice treated with CdTe QDs, while remaining within normal ranges in both the

CdTe and SiO_2 NPs and control group. The results of histopathology showed that the CdTe QDs caused mild nephrotoxicity, while other organs were normal, and no abnormalities were detected in control and CdTe- and SiO_2-treated group. These findings suggest that the nephrotoxicity could be minimized by silica coating, which would be useful for many biomedical applications [125].

Recent studies on the synthesis of aqueous cadmium sulfide (CdS) quantum dots (QDs) and silica-encapsulated CdS QDs by reverse microemulsion method and utilized as targeted bio-optical probes have reported the role of CdS as an efficient cell tag with fluorescence on par with previously documented cadmium telluride and cadmium selenide QDs, which have been considered to impart high levels of toxicity. In the study, the toxicity of bare QDs was efficiently quenched by encapsulating them in a biocompatible coat of silica. The toxicity profile and uptake of bare CdS QDs and silica-coated QDs, along with the CD31-labeled, silica-coated CdS QDs on human umbilical vein endothelial cells and glioma cells, were investigated. The effect of size, along with the time-dependent cellular uptake of the nanomaterials, has also been emphasized. Enhanced, high-specificity imaging toward endothelial cell lines in comparison with glioma cells was achieved with CD31 antibody-conjugated nanoparticles. The silica-coated nanomaterials exhibited excellent biocompatibility and greater photostability inside live cells, in addition to possessing an extended shelf life. In vivo biocompatibility and localization study of silica-coated CdS QDs in medaka fish embryos, following direct nanoparticle exposure for 24 h, authenticated the nanomaterials' high potential for in vivo imaging, augmented with superior biocompatibility. As expected, CdS QD-treated embryos showed 100 % mortality, whereas the silica-coated QD-treated embryos stayed viable and healthy throughout and after the experiments, devoid of any deformities. The study provided a highly cogent and convincing evidence for such silica-coated QDs as a model nanoparticle in practice, to achieve in vitro and in vivo precision targeted imaging [175]. In another study, the dependence of CdSe QD toxicity on surface coating was clearly demonstrated by the influence of in-house synthesized QDs on cell viability in comparison to commercially available coated QDs. Sensitivity to gastric fluid treatment suggests that toxicity of CdSe QDs can depend on the route of exposure. Specifically, the acidic gastric fluid may damage QDs' protective coating and lead to direct contact of the CdSe core with cells, resulting in cell death. On the other hand, an increase in cell attachment and viability was observed after treatment of QDs with simulated gastric fluid in the case of in-house synthesized CdSe QD preparations containing free Cd^{2+}, possibly due to the formation of a cadmium carbonate precipitate removing free Cd^{2+} from the QD preparation. This suggests that the secretion of sodium carbonate to neutralize gastric acid during the digestion process in the human GI tract may help to reduce free Cd^{2+} released by CdSe QDs through formation of a cadmium carbonate precipitate. The removal of the free Cd^{2+} ion through dialysis greatly decreased the toxic effect of in-house synthesized QDs, indicating that the release of $Cd^{2+}$is one of the main mechanisms of CdSe QD cytotoxicity. In general, the results have shown that CdSe-core QD toxicity can vary depending on coating and treatment with acid, highlighting the importance of considering exposure route in evaluating nanomaterial toxicity [159].

22.6 Nanotoxicity of Nanobiomaterials: In Vitro–In Vivo Evaluation Methods

One important aspect of nanobiomaterials is their potential interaction or their interference with the fluid in the ocular area. Some nanobiomaterials have proved to be toxic to other tissues of the body resulting in increase in oxidative stress, producing inflammations, etc.

22.7 Nanotoxicity Assessment: In Vitro Assays

22.7.1 Cell Viability

Cell viability is one of the most commonly investigated and utilized methodologies in cytotoxicity testing. There are various measurable cellular processes which indicate cell viability and also many endpoints currently investigated to depict an actual state of cultured cells in vitro. With the increasing concern of investigating potential toxicity by nanoparticles or nanomaterials, these methods are constantly under challenge in terms of their efficacy in this new arena of nanotoxicity assessment. The following are the methods utilized for the assessment of nanotoxicity in vitro.

22.7.2 MTT Assay Detecting Mitochondrial Activity

This is a colorimetric assay using a yellow tetrazolium dye 3-(4,5-dimethylthiazol-2-yl)-2,5-diphenyltetrazolium bromide (MTT). It is one of the most popular assay methods utilized for measuring cell viability. The assay is based on the reduction of the yellow tetrazolium dye (MTT) to a purple water-insoluble formazan complex in cells bearing intact mitochondria [97, 142]. It is a very popular method continuously improving with time and has also been applied in numerous cytotoxicity studies [13]. The MTT assay has been used to validate other methods and to determine nanoparticle toxicity [145]. Several studies reported utilization of MTT assay method for nanotoxicity assessment of nanoparticles composed of titanium dioxide [164], iron oxide [59], zinc oxide [132], chitosan [58], and silica [111] and for fullerenes ([172]) and for naked or chitosan-coated quantum dots [149].

22.7.3 Detection of LDH Release Upon Necrosis

The colorimetric lactate dehydrogenase (LDH) assay is a very popular assay method used to evaluate cell or tissue damage [77, 167]. The LDH assay method is based on the oxidation of the yellow tetrazolium salt INT

(2-(p-iodophenyl)-3-(p-nitrophenyl)-5-phenyltetrazolium chloride) to a red formazan complex. Upon cellular necrosis, a sufficient amount of LDH is leaked out of the cell [80]. Therefore, detection and measurement of LDH can be used a marker for cellular necrosis. LDH activity is measured in the cell culture supernatant. In cellular metabolic activity, LDH uses pyruvate as a substrate to produce lactate releasing NADH + H^+ [80]. For assay, INT is utilized as a substitute substrate for the original substrate pyruvate LDH [102]. The cytotoxic potential of various nanoparticles or nanomaterials has been investigated using INT as substrate. The reported studies included cytotoxicity of silica [86], iron oxide [59], TiO_2 [64], ZnO [132], SWCNTs [170], and fullerenes [122].

22.7.4 Annexin V/Propidium Iodide Staining for Apoptotic and Necrotic Cells

Annexin V [vascular anticoagulant alpha (VAC) alpha] is a member of the family of calcium- and phospholipid-binding proteins, which is generally used to detect apoptotic cells [79]. In a calcium-dependent manner, it binds strongly to phosphatidylserine [154]. In the method, fluorescently labeled annexin V is used to detect apoptotic cells [38]. Normally annexin V binds to phosphatidylserine present inside plasma membrane of necrotic cells, producing false-negative results upon cell disintegration. Therefore, propidium iodide (3,8-diamino-5-[3-(diethylmethylammonio) propyl]-6-phenylphenanthridinium diiodide) has to be used as a co-stain to selectively stain the necrotizing cells [9, 11]. Several studies reported utilization of this staining method along with flow cytometry for the evaluation of apoptosis for TiO_2 nanoparticles [6], for pure and polyhydroxylated fullerenes [60], for SWCNTs [165], and for quantum dots [22].

22.7.5 Detection of Intact Lysosomes Via Neutral Red Uptake

Neutral red (3-amino-7-dimethylamino-2-methylphenazine hydrochloride) is a very renowned viability stain that has found its use in numerous cytotoxicity studies including cell proliferation and adhesion assays [14, 83]. Neutral red is a weak cationic dye. It is thought to be engulfed into the cytosol by nonionic diffusion through the cell membrane. Then the dye gets accumulated in the lysosomes of viable cells, where it is excluded from dead cells [105]. Fluorimetry or spectrophotometry can be used to study the uptake of neutral red. REACH (Registration, Evaluation, and Authorization of Chemicals directive by the European Commission) included this neutral red uptake method utilizing NIH3T3 mouse fibroblasts for the in vitro toxicity assessment of chemicals [28]. This method is validated only in NIH3T3 mouse fibroblasts for toxicity testing [36]. Various studies reported the use of neutral red uptake assay for a number of different nanoparticles such as TiO_2 [120], SWCNTs [30], multiwalled carbon nanotubes (MWCNTs) [96], and chitosan nanoparticles [58].

22.7.6 Detection of the Apoptosis Marker Caspase-3

Activation of the caspase-3 pathway is a distinctive characteristic of apoptosis that can be used to quantify activators and inhibitors of the "death cascade" in cellular assays. The detection of active caspase-3 is a very popular and important assay method used in apoptosis assays. Upon activation of caspase-3, it is inevitable that there will be cell death. Activated caspase-3 is detected by measuring the cleavage of a chromophore-linked (pNA) or fluorophore-linked (AFC, AMC) caspase-3 substrate (preferably the amino acids DEVD) (H-Asp-Glu-Val-Asp-OH = DEVD), which absorbs or emits light when separated from the substrate [146]. Caspase-3 assay has been employed to investigate and examine apoptosis in cell culture upon exposure to fullerenes [54], SWCNTs [176], silica nanoparticles [152], quantum dots [23], and to TiO_2 nanoparticles [23].

22.7.7 AlamarBlue Redox System

AlamarBlue is a nontoxic, nonradioactive, eco-friendly reagent used for assessing cellular health. The active ingredient of alamarBlue is a nonfluorescent blue-colored, nontoxic, and cell-permeable compound named as resazurin. The assay utilizes the natural reducing power of living cells. Upon entering the cells, resazurin is reduced to resorufin, which is a fluorescent compound that produces a very bright red fluorescence. Normal viable cells with normal metabolic activity can continuously convert resazurin to resorufin, which generates a quantitative amount of fluorescence, thereby indicating cellular viability – and cytotoxicity [5]. It is a redox indicator and nontoxic to the cells as well as the users. The application of this assay method in nanotoxicity assessment is relatively new.

22.7.8 Incorporation of [3H] Thymidine into the DNA (Deoxyribonucleic Acid)

Incorporation of [3H] thymidine into the DNA is a sensitive and accurate measurement of cell proliferation. This method utilizes radioactive thymidine incorporation into the DNA upon incubation and further measuring the radioactivity by gamma or related counters to assess cellular status. The possible in vitro toxicity, significantly high cost of radioactive material, and requirement of special training and facilities make the utilization of this assay method unadoptable. Moreover, this technique generally requires a longer incubation period (24–48 h) with [3H] thymidine [123]. The use of this method in nanotoxicity assessment is scanty.

22.7.9 Clonogenic Cell Survival Assays

The clonogenic cell survival assay is an old technique that determines the ability of a cell to proliferate indefinitely, thereby retaining its reproductive ability to form a large colony or a clone. This cell is then said to be clonogenic. A cell survival curve can be defined as a relationship between the dose of the agent used to produce an insult and the fraction of cells retaining their ability to reproduce. This particular assay can be used to determine the effectiveness of various agents of concern upon the survival and proliferation of cells [69, 117]. In the assay, cells can be seeded at sufficiently low density and appropriated for growing till colonies are formed. Cells can be pretreated with test sample before plating or can be treated after plating. Colonies formed from single cells as assumed and can be stained with crystal violet or nuclear stains. Colonies with highly proliferating cells can be visualized by visual inspection.

22.7.10 DNA Laddering

DNA laddering is one of the oldest techniques used for DNA damage assay. The technique involves fragmentation of DNA by potential toxicants exposed to the culture followed by fluorescent labeling of the DNA and detection of damage can be done by gel electrophoresis [74].

22.7.11 The Comet Assay

The comet assay is also known as single-cell gel electrophoresis. DNA damage in individual cells can be quantified and analyzed by this technique with excellent sensitivity and rapidity. Agarose gel electrophoresis is used for separating DNA fragments followed by detection through DNA-specific fluorescent dye such as ethidium bromide or propidium iodide. A comet is formed in agarose gel slides upon electrophoresis, demonstrating a head and a tail. The extent of DNA liberated, as suggested by the amount of fluorescent from the head of the comet, is directly proportional to the amount of DNA damage [140].

22.7.12 TUNEL Assay

TUNEL assay is terminal deoxynucleotidyl transferase dUTP(deoxyuridine triphosphate) nick end labeling assay. During apoptosis, DNA fragments are formed, and the assay is based on the incorporation of biotinylated nucleotides conjugated to bromodeoxyuridine (BrdU) at the 3' OH ends of these DNA fragments. The detection system in the assay utilizes a biotin-conjugated anti-BrdU antibody and streptavidin–horseradish peroxidase [43].

22.7.13 Trypan Blue Assay

Trypan blue assay is based on the principle of selective uptake of trypan blue dye by dead cells. The assay involves the treatment of cells with toxicant sample, followed by trypsinization, and subsequent staining with trypan blue, a diazo dye. The dye is only taken up by dead cells, but viable cells remain intact without staining. This method is very advantageous and reflects the actual number of viable cells as compared to control, untreated cells [148].

22.7.14 Stress Response

22.7.14.1 Detection of Reactive Oxygen Species

Molecular probes extend derivatives of reduced fluorescein and calcein as cell-permeant indicators for detecting reactive oxygen species (ROS). Reactive oxygen species can be good indicator of cellular stress or oxidative stress. H_2DCF-DA (2′, 7′-dichlorodihydrofluorescein diacetate) is often used to investigate cellular stress response. It is an extensively used molecular probe for the in vitro detection of intracellular reactive oxygen species (ROS) [56, 63]. In cell culture, the acetylated (non-fluorescent) molecule is taken up by the cells followed by getting trapped inside cytosol through deacetylation. Then intracellular oxidation occurs, and the molecule becomes fluorescent [137, 151]. The fluorescence can be quantified to investigate the cellular oxidative stress level. Generally, the technique involving the DCF is advocated to be a more generalized marker for cellular oxidative stress [151]. The increase in the fluorescence was investigated in cell culture after getting exposed to various nanoparticles including TiO_2 [137], SWCNTs [118], fullerenes [90], and silica particles [32], as hallmark sign of nanotoxicity.

Another method for the detection of oxidative stress induced by nanoparticles or nanomaterials is electroparamagnetic resonance (EPR). *EPR* is, also, a very sensitive technique that has been widely used to assess ROS generated by nanoparticle exposure. An adduct-forming, spin-trapping agent (5,5-dimethyl-l-pyrroline N-oxide, DMPO) for hydroxide (OH-) or superoxide (O^{2-}) radicals or a radical-consuming spin probe (4-hydroxy-2,2,6,6-tetramethylpiperidine-1-oxyl) is used for the detection of radicals by EPR [133, 141].

During oxidative damage or degradation, peroxidation of lipids of cellular membrane occurs, releasing lipid peroxidation products like malondialdehyde as hallmark marker of damage. Reactive oxygen species (ROS) and reactive nitrogen species (RNS) are the culprits responsible for this kind of damage. The presence of malondialdehyde or other thiobarbituric acid-reactive substances ([19]; C. F. [171]) can be assayed for measurement of oxidative degradation of cell membranes. The method has been used widely to demonstrate the power of various nanomaterials to elicit lipid peroxidation in multiple cell types, which include fullerenes in human dermal fibroblast (HDF) and human liver carcinoma (HepG2) cells [129].

22.7.14.2 The Plasmid Assay

The plasmid assay is another technique that has been widely used to detect ROS production [45]. In this assay, unwinding and linearization of a coiled bacterial DNA plasmid is implicated to estimate free radical and ROS exposure and the possible damage by these.

22.8 Inflammatory Response

Detection of markers of inflammation is one of the widely applied techniques to estimate toxicity associated with nanomaterials or nanoparticles. There are various markers identified which serve as hallmark for any kind of inflammatory response apart from the five cardinal signs of inflammation. These markers can be identified and detected in vitro by using enzyme-linked immunosorbent assay (ELISA). The ELISA technique is a very simple, time-tested, and accurate method which utilizes antigen–antibody reaction with enzyme-linked detection system for quantification of inflammatory markers in cell culture supernatants [84]. A growing body of published literature involves assessment of the effect of nanoparticles of different composition and origin, e.g., titanium dioxide [150], iron oxide [166], zinc oxide [132], carbon black [[35], [96]], carbon nanotubes [41], fullerenes [131], silica [121], and quantum dots [124] on living cells as a measure of risk assessment. The most commonly tested inflammatory markers of human and murine origin are the different chemokines including interleukin-8 (IL-8), followed by TNF-α and IL-6 and IL-1b, as well as a few other cytokines and stimulating factors. The chemokine MIP-2 is also usually quantified in rat model systems together with TNF-α and/or IL-6.

22.9 Nanotoxicity Assessment: In Vivo Assays

As of now, there are some models for testing nanotoxicities in the eye. Choosing the best animal model is one of the important criteria. Mice and rats are not the most common model. Rabbit is by far the most common model for toxicity testing. However, there are hindrances as the retinal vasculature of rabbit and humans is different. As compared to humans, the blood vessels of rabbit are rather large and radiate laterally from the optic disk on top of the retina. The lens of both rabbits and mice are very large, and intraocular injections are rather difficult. However, toxicity is being evaluated with the aid of analysis of parameters like behavior, weight change, and mortality. Hematologic toxicity assessment by complete blood count of blood sample from animals exposed to nanoparticles is another method in the battery. Evaluations of immunogenicity, hepatotoxicity, and nephrotoxicity induced by nanoparticles have also been performed in animals preferably in rats and mice. These studies have been supported by serum biochemical markers and histopathology. Additionally, chronic toxicity and the testing of carcinogenicity studies in animals are quite challenging. In some cases, MRI (magnetic resonance imaging) and PET (positron emission tomography) can be used to support the animal toxicological studies. Since nanoparticles possess unique physicochemical properties that are significantly different from traditionally studied particles discussed initially, the traditional animal models of toxicity screening cannot be successfully implicated to nanoparticles. Therefore, there is an urgent need for newer models especially for nanolevel particles or structures.

Nanotoxicology is the rising subdiscipline of nanotechnology and is boosting up the controversial issue of nanotoxicity concerns of nanomaterials. Most of the growing bodies of research on nanotoxicity are focused on the in vitro assays utilizing cell culture systems. The data from this research could be misleading due to the lack of in vitro–in vivo correlation. Keeping in mind the extreme complexity of in vivo systems, it is quite impossible to predict the interactions of the nanostructures with biological components such as proteins and cells. The interaction of nanomaterials could lead to unique biodistribution, clearance, immune response, and metabolism. It becomes necessary to understand the in vivo behavior of nanomaterials for assessing their possible toxic effect and also to establish some predictive models. New toxicity testing systems that rely on predictive, high-throughput technologies are the ultimate goal of testing nanobiomaterials for evaluating their potential hazards. Both in vitro and in vivo assay system and comparison of their data with those of the reference materials in the specific class/subclass are important. Well-designed experiments, determining the physicochemical parameters, may be able to identify the factors responsible for contributing for the toxicities.

One of the promising models of nanotoxicity assessment as reported is the zebra fish (*Danio rerio*). The zebra fish has been used in biotoxicity screening due to the close homology of zebra fish with the human genome. The zebra fish has been found to be physiologically and anatomically similar to human in terms of blood–brain barrier, endothelial cells, and immunogenic responses. The unique physiology and optical clarity of zebra fish embryos also allow testing at all stages of embryonic development. Additionally, expenses are significantly lowered, as zebra fish are much smaller and require less expensive husbandry/housing than do mammals. The time period of study is also reduced due to their high fecundity along with the capability of fertilizing 200–300 eggs per day, every 5–7 days, and has an equivalent longevity and generation time to mice (3–5 months). Moreover, post-fertilization, zebra fish embryos rapidly develop ex utero, and less than 1 week is required for the development of the digestive, nervous, and cardiovascular organ systems. An interesting consideration in the use of zebra fish embryos is the presence of the chorion that surrounds the embryo during development. The chorion comprises pores or channels ca. 0.5–0.7 μm in diameter. Therefore, it can

Table 22.3 Nanomaterial toxicity as determined by zebra fish testing

Nanomaterial class	Zebra fish test	Toxicity reported	References
Fullerenes	Zebra fish embryos; nC_{60} and fullerol ($C_{60}(OH)_{16–18}$) within 96 hpf	• Fullerol – nontoxic at 50 mg/L [70] • nC_{60} – severely toxic at 1.5 mg/L	[10, 178]
	Zebra fish embryos; toxic potential of C60 following light activation in an antioxidant environment to determine toxicity mechanism	• Toxicity determined to be induced by oxidative stress	[157]
Carbon nanotubes	Zebra fish embryos; SWNT and DWNT	• SWNT – not significantly toxic up to 360 mg/L • DWNT – delayed hatching	[24]
Dendrimers	Zebra fish embryos; up to 220 μM G 3.5 and G4±RGD PAMAM dendrimers	• G3.5 – nontoxic • RGD G 3.5 – nontoxic • G4 – dose and time-dependent toxicity • RGD G4 – less potent dose and time-dependent toxicity than G4 dendrimers	[55]
Metal oxides	96 h zebra fish embryo–larval assay; nZnO, $nTiO_2$, nAl_2O_3, and bulk counterparts	• nZnO and bulk – delayed hatching rate and development • $nTiO_2$, nAl_2O_3, and bulk counterparts – nontoxic	[168, 177]
Copper nanoparticles	Adult zebra fish; 80-nm CuNPs and soluble copper ($CuSO_4$)	• CuNP – 48 h LC_{50} of 1.5 mg/L; only 16 % of toxicity is caused by dissolved silver • $CuSO_4$ – 48 h LC_{50} of 0.25 mg/L	[48]

be assumed that the size of common NPs alone should not preclude passage through the chorion, though the specific and/or nonspecific interactions with the chorion are still unknown. All these characteristics allow examining the in vivo course of pathological processes real time, such as organogenesis, vasculogenesis, and deleterious effects during anatomical development. Therefore, the robust zebra fish model allows for nanotoxicity screens of a wide array of nanomaterials including precious metal nanoparticles, carbon-based nanomaterials, and polymers. To date, these screens have not been focused toward facilitating rapid preclinical development of novel pharmaceuticals but rather conservative, parametric benchmarks of toxicity. Some of the reports involving nanotoxicity studies in zebra fish are summarized in the Table 22.3.

22.10 Challenges and Limitations of Nanotoxicity Assessment

Nanotoxicity to the ocular system provokes a great deal of challenges and hurdles. The overall safety of the nanoparticulate system is the most important and unique consideration in the production of nanomedicines, as predictors of nanoparticle toxicity assessment international standard-setting bodies have set up a minimum set of measurement of size, zeta potential (surface charge), and solubility of nanoparticles [31]. Toxicity assessment of nanoparticles is an important topic for researchers, both in material science and biomedical fields. Toxicity assessment methods are not on par with the development of technology, specifically in the field of biological applications of nanoparticles. The lacunae found in the in vitro assays' results are their inconsistencies. In in vivo application, well-defined, consistent assay protocol and techniques should be framed so that investigators could reveal the lead to the unknown, "black box" of particle toxicity. The immediate concern, in this regard, will be the standardization of assessment protocols for nanoparticle toxicity. Government, academics, and worldwide cooperations are desirable to facilitate this process for standardization of assays. In vitro findings should be carefully integrated to the in vivo behavior of nanoparticles since it is a fairly different environment that nanoparticles will experience. For in vivo applications, therefore, extra care should be taken in prediction of potential toxicity of nanoparticles before their actual implementation [25]. Current preclinical investigations seem to provide new approaches to assess the toxicities of the nanobiomaterials in the ocular system.

22.11 Summary and Conclusion

The present review indicates the potential toxicity issues related to various therapeutically implicated nanoparticles that were previously considered biocompatible and safe and their evaluation. The unique properties of nanoparticles like size, shape, surface chemistry, and degree of aggregation influence the production of free radicals and subsequent oxidative stress. Most of the published reports have emphasized the acute toxicity of nanoparticles in traditional animal models only. Due to significantly modified bulk physiochemical properties possessed by nanodimension particles or structures, the traditional methods of toxicity evaluation have failed significantly. Moreover, long-term toxicity issues have been overlooked. Ultimately, appraisal of toxicity of nanoparticles has proven to be challenging, as many factors are associated. It is also true that the possible interaction of nanoparticles or nanomaterials with the biological components is highly unpredictable. Therefore, it has become necessary to design and establish newer in vivo models for nanotoxicity assessment so that nanomaterials can be safely used as therapeutics and as diagnostic tools in ophthalmic drug delivery.

Acknowledgments The author (BBK) gratefully acknowledges the Department of Biotechnology (DBT) of the government of India for providing the Overseas Associate Fellowship for NE Region 2014 and Dibrugarh University, Dibrugarh, Assam, India, for permitting to undertake the proposed study at the University of South Florida, USA.

References

1. Abdelbary G (2011) Ocular ciprofloxacin hydrochloride mucoadhesive chitosan-coated liposomes. Pharm Dev Technol 16(1):44–56. doi:10.3109/10837450903479988
2. Abrishami M, Ganavati SZ, Soroush D, Rouhbakhsh M, Jaafari MR, Malaekeh-Nikouei B (2009) Preparation, characterization, and in vivo evaluation of nanoliposomes-encapsulated bevacizumab (avastin) for intravitreal administration. Retina 29(5):699–703
3. Aksungur P, Demirbilek M, Denkbas EB, Vandervoort J, Ludwig A, Unlu N (2011) Development and characterization of Cyclosporine A loaded nanoparticles for ocular drug delivery: cellular toxicity, uptake, and kinetic studies. J Control Release 151(3):286–294. doi:10.1016/j.jconrel.2011.01.010
4. Al-Halafi AM (2014) Nanocarriers of nanotechnology in retinal diseases. Saudi J Opthalmol. doi:10.1016/j.sjopt.2014.02.009
5. Al-Nasiry S, Geusens N, Hanssens M, Luyten C, Pijnenborg R. The use of Alamar Blue assay for quantitative analysis of viability, migration and invasion of choriocarcinoma cells. (0268–1161 (Print))
6. Aljandali A, Pollack H, Yeldandi A, Li Y, Weitzman SA, Kamp DW (2001) Asbestos causes apoptosis in alveolar epithelial cells: role of iron-induced free radicals. J Lab Clin Med 137:330–339
7. Alonso MJ, Sánchez A (2003) The potential of chitosan in ocular drug delivery. J Pharm Pharmacol 55:1451–1463
8. Amo EMd, Urtti A (2008) Current and future ophthalmic drug delivery systems. A shift to the posterior segment. Drug Discov Today 13(3–4):135–143
9. Aubry JP, Blaecke A, Lecoanet-Henchoz S, Jeannin P, Herbault N, Caron G, . . . Bonnefoy JY (1999) Annexin V used for measuring apoptosis in the early events of cellular cytotoxicity. Cytometry 37:197–204
10. Azevedo Costa CL, Chaves IS, Ventura-Lima J, Ferreira JLR, Ferraz L, de Carvalho LM, Monserrat JM (2012) In vitro evaluation of co-exposure of arsenium and an organic nanomaterial (fullerene, C_{60}) in zebrafish hepatocytes. Comp Biochem Physiol C Toxicol Pharmacol 155(2):206–212. doi:10.1016/j.cbpc.2011.08.005
11. Bartkowiak D, Hogner S, Baust H, Nothdurft W, Rottinger EM (1999) Comparative analysis of apoptosis in HL60 detected by annexin-V and fluorescein diacetate. Cytometry 37:191–196
12. Bejjani RA, BenEzra D, Cohen H, Rieger J, Andrieu C, Jeaanny JC, . . . Behar-Cohen FF (2005) Nanoparticles for gene delivery to retinal pigment epithelial cells. Mol Vis 11:124–132
13. Berridge MV, Herst PM, Tan AS (2005) Tetrazolium dyes as tools in cell biology: new insights into their cellular reduction. Biotechnol Annu Rev 11:127–152
14. Borenfreund E, Puerner JA (1985) Toxicity determined in vitro by morphological alterations and neutral red absorption. Toxicol Lett 24:119–124
15. Bottini M, Bruckner S, Nika K, Bottini N, Bellucci S, Magrini A et al (2006) Multi-walled carbon nanotubes induce T lymphocyte apoptosis. Toxicol Lett 160:121–126
16. Bourges JL, Gautier SE, Delie F, Bejjani RA, Jeanny JC, Gurny R, . . . Behar-Cohen FF (2003) Ocular drug delivery targeting the retina and retinal pigment epithelium using polylactide nanoparticles. Investig Ophthalmol Vis Sci 44:3562–3569

17. Braydich-Stolle L, Hussain SM, Schlager J, Hofmann MC (2005) In vitro cytotoxicity of nanoparticles in mammalian germline stem cells. Toxicol Sci 88:412–419
18. Browning LM, Kerry J, Lee KJ, Huang T, Nallathamby PD, Lowman JE, Xu NXH (2009) Random walk of single gold nanoparticles in zebrafish embryos leading to stochastic toxic effects on embryonic developments. Nanoscale 1:138–152
19. Buege JA, Aust SD (1978) Microsomal lipid peroxidation. Methods Enzymol 52(0076–6879 (Print)):302–310
20. Calvo P, Sánchez A, Martínez J, López MI, Calonge M, Pastor JC, Alonso MJ (1996) Polyester nanocapsules as new topical ocular delivery systems for cyclosporin A. Pharm Res 13:311–315
21. Cardillo JA, Oliveira AAS-F AG (2006) Intravitreal bioerudivel sustained-release triamcinolone microspheres system (RETAAC). Preliminary report of its potential usefulnes for the treatment of diabetic macular edema. Arch Soc Esp Oftalmol 81(12). doi:10.4321/S0365-66912006001200002
22. Chan WH, Shiao NH (2008) Cytotoxic effect of CdSe quantum dots on mouse embryonic development. Acta Pharmacol Sin 29:259–266
23. Chan WH, Shiao NH, Lu PZ (2006) CdSe quantum dots induce apoptosis in human neuroblastoma cells via mitochondrial-dependent pathways and inhibition of survival signals. Toxicol Lett 167:191–200
24. Cheng J, Flahaut E, Cheng S (2007) Effect of carbon nanotubes on developing zebrafish (*Danio rerio*) embryos. Environ Toxicol Chem 26:708–716
25. Choi J, Wang NS (2011) Nanoparticles in biomedical applications and their safety concerns. In: Fazel R (ed) Biomedical engineering – from theory to applications. 51000 Rijeka, Croatia-European Union. InTech, pp 300–314
26. Chou CC, Hsiao HY, Hong QS, Chen CH, Peng YW, Chen HW et al (2008) Single-walled carbon nanotubes can induce pulmonary injury in mouse model. Nano Lett 8:437–445
27. Civiale C, Licciardi M, Cavallaro G, Giammona G, Mazzone MG (2009) Polyhydroxyethylaspartamide-based micelles for ocular drug delivery. Int J Pharm 378 (1–2):177–186. doi:10.1016/j.ijpharm.2009.05.028
28. European Commission (2004) Methods for the determination of physico-chemical properties, toxicity and ecotoxicity. B-1049 Brussels. European Commission
29. Cui DX (2005) Effect of single wall carbon nanotubes on human HEK293 cells. Toxicol Lett 155:73–85
30. Davoren M, Herzog E, Casey A, Cottineau B, Chambers G, Byrne HJ, Lyng FM (2007) In vitro toxicity evaluation of single walled carbon nanotubes on human A549 lung cells. Toxicol In Vitro 21:438–448
31. Desai N (2012) Challenges in development of nanoparticle-based therapeutics. AAPS J 14(2):282–295
32. Deshpande A, Narayanan K, Lehnert BE (2002) Silica-induced generation of extracellular factor(s) increases reactive oxygen species in human bronchial epithelial cells. Toxicol Sci 67:275–283
33. Di Tommaso C, Torriglia A, Furrer P, Behar-Cohen F, Gurny R, Möller M (2011) Ocular biocompatibility of novel Cyclosporin A formulations based on methoxy poly(ethylene glycol)-hexylsubstituted poly(lactide) micelle carriers. Int J Pharm 416(2):515–524. doi:10.1016/j.ijpharm.2011.01.004
34. Diebold Y, Clonge M (2010) Applications of nanoparticles in ophthalmology. Prog Retin Eye Res 29:596–609
35. Duffin R, Tran L, Brown D, Stone V, Donaldson K (2007) Proinflammogenic effects of low-toxicity and metal nanoparticles in vivo and in vitro: highlighting the role of particle surface area and surface reactivity. Inhal Toxicol 19:849–856
36. ECVAM (1997) Statement on the scientific validity of the 3T3 NRU PT test (an in vitro test for phototoxic potential). European Centre for the Validation of Alternative Methods. JRC Environment Institute 21020 Ispra (VA) Italy

37. Edwards A, Prausnitz MR (2001) Predicted permeability of the corneal to topical drugs. Pharm Res 18:1497–1508
38. Engeland M v, Nieland LJW, Ramaekers FCS, Schutte B, Rtelingsperger CPM (1998) Annexin V-affinity assay: a review on an apoptosis detection system based on phosphatidylserine exposure. Cytometry 31:1–9
39. Espuelas MMMS, Mirshahi M, Arnedo A, Irache JM (2002) Efficacy of ganciclovir-loaded nanoparticles in human cytomegalovirus (HCMV)-infected cells. J Drug Target 10:231–238
40. Filion MC, Phillips NC (1998) Major limitations in the use of cationic liposomes for DNA delivery. Int J Pharm 162:159–170
41. Firme Iii CP, Bandaru PR (2010) Toxicity issues in the application of carbon nanotubes to biological systems. Nanomedicine 6(2):245–256. doi:10.1016/j.nano.2009.07.003
42. Fujii NKS (2011) Recent advances in ocular drug delivery systems. Polymers 3:193–221
43. Gavrieli Y, Sherman Y, Ben-Sasson SA (1992) Identification of programmed cell death in situ via specific labeling of nuclear DNA fragmentation (0021–9525 (Print))
44. Ghaderi S, Ramesh B, Seifalian AM (2011) Fluorescence nanoparticles "quantum dots" as drug delivery system and their toxicity: a review. J Drug Target 19(7):475–486
45. Gilmour PS, Brown DM, Beswick PH, MacNee W, Rahman I, Donaldson K (1997) Free radical activity of industrial fibers: role of iron in oxidative stress and activation of transcription factors. Environ Health Perspect 105(Suppl 5):1313–1317
46. Gonzalez-Mira E, Egea MA, Garcia ML, Souto EB (2010) Design and ocular tolerance of flurbiprofen loaded ultrasound-engineered NLC. Colloids Surf B Biointerfaces 81(2):412–421. doi:10.1016/j.colsurfb.2010.07.029
47. Goodman CM, McCusker CD, Yilmaz T, Rotello VM (2004) Toxicity of gold nanoparticles functionalized with cationic and anionic side chains. Bioconjug Chem 15:897–900
48. Griffitt R, Weil R, Hyndman K, Denslow N, Powers K, Taylor D, Barber D (2007) Exposure to copper nanoparticles causes gill injury and acute lethality in zebrafish (*Danio rerio*). Environ Sci Technol 41:8178–8186
49. Guđmundsdóttir E, Stefánsson E, Bjarnadóttir G, Sigurjónsdóttir JF, Guđmundsdóttir G, Masson M, Loftsson T (2000) Methazolamide 1 % in cyclodextrin solution lowers IOP in human ocular hypertension. Invest Ophthalmol Vis Sci 41(11):3552–3554
50. Gupta H, Aqil M, Khar RK, Ali A, Bhatnagar A, Mittal G (2010) Sparfloxacin-loaded PLGA nanoparticles for sustained ocular drug delivery. Nanomedicine 6(2):324–333. doi:10.1016/j.nano.2009.10.004
51. Habib FS, Fouad EA, Abdel-Rhaman MS, Fathalla D (2010) Liposomes as an ocular delivery system of fluconazole: in-vitro studies. Acta Ophthalmol (Copenh) 88(8):901–904. doi:10.1111/j.1755-3768.2009.01584.x
52. Hämäläinen KM, Kananen K, Auriola S, Kontturi K, Urtti A (1997) Characterization of paracellular and aqueous penetration routes in cornea, conjunctiva, and sclera. Invest Ophthalmol Vis Sci 38:627–634
53. Han SG, Andrews R, Gairola C, Bhalla DK (2008) Acute pulmonary effects of combined exposure to carbon nanotubes and ozone in mice. Inhal Toxicol 20:391–398
54. Harhaji L, Isakovic A, Vucicevic L, Janjetovic K, Misirkic M, Markovic Z, . . . Trajkovic V (2007) Modulation of tumor necrosis factor-mediated cell death by fullerenes. Pharm Res 25:1365–1376
55. Heiden T, Dengler E, Kao W, Heideman W, Peterson R (2007) Developmental toxicity of low generation PAMAM dendrimers in zebrafish. Toxicol Appl Pharmacol 225:70–79
56. Hempel SL, Buettner GR, O'Malley YQ, Wessels DA, Flaherty DM (1999) Dihydrofluorescein diacetate is superior for detecting intracellular oxidants: comparison with 20,70-dichlorodih ydrofluorescein diacetate, 5(and 6)- carboxy-20,70 dichlorodihydrofluorescein diacetate, and dihydrorhodamine 123. Free Radic Biol Med 27:146–159
57. Hillyer JF, Albrecht RM (2001) Gastrointestinal persorption and tissue distribution of differently sized colloidal gold nanoparticles. J Pharm Sci 90:1927–1936

58. Huang M, Khor E, Lim LY (2004) Uptake and cytotoxicity of chitosan molecules and nanoparticles: effects of molecular weight and degree of deacetylation. Pharm Res 21:344–353
59. Hussain SM, Hess KL, Gearhart JM, Geiss KT, Schlager JJ (2005) In vitro toxicity of nanoparticles in BRL 3A rat liver cells. Toxicol In Vitro 19:975–983
60. Isakovic A, Markovic Z, Todorovic-Markovic B, Nikolic N, Vranjes-Djuric S, Mirkovic M, . . . Trajkovic V (2006) Distinct cytotoxic mechanisms of pristine versus hydroxylated fullerene. Toxicol Sci 91:173–183
61. Jackson BP, Bugge D, Ranville JF, Chen CY (2012) Bioavailability, toxicity, and bioaccumulation of quantum dot nanoparticles to the amphipod *Leptocheirus plumulosus*. Environ Sci Technol 46(10):5550–5556
62. Jacobsen NR, Moller P, Jensen KA, Vogel U, Ladefogel O, Loft S et al (2009) Lung inflammation and genotoxicity following pulmonary exposure to nanoparticles in $ApoE^{-/-}$ mice. Toxicology 6:1–17
63. Jakubowski W, Bartosz G (2000) 2,7-Dichlorofluorescin oxidation and reactive oxygen species: what does it measure? Cell Biol Int 24:757–760
64. Jeng HA, Swanson J (2006) Toxicity of metal oxide nanoparticles in mammalian cells. J Environ Sci Health A Tox Hazard Subst Environ Eng 41:2699–2711
65. Jia G (2005) Cytotoxicity of carbon nanomaterials: single-wall nanotube, multiwall nanotube, and fullerene. Environ Sci Technol 39:1378–1383
66. Jong WHD, Borm P (2008) Drug delivery and nanoparticles: applications and hazards. Int J Nanomedicine 3:133–149
67. Juberías JR, Calonge M, Gómez S, López MI, Calvo P, Herreras JM, Alonso MJ (1998) Efficacy of topical cyclosporine-loaded nanocapsules on keratoplasty rejection model in the rat. Curr Eye Res 17:39–46
68. Kalbacova M, Kalbac M, Dunsch L, Kataura H, Hempel U (2006) The study of the interaction of human mesenchymal stem cells and monocytes/macrophages with single-walled carbon nanotube films. Phys Status Solidi B 243:3514–3518
69. Kapadia MR, Chow LW, Tsihlis ND, Ahanchi SS, Eng JW, Murar J, . . . Kibbe MR (2008) Nitric oxide and nanotechnology: a novel approach to inhibit neointimal hyperplasia. J Vasc Surg 47(1):173–182
70. Kassem MA, AbdelRahman AA, Ghorab MM, Ahmed MB, Khalil RM (2007) Nanosuspensions as an ophthalmic delivery system for certain glucocorticoid drugs. Int J Pharm 340:126–133
71. Kaur IP, Garg A, Singla AK, Aggarwal D (2004) Vesicular systems in ocular drug delivery: an overview. Int J Pharm 269:1–14
72. Kaur IP, Kanwar M (2002) Ocular preparations: the formulation approach. Drug Dev Ind Pharm 28:473–493
73. Kayat J, Gajbhiye V, Tekade RK, Jain NK (2011) Pulmonary toxicity of carbon nanotubes: a systematic report. Nanomedicine 7(1):40–49. doi:10.1016/j.nano.2010.06.008
74. Keane RW, Srinivasan A, Foster LM, Testa M-P, Örd T, Nonner D, . . . Kayalar C (1997) Activation of CPP32 during apoptosis of neurons and astrocytes. J Neurosci Res 48(2):168–180. doi:10.1002/(sici)1097-4547(19970415)48:2<168::aid-jnr9>3.0.co;2-a
75. Keister JC, Cooper ER, Missel PJ, Lang JC, Hager DF (1991) Limits on optimizing ocular drug delivery. J Pharm Sci 80:50–53
76. King Heiden TC, Dengler E, Kao WJ, Heideman W, Peterson RE (2007) Developmental toxicity of low generation PAMAM dendrimers in zebrafish. Toxicol Appl Pharmacol 225(1):70–79. doi:10.1016/j.taap.2007.07.009
77. Koh JY, Choi DW (1987) Quantitative determination of glutamate mediated cortical neuronal injury in cell culture by lactate dehydrogenase efflux assay. J Neurosci Methods 20:83–90
78. Kompella UB, Amrite AC, Ravi RP, Durazo SA (2013) Nanomedicines for back of the eye drug delivery, gene delivery, and imaging. Prog Retin Eye Res 36:172–198

79. Koopman G, Reutelingsperger CP, Kuijten GA, Keehnen RM, Pals ST, Oers MH (1994) Annexin V for flow cytometric detection of phosphatidylserine expression on B cells undergoing apoptosis. Blood 84:1415–1420
80. Korzeniewski C, Callewaert DM (1983) An enzyme-release assay for natural cytotoxicity. J Immunol Methods 64:313
81. Lam CW, James JT, McCluskey R, Hunter RL (2004) Pulmonary toxicity of single-wall carbon nanotubes in mice 7 and 90 days after intratracheal instillation. Toxicol Sci 77:126–134
82. Langer K, Zimmer A, Kreuter J (1997) Acrylic nanoparticles for ocular drug delivery. S.T.P Pharm Sci 7(6): 445–451
83. Lee JK, Kim DB, Kim JI, Kim PY (2000) In vitro cytotoxicity tests on cultured human skin fibroblasts to predict skin irritation potential of surfactants. Toxicol In Vitro 14:345–349
84. Lequin RM (2005) Enzyme immunoassay (EIA)/enzyme-linked immunosorbent assay (ELISA). Clin Chem 51:2415–2418
85. Li JJ, Zou L, Hartano D, Ong CN, Bay BH, Yung LYL (2008) Gold nanoparticles induce oxidative damage in lung fibroblasts in vitro. Adv Mater 20:138–142
86. Lison D, Thomassen LC, Rabolli V, Gonzalez L, Napierska D, Seo JW, . . . Martens JA (2008) Nominal and effective dosimetry of silica nanoparticles in cytotoxicity assays. Toxicol Sci 104:155–162
87. Liu M, Fréchet JMJ (1999) Designing dendrimers for drug delivery. Pharm Sci Technol Today 2(10):393–401. doi:10.1016/s1461-5347(99)00203-5
88. Losa C, Marchal-Heussler L, Orallo F, Vila-Jato JL, Alonso MJ (1993) Design of new formulations for topical ocular administration: polymeric nanocapsules containing metipranolol. Pharm Res 10:80–87
89. M, MP, Yin XJ, Zhao J, Ding M, Leonard S, Schwegler-Berry D, et al (2008) Raw single-wall carbon nanotubes induce oxidative stress and activate MAPKs, AP-1, NF-kB, and Akt in normal and malignant human mesothelial cells. Environ Health Perspect 116:1211–1217
90. Maeda R, Noiri E, Isobe H, Homma T, Tanaka T, Negishi K, . . . Nakamura E (2008) A water-soluble fullerene vesicle alleviates angiotensin II-induced oxidative stress in human umbilical venous endothelial cells. Hypertens Res 31:141–151
91. Magrez A (2006) Cellular toxicity of carbon-based nanomaterials. Nano Lett 6:1121–1125
92. Mannermaa E, Vellonen KS, Urtti A (2006) Drug transport in corneal epithelium and blood-retina barrier: emerging role of transporters in ocular pharmacokinetics. Adv Drug Deliv Rev 58:1136–1163
93. Maynard AD, Aitken JR, Butz T, Colvin V, Donaldson K, Oberdörster G et al (2006) Safe handling of nanotechnology. Nature 444:267–269
94. Mehanna MM, Elmaradny HA, Samaha MW (2010) Mucoadhesive liposomes as ocular delivery system: physical, microbiological, and in vivo assessment. Drug Dev Ind Pharm 36(1):108–118. doi:10.3109/03639040903099751
95. Merodio M, Arnedo A, Renedo MJ, Irache JM (2001) Ganciclovir-loaded nanoparticles: characterization and in vitro release properties. Eur J Pharm Sci 12:251–259
96. Monteiro-Riviere NA, Inman AO (2006) Challenges for assessing carbon nanomaterial toxicity to the skin. Carbon 44:1070–1078
97. Mosmann T (1983) Rapid colorimetric assay for cellular growth and survival: application to proliferation and cytotoxicity assays. J Immunol Methods 65:55–63
98. Motwani SK, Chopra S, Talegaonkar S, Kohli K, Ahmad FJ, Khar RK (2008) Chitosan-sodium alginate nanoparticles as submicroscopic reservoirs for ocular delivery: formulation, optimisation and in vitro characterisation. Eur J Pharm Biopharm 68:513–525
99. Mukherjee SP, Davoren M, Byrne HJ (2010) In vitro mammalian cytotoxicological study of PAMAM dendrimers – towards quantitative structure activity relationships. Toxicol In Vitro 24(1):169–177. doi:10.1016/j.tiv.2009.09.014
100. Mukherjee SP, Lyng FM, Garcia A, Davoren M, Byrne HJ (2010) Mechanistic studies of in vitro cytotoxicity of poly(amidoamine) dendrimers in mammalian cells. Toxicol Appl Pharmacol 248(3):259–268. doi:10.1016/j.taap.2010.08.016

101. Muller J, Huauxa F, Moreaub N, Missona P, Heiliera JF, Delosc M et al (2005) Respiratory toxicity of multi-wall carbon nanotubes. Toxicol Appl Pharmacol 207:221–231
102. Nachlas MM, Margulies SI, Goldberg JD, Seligman AM (1960) The determination of lactic dehydrogenase with a tetrazolium salt. Anal Biochem 1:317
103. Nagarwal RC, Kant S, Singh PN, Maiti P, Pandit JK (2009) Polymeric nanoparticulate system: a potential approach for ocular drug delivery. J Control Release 136:2–13
104. Naha PC, Davoren M, Lyng FM, Byrne HJ (2010) Reactive oxygen species (ROS) induced cytokine production and cytotoxicity of PAMAM dendrimers in J774A.1 cells. Toxicol Appl Pharmacol 246(1–2):91–99. doi:10.1016/j.taap.2010.04.014
105. Nemes Z, Dietz R, Lüth JB, Gomba S, Hackenthal E, Gross F (1979) The pharmacological relevance of vital staining with neutral red. Cell Mol Life Sci 35:1475
106. Normand N, Valamanesh F, Savoldelli M, Mascarelli F, BenEzra D, Courtois Y, Behar-Cohen FF (2005) VP22 light controlled delivery of oligonucleotides to ocular cells in vitro and in vivo. Mol Vis 11:184–191
107. Jacobsen NR, Pojana G, White P, Moller P, Cohn CA, Korsholm KS, et al (2008) Genotoxicity, cytotoxicity, and reactive oxygen species induced by single-walled carbon nanotubes and C_{60} fullerenes in the FE1-Muta mouse lung epithelial cells. Environ Mol Mutagen 49:476–487
108. Pan Y, Neuss S, Leifert A, Fischler M, Wen F, Simon U, . . . Jahnen-Dechent W (2007) Size-dependent cytotoxicity of gold nanoparticles. Small 3:1941–1949
109. Paolicelli P, Fuente M d l, Sánchez A, Seijo B, Alonso MJ (2009) Chitosan nanoparticles for drug delivery to the eye. Expert Opin Drug Deliv 6:239–253
110. Parkinson TM, Ferguson E, Febbraro S, Bakhtyari A, King M, Mundasad M (2003) Tolerance of ocular iontophoresis in healthy volunteers. J Ocul Pharmacol Ther 19(2):145–151. doi:10.1089/108076803321637672
111. Peng J, He X, Wang K, Tan W, Li H, Xing X, Wang Y (2006) An antisense oligonucleotide carrier based on amino silica nanoparticles for antisense inhibition of cancer cells. Nanomedicine 2:113–120
112. Pepić I, Hafner A, Lovrić J, Pirkić B, Filipović-Grčić J (2010) A nonionic surfactant/chitosan micelle system in an innovative eye drop formulation. J Pharm Sci 99(10):4317–4325. doi:10.1002/jps.22137
113. Pernodet N, Fang X, Sun Y, Bakhtina A, Ramakrishnan A, Sokolov J, . . . Rafailovich M (2006) Adverse effects of citrate/gold nanoparticles on human dermal fibroblasts. Small 2:766–773
114. Petit A-N, Eullaffroy P, Debenest T, Gagné F (2010) Toxicity of PAMAM dendrimers to Chlamydomonas reinhardtii. Aquat Toxicol 100(2):187–193. doi:10.1016/j.aquatox.2010.01.019
115. Pignatello R, Bucolo C, Ferrara P, Maltese A, Puleo A, Puglisi G (2002) Eudragit RS100 nanosuspensions for the ophthalmic controlled delivery of ibuprofen. Eur J Pharm Sci 16:53–61
116. Poland CA, Duffin R, Kinloch I, Maynard A, Wallace WA, Seaton A et al (2008) Carbon nanotubes introduced into the abdominal cavity of mice show asbestos-like pathogenicity in a pilot study. Nat Nanotechnol 3:423–428
117. Puck TT, Marcus PI (1956) Action of X-rays on mammalian cells. J Exp Med 103(5):653–666. doi:10.1084/jem.103.5.653
118. Pulskamp K, Diabate S, Krug HF (2007) Carbon nanotubes show no sign of acute toxicity but induce intracellular reactive oxygen species in dependence on contaminants. Toxicol Lett 168:58–74
119. Rafie F, Javadzadeh Y, Javadzadeh AR, Ghavidel LA, Jafari B, Moogooee M, Davaran S (2010) In vivo evaluation of novel nanoparticles containing dexamethasone for ocular drug delivery on rabbit eye. Curr Eye Res 35(12):1081–1089. doi:10.3109/02713683.2010.508867
120. Ramires PA, Romito A, Cosentino F, Milell E (2001) The influence of titania/hydroxyapatite composite coatings on in vitro osteoblasts behaviour. Biomaterials 22:1467–1474
121. Rao KM, Porter DW, Meighan T, Castranova V (2004) The sources of inflammatory mediators in the lung after silica exposure. Environ Health Perspect 112:1679–1686

122. Roberts JE, Wielgus AR, Boyes WK, Andley U, Chignell CF (2008) Phototoxicity and cytotoxicity of fullerol in human lens epithelial cells. Toxicol Appl Pharmacol 228:49–58
123. Robledo RF, Buder-Hoffmann SA, Cummins AB, Walsh ES, Taatjes DJ, Mossman BT (2000) Increased phosphorylated extracellular signal-regulated kinase immunoreactivity associated with proliferative and morphologic lung alterations after chrysotile asbestos inhalation in mice. Am J Pathol (0002–9440 (Print)). 156(4):1307–1316.
124. Ryman-Rasmussen JP, Riviere JE, Monteiro-Riviere NA (2007) Surface coatings determine cytotoxicity and irritation potential of quantum dot nanoparticles in epidermal keratinocytes. J Invest Dermatol 127:143–153
125. Sadaf A, Zeshan B, Wang Z, Zhang R, Xu S, Wang C, Cui Y (2012) Toxicity evaluation of hydrophilic CdTe quantum dots and CdTe@SiO_2 nanoparticles in mice. J Nanosci Nanotechnol 12(11):8287–8292
126. Sahoo SK, Dilnawaza F, Krishnakumar S (2008) Nanotechnology in ocular drug delivery. Drug Discov Today 13(3–4):144–151
127. Sakai T, Kohno H, Higaki TIM, Saito S, Matsushima M, Mizushima Y, Kitahara K (2006) Treatment of experimental autoimmune uveoretinitis with poly(lactic acid) nanoparticles encapsulating betamethasone phosphate. Exp Eye Res 82:657–663
128. Sasaki H, Yamamura K, Mukai T, Nishida K, Nakamura J, Nakashima M, Ichikawam M (1999) Enhancement of ocular drug penetration. Crit Rev Ther Drug Carrier Syst 16:85–146
129. Sayes CM, Fortner JD, Guo W, Lyon D, Boyd AM, Ausman KD, . . . Colvin VL (2004) The differential cytotoxicity of water-soluble fullerenes. Nano Lett 4(10):1881–1887. doi:10.1021/nl0489586
130. Sayes CM, Gobin AM, Ausman KD, Mendez J, West JL, Colvin VL (2005) Nano-C_{60} cytotoxicity is due to lipid peroxidation. Biomaterials 26(36):7587–7595. doi:10.1016/j.biomaterials.2005.05.027
131. Sayes CM, Marchione AA, Reed KL, Warheit DB (2007) Comparative pulmonary toxicity assessments of C_{60} water suspensions in rats: few differences in fullerene toxicity in vivo in contrast to in vitro profiles. Nano Lett 7:2399–2406
132. Sayes CM, Reed KL, Warheit DB (2007) Assessing toxicity of fine and nanoparticles: comparing in vitro measurements to in vivo pulmonary toxicity profiles. Toxicol Sci 97:163–180
133. Schins RP, Duffin R, Hohr D, Knaapen AM, Shi T, Weishaupt C, Stone V, . . . Borm PJ (2002) Surface modification of quartz inhibits toxicity, particle uptake, and oxidative DNA damage in human lung epithelial cells. Chem Res Toxicol 15(9):1116–1173
134. Schoenwald RD (1990) Ocular drug delivery: pharmacokinetics considerations. Clin Pharmacokinet 18:255–269
135. Shafie MAA, Fayek HHM (2013) Formulation and evaluation of betamethasone sodium phosphate loaded nanoparticles for ophthalmic delivery. J Clin Exp Ophthalmol 4:273. doi:10.4172/2155-9570.1000273
136. Sharma A, Tandon A, Tovey JCK, Gupta R, Robertson JD, Fortune JA, . . . Mohan RR (2011) Polyethylenimine-conjugated gold nanoparticles: gene transfer potential and low toxicity in the cornea. Nanomedicine: Nanotechnol Biol Med 7(4):505–513
137. Shen B, Scaiano JC, English AM (2006) Zeolite encapsulation decreases TiO_2- photosensitized ROS generation in cultured human skin fibroblasts. Photochem Photobiol 82:5–12
138. Shvedova AA (2008) Inhalation vs. aspiration of single-walled carbon nanotubes in C57BL/6 mice: inflammation, fibrosis, oxidative stress, and mutagenesis. Am J Physiol Lung Cell Mol Physiol 295:552–565
139. Shvedova AA, Castranova V, Kisin KR, Schwegler-Berry D, Murray AR, Gandelsman VZ (2003) Exposure to carbon nanotube material: assessment of nanotube cytotoxicity using human keratinocyte cells. J Toxicol Environ Health A 66:1909–1926
140. Singh NP, McCoy MT, Tice RR, Schneider EL (1988) A simple technique for quantitation of low levels of DNA damage in individual cells. (0014–4827 (Print))

141. Singh S, Shi T, Duffin R, Albrecht C, van Berlo D, Höhr D, . . . Schins RPF (2007) Endocytosis, oxidative stress and IL-8 expression in human lung epithelial cells upon treatment with fine and ultrafine TiO_2: role of the specific surface area and of surface methylation of the particles. Toxicol Appl Pharmacol 222(2):141–151. doi:10.1016/j.taap.2007.05.001
142. Slater TF, Sawyer B, Straeuli U (1963) Studies on succinate-tetrazolium reductase systems. III. Points of coupling of four different tetrazolium salts. Biochim Biophys Acta 77:383–393
143. Smart SK, Cassady A, Lu GQ, Martin DJ (2006) The biocompatibility of carbon nanotubes. Carbon 44:1034–1047
144. Spataro G, Malecaze F, Turrin C-O, Soler V, Duhayon C, Elena P-P, . . . Caminade A-M (2010) Designing dendrimers for ocular drug delivery. Eur J Med Chem 45(1):326–334. doi:10.1016/j.ejmech.2009.10.017
145. Spielmann H, Hoffmann S, Liebsch M, Botham P, Fentem JH, Eskes C, . . . Zuang V (2007) The ECVAM international validation study on in vitro tests for acute skin irritation: report on the validity of the EPISKIN and EpiDerm assays and on the Skin Integrity Function Test. Altern Lab Anim 35:559–601
146. Srinivasula SM, Saleh A, Ahmad M, Fernandes-Alnemri T, Alnemri ES (2001) Isolation and assay of caspases. Methods Cell Biol 66:1–27
147. Stoker E, Purser F, Kwon S, Park YB, Lee JS (2008) Alternative estimation of human exposure of single-walled carbon nanotubes using three-dimensional tissue-engineered human lung. Int J Toxicol 27:441–448
148. Strober W (2001) Trypan blue exclusion test of cell viability. Curr Protoc Immunol (1934-368X (Electronic))
149. Tan WB, Zhang Y (2005) Surface modification of gold and quantum dot nanoparticles with chitosan for bioapplications. J Biomed Mater Res A 75:56–62
150. Tao F, Kobzik L (2002) Lung macrophage-epithelial cell interactions amplify particle-mediated cytokine release. Am J Respir Cell Mol Biol 26:499–505
151. Tarpey MM, Wink DA, Grisham MB (2004) Methods for detection of reactive metabolites of oxygen and nitrogen: in vitro and in vivo considerations. Am J Physiol Regul Integr Comp Physiol 286:R431–R444
152. Thibodeau M, Giardina C, Hubbard AK (2003) Silica-induced caspase activation in mouse alveolar macrophages is dependent upon mitochondrial integrity and aspartic proteolysis. Toxicol Sci 76:91–101
153. Torchilin VP (2005) Recent advances with liposomes as pharmaceutical carriers. Nat Rev Drug Discov 4:145–160
154. Trotter PJ, Orchard MA, Walker JH (1995) Ca^{2+} concentration during binding determines the manner in which annexin V binds to membranes. Biochem J 308(Pt 2):591–598
155. Ulrich KE, Cannizzaro SM, Langer RS, Shakeshelf KM (1999) Polymeric systems for controlled drug release. Chem Rev 99:3181–3198
156. Urtti A (2006) Challenges and obstacles of ocular pharmacokinetics and drug delivery. Adv Drug Deliv Rev 58:1131–1135
157. Usenko C, Harper S, Tanguay R (2008) Fullerene C_{60} exposure elicits an oxidative stress response in embryonic zebrafish. Toxicol Appl Pharmacol 229:44–55
158. Vandamme TF, Brobeck L (2005) Poly(amidoamine) dendrimers as ophthalmic vehicles for ocular delivery of pilocarpine nitrate and tropicamide. J Control Release 102(1):23–38. doi:10.1016/j.jconrel.2004.09.015
159. Veeranarayanan S, P A, Mohamed MS, Nagaoka Y, Iwai S, Nakagame Y, Kashiwada S, Yoshida Y, Maekawa T, Kumar DS (2012) Synthesis and application of luminescent single CdS quantum dot encapsulated silica nanoparticles directed for precision optical bioimaging. Int J Nanomedicine 7:3769–3786
160. Vega E, Egea MA, Valls O, Espina M, García ML (2006) Flurbiprofen loaded biodegradable nanoparticles for ophthalmic administration. J Pharm Sci 95:2393–2405

161. Volotinen M, Mäenpää J, Kautiainen H, Tolonen A, Uusitalo J, Ropo A, . . . Aine E (2009) Ophthalmic timolol in a hydrogel vehicle leads to minor inter-individual variation in timolol concentration in aqueous humor. Eur J Pharm Sci 36(2–3):292–296. doi:10.1016/j.ejps.2008.10.004
162. Wagner AJ, Bleckmann CA, Murdock RC, Schrand AM, Schlager JJ, Hussain SM (2007) Cellular interaction of different forms of aluminum nanoparticles in rat alveolar macrophages. J Phys Chem B 111:7353–7359
163. Wang SG, Lu WT, Tovmachenko O, Rai US, Yu HT, Ray PC (2008) Challenge in understanding size and shape dependent toxicity of gold nanomaterials in human skin keratinocytes. Chem Phys Lett 463:145–149
164. Wilson MR, Stone V, Cullen RT, Searl A, Maynard RL, Donaldson K (2000) In vitro toxicology of respirable Montserrat volcanic ash. Occup Environ Med 57:727–733
165. Worle-Knirsch JM, Pulskamp K, Krug HF (2006) Oops they did it again! Carbon nanotubes hoax scientists in viability assays. Nano Lett 6:1261–1268
166. Wottrich R, Diabate S, Krug HF (2004) Biological effects of ultrafine model particles in human macrophages and epithelial cells in mono- and co-culture. Int J Hyg Environ Health 207:353–361
167. Wroblewski F, Ladue JS (1955) Lactic dehydrogenase activity in blood. Proc Soc Exp Biol Med 90:210–213
168. Xiong D, Fang T, Yu L, Sima X, Zhu W (2011) Effects of nano-scale TiO_2, ZnO and their bulk counterparts on zebrafish: acute toxicity, oxidative stress and oxidative damage. Sci Total Environ 409(8):1444–1452. doi:10.1016/j.scitotenv.2011.01.015
169. Xu J, Li X, Sun F (2011) In vitro and in vivo evaluation of ketotifen fumarate-loaded silicone hydrogel contact lenses for ocular drug delivery. Drug Deliv 18(2):150–158. doi:10.3109/10717544.2010.522612
170. Yacobi NR, Phuleria HC, Demaio L, Liang CH, Peng CA, Sioutas C, . . . Crandall ED (2007) Nanoparticle effects on rat alveolar epithelial cell monolayer barrier properties. Toxicol In Vitro 21:1373–1381
171. Yang CF, Shen HM, Shen Y, Zhuang ZX, Ong CN (1997) Cadmium-induced oxidative cellular damage in human fetal lung fibroblasts (MRC-5 cells). Environ Health Perspect 105(7):712–716
172. Yang XL, Fan CH, Zhu HS (2002) Photo-induced cytotoxicity of malonic acid [$C_{(60)}$]fullerene derivatives and its mechanism. Toxicol In Vitro 16:41–46
173. Yao W, Sun K, Mu H, Liang N, Liu Y, Yao C, . . . Wang A (2010) Preparation and characterization of puerarin-dendrimer complexes as an ocular drug delivery system. Drug Dev Ind Pharm 36(9):1027–1035. doi:10.3109/03639041003610799
174. Yen HJ, Hsu SH, Tsai CL (2009) Cytotoxicity and immunological response of gold and silver nanoparticles of different sizes. Small 5:1553–1561
175. Yong K-T, Wang Y, Roy I, Rui H, Swihart MT, Law W-C, . . . Reynolds JL (2012) Preparation of quantum dot/drug nanoparticle formulations for traceable targeted delivery and therapy. Theranostics 2(7):681–694
176. Zeni O, Palumbo R, Bernini R, Zeni P, Sarti M, Scarfì MR (2008) Cytotoxicity investigation on cultured human blood cells treated with single-wall carbon nanotubes. Sensors 8:500–519
177. Zhu X, Zhu L, Duan Z, Qi R, Li Y, Lang Y (2008) Comparative toxicity of several metal oxide nanoparticle aqueous suspensions to zebrafish (*Danio rerio*) early developmental stage. J Environ Sci Health A Tox Hazard Subst Environ Eng 43:278–284
178. Zhu X, Zhu L, Li Y, Duan Z, Chen W, Alvarez P (2007) Developmental toxicity in zebrafish (*Danio rerio*) embryos after exposure to manufactured nanomaterials: buckminsterfullerene aggregates (nC_{60}) and fullerol. Environ Toxicol Chem 26:976–979

Chapter 23
Nanosystem in Ocular Bioenvironment

Fahima Dilnawaz and Sanjeeb Kumar Sahoo

Abstract Delivering drugs at effective therapeutic concentrations to the ocular tissue with minimal side effects is a challenging task. Various physiological and anatomical barriers impede effective ocular delivery. Numerous approaches have been adapted to increase the bioavailability and the duration of drug action, but the effectiveness of drugs still have their limitations. Recently, nanotechnology-based drug delivery approaches have emerged as promising strategies for the delivery of water-soluble/water-insoluble drugs due to improved targeting capability, solubility, efficacy, and safety in administration. Here in this chapter, we have presented an overview of different biological barriers of ocular drug delivery and have reported how different nanocarriers are effective in ocular therapy with improved patient compliance.

Keywords Ocular drug delivery • Nanotechnology • Nanoparticles • Ocular transport • Ocular barriers

23.1 Introduction

Eyes have a complex anatomical structure that make it a challenge to deliver therapeutic agents when affected by disease. The anatomy, physiology, and biochemistry restrict the entry of drug molecules at the required site of action. Clinically, the anterior segment of the eye (cornea, conjunctiva, sclera, anterior uvea) is mostly treated with topical eyedrops. However, the administered eyedrops are drained out rapidly from the ocular surface with less absorption, and as a result, the bioavailability of the drug is very low, typically less than 5 % [1, 2]. Bioavailability and duration of drug activity has been increased modestly by prolonged action of different dosage forms, but they have not resulted in wide acceptance by the patients [2]. Even with modified formulations, there is limited ocular drug absorption by the corneal and conjunctival epithelial barriers of the eye [2], whereas the posterior

F. Dilnawaz • S.K. Sahoo (✉)
Laboratory for Nanomedicine, Institute of Life Sciences,
Nalco Square, Chandrasekharpur, Bhubaneswar, Orissa, India
e-mail: sanjeebsahoo2005@gmail.com

© Springer International Publishing Switzerland 2016

Y. Pathak et al. (eds.), *Nano-Biomaterials For Ophthalmic Drug Delivery*,
DOI 10.1007/978-3-319-29346-2_23

segment (retina, vitreous, choroid) is mainly treated with high doses of drug either intravenously or by intravitreal administration [3]. These invasive methods may cause endophthalmitis. As a result, these routes of administration are not considered to be ideal to a large number of patients. There is an increasing need to expand new prolonged action of dosage forms for subconjunctival and periocular administration. Alternatively, another way of efficient drug targeting is from the bloodstream to the retinal pigment epithelium or choroidal vasculature. Advance studies in this approach should facilitate safer and more efficient treatment of posterior segment diseases with antibodies, genes, oligonucleotides, and growth factors. After the drug has been administered, there are two major obstacles: poor stability and limited transport across epithelia, which prevent the drugs from reaching the ideal biological compartments. The abovementioned problems become acute when biomolecules such as peptides, proteins, and genes face restrictive steps for their adequate clinical exploitation. Hence, an appropriate delivery system that can protect them from the biological barriers and facilitate its movement through the biological environment is extremely necessary [4]. Therefore, ocular drug delivery is one of the most challenging and interesting endeavors for pharmaceutical scientists. In this chapter, various nanocarriers for ocular therapy are discussed along with different biological barriers, routes of administration, and types of nanocarrier disposition to the ocular tissues.

23.2 Biological Barriers to Ocular Drug Delivery

The eye contains numerous barriers such as tear film, cornea, conjunctiva, sclera, and the blood–ocular barrier that impede the efficient passage of many drugs [5].

23.2.1 Tear Film

It is the muco-aqueous layer of the tear film that gives protection to the anterior surface of the eye. The thickness is around ~3–10 μm [6]. It is a three-layered structure consisting of: a, b, and c. (a) There is lipid layer that provides a smooth tear surface and retards the rate of tear evaporation from the cornea. It also mechanically traps and flushes out foreign particles and chemicals. Apart from that, it contains bacteriostatic substances that inhibit the growth of microorganisms. (b) Next is an aqueous layer produced by the lacrimal gland that is composed of water, proteins, and other substances, such as lipocalin, lactoferrin, lysozyme, and lacritin, which are responsible for control of osmotic balance, infection, and water-promoted spreading of the tear film. (c)The mucus layer over the corneal epithelium is the innermost layer and is hydrophilic and serves as an anchor for the tear film. The tear film creates a smooth surface for the light to pass, protects the eye from infection, and nourishes the eye [7, 8].

23.2.2 Cornea

It is a transparent, dome-shaped window covering the front of the eye, which has mostly dense innervated tissues [9, 10]. Its powerful refracting surface provides two-thirds of the eye's focusing power. The cornea constitutes of three layers: epithelium, stroma, and endothelium–all three act as mechanical barriers for foreign substances. Each layer has a different polarity and rate-limiting structure for drug permeation. The lipophilic nature of the corneal epithelium and the tight junctions restrict the paracellular drug permeation from ocular tear film and permeate the lipophilic drugs by transcellular pathways. The hydrophilic charged cationic compounds infuse easily through the cornea than anionic compounds, as the corneal epithelium is negatively charged [11]. The middle stroma layer is highly hydrated and acts as a barrier for the permeation of lipophilic drugs. The corneal endothelium is the innermost monolayer of hexagonal-shaped cells, and it acts as a separating barrier between the stroma and aqueous humor. The endothelial junctions are leaky and facilitate the passage of macromolecules between the aqueous humor and stroma [11, 12].

23.2.3 Conjunctiva

The conjunctiva is the thin, transparent, vascularized mucus membrane that covers the outer surface of the eye. It is involved in the formation and maintenance of the tear film, and it has a rich supply of capillaries and lymphatics. Therefore, the drugs are cleared through blood and lymph [13]. There is no tight junction barrier in the conjunctival blood vessel, so the drug molecules enter the blood circulation by pinocytosis and through the paracellular pores present in the vascular endothelial layer. The conjunctival lymphatics act as an efflux system for the efficient elimination of the drug molecules from the conjunctival space.

23.2.4 Sclera

The sclera (white of the eye) is a fibrous collagen tissue consisting of three layers, starting from the anterior to the posterior: episclera, sclera stroma, and lamina fusca, and continues to the cornea anteriorly. It provides structural integrity and defines the shape and length of the eye [14]. The posterior sclera consists of more loosely weaved collagen fibers than that of the anterior sclera [15]. The human sclera is thick near the limbus, thin at the equator, and thicker near the optic nerve. Hydrophobic drugs are impermeable; the lipophilicity of the drug eliminates permeability. Hydrophilic drugs mostly diffuse by the aqueous medium of proteoglycans in the fiber matrix pores and thus, more easily than lipophilic drugs [16].

23.2.5 Blood–Ocular Barrier

The blood–ocular barrier is formed by two main barriers, i.e., blood–aqueous barrier (BAB) and the blood–retinal barrier (BRB). These physical barriers are mainly created by the endothelium of the retinal and iris capillaries, ciliary epithelium, and retinal pigment epithelium (RPE), and these protect the eye by preventing the entry of toxic substances and maintaining the homeostatic control that underpins the ocular physiology [17]. However, inflammation can break down this barrier, allowing the drugs and large molecules to penetrate into the eye. The permeability of the BAB depends on the ionic concentration gradients. The BAB is composed of iris capillaries and pigmentary epithelium, which allows the transcellular transport (i.e., transport mediated by cell to cell) by means of vesicles, whereas the paracellular transport (i.e., transport across the epithelium passing through the intercellular space between the cells) is controlled by the extension of the tight junctions. Along with the ciliary and retinal pigment epithelium, the iris pigment epithelium acts as an obstacle to the passage of type T activated lymphocytes. It allows small lipophilic drugs to enter the uveal blood circulation and facilitates the rapid elimination from the anterior chamber, whereas the larger hydrophilic drugs are merely eliminated by aqueous humor turnover across the BAB [18]. Apart from that, the physiological process, such as blinking and tears, also lessens the residence time of topical application in less than 30 s [1, 19]. The corneal epithelium is impermeable to any substance larger than 500 Da. Instead of crossing the cornea, these larger substances permeate throughout the conjunctiva and the underlying sclera. This is known as "nonproductive passage"; in this manner, less than 5 % of topically administered drugs reach the intraocular tissues [1, 11]. In this regard, physiological processes and cellular barriers impede the efficient passage of various drugs that led to a minimum dose absorption into the eye. Therefore, the development of an appropriate delivery system that protects them while facilitating their transport through biological barriers is desired.

23.3 Transporters for Ocular Drug Delivery

The membrane-bound proteins present on various ocular tissues, i.e., epithelia of the cornea, conjunctiva, and retina, play an important role in active transport of drugs across biological membranes. There are mainly two types of transport systems in ocular drug delivery, i.e., efflux transporters and influx transporters. The efflux transporters belong to the ATP-binding cassette superfamily. Prominent efflux transporters identified on ocular tissues include multidrug resistance protein (MRP), P-glycoprotein (P-gp), and breast cancer resistance protein (BCRP). The P-gp has an affinity to efflux lipophilic compounds in normal as well as in cancerous cells, which possibly leads to emergence of drug resistance [20]. There is a functional activity of P-gp on various ocular cell lines and tissues such as the cornea, conjunctiva, and retinal pigmented epithelium (RPE) [21, 22]. MRP works in a similar manner but effluxes out organic anions and conjugated compounds. These efflux transporters lower the bioavailability of the molecule at the site of action. The influx transporters belong to the solute carrier superfamily, which facilitate the translocation of essential nutrients and xenobiotics (amino acids, peptides, vitamins, glucose, lactate, and nucleoside/nucleobases) across biological membranes. The most common influx transporters for ocular drug delivery are amino acid and peptide transporters. These proteins may have a physiological role in transporting various amino acids and nutrients into ocular tissues.

23.4 Challenges Faced in Ocular Transport

The occurrence of various ocular disorders may, in the long run, be vision threatening. The principal reasons for this relate back to the natural characteristics of the eye that restrict drug bioavailability. Existence of multiple ocular tissue layers, cellular barriers, and various physiological processes impede the effective passage of many drugs, which lead to a minimal dose absorption into the eye. Various approaches have been taken to develop more efficient therapeutic systems to fight against numerous ocular diseases. Recently, much attention has been given to tailor or personalize effective safe delivery vehicles that can improve the ocular bioavailability of existing therapeutic molecules. Nanotechnology is the science of manipulating matter to a nanometer scale, which holds enough promise to revolutionize therapy. In nanotechnology-based drug delivery systems, the drug is protected from degradation, released in a controlled manner along with targeted delivery, and has a reduced number of side effects. With this approach, various limitations imposed by the conventional formulations are mostly surmounted. Using different nanoformulations, the efflux transporter may be inhibited and the influx transporter may be transported to the targeted place by crossing different ocular barriers [23].

23.5 Colloidal Carriers for Ocular Delivery

Colloidal carriers were developed with an aim to increase the bioavailability of ophthalmic drugs by prolonging corneal/conjunctival epithelium residence time. These new delivery systems have many advantages such as improved residence time, controlled and sustained drug release for prolonged periods, reduction of side effects, and better patient compliance, in comparison to the conventional formulation for ocular delivery (Fig. 23.1). This is possible due to the use of compatible, biodegradable, synthetic/natural polymers for the preparation of different colloidal carriers [1, 23].

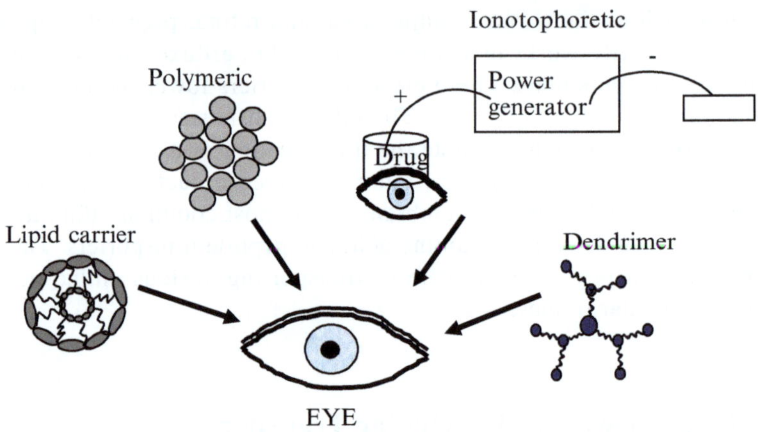

Fig. 23.1 Different colloidal carriers for ocular drug delivery

23.5.1 Polymeric

In polymer-based colloidal carrier systems, the drug is either entrapped or adsorbed in the matrix. Different biodegradable polymers (synthetic or natural) such as poly (lactic-co-glycolic acid) (PLGA), poly-e-caprolactone, polyacrylamide, polycyanoacrylate, poly(methyl methacrylate), etc. [24, 25], and natural polymers like chitosan, gelatin, sodium alginate, albumin, and tamarind kernel polysaccharide [26–29] are effectively used as drug delivery vehicles for the ocular tissues. Among them, PLGA is the most suitable candidate because of ease of formulation and FDA approval. Jwala et al. studied the efficacy of acyclovir-loaded PLGA nanoparticles for the treatment of ocular herpes keratitis which demonstrated sustained release activity, and thus, retarded the degradation of prodrug in deep corneal tissues [30]. Agnihotri and Vavi developed and evaluated diclofenac sodium-loaded PLGA-NPs for ocular use and found good biocompatibility with the eye [31]. Chennamaneni et al. developed dexamethasone-loaded–PLGA–hydroxypropyl methyl cellulose (HPMC) microparticle for the treatment of uveitis and cataract inflammation. In vivo studies in rabbit, where the intracapsular bag was implanted, demonstrated dexamethasone flow from endocapsular space, and significant concentration was present in the anterior and posterior chambers [32]. Basaran et al. reported the efficacy of cyclosporine-loaded chitosan nanoparticles, and the in vivo study in sheep demonstrated prolonged release of drugs due to enhanced residence time at the corneal and conjunctival surfaces [33]. Sabzevari et al. studied the anti-inflammatory effects using polymeric mucoadhesive nanoparticles of triamcinolone acetonide-loaded-poly β-amino ester in rabbit eyes, which demonstrated anti-inflammatory effects, and compared it to microparticles of prednisolone acetate and triamcinolone acetonide, concluding that it is as good as a subconjunctival injection method [34]. The mucoadhesive nature of chitosan along with hyaluronic acid (chief component of extracellular matrix) was used in ocular delivery formulation. Wadhwa et al. reported that the addition of hyaluronic acid (HS) to the timolol maleate-loaded or dorzolamide hydrochloride chitosan nanoparticles gave a synergistic effect for mucoadhesion and resulted in a higher reduction of intraocular pressure level as compared to pure drug solution [35]. Tayel et al. formulated terbinafine hydrochloride nanoparticles with Eudragit RS100 and chitosan polymers using Pluronic F68 surfactant, which showed increased drug mean residence time and improved its ocular bioavailability, as compared to terbinafine hydrochloride eyedrops [36]. Singh et al. prepared acetazolamide-loaded pH-triggered polymeric nanoparticulate in situ gel (NP-ISG) for ophthalmic delivery. The Draize test and corneal toxicity study indicated non-irritant property of (NP-ISG) as well as no tissue damage effect. The developed formulation on an in vivo study of rabbits showed prolonged precorneal residence time, sustained drug release, and significant decrease of IOP compared to other eyedrops [37]. Du Toit et al. developed a hybrid polymeric system with lipoidal-chitosan-poly (e-caprolactone) for the treatment of inflammatory disorders (specifically ocular afflictions), which demonstrated enhanced efficacy in tissue permeation, cell uptake, and anti-inflammatory activity [38]. Mohammed et al. developed fluconazole-loaded chitin nanogels (Flu-CNGs) for the treatment of corneal fungal infections in ocular inflammation studies [39]. Jayaraman et al. developed a nanochitosan peptide having serine–threonine–tyrosine sequence that will transduce the signaling agent within the retinal pigmented epithelium cells that could potentially treat age-related macular degeneration [40].

23.5.2 Iontophoretic

Iontophoresis is a physical method by which drug permeation is enhanced across the biological barrier with the application of an electrical current. This technique is widely exploited for disorders affecting the posterior segment of the eye such as diabetic retinopathy, age-related macular degeneration (AMD), retinitis pigmentosa, and glaucoma, which account for irreversible blindness worldwide [41]. Tratta et al. used anodal transscleral iontophoresis for the penetration of cytochrome C across different ocular tissues [42]. The process of cathodal iontophoresis executed the permeation of negatively charged compounds or macromolecules by electrorepulsion [43]. Application of antibiotics through the ionophoretic method increased the bactericidal activity and reduced the severity of the disease in an in vivo study of the rabbit model [44].

23.5.3 Lipid Carrier

Lipid carriers have a lipidic core surrounded by a coating of different surfactants such as Span 20, Span 40, Span 60 Span 80, Tween 80, etc. Using the above carrier, a hydrophilic or a lipophilic drug can be encapsulated, resulting in advantages like long-term stability, high drug loading, and negligent biotoxicity due to the use of physiological lipids during the preparation. Moreover, lipidic carriers enhance corneal absorption and ocular bioavailability, and also prolong residence time [45]. Araújo et al. developed a nanolipid carrier that demonstrated prolonged retention in the posterior segment of the eye [3]. Delgado et al. complexed DNA along with protamine (helps in nuclear localization) and adsorbed it to solid lipid nanoparticles (SLN) that demonstrated higher transfection efficiency as compared to DNA-SLN without protamine [46]. Similarly, in another study, Gasco et al. used polysaccharides for the modification of DNA-SLN, which significantly increased the transfection of RPE [47]. Attama et al. formulated surface-modified SLN containing timolol hydrogen maleate for the treatment of glaucoma and evaluated its efficacy in a human corneal construct, which demonstrated enhanced ocular bioavailability [48]. Diebold et al. developed liposome–chitosan nanoparticle complexes (LCS-NP) and studied its acute tolerance in eyeball and lid tissue in rabbits, which turned out to be very low, establishing it as a potential drug carrier for ocular delivery [49]. Niosomes are nonionic surfactants, a bilayered vesicular system formed by the self-assembly of hydrated surfactants. These are water-based vesicle suspensions that could offer high patient compliance in comparison with oily dosage forms. Their infrastructure mostly consists of hydrophilic, amphiphilic, and lipophilic moieties that can entrap both hydrophilic and lipophilic drugs with a broad range of solubilities. The bilayer of niosomes protects the encapsulated drug from inside and outside of the body; hence, the labile and sensitive drugs can be protected [50]. Basha et al. formulated nanovesicles made from surfactants (Span 60 and Tween 80) for ocular delivery of clotrimazole, which demonstrated sustained antifungal effect for more than 12 h against *Candida albicans* [51]. Similarly, Abdelkader et al. developed niosomes from Span 60 that were loaded with naltrexone hydrochloride that permeated to the transcorneal region and non-irritant [52]. Aggarwal et al. developed acetazolamide-loaded niosomes using Span 60 cholesterol, dicetyl phosphate, and stearylamine for the topical treatment of glaucoma. The developed formulation was more effective at lowering IOP as compared to marketed formulation dorzolamide [53]. Aggarwal and Kaur formulated chitosan–carbopol-coated niosomes and entrapped timolol maleate for antiglaucoma therapy, which extended the drug release up to 10 h [54].

23.5.4 Dendrimers

Dendrimers, the new class of polymeric materials, have a unique architecture, offering a number of advantages as a carrier system. Dendrimers enhance the water solubility, biocompatibility, and bioavailability of different drugs, either in the form of encapsulation or by conjugation. Hence, they have been recognized as a novel delivery system for diverse ocular administration. Vandamme and Brobeck developed a series of poly (amidoamine) (PAMAM) dendrimers where the anionic dendrimer solution demonstrated longer residence time of pilocarpine nitrate and tropicamide along with the reduction of IOP [55]. Durairaj et al. demonstrated the efficacy of dendrimeric polyguanidilyated translocators (DPTs) that could efficiently translocate the molecules across the biological barrier. For conjunctivitis treatment, an eyedrop formulation of gatifloxacin-loaded (DPTs) was developed, and the in vivo studies in New Zealand white rabbits showed excellent efficacy with once-a-day dosing [56]. A synergistic approach was conceptualized to obtain immunomodulatory and antiangiogenic effects for the treatment of scar tissue after surgery. For this, water-soluble conjugates of D(+)-glucosamine and D(+)-glucosamine 6-sulfate with anionic PAMAM (G3.5) dendrimers were used in clinically relevant scar tissue rabbit models after glaucoma filtration surgery, where long-term success was observed with fewer clinical microbial infection [57]. PAMAM dendrimers with carboxyl end groups (G3.5-COOH) for extended half-life and sustained delivery were used for intraocular tumors such as retinoblastoma. Carboplatin-loaded PAMAM demonstrated better efficacy for an extended time in tumor vasculature in transgenic murine retinoblastoma model following subconjunctival administration [58]. To treat choroidal neovascularization (CNV), a dendrimer-based approach was used for antivascular endothelial growth factor oligonucleotide (VEGF-ODN) delivery. The overexpression of VEGF has been known to be associated with numerous diseases with angiogenesis. Marano et al. reported the efficacy of anti-VEGF-ODN-1 administered to rat eyes with CNV, which demonstrated significant suppression of VEGF expression in the cell [59]. Holden and co-workers reported the efficacy of a PAMAM dendrimer hydrogel that is made from "ultraviolet-cured PAMAM dendrimer" linked with PEG-acrylate chains for delivery of two antiglaucoma drugs: timolol maleate and brimonidine. Compared to other eyedrops, there was significant uptake of drug-loaded dendrimers in "human corneal epithelial cells" and "bovine corneal transport" [60]. This work was further extended to develop novel "hybrid PAMAM dendrimer hydrogel/poly(lactic-co-glycolic acid) (PLGA) nanoparticle platform (HDNP)," for the codelivery of timolol maleate and brimonidine. After topical administration of the HDNP in adult "normotensive Dutch-belted male rabbits," the formulation maintained significantly higher concentration for up to a week in aqueous humor and cornea when compared to saline. These hybrid NPs increases the drug's bioavailability and sustained drug activity [60]. For the treatment of retinoblastoma, photodynamic therapy is a potentially efficient approach. Makky et al. reported the efficacy of photosensitizers, porphyrin-based glycodendrimers–surfaces conjugated with the mannose-specific ligand protein concanavalin A for targeted delivery to tumor cells in the retina. This dendrimer complex demonstrated specific interactions with the receptors in the lipid bilayer along with enhanced accumulation in malignant ocular tissues [61]. Similarly, another study conducted by Nishiyama et al. reported the efficacy of porphyrin-based dendrimers micelles, which demonstrated selective accumulation within 24 h in a CNV rat model [62].

23.6 Disposition of Nanocarriers in Ocular Tissues

After its administration, the fate of nanoparticles is governed by numerous factors including routes of administration, i.e., topical, periocular, intravitreal, etc. The possible pathway of a nanocarrier is depicted in Fig. 23.2.

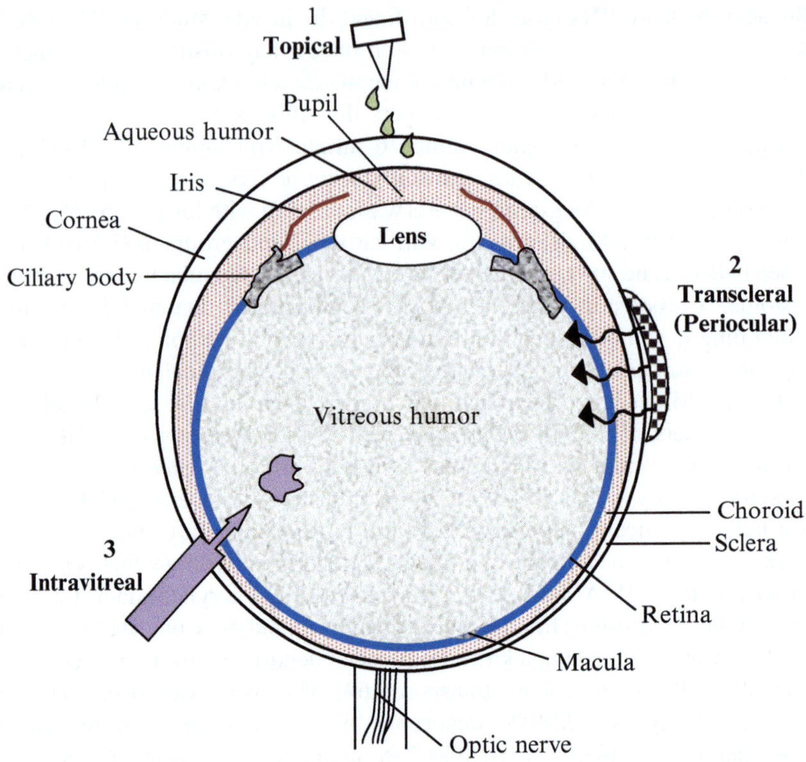

Fig. 23.2 Schematic model of eye anatomy and its mode of drug delivery. (*1*) Topical delivery diffuse across the tear film and cornea and quickly cleared through the aqueous humor (often used for anterior segment disease) (*2*) Transscleral/periocular, (*3*) Intravitreal are more direct approach for treating posterior segment disease

23.6.1 Topical

For various pathogenic ocular diseases, the topical route of administration remains the route of choice for drug delivery to the anterior segment of the eye [63]. Most of the developed formulations for ocular therapeutics are administered in topical dosage form. The fate of nanoparticles following the topical administration is comparable to other topical, aqueous ophthalmic products. For various eye disorders such as conjunctivitis, episcleritis, dacryocystitis, keratitis, and anterior uveitis, steroid hormonelike drugs such as corticosteroids are applied as primary choice [64]. The topical administration has several advantages such as direct delivery of the drugs at the site of action, direct drug penetration, and patient convenience, as this route bypasses the side effects associated with systemic absorption [65]. Moreover, the nanoparticulate carrier mostly gains entry either into the cornea or conjunctiva or passes through the nasolacrimal duct and enters the gastrointestinal tract. Nanoparticulate carriers increase the drug level in different ocular tissues, as they increase the precorneal residence time, boost the cellular uptake and accumulation, and maintain prolonged duration of the drug at site [66].

23.6.2 Intravitreal Disposition

Following the intravitreal administration, within hours, the nanoparticulate carrier mostly settles onto the inner limiting membrane. After this, it infiltrates into various retinal layers, and from there, a fraction may reach the systemic circulation through retinal or choroid vasculature [27]. Studies conducted also revealed that a fraction of the particles also settles and penetrates into the iris and ciliary. The settlement of different particles is primarily governed by the size, density, and property of the vitreous, i.e., viscosity, density, convection currents, and fluid clearance [67]. De Kozak et al. reported the duration and fate of nanoparticulate formulation after intravitreal administration in an experimental autoimmune uveitis rat model. In an inflammatory condition, the NPs penetrated after 24 h into the astrocytes RPE, into the anterior chamber of the eye and were captured by the infiltrating macrophages. After 3 days, NPs were spotted in cervical lymph nodes and negligible amounts of NPs were detected in the liver and spleen. Whereas, in normal eyes, the NPs infiltrated within 24 h in cells of the iris and astrocytes within the inner limiting membrane of the retina, and in 3 days, they penetrated to the RPE, choroid, and ciliary body. Possibly in this way, enhanced uptake of NPs by the RPE takes place [68].

23.6.3 Periocular Disposition

The fate of NPs administered through the periocular route is quite distinctive, as it involves the route of subconjunctiva, subtenon, and retrobulbar administrations. After periocular administration, the released drug can reach the conjunctiva, tear film, sclera, and choroid; it could also appear at the retina and vitreous and can reside for a longer period of time, making it useful for sustained drug delivery application [66].

23.7 Ocular Gene Therapy

Delivery of genetic material to the eye is quite a challenge from a pharmaceutical point of view. However, it is an excellent candidate for gene therapy, as it is immune-privileged and affected by genetic diseases [69]. Both environmental trauma and genetic disorders can cause varying degrees of ocular diseases such as cataracts, glaucoma, retinitis pigmentosa, etc. Several treatment strategies have been attempted for overcoming these genetic deficits in various animal models. From a pharmaceutical point of view, delivery of genetic material is a challenging task, as it is unstable in biological fluids and has poor cell penetration due to its size or charge. For these, a suitable carrier is needed to deliver it to ocular tissues. Correcting the genetic deficit, gene delivery using non-viral carrier systems holds promise for an effective carrier system, as the above-cited limiting factors do not influence them strongly. Farjo et al. reported the feasibility of using compacted DNA nanoparticles as an efficacious platform for non-viral gene transfer to the eye, which appears to be an excellent strategy for gene transfer to various ocular tissues. These nanoparticles consist of a neutrally charged complex containing a single molecule of plasmid DNA-enhanced green fluorescent protein (pEGFP) compacted with polyethylene glycol (PEG)-substituted lysine peptides. After intravitreal injection two days later, the fluorescence was present in the lens, cornea, trabecular meshwork, sclera, choroid, RPE, and other retinal cells [69]. Klausner et al. reported the efficacy of NOVAFECT chitosans for gene delivery. NOVAFECT chitosans are ultrapure chitosan oligomers that were recently marketed as carriers for non-viral gene therapy. The group designed the oligomeric chitosan–DNA nanoparticles and evaluated them in rat corneas. The luciferase gene expression (green fluorescent protein) was 5.4 times greater than that of the control (polyethylenimine–DNA nanoparticles). The keratocytes (corneal fibroblasts) expressed the transgene. Hence, the oligomeric chitosan–DNA nanoparticles can be a promising approach for the treatment of acquired and inherited corneal diseases [70]. The precise local delivery of oligonucleotides (ODN) can be controlled with an illumination using lasers for therapeutic purpose. This technology is mostly used for tumor cells. Normand et al. developed light-sensitive NPs using the structural protein of the Herpes simplex virus, VP22, that was expressed in *E. coli*. The VP22 is bound through the C-terminal of the viral protein that led to form a light-sensitive particle (vectosome). This vectosome leads to intracellular internalization in the studied human retinal pigment epithelial cells (ARPE-19). With light or laser beam illumination, there was disruption of the vectosome and release of complexed ODN, whereas without illumination, the internalized vectosome remained stable in the cell cytoplasm. The in vivo study (rat eye) with a single intravitreal injection of vectosomes supplementing with transscleral illumination released the ODN to retina and RPE cells [71]. Ding et al. reported the safety and efficacy of polyethylene glycol-substituted lysine peptide (CK30PEG)-compacted DNA nanoparticles in retinal cells. The NPs formed were injected in mice and transfection was observed. With administration, there was no sign of local inflammation or chemokine marker expression after 2 days of observation. This indicated that the retina may be the suitable target for nanoparticle-based clinical involvement [72]. Alqawlaq et al. formulated non-viral gemini surfactant–phospholipid nanoparticles (GL-NPs) for intravitreal and topical administration of glaucoma gene therapy [73]. GL-NPs carrying Cy5-labeled plasmid DNA demonstrated a distinct trafficking performance and biodisposition after intravitreal injection in mice. These particles were located within the retinal nerve fiber layer, whereas in topical application, GL-NPs were located in various anterior chambers like the limbus, iris, and conjunctiva [73]. In another approach, Apaolaza et al. developed hyaluronic acid–SLNPs for the treatment of ocular disease by gene therapy. The developed NPs consisted of SLNs, protamine (P), and hyaluronic acid

(HA), as carrier for gene therapy. With the incorporation of NPs complex to the APRE-19 cells, the transfection capacity increased sevenfold. The plasmid that encoded the protein retinoschisin was incorporated, which led to high transfection. This novel approach illustrated a non-viral mode of treatment for X-linked juvenile retinoschisis by gene therapy [74]. Mitra et al. developed glycol chitosan (GCS) and plasmid DNA (pDNA) for a number of non-monogenic degenerative ocular diseases. The efficacy was determined using a pDNA carrying ubiquitously expressed CBA-eGFP cassette that was subretinally injected into adult wild-type albino mice. After 14 days postinjection, the GFP was observed in the RPE in GCS NP-treated mice, whereas no expression was observed in saline or uncompacted pDNA. This vehicle can be potentially explored for ocular gene therapy [75].

23.8 Ocular Microrobots

The anterior and posterior segment of the eyes can be possibly treated with ocular microrobots. Microrobotics primarily deals with wireless navigation and control to produce wireless millimeter and submillimeter-sized platform for minimally invasive targeted therapy. However, the ocular delivery through microrobots is still in infancy due to many challenges related to fabrication and control in a complex biological environment. The exercising magnetic field for maneuvering microrobots in the human body is an advantageous approach due to their safe operation [76]. The retina requires sufficient oxygen and other essential nutrients to execute its primary visual function. Retinal hypoxia (inadequate oxygen supply) is mostly observed at the end stage of retinal disease in diabetic retinopathy, glaucoma, age-related macular degeneration, retinopathy of prematurity, and retinal vein occlusions. However, the influence of oxygen is not well understood, making development of an in vivo oxygen measurement essential for better diagnosis and treatment. Measurement of oxygen tensions, both in aqueous humor and vitreous humor, and particularly in the preretinal area, can be useful. In this regard, Ergeneman et al. developed an intraocular optical oxygen sensor utilizing a luminescence coating. The sensor works by quenching luminescence in the presence of oxygen; therefore, this iridium phosphorescent complex was designed and synthesized to be used as an oxygen probe. Iridium has a longer lifetime, stronger absorption band in the visible region, higher luminescence quantum yield, higher photostability, and larger Stokes shift. The sensor was dip coated with biocompatible polystyrene film containing luminescence dye. A blue LED light is used as a light source, and a photodiode is used for detection, and with the help of optical filters, the emission signal is separated from excitation signal, and the location of sensor in the setup is maintained with a magnet. This sensor was demonstrated as a concept of in vivo application of intraocular treatment application [77]. Retinal vein occlusion (RVO) is caused by obstruction of blood flow due to clot formation and is the common cause for vision loss. Different methods of treatments have been proposed for RVO, but alternative methods have been adopted to avoid postoperative complication. Currently, the most promising treatment for RVO-prolonged local intravenous thrombolysis (i.e., clot dissolution) method with tissue plasminogen activator (t-PA) injection is employed [78]. A substitute approach for RVO treatment was developed and proposed with a wireless magnetic microrobot coated with clot-dissolving t-PA. Here, the microrobot will be steered through externally applied magnetic fields at the thrombus site and can be traced visually through the pupil and will be immobilized in close proximity to the retinal vein. There will be diffusion of t-PA from the surface coating of the microrobot, and the dissolution of clot will start. The developed method will prolong drug release without any micropump and is potentially less invasive than other methods [79]. Kummer et al. developed an ellipsoid microrobotic device using electroplated nickel for the implementation of unrestrained magnetic control. The system consists of eight electromagnets placed in a hemispherical arrangement that accommodates the geometry of a small animal head, neck, and shoulders, i.e., OctoMag [80]. The OctoMag is designed for an ophthalmoscope equipped with a camera that is located along the central axis for imaging the microrobot in eye. With the adjustment of the magnetic fields and magnetic field gradients, the wireless soft magnetic microrobots can experience mobility and controllability, respectively. In an ex vivo system (different media, vitreous, balanced salt solution, silicone oil) and in vivo animal experiment at the posterior eye segment, the developed electromagnetic system facilitates precise control of the microrobots in and around different axes [81].

23.9 Ocular Implants

An ocular implant is an artificial device designed in such a way to replace a damaged or missing visual organ. Retinitis pigmentosa is an inherited retinal degenerative disease that results in progressive vision loss because of gradual loss of light-sensitive retinal cells (rods and cones). To treat this disease, the Argus® II Retinal Prosthesis System has been developed by Second Sight® Medical Products Inc. of Sylmar, California, and it is the first and only retinal prosthesis approved in the USA. The Argus® II (first bionic eye) can be surgically implanted in a patient's eye, and it works by converting video image that was captured in a miniature camera housed, which could transmit wirelessly to an array of electrodes on the surface of the retina. The pulses generated then in turn are proposed to stimulate the retina's remaining cells, for the perception of patterns of light in the brain. With this aid, the patient can interpret the visual patterns and gain some visual function (da Cruz et al 2013). For stimulation of the visual cortex, a silicon microchip can be implanted on a network of electrodes made of biocompatible material. Each electrode can inject a stimulating electrical current that can provoke a series of luminous points to appear (an array of pixels) in the field of vision of the blind person. The implant is a two-part system: one, the implant and the other, external controller. The implant can be lodged in the visual cortex that could wirelessly receive the data and energy from the external controller. The device contains all the circuits necessary to generate the electrical stimuli and to monitor the changing in the biological tissue interface [82].

Intracortically, visual prosthesis is to provide a blind patient some amount of useful vision so that he can be able to perform simple task, i.e., to recognize basic shapes and perform simple eye–hand coordination tasks, navigation, and reading large printed text. In this regard, the cortical visual prosthesis device is designed so that it can be implanted as arrays of activated iridium oxide-coated electrodes in the visual cortex located on the occipital lobe of the brain, which will image, process it, and then deliver the consequent signals to the implanted chips, which will eventually stimulate the implanted electrodes. If such a cortical prosthesis can be implanted in blind/low vision human subjects, they might be able to perform tasks and, with practice, should be able to adapt to dotted images even with a low resolution of 325 dots of phosphene (spots of light). In prosthetic vision, a visual scene is elicited in the visual field of a person whose visual pathway is being electrically excited, that is, composed of large, isolated spots of light, which look alike, like a magnified pictorial print [83]. Mathieson et al. designed electronic retinal prostheses that restore sights by electrically stimulating surviving neurons. The developed photovoltaic subretinal prosthesis where the silicon photodiodes in each pixel receive power and the data is generated through pulsed near-infrared illumination and that, in turn, electrically stimulate neurons [84]. Patients with persistent diabetic macular edema (DME) are clinically administered with fluocinolone acetonide vitreous implants for sustained delivery of drug so that it can have long-term effect [85].

23.10 Conclusions and Future Perspectives

Effectual treatment of ocular diseases is a formidable challenge due to the presence of ocular barriers. Considerable research has focused on the development of nanotechnology-based drug delivery systems for ocular therapy because of their controlled release, biocompatibility, avoidance of systemic side effects, improved pharmacokinetics, enhanced stability, etc. The nanosized formulations provide enormous ease for the convenience of application in the form of eyedrops for topical, intravitreal, transscleral, or implantable administration. With the aid of various nanovehicles, complex eye disorders like glaucoma, cataract, and retinal diseases can be treated. Robust gene therapy as well nanodevice application were successful for the treatment of genetically based blinding diseases. Future research will focus on designing microrobots that can be efficiently monitored for real-time diagnosis as well as for minimally invasive eye surgery through a single wirelessly controlled platform.

References

1. Sahoo S, Dilnawaz F, Krishnakumar S (2008) Nanotechnology in ocular drug delivery. Drug Discov Today 13:144–151
2. Urtti A (2006) Challenges and obstacles of ocular pharmacokinetics and drug delivery. Adv Drug Deliv Rev 58:1131–1135

3. Araújo J, Nikolic S, Egea MA, Souto EB, Garcia ML (2011) Nanostructured lipid carriers for triamcinolone acetonide delivery to the posterior segment of the eye. Colloids Surf B: Biointerfaces 88:150–157
4. Kang-Mieler J, Osswald C, Mieler W (2014) Advances in ocular drug delivery: emphasis on the posterior segment. Expert Opin Drug Deliv 11:1647–1660
5. Ali J, Fazil M, Qumbar M, Khan N, Ali A (2014) Colloidal drug delivery system: amplify the ocular delivery. Drug deliv 1–17
6. Szczesna I, Iskander DR (2012) Future directions in noninvasive measurements of tear film surface kinetics. Optom Vis Sci 89:749–759
7. Montes-Mico R, Cervino A, Ferrer-Blasco T, García-Lázaro S, Madrid-Costa D (2010) The tear film and the optical quality of the eye. Ocul Surf 8:185–192
8. Stahl U, Willcox M, Stapleton F (2012) Osmolality and tear film dynamics. Clin Exp Optom 95:3–11
9. Baspinar Y, Bertelmann E, Pleyer U, Buech G, Siebenbrodt I, Borchert HH (2008) Corneal permeation studies of everolimus microemulsion. J Ocul Pharmacol Ther 24:399–402
10. Muller L, Marfurt CF, Kruse F, Tervo TM (2003) Corneal nerves: structure, contents and function. Exp Eye Res 76:521–542
11. Araujo J, Gonzalez E, Egea MA, Garcia ML, Souto EB (2009) Nanomedicines for ocular NSAIDs: safety on drug delivery. Nanomedicine 5:394–401
12. Fischbarg J, da Silva-Cunha (2006) The corneal endothelium. In: Fischbarg J (ed) The biology of eye. Academic, New York, pp 113–125
13. Singh D (2003) Conjunctival lymphatic system. J Cataract Refract Surg 29:632–633
14. Pescina S, Santi P, Ferrari G, Nicoli S (2011) Trans-scleral delivery of macromolecules. Ther Deliv 2:1331–1349
15. Miao H, Wu BD, Tao Y, Li XX (2013) Diffusion of macromolecules through sclera. Acta Ophthalmol Scand 91:e1–e6
16. Wen H, Hao J, Li SK (2013) Characterization of human sclera barrier properties for transscleral delivery of bevacizumab and ranibizumab. J Pharm Sci 102:892–903
17. Singh V, Ahmad R, Heming T (2011) The challenges of ophthalmic drug delivery: a review. Int J Drug Discov 3:56–62
18. Freddo TF (2013) A contemporary concept of the blood-aqueous barrier. Prog Retin Eye Res 32:181–195
19. Kaur I, Kanwar M (2002) Ocular preparations: the formulation approach. Drug Dev Ind Pharm 28:473–493
20. Dey S, Gunda S, Mitra AK (2004) Pharmacokinetics of erythromycin in rabbit corneas after single-dose infusion: role of P-glycoprotein as a barrier to in vivo ocular drug absorption. J Pharmacol Exp Ther 311:246–255
21. Constable P, Lawrenson JG, Dolman DE, Arden GB, Abbott NJ (2006) P-Glycoprotein expression in human retinal pigment epithelium cell lines. Exp Eye Res 83:24–30
22. Dey S, Patel J, Anand BS, Jain-Vakkalagadda B, Kaliki P, Pal D, Ganapathy V, Mitra AK (2003) Molecular evidence and functional expression of P-glycoprotein (MDR1) in human and rabbit cornea and corneal epithelial cell lines. Invest Ophthalmol Vis Sci 44:2909–2918
23. Gaudana R, Ananthula HK, Parenky A, Mitra AK (2010) Ocular drug delivery. AAPS J 12:348–360
24. Aksungur P, Demirbilek M, Denkbas EB, Vandervoot J, Ludwik A, Unlu N (2011) Development and characterization of cyclosporine a loaded NPs for ocular drug delivery: cellular toxicity, uptake, and kinetic studies. J Control Release 151:286–294
25. Sahoo S, Labhasetwar V (2003) Nanotech approaches to drug delivery and imaging. Drug Discov Today 8:1112–1120
26. Jain K, Kumar RS, Sood S, Dhyanandhan G (2013) Betaxolol hydrochloride loaded chitosan NPs for ocular delivery and their anti-glaucoma efficacy. Curr Drug Deliv 10:193–199
27. Merodio M, Irache JM, Valamanesh F, Mirshahi M (2002) Ocular disposition and tolerance of ganciclovir-loaded albumin nanoparticles after intravitreal injection in rats. Biomaterials 23:1587–1594

28. Vandervoort J, Ludwig A (2007) Ocular drug delivery: nanomedicine applications. Nanomedicine (Lond) 2:11–21
29. Zhu X, Su M, Tang S, Wang L, Liang X, Meng F, Hong Y, Xu Z (2012) Synthesis of thiolated chitosan and preparation NPs with sodium alginate for ocular drug delivery. Mol Vis 18:1973–1982
30. Jwala J, Boddu SH, Shah S, Sirimulla S, Pal D, Mitra AK (2011) Ocular sustained release NPs containing stereoisomeric dipeptide prodrugs of acyclovir. J Ocul Pharmacol Ther 27:163–172
31. Agnihotri S, Vavia PR (2009) Diclofenac loaded biopolymeric nanosuspension for ophthalmic application. Nanomedicine 5:90–95
32. Chennamaneni S, Mamalis C, Archer B, Oakey Z, Ambati BK (2013) Development of a novel bioerodible dexamethasone implant for uveitis and postoperative cataract inflammation. J Control Release 167(1):53–59
33. Başaran E, Yenilmez E, Berkman MS, Büyükköroğlu G, Yazan Y (2013) Chitosan NPs for ocular delivery of cyclosporine A. J Microencapsul 41:49–57
34. Sabzevari A, Adibkia K, Hashemi H, De- Geest BG, Mohsenzadeh N, Atyabi F, Ghahremani MH, Khoshayand MR, Dinarvand R (2013) Improved antiinflammatory effects in rabbit eye model using biodegradable poly beta-amino ester NPs of triamcinolone acetonide. Invest Ophthalmol Vis Sci 54:5520–5526
35. Wadhwa S, Paliwal R, Paliwal SR, Vyas SP (2009) Nanocarriers in ocular drug delivery: an update review. Curr Pharm Des 15:2724–2750
36. Tayel S, El-Nabarawi MA, Tadros MI, Abd-Elsalam WH (2013) Positively charged polymeric nanoparticle reservoirs of terbinafine hydrochloride: preclinical implications for controlled drug delivery in the aqueous humor of rabbits. AAPS PharmSciTech 14:782–793
37. Singh J, Chhabra G, Pathak K (2013) Development of acetazolamide-loaded, pH-triggered polymeric nanoparticulate in situ gel for sustained ocular delivery: in vitro ex vivo evaluation and pharmacodynamic study. Drug Dev Ind Pharm 40:1223–1232
38. Du Toit LC, Govender T, Carmichael T, Kumar P, Choonara YE, Pillay V (2013) Design of an antiinflammatory composite nanosystem and evaluation of its potential for ocular drug delivery. J Pharm Sci 102:2780–2805
39. Mohammed N, Rejinold NS, Mangalathillam S, Biswas R, Nair SV, Jayakumar R (2013) Fluconazole loaded chitin nanogels as a topical ocular drug delivery agent for corneal fungal infections. J Biomed Nanotechnol 9:1521–1531
40. Jayaraman M, Bharali DJ, Sudha T, Mousa SA (2012) Nano chitosan peptide as a potential therapeutic carrier for retinal delivery to treat age-related macular degeneration. Mol Vis 18:2300–2308
41. Pascolini D, Mariotti SP (2012) Global estimates of visual impairment: 2010. Brit J Ophthalmol 96:614–618
42. Tratta E, Pescina S, Padula C, Santi P, Nicoli S (2014) In vitro permeability of a model protein across ocular tissues and effect of iontophoresis on the transscleral delivery. Eur J Pharm Biopharm
43. Pescina S, Antopolsky M, Santi P, Nicoli S, Murtomäki L (2013) Effect of iontophoresis on the in vitro trans-scleral transport of three single stranded oligonucleotides. Eur J Pharm Sci 49:142–147
44. Rootman D, Jantzen JA, Gonzalez JR, Fischer MJ, Beuerman R, Hill JM (1988) Pharmacokinetics and safety of transcorneal iontophoresis of tobramycin in the rabbit. Invest Ophthalmol Vis Sci 29:1397–1401
45. Seyfoddin A, Shaw J, Al-Kassas R (2010) Solid lipid nanoparticles for ocular drug delivery. Drug Deliv 17:1–23
46. Delgado D, del Pozo-Rodríguez A, Solinís MA, Rodríguez-Gascón A (2011) Understanding the mechanism of protamine in solid lipid nanoparticle-based lipofection: the importance of the entry pathway. Eur J Pharm Biopharm 79:495–502
47. Gasco'n AR, Solinı's MA, del Pozo-Rodrı'guez A, Delgado D, Pedraz JL (2012) Lipid nanoparticles for gene therapy (EP 2 460 516 A2)

48. Attama AA, Reichl S, Müller-Goymann CC (2009) Sustained release and permeation of timolol from surface-modified solid lipid nanoparticles through bioengineered human cornea. Curr Eye Res 34:698–705
49. Diebold Y, Jarrin M, Saiez V, Carvalho EL, Orea M, Calonge M, Seijo B, Alonso MJ (2007) Ocular drug delivery by liposome-chitosan nanoparticle complexes (LCS-NP). Biomaterials 28:1553–1564
50. Khan ASP, Visht S, Malviya R (2011) Niosomes as colloidal drug delivery system: a review. J Chronother Drug Del 2:15–21
51. Basha M, Abd El-Alim SH, Shamma RN, Awad GE (2013) Design and optimization of surfactant based nanovesicles for ocular delivery of Clotrimazole. J Liposome Res 23:203–210
52. Abdelkader H, Wu Z, Al-Kassas R, Alany R (2012) Niosomes and discomes for ocular delivery of naltrexone hydrochloride: morphological, rheological, spreading properties and photoprotective effects. Int J Pharmaceu 433:142–148
53. Aggarwal D, Garg A, Kaur I (2004) Development of a topical niosomal preparation of acetazolamide: preparation and evaluation. J Pharm Pharmacol 1–9
54. Aggarwal D, Kaur IP (2005) Improved pharmacodynamics of timolol maleate from a mucoadhesive niosomal ophthalmic drug delivery system. Int J Pharm 290:155–159
55. Vandamme TF, Brobeck L (2005) Poly (amidoamine) dendrimers as ophthalmic vehicles for ocular delivery of pilocarpine nitrate and tropicamide. J Control Release 102:23–38
56. Durairaj C, Kadam RS, Chandler JW, Hutcherson SL, Kompella UB (2010) Nanosized dendritic polyguanidilyated translocators for enhanced solubility, permeability and delivery of gatifloxacin. Invest Ophthalmol Vis Sci 51:5804–5816
57. Sugisaki K, Usui T, Nishiyama N, Jang WD, Yanagi Y, Yamagami S, Amano S, Kataoka K (2008) Photodynamic therapy for corneal neovascularization using polymeric micelles encapsulating dendrimer porphyrins. Invest Ophthalmol Vis Sci 49:894–899
58. Kang SJ, Durairaj C, Kompella UB, O'Brien JM, Grossniklaus H (2009) Subconjunctival nanoparticle carboplatin in the treatment of murine retinoblastoma. Arch Ophthalmol 127:1043–1047
59. Marano RJ, Toth I, Wimmer N, Brankov M, Rakoczy PE (2005) Dendrimer delivery of an anti-vegf oligonucleotide into the eye: a long-term study into inhibition of laser-induced cnv, distribution, uptake and toxicity. Gene Therapy 12:1544–1550
60. Holden CA, Tyagi P, Thakur A, Kadam R, Jadhav G, Kompella UB, Yang H (2012) Polyamidoamine dendrimer hydrogel for enhanced delivery of antiglaucoma drugs. Nanomedicine 8:776–783
61. Makky A, Michel JP, Maillard P, Rosilio V (1808) Biomimetic liposomes and planar supported bilayers for the assessment of glycodendrimeric porphyrins interaction with an immobilized lectin. Biochim Biophys Acta 2011:656–666
62. Nishiyama N, Stapert HR, Zhang GD, Takasu D, Jiang D-L, Nagano T, Aida T, Kataoka K (2003) Light-harvesting ionic dendrimer porphyrins as new photosensitizers for photodynamic therapy. Bioconjug Chem 14:58–66
63. Lallemanda F, Felt-Baeyensa O, Besseghirb K, Behar-Cohen F, Gurny R (2003) Cyclosporine A delivery to the eye: a pharmaceutical challenge. Eur J Pharm Biopharm 56:307–318
64. Suresh P, Dewangan D (2011) Ophthalmic delivery system for dexamethasone: an overview. Int J Inno Pharm Res 2:161–165
65. Yeh S, Nussenblatt RB (2008) Fluocinolone acetonide for the treatment of uveitis: weighing the balance between local and systemic immunosuppression. Arch Ophthalmol 126:1287–1289
66. Amrite A, Kompella UB (2006) Nanoparticles for ocular drug delivery. In: Gupta RB, Kompella UB (eds) Nanoparticle technology for drug delivery,vol 159, pp 319–353
67. Bourges JL, Gautier SE, Delie F, Bejjani RA, Jeanny JC, Gurny R, BenEzra D, Behar-Cohen FF (2003) Ocular drug delivery targeting the retina and retinal pigment epithelium using polylactide nanoparticles. Invest Ophthalmol Vis Sci 44:3562–3569
68. De Kozak Y, Andrieux K, Villarroya H, Klein C, Thillaye-Goldenberg B, Naud MC, Garcia E, Couvreur P (2004) Intraocular injection of tamoxifen-loaded nanoparticles: a new treatment of experimental autoimmune uveoretinitis. Eur J Immunol 34:3702–3712

69. Farjo R, Skaggs J, Quiambao AB, Cooper MK, Naash MI (2006) Efficient Non-viral ocular gene transfer with compacted DNA nanoparticles. PLoS One 1:e38
70. Klausner EA, Zhang Z, Chapman RL, Multack RF, Volin MV (2010) Ultrapure chitosan oligomers as carriers for corneal gene transfer. Biomaterials 31:1814–1820
71. Normand N, Valamanesh F, Savoldelli M, Mascarelli F, BenEzra D, Courtois Y, Behar-Cohen F (2005) VP22 light controlled delivery of oligonucleotides to ocular cells in vitro and in vivo. Mol Vis 11:184–191
72. Ding X, Quiambao AB, Fitzgerald JB, Cooper MJ, Conley SM, Naash MI (2009) Ocular delivery of compacted DNA-nanoparticles does not elicit toxicity in the mouse retina. PLoS One 4:e7410
73. Alqawlaq S, Sivak JM, Huzil JT, Ivanova MV, Flanagan JG, Beazely MA, Foldvari M (2014) Preclinical development and ocular biodistribution of Gemini-DNA nanoparticles after intravitreal and topical administration: towards non-invasive glaucoma gene therapy. Nanomedicine 10(8):1637–1647, S1549-9634(14)00223-8
74. Apaolaza P, Delgado D, del Pozo-Rodríguez A, Gascón AR, Solinís MÁ (2014) A novel gene therapy vector based on hyaluronic acid and solid lipid nanoparticles for ocular diseases. Int J Pharm 465:413–426
75. Mitra R, Han Z, Merwin M, Al Taai M, Conley SM, Naash MI (2014) Synthesis and characterization of glycol chitosan DNA nanoparticles for retinal gene delivery. ChemMedChem 9:189–196
76. Fusco S, Ullrich F, Pokki J, Chatzipirpiridis G, Ozkale B, Sivaraman KM, Ergeneman O, Pané S, Nelson BJ (2014) Microrobots: a new era in ocular drug delivery. Expert Opin Drug Deliv:1–12
77. Ergeneman O, Dogangil G, Kummer MP, Abbott JJ, Nazeeruddin MK, Nelson BJ (2008) A magnetically controlled wireless optical oxygen sensor for intraocular measurements. IEEE Sens J 8:29–37
78. Shahid H, Hossain P, Amoaku WM (2006) The management of retinal vein occlusion: is interventional ophthalmology the way forward? British J Ophthalmol 90:627–639
79. Dogangil G, Ergeneman O, Abbott JJ, Pane S, Hall H, Muntwyler S, Nelson BJ (2008) Toward targeted retinal drug delivery with wireless magnetic microrobots. In: Proceedings of IEEE/RSJ international conference on intelligent robots and systems, pp 1921–1926
80. Kummer M, Abbott JJ, Kratochvil BE et al (2010) OctoMag: an electromagnetic system for 5-DOF wireless micromanipulation. IEEE Trans Robot 26:1006–1017
81. Ullrich F, Bergeles C, Pokki J, Ergeneman O, Erni S, Chatzipirpiridis G, Pané S, Framme C, Nelson BJ (2013) Mobility experiments with microrobots for minimally invasive intraocular surgery. Invest Ophthalmol Vis Sci 54:2853–2863
82. Sawan (2011) Intra-cortical visual prosthesis. Montréal polytechnical
83. Srivastava N, Troyk P, Dagnelie G (2009) Detection, eye-hand coordination and virtual mobility performance in simulated vision for a cortical visual prosthesis device. J Neural Eng 6:035008
84. Mathieson K, Loudin J, Goetz G, Huie P, Wang L, Kamins TI, Galambos L, Smith R, Harris JS, Sher A, Palanker D (2012) Photovoltaic retinal prosthesis with high pixel density. Nat Photonics 6:391–397
85. Cunha-Vaz J, Ashton P, Iezzi R, Campochiaro P, Dugel PU, Holz FG, Weber M, Danis RP, Kuppermann BD, Bailey C, Billman K, Kapik B, Kane F, Green K, FAME Study Group (2014) Sustained delivery fluocinolone acetonide vitreous implants: long-term benefit in patients with chronic diabetic macular edema. Ophthalmology 121(10):1892–1904, pii: S0161-6420(14)00366-2

Chapter 24
Excipients Utilized for Ophthalmic Drug Delivery Systems

Dev Prasad and Harsh Chauhan

Abstract The chapter lists and discusses different types of excipients used in ophthalmic drug delivery systems, along with their important functions like solubility, penetration enhancers, etc. Further, the focus on the chapter is to list excipients and their maximum potency amount used in FDA-approved ophthalmic formulations.

Keywords Excipients • Cyclodextrins • Polymers • Bio adhesives • FDA-approved excipients • Ophthalmic

Chapter Objectives

1. Introduce excipients used in ophthalmic drug delivery systems
2. Classify excipients used in ophthalmic drug delivery systems
3. List excipients and their applications in ophthalmic drug delivery systems
4. List FDA-approved excipients in ophthalmic formulations

24.1 Introduction

Drug delivery to the eye has always posed a challenge to researchers, primarily due to the physiological and anatomical barriers present in the eye. The sensitive nature of the eye and toxicity associated with drugs/excipients have further added challenges for ophthalmic drug delivery. Conventional drug delivery to the eye suffers from many disadvantages such as poor bioavailability due to rapid lacrimal

D. Prasad
School of Pharmacy, Massachusetts College of Pharmacy and Health Sciences, Boston, MA, USA
e-mail: dev.prasad@my.mcphs.edu

H. Chauhan (✉)
Pharmacy Sciences, Creighton University, Omaha, NE, USA
e-mail: harshchauhan@creighton.edu

© Springer International Publishing Switzerland 2016

Y. Pathak et al. (eds.), *Nano-Biomaterials For Ophthalmic Drug Delivery*,
DOI 10.1007/978-3-319-29346-2_24

elimination, reduced contact time, reduced patient acceptability, and burst release of the drug. Thus, in recent times, there have been numerous studies for developing novel ophthalmic delivery systems with the primary aim of obtaining an optimum concentration of the drug at the desired site with a sustained drug release profile without toxicity. Novel drug delivery systems such as nanoparticles, liposomes, etc., were designed to overcome these challenges. Excipients play an important role in both conventional and novel drug delivery systems. They are an integral part of drug delivery systems, with functions ranging from inert diluents to controlling the release rate of the drug. The chapter lists and discusses different types of excipients used in ophthalmic drug delivery systems. Further, the focus on the chapter is to list excipients used in FDA-approved ophthalmic formulations.

24.2 Classification of Excipients for Ophthalmic Drug Delivery System

Excipients in ophthalmic drug delivery systems can be classified according to their function in the drug delivery systems. Various excipients are utilized in drug delivery systems to prepare/modify the formulation or to develop specific drug delivery systems such as inserts or liposomes [1]. Figure 24.1 lists various drug delivery systems used for ophthalmic drug delivery. These drug delivery systems work on different principles, and excipients need to be compatible with the specific processes required for the preparation and evaluation of these products. Each excipient is evaluated based on the advantages it offers in the specific drug delivery system. However, some specific properties of excipients are desired from these excipients so that they can be utilized for ophthalmic delivery.

Desired physicochemical properties of excipients for ophthalmic drug delivery system:

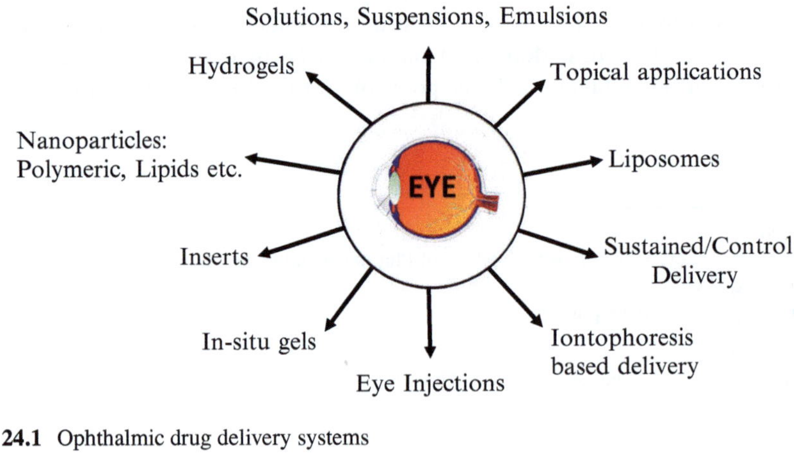

Fig. 24.1 Ophthalmic drug delivery systems

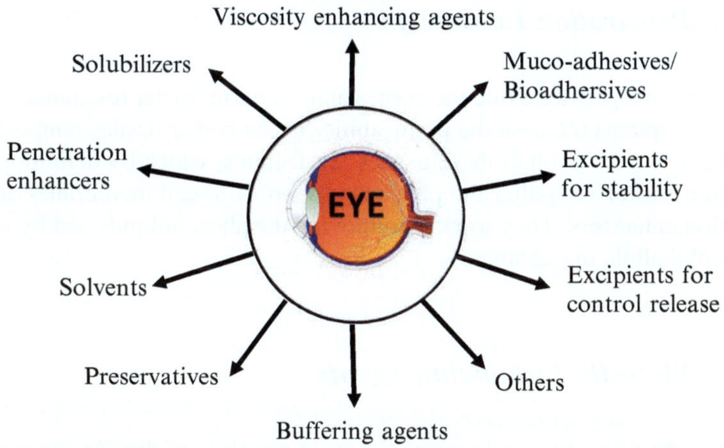

Fig. 24.2 Excipients utilized in ophthalmic drug delivery systems

1. Safe with no local and systemic side effects.
2. Increase the ocular residence time of the drug instilled in the ocular tissues.
3. Act as reservoirs for drugs to prolong the residence time of the active molecule.
4. Control drug release and thus, decrease the frequency of administration.
5. Excipients should be cheap, easy to handle, and stable.
6. Compatible with various drugs utilized in ophthalmic drug delivery systems.
7. Biodegradable and biocompatible.

24.3 Application of Excipients for Ophthalmic Drug Delivery Systems

Figure 24.2 shows various functions of excipients in ophthalmic dosage form. Excipients can be classified into the following categories:

24.3.1 Solubility Enhancers

Drug formulation prepared as solution is commonly used in ophthalmic delivery. It is very important to effectively dissolve the drug to form a solution so that it can permeate through the membrane. A lot of potent lipophilic drugs have shown pharmacological benefits in various eye conditions, but their solubility is an issue. Various cyclodextrins (CDs) and polymers are utilized to solubilize these drugs so that they can be formulated for ophthalmic delivery.

24.3.2 Penetration Enhancers

These excipients penetrate into the cornea and decrease barrier resistance. By doing so, these excipients increase the permeability of the ocular tissues temporarily and allow drugs to pass through these tissues. Surfactants, which by definition, reduce the surface tension, can alter the physical properties of cell membranes and act as penetration enhancers. They work by removing the phospholipids and by solubilizing the ophthalmic membranes.

24.3.3 Viscosity-Enhancing Agents

Enhancing the viscosity can increase the residence time of drug in the eyes which can lead to increase diffusion through the cornea. Gels are commonly used for the ophthalmic preparation due to their high viscosity. Classical preformed gels containing viscosity-enhancing agents as excipients can be directly applied to the eyes, whereas other gelling excipients can be used for in situ gel formation. In situ gel formation uses excipients, mainly polymers, to form viscous liquids which undergoes sol–gel phase transitions. They usually need exposure to specific physiological conditions like temperature, pH, etc., for the formation of in situ gels. The disadvantage of using viscous formulation is that they can result in blurred vision during the application. Also, dose variability can be an issue since it's difficult to administer accurate dose of high viscosity formulation.

24.3.4 Excipients Used for In Situ Gelation

To overcome challenges associated with viscous ophthalmic drug delivery systems, in situ gelling is achieved by changing temperature, ion concentration, or pH. There are several polymers with such properties [2]. Sustained drug delivery in the eye can be achieved by using excipients such as poloxamers which changes from solution to gel at the temperature of the eye. However, these thermally gelling systems have many disadvantages, including the risk of gelling before administration by an increase in the ambient temperature during packing or storage, for example. Other excipients like gellan gum have been utilized to form osmotic-induced gelation for ophthalmic delivery.

24.3.5 Bioadhesives/Mucoadhesives

Excipients adhering to biological membranes, like epithelial tissues or the mucous coat can improve the contact time of formulations. Hydrophilic polymers are commonly used as bioadhesives. These bioadhesives include polymers belonging to cellulose derivatives like HPMC, poly-acrylic like carbomer, polyvinylpyrrolidone (PVP), polyvinyl alcohol (PVA), sodium hyaluronate, sodium alginate, etc. The major disadvantage of using bioadhesives/mucoadhesives is patient compliance due to the stickiness of the formulation containing adhesive to the eye.

24.3.6 Excipients for Controlling Drug Release

Drug release rate needs to be controlled so as to maintain optimal drug concentrations in the targeted areas of the eyes. A matrix containing various excipients is used to control the release of drug from the formulations.

24.3.7 Solvents

Solvents are used to incorporate drugs, and sometimes excipients, for the formulation. In many cases, like ophthalmic solutions and suspensions, they form the bulk of the drug delivery systems. They may not be classified as excipients, but they play an important role in the formulations. It's very important to study the stability of drug and excipients in the solvent systems utilized for ophthalmic drug delivery.

24.3.8 Preservatives

Preservatives are added to inhibit the growth of microbial organism or to prevent any undesirable physical/chemical reactions. Since the eyes are very sensitive, it's very important to add preservative to various ophthalmic formulations. Excipients like benzalkonium chloride are commonly used as preservative in eye formulations.

24.3.9 Buffering Agents

A buffering agent is a weak acid or base used to maintain the pH of ophthalmic formulations. Buffering excipients prevent a rapid change in pH which can result in the degradation of the drug or harmful side effects to the eyes.

24.3.10 Stability Enhancers

Various excipients are used to enhance the physical and chemical stability of ophthalmic formulations.

24.4 Excipients Utilized in Ophthalmic Drug Delivery Systems

24.4.1 Cyclodextrins (CDs)

CDs have been used frequently to solubilize and stabilize the drugs in ophthalmic formulations. CDs belong to a group of cyclic oligosaccharides and improve the solubility of lipophilic drugs by forming complexes. Interactions of CDs with the biological membrane have also been found to play a role in their solubility enhancement efficiency. Amount of CDs utilized for the purpose of solubility enhancement is very critical. Ideally, 15 % or less should be added; large quantities can decrease bioavailability by retaining drug in tears [3, 4]. Eyedrop products containing CDs registered in Europe include: chloramphenicol (Clorocil®: Edol), diclofenac (Voltaren Ophthalmic®: Novartis), and indomethacin (Indocid®: Merck Sharp & Dohme-Chibret).

24.4.2 Carbopol

Carbopol or carbomer or acrylic acid polymers are synthetically high molecular weight polymers of acrylic acid with allyl sucrose or allyl ethers of pentaerythritol [5]. Each can be used as a bioadhesive material, controlled release agent, emulsifying agent, emulsion stabilizer, rheology modifier, or as a stabilizing agent in various ophthalmic formulations [6–8]. A formulation containing β-blockers timolol, betaxolol, carteolol, and metipranolol was prepared with carbopol and was shown to be very effective in reducing intraocular pressure [9]. Carbopol-containing hydrogels were found to incorporate additional drugs like pilocarpine, timolol, antibiotics, betaxolol, ribozymes, and topical steroids, in comparison to other polymers. Carbopol has also been used in the treatment of dry eye syndrome in Gel-Larmes-Thea formulation [10]. It undergoes a sol-to-gel transition in aqueous solution when the pH increases above its pKa and exhibits Newtonian time-dependent response to shear strain. Structurally, the carbopol resins (910, 934, 940, 941, and 962) are synthetic, high molecular weight, nonlinear polymers of predominantly acrylic acid, cross-linked with a polyalkenyl polyether. Johnson et al. confirmed the efficacy of carbopol in enhancing residence time in the cornea by instilling pilocarpine nitrate in rabbits' eyes [11]. Another advantage of carbopol is that it exhibits bioadhesive properties which can increase viscosity and thus, residence time. Furthermore, they are capable of forming strong non-covalent bonds with the mucin coating biological membranes and remain in place, for comparably as long as the mucin [12].

24.4.3 Poloxamers

Poloxamers are nonionic tri-block copolymers composed of a central hydrophobic chain of polypropylene oxide flanked by two hydrophilic chains of polyethylene oxide [5]. Miller and Donovan examined a temperature-sensitive solution of poloxamer for the delivery of pilocarpine. An increase in meiosis was found in the formulation containing Pluronic F127 when compared to the aqueous solution of the drug [12].

24.4.4 Xyloglucan

Xyloglucan is a polysaccharide derived from tamarind seeds. It was partially degraded by β-galactosidase and exhibits thermally reversible gelation in dilute aqueous solution. Xyloglucan formulations were used as sustained release vehicles for pilocarpine hydrochloride delivery; very low concentrations of xyloglucan were found to enhance the miotic response, an effect similar to high concentrations of poloxamer 407 gel [13].

24.4.5 Methylcellulose (MC)

It is a water-soluble derivative of cellulose and turns into gel at around 40–50 °C because of hydrophobic interactions. Gels prepared from MC are thermo-reversible, i.e., they gel upon heating and liquefy upon cooling. Additions of salts like NaCl decrease the transition temperature to 32–34 °C.

24.4.6 Hydroxypropyl Methylcellulose (HPMC)

It shows sol–gel transitions between 75 and 90 °C. Chemical or physical modifications are used to lower this transition temperature to appropriate temperatures.

24.4.7 Ethyl (Hydroxyethyl) Cellulose (EHEC)

In aqueous solutions, EHEC also exhibits thermo-sensitive behavior. EHEC is liquid at room temperature (20–25 °C) and undergoes gelation at 30–40 °C.

24.4.8 Pseudolatexes

These are artificial latexes prepared by dispersing polymers in aqueous medium. Pseudolatex-based ocular formulations of pilocarpine have shown promise for controlled ocular delivery [14].

24.4.9 Cellulose Acetate Phthalate (CAP)

CAP is a cellulose polymer. It undergoes coagulation when the pH of the solution is raised to 7.4 by the presence of tears.

24.4.10 Gellan Gum

It is a natural polymer extracted from cultures of *Pseudomonas elodea*. In the presence of mono- and divalent cations, gellan gum forms a clear solution. The aqueous solution of gellan gum forms high viscosity hydrogels in the presence of tear fluid. Various studies used gellan gum for the ophthalmic formulations and found it to be superior to other ophthalmic formulations [15–17].

24.4.11 Alginate

Alginate is a natural block copolymer of β-D mannuronic acid and α-guluronic acid. Ionotropic hydrogels are formed by the interaction of calcium ions with guluronic acid monomers [18]. Alginate has been successfully used in ophthalmic formulations including control release formulations [2, 19–21].

24.4.12 Carrageenans

Carrageenans are a group of natural, water-soluble, sulfated galactans extracted from red seaweed. They also undergo gelation in the presence of tears. Lang et al. developed a topical for the treatment of dry eye using carrageenans [21].

24.4.13 Hyaluronic Acid (HA)

To treat dry eye syndrome, hydrogels based on HA are increasingly being used. HA is a high molecular weight polymer consisting of linear polysaccharides. It forms hydrogels and have been used in various eye formulations, especially for dry eye. HA offers desired properties, such as: high water binding capacity; rheological properties, similar to those of mucus; and is comparably safe. Johnson et al. showed that sodium salt of hyaluronate at 0.1 % and 0.3 % concentrations significantly reduces the symptoms of ocular irritation [22]. Antibiotic gentamicin formulated with 0.25 % HA was found to increase precorneal drug residence time [22].

24.4.14 *Combination of Excipients*

Excipients can be used in combination to take advantage of their pharmaceutical functions for the preparation and optimization of ophthalmic formulations. Conventional penetration enhancers such as benzalkonium chloride (BAC) and solubility enhancers and CDs can be used to increase the drug availability at the lipophilic eye surface. Similarly, CDs can be combined with hydrophilic polymers. Recently, Ito et al. examined the effect of hydroxypropyl-beta-cyclodextrins (HP-β-CD) and hydroxypropyl methylcellulose (HPMC) on disulfiram formulation for intraocular pressure in experimentally induced ocular hypertension in rabbits [23]. They found that a combination of HP-β-CD and HPMC increases the solubility of disulfiram and resulted in better pharmacological benefits. Combining viscosity enhancers and in situ gelling properties of polymers have shown to significantly decrease the concentrations of polymers without compromising its gelling properties. For example, in vivo studies for timolol maleate showed that ocular bioavailability increases by approximately 2.5 times by adding Pluronic with methylcellulose, compared to timolol maleate solutions [24].

Combinations of excipients have been used by various researchers to effectively enhance the efficacy of many ophthalmic formulations [2, 25]. Most of these excipients when used alone do not provide the ideal platform for the ophthalmic delivery. However, combining these excipients was found to significantly improve the formulation characteristics, patient compliance, and therapeutic efficacy [26–29].

24.5 FDA-Approved Excipients

As discussed earlier, the conventional formulations for ophthalmic formulation consist of several categories such as solutions, suspensions, emulsions, gels, and ointments. A number of previous chapters have covered several aspects of these dosage forms; therefore, the focus of this section is to provide a comprehensive list of FDA-approved ophthalmic drug products along with extensive information about those products. Table 24.1 is a list of selected, commercially available ophthalmic formulations with the drug name, its chemical structure, clinical indication, and the list of excipients used in formulations.

Regulatory bodies in different parts of the world have set strict requirements and guidelines for the development and approval of drug products. The excipients and the concentration used in the formulation are important information requested by regulatory bodies. The excipients used to formulate ophthalmic formulations include pH modifiers, isotonic agents, penetration enhancers, solubilizers, solvents, mucoadhesives, and preservatives. Therefore, Table 24.2 focuses on excipients utilized in the drug products approved by FDA with the maximum strength used in the various dosage forms.

Table 24.1 List of selected commercially available FDA-approved formulations for ophthalmic delivery (Compiled from FDA website)

Drug product	API structure	MoA/indication	API	Excipients
Apraclonidine ophthalmic solution	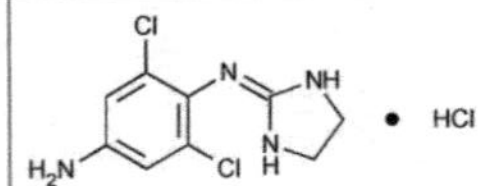	Selective alpha-2-adrenergic agonist/ indicated for the reduction of elevated intraocular pressure (IOP)	Apraclonidine – 0.5 %	Benzalkonium chloride – 0.01 %; Sodium acetate Sodium chloride
Atropine sulfate ophthalmic solution	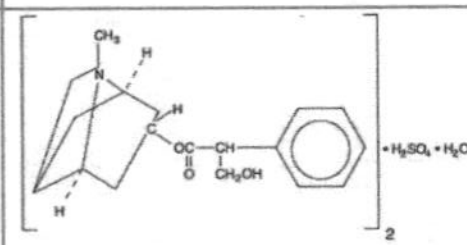	Antimuscarinic agent/indicated for cycloplegia, mydriasis, penalization of the healthy eye in the treatment of amblyopia	Atropine sulfate – 1 %	Benzalkonium chloride – 0.01 %; Edetate disodium Hypromellose (2910)
Azelastine hydrochloride solution	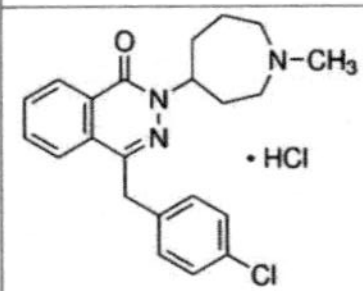	Selective histamine H1 antagonist/ treatment of itching of the eye associated with allergic conjunctivitis	Azelastine hydrochloride – 0.5 mg/ml	Benzalkonium chloride – 0.125 mg/ml; Disodium edetate dihydrate Hydroxypropyl methylcellulose Sorbitol solution
Azithromycin ophthalmic solution	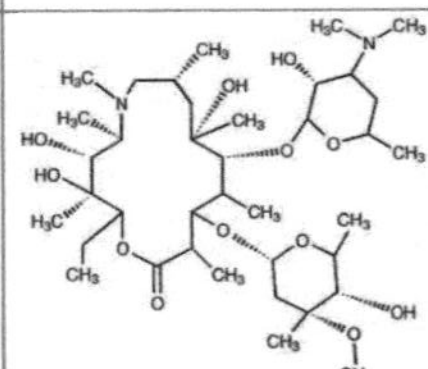	Macrolide antibiotic/treatment of bacterial conjunctivitis	Azithromycin – 1 %	Mannitol Citric acid monohydrate Sodium citrate Poloxamer 407 Polycarbophil Sodium chloride Edetate disodium Benzalkonium chloride – 0.003 %
Bacitracin zinc and polymyxin B sulfate ophthalmic ointment	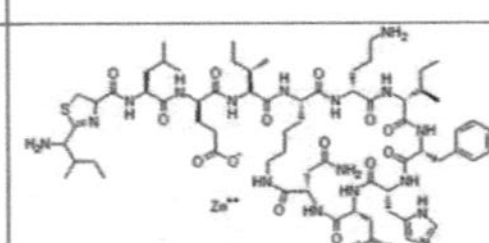 Bacitracin zinc 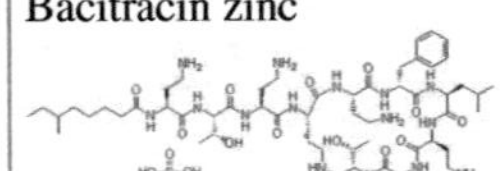 Polymyxin B sulfate	Antibacterial/infections of the external eye and its adnexa caused by susceptible bacteria	Bacitracin zinc – 500 units in 1 g Polymyxin B sulfate – 10,000 units in 1 g	Petrolatum Mineral oil

Bepotastine besilate solution	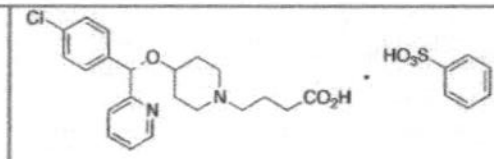	Histamine H1 receptor antagonist/ indicated for the treatment of itching associated with signs and symptoms of allergic conjunctivitis	Bepotastine besilate – 15 mg/ml	Benzalkonium chloride – 0.05 mg/ml Sodium phosphate monobasic Dihydrate sodium chloride
Besifloxacin ophthalmic suspension	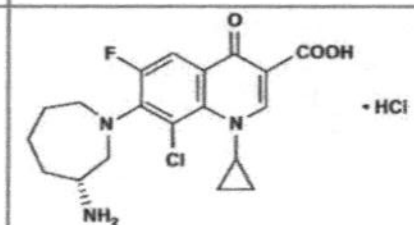	Anti-infective	Besifloxacin – 0.6 %	Benzalkonium chloride – 0.01 % Polycarbophil Mannitol Poloxamer 407 Sodium chloride Edetate disodium dihydrate
Brimonidine tartrate ophthalmic solution	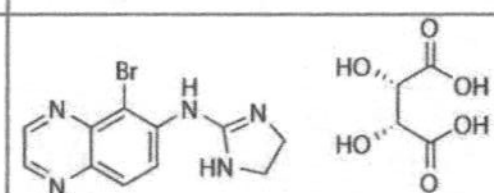	Selective alpha-2 adrenergic receptor agonist/indicated for the reduction of elevated intraocular pressure (IOP) in patients with open-angle glaucoma or ocular hypertension	Brimonidine tartrate – 0.1 % or 0.15 %	Boric acid Calcium chloride Magnesium chloride Potassium chloride Sodium borate Carboxymethylcellulose sodium Sodium chloride
Brinzolamide suspension	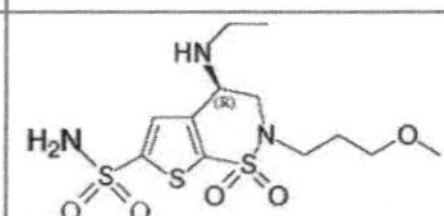	Carbonic anhydrase inhibitor/ treatment of elevated intraocular pressure in patients with ocular hypertension or open-angle glaucoma	Brinzolamide – 10 mg/ml	Benzalkonium chloride – 0.1 mg/ml Mannitol Carbomer 974P Tyloxapol Edetate disodium
Brinzolamide/ brimonidine tartrate ophthalmic suspension	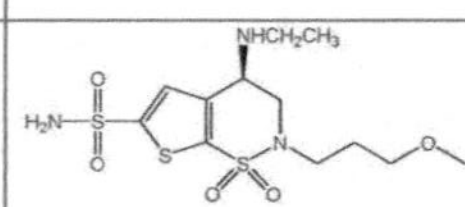	Carbonic anhydrase inhibitor/ indicated for the reduction of elevated intraocular pressure (IOP) in patients with open-angle glaucoma or ocular hypertension	Brinzolamide – 10 mg/ml Brimonidine tartrate 2 mg/ ml (equivalent to 1.32 mg/ ml as brimonidine free base)	Benzalkonium chloride 0.03 mg/ml Propylene glycol Carbomer 974P Boric acid Mannitol Sodium chloride Tyloxapol

(continued)

Table 24.1 (continued)

Drug product	API structure	MoA/indication	API	Excipients
Bromfenac ophthalmic solution	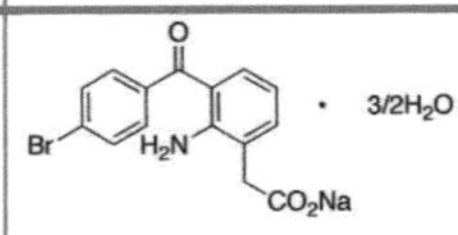	Nonsteroidal anti-inflammatory drug (NSAID)/treatment of postoperative inflammation and reduction of ocular pain	Bromfenac sodium hydrate – 0.1035 %	Benzalkonium chloride – 0.05 mg/ml Boric acid Disodium edetate – 0.2 mg/ml Polysorbate 80–1.5 mg/ml Povidone – 20 mg/ml Sodium borate Sodium sulfite anhydrous – 2 mg/ml
Carbachol intraocular solution	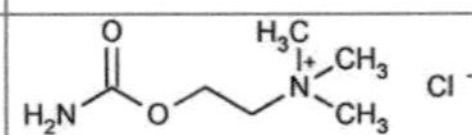	Potent cholinergic agent/reduces the intensity of intraocular pressure elevation in the first 24 h after cataract surgery	Carbachol – 0.01 %	Sodium chloride – 0.64 % Potassium chloride – 0.075 % Calcium chloride dihydrate – 0.048 % Magnesium chloride hexahydrate – 0.03 % Sodium acetate trihydrate – 0.39 % Sodium citrate dehydrate – 0.17 %
Chloromycetin ophthalmic ointment	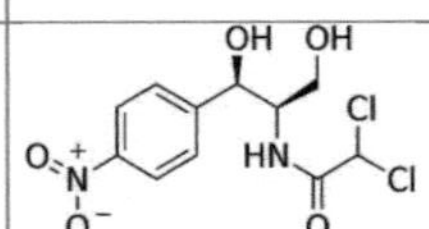	Broad-spectrum antibiotic	Chloromycetin – 1 %	Base of liquid petrolatum and polyethylene
Ciprofloxacin ophthalmic solution	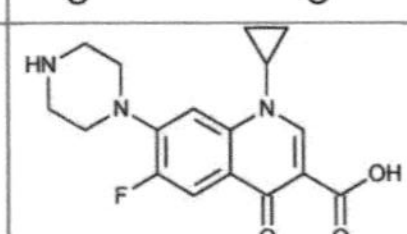	Anti-infective	Ciprofloxacin HCl – 3.5 mg/ml equivalent to 3 mg/ml base	Benzalkonium chloride – 0.006 % Acetic acid Edetate disodium – 0.05 % Mannitol – 4.6 % Sodium acetate
Cyclosporine ophthalmic emulsion	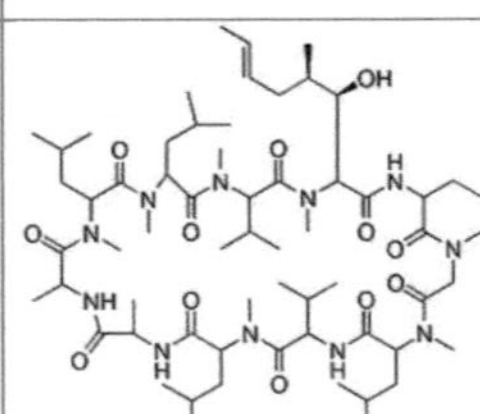	Immunosuppressant/indicated to increase tear production in patients whose tear production is presumed to be suppressed due to ocular inflammation associated with keratoconjunctivitis sicca	Cyclosporine – 0.05 %	Glycerin Castor oil Polysorbate 80 Carbomer copolymer type A

Dexamethasone sodium phosphate ophthalmic solution	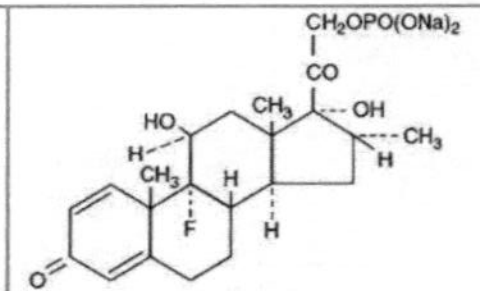	Anti-inflammatory	Dexamethasone sodium phosphate, [equivalent To 0.1 % of dexamethasone phosphate]	Sodium citrate Sodium borate Creatinine Polysorbate 80 Edetate disodium dihydrate Sodium bisulfite – 0.1 % Phenylethyl alcohol – 0.25 % Benzalkonium chloride – 0.02 %
Diclofenac sodium solution	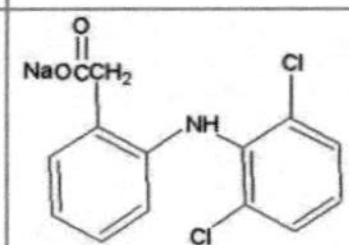	Anti-inflammatory and analgesic/ postoperative inflammation	Diclofenac sodium – 0.1 %	Polyoxyl 35 castor oil Boric acid Tromethamine Sorbic acid – 0.2 % Edetate disodium – 0.1 %
Difluprednate ophthalmic emulsion	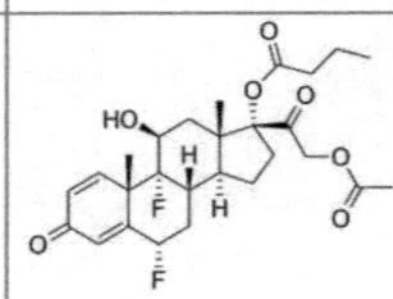	Corticosteroids/treatment of inflammation and pain associated with ocular surgery	Difluprednate – 0.05 %	Boric acid Castor oil Glycerin Polysorbate 80 Sodium acetate Sodium EDTA Sorbic acid – 0.1 %
Dorzolamide hydrochloride ophthalmic solution	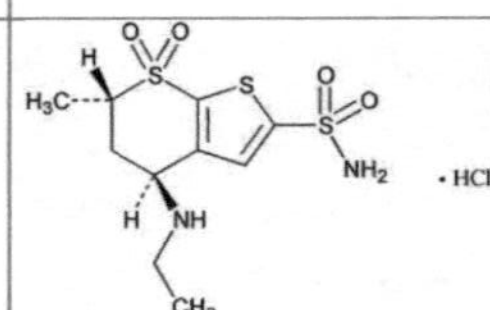	Carbonic anhydrase/treatment of elevated intraocular pressure in patients with ocular hypertension or open-angle glaucoma	Dorzolamide – 2 %	Hydroxyethyl cellulose (2000cps) – 1 % Mannitol Sodium citrate dihydrate Benzalkonium chloride – 0.0075 %
Emedastine difumarate ophthalmic solution	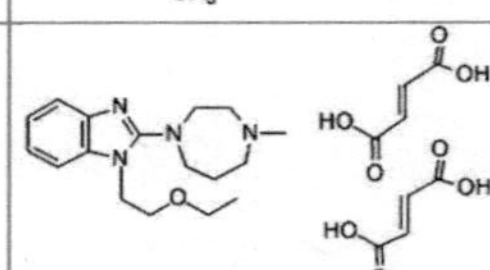	Histamine H1 antagonist/temporary relief of the signs and symptoms of allergic conjunctivitis	Emedastine difumarate equivalent to 0.5 mg/ml emedastine	Benzalkonium chloride – 0.01 % Tromethamine Sodium chloride Hypromellose

(continued)

Table 24.1 (continued)

Drug product	API structure	MoA/indication	API	Excipients
Fluorometholone ophthalmic suspension	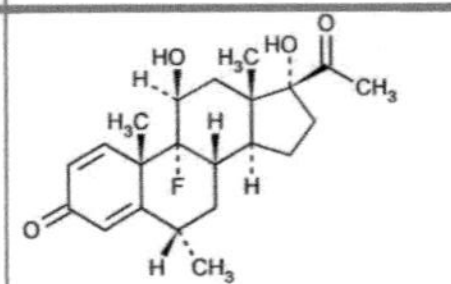	Corticosteroids	Fluorometholone – 0.1 %	Benzalkonium chloride – 0.004 % Edetate disodium Polysorbate 80 Polyvinyl alcohol – 1.4 % Sodium chloride
Flurbiprofen sodium ophthalmic solution	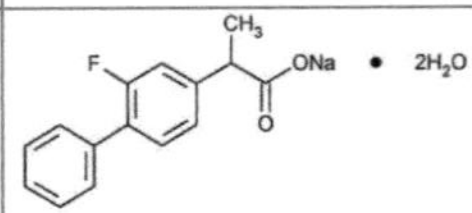	Analgesic, antipyretic, and anti-inflammatory activity/indicated for the inhibition of intraoperative miosis	Flurbiprofen sodium – 0.03 %	Sodium citrate Polyvinyl alcohol – 1.4 % Edetate disodium Potassium chloride Sodium chloride Citric acid Thimerosal – 0.005 %
Ganciclovir ophthalmic gel	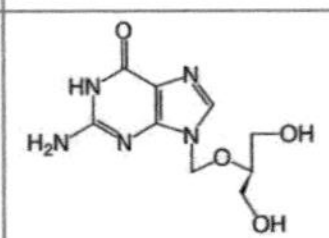	Inhibits DNA replication by Herpes simplex viruses/treatment of acute herpetic keratitis	Ganciclovir – 0.15 %	Carbopol Mannitol Benzalkonium chloride – 0.075 mg/ml
Gentamicin sulfate ointment	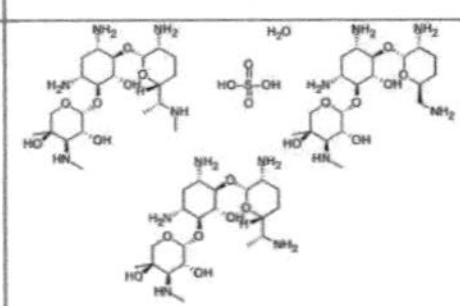	Antibacterial	Gentamicin sulfate – 3 mg/g of base	Methylparaben – 0.5 mg/g Propylparaben – 0.1 mg/g Petrolatum Mineral oil
Hypromellose 2910 (4000 MPAs) solution	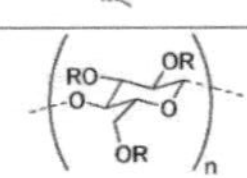 R = H or CH_3 or $CH_2CH(OH)CH_3$	For use as a lubricant to prevent further irritation or to relieve dryness of the eye	Hypromellose 2910 (4000 MPAs)	Boric acid Edetate disodium Sodium borate Benzalkonium chloride

Ketorolac tromethamine ophthalmic solution	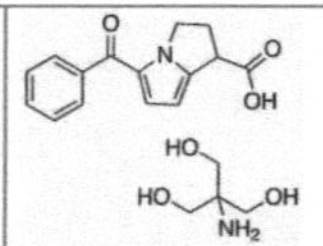	Nonsteroidal anti-inflammatory/relief of ocular itching due to seasonal allergic conjunctivitis	Ketorolac tromethamine – 0.5 %	Benzalkonium chloride – 0.01 % Edetate disodium – 0.1 % Octoxynol 40 Sodium chloride
Levobunolol hydrochloride ophthalmic solution	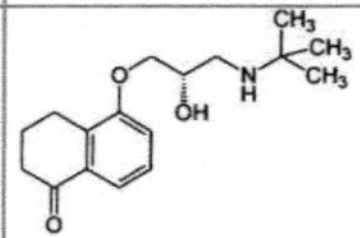	Non-cardioselective beta-adrenoceptor blocking agent/lowering intraocular pressure and may be used in patients with chronic open-angle glaucoma or ocular hypertension	Levobunolol HCl – 0.5 %	Benzalkonium chloride – 0.004 % Polyvinyl alcohol – 1.4 % Edetate disodium Sodium metabisulfite Sodium phosphate, dibasic Potassium phosphate, monobasic Sodium chloride
Levocabastine hydrochloride ophthalmic suspension	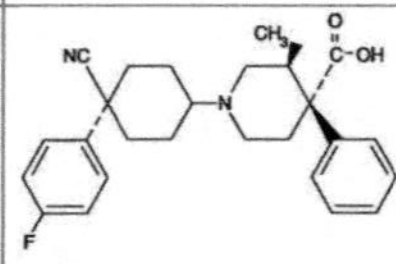	Selective histamine H1 antagonist/relief of the signs and symptoms of seasonal allergic conjunctivitis	Levocabastine hydrochloride equivalent to 0.5 mg/ml of levocabastine	Benzalkonium chloride – 0.15 mg/ml Propylene glycol Polysorbate 80 Disodium edetate Hydroxypropyl methylcellulose
Lodoxamide tromethamine ophthalmic solution	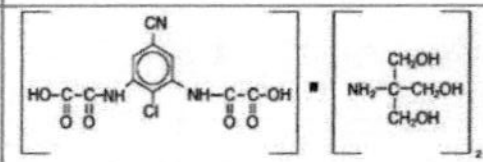	Mast cell stabilizer/indicated in the treatment of the ocular disorders such as vernal keratoconjunctivitis, vernal conjunctivitis, and vernal keratitis	Lodoxamide tromethamine – 0.178 %	Benzalkonium chloride – 0.01 % Mannitol Hypromellose 2910 (4000 MPAs) Sodium citrate Citric acid monohydrate Edetate disodium Tyloxapol

(continued)

Table 24.1 (continued)

Drug product	API structure	MoA/indication	API	Excipients
Loteprednol etabonate and tobramycin ophthalmic suspension	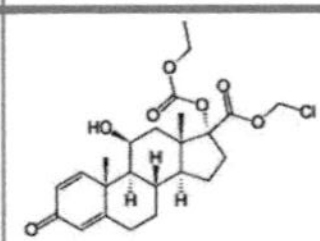 Loteprednol etabonate 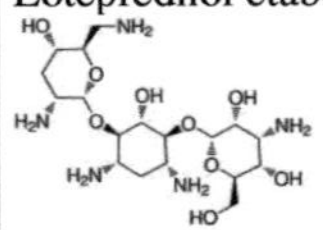 Tobramycin	Topical anti-infective and corticosteroid combination	Loteprednol etabonate – 0.5 % and tobramycin – 0.3 %	Edetate disodium Glycerin Povidone Tyloxapol
Loteprednol etabonate ophthalmic gel	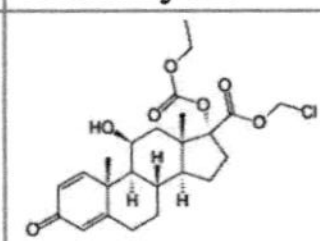	Corticosteroids	Loteprednol etabonate – 0.5 %	Boric acid Edetate disodium dehydrate Glycerin Polycarbophil Propylene glycol Sodium chloride Tyloxapol Benzalkonium chloride – 0.003 %
Loteprednol etabonate ophthalmic suspension	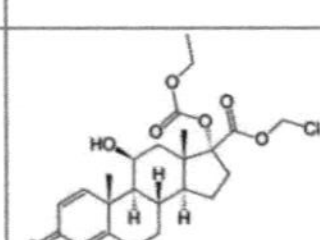	Topical anti-inflammatory corticosteroid/indicated for the temporary relief of the signs and symptoms of seasonal allergic conjunctivitis	Loteprednol etabonate – 0.5 %	Benzalkonium chloride – 0.01 % Edetate disodium Glycerin Povidones Tyloxapol
Medrysone ophthalmic suspension	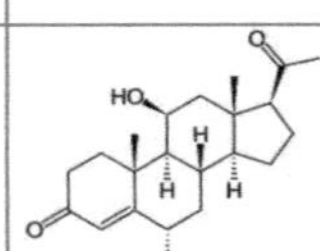	Anti-inflammatory agent/treatment of allergic conjunctivitis, vernal conjunctivitis, episcleritis, and epinephrine sensitivity	Medrysone – 1 %	Edetate disodium Hypromellose Polyvinyl alcohol – 1.4 % Potassium chloride Sodium chloride Sodium phosphate, dibasic Sodium phosphate, monobasic

Metipranolol ophthalmic solution	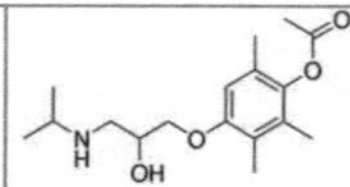	Blocks beta1 and beta2 (nonselective) adrenergic receptors/reduction of elevated intraocular pressure	Metipranolol – 3 mg/ml	Povidone Glycerol Sodium chloride Edetate disodium Benzalkonium chloride – 0.004 %
Moxifloxacin hydrochloride solution	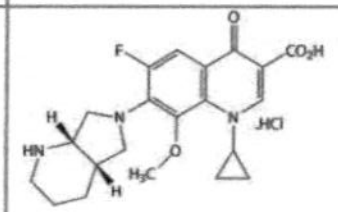	Anti-infective drugs	Moxifloxacin hydrochloride, equivalent to 5 mg/ml moxifloxacin base	Sodium chloride Xanthan gum Boric acid Sorbitol Tyloxapol
Nedocromil sodium ophthalmic solution	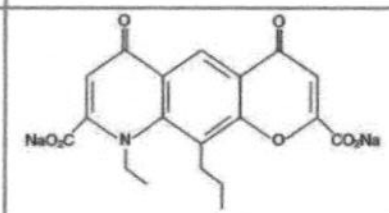	Mast cell stabilizer/treatment of itching associated with allergic conjunctivitis	Nedocromil sodium – 2 %	Edetate disodium – 0.05 % Sodium chloride – 0.50 % Benzalkonium chloride – 0.01 %
Neomycin and polymyxin B sulfates and hydrocortisone ophthalmic suspension	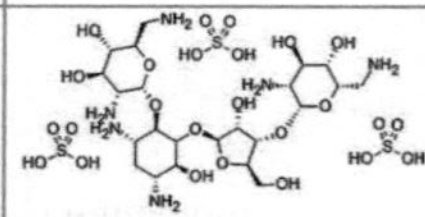 Neomycin sulfate	Topical anti-infective and corticosteroid combination	Neomycin sulfate (equivalent to 3.5 mg neomycin base), polymyxin B sulfate equal to 10,000 polymyxin B units, hydrocortisone 10 mg	Thimerosal – 0.001 % Cetyl alcohol Glyceryl monostearate Mineral oil Polyoxyl 40 stearate Propylene glycol
Neomycin and polymyxin B sulfates ophthalmic ointment	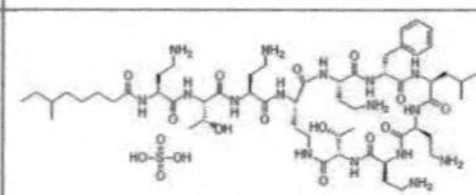 Polymyxin B sulfate	Anti-infective	Neomycin sulfate and polymyxin B sulfate same amount as above	Methylparaben – 0.05 % Propylparaben – 0.01 % White petrolatum Anhydrous liquid lanolin
Neomycin and polymyxin B sulfates ophthalmic solution	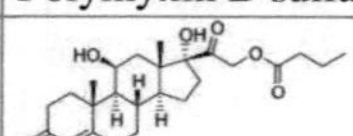 Hydrocortisone	Anti-infective	Neomycin sulfate equivalent to 3.5 mg/g neomycin base, polymyxin B sulfate equal to 16,250 polymyxin B units/g	Benzalkonium chloride – 0.004 % Hydroxypropyl methylcellulose 2910–0.5 % Boric acid Sodium chloride

(continued)

Table 24.1 (continued)

Drug product	API structure	MoA/indication	API	Excipients
Neomycin and polymyxin B sulfates and dexamethasone ophthalmic suspension	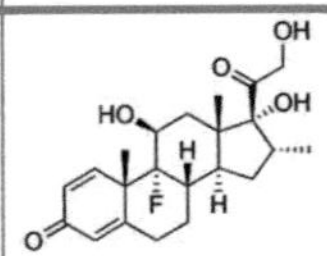 Dexamethasone	Topical anti-infective and corticosteroid combination	Neomycin sulfate equivalent to neomycin 3.5 mg/ml, polymyxin B sulfate 10,000 units/ml, dexamethasone 0.1 %	Hypromellose 2910–0.5 % Sodium chloride Polysorbate 20 Benzalkonium chloride – 0.004 %
Neomycin sulfate, polymyxin B sulfate, and gramicidin solution	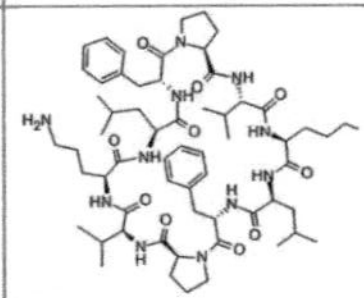 Gramicidin	Antimicrobial solution	Each mL contains neomycin sulfate equivalent to 1.75 mg neomycin base, polymyxin B sulfate equivalent to 10,000 polymyxin B units, and gramicidin 0.025 mg	Sodium chloride Alcohol – 0.5 % (v/v) Poloxamer 188 Propylene glycol Thimerosal – 0.001 %
Nepafenac ophthalmic suspension	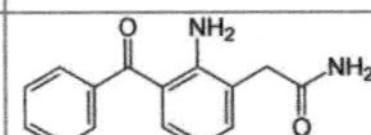	NSAIDs/indicated for the treatment of pain and inflammation associated with cataract surgery	Nepafenac – 0.1 %	Mannitol Carbomer 974P Sodium chloride Tyloxapol Edetate disodium Benzalkonium chloride – 0.005 %
Nepafenac ophthalmic suspension			Nepafenac – 0.3 %	Boric acid Propylene glycol Carbomer 974P Sodium chloride Guar gum Carboxymethylcellulose sodium Edetate disodium Benzalkonium chloride – 0.005 %
Olopatadine hydrochloride ophthalmic solution	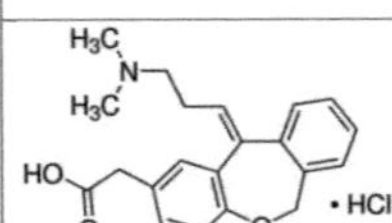	Mast cell stabilizer and a histamine H1 antagonist/treatment of ocular itching associated with allergic conjunctivitis	Olopatadine hydrochloride equal to 2 mg/ml of olopatadine	Povidone Dibasic sodium phosphate Sodium chloride Edentate disodium Benzalkonium chloride – 0.01 %

Pemirolast potassium ophthalmic solution	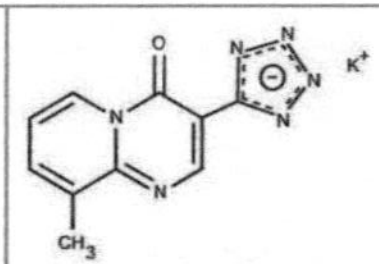	Mast cell stabilizer/prevention of itching of the eye due to allergic conjunctivitis	Pemirolast potassium – 0.1 %	Benzododecinium chloride – 0.005 % Glycerin – 1.8 %
Petrolatum and mineral oil ointment		Lubricant to prevent irritation to relieve dryness of the eye(s)	Mineral oil – 15 % White petrolatum – 85 %	
Echothiophate iodide solution	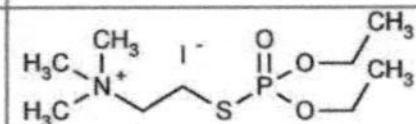	Long-acting cholinesterase inhibitor/ chronic open-angle glaucoma	Echothiophate iodide – 0.03 %, 0.06 %, 0.125 %, or 0.25 % with 40 mg potassium acetate in each case	Chlorobutanol – 0.55 % Mannitol – 1.2 % Boric acid – 0.06 % Sodium phosphate – 0.026 %
Pilocarpine hydrochloride ophthalmic solution	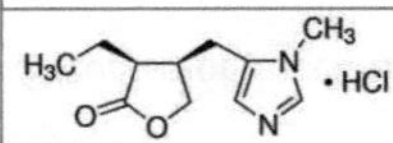	Direct-acting cholinergic parasympathomimetic agent/reduction of elevated intraocular pressure	Pilocarpine hydrochloride – 1 %, 2 %, or 4 %	Benzalkonium chloride – 0.01 % Hypromellose 2910 Boric acid Sodium citrate Sodium chloride – 1 %
Prednisolone acetate suspension	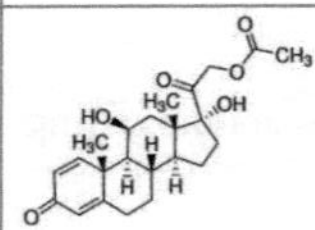	Corticosteroids	Prednisolone acetate – 1.0 %	Hypromellose Dibasic sodium phosphate Polysorbate 80 Edetate disodium Glycerin Citric acid Benzalkonium chloride – 0.01 %
Proparacaine hydrochloride solution	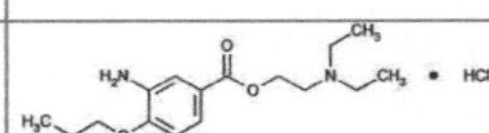	Rapidly acting topical anesthetic/ corneal anesthesia of short duration	Proparacaine hydrochloride – 0.5 %	Benzalkonium chloride – 0.01 % Glycerin
Retapamulin ointment	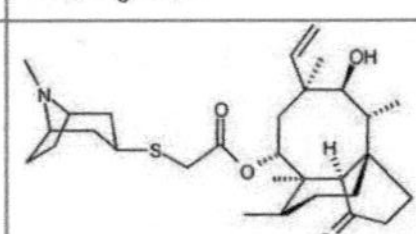	Antibacterial agent/topical treatment of impetigo	Retapamulin – 10 mg/g of white petrolatum	White petrolatum

(continued)

Table 24.1 (continued)

Drug product	API structure	MoA/indication	API	Excipients
Rimexolone suspension	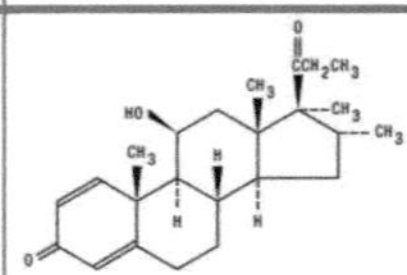	Corticosteroids/treatment of postoperative inflammation following ocular surgery and in the treatment of anterior uveitis	Rimexolone – 1 %	Benzalkonium chloride 0.01 % Carbomer 974P Polysorbate 80 Sodium chloride Edetate disodium
Sulfacetamide sodium and prednisolone acetate ophthalmic ointment	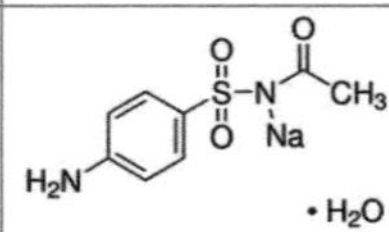 Sulfacetamide sodium	Corticosteroids/suppress the inflammatory response	Sulfacetamide sodium – 10 % and prednisolone acetate – 0.2 %	Phenylmercuric acetate 0.0008 % Mineral oil Petrolatum Lanolin alcohol White petrolatum
Sulfacetamide sodium and prednisolone acetate suspension	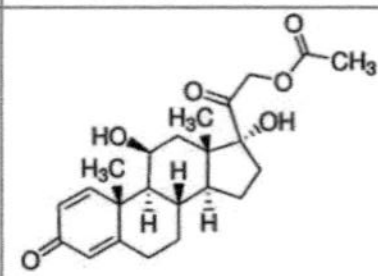 Prednisolone acetate	Anti-inflammatory/anti-infective combination	Sulfacetamide sodium – 10 % Prednisolone acetate (microfine suspension) – 0.2 %	Benzalkonium chloride – 0.004 % Edetate disodium Polysorbate 80 Polyvinyl alcohol 1.4 % Potassium phosphate, monobasic Sodium phosphate, dibasic Sodium thiosulfate
Sulfacetamide sodium ophthalmic solution USP		Topical antibacterial agents	Sulfacetamide sodium – 10 %	Methylparaben – 0.05 % Propylparaben – 0.01 % White petrolatum Anhydrous liquid lanolin Mineral oil
Suprofen solution	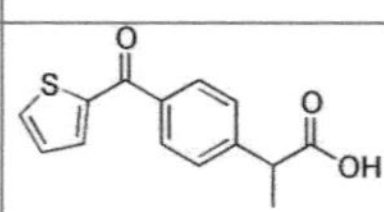	Analgesic, antipyretic, and anti-inflammatory	Suprofen – 1.0 %	Thimerosal – 0.005 % Caffeine – 2 % Edetate disodium Dibasic sodium phosphate Monobasic sodium phosphate Sodium chloride

Tafluprost solution	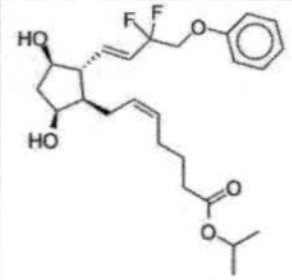	Selective FP prostanoid receptor agonist/indicated to elevate intraocular pressure	Tafluprost – 0.015 mg/ml	Glycerol Sodium dihydrogen phosphate dihydrate Disodium edetate Polysorbate 80
Tetracaine hydrochloride ophthalmic solution	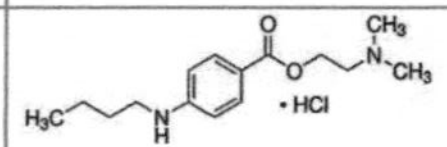	Local anesthetic	Tetracaine hydrochloride – 0.5 %	Chlorobutanol Boric acid Edetate disodium Potassium chloride
Tobramycin and dexamethasone ophthalmic suspension	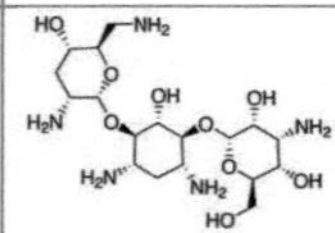 Tobramycin 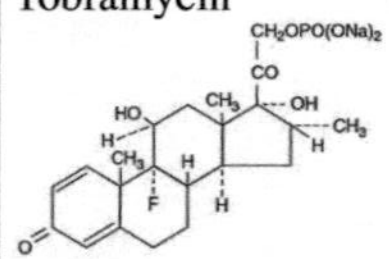 Dexamethasone	Corticosteroid and antibacterial combination	Tobramycin – 0.3 % and dexamethasone – 0.1 %	Edetate disodium Hydroxyethyl cellulose (MPAs) – 1 % Sodium chloride Sodium sulfate Tyloxapol Benzalkonium chloride – 0.01 %
Travoprost solution	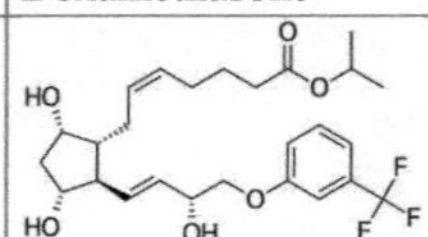	Prostaglandin analog is a selective FP prostanoid receptor agonist which is believed to reduce intraocular pressure by increasing uveoscleral outflow	Travoprost – 0.04 mg/ml	Polyoxyl 40 hydrogenated castor oil Boric acid Propylene glycol Sorbitol Zinc chloride
Travoprost solution			Travoprost – 0.03 mg/ml	Polyquaternium 1 – 0.01 mg/ml Polyoxyethylene hydrogenated castor oil 40 Propylene glycol Boric acid Mannitol Sodium chloride

(continued)

Table 24.1 (continued)

Drug product	API structure	MoA/indication	API	Excipients
Triamcinolone acetonide injectable suspension	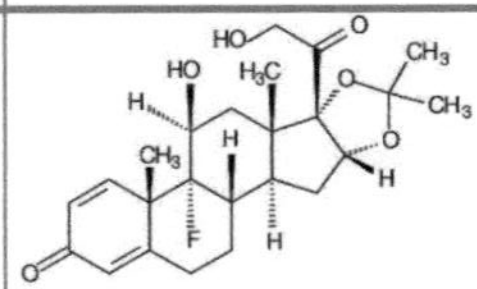	Glucocorticoids/indicated in sympathetic ophthalmia, temporal arteritis, uveitis, and ocular inflammatory conditions unresponsive to topical corticosteroids	Triamcinolone acetonide – 40 mg/ml	Sodium chloride – 0.5 % Carboxymethylcellulose sodium Polysorbate 80 – 0.015 % Potassium chloride Calcium chloride (dihydrate) Magnesium chloride (hexahydrate) Sodium acetate (trihydrate) Sodium citrate (dihydrate)

Table 24.2 List of excipients and maximum potency amount used in FDA-approved ophthalmic drug delivery systems (Compiled from FDA inactive ingredient list)

Excipient	Ophthalmic dosage form	Maximum potency amount (%)
Acetic acid	Solution	0.2
Alcohol	Solution	1.4
Alcohol, dehydrated	Solution	0.5
Alginic acid	Insert, controlled release	0.01
Amerchol-cab	Ointment	0.002
Antipyrine	Solution	0.1
Benzalkonium chloride	Gel	0.008
	Ointment	0.025
	Solution	2
	Suspension	0.025
Benzododecinium bromide	Solution	0.012
Boric acid	Emulsion	0.1
	Solution	37.2
	Suspension	1
	Solution	1.9
Caffeine	Solution	2
Calcium chloride	Solution	0.02
Carbomer 1342	Emulsion	0.05
Carbomer 934P	Suspension	0.45
Carbomer 940	Gel	4
Carbomer homopolymer type B (allyl pentaerythritol cross-linked)	Suspension	0.5
Carboxymethylcellulose sodium	Solution	0.5
Castor oil	Emulsion	5
Cetyl alcohol	Suspension	0.5
Chlorobutanol	Ointment	0.65
	Solution	0.5
Chlorobutanol, anhydrous	Solution	0.5
Citric acid	Solution	0.2
Citric acid monohydrate	Solution	0.05
Creatinine	Solution	0.5
Divinylbenzene styrene copolymer	Suspension	0.75
Edetate disodium	Solution	10
	Solution	0.1
	Suspension	0.13
Edetate sodium	Emulsion	0.02
	Solution	0.1
	Suspension	0.02
Ethylene vinyl acetate copolymer	Insert, controlled release	0.14

(continued)

Table 24.2 (continued)

Excipient	Ophthalmic dosage form	Maximum potency amount (%)
Gellan gum (low acyl)	Solution	0.6
Glycerin	Emulsion	2.2
	Suspension	2.5
	Solution	3
Glyceryl stearate	Suspension	0.5
Hydroxyethyl cellulose	Solution	0.5
	Suspension	0.35
Hydroxypropyl methylcellulose 2906	Solution	0.5
Hypromellose 2910 (15,000 MPAs)	Solution	0.5
	Suspension	0.5
Hypromelloses	Gel	2.25
	Solution	50
	Suspension	0.6
Lanolin	Ointment	2
Lanolin alcohols	Ointment	10
Lanolin anhydrous	Ointment	3
Lauralkonium chloride	Solution	0.005
Lauroyl sarcosine	Suspension	0.03
Magnesium chloride	Solution	0.0065
Mannitol	Solution	23
	Suspension	4
Methylcellulose (4000 MPAs)	Solution	0.5
Methylparaben	Ointment	0.05
	Solution	0.05
	Suspension	0.05
Mineral oil	Ointment	59.5
	Suspension	0.1
Nonoxynol-9	Solution	0.125
Octoxynol-40	Solution	0.05
Petrolatum	Ointment	85
Petrolatum, white	Ointment	89
Phenylethyl alcohol	Solution	0.5
Phenylmercuric acetate	Ointment	0.0008
Phenylmercuric nitrate	Ointment	0.002
	Solution	0.002
Polidronium chloride	Solution	0.0005
Poloxamer 188	Solution	0.1
Poloxamer 407	Solution	0.2
	Suspension	0.101
Polycarbophil	Solution	0.9
	Suspension	0.859
Polyethylene glycol 400	Ointment	4.997
Polyethylene glycol 8000	Solution	2

(continued)

Table 24.2 (continued)

Excipient	Ophthalmic dosage form	Maximum potency amount (%)
Polyoxyl 35 castor oil	Solution	5
Polyoxyl 40 hydrogenated castor oil	Solution	0.5
Polyoxyl 40 stearate	Solution	7
	Suspension	0.5
Polypropylene glycol	Solution	15
Polysorbate 20	Ointment	15
	Emulsion	4
	Solution	1
	Suspension	0.1
Polyvinyl alcohol	Solution	1.4
	Suspension	1.4
Potassium chloride	Solution	22.2
Potassium sorbate	Solution	0.47
Povidone K29/32	Solution	1.8
Povidone K30	Solution	2
	Suspension	0.6
Povidone K90	Solution	1.2
Povidones	Solution	1.8
	Suspension	1.8
Propylene glycol	Solution	10
	Suspension	5
Propylparaben	Solution	0.015
	Suspension	0.01
Soda ash	Solution	0.005
Sodium acetate	Emulsion	0.05
	Solution	1.279
Sodium bisulfate	Solution	0.1
	Suspension	0.06
Sodium borate	Solution	1.1
	Suspension	0.0285
Sodium borate decahydrate	Solution	0.15
Sodium carbonate	Solution	1
Sodium chloride	Solution	55
	Suspension	0.85
	Solution	0.9
Sodium citrate	Solution	2.2
	Suspension	0.45
Sodium metabisulfite	Solution	0.25
Sodium nitrate	Solution	1.18
Sodium sulfate	Solution	0.226
	Suspension	1.2
Sodium sulfite	Solution	0.2

(continued)

Table 24.2 (continued)

Excipient	Ophthalmic dosage form	Maximum potency amount (%)
Sodium thiosulfate	Solution	5
	Suspension	0.32
Sorbic acid	Emulsion	0.1
	Solution	0.2
Sorbitol	Solution	6.6666
Stabilized oxychloro complex	Solution	0.005
Sulfuric acid	Solution	0.02
Thimerosal	Solution	0.01
	Suspension	1
Titanium dioxide	Insert, controlled release	0.004
Tocophersolan	Solution	0.5
Trisodium citrate dihydrate	Solution	0.3
Tromethamine	Solution	0.936
Tyloxapol	Solution	0.1
	Suspension	0.3
Xanthan gum	Solution	0.6
Zinc chloride	Solution	0.0025

In order to secure a prompt development process and the shortest possible approval time, the formulator should be familiar with the current marketed products, and the extensive information provided here should help achieve those goals. These tables will also be useful for researchers as a quick guide to know which excipients are used in current ophthalmic products and the concentrations of these administrated in clinical practice.

24.6 Toxicological Considerations of Excipients for Ophthalmic Drug Delivery System

The sensitive nature of the corneal conjunctival tissues imposes great caution in the selection of excipients because they can diffuse/penetrate into the eye and therefore can lead to unknown toxicological complications, e.g., BAC can get accumulated in the cornea for days. Similarly, EDTA was found to reach the iris–ciliary body in high concentrations, enough to significantly change the permeability of the blood vessels, and thus, indirectly increasing drug removal from the aqueous humor [2]. Also, some surfactants used in ophthalmic formulations were found to cause irritation of the eye mucosa [4]. Penetration enhancers offer numerous advantages for improvement of bioavailability but can easily cause irritation and damage to eye tissue if used in high concentrations. Hence, it's very important to look at the toxicity profile of excipients at the concentrations desired for the preparation of ophthalmic drug delivery systems.

References

1. Achouri D, Alhanout K, Piccerelle P, Andrieu V (2013) Recent advances in ocular drug delivery. Drug Dev Ind Pharm 39(11):1599–617
2. Nanjawade BK, Manvi FV, Manjappa AS (2007) In situ-forming hydrogels for sustained ophthalmic drug delivery. J Control Release 122(2):119–34
3. Kaur IP, Kanwar M (2002) Ocular preparations: the formulation approach. Drug Dev Ind Pharm 28(5):473–93
4. Kaur IP, Chhabra S, Aggarwal D (2004) Role of cyclodextrins in ophthalmics. Curr Drug Deliv 1(4):351–60
5. Rowe RC, Sheskey PJ, Owen SC (2006) Handbook of pharmaceutical excipients, 6th edn. Pharmaceutical Press/American Pharmacists Association, London/Washington, DC
6. Amin P, Bhogte C, Deshpande M (1996) Studies on gel tears. Drug Dev Ind Pharm 22(7):735–9
7. Unlü N, Ludwig A, Van Ooteghem M, Hincal A (1991) Formulation of Carbopol 940 ophthalmic vehicles, and in vitro evaluation of the influence of simulated lacrimal fluid on their physico-chemical properties. Pharmazie 46(11):784–8
8. Deshpande S, Shirolkar S (1989) Sustained release ophthalmic formulations of pilocarpine. J Pharm Pharmacol 41(3):197–200
9. Lu GW (2010) Recent advances in developing ophthalmic formulations: a patent review. Recent Pat Drug Deliv Formul 4(1):49–57
10. Laroche L, Arrata M, Brasseur G, Lagoutte F, Le Mer Y, Lumbroso P et al (1991) Treatment of keratoconjunctivitis sicca with lacrimal gel: a multicentre randomized study. J Fr Ophtalmol 14(5):321–6
11. Johnson ME, Murphy PJ, Boulton M (2008) Carbomer and sodium hyaluronate eyedrops for moderate dry eye treatment. Optom Vis Sci 85(8):750–7
12. Le Bourlais C, Acar L, Zia H, Sado PA, Needham T, Leverge R (1998) Ophthalmic drug delivery systems – recent advances. Prog Retin Eye Res 17(1):33–58
13. Miyazaki S, Suzuki S, Kawasaki N, Endo K, Takahashi A, Attwood D (2001) In situ gelling xyloglucan formulations for sustained release ocular delivery of pilocarpine hydrochloride. Int J Pharm 229(1–2):29–36
14. Vyasd SP, Ramchandraiah S, Jain CP, Jain SK (1992) Polymeric pseudolatices bearing pilocarpine for controlled ocular delivery. J Microencapsul 9(3):347–55
15. Saari KM, Nelimarkka L, Ahola V, Loftsson T, Stefánsson E (2006) Comparison of topical 0.7% dexamethasone-cyclodextrin with 0.1% dexamethasone sodium phosphate for postcataract inflammation. Graefes Arch Clin Exp Ophthalmol 244(5):620–6
16. Sigurdsson HH, Konrádsdóttir F, Loftsson T, Stefánsson E (2007) Topical and systemic absorption in delivery of dexamethasone to the anterior and posterior segments of the eye. Acta Ophthalmol Scand 85(6):598–602
17. Singh SK, Bandyopadhyay P, Hasan S (2004) Ophthalmic antibiotic drug formulations containing a cyclodextrin compound and cetyl pyridinium chloride. US patent 20,040,019,012
18. Ludwig A (2005) The use of mucoadhesive polymers in ocular drug delivery. Adv Drug Deliv Rev 57(11):1595–639
19. Séchoy O, Tissié G, Sébastian C, Maurin F, Driot JY, Trinquand C (2000) A new long acting ophthalmic formulation of Carteolol containing alginic acid. Int J Pharm 207(1–2):109–16
20. Demailly P, Allaire C, Trinquand C (2001) Ocular hypotensive efficacy and safety of once daily carteolol alginate. Br J Ophthalmol 85(8):921–4
21. Keister JC, Lang JC, Missel PJ, Stancioff DJ (1995) Use of carrageenans in topical ophthalmic compositions. Google patents
22. Johnson ME, Murphy PJ, Boulton M (2006) Effectiveness of sodium hyaluronate eyedrops in the treatment of dry eye. Graefes Arch Clin Exp Ophthalmol 244(1):109–12
23. Ito Y, Nagai N, Shimomura Y (2010) Reduction in intraocular pressure by the instillation of eye drops containing disulfiram included with 2-hydroxypropyl-β-cyclodextrin in rabbit. Biol Pharm Bull 33(9):1574–8

24. El-Kamel AH (2002) In vitro and in vivo evaluation of Pluronic F127-based ocular delivery system for timolol maleate. Int J Pharm 241(1):47–55
25. Shastri DH, Patel LD, Parikh RK (2010) Studies on in situ hydrogel: a smart way for safe and sustained ocular drug delivery. J Young Pharm 2(2):116–20
26. Srividya B, Cardoza RM, Amin PD (2001) Sustained ophthalmic delivery of ofloxacin from a pH triggered in situ gelling system. J Control Release 73(2–3):205–11
27. Liu Z, Pan W, Nie S, Zhang L, Yang X, Li J (2005) Preparation and evaluation of sustained ophthalmic gel of enoxacin. Drug Dev Ind Pharm 31(10):969–75
28. Liu Z, Li J, Nie S, Liu H, Ding P, Pan W (2006) Study of an alginate/HPMC-based in situ gelling ophthalmic delivery system for gatifloxacin. Int J Pharm 315(1–2):12–7
29. Singh V, Bushetti SS, Appala R, Shareef A, Imam SS, Singh M (2010) Stimuli-sensitive hydrogels: a novel ophthalmic drug delivery system. Indian J Ophthalmol 58(6):477–81

Chapter 25
Ophthalmic Drug Delivery Systems for the Treatment of Corneal Diseases

Ankit Shah

Abstract The role of nanotherapeutics as it applies to the treatment and management of current corneal pathology is evolving to provide increased therapeutic concentration, targeted therapy, and reduced toxicity. The current limitations to such novel therapies are based on structural and functional barriers of a lipophilic corneal epithelium and a hydrophilic stroma. This chapter provides a detailed review of the corneal anatomy to help facilitate discussion into how nanotherapeutics may be applied in certain clinical conditions. Current modes of ophthalmic drug delivery to the cornea such as solutions, suspensions, and ointments have their own inherent flaws including lower bioavailability, lower contact time, and bypass of first-pass metabolism. In contrast, the development of colloids such as liposomes, microemulsions, and niosomes overcomes many of these limitations by providing a biphasic environment with lipophilic and hydrophilic properties. Furthermore, they allow for sustained release of pharmacotherapy and higher bioavailability. Current innovative applications include using voriconazole microemulsions and cyclosporine micelles in the treatment of fungal keratitis and dry eye syndrome, respectively. This chapter further details the use of transporter proteins, by manipulating their influx and efflux transport properties, it becomes possible to deliver antiviral medications for the treatment of herpetic corneal disease. The use of an electrical stimulus, via iontophoresis or ultrasound, via sonophoresis to drive pharmacologic agents through tissue planes is also discussed in this chapter with several examples of current experimental models. Finally, we describe the use of microneedles to introduce gene therapy to potentially treat corneal disease.

Keywords Suspensions • Colloids • Microemulsions • Microneedles • Sonophoresis • Iontophoresis • Dendrimers • Transporter proteins • Liposomes

---

A. Shah (✉)
Department of Ophthalmology, University of South Florida, Tampa, FL, USA
e-mail: shah.aa7@gmail.com

© Springer International Publishing Switzerland 2016

Y. Pathak et al. (eds.), *Nano-Biomaterials For Ophthalmic Drug Delivery*,
DOI 10.1007/978-3-319-29346-2_25

This chapter briefly revisits the corneal anatomy and physiology to better understand the barriers that pharmacologic agents have to overcome in order to reach a therapeutic threshold. Furthermore, we discuss several common corneal pathologic processes and their current pharmacologic treatments. Finally, we touch upon current and future applications of nanotherapeutics and ophthalmic drug delivery in the treatment of corneal disease.

25.1 Anatomy

The cornea of the eye is the anterior most avascular clear tissue segment of the eye, which functions predominantly as the main refractive medium of the eye (not the lens, contrary to popular belief). It also serves as a protective barrier against ocular penetration from pathogenic microbes, foreign bodies, and pharmacologic agents. Familiarity of the corneal anatomy is important to understand how the cornea maintains clarity and also how nanotherapeutic agents can overcome structural barriers to treat pathological corneal conditions.

There are five distinct corneal layers. These include (from anterior to posterior): the corneal epithelium, Bowman's layer, corneal stroma, Descemet's layer, and the corneal endothelium. Each is unique in structure and function and serves an important role in maintaining corneal integrity.

The corneal epithelium is composed of stratified squamous cells and reflects the anterior most cellular surface of the cornea. It measures approximately 50 μm in width ([21], 197). This layer is further segregated into the basal most columnar cells, slightly anterior wing cells, and surface cells. The surface cells, at their apical edge contain microvilli that intimately interact with the tear film to provide a lubricating medium. These same surface cells and their microvilli are also involved in increasing contact time between pharmacologic ophthalmic agents and corneal epithelial cells. Furthermore, the desmosome helps form tight bonds or adhesions between adjacent surface cells and helps to form an impermeable barrier against bacterial entry. Epithelial cells are shed routinely and regenerate from corneal limbal stem cells. Basal cells are fixed to the epithelium basement membrane through hemidesmosomes. The corneal epithelium, through its tight junctions, is a lipophilic agent.

Together, the corneal epithelium basal cells and the stromal keratocytes help produce an acellular basement membrane-like layer of collagen fibrils termed the Bowman's layer. This layer does not regenerate if destroyed and, as a result, may cause corneal edema and scarring if damaged.

The corneal stroma is the thickest layer, measuring approximately 500 μm and composed of keratocytes, type I collagen, and glycosaminoglycans (specifically keratin sulfate) [22]. The cornea maintains its clarity because stromal proteins and cells are arranged in a parallel plane helping preserve an optically clear corneal media. The regular arrangement of the corneal stroma also allows for easy dissection during lamellar surgery. In contrast to the corneal epithelium, the stroma is a hydrophilic medium.

Descemet's membrane is a true basement membrane produced by the corneal endothelium and measures approximately 10 μm [22]. Descemet's membrane is composed of type IV collagen fibrils and, unlike Bowman's layer, may be synthesized if damaged. It serves to anchor the underlying endothelial cells to the corneal stroma.

The corneal endothelium is the posterior most cellular layer of the cornea, separating the corneal stroma from the anterior chamber. More importantly, it serves an important role in maintaining corneal clarity through energy-dependent channel proteins that help keep the cornea dehydrated. Composed of a single layer of hexagonal cells, the corneal endothelial cell density is fixed from birth and, if damaged, does not regenerate. Fuchs' dystrophy is a pathological corneal condition associated with the loss of corneal endothelial cells. If enough endothelial cells are damaged, the cornea begins to swell, and edema ensues.

It is important to note that the cornea itself is completely devoid of any vascular supply and receives oxygen through the air/tear film complex and glucose from the aqueous fluid. In fact, sub-threshold therapeutic concentrations of topical ophthalmic medications are partly a result of lymphatic and vascular flow of pre-corneal space. By the time ophthalmic drugs are delivered to the cornea, therapeutic concentrations have decreased substantially. Other limitations to drug delivery to the front of the eye include unintended dilution of medications through contact with the tear film and drainage through the nasolacrimal system [1]. The natural blinking phenomenon and normal tear drainage in patients greatly reduce the total concentration of medication which eventually reaches the pre-corneal space.

Once it reaches the cornea, there are static structural barriers that must be overcome including tight junctions between corneal epithelial cells and inherent molecular and electrical charges of the epithelium and the stromal tissue [20].

25.2 Current Modes of Ophthalmic Drug Delivery

In general, there are three main mechanisms of ophthalmic drug delivery: solutions, suspensions, and ointments. We discuss benefits and disadvantages of each mode of delivery in the following section.

Ophthalmic topical solutions have distinct benefits over other systemic formulations (intravenous, oral, and even intravitreal) but also have some major disadvantages. The benefits lie in the fact that topical medications, in general, are poorly absorbed systemically and, as a result, systemic toxicity, such as renal failure or hepatic dysfunction is less likely. At the same time, if a topical preparation of a drug does reach the systemic circulation, the likelihood of an adverse pharmacologic event may be higher because it bypasses first-pass metabolism from the liver. As mentioned previously, ophthalmic solutions as a means of drug delivery are not an ideal mechanism because of its poor bioavailability, minimal corneal contact time, and dependence on preservatives.

Ophthalmic ointments are another means of ophthalmic drug delivery, which in itself have certain benefits over ophthalmic solutions. For instance, because ointments, by definition, have a higher viscosity than their solution counterparts, they promote longer surface contact with the cornea and allow for an improved therapeutic effect. This benefit is even more significant in pediatric populations where sufficient topical drop administration is limited by patient cooperation and excessive tearing. A major disadvantage to ophthalmic ointments is that it limits accessibility of other ophthalmic medications to the ocular surface. Additionally, ointments cause blurry vision and are not recommended for daytime use [3].

25.3 Future Ophthalmic Drug Delivery Methods: Nanotherapeutics

Recent focus on ophthalmic drug delivery has shifted to include the use of nanotherapeutics to improve therapeutic concentrations, corneal penetration, and reduce toxicity. Nanotherapeutics refers to the use of nanoparticles or micromolecules, ranging from 1 to 100 nm in size, that are genetically or pharmacologically manufactured to provide a means of drug delivery at the cellular level. There are several subgroups of nanoparticles that have been implicated for its use specifically with corneal pathology [2].

The use of nanotherapeutics for ophthalmic drug delivery has become a very hot topic of interest in recent years as physicians, scientists, and pharmacists find ways to overcome static and dynamic barriers to ophthalmic delivery. Current modalities of delivery include solutions, suspensions, and ointments. While each has its own advantages and disadvantages, overall, their effectiveness is limited by the tear film, the nasolacrimal drainage system, the blink reflex, the scleral/conjunctival vascular and lymphatic flow, and the tight junctions between corneal epithelial cells. The majority of nanotherapeutic research involves local therapy against macular degeneration, retinitis pigmentosa, and glaucoma management and treatments. The following sections describe recent nanotherapeutic developments in relationship to corneal pathology.

25.4 Colloids

Recent progress has been made in identifying unique modalities for ophthalmic drug delivery, specifically in the scope of corneal pathology. Current advances include reliance on colloids through delivery of medications via liposomes, microemulsions, and nanoemulsions. Colloids are small solid particles that do not dissolve in a solution and, unlike other solid particles, have an ability to scatter light [11].

25.5 Liposomes

Liposomes are microvesicles made of a surrounding lipid bilayer coat and a central aqueous center. The aqueous center allows for incorporation of water-soluble drugs, while lipid soluble drugs can be attached to the lipid coating bilayer membrane [13]. This amphiphilic property of liposomes makes them an ideal drug delivery mechanism because it improves corneal penetration through the hydrophobic epithelium and the lipophilic corneal stroma. Additionally, due to the inherent properties of liposomes, they provide a longer half-life and a persistent steady release of medication than topical applications of the same medication. In fact, liposomes have been manipulated to include vitamins, specifically vitamin A and C, to promote corneal hydration in the treatment of dry eye syndrome and also in the delivery of idoxuridine to treat herpetic keratitis to prevent epithelial toxicity [4].

Unlike ophthalmic suspensions and solutions, liposomes and microemulsions provide a means for greater bioavailability, sustained and controlled release of medications, and allow for bypass of the blood ocular barriers. By exchanging different phospholipids on the cell membrane, physician-scientists are able to selectively modify and assemble colloids which would target specific cells. It is important to note that liposomes may be manipulated to overcome the net negative charge of the corneal epithelium. Catatonic liposomes have better permeability and penetration into the corneal stroma, a characteristic which is useful when targeting medications to the stroma [5]. Unfortunately, liposomes are easily biodegraded and are relatively unstable. Historical studies centered on liposomal carrier systems, specifically penicillin G and indoxole in rat corneas, have established liposomes as adequate alternative carrier mechanism with impressive corneal and aqueous penetration and permeability [19].

In contrast, niosomes are a cheaper and more stable alternative to liposomes because they are made up of nonionic surfactant vesicles yet maintain a bilayer amphiphilic structure. Their lipid surfactant membrane and aqueous center make niosomes biodegradable, stable, and flexible, while reducing toxicity and cost. Furthermore, the nonionic surfactant makes niosomes less likely to promote an inflammatory reaction. Current studies in the use of niosomes revolve around the delivery of ophthalmic mydriatics including cyclopentolate and anti-ocular-hypertensive agents such as timolol. These studies show promising results in in vivo animal models by establishing increased intracameral concentrations than their solution and suspension counterparts [15].

25.6 Microemulsions

In contrast, microemulsions are nanoparticles made up of two immiscible liquid moeities: usually, a hydrophilic agent such as water and a hydrophobic oil with adjunct surfactant. Microemulsions are easily created in the lab, requiring little to no energy expenditure. The amphiphilic property of microemulsions, in addition to their small size, allow for easy penetration, accessibility, and prolonged release of therapeutic drugs to the target tissue through the formation of intracorneal depots. Microemulsions exist through the formation of micelles [14]. Current experimental therapeutic microemulsions that have been developed incorporate anti-inflammatory nonsteroidal agents, steroids, and ocular anti-hypertensives to treat difficult ophthalmic disease entities. Their benefits reside in the fact that they are not subject to precorneal barriers of the conjunctival lymphatic and vascular system and are able to circumvent structural impediments of the cornea [6].

The creation and use of voriconazole microemulsions in the treatment of fungal corneal ulcers has been researched, and early studies ex vivo show a clear increase in voriconazole concentrations after 12 h in the corneal tissue when compared to suspensions. In vivo studies are being planned [7].

Similarly, dry eye syndrome is another corneal pathological entity which has benefited from the use of nanotherapeutics for treatment. Colloidal application of topical cyclosporine A, through the creation of catatonic micelles, has been widely successful in treating patients with refractory dry eyes and has resulted in improved symptoms, patient cooperation, and improved bioavailability, more so than their oil-based microemulsion counterparts [8].

25.7 Transporter Proteins

Another pattern of how nanotherapeutics is shaping the way corneal disease is managed is through the advent of transporter proteins. These membrane transport proteins can selectively target and bind ligands and allow for passage into and out of the cells. Recent advances to transporter proteins have enabled physician-scientists to manipulate this system so that certain drug moieties may be combined to the ligand to allow intracellular diffusion of the medication [16]. There are two specific types of transporter proteins: efflux and influx proteins. Efflux transport proteins are energy-dependent proteins that enable for targeted ligands to be removed extracellularly. In contrast, influx transport proteins rely on a solute transport mechanism down concentration gradients without requiring energy. Transporter proteins allow for selective drug delivery of medications to the epithelial of the cornea, conjunctiva, and sclera.

These transport proteins have been vital in the delivery of antiviral agents for the treatment of herpetic epithelial/stromal keratopathy. By binding pro-drugs of acyclovir and ganciclovir to certain ligands, scientists have been able to manipulate transport proteins to allow for influx and efflux of these agents and increase the intracellular concentration of these antivirals (ocular drug delivery). Furthermore, transport proteins are being tested in animal models for adjunct vitamin C delivery to the cornea [16].

Efflux transport proteins appear to serve as an inherent protective mechanism to the cell by ridding itself of molecules that the cell believes are toxic [9]. Future research may evolve into selectively targeting these genes in hopes to increase intracellular concentrations of certain medications by limiting its exit out of the cell [16].

25.8 Dendrimers

Dendrimers are polymers of repetitive branching molecules that form structurally symmetric spherical nanomolecules in a concentric fashion, likewise to the layers of an onion. Each tier of repetitive monomeric subgroups encompasses a core, and it is through alterations of the functional subgroups that allow specific interactions between different drugs, target cells, and pathogens [4]. Dendrimers essentially serve as transporters for nanoparticles, by either incorporating the target drug in the core or by conjugating it through one of its subgroups. Current use of dendrimers in ophthalmic delivery has only been shown to be effective in gene transplant, and through instillation of ophthalmic miotic and mydriatic agents.

25.9 Ocular Iontophoresis

Ocular iontophoresis involves using an electrical stimulus to drive pharmacologic agents through the tissue planes by taking advantage of the fact that similarly charged agents repel each other and opposite charged agents attract. This system enables scientists and ophthalmologists to manipulate medications by introducing a net charge, and then, applying an electrical stimulus so that the medication can be pushed or pulled through a transcorneal approach. Furthermore, net migration of medications is also assisted through electroosmosis, which relies on particles moving with a current from the anode to the cathode [10].

Introduction of antibiotics, antivirals, antifungals, and anti-inflammatory medications through the use of ocular iontophoresis has proved to be successful in the experimental stage. Every class of antibiotics has been studied and found to successfully penetrate the cornea for the treatment of bacterial keratitis. In fact, recent advances of ocular iontophoresis have introduced anesthetics such as lidocaine and epinephrine as viable medications that can successfully be concentrated into the eyelids for oculoplastic procedures. The use of an electric probe to electrically stimulate the transportation of medications through the corneal planes appears to be a safe, painless, effective, and fast alternative to current invasive measures [11].

The idea of reverse iontophoresis is another up-and-coming nanotherapeutic procedure, which serves to produce a unique avenue for ophthalmic drug delivery. In vitro studies with hemosiderin-stained corneas illustrate the successful extraction of hemosiderin deposits from the corneal stroma and endothelium. By altering the net charge of iron, an electrical stimulus was used to remove the residual hemosiderin granules from the corneal tissue [17]. This is a particularly useful strategy in the treatment of complicated hyphema after trauma with resulting endothelium blood staining. Current treatments consist of emergent anterior chamber washout and, if staining is severe, corneal transplantation.

The use of the above nanoparticles, specifically liposomes and microemulsions, can be concurrently combined with the current uses of contact lenses to provide a unique combination of ophthalmic drug delivery. Current contact lenses are composed of a polymeric material that mimics the ocular surface shape and comfortably rests on the ocular surface. While pre-soaking contact lenses with a drug may initially prolong the duration of a therapeutic effect, contact lenses are still subject to the same limitations as ophthalmic solutions, specifically, the amount of drug that can be loaded to the contact lenses and a half-life only of hours. Soaking a pair of contact lenses after it has been placed on the eye does not fare much better. However, combining nanoparticles, such as microemulsions and liposomes onto contact lenses, not only provides an immediate release of the drug lasting several hours, but also provides a sustained release for several days by taking advantage of the hydrophobic and hydrophilic properties of oil and water emulsions. Drugs mixed with aqueous agents tends to diffuse immediately, while drugs soaked in oil takes days to diffuse [4].

25.10 Sonophoresis

Ultrasound has been used for generations in the evaluation and diagnosis of ophthalmic pathology when the visualization of the posterior pole is difficult or obstructed as in the case of an evaluation of the retina during vitreous hemorrhage or mature cataracts. Recent advances have enabled scientists to link high-frequency sound waves to ophthalmic medications in efforts to facilitate drug delivery into the cornea and even the retina. Sonophoresis or ultrasound-mediated delivery of ophthalmic medications relies on the creation of cavitation or gas bubbles that can penetrate through intact ocular tissue. The use of an ultrasound, often at a frequency of greater than 800 kHz in conjunction with a coupling medium (often the medication to be infused), creates micro-bubbles which oscillate to form pores that enable cellular and vascular permeability. The use of sonophoresis in rabbit models to deliver ophthalmic dexamethasone was studied and found that ultrasound frequency at 800 kHz enhanced and optimized corneal drug delivery. However, limitations to the widespread use of sonophoresis stem from the concerns of thermal and mechanical damage to the avascular cornea, lens, and vitreous from the high-frequency and prolonged ultrasound energy [18].

25.11 Microneedles

Microneedle therapy will be briefly introduced as a potential new ophthalmic drug delivery option in the treatment of anterior segment and corneal disease processes.

The use of microneedles as a therapeutic means has gained traction in recent years. Microneedle therapy has been used for some time in the delivery of vaccines and medications intradermally, with overall good effect. However, recent use of microneedles, which are customized needles, ranging in size from 25 to 2000 μm in length, provides a means of localized delivery without causing pain and extensive tissue damage. While microneedles are very effective in the treatment of posterior segment disease, there are novel applications for their use in corneal pathology. Current innovation for microneedle therapy is used for the treatment of cystoid macular edema and posterior uveitis via intrascleral injection of steroids directly into the suprachoroidal space. This provides faster entry and higher concentrations in the retinal space [12]. Current corneal applications of microneedle therapy are limited to the introduction of gene vectors into the epithelial and stromal cells. Further studies which link microneedles with coated antibiotics for the treatment of infectious keratitis are underway; however, extensive corneal damage from the underlying infection makes microneedles difficult to use.

References

1. Zielinski WL, Sullivan TR (2007) Ophthalmic drug therapy – challenges and advances in front-of-the-eye delivery. Mystic Pharmaceuticals Inc., Special Report. Ocul Deliv:44–45
2. Bhargavi, Ch.Anil, Bhowmik D, Desale P, Kumar KPS (2013) Nanotherapeutics – an era of drug delivery system in nanoscience. Indian J Res Pharm Biotechnol 1(2):210
3. Ophthalmic Fluoroquinolones Review, Provider Synergies 1(3) (2004–2008)
4. Souza JG, Dias K, Pereira TA, Bernardi DS (2013) Topical delivery of ocular therapeutics: carrier systems and physical methods. J Pharm Pharmacol 66:507–530
5. Kaur IP, Kanwar M (2002) Ocular preparations: The formulations approach. Drug Dev Ind Pharm 28(5):473–493
6. Hegde RR, Verma A, Ghosh A (2013) Microemulsion: new insights into the ocular drug delivery. ISRN Pharmaceut 2013, 826798. doi:10.1155/2013/826798, 11 pages
7. Kumar R, Sinha VR (2014) Preparation and optimization of voriconazole microemulsion for ocular delivery. Colloids Surf B: Biointerfaces (117):84
8. Luschmann C, Tessmar J, Schoeberl S, Strau O, Luschmann K (2014) Self assembling colloidal system for the ocular administration of cyclosporine A. Cornea 33(1):78
9. Barot M, Gokulgandhi MR, Pal D, Mitra AK (2013) Mitochondrial localization of P-glycoprotein and peptide transporters in corneal epithelial cells – novel strategies for intracellular drug targeting. Exp Eye Res 106:52
10. Myles ME, Loutsch JM, Higaki S, Hill JM. Ocular Iontophoresis. In: Mitra AK, editor. Ophthalmic Drug Delivery Systems. 2nd ed. New York: Marcel Dekker Inc; 2003. pp. 365–408
11. Bourlais CL, Acar L, Zia H, Sado RA (1998) Ophthalmic drug delivery systems – recent advances. Prog Retin Eye Res 17(1):42
12. Jiang J, Gill HS, McCarey BE, Patel SR, Edelhauser HF (2007) Coated microneedles for drug delivery to the eye. Invest Ophthalmol Vis Sci 48(9):4039–4040

13. Meisner D, Mezei M (1995) Liposome ocular delivery system. Adv Drug Deliv Rev 16:89
14. Vandamme TF (2002) Microemulsions as ocular drug delivery systems: recent developments and future challenges. Prog Retin Eye Res 21:17
15. Kaur IP, Garg A, Singla AK, Aggarwal D (2004) Vesicular systems in ocular drug delivery: an overview. Int J Pharm 269:9–10
16. Gaudana R, Jwala J, Boddu SHS, Mitra AK (2009) Recent perspectives in ocular drug delivery. Pharm Res 26(5):1198–1200
17. Bai JH, Su S, Huang L, Xhang YY, Wang YS, Guo MH, et al (2014) In vitro extraction of intracorneal iron using reverse iontophoresis and vitamin C. Graefes Arch Clin Exp Ophthalmol. 2014 Aug;252(8):1245–58
18. Nabili M, Shenoy A, Chawla S, Mahesh S, Liu J, et al (2014) Ultrasound-enhanced ocular delivery of dexamethasone sodium phosphate: an in vivo study. J Ther Ultrasound 2(6):8–9
19. Schaeffer HE, Krohn DL (1982) Liposomes in topical drug delivery. Invest Ophthalmol Vis Sci 22(2):221–224
20. Kim YC, Chiang B, Wu X, Prausnitz MR (2014) Ocular drug delivery of macromolecules. J Control Release. 2014 Sep 28;190:172–81
21. Friedman NJ, Kaiser PK, Trattler WB (2005) Review of ophthalmology. Elsevier Saunders, Philadelphia
22. American Academy of Ophthalmology (2013) Basic and clinical science course – external disease and cornea, section 8. EB, San Francisco

Chapter 26
Challenges in Ocular Pharmacokinetics and Drug Delivery

Joyce S. Macwan, Anjali Hirani, and Yashwant Pathak

Abstract The ideal ophthalmic drug delivery system must be able to deliver the effective therapeutic concentrations in the target tissues for the intended duration with minimal adverse effects. Successful tissue-targeted delivery of ocular drugs is a challenging task for pharmaceutical scientists. The principal reasons are inherent and unique anatomical and physiological differences among ocular tissues. Ocular drug delivery is restricted by the barriers protecting the eye. The chief ocular membrane barriers are located in the cornea and conjunctiva of the eye which can significantly affect drug entry into the eye.

Conventional routes for ophthalmic drug delivery include topical, systemic, intravitreal, and periocular. The topical route is the most common, mainly for the treatment of anterior segment diseases. However, the tear film drainage and permeation through the cornea and conjunctiva are the key barriers encountered by topically administered drugs resulting in poor ocular bioavailability. Following systemic administration of ocular drugs, the blood-aqueous barrier and blood-retinal barrier prevent entry of substances from the systemic circulation into the intraocular compartments of the anterior and posterior segments of the eye, respectively. Intravitreal injections are good treatment option for posterior segment diseases; however; it is an invasive approach and associated with severe complications. The periocular/transscleral route is a safer substitute to the intravitreal route, and therefore, it has gained significant interest in recent years.

J.S. Macwan
Simulations Plus, Inc., Lancaster, CA, USA

A. Hirani
Department of Pharmaceutical Sciences, USF College of Pharmacy, University of South Florida, Tampa, FL 33612, USA

School of Biomedical Engineering and Sciences, Virginia Tech-Wake Forest University, Blacksburg, VA 24061, USA

Y. Pathak (✉)
Department of Pharmaceutical Sciences, USF College of Pharmacy, University of South Florida, Tampa, FL 33612, USA
e-mail: ypathak1@health.usf.edu

© Springer International Publishing Switzerland 2016

Y. Pathak et al. (eds.), *Nano-Biomaterials For Ophthalmic Drug Delivery*,
DOI 10.1007/978-3-319-29346-2_26

This chapter provides an overview of the main aspects of the primary ocular barriers and challenges associated with various routes of ophthalmic drug administration. Advances in understanding the significance of challenges and barriers will facilitate the development of safe and effective novel ocular drug delivery systems.

Keywords Ocular pharmacokinetics • Drug delivery • Barriers • Ophthalmic • Eye • Challenges • Routes of ocular drug administration • Ocular membranes

26.1 Introduction

An optimal delivery of ocular therapeutics is a challenging task due to unique anatomical and physiological constraints posed by the ocular barriers. The primary membrane barriers are located in the cornea, conjunctiva, sclera, iris-ciliary body, and retinal tissues of the eye [1] (Fig. 26.1). These defensive barriers also hinder the

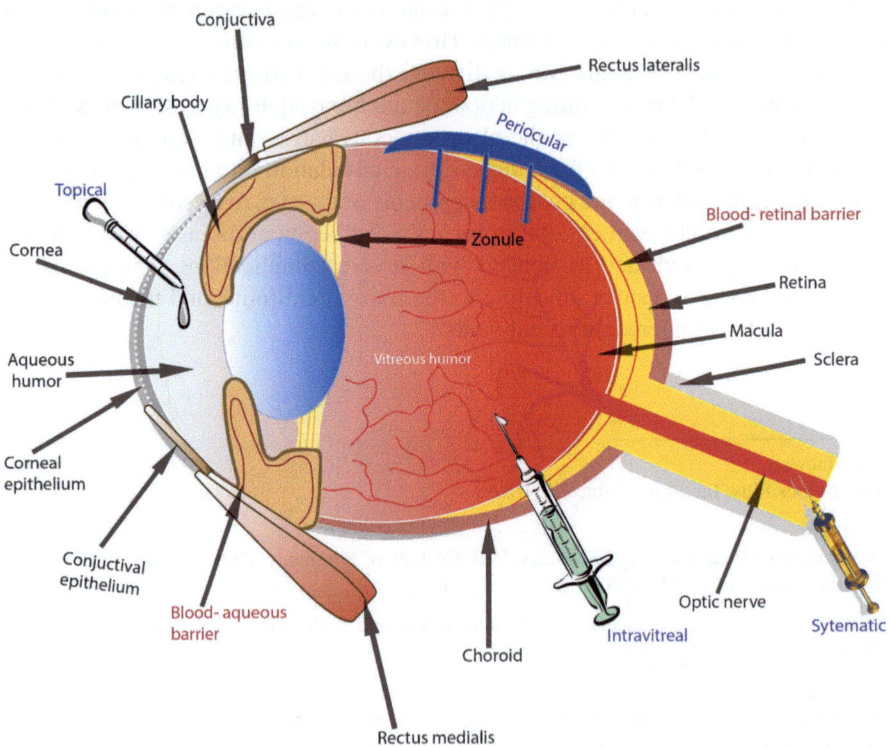

Fig. 26.1 Schematic presentation of the eye and its biological barriers as well as drug delivery routes

delivery and site-specific targeting of ophthalmic drugs. In recent decades, several studies have been focused to develop effective ocular drug delivery approaches to overcoming such biological obstacles.

Various ocular drug delivery modalities for treating anterior and posterior segment disorders are available. Traditionally, topical ocular eye drops are the most commonly utilized dosage form to treat anterior segment disorders (e.g., conjunctivitis, chronic dry eye, and glaucoma). However, the barrier exerted by the corneal and precorneal factors including tear turnover, solution drainage, nasolacrimal drainage, and low drug residence time limit ocular drug bioavailability below 5 % [2, 3]. Several drug delivery strategies including long-acting and modified formulations have been developed to overcome significant challenges faced during topical ocular drug administration; however, they are still not entirely successful to overcome the issues associated with this route.

The topical route fails to deliver appropriate therapeutic drug levels to the posterior segment of the eye, whereas the systemic route requires a large dose and often results in systemic toxicity [4]. The inability or inefficiency of the topical and systemic drug administration to the posterior portion of the eye resulted in rapidly growing interest in alternative routes of drug delivery such as intravitreal and periocular. These routes are usually preferred for the treatment of the posterior segment diseases that include retinitis pigmentosa, age-related macular degeneration, glaucoma, and diabetic retinopathies. Intravitreal administration delivers high local concentrations to the posterior tissues including the vitreous body, neural retina, retinal pigment epithelium (RPE), and choroid. However, it is the most invasive mode of drug delivery to the posterior portion of the eye and associated with risk of adverse effects. On the other hand, the periocular route is considered less invasive relative to intravitreal administration [5]. Successful posterior segment drug delivery of small or large molecules such as proteins, virus vectors, DNA, and other novel products of biotechnology is still challenging.

Recent advancement in ocular drug delivery systems approaches the obstacles of drug entry into the eye through numerous perspectives. However, due to the inherent complexity of the eye and complex pathophysiological processes of ophthalmic disorders, continued research efforts are still needed to develop safe and effective ocular drug therapies. This chapter provides a general review of challenges in ocular pharmacokinetics and drug delivery.

26.2 Ocular Membrane and Barriers in Ophthalmic Therapeuticss

Depending upon the route of administration, drug molecules traverse various membrane barriers in order to reach the preferred target sites. The corneal route is the main route of entry for the most of topically applied drugs into the eye. On the other hand, the non-corneal route which involves conjunctival/scleral pathway is the minor ocular absorption route. However, it may serve as the major absorption pathway for some of the compounds.

26.2.1 Corneal Route

26.2.1.1 Cornea

The cornea is the outermost avascular transparent layer of the anterior segment of the eye. It is considered the principal route for the entry of the most of the ocular drugs into the eye. The cornea is covered with the three-layered tear film that serves as a protective barrier to the eye. It is composed of five distinct layers including epithelium (the outermost), Bowman's membrane, stroma (middle), Descemet's membrane, and endothelium (the innermost) [6]. Each layer affects drug permeation differently due to structural and compositional differences [7]. The individual permeability of the epithelium, stroma, and endothelium determines the overall permeability of the cornea.

Two routes are obtainable for solutes diffusing across the epithelium: a paracellular pathway and a transcellular pathway. The epithelium is lipophilic in nature and thereby greatly favors penetration of lipophilic substances and impedes permeation of hydrophilic and ionized drugs [8]. The superficial corneal epithelium layer has tight junctions, and that diminish the paracellular transport of drugs into deeper layers of the cornea where larger gap junctions are present [9]. The extent of passive drug penetration and thereby tissue bioavailability are dependent on physicochemical properties of drugs such as lipophilicity, molecular weight, solubility, charge, polarity, molecular size, and shape. Tissue properties including integrity of tight/leaky junctions, presence of metabolic enzymes, and transporters also play a crucial role. Parabolic and sigmoidal relationships were established to illustrate the impact of drug lipophilicity on the corneal permeation of drugs such as beta-blockers, alkyl p-amino-benzoates, aniline derivatives, and steroids described previously [10–14]. The transcellular permeability of lipophilic drug betaxolol (permeability coefficient is 25 times higher than that of atenolol) was greater than the same molecular size hydrophilic atenolol in the cornea of pigmented rabbit [14, 15].

Hamalainen and coworkers determined the corneal permeability of mixture of 17 small hydrophilic polyethylene glycols (PEG) of different sizes, which diffuse largely through paracellular route. The study concluded that the rate-limiting step for the corneal permeability through paracellular pathway is the likelihood of finding hydrophilic pores [16]. These authors also demonstrated that paracellular permeability of PEG oligomers reduces with increasing molecular weight. Paracellular permeation of small molecules with an average diameter of 2.0 nm has been reported in the rabbit corneal epithelium [17]. Palmgren et al. assessed the paracellular permeability of neutral PEG and positively charged amino PEG across the corneal epithelial characterizing the interaction of size and charge with paracellular pores [18]. This study indicated that at physiological pH, paracellular pores of the corneal epithelial layer are negatively charged; consequently, the permeability of positively and neutral PEG are higher than that of negatively charged.

Unlike the epithelial layer, the stroma possesses a high content of hydrated collagen, which renders it hydrophilic [19]. Therefore, it permits permeation of hydrophilic molecules and restricts entry of hydrophobic molecules that can cross lipophilic epithelium further into deep ocular tissues. Moreover, it also acts as a barrier for macromolecules; only molecules which are less than 50 kilodalton (kDa) sizes can permeate through the stroma [20]. Like the epithelium layer, the endothelium also offers both paracellular (preferred by hydrophilic molecules) and transcellular (preferred by hydrophobic molecules) routes [21]. Monolayered endothelium that lies between the aqueous humor and stroma represents leaky structure that allows easy entry of macromolecules between the stroma and aqueous humor [6, 8]. The endothelium is primarily composed of phospholipids and therefore permeable to lipophilic molecules. Ideally, the molecules that have amphiphilic characteristics can be considered suitable candidate for drug delivery via the corneal route [20].

Typically, molecules that are successful crossing the corneal barriers can diffuse into the anterior uvea and aqueous humor; however, they fail to reach the posterior site of the eye at adequate, effective concentrations [4]. In addition, the presence of drug-metabolizing enzymes mainly esterase, aldehyde, and ketone reductases and transmembrane efflux transporters including P-glycoprotein (P-gp), multidrug resistance protein (MRP), and breast cancer resistance protein in the cornea [6, 22] influence overall disposition of the drug in the ocular tissues.

26.2.2 Non-corneal Routes

26.2.2.1 Conjunctiva

The conjunctiva is a transparent, thin, and elastic tissue covering nearly 80 % of the ocular surface, offering a large surface area for absorption compared to the cornea [6]. The part of the conjunctiva which lines the eyelids and covers anterior one-third of the eyeball is called the palpebral and bulbar conjunctiva, respectively. The junction between the palpebral and bulbar conjunctiva is called the fornix conjunctiva [23]. The conjunctiva provides lubrication to the eye by generating mucus and protection by producing antimicrobial peptides [24]. The main function of the conjunctiva is the formation of the tear film.

The conjunctiva is composed of two layers: an outer stratified epithelium and its underlying stroma [6]. The outer apical epithelial layer has tight junctions similar to the cornea but weaker. This layer restricts paracellular drug permeation across layers [25]. On the other hand, the stroma is furnished with blood, nerves, and lymph vessels. As a result, it can significantly decrease drug levels through considerable loss in the lymphatic and systemic circulation. Therefore, the conjunctival drug absorption is considered to be nonproductive than the cornea [6].

The density and size of pores are two and sixteen times higher, respectively, in the conjunctival epithelium of rabbit than the cornea. In addition, the estimated total paracellular space is 230 times larger in the conjunctiva than the cornea [17]. The conjunctiva is fairly permeable to hydrophilic drugs with less than 20 kDa molecular weights. As an example, the conjunctiva was permeable to large molecular weight substances such as inulin and fluorescein isothiocyanate (FITC)-dextran, whereas the cornea was permeable to only mannitol (molecular weight 182) but not to inulin or dextran [25]. For this reason, it is a favorable route for delivery of proteins and peptide molecules [6, 26]. Huang and associates evaluated paracellular permeability of the normal rabbit cornea and conjunctiva. Horseradish peroxidase was able to leach through the intercellular space of the conjunctival epithelia upon intravenous administration and was hindered by tight junctions of the superficial epithelium. The study confirmed that the superficial epithelial layer serves as a major barrier to the corneal and conjunctival permeability [25].

Similar to the cornea, the conjunctival tissues express esterase enzymes and efflux drug transporters (P-gp and MRP1) that can further reduce ocular drug absorption [27, 28]. The conjunctival epithelium transporters including neutral and cationic amino acids transporter, nucleoside transporter, and peptide transporter play a considerable role in the transport of drugs such as acyclovir [23, 29, 30].

26.2.2.2 Sclera

The sclera is a firm, opaque white, and fibrous tissue which is a continuation of the cornea to the posterior part of the eye. It offers large accessible surface area (~95 % of the total ocular surface area) and excessive hydration which makes it permeable to hydrophilic compounds. Owing to its hypocellular nature, it has less proteolytic enzymes and few protein-binding sites; in addition, the permeability does not decrease sustainably with age [31–33]. Scleral permeability is approximately ten times more and two times less to that of the cornea and conjunctiva, respectively [17]. Therefore, the sclera is not considered as a major barrier mainly due to higher permeation of drug in the sclera than nearby tissues like the cornea. The previously published study reported five times higher scleral permeation of radiolabeled hydrocortisone and mannitol compared to the corneal permeation in an isolated rabbit scleral membrane [34]. The permeability of the sclera is considered equivalent to the stroma layer of the cornea and thus likely favors the transport of hydrophilic substances. The transscleral drug transport occurs primarily by passive diffusion through the intercellular aqueous pore pathway. Consequently, the extent of drug permeation is independent of lipophilicity but considerably dependent on molecular radius, geometry, and charge of the molecule [20]. The results of in vitro studies indicated that large molecules (120–150 kDa) such as FITC-conjugated dextrans and immunoglobulins could diffuse through the sclera [35, 36]. Ambati et al. studied scleral permeability of sodium fluorescein, FITC-conjugated bovine serum albumin, FITC-IgG, and FITC dextrans ranging in molecular weight from 4 to 150 kDa across the fresh rabbit sclera. They observed reduction in the scleral permeability with increased molecular weight and molecular radius; furthermore, the sclera displayed better permeability to globular proteins compared to linear dextrans of similar molecular weight. The study concluded that molecular radius was a better predictor of the scleral permeability than molecular weight [37]. The sclera is a poorly vascularized tissue and mainly composed of collagen fibrils and negatively charged glycoproteins. Thus, it greatly favors permeability of negatively charged solutes [20]. Moreover, some fraction of topically applied drugs may reach the posterior segment by diffusion across the sclera [19]. Usually substances are more permeable to the posterior sclera than that of the anterior sclera as a result of loose meshwork of collagen fibers [22].

26.2.3 Blood-Ocular Barriers

The blood-aqueous and blood-retinal barriers are collectively named the blood-ocular barriers.

26.2.3.1 Blood-Aqueous Barrier

The blood-aqueous barrier lies in the anterior part of the eye. It obstructs the access of systemic drug into internal ocular compartments like the aqueous humor. It constitutes the tight junctional complexes of the iris-ciliary blood vessels endothelia and nonpigmented cell layer of the ciliary epithelium [38, 39]. The blood-aqueous barrier cannot offer barrier functionalities to the greatest extent even when it is intact. As an example, intravenously administered horse radish peroxidase (molecular mass 40 kDa) could reach the aqueous humor via penetration through the fenestrated capillaries of the ciliary processes but not via the iris blood vessels which control the passage of plasma proteins into the aqueous humor [40, 41]. Small and lipophilic molecules can enter the anterior uveal blood circulation via the blood-aqueous barrier resulting in quick clearance from the anterior chamber. On the contrary, large and fairly hydrophilic molecules are removed from the anterior chamber only by the aqueous humor turnover. As a consequence, large and fairly hydrophilic molecules exhibit relatively slow clearance than small and lipophilic molecules [42]. For instance, moderately lipophilic pilocarpine clearance was 13 μL/min, while the clearance of hydrophilic insulin was similar to the rate of aqueous humor turnover (2–3 μL/min in the humans and 3.0–4.7 μL/min in the rabbits) [43, 44].

26.2.3.2 Blood-Retinal Barrier

The blood-retinal barrier is located in the posterior segment of the eye. It is formed by nonleaky tight junctions between the retinal capillary endothelial cells (the inner blood-retinal barrier) and retinal pigment epithelial cells (the outer blood-retinal barrier) [20, 45]. They obstruct the entry of drug from blood into the posterior portion of the eye. The blood-retinal barriers are particularly permeable to more lipophilic substances, while lack of cellular components limits a convection of molecules [1, 46].

The retinal capillary endothelial cells have the intercellular tight junctions and exhibit poor permeability to small hydrophilic molecules (e.g., sodium fluorescein) and proteins (e.g., horse radish peroxidase) due to the restricted paracellular transport [47, 48]. The RPE exhibits tight junctions and separates outer surface of the multilayer neural retina from the choroid [49]. It allows the exchange of only selected nutrients, and the passage of other molecules through transcellular and paracellular pathways across the retina is prohibited. The concentrations of systemically administered drugs in the vitreous body are usually ~1–2 % that of plasma levels due to the blood-retinal barriers [4].

Recent studies have reported expression of P-gp and MRP efflux transporters [50, 51] and the presence of drug metabolic enzymes in the RPE that may play a significant role in the clearance of drugs. These efflux pumps diminish permeability of various compounds into the vitreous humor.

26.3 Routes of Ocular Drug Administration

Various routes of administration are available for delivery of ocular drug therapies. Commonly, topical, intravitreal, transscleral, and systemic routes are used for ocular drug delivery. Each route has its challenges and benefits which are summarized in Table 26.1 [7]

26.3.1 Topical Administration

Topical administration is the most conventional route of ophthalmic drug delivery to the anterior segment of the eye (cornea, aqueous humor, lens, iris-ciliary body). Commonly used topical formulations, such as eye drops, suspensions, solutions, gels, or ointments, each has its pros and cons. Topical administration is the most preferred route to treat common diseases in the anterior segment of the eye. Topically applied drug cannot reach the posterior segment of the eye at adequate therapeutic concentrations through the corneal route. As an example, after topical administration of eye drops of timolol, the concentration in the vitreous humor was 26 times less than the aqueous humor after one half hour [52].

Absorption of topical ophthalmic drugs occur via two main pathways: (1) the corneal route which involves traverse of drugs into intraocular tissues via the cornea and aqueous humor and (2) the non-corneal route that is absorption of drugs from the conjunctiva and that of the sclera to the choroid/RPE [53]. The compound's corneal permeability is the determinant of the most favorable route. Precorneal factors including but are not limited to tear turnover, the tear film, induced lacrimation, conjunctival absorption, tear drainage, blink reflex, tear dilution, solution drainage, and nasolacrimal drainage are responsible for less than 5 % ocular bioavailability as a consequence of significant drug loss. In addition, precorneal factors limit residence time of drugs on the ocular surface. Consequently, the topical route fails to maintain therapeutic levels and demands the necessity of frequent administration that leads to poor patient compliance. Several formulations with viscous

Table 26.1 Summary of benefits and challenges of ocular routes of drug delivery [7]

Routes	Benefits	Challenges
Topical	Easy application	Nasolacrimal and solution drainage
	High patient compliance	Higher tear dilution and turnover rate
	Noninvasive	Corneal barrier and efflux pump (<5 % bioavailability)
		Inefficient for the treatment of posterior segment disorders
		Short residence time on the ocular surface
Intravitreal	Allows direct delivery of drug to the vitreous and retina	Invasive method
	Improves drug absorption over the systemic and topical routes	Injections require repeated administration
	Controlled drug delivery for a long period of time with the use of nonbiodegradable implants	Associated with complications such as astigmatism, cataract, vitreous hemorrhage, retinal detachment, and endophthalmitis
	Biodegradable implants do not require surgical removal procedures	Poor patient acceptance due to pain and discomfort cause by frequent injections
	Implants may improve patient acceptance and adherence	Insertion and removal of nonbiodegradable device require surgery
Periocular/transscleral	dequate anterior and vitreous drug levels can be obtained	Fail to deliver drugs to the retina due to the presence of retinal pigment epithelium
	Controlled drug delivery for a long period of time with the use of nonbiodegradable implants	Insertion and removal of nonbiodegradable device require surgery
	Biodegradable implants do not require surgical removal procedures	
	Implants may improve patient acceptance and adherence	
al/systemic	Good patient acceptance	Encounters by the blood-ocular barriers
	More effective to treat posterior segment diseases compared to topically applied eye drops	Need of high dosing results in systemic toxicity

vehicles including gels and ointment are used to increase precorneal residence time to improve ocular bioavailability; nevertheless, they may cause patient intolerance and blurred vision [22, 54].

Topical formulations applied through the non-corneal route encounter the bulbar conjunctiva as the first main barrier which favors entry of macromolecules and hydrophilic substances into intraocular regions [55]. Quick clearance of drugs through lymphatic and vascularate structure of the conjunctiva and episclera contribute to lowering ocular bioavailability. Hydrophilic substances administered through the topical route are eliminated primarily via the anterior route which involves passive diffusion to the anterior chamber and clearance via the trabecular meshwork and Schlemm's canal. From the aqueous humor, drug molecules move to the iris and ciliary body where they bind to melanin in these tissues. Melanin-bound drug may act as a reservoir and thereof provide sustained release of drug [19].

Topically applied drugs are cleared from the aqueous humor by both the conventional and nonconventional pathways. The conventional pathway consists of the aqueous humor moving through the trabecular meshwork, across the inner wall of Schlemm's canal, into its lumen, draining collector channels, aqueous veins, and episcleral veins [56, 57]. The nonconventional route involves passing through the uveal meshwork and the anterior portion of the ciliary muscle [58, 59]. The elimination through later pathway is dependent on drug's permeability characteristics; on the contrary, the convective flow of the conventional clearance route is independent of drug properties. As a result, clearance rate of lipophilic drugs is higher than that of hydrophilic drugs from the anterior chamber [60].

Tears are secreted from the lacrimal gland and pose a significant barrier to topically administered drugs. The typical volume and turnover rate of the human tear film is around 7 μL and 0.5–2.2 μL/min, respectively. The cul-de-sac can momentarily hold ~30 μL of an administered eye drops; however, most of eye drop containers dispense ~50 μL volume, and as a result, drug loss occurs due to overflow and/or nasolacrimal drainage. The tear film is composed of three layers: the outermost lipid layer, a middle aqueous layer, and the innermost mucus layer [20, 54]. The tear film exhibits a fast restoration time of approximately 2–3 min, and 80 % or more topically applied eye drops can wash away within ~15–30 s after instillation [61]. Typically, less than 1 % of topically administered drugs get absorbed into the aqueous humor [62]. Further loss of drugs occurs due to metabolic enzymes in the anterior chamber [4]. The tear film is comprised of a buffer system with approximately between 7.2 and 7.5 pH ranges [61]. Eye drops with pH 6–9 are well tolerated, but formulations below or above this pH range may cause irritation and hence promote dilution and drug loss mediated by induced lacrimation [54]. Proteins and mucin content of tear can bind to drug molecules and reduce effective drug concentrations [20]. Despite these drawbacks, administration of topical eye drops is a preferable ocular route for all age groups due to easy application, noninvasive nature, and cost-effectiveness.

26.3.2 Intravitreal Administration

The transport of drug at therapeutic levels from the anterior segment to the posterior segment is difficult due to continuous drainage of the aqueous humor and longer diffusion distance. Ophthalmic drug delivery in the posterior segment of the eye is still a major challenge to drug delivery scientists. The intravitreal route is the most direct drug delivery modality that involves direct injection of various drug formulations including solution, suspension, and implants through the pars plana into the posterior segment of the eye [63]. The possibility and occurrence of severe complications with this treatment method is low and can be further reduced with proper procedures [64]. However, careful monitoring of patients is imperative due to associated adverse effects following intravitreal administration, the most invasive approach.

The area between the lens and retina is occupied with gel-like avascular connective tissue of the vitreous body. The motion of drug in the vitreous humor is predominantly controlled by diffusion rate [65]. Diffusivity of large molecules mainly positively charged are hindered, but small molecules can easily diffuse in the vitreous body [66]. Macha and associates conducted permeation study of paracellular marker, fluorescein, into the aqueous and vitreous humor of rabbits. They indicated that the vitreous humor exhibits tighter barrier than the aqueous humor that result in very low levels of fluorescein (~1–2 %) in the vitreous chamber to that of plasma [67]. High-strength oral formulations or rapid intravenous infusions were administered to obtain higher plasma levels in order to attain adequate elevated drug levels in the vitreous body [68, 69] but that led to inadvertent toxicity [70, 71]. The lens poses as an additional hindrance and limits the entry of drug from the aqueous humor to the vitreous humor. The cell layer between the retina and vitreous humor is called the inner limiting membrane that acts as a barrier for gene therapy delivery through intravitreal administration [72]. Furthermore, the RPE contributes significantly to diminish the transport of drug from the vitreous body to the outer segments of retina and choroid [7].

Elimination of drugs occurs through the anterior and posterior routes after intravitreal administration. The anterior route involves clearance of drugs via the aqueous humor turnover; on the other hand, the posterior pathway involves elimination of drugs with optimum passive permeability and/or substrates of active transporters to the systemic circulation via the blood-retinal barriers [60]. As a consequence, extended half-life in the vitreous humor can be expected for hydrophilic and large molecular weight compounds. Moreover, drug may display a long half-life in the vitreous body, if instead of clearing through the blood-retinal barrier; it diffuses into the anterior chamber where it is eliminated either by aqueous blood flow or after diffusion across the iris surface. For instance, gentamicin, which was cleared from the vitreous body via the anterior chamber after intravitreal injection, demonstrated five times longer half-life compared to penicillin that was eliminated through the retinal route in the rabbit eye [60].

26.3.2.1 Intravitreal Injections

Intravitreal injections have an advantage over topical and systemic administrations as they bypass corneoscleral barriers and thereby have gained significant attention for successful delivery of ophthalmic formulations to treat vitreoretinal diseases. With this approach, it is likely to achieve sustained therapeutic drug levels in the vitreous body through lower doses and thus reduce the long-term adverse effects associated with high dose or drug administered through rapid intravenous infusions [4]. The major disadvantage with intravitreal injections are patient noncompliance and associated side effects including astigmatism, cataract, vitreous hemorrhage, retinal detachment, and endophthalmitis due to repeated injections of drug formulations that cannot maintain effective drug concentration over the treatment period [73]. Development of modified formulations including prodrug approach was explored to overcome the need of frequent injections. Kawakami et al. formulated lipophilic prodrug liposomes to prolong the residence time of drug in the vitreous body in order to reduce frequent necessity of intravitreal injections. They observed significantly high levels of tilisolol in the vitreous body of rabbit eye following intravitreal injection of liposomes of O-palmitoyl prodrug of tilisolol compared to tilisolol liposomes [74]. Nevertheless, there are still challenges remaining with this route.

26.3.2.2 Intravitreal Implants

Intravitreal implants have been developed to overcome shortcomings associated with multiple intravitreal injections. Intravitreal implants compromised of either biodegradable or nonbiodegradable polymeric matrices and are injected into the vitreous body or surgically placed in the pars plana area. Intravitreal implants are advantageous over the conventional drug delivery methods because they deliver sustained therapeutic levels of drugs directly to the target tissues in the posterior segment of the eye and hence diminish systemic toxicity associated with high concentrations of drugs. The other advantages are dose reduction as a result of targeted drug delivery and fewer serious complications attributed to less need of repeated dosing [75, 76]. There is a minimal risk of side effects with sustained release drug delivery approach as a result of infrequent administration and lack of posttreatment removal of biodegradable implants [1]. Furthermore, they need to be inserted once every few years and are convenient to patients. However, precise drug release control is more challenging with biodegradable implants. On the other hand, nonbiodegradable implants produce more accurate sustained release over the treatment period; however, they require invasive post-therapy removal [75].

The dexamethasone drug delivery system (Ozurdex®; Allergan Inc., Irvine, CA) is a biodegradable intravitreal implant and the first drug therapy for macular edema and noninfectious posterior uveitis. It provides persistent therapeutic effects on target tissue up to six months with minimal systemic absorption after implantation. Nonetheless, intravitreal implants are not completely successful in elimination of side effects associated with intravitreal injections. Side effects reported for Ozurdex® include endophthalmitis, cataract, eye inflammation, increased eye pressure, and retinal detachments [77].

26.3.3 *Periocular/Transscleral Administration*

Upon topical administration, significant drug absorption into the systemic circulation occurs through the nasolacrimal duct; as a result, therapeutic goals are not attained in the posterior segment of the eye. Therefore, the subconjunctival route of drug administration has been preferred for locally administered drugs [24]. Drug delivery to the posterior chamber of the eye via the sclera is relatively less invasive and an alternative approach to intravitreal administration. This approach can overcome the surgery-related risk of retinal detachment, a major shortcoming of intravitreal injections and implants [78].

Periocular administration consists of the peribulbar, posterior juxtascleral, retrobulbar, subtenon and subconjunctival routes. Concentration of drugs in the eye varies accordingly with the various pathways in the periocular route. For example, maximum and controlled concentrations of sodium fluorescein were achieved via subtenon injection, relative to retrobulbar, and subconjunctival routes after periocular administration of sodium fluorescein in the rabbit eye [79]. The fibrous nature of sclera results in less resistance to drug diffusion via the scleral route compared to the topical or systemic route. Following administration through periocular routes, drug arrives at the posterior site via three pathways: transscleral, systemic circulation (through the choroidal vasculature), and the anterior (through the tear film, cornea, aqueous humor, and vitreous humor) [79]. After subconjunctival administration, drug clearance rapidly occurs through the dynamic barriers that include the conjunctival blood and lymphatic circulation. Thus, drug clearance from the subconjunctival space determines drug concentrations in the vitreous body during subconjunctival drug delivery. Molecules that circumvent elimination through the conjunctival blood and lymphatic circulation pathways may able to cross the sclera and retina and can reach the inner retinal layers [80, 81]. However, the choroidal vasculature poses a significant barrier, and substantial loss of drug may occur before it can reach the neural retina. Additionally, the blood-retinal barriers restrict drug availability to the photoreceptor cells.

Lipophilic drug molecules can substantially bind to the choroid-Bruch's membrane and exhibit higher partitioning into the choroid RPE compared to the retina. Despite overcoming the epithelial barriers, drug permeation to the neural retina is impeded due to this binding following transscleral administration [82, 83]. Moreover, periocular routes are also associated with potential side effects including cataract development, strabismus, hyphema, intraocular pressure, and corneal decompensation in some patients [84].

26.3.3.1 Subconjunctival Injections

Drugs slowly diffuse from the sclera and conjunctiva to the retina and choroid following subconjunctival injections and, as a result, are more likely to escape epithelial barriers. Higher levels of dacarbazine, an anticancer drug, were obtained in the aqueous and vitreous humor of rabbit eye following subconjunctival administration relatively to intravenous administration [85]. Lee and coworkers assessed elimination pathways of hydrophilic sodium fluorescein after subconjunctival administration in live and postmortem rat eyes. The study results suggested that, after subconjunctival administration, hydrophilic small drugs and macromolecules were predominantly eliminated through the conjunctival blood and lymphatic vessels, respectively [86].

Higher intraocular and lower plasma levels are achieved through the scleral route, while the opposite is true for the conjunctival application as a result of drainage of drug to the systemic circulation from the conjunctiva. For instance, elevated levels of tilisolol, a beta-blocker drug, were observed in ocular tissues following the scleral than conjunctival application; the opposite was true for plasma levels [87].

26.3.3.2 Scleral Plugs and Implants

Scleral plugs and implant therapies involve invasive methods. They are inserted onto the sclera with careful surgical procedure for sustained drug release [88]. Intravitreal administration has gained considerable attention for the treatment of cytomegalovirus retinitis as a result of risk of side effects associated with administration of oral or intravenous ganciclovir. However, frequent administration is necessary to sustain intravitreal therapeutic drug concentrations. To conquer this problem, Sanborn and colleagues developed a nonbiodegradable ganciclovir scleral implant for prolong drug delivery [89]. However, it also requires a surgical procedure for removal and implantation. To overcome the limitation associated with nonbiodegradable ganciclovir scleral implant, Yasukawa et al. developed a ganciclovir scleral implant using various blends of polylactide (PLA) and polylactide-co-glycolide biodegradable polymers exhibiting a triphasic release pattern. The implants can be easily inserted through a small scleral incision at the pars plana region of the eye. The scleral plugs of PLA-70,000 and PLA-5000 containing 25 % of ganciclovir provided effectively sustained drug release over the period of 24 weeks [46].

26.3.4 Systemic Administration

This route has also been used for drug delivery to the posterior segment of the eye via oral or intravenous administration. Poor ocular bioavailability and elevated risk of systemic side effects are major drawbacks associated with this route. A large systemic dose (only drugs with wide therapeutic index can be administered, e.g., antibiotics) and frequent administration of drugs are often necessary to surmount this issue. This encourages poor patient compliance along with increased risk of systemic toxicity. The transport of systemically administered drugs into the anterior and posterior segment of the eye is significantly limited by the blood-aqueous and blood-retinal barriers, respectively. The detailed functionality of these barriers is discussed under Sect. 26.2.3. Following intravenous administration, drugs are distributed in ocular tissues. However, due to systemic toxicity and delivery concerns, it is not the preferred route for the treatment of eye diseases. Drug delivery scientists encounter significant challenges to developing efficient, targeted ocular drug delivery approaches in order to overcome limitations associated with systemic administration.

In summary, the unique anatomical and physiological constraints of ocular structure pose challenging problems for pharmaceutical scientists to develop safe and efficient ocular therapeutics. Increased knowledge of dynamic ocular barriers and thorough understanding of ophthalmic diseases would enable formulation scientists to improve ophthalmic drug delivery and target their approaches rendering new formulations safe and effective.

New delivery technologies including nanoparticulate systems like liposomes, microemulsions, nanoparticles, nanosuspensions, dendrimers, niosomes, and cyclodextrins have gained significant attention to overcome various constraints related to ocular drug delivery. These aspects are discussed in depth in other chapters of this book.

References

1. Kang-Mieler JJ, Osswald CR, Mieler WF (2014) Advances in ocular drug delivery: emphasis on the posterior segment. Expert Opin Drug Deliv 30:1–14
2. Hughes PM, Olejnik O, Chang-Lin JE, Wilson CG (2005) Topical and systemic drug delivery to the posterior segments. Adv Drug Deliv Rev 57(14):2010–2032
3. Lee VH, Robinson JR (1986) Topical ocular drug delivery: recent developments and future challenges. J Ocul Pharmacol 2(1):67–108
4. Duvvuri S, Majumdar S, Mitra AK (2003) Drug delivery to the retina: challenges and opportunities. Expert Opin Biol Ther 3(1):45–56
5. Ranta VP, Mannermaa E, Lummepuro K, Subrizi A, Laukkanen A, Antopolsky M et al (2010) Barrier analysis of periocular drug delivery to the posterior segment. J Control Release 148(1):42–48
6. Cholkar K, Patel SP, Vadlapudi AD, Mitra AK (2013) Novel strategies for anterior segment ocular drug delivery. J Ocul Pharmacol Ther 29(2):106–123
7. Gaudana R, Ananthula HK, Parenky A, Mitra AK (2010) Ocular drug delivery. AAPS J 12(3):348–360
8. Gaudana R, Jwala J, Boddu SH, Mitra AK (2009) Recent perspectives in ocular drug delivery. Pharm Res 26(5):1197–1216
9. Tanaka M, Ohnishi Y, Kuwabara T (1983) Membrane structure of corneal epithelium: freeze-fracture observation. Jpn J Ophthalmol 27(3):434–443
10. Chien DS, Sasaki H, Bundgaard H, Buur A, Lee VH (1991) Role of enzymatic lability in the corneal and conjunctival penetration of timolol ester prodrugs in the pigmented rabbit. Pharm Res 8(6):728–733
11. Kishida K, Otori T (1980) A quantitative study on the relationship between transcorneal permeability of drugs and their hydrophobicity. Jpn J Ophthalmol 24:251–259
12. Mosher GL, Mikkelson TJ (1979) Permeability of the n-alkyl p-aminobenzoate esters across the isolated corneal membrane of the rabbit. Int J Pharm 2(3–4):239–243
13. Schoenwald RD, Huang HS (1983) Corneal penetration behavior of beta-blocking agents I: physiochemical factors. J Pharm Sci 72(11):1266–1272
14. Wang W, Sasaki H, Chien DS, Lee VH (1991) Lipophilicity influence on conjunctival drug penetration in the pigmented rabbit: a comparison with corneal penetration. Curr Eye Res 10(6):571–579

15. Prausnitz MR, Noonan JS (1998) Permeability of cornea, sclera, and conjunctiva: a literature analysis for drug delivery to the eye. J Pharm Sci 87(12):1479–1488
16. Hamalainen KM, Kontturi K, Auriola S, Murtomäki L, Urtti A (1997) Estimation of pore size and pore density of biomembranes from permeability measurements of polyethylene glycols using an effusion-like approach. J Control Release 49(2–2):97–104
17. Hamalainen KM, Kananen K, Auriola S, Kontturi K, Urtti A (1997) Characterization of paracellular and aqueous penetration routes in cornea, conjunctiva, and sclera. Invest Ophthalmol Vis Sci 38(3):627–634
18. Palmgren JJ, Toropainen E, Auriola S, Urtti A (2002) Liquid chromatographic-electrospray ionization mass spectrometric analysis of neutral and charged polyethylene glycols. J Chromatogr A 976(1–2):165–170
19. Tangri P, Khurana S (2011) Basics of ocular drug delivery systems. Int J Res Pharm Biomed Sci 2(4):1541–1552
20. Achouri D, Alhanout K, Piccerelle P, Andrieu V (2013) Recent advances in ocular drug delivery. Drug Dev Ind Pharm 39(11):1599–1617
21. Edward A, Prausnitz MR (2001) Predicted permeability of the cornea to topical drugs. Pharm Res 18(11):1497–1508
22. Molokhia SA, Thomas SC, Garff KJ, Mandell KJ, Wirostko BM (2013) Anterior eye segment drug delivery systems: current treatments and future challenges. J Ocul Pharmacol Ther 29(2):92–105
23. Hosoya K, Lee VH, Kim KJ (2005) Roles of the conjunctiva in ocular drug delivery: a review of conjunctival transport mechanisms and their regulation. Eur J Pharm Biopharm 60(2):227–240
24. Stjernschantz J, Astin M (1993) Anatomy and physiology of the eye. Physiological aspects of ocular drug therapy. In: Edman P (ed) Biopharmaceutics of ocular drug delivery. CRC Press, Boca Raton
25. Huang AJ, Tseng SC, Kenyon KR (1989) Paracellular permeability of corneal and conjunctival epithelia. Invest Ophthalmol Vis Sci 30(4):684–689
26. Urtti A (2006) Challenges and obstacles of ocular pharmacokinetics and drug delivery. Adv Drug Deliv Rev 58(11):1131–1135
27. Gukasyan HJ, Yerxa BR, Pendergast W, Lee VH (2002) Metabolism and transport of purinergic receptor agonists in rabbit conjunctival epithelial cells. Adv Exp Med Biol 506(Pt A):255–259
28. Yang JJ, Kim KJ, Lee VH (2000) Role of P-glycoprotein in restricting propranolol transport in cultured rabbit conjunctival epithelial cell layers. Pharm Res 17(5):533–538
29. Ganapathy ME, Ganapathy V (2005) Amino acid transporter ATB0,+ as a delivery system for drugs and prodrugs. Curr Drug Targets Immune Endocr Metabol Disord 5(4):357–364
30. Hosoya K, Horibe Y, Kim KJ, Lee VH (1998) Nucleoside transport mechanisms in the pigmented rabbit conjunctiva. Invest Ophthalmol Vis Sci 39(2):372–377
31. Foster CS, Sainz de la Maza M (1994) The sclera. Springer, New York
32. Olsen TW, Aaberg SY, Geroski DH, Edelhauser HF (1998) Human sclera: thickness and surface area. Am J Ophthalmol 125(2):237–241
33. Olsen TW, Edelhauser HF, Lim JI, Geroski DH (1995) Human scleral permeability. Effects of age, cryotherapy, transscleral diode laser, and surgical thinning. Invest Ophthalmol Vis Sci 36(9):1893–1903
34. Unlu N, Robinson JR (1998) Scleral permeability to hydrocortisone and mannitol in the albino rabbit eye. J Ocul Pharmacol Ther 14(3):273–281
35. Ambati J, Gragoudas ES, Miller JW, You TT, Miyamoto K, Delori FC et al (2000) Transscleral delivery of bioactive protein to the choroid and retina. Invest Ophthalmol Vis Sci 41(5):1186–1191
36. Nicoli S, Ferrari G, Quarta M, Macaluso C, Govoni P, Dallatana D et al (2009) Porcine sclera as a model of human sclera for in vitro transport experiments: histology, SEM, and comparative permeability. Mol Vis 15:259–266
37. Ambati J, Canakis CS, Miller JW, Gragoudas ES, Edwards A, Weissgold DJ et al (2000) Diffusion of high molecular weight compounds through sclera. Invest Ophthalmol Vis Sci 41(5):1181–1185

38. Bill A (1986) The blood-aqueous barrier. Trans Ophthalmol Soc UK 105(Pt 2):149–155
39. Freddo TF (2001) Shifting the paradigm of the blood-aqueous barrier. Exp Eye Res 73(5):581–592
40. Barsotti MF, Bartels SP, Freddo TF, Kamm RD (1992) The source of protein in the aqueous humor of the normal monkey eye. Invest Ophthalmol Vis Sci 33(3):581–595
41. Schlingemann RO, Hofman P, Klooster J, Blaauwgeers HG, Van der Gaag R, Vrensen GF (1998) Ciliary muscle capillaries have blood-tissue barrier characteristics. Exp Eye Res 66(6):747–754
42. Urtti A, Salminen L (1993) Animal pharmacokinetic studies. In: Mitra AK (ed) Ophthalmic drug delivery systems. Marcel Dekker, New York
43. Conrad JM, Robinson JR (1977) Aqueous chamber drug distribution volume measurement in rabbits. J Pharm Sci 66(2):219–224
44. Schoenwald RD (2003) Ocular pharmacokinetics and pharmacodynamics. In: Mitra AK (ed) Ophthalmic drug delivery systems. Marcel Dekker, New York
45. Occhiutto ML, Freitas FR, Maranhao RC, Costa VP (2012) Breakdown of the blood-ocular barrier as a strategy for the systemic use of nanosystems. Pharmaceutics 4(2):252–275
46. Yasukawa T, Kimura H, Tabata Y, Ogura Y (2001) Biodegradable scleral plugs for vitreoretinal drug delivery. Adv Drug Deliv Rev 52(1):25–36
47. Gardner TW, Antonetti DA, Barber AJ, Lieth E, Tarbell JA (1999) The molecular structure and function of the inner blood-retinal barrier. Penn State Retina Research Group. Doc Ophthalmol 97(3–4):229–237
48. Sunkara G, Kompella UB (2003) Membrane transport processes in the eye. In: Mitra AK (ed) Ophthalmic drug delivery systems. Marcel Dekker, Inc, New York
49. Cunha-Vaz JG (1976) The blood-retinal barriers. Doc Ophthalmol 41(2):287–327
50. Aukunuru JV, Sunkara G, Bandi N, Thoreson WB, Kompella UB (2001) Expression of multidrug resistance-associated protein (MRP) in human retinal pigment epithelial cells and its interaction with BAPSG, a novel aldose reductase inhibitor. Pharm Res 18(5):565–572
51. Kennedy B, Mangini N (2002) Native and cultured human RPE express P-glycoprotein. Invest Ophthalmol Vis Sci 43:903, E-Abstract
52. Urtti A, Pipkin JD, Rork G, Sendo T, Finne U, Repta AJ (1990) Controlled drug delivery devices for experimental ocular studies with timolol 2. Ocular and systemic absorption in rabbits. Int J Pharm Sci 61(3):241–249
53. Maurice DM (2002) Drug delivery to the posterior segment from drops. Surv Ophthalmol 47(Suppl 1):S41–S52
54. Koevary SB (2003) Pharmacokinetics of topical ocular drug delivery: potential uses for the treatment of diseases of the posterior segment and beyond. Curr Drug Metab 4(3):213–222
55. Ahmed I, Patton TF (1985) Importance of the noncorneal absorption route in topical ophthalmic drug delivery. Invest Ophthalmol Vis Sci 26(4):584–587
56. Ascher KW (1954) Veins of the aqueous humor in glaucoma. Boll Ocul 33(3):129–144
57. Goldmann H (1950) Minute volume of the aqueous in the anterior chamber of the human eye in normal state and in primary glaucoma. Ophthalmologica 120(1–2):19–21
58. Bill A, Hellsing K (1965) Production and drainage of aqueous humor in the cynomolgus monkey (Macaca irus). Invest Ophthalmol 4(5):920–926
59. Johnson M, Erickson K (2000) Mechanisms and routes of aqueous humor drainage. In: Albert DM, Jakobiec FA (eds) Principles and practice of ophthalmology. WB Saunders, Philadelphia
60. Maurice DM, Mishima S (1984) Ocular pharmacokinetics. In: Sears ML (ed) Handbook of experimental pharmacology. Springer, Berlin/Heidelberg
61. Maurice DM (1967) The use of fluorescein in ophthalmological research. Invest Ophthalmol 6(5):464–477
62. Hughes PM, Mitra AK (1993) Overview of ocular drug delivery. In: Mitra AK (ed) Drug delivery systems. Marcell Dekker, Inc, New York
63. Thrimawithana TR, Young S, Bunt CR, Green C, Alany RG (2011) Drug delivery to the posterior segment of the eye. Drug Discov Today 16(5–6):270–277

64. Peyman GA, Lad EM, Moshfeghi DM (2009) Intravitreal injection of therapeutic agents. Retina 29(7):875–912
65. Mitra AK, Anand BS, Duvvuri S (2006) Drug delivery to the eye. In: Fischbarg J (ed) The biology of eye. Academic, New York
66. Pitkanen L, Ruponen M, Nieminen J, Urtti A (2003) Vitreous is a barrier in nonviral gene transfer by cationic lipids and polymers. Pharm Res 20(4):576–583
67. Macha S, Mitra AK (2001) Ocular pharmacokinetics in rabbits using a novel dual probe microdialysis technique. Exp Eye Res 72(3):289–299
68. Lalezari JP, Friedberg DN, Bissett J, Giordano MF, Hardy WD, Drew WL et al (2002) High dose oral ganciclovir treatment for cytomegalovirus retinitis. J Clin Virol 24(1–2):67–77
69. Lopez-Cortes LF, Ruiz-Valderas R, Lucero-Munoz MJ, Cordero E, Pastor-Ramos MT, Marquez J (2000) Intravitreal, retinal, and central nervous system foscarnet concentrations after rapid intravenous administration to rabbits. Antimicrob Agents Chemother 44(3):756–759
70. Henderly DE, Freeman WR, Causey DM, Rao NA (1987) Cytomegalovirus retinitis and response to therapy with ganciclovir. Ophthalmology 94(4):425–434
71. Jabs DA, Newman C, De Bustros S, Polk BF (1987) Treatment of cytomegalovirus retinitis with ganciclovir. Ophthalmology 94(7):824–830
72. Dalkara D, Kolstad KD, Caporale N, Visel M, Klimczak RR, Schaffer DV et al (2009) Inner limiting membrane barriers to AAV-mediated retinal transduction from the vitreous. Mol Ther 17(12):2096–2102
73. Myles ME, Neumann DM, Hill JM (2005) Recent progress in ocular drug delivery for posterior segment disease: emphasis on transscleral iontophoresis. Adv Drug Deliv Rev 57(14):2063–2079
74. Kawakami S, Yamamura K, Mukai T, Nishida K, Nakamura J, Sakaeda T et al (2001) Sustained ocular delivery of tilisolol to rabbits after topical administration or intravitreal injection of lipophilic prodrug incorporated in liposomes. J Pharm Pharmacol 53(8):1157–1161
75. Bourges JL, Bloquel C, Thomas A, Froussart F, Bochot A, Azan F et al (2006) Intraocular implants for extended drug delivery: therapeutic applications. Adv Drug Deliv Rev 58(11):1182–1202
76. Choonara YE, Pillay V, Danckwerts MP, Carmichael TR, du Toit LC (2010) A review of implantable intravitreal drug delivery technologies for the treatment of posterior segment eye diseases. J Pharm Sci 99(5):2219–2239
77. Hunter RS, Lobo AM (2011) Dexamethasone intravitreal implant for the treatment of noninfectious uveitis. Clin Ophthalmol 5:1613–1621
78. Raghava S, Hammond M, Kompella UB (2004) Periocular routes for retinal drug delivery. Expert Opin Drug Deliv 1(1):99–114
79. Ghate D, Edelhauser HF (2006) Ocular drug delivery. Expert Opin Drug Deliv 3(2):275–287
80. Kim SH, Csaky KG, Wang NS, Lutz RJ (2008) Drug elimination kinetics following subconjunctival injection using dynamic contrast-enhanced magnetic resonance imaging. Pharm Res 25(3):512–520
81. Weijtens O, Feron EJ, Schoemaker RC, Cohen AF, Lentjes EG, Romijn FP et al (1999) High concentration of dexamethasone in aqueous and vitreous after subconjunctival injection. Am J Ophthalmol 128(2):192–197
82. Cheruvu NP, Kompella UB (2006) Bovine and porcine transscleral solute transport: influence of lipophilicity and the Choroid-Bruch's layer. Invest Ophthalmol Vis Sci 47(10):4513–4522
83. Kadam RS, Kompella UB (2010) Influence of lipophilicity on drug partitioning into sclera, choroid-retinal pigment epithelium, retina, trabecular meshwork, and optic nerve. J Pharmacol Exp Ther 332(3):1107–1120
84. Castellarin A, Pieramici DJ (2004) Anterior segment complications following periocular and intraocular injections. Ophthalmol Clin North Am 17(4):583–590, vii
85. Kalsi GS, Silver HK, Rootman J (1991) Ocular pharmacokinetics of dacarbazine following subconjunctival versus intravenous administration in the rabbit. Can J Ophthalmol 26(5):247–251
86. Lee SJ, He W, Robinson SB, Robinson MR, Csaky KG, Kim H (2010) Evaluation of clearance mechanisms with transscleral drug delivery. Invest Ophthalmol Vis Sci 51(10):5205–5212

87. Sasaki H, Ichikawa M, Kawakami S, Yamamura K, Nishida K, Nakamura J (1996) In situ ocular absorption of tilisolol through ocular membranes in albino rabbits. J Pharm Sci 85(9):940–943
88. Ali M, Byrne ME (2008) Challenges and solutions in topical ocular drug-delivery systems. Expert Rev Clin Pharmacol 1(1):145–161
89. Sanborn GE, Anand R, Torti RE, Nightingale SD, Cal SX, Yates B et al (1992) Sustained-release ganciclovir therapy for treatment of cytomegalovirus retinitis. Use of an intravitreal device. Arch Ophthalmol 110(2):188–195

Index

© Springer International Publishing Switzerland 2016

Y. Pathak et al. (eds.), *Nano-Biomaterials For Ophthalmic Drug Delivery*,
DOI 10.1007/978-3-319-29346-2

O

P

S

T

U

V

MIX
Papier aus verantwortungsvollen Quellen
Paper from responsible sources
FSC® C105338

If you have any concerns about our products,
you can contact us on
ProductSafety@springernature.com

In case Publisher is established outside the EU,
the EU authorized representative is:
Springer Nature Customer Service Center GmbH
Europaplatz 3, 69115 Heidelberg, Germany

Printed by Libri Plureos GmbH
in Hamburg, Germany